Periodic Table of Elements

1 1A	2 2A	3	4	5	6	7	8	9	10	11	12	13 3A	14 4A	15 5A	16 6A	17 7A	18 8A
1 H																	2 He
3 Li	4 Be											5 B	6 C	7 N	8 O	9 F	10 Ne
11 Na	12 Mg											13 Al	14 Si	15 P	16 S	17 Cl	18 Ar
19 K	20 Ca	21 Sc	22 Ti	23 V	24 Cr	25 Mn	26 Fe	27 Co	28 Ni	29 Cu	30 Zn	31 Ga	32 Ge	33 As	34 Se	35 Br	36 Kr
37 Rb	38 Sr	39 Y	40 Zr	41 Nb	42 Mo	43 Tc	44 Ru	45 Rh	46 Pd	47 Ag	48 Cd	49 In	50 Sn	51 Sb	52 Tc	53 I	54 Xc
55 Cs	56 Ba	57 La*	72 Hf	73 Ta	74 W	75 Re	76 Os	77 Ir	78 Pt	79 Au	80 Hg	81 Tl	82 Pb	83 Bi	84 Po	85 At	86 Rn
87 Fr	88 Ra	89 Ac†	104 Rf	105 Db	106 Sg	107 Bh	108 Hs	109 Mt	110 Ds	111 Rg	112 Cn	113 Uut	114 Fl	115 Uup	116 Lv	117 Uus	118 Uuo

Labels: Alkali metals (group 1, rows 3–87); Alkaline earth metals (group 2); Transition metals (groups 3–12); Halogens (group 17); Noble gases (group 18)

*Lanthanides	58 Ce	59 Pr	60 Nd	61 Pm	62 Sm	63 Eu	64 Gd	65 Tb	66 Dy	67 Ho	68 Er	69 Tm	70 Yb	71 Lu
†Actinides	90 Th	91 Pa	92 U	93 Np	94 Pu	95 Am	96 Cm	97 Bk	98 Cf	99 Es	100 Fm	101 Md	102 No	103 Lr

Basic Chemistry

EIGHTH EDITION

Basic Chemistry

Steven S. Zumdahl
University of Illinois

Donald J. DeCoste
University of Illinois

Australia • Brazil • Mexico • Singapore • United Kingdom • United States

***Basic Chemistry,* Eighth Edition**
Steven S. Zumdahl and Donald J. DeCoste

Product Director: Mary Finch

Senior Product Team Manager: Lisa Lockwood

Product Manager: Thomas Martin

Content Coordinator: Brendan Killion

Product Assistant: Karolina Kiwak

Media Developer: Lisa Weber

Executive Brand Manager: Nicole Hamm

Market Development Manager: Janet Del Mundo

Content Project Manager: Teresa L. Trego

Art Director: Maria Epes

Manufacturing Planner: Judy Inouye

Rights Acquisitions Specialist: Thomas McDonough

Production and Composition: Graphic World Inc.

Photo Researcher: Sharon Donahue/PreMedia Global

Text Researcher: PreMedia Global

Copy Editor: Graphic World Inc.

Illustrator: Graphic World Inc.

Text Designer: Ellen Pettengell Designs

Cover Designer: Irene Morris

Cover Image: "Emptyful" sculpture by Bill Pechet with co-designer Chris Pekar of Lightworks and Lumenppulse. Photo by Gerry Kopelow.

Title Page Photo: Yamada Taro/Getty Images

Library of Congress Control Number: 2013942838

Student Edition:

ISBN-13: 978-1-285-45314-9

ISBN-10: 1-285-45314-X

Cengage Learning
200 First Stamford Place, 4th Floor
Stamford, CT 06902
USA

Cengage Learning is a leading provider of customized learning solutions with office locations around the globe, including Singapore, the United Kingdom, Australia, Mexico, Brazil, and Japan. Locate your local office at **www.cengage.com/global.**

Cengage Learning products are represented in Canada by Nelson Education, Ltd.

To learn more about Cengage Learning Solutions, visit **www.cengage.com.**

Purchase any of our products at your local college store or at our preferred online store **www.cengagebrain.com.**

Printed in the United States of America
1 2 3 4 5 6 7 17 16 15 14 13

Brief Contents

Contents

Dr. John Brackenbury/Science Photo Library/Photo Researchers, Inc.

Richard Cummins/Lonely Planet Images/Getty Images

View Stock/Getty Images

Image by © Raven Regan/Design Pics/Corbis

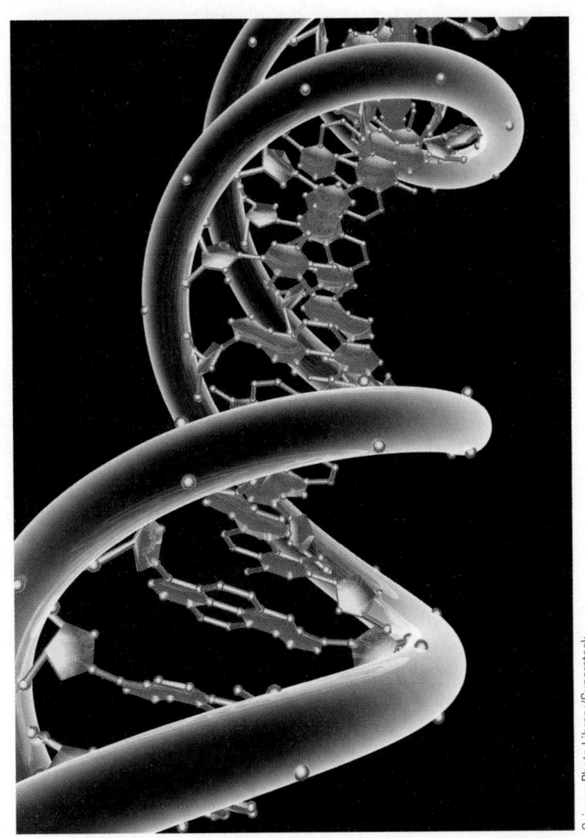
Science Photo Library/Superstock

Gregg Epperson/Shutterstock.com

Preface

The eighth edition of *Introductory Chemistry* continues toward the goals we have pursued for the first seven editions: to make chemistry interesting, accessible, and understandable to the beginning student. For this edition, we have included additional support for instructors and students to help achieve these goals.

Learning chemistry can be very rewarding. And even the novice, we believe, can relate the macroscopic world of chemistry—the observation of color changes and precipitate formation—to the microscopic world of ions and molecules. To achieve that goal, instructors are making a sincere attempt to provide more interesting and more effective ways to learn chemistry, and we hope that *Introductory Chemistry* will be perceived as a part of that effort. In this text we have presented concepts in a clear and sensible manner using language and analogies that students can relate to. We have also written the book in a way that supports active learning. In particular, the Active Learning Questions, found at the end of each chapter, provide excellent material for collaborative work by students. In addition, we have connected chemistry to real-life experience at every opportunity, from chapter opening discussions of chemical applications to "Chemistry in Focus" features throughout the book. We are convinced that this approach will foster enthusiasm and real understanding as the student uses this text. Highlights of the *Introductory Chemistry* program are described below.

New to This Edition

Building on the success of previous editions of *Introductory Chemistry,* the following changes have been made to further enhance the text:

Updates to the Student Text and Instructor's Annotated Edition

Changes to the student text and the accompanying Instructor's Annotated Edition are outlined below:

Critical Thinking Questions We have added these questions throughout the text to emphasize the importance of conceptual learning. These are particularly useful for generating class discussion.

Interactive Examples We have included computer-based examples throughout the text that require students to think through the Example step-by-step rather than simply scan the written Example in the text as many students do.

ChemWork Problems We have added these to the end-of-chapter problems throughout the text. These problems test students' understanding of core concepts from each chapter. Students who solve a particular problem with no assistance can proceed directly to the answer. However, students who need help can get assistance through a series of online hints. The online procedure for assisting students is modeled after the way a teacher would help with homework problems in his or her office. The hints are usually in the form of interactive questions that guide students through the problem-solving process. Students cannot receive the correct answer from the computer; rather, it encourages students to continue working through the hints to arrive at the answer. ChemWork problems in the text can be worked using the online system or as pencil-and-paper problems.

Limiting Reactant Approach In Chapter 9, we have enhanced the treatment of stoichiometry by adding a new section on limiting reactants, which emphasizes calculating the amounts of products that can be obtained from each reactant. Now, students are taught how to select a limiting reactant both by comparing the amounts of reactants present and by calculating the amounts of product that can be formed by complete consumption of each reactant.

Art Program We have revised, modified, and updated the figures in the textbook as needed to better serve visual learners.

"Chemistry in Focus" Boxes We have revised many of the "Chemistry in Focus" boxes and added new ones with up-to-date topics such as Gorilla glass, nanocars, and use of breath analysis to diagnose disease.

End-of-Chapter Exercises We replaced over 10% of the end-of-chapter questions and problems as well as added new problems using the new "Chemistry in Focus" boxes and ChemWork problems. As before, the margin of the Instructor's Annotated Edition includes answers to all of the Self Check and end-of-chapter exercises, along with additional examples for all Example problems. In the student edition, answers to Self Checks and to even-numbered exercises are provided at the back of the book.

Glossary Terms/Key Terms We have updated all glossary and key terms and have added many new definitions.

Emphasis on Reaction Chemistry

We continue to emphasize chemical reactions early in the book, leaving the more abstract material on orbitals for later chapters. In a course in which many students encounter chemistry for the first time, it seems especially important that we present the chemical nature of matter before we discuss the theoretical intricacies of atoms and orbitals. Reactions are inherently interesting to students and can help us draw them to chemistry. In particular, reactions can form the basis for fascinating classroom demonstrations and laboratory experiments.

We have therefore chosen to emphasize reactions before going on to the details of atomic structure. Relying only on very simple ideas about the atom, Chapters 6 and 7 represent a thorough treatment of chemical reactions, including how to recognize a chemical change and what a chemical equation means. The properties of aqueous solutions are discussed in detail, and careful attention is given to precipitation and acid–base reactions. In addition, a simple treatment of oxidation–reduction reactions is given. These chapters should provide a solid foundation, relatively early in the course, for reaction-based laboratory experiments.

For instructors who feel that it is desirable to introduce orbitals early in the course, prior to chemical reactions, the chapters on atomic theory and bonding (Chapters 11 and 12) can be covered directly after Chapter 4. Chapter 5 deals solely with nomenclature and can be used wherever it is needed in a particular course.

Development of Problem-Solving Skills

Problem solving is a high priority in chemical education. We all want our students to acquire problem-solving skills. Fostering the development of such skills has been a central focus of the earlier editions of this text, and we have maintained this approach in this edition.

In the first chapters we spend considerable time guiding students to an understanding of the importance of learning chemistry. At the same time, we explain that the complexities that can make chemistry frustrating at times can also provide the opportunity to develop the problem-solving skills that are beneficial in any profession. Learning to think like a chemist is useful to everyone. To emphasize this idea, we apply scientific thinking to some real-life problems in Chapter 1.

One reason chemistry can be challenging for beginning students is that they often do not possess the required mathematical skills. Thus we have paid careful attention to such fundamental mathematical skills as using scientific notation, rounding off to the correct number of significant figures, and rearranging equations to solve for a particular quantity. And we have meticulously followed the rules we have set down so as not to confuse students.

Attitude plays a crucial role in achieving success in problem solving. Students must learn that a systematic, thoughtful approach to problems is better than brute force memorization. We foster this attitude early in the book, using temperature conversions as a vehicle in Chapter 2. Throughout the book we encourage an approach that starts with trying to represent the essence of the problem using symbols and/or diagrams and ends with thinking about whether the answer makes sense. We approach new concepts by carefully working through the material before we give mathematical formulas or overall strategies. We encourage a thoughtful step-by-step approach rather than the premature use of algorithms. Once we have provided the necessary foundation, we highlight important rules and processes in skill development boxes so that students can locate them easily.

Section 8-4: Learning to Solve Problems is written specifically to help students better understand how to think their way through a problem. We discuss how to solve problems in a flexible, creative way based on understanding the fundamental ideas of chemistry and asking and answering key questions. We model this approach in the in-text Examples throughout the text.

Many of the worked examples are followed by Self Check Exercises, which provide additional practice. The Self Check Exercises are keyed to end-of-chapter exercises to offer another opportunity for students to practice a particular problem-solving skill or understand a particular concept.

We have expanded the number of end-of-chapter exercises. As in the first seven editions, the end-of-chapter exercises are arranged in "matched pairs," meaning that both problems in the pair explore similar topics. An Additional Problems section includes further practice in chapter concepts as well as more challenging problems. Cumulative Reviews, which appear after every few chapters, test concepts from the preceding chapter block. Answers for all even-numbered exercises appear in a special section at the end of the student edition.

Handling the Language of Chemistry and Applications

We have gone to great lengths to make this book "student friendly" and have received enthusiastic feedback from students who have used it.

As in the earlier editions, we present a systematic and thorough treatment of chemical nomenclature. Once this framework is established, students can progress through the book comfortably.

Along with chemical reactions, applications form an important part of descriptive chemistry. Because students are interested in chemistry's impact on their lives, we have included many new "Chemistry in Focus" boxes, which describe current applications of chemistry. These special-interest boxes cover such topics as hybrid cars, artificial sweeteners, and Gorilla Glass.

Visual Impact of Chemistry

In response to instructors' requests to include graphic illustrations of chemical reactions, phenomena, and processes, we use a full-color design that enables color to be used functionally, thoughtfully, and consistently to help students understand chemistry and to make the subject more inviting to them. We have included only those photos that illustrate a chemical reaction or phenomenon or that make a connection between chemistry and the real world. Many new photos enhance the eighth edition.

Choices of Coverage

For the convenience of instructors, four versions of the eighth edition are available: two paperback versions and two hardbound versions. *Basic Chemistry,* Eighth Edition, a paperback text, provides basic coverage of chemical concepts and applications through acid–base chemistry and has 16 chapters. *Introductory Chemistry,* Eighth Edition, available in hardcover and paperback, expands the coverage to 19 chapters with the addition of equilibrium, oxidation–reduction reactions and electrochemistry, radioactivity, and nuclear energy. Finally, *Introductory Chemistry: A Foundation,* Eighth Edition, a hardbound text, has 21 chapters, with the final two chapters providing a brief introduction to organic and biological chemistry.

About the Annotated Instructor's Edition

The Annotated Instructor's Edition (AIE) gathers a wealth of teaching support in one convenient package. The AIE contains all 21 chapters (the full contents of *Introductory Chemistry: A Foundation,* Eighth Edition). Annotations in the wraparound margins of the AIE include:

- Answers to Self Check Exercises, at point of use.
- Answers to all end-of-chapter questions and exercises, at point of use.
- Additional Examples with answers to supplemental worked-out Examples in the text.
- Teaching Support Suggestions for specific lecture/instruction methods, activities, and in-class demonstrations to help convey concepts.
- An Overview of the chapter's learning objectives.
- Teaching Tips: Guidelines for highlighting critical information in the chapter.
- Misconceptions: Tips on where students may have trouble or be confused with a topic.
- Demonstrations: Detailed instructions for in-class demonstrations and activities. (These are similar to material in Teaching Support and may be referenced in Teaching Support annotations.)
- Laboratory Experiments: Information on which labs in the Laboratory Manual are relevant to chapter content.
- Background Information: Explanations of conventions used in the text.
- Icons mark material correlations between the main text and the electronic support materials, the Test Bank, and the Laboratory Manual.

- Historical Notes: Biographical or other historical information about science and scientists.

To get access, visit www.cengage.com/chemistry/zumdahl/introchem8e

Supporting Materials

Please visit www.cengage.com/chemistry/zumdahl/introchem8e for information about student and instructor resources for this text.

Acknowledgments

The successful completion of this book is due to the efforts of many talented and dedicated people. Mary Finch, product director, and Lisa Lockwood, senior product team manager, were extremely supportive of the revision. We also are glad to have worked again with Thomas Martin, product manager. Tom never fails to have good ideas, is extremely well organized, and has an eye for detail that is indispensable. We are also appreciative of Teresa Trego, content project manager, who managed the project with her usual grace and professionalism. We also very much appreciate the work of Sharon Donahue, photo researcher, who has a knack for finding the perfect photo.

We greatly appreciate the efforts of Gretchen Adams from the University of Illinois, who managed the revision of the end-of-chapter exercises and problems (including the ChemWork problems) and the solutions manuals. Gretchen also created the interactive examples and worked on the *Annotated Instructor's Guide.* Gretchen is extremely creative and a tireless worker. Thanks as well to John Little, who contributed to the work James Hall has done to *Introductory Chemistry in the Laboratory;* to Nicole Hamm, executive brand manager, who knows the market and works very hard in support of this book; and to Simon Bott, who revised the test banks.

Thanks to others who provided valuable assistance on this revision: Brendan Killion, content coordinator; Karolina Kiwak, product assistant; Lisa Weber, media developer; Janet Del Mundo, market development manager; Maria Epes, art director; Mallory Skinner, production editor (Graphic World); and Michael Burand, who checked the textbook and solutions for accuracy.

Our sincerest appreciation goes to all of the reviewers whose feedback and suggestions contributed to the success of this project.

Angela Bickford
Northwest Missouri State University

Simon Bott
University of Houston

Jabe Breland
St. Petersburg College

Michael Burand
Oregon State University

Frank Calvagna
Rock Valley College

Jing-Yi Chin
Suffolk County Community College

Carl David
University of Connecticut

Cory DiCarlo
Grand Valley State University

Cathie Keenan
Chaffey College

Pamela Kimbrough
Crafton Hills College

Wendy Lewis
Stark State College of Technology

Guillermo Muhlmann
Capital Community College

Lydia Martinez Rivera
University of Texas at San Antonio

Sharadha Sambasivan
Suffolk County Community College

Perminder Sandhu
Bellevue Community College

Lois Schadewald
Normandale Community College

Marie Villarba
Seattle Central Community College

Basic Chemistry

Chemistry: An Introduction

1

CHAPTER 1

Chemistry deals with the natural world. Dr. John Brackenbury/Science Photo Library/Photo Researchers, Inc.

Did you ever see a fireworks display on July Fourth and wonder how it's possible to produce those beautiful, intricate designs in the air? Have you read about dinosaurs—how they ruled the earth for millions of years and then suddenly disappeared? Although the extinction happened 65 million years ago and may seem unimportant, could the same thing happen to us? Have you ever wondered why an ice cube (pure water) floats in a glass of water (also pure water)? Did you know that the "lead" in your pencil is made of the same substance (carbon) as the diamond in an engagement ring? Did you ever wonder how a corn plant or a palm tree grows seemingly by magic, or why leaves turn beautiful colors in autumn? Do you know how the battery works to start your car or run your calculator? Surely some of these things and many others in the world around you have intrigued you. The fact is that we can explain all of these things in convincing ways using the models of chemistry and the related physical and life sciences.

Fireworks are a beautiful illustration of chemistry in action. PhotoDisc/Getty Images

1-1 Chemistry: An Introduction

OBJECTIVE **To understand the importance of learning chemistry.**

Although chemistry might seem to have little to do with dinosaurs, knowledge of chemistry was the tool that enabled paleontologist Luis W. Alvarez and his coworkers from the University of California at Berkeley to "crack the case" of the disappearing dinosaurs. The key was the relatively high level of iridium found in the sediment that represents the boundary between the earth's Cretaceous (K) and Tertiary (T) periods—the time when the dinosaurs disappeared virtually overnight (on the geologic scale). The Berkeley researchers knew that meteorites also have unusually high iridium content (relative to the earth's composition), which led them to suggest that a large meteorite impacted the earth 65 million years ago, causing the climatic changes that wiped out the dinosaurs.

A knowledge of chemistry is useful to almost everyone—chemistry occurs all around us all of the time, and an understanding of chemistry is useful to doctors, lawyers, mechanics, business people, firefighters, and poets among others. Chemistry is important—there is no doubt about that. It lies at the heart of our efforts to produce new materials that make our lives safer and easier, to produce new sources of energy that are abundant and nonpolluting, and to understand and control the many diseases that threaten us and our food supplies. Even if your future career does not require the daily use of chemical principles, your life will be greatly influenced by chemistry.

A strong case can be made that the use of chemistry has greatly enriched all of our lives. However, it is important to understand that the principles of chemistry are inherently neither good nor bad—it's what we do with this knowledge that really matters. Although humans are clever, resourceful, and concerned about others, they also can be greedy, selfish, and ignorant. In addition, we tend to be shortsighted; we concentrate too much on the present and do not think enough about the long-range implications of our actions. This type of thinking has already caused us a great deal of trouble—severe environmental damage has occurred on many fronts. We cannot place all the responsibility on the chemical companies, because everyone has contributed to these prob-

Bart Eklund checking air quality at a hazardous waste site.

lems. However, it is less important to lay blame than to figure out how to solve these problems. An important part of the answer must rely on chemistry.

One of the "hottest" fields in the chemical sciences is environmental chemistry—an area that involves studying our environmental ills and finding creative ways to address them. For example, meet Bart Eklund, who works in the atmospheric chemistry field for Radian Corporation in Austin, Texas. Bart's interest in a career in environmental science was fostered by two environmental chemistry courses and two ecology courses he took as an undergraduate. His original plan to gain several years of industrial experience and then to return to school for a graduate degree changed when he discovered that professional advancement with a B.S. degree was possible in the environmental research field. The multidisciplinary nature of environmental problems has allowed Bart to pursue his interest in several fields at the same time. You might say that he specializes in being a generalist.

The environmental consulting field appeals to Bart for a number of reasons: the chance to define and solve a number of research problems; the simultaneous work on a number of diverse projects; the mix of desk, field, and laboratory work; the travel; and the opportunity to perform rewarding work that has a positive effect on people's lives.

Among his career highlights are the following:

- Spending a winter month doing air sampling in the Grand Tetons, where he also met his wife and learned to ski;
- Driving sampling pipes by hand into the rocky ground of Death Valley Monument in California;
- Working regularly with experts in their fields and with people who enjoy what they do;
- Doing vigorous work in 100 °F weather while wearing a rubberized suit, double gloves, and a respirator; and
- Getting to work in and see Alaska, Yosemite Park, Niagara Falls, Hong Kong, the People's Republic of China, Mesa Verde, New York City, and dozens of other interesting places.

Bart Eklund's career demonstrates how chemists are helping to solve our environmental problems. It is how we use our chemical knowledge that makes all the difference.

An example that shows how technical knowledge can be a "double-edged sword" is the case of chlorofluorocarbons (CFCs). When the compound CCl_2F_2 (originally called Freon-12) was first synthesized, it was hailed as a near-miracle substance. Because of its noncorrosive nature and its unusual ability to resist decomposition, Freon-12 was rapidly applied in refrigeration and air-conditioning systems, cleaning applications, the blowing of foams used for insulation and packing materials, and many other ways. For years everything seemed fine—the CFCs actually replaced more dangerous materials, such as the ammonia formerly used in refrigeration systems. The CFCs were definitely viewed as "good guys." But then a problem was discovered—the ozone in the upper atmosphere that protects us from the high-energy radiation of the sun began to decline. What was happening to cause the destruction of the vital ozone?

Much to everyone's amazement, the culprits turned out to be the seemingly beneficial CFCs. Inevitably, large quantities of CFCs had leaked into the atmosphere but nobody was very worried about this development because these compounds seemed totally benign. In fact, the great stability of the CFCs (a tremendous advantage for their various applications) was in the end a great disadvantage when they were released into the environment. Professor F. S. Rowland and his colleagues at the University of California at Irvine demonstrated that the CFCs eventually drifted to high altitudes in the atmosphere, where the energy of the sun stripped off chlorine atoms. These chlorine atoms in turn promoted the decomposition of the ozone in the upper

Dr. Ruth—Cotton Hero

Dr. Ruth Rogan Benerito may have saved the cotton industry in the United States. In the 1960s, synthetic fibers posed a serious competitive threat to cotton, primarily because of wrinkling. Synthetic fibers such as polyester can be formulated to be highly resistant to wrinkles both in the laundering process and in wearing. On the other hand, 1960s' cotton fabrics wrinkled easily—white cotton shirts had to be ironed to look good. This requirement put cotton at a serious disadvantage and endangered an industry very important to the economic health of the South.

AP Photo/Eric Risberg

Ruth Benerito, the inventor of easy-care cotton.

During the 1960s Ruth Benerito worked as a scientist for the Department of Agriculture, where she was instrumental in developing the chemical treatment of cotton to make it wrinkle resistant. In so doing she enabled cotton to remain a preeminent fiber in the market—a place it continues to hold today. She was honored with the Lemelson–MIT Lifetime Achievement Award for Inventions in 2002 when she was 86 years old.

Dr. Benerito, who holds 55 patents, including the one for wrinkle-free cotton awarded in 1969, began her career when women were not expected to enter scientific fields. However, her mother, who was an artist, adamantly encouraged her to be anything she wanted to be.

Dr. Benerito graduated from high school at age 14 and attended Newcomb College, the women's college associated with Tulane University. She majored in chemistry with minors in physics and math. At that time she was one of only two women allowed to take the physical chemistry course at Tulane. She earned her B.S. degree in 1935 at age 19 and subsequently earned a master's degree at Tulane and a Ph.D. at the University of Chicago.

In 1953 Dr. Benerito began working in the Agriculture Department's Southern Regional Research Center in New Orleans, where she mainly worked on cotton and cotton-related products. She also invented a special method for intravenous feeding in long-term medical patients.

Since her retirement in 1986, she has continued to tutor science students to keep busy. Everyone who knows Dr. Benerito describes her as a class act.

atmosphere. (We will discuss this in more detail in Chapter 13.) Thus a substance that possessed many advantages in earth-bound applications turned against us in the atmosphere. Who could have guessed it would turn out this way?

The good news is that the U.S. chemical industry is leading the way to find environmentally safe alternatives to CFCs, and the levels of CFCs in the atmosphere are already dropping.

The saga of the CFCs demonstrates that we can respond relatively quickly to a serious environmental problem if we decide to do so. Also, it is important to understand that chemical manufacturers have a new attitude about the environment—they are now among the leaders in finding ways to address our environmental ills. The industries that apply the chemical sciences are now determined to be part of the solution rather than part of the problem.

As you can see, learning chemistry is both interesting and important. A chemistry course can do more than simply help you learn the principles of chemistry, however. A major by-product of your study of chemistry is that you will become a better problem solver. One reason chemistry has the reputation of being "tough" is that it often deals with rather complicated systems that require some effort to figure out. Although this might at first seem like a disadvantage, you can turn it to your advantage if you

© Cengage Learning

A chemist in the laboratory.

have the right attitude. Recruiters for companies of all types maintain that one of the first things they look for in a prospective employee is the ability to solve problems. We will spend a good deal of time solving various types of problems in this book by using a systematic, logical approach that will serve you well in solving any kind of problem in any field. Keep this broader goal in mind as you learn to solve the specific problems connected with chemistry.

Although learning chemistry is often not easy, it's never impossible. In fact, anyone who is interested, patient, and willing to work can learn the fundamentals of chemistry. In this book we will try very hard to help you understand what chemistry is and how it works and to point out how chemistry applies to the things going on in your life.

Our sincere hope is that this text will motivate you to learn chemistry, make its concepts understandable to you, and demonstrate how interesting and vital the study of chemistry is.

1-2 What Is Chemistry?

OBJECTIVE To define chemistry.

Chemical and physical changes will be discussed in Chapter 3.

Chemistry can be defined as *the science that deals with the materials of the universe and the changes that these materials undergo.* Chemists are involved in activities as diverse as examining the fundamental particles of matter, looking for molecules in space, synthesizing and formulating new materials of all types, using bacteria to produce such chemicals as insulin, and inventing new diagnostic methods for early detection of disease.

Chemistry is often called the central science—and with good reason. Most of the phenomena that occur in the world around us involve chemical changes, changes where one or more substances become different substances. Here are some examples of chemical changes:

Wood burns in air, forming water, carbon dioxide, and other substances.

A plant grows by assembling simple substances into more complex substances.

The steel in a car rusts.

Eggs, flour, sugar, and baking powder are mixed and baked to yield a cake.

The definition of the term *chemistry* is learned and stored in the brain.

Emissions from a power plant lead to the formation of acid rain.

The launch of the space shuttle gives clear indications that chemical reactions are occurring. NASA

As we proceed, you will see how the concepts of chemistry allow us to understand the nature of these and other changes and thus help us manipulate natural materials to our benefit.

1-3 Solving Problems Using a Scientific Approach

OBJECTIVE To understand scientific thinking.

One of the most important things we do in everyday life is solve problems. In fact, most of the decisions you make each day can be described as solving problems.

It's 8:30 a.m. on Friday. Which is the best way to drive to school to avoid traffic congestion?

CHEMISTRY IN FOCUS

A Mystifying Problem

To illustrate how science helps us solve problems, consider a true story about two people, David and Susan (not their real names). David and Susan were healthy 40-year-olds living in California, where David was serving in the Air Force. Gradually Susan became quite ill, showing flu-like symptoms including nausea and severe muscle pains. Even her personality changed: she became uncharacteristically grumpy. She seemed like a totally different person from the healthy, happy woman of a few months earlier. Following her doctor's orders, she rested and drank a lot of fluids, including large quantities of coffee and orange juice from her favorite mug, part of a 200-piece set of pottery dishes recently purchased in Italy. However, she just got sicker, developing extreme abdominal cramps and severe anemia.

During this time David also became ill and exhibited symptoms much like Susan's: weight loss, excruciating pain in his back and arms, and uncharacteristic fits of temper. The disease became so debilitating that he retired early from the Air Force and the couple moved to Seattle. For a short time their health improved, but after they unpacked all their belongings (including those pottery dishes), their health began to deteriorate again. Susan's body became so sensitive that she could not tolerate the weight of a blanket. She was near death. What was wrong? The doctors didn't know, but one suggested she might have porphyria, a rare blood disease.

Desperate, David began to search the medical literature himself. One day while he was reading about porphyria, a phrase jumped off the page: "Lead poisoning can sometimes be confused with porphyria." Could the problem be lead poisoning?

We have described a very serious problem with life-or-death implications. What should David do next? Overlooking for a moment the obvious response of calling the couple's doctor immediately to discuss the possibility of lead poisoning, could David solve the problem via scientific thinking? Let's use the three steps described in Section 1.3 to attack the problem one part at a time. This is important: usually we solve complex problems by breaking them down into manageable parts. We can then assemble the solution to the overall problem from the answers we have found "piecemeal."

In this case there are many parts to the overall problem:

What is the disease?

Where is it coming from?

Can it be cured?

Let's attack "What is the disease?" first.

Observation: David and Susan are ill with the symptoms described. Is the disease lead poisoning?

Hypothesis: The disease is lead poisoning.

Experiment: If the disease is lead poisoning, the symptoms must match those known to characterize lead poisoning. Look up the symptoms of lead poisoning. David did this and found that they matched the couple's symptoms almost exactly.

This discovery points to lead poisoning as the source of their problem, but David needed more evidence.

Observation: Lead poisoning results from high levels of lead in the bloodstream.

Hypothesis: The couple have high levels of lead in their blood.

You have two tests on Monday. Should you divide your study time equally or allot more time to one than to the other?

Your car stalls at a busy intersection and your little brother is with you. What should you do next?

These are everyday problems of the type we all face. What process do we use to solve them? You may not have thought about it before, but there are several steps that almost everyone uses to solve problems:

1. Recognize the problem and state it clearly. Some information becomes known, or something happens that requires action. In science we call this step *making an observation.*

Experiment: Perform a blood analysis. Susan arranged for such an analysis, and the results showed high lead levels for both David and Susan.

This confirms that lead poisoning is probably the cause of the trouble, but the overall problem is still not solved. David and Susan are likely to die unless they find out where the lead is coming from.

Observation: There is lead in the couple's blood.

Hypothesis: The lead is in their food or drink when they buy it.

Experiment: Find out whether anyone else who shopped at the same store was getting sick (no one was). Also note that moving to a new area did not solve the problem.

Observation: The food they buy is free of lead.

Hypothesis: The dishes they use are the source of the lead poisoning.

Experiment: Find out whether their dishes contain lead. David and Susan learned that lead compounds are often used to put a shiny finish on pottery objects. And laboratory analysis of their Italian pottery dishes showed that lead was present in the glaze.

Observation: Lead is present in their dishes, so the dishes are a possible source of their lead poisoning.

Hypothesis: The lead is leaching into their food.

Experiment: Place a beverage, such as orange juice, in one of the cups and then analyze the beverage for lead. The results showed high levels of lead in drinks that had been in contact with the pottery cups.

After many applications of the scientific method, the problem is solved. We can summarize the answer to the problem (David and Susan's illness) as follows: the Italian pottery they used for everyday dishes contained a lead glaze that contaminated their food and drink with lead. This lead accumulated in their bodies to the point where it interfered seriously with normal functions and produced severe symptoms. This overall explanation, which summarizes the hypotheses that agree with the experimental results, is called a *theory* in science. This explanation accounts for the results of all the experiments performed.*

Photo by Ken O'Donoghue © Cengage Learning

Italian pottery.

We could continue to use the scientific method to study other aspects of this problem, such as

What types of food or drink leach the most lead from the dishes?

Do all pottery dishes with lead glazes produce lead poisoning?

As we answer questions using the scientific method, other questions naturally arise. By repeating the three steps over and over, we can come to understand a given phenomenon thoroughly.

*"David" and "Susan" recovered from their lead poisoning and are now publicizing the dangers of using lead-glazed pottery. This happy outcome is the answer to the third part of their overall problem, "Can the disease be cured?" They simply stopped eating from that pottery!

2. Propose *possible* solutions to the problem or *possible* explanations for the observation. In scientific language, suggesting such a possibility is called *formulating a hypothesis.*
3. Decide which of the solutions is the best or decide whether the explanation proposed is reasonable. To do this we search our memory for any pertinent information or we seek new information. In science we call searching for new information *performing an experiment.*

As we will discover in the next section, scientists use these same procedures to study what happens in the world around us. The important point here is that scientific thinking can help you in all parts of your life. It's worthwhile to learn how to think scientifically—whether you want to be a scientist, an auto mechanic, a doctor, a politician, or a poet!

1-4 The Scientific Method

OBJECTIVE **To describe the method scientists use to study nature.**

In the last section we began to see how the methods of science are used to solve problems. In this section we will further examine this approach.

Science is a framework for gaining and organizing knowledge. Science is not simply a set of facts but also a plan of action—a *procedure* for processing and understanding certain types of information. Although scientific thinking is useful in all aspects of life, in this text we will use it to understand how the natural world operates. The process that lies at the center of scientific inquiry is called the **scientific method.** As we saw in the previous section, it consists of the following steps:

Steps in the Scientific Method

1. *State the problem and collect data (make observations).* Observations may be *qualitative* (the sky is blue; water is a liquid) or *quantitative* (water boils at 100 °C; a certain chemistry book weighs 4.5 pounds). A qualitative observation does not involve a number. A quantitative observation is called a **measurement** and does involve a number (and a unit, such as pounds or inches). We will discuss measurements in detail in Chapter 2.
2. *Formulate hypotheses.* A hypothesis is a *possible* explanation for the observation.
3. *Perform experiments.* An experiment is something we do to test the hypothesis. We gather new information that allows us to decide whether the hypothesis is supported by the new information we have learned from the experiment. Experiments always produce new observations, and this brings us back to the beginning of the process again.

To explain the behavior of a given part of nature, we repeat these steps many times. Gradually we accumulate the knowledge necessary to understand what is going on.

Once we have a set of hypotheses that agrees with our various observations, we assemble them into a theory that is often called a *model.* A **theory** (model) is a set of tested hypotheses that gives an overall explanation of some part of nature (Fig. 1.1).

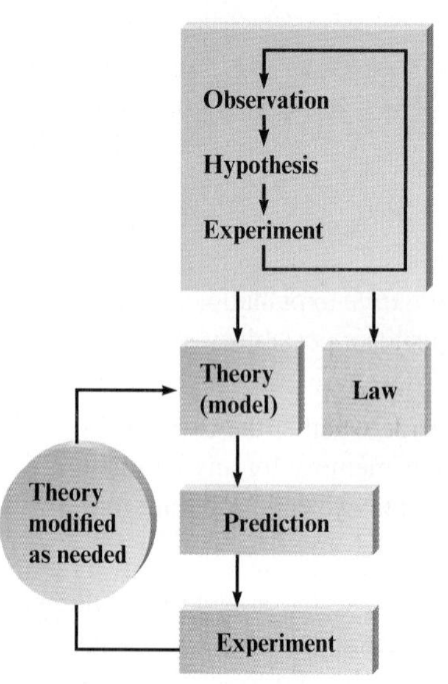

Figure 1.1 ▸ The various parts of the scientific method.

It is important to distinguish between observations and theories. An observation is something that is witnessed and can be recorded. A theory is an *interpretation*—a possible explanation of *why* nature behaves in a particular way. Theories inevitably change as more information becomes available. For example, the motions of the sun and stars have remained virtually the same over the thousands of years during which humans have been observing them, but our explanations—our theories—have changed greatly since ancient times.

The point is that we don't stop asking questions just because we have devised a theory that seems to account satisfactorily for some aspect of natural behavior. We continue doing experiments to refine our theories. We generally do this by using the theory to make a prediction and then doing an experiment (making a new observation) to see whether the results bear out this prediction.

Always remember that theories (models) are human inventions. They represent our attempts to explain observed natural behavior in terms of our human experiences. We must continue to do experiments and refine our theories to be consistent with new knowledge if we hope to approach a more nearly complete understanding of nature.

As we observe nature, we often see that the same observation applies to many different systems. For example, studies of innumerable chemical changes have shown that the total mass of the materials involved is the same before and after the change. We often formulate such generally observed behavior into a statement called a

Critical Thinking

What if everyone in the government used the scientific method to analyze and solve society's problems, and politics were never involved in the solutions? How would this be different from the present situation, and would it be better or worse?

natural law. The observation that the total mass of materials is not affected by a chemical change in those materials is called the law of conservation of mass.

You must recognize the difference between a law and a theory. A law is a summary of observed (measurable) behavior, whereas a theory is an explanation of behavior. *A law tells what happens; a theory (model) is our attempt to explain why it happens.*

In this section, we have described the scientific method (which is summarized in Fig. 1.1) as it might ideally be applied. However, it is important to remember that science does not always progress smoothly and efficiently. Scientists are human. They have prejudices; they misinterpret data; they can become emotionally attached to their theories and thus lose objectivity; and they play politics. Science is affected by profit motives, budgets, fads, wars, and religious beliefs. Galileo, for example, was forced to recant his astronomical observations in the face of strong religious resistance. Lavoisier, the father of modern chemistry, was beheaded because of his political affiliations. And great progress in the chemistry of nitrogen fertilizers resulted from the desire to produce explosives to fight wars. The progress of science is often slowed more by the frailties of humans and their institutions than by the limitations of scientific measuring devices. The scientific method is only as effective as the humans using it. It does not automatically lead to progress.

1-5 Learning Chemistry

OBJECTIVE To develop successful strategies for learning chemistry.

Chemistry courses have a universal reputation for being difficult. There are some good reasons for this. For one thing, the language of chemistry is unfamiliar in the beginning; many terms and definitions need to be memorized. As with any language, *you must know the vocabulary* before you can communicate effectively. We will try to help you by pointing out those things that need to be memorized.

But memorization is only the beginning. Don't stop there or your experience with chemistry will be frustrating. Be willing to do some thinking, and learn to trust yourself to figure things out. To solve a typical chemistry problem, you must sort through the given information and decide what is really crucial.

It is important to realize that chemical systems tend to be complicated—there are typically many components—and we must make approximations in describing them. Therefore, trial and error play a major role in solving chemical problems. In tackling a complicated system, a practicing chemist really does not expect to be right the first time he or she analyzes the problem. The usual practice is to make several simplifying assumptions and then give it a try. If the answer obtained doesn't make sense, the chemist adjusts the assumptions, using feedback from the first attempt, and tries again. The point is this: in dealing with chemical systems, do not expect to understand immediately everything that is going on. In fact, it is typical (even for an experienced chemist) *not* to understand at first. Make an attempt to solve the problem and then analyze the feedback. *It is no disaster to make a mistake as long as you learn from it.*

The only way to develop your confidence as a problem solver is to practice solving problems. To help you, this book contains examples worked out in detail. Follow these through carefully, making sure you understand each step. These examples are usually followed by a similar exercise (called a self check exercise) that you should try on your own (detailed solutions of the self check exercises are given at the end of each chapter). Use the self check exercises to test whether you are understanding the material as you go along.

There are questions and problems at the end of each chapter. The questions review the basic concepts of the chapter and give you an opportunity to check whether you properly understand the vocabulary introduced. Some of the problems are really just exercises that are very similar to examples done in the chapter. If you understand the material in the chapter, you should be able to do these exercises in a straightforward

Chemistry: An Important Component of Your Education

What is the purpose of education? Because you are spending considerable time, energy, and money to pursue an education, this is an important question.

Some people seem to equate education with the storage of facts in the brain. These people apparently believe that education simply means memorizing the answers to all of life's present and future problems. Although this is clearly unreasonable, many students seem to behave as though this were their guiding principle. These students want to memorize lists of facts and to reproduce them on tests. They regard as unfair any exam questions that require some original thought or some processing of information. Indeed, it might be tempting to reduce education to a simple filling up with facts, because that approach can produce short-term satisfaction for both student and teacher. And of course, storing facts in the brain *is* important. You cannot function without knowing that red means stop, electricity is hazardous, ice is slippery, and so on.

Blend Images/Superstock

Students pondering the structure of a molecule.

However, mere recall of abstract information, without the ability to process it, makes you little better than a talking encyclopedia. Former students always seem to bring the same message when they return to campus. The characteristics that are most important to their success are a knowledge of the fundamentals of their fields, the ability to recognize and solve problems, and the ability to communicate effectively. They also emphasize the importance of a high level of motivation.

How does studying chemistry help you achieve these characteristics? The fact that chemical systems are complicated is really a blessing, though one that is well disguised. Studying chemistry will not by itself make you a good problem solver, but it can help you develop a positive, aggressive attitude toward problem solving and can help boost your confidence. Learning to "think like a chemist" can be valuable to anyone in any field. In fact, the chemical industry is heavily populated at all levels and in all areas by chemists and chemical engineers. People who were trained as chemical professionals often excel not only in chemical research and production but also in the areas of personnel, marketing, sales, development, finance, and management. The point is that much of what you learn in this course can be applied to any field of endeavor. So be careful not to take too narrow a view of this course. Try to look beyond short-term frustration to long-term benefits. It may not be easy to learn to be a good problem solver, but it's well worth the effort.

way. Other problems require more creativity. These contain a knowledge gap—some unfamiliar territory that you must cross—and call for thought and patience on your part. For this course to be really useful to you, it is important to go beyond the questions and exercises. Life offers us many exercises, routine events that we deal with rather automatically, but the real challenges in life are true problems. This course can help you become a more creative problem solver.

As you do homework, be sure to use the problems correctly. If you cannot do a particular problem, do not immediately look at the solution. Review the relevant material in the text and then try the problem again. Don't be afraid to struggle with a problem. Looking at the solution as soon as you get stuck short-circuits the learning process.

Learning chemistry takes time. Use all the resources available to you and study on a regular basis. Don't expect too much of yourself too soon. You may not understand everything at first, and you may not be able to do many of the problems the first time you try them. This is normal. It doesn't mean you can't learn chemistry. Just remember to keep working and to keep learning from your mistakes, and you will make steady progress.

CHAPTER 1 REVIEW

F directs you to the *Chemistry in Focus* feature in the chapter

Key Terms

chemistry (1-2)
scientific method (1-4)
measurement (1-4)
theory (1-4)
natural law (1-4)

For Review

- Chemistry is important to everyone because chemistry occurs all around us in our daily lives.
- Chemistry is the science that deals with the materials of the universe and the changes that these materials undergo.
- Scientific thinking helps us solve all types of problems that we confront in our lives.
- Scientific thinking involves observations that enable us to clearly define both a problem and the construction and evaluation of possible explanations or solutions to the problem.
- The scientific method is a procedure for processing the information that flows from the world around us in which we
  - Make observations.
  - Formulate hypotheses.
  - Perform experiments.

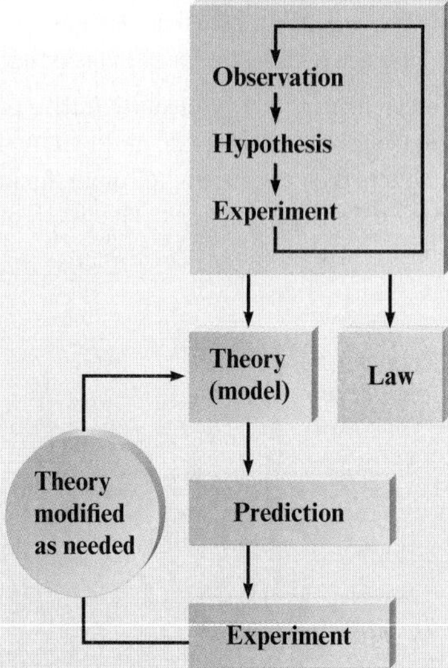

- Models represent our attempt to understand the world around us.
  - Models are not the same as "reality."
  - Elementary models are based on the properties of atoms and molecules.
- Understanding chemistry takes hard work and patience.
- As you learn chemistry, you should be able to understand, explain, and predict phenomena in the macroscopic world by using models based in the microscopic world.
- Understanding is different from memorizing.
- It is acceptable to make mistakes as long as you learn from them.

All even-numbered Questions and Problems have answers in the back of this book and solutions in the *Student Solutions Guide.*

Active Learning Questions

These questions are designed to be considered by groups of students in class. Often these questions work well for introducing a particular topic in class.

1. Discuss how a hypothesis can become a theory. Can a theory become a law? Explain.
2. Make five qualitative and five quantitative observations about the room in which you now sit.
3. List as many chemical reactions you can think of that are part of your everyday life. Explain.
4. Differentiate between a "theory" and a "scientific theory."
5. Describe three situations when you used the scientific method (outside of school) in the past month.
6. Scientific models do not describe reality. They are simplifications and therefore incorrect at some level. So why are models useful?
7. Theories should inspire questions. Discuss a scientific theory you know and the questions it brings up.
8. Describe how you would set up an experiment to test the relationship between completion of assigned homework and the final grade you receive in the course.
9. If all scientists use the scientific method to try to arrive at a better understanding of the world, why do so many debates arise among scientists?
10. As stated in the text, there is no one scientific method. However, making observations, formulating hypotheses, and performing experiments are generally components of "doing science." Read the following passage, and list any observations, hypotheses, and experiments. Support your answer.

    Joyce and Frank are eating raisins and drinking ginger ale. Frank accidentally drops a raisin into his ginger ale. They both notice that the raisin falls to the bottom of the glass. Soon, the raisin rises to the surface of the ginger ale, and then sinks. Within a couple of minutes, it rises and sinks again. Joyce asks, "I wonder why that happened?" Frank says, "I don't know, but let's see if it works in water." Joyce fills a glass with water and drops the raisin into the glass. After a few minutes, Frank says, "No, it doesn't go up and down in the water." Joyce closely observes the raisins in the two glasses and states, "Look, there are bubbles on the raisins in the ginger ale but not on the raisins in the water." Frank says, "It must be the bubbles that make the raisin rise." Joyce asks, "OK, but then why do they sink again?"

11. In Section 1-3 the statement is made that it is worthwhile for scientists, auto mechanics, doctors, politicians, and poets to take a scientific approach to their professions. Discuss how each of these people could use a scientific approach in his or her profession.

12. As part of a science project, you study traffic patterns in your city at an intersection in the middle of downtown. You set up a device that counts the cars passing through this intersection for a 24-hour period during a weekday. The graph of hourly traffic looks like this.

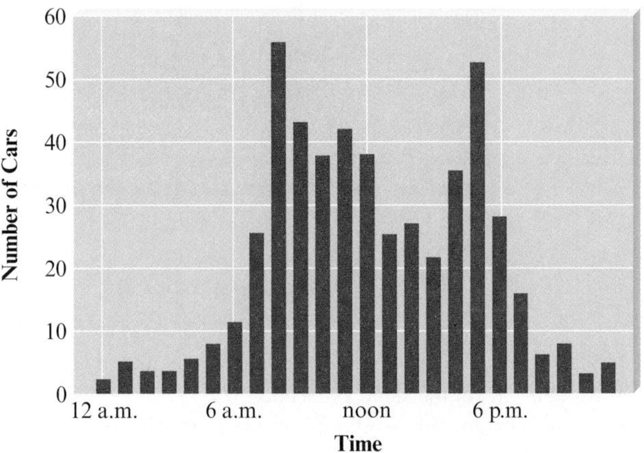

a. At what time(s) does the highest number of cars pass through the intersection?
b. At what time(s) does the lowest number of cars pass through the intersection?
c. Briefly describe the trend in numbers of cars over the course of the day.
d. Provide a hypothesis explaining the trend in numbers of cars over the course of the day.
e. Provide a possible experiment that could test your hypothesis.

13. Confronted with the box shown in the diagram, you wish to discover something about its internal workings. You have no tools and cannot open the box. You pull on rope B, and it moves rather freely. When you pull on rope A, rope C appears to be pulled slightly into the box. When you pull on rope C, rope A almost disappears into the box.*

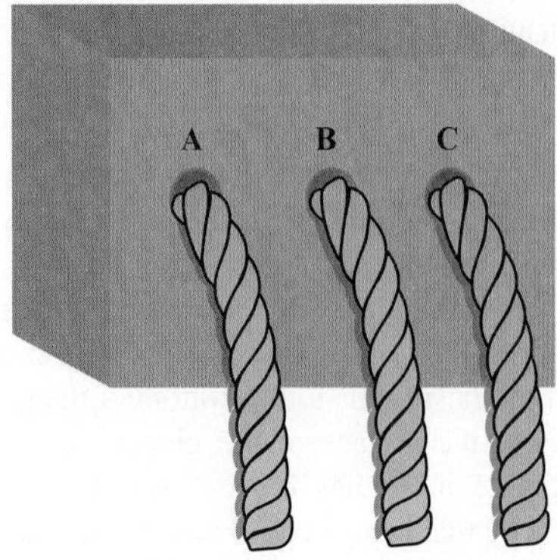

*From Yoder, Suydam, and Snavely, *Chemistry* (New York: Harcourt Brace Jovanovich, 1975), pp. 9–11.

a. Based on these observations, construct a model for the interior mechanism of the box.
b. What further experiments could you do to refine your model?

Questions and Problems

1-1 Chemistry: An Introduction

Questions

1. Chemistry is an intimidating academic subject for many students. You are not alone if you are afraid of not doing well in this course! Why do you suppose the study of chemistry is so intimidating for many students? What about having to take a chemistry course bothers you? Make a list of your concerns and bring them to class for discussion with your fellow students and your instructor.

2. The first paragraphs in this chapter ask you if you have ever wondered how and why various things in our everyday lives happen the way they do. For your next class meeting, make a list of five similar chemistry-related things for discussion with your instructor and the other students in your class.

3. This section presents several ways our day-to-day lives have been enriched by chemistry. List three materials or processes involving chemistry that you feel have contributed to such an enrichment and explain your choices.

F 4. The "Chemistry in Focus" segment titled *Dr. Ruth—Cotton Hero* discusses the enormous contribution of Dr. Ruth Rogan Benerito to the survival of the cotton fabric industry in the United States. In the discussion, it was mentioned that Dr. Benerito became a chemist when women were not expected to be interested in, or good at, scientific subjects. Has this attitude changed? Among your own friends, approximately how many of your female friends are studying a science? How many plan to pursue a career in science? Discuss.

1-2 What Is Chemistry?

Questions

5. This textbook provides a specific definition of chemistry: the study of the materials of which the universe is made and the transformations that these materials undergo. Obviously, such a general definition has to be very broad and nonspecific. From your point of view at this time, how would *you* define chemistry? In your mind, what are "chemicals"? What do "chemists" do?

6. We use chemical reactions in our everyday lives, too, not just in the science laboratory. Give at least five examples of chemical transformations that you use in your daily activities. Indicate what the "chemical" is in each of your examples and how you recognize that a chemical change has taken place.

1-3 Solving Problems Using a Scientific Approach

Questions

F 7. Read the "Chemistry in Focus" segment *A Mystifying Problem* and discuss how David and Susan analyzed the situation, arriving at the theory that the lead glaze on the pottery was responsible for their symptoms.

All even-numbered Questions and Problems have answers in the back of this book and solutions in the *Student Solutions Guide*.

8. Being a scientist is very much like being a detective. Detectives such as Sherlock Holmes or Miss Marple perform a very systematic analysis of a crime to solve it, much like a scientist does when addressing a scientific investigation. What are the steps that scientists (or detectives) use to solve problems?

1-4 The Scientific Method

Questions

9. A ________ is a summary of observed behavior, whereas a ________ is an explanation of behavior.

10. Observations may be either qualitative or quantitative. Quantitative observations are usually referred to as *measurements.* List five examples of *qualitative observations* you might make around your home or school. List five examples of *measurements* you might make in everyday life.

11. True or false? Once we have a set of hypotheses that agrees with our various observations, we assemble them into a theory that is often called a model.

12. True or false? If a theory is disproven, then all of the observations that support that theory are also disproven. Explain.

13. Although, in general, science has advanced our standard of living tremendously, there is sometimes a "dark side" to science. Give an example of the misuse of science and explain how this has had an adverse effect on our lives.

14. Discuss several political, social, or personal considerations that might affect a scientist's evaluation of a theory. Give examples of how such external forces have influenced scientists in the past. Discuss methods by which such bias might be excluded from future scientific investigations.

1-5 Learning Chemistry

Questions

15. Although reviewing your lecture notes and reading your textbook are important, why does the study of chemistry depend so much on problem solving? Can you learn to solve problems yourself just by looking at the solved examples in your textbook or study guide? Discuss.

16. Why is the ability to solve problems important in the study of chemistry? Why is it that the *method* used to attack a problem is as important as the answer to the problem itself?

17. Students approaching the study of chemistry must learn certain basic facts (such as the names and symbols of the most common elements), but it is much more important that they learn to think critically and to go beyond the specific examples discussed in class or in the textbook. Explain how learning to do this might be helpful in any career, even one far removed from chemistry.

F 18. The "Chemistry in Focus" segment *Chemistry: An Important Component of Your Education* discusses how studying chemistry can be beneficial not only in your chemistry courses but in your studies in general. What are some characteristics of a good student, and how does studying chemistry help achieve these characteristics?

All even-numbered Questions and Problems have answers in the back of this book and solutions in the *Student Solutions Guide.*

2 Measurements and Calculations

CHAPTER 2

An enlarged view of a graduated cylinder Masterfile

As we pointed out in Chapter 1, making observations is a key part of the scientific process. Sometimes observations are *qualitative* ("the substance is a yellow solid") and sometimes they are *quantitative* ("the substance weighs 4.3 grams"). A quantitative observation is called a measurement. Measurements are very important in our daily lives. For example, we pay for gasoline by the gallon, so the gas pump must accurately measure the gas delivered to our fuel tank. The efficiency of the modern automobile engine depends on various measurements, including the amount of oxygen in the exhaust gases, the temperature of the coolant, and the pressure of the lubricating oil. In addition, cars with traction control systems have devices to measure and compare the rates of rotation of all four wheels. As we will see in the "Chemistry in Focus" discussion in this chapter, measuring devices have become very sophisticated in dealing with our fast-moving and complicated society.

As we will discuss in this chapter, a measurement always consists of two parts: a number and a unit. Both parts are necessary to make the measurement meaningful. For example, suppose a friend tells you that she saw a bug 5 long. This statement is meaningless as it stands. Five what? If it's 5 millimeters, the bug is quite small. If it's 5 centimeters, the bug is quite large. If it's 5 meters, run for cover!

The point is that for a measurement to be meaningful, it must consist of both a number and a unit that tells us the scale being used.

In this chapter we will consider the characteristics of measurements and the calculations that involve measurements.

A gas pump measures the amount of gasoline delivered. Shannon Fagan/The Image Bank/Getty Images

2-1 Scientific Notation

OBJECTIVE **To show how very large or very small numbers can be expressed as the product of a number between 1 and 10 and a power of 10.**

A measurement must always consist of a number *and* a unit.

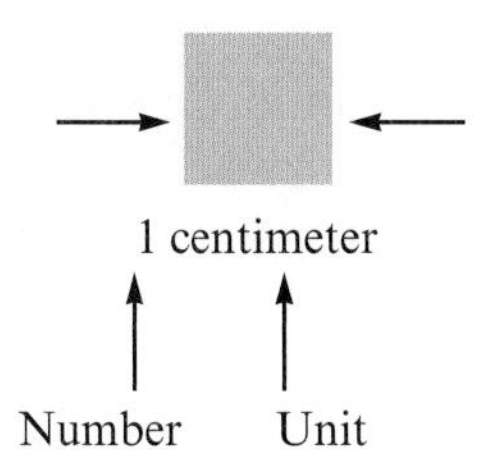

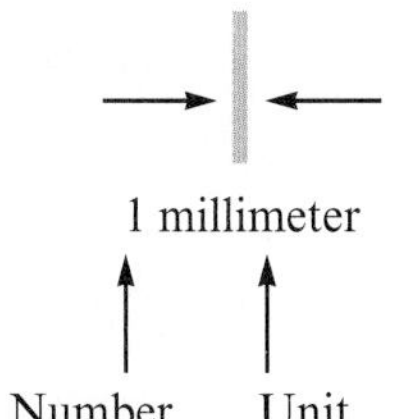

The numbers associated with scientific measurements are often very large or very small. For example, the distance from the earth to the sun is approximately 93,000,000 (93 million) miles. Written out, this number is rather bulky. Scientific notation is a method for making very large or very small numbers more compact and easier to write.

To see how this is done, consider the number 125, which can be written as the product

$$125 = 1.25 \times 100$$

Because $100 = 10 \times 10 = 10^2$, we can write

$$125 = 1.25 \times 100 = 1.25 \times 10^2$$

Similarly, the number 1700 can be written

$$1700 = 1.7 \times 1000$$

and because $1000 = 10 \times 10 \times 10 = 10^3$, we can write

$$1700 = 1.7 \times 1000 = 1.7 \times 10^3$$

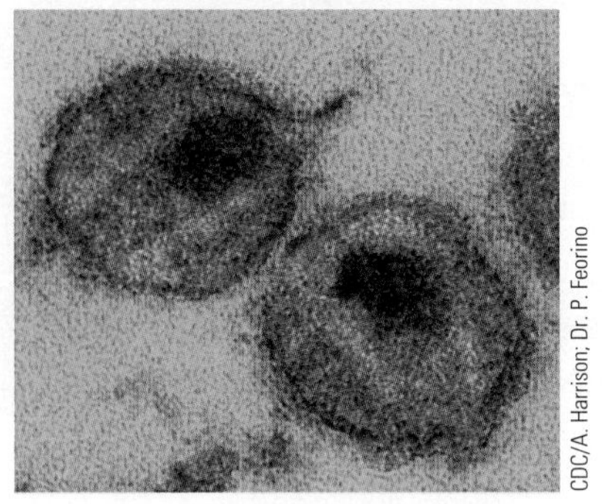
CDC/A. Harrison; Dr. P. Feorino

When describing very small distances, such as the diameter of these HIV virions, it is convenient to use scientific notation.

Scientific notation simply expresses a number as *a product of a number between 1 and 10 and the appropriate power of 10.* For example, the number 93,000,000 can be expressed as

$$93{,}000{,}000 = 9.3 \times 10{,}000{,}000 = 9.3 \times 10^7$$

Number between 1 and 10 — Appropriate power of 10 ($10{,}000{,}000 = 10^7$)

The easiest way to determine the appropriate power of 10 for scientific notation is to start with the number being represented and count the number of places the decimal point must be moved to obtain a number between 1 and 10. For example, for the number

9 3 0 0 0 0 0 0
7 6 5 4 3 2 1

Math skill builder
Keep one digit to the left of the decimal point.

we must move the decimal point seven places to the left to get 9.3 (a number between 1 and 10). To compensate for every move of the decimal point to the left, we must multiply by 10. That is, each time we move the decimal point to the left, we make the number smaller by one power of 10. So for each move of the decimal point to the left, we must multiply by 10 to restore the number to its original magnitude. Thus moving the decimal point seven places to the left means we must multiply 9.3 by 10 seven times, which equals 10^7:

$$93{,}000{,}000 = 9.3 \times 10^7$$

We moved the decimal point seven places to the left, so we need 10^7 to compensate.

Math skill builder
Moving the decimal point to the left requires a positive exponent.

Remember: whenever the decimal point is moved to the *left,* the exponent of 10 is *positive.*

We can represent numbers smaller than 1 by using the same convention, but in this case the power of 10 is negative. For example, for the number 0.010 we must move the decimal point two places to the right to obtain a number between 1 and 10:

0 . 0 1 0
1 2

Math skill builder
Moving the decimal point to the right requires a negative exponent.

This requires an exponent of -2, so $0.010 = 1.0 \times 10^{-2}$. Remember: whenever the decimal point is moved to the *right,* the exponent of 10 is *negative.*

Next consider the number 0.000167. In this case we must move the decimal point four places to the right to obtain 1.67 (a number between 1 and 10):

0 . 0 0 0 1 6 7
1 2 3 4

Math skill builder
Read the Appendix if you need a further discussion of exponents and scientific notation.

Moving the decimal point four places to the right requires an exponent of -4. Therefore,

$$0.000167 = 1.67 \times 10^{-4}$$

We moved the decimal point four places to the right.

We summarize these procedures below.

Math skill builder

$100 = 1.0 \times 10^2$
$0.010 = 1.0 \times 10^{-2}$

Math skill builder

Left Is Positive; remember LIP.

Using Scientific Notation

- Any number can be represented as the product of a number between 1 and 10 and a power of 10 (either positive or negative).
- The power of 10 depends on the number of places the decimal point is moved and in which direction. The *number of places* the decimal point is moved determines the *power of 10*. The *direction* of the move determines whether the power of 10 is *positive* or *negative*. If the decimal point is moved to the left, the power of 10 is positive; if the decimal point is moved to the right, the power of 10 is negative.

Interactive Example 2.1 Scientific Notation: Powers of 10 (Positive)

Math skill builder

A number that is greater than 1 will always have a positive exponent when written in scientific notation.

Represent the following numbers in scientific notation.

a. 238,000
b. 1,500,000

SOLUTION

a. First we move the decimal point until we have a number between 1 and 10, in this case 2.38.

2 3 8 0 0 0
5 4 3 2 1 The decimal point was moved five places to the left.

Because we moved the decimal point five places to the left, the power of 10 is positive 5. Thus $238{,}000 = 2.38 \times 10^5$.

b. 1 5 0 0 0 0 0
6 5 4 3 2 1 The decimal point was moved six places to the left, so the power of 10 is 6.

Thus $1{,}500{,}000 = 1.5 \times 10^6$. ■

Interactive Example 2.2 Scientific Notation: Powers of 10 (Negative)

Math skill builder

A number that is less than 1 will always have a negative exponent when written in scientific notation.

Represent the following numbers in scientific notation.

a. 0.00043
b. 0.089

SOLUTION

a. First we move the decimal point until we have a number between 1 and 10, in this case 4.3.

0 . 0 0 0 4 3
1 2 3 4 The decimal point was moved four places to the right.

Because we moved the decimal point four places to the right, the power of 10 is negative 4. Thus $0.00043 = 4.3 \times 10^{-4}$.

b. 0 . 0 8 9
1 2 The power of 10 is negative 2 because the decimal point was moved two places to the right.

Thus $0.089 = 8.9 \times 10^{-2}$.

SE

Critical Thinkin

What if you were no
use units for one da
would this affect yo
that day?

Table 2.3 ▸ The Metric System for Measuring Length

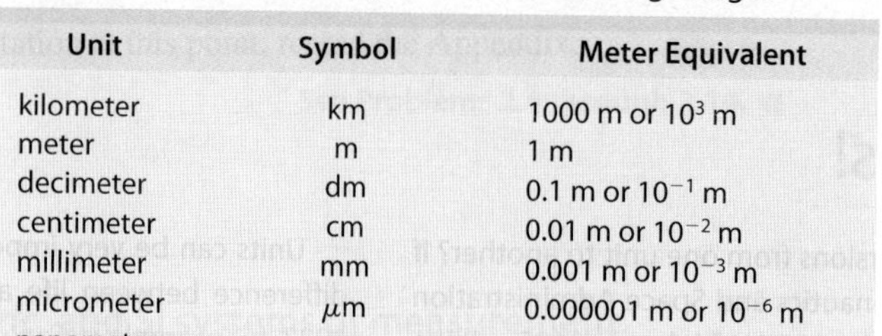

Unit	Symbol	Meter Equivalent
kilometer	km	1000 m or 10^3 m
meter	m	1 m
decimeter	dm	0.1 m or 10^{-1} m
centimeter	cm	0.01 m or 10^{-2} m
millimeter	mm	0.001 m or 10^{-3} m
micrometer	μm	0.000001 m or 10^{-6} m
nanometer	nm	0.000000001 m or 10^{-9} m

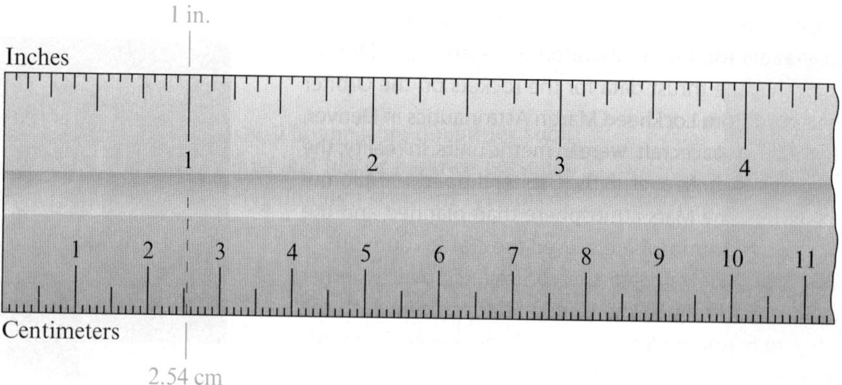

Figure 2.1 ▸ Comparison of English and metric units for length on a ruler.

Volume is the amount of three-dimensional space occupied by a substance. The fundamental unit of volume in the SI system is based on the volume of a cube that measures 1 meter in each of the three directions. That is, each edge of the cube is 1 meter in length. The volume of this cube is

$$1 \text{ m} \times 1 \text{ m} \times 1 \text{ m} = (1 \text{ m})^3 = 1 \text{ m}^3$$

or, in words, one cubic meter.

In Fig. 2.2 this cube is divided into 1000 smaller cubes. Each of these small cubes represents a volume of 1 dm^3, which is commonly called the **liter** (rhymes with "meter" and is slightly larger than a quart) and abbreviated L.

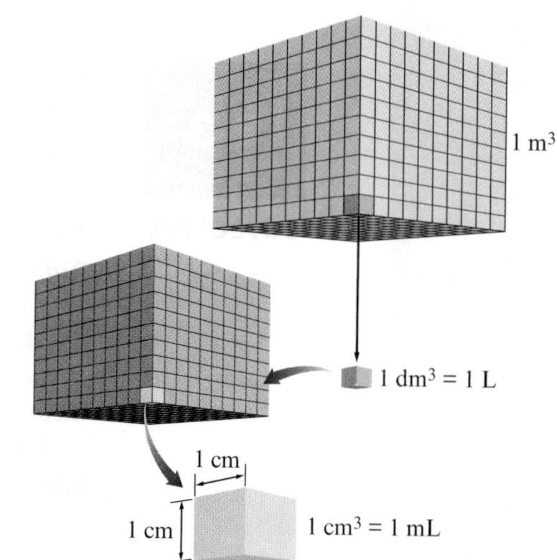

Figure 2.2 ▸ The largest drawing represents a cube that has sides 1 m in length and a volume of 1 m^3. The middle-size cube has sides 1 dm in length and a volume of 1 dm^3, or 1 L. The smallest cube has sides 1 cm in length and a volume of 1 cm^3, or 1 mL.

CHEMISTRY IN FOCUS

Measurement: Past, Present, and Future

Measurement lies at the heart of doing science. We obtain the data for formulating laws and testing theories by doing measurements. Measurements also have very practical importance; they tell us if our drinking water is safe, whether we are anemic, and the exact amount of gasoline we put in our cars at the filling station.

Although the fundamental measuring devices we consider in this chapter are still widely used, new measuring techniques are being developed every day to meet the challenges of our increasingly sophisticated world. For example, engines in modern automobiles have oxygen sensors that analyze the oxygen content in the exhaust gases. This information is sent to the computer that controls the engine functions so that instantaneous adjustments can be made in spark timing and air–fuel mixtures to provide efficient power with minimum air pollution.

A very recent area of research involves the development of paper-based measuring devices. For example, the nonprofit organization Diagnostics for All (DFA) based in Cambridge, Massachusetts, has invented a paper-based device to detect proper liver function. In this test a drop of the patient's blood is placed on the paper and the resulting color that develops can be used to determine whether the person's liver function is normal, worrisome, or requires immediate action. Other types of paper-based measuring devices include one used to detect counterfeit pharmaceuticals and another used to detect whether someone has been immunized against a specific disease. Because these paper-based devices are cheap, disposable, and easily transportable, they are especially useful in developing countries.

Scientists are also examining the natural world to find supersensitive detectors because many organisms are sensitive to tiny amounts of chemicals in their environments—recall, for example, the sensitive noses of bloodhounds. One of these natural measuring devices uses the sensory hairs from Hawaiian red swimming crabs, which are connected to electrical analyzers and used to detect hormones down to levels of 10^{-8} g/L. Likewise, tissues from pineapple cores can be used to detect tiny amounts of hydrogen peroxide.

These types of advances in measuring devices have led to an unexpected problem: detecting all kinds of substances in our food and drinking water scares us. Although these substances were always there, we didn't worry so much when we couldn't detect them. Now that we know they are present, what should we do about them? How can we assess whether these trace substances are harmful or benign? Risk assessment has become much more complicated as our sophistication in taking measurements has increased.

Ben Osbourne/Stone/Getty Images

A pollution control officer measuring the oxygen content of river water.

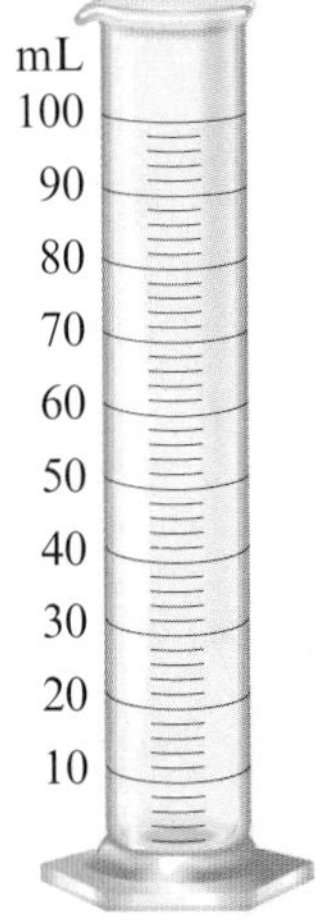

Figure 2.3 A 100-mL graduated cylinder.

The cube with a volume of 1 dm^3 (1 liter) can in turn be broken into 1000 smaller cubes, each representing a volume of 1 cm^3. This means that each liter contains 1000 cm^3. One cubic centimeter is called a **milliliter** (abbreviated mL), a unit of volume used very commonly in chemistry. This relationship is summarized in Table 2.4.

The *graduated cylinder* (Fig. 2.3), commonly used in chemical laboratories for measuring the volumes of liquids, is marked off in convenient units of volume (usually milliliters). The graduated cylinder is filled to the desired volume with the liquid, which then can be poured out.

Another important measurable quantity is **mass,** which can be defined as the quantity of matter present in an object. The fundamental SI unit of mass is the **kilogram.**

Table 2.4 The Relationship of the Liter and Milliliter

Unit	Symbol	Equivalence
liter	L	1 L = 1000 mL
milliliter	mL	$\frac{1}{1000}$ L = 10^{-3} L = 1 mL

METTLER TOLEDO

Figure 2.4 ▸ An electronic analytical balance used in chemistry labs.

Table 2.5 ▸ The Most Commonly Used Metric Units for Mass

Unit	Symbol	Gram Equivalent
kilogram	kg	1000 g = 10^3 g = 1 kg
gram	g	1 g
milligram	mg	0.001 g = 10^{-3} g = 1 mg

Table 2.6 ▸ Some Examples of Commonly Used Units

length	A dime is 1 mm thick. A quarter is 2.5 cm in diameter. The average height of an adult man is 1.8 m.
mass	A nickel has a mass of about 5 g. A 120-lb woman has a mass of about 55 kg.
volume	A 12-oz can of soda has a volume of about 360 mL. A half gallon of milk is equal to about 2 L of milk.

Because the metric system, which existed before the SI system, used the gram as the fundamental unit, the prefixes for the various mass units are based on the **gram,** as shown in Table 2.5.

In the laboratory we determine the mass of an object by using a balance. A balance compares the mass of the object to a set of standard masses ("weights"). For example, the mass of an object can be determined by using a single-pan balance (Fig. 2.4).

To help you get a feeling for the common units of length, volume, and mass, some familiar objects are described in Table 2.6.

2-4 Uncertainty in Measurement

OBJECTIVES

- ▸ **To understand how uncertainty in a measurement arises.**
- ▸ **To learn to indicate a measurement's uncertainty by using significant figures.**

When you measure the amount of something by counting, the measurement is exact. For example, if you asked your friend to buy four apples from the store and she came back with three or five apples, you would be surprised. However, measurements are not always exact. For example, whenever a measurement is made with a device such as a ruler or a graduated cylinder, an estimate is required. We can illustrate this by measuring the pin shown in Fig. 2.5(a). We can see from the ruler that the pin is a little longer than 2.8 cm and a little shorter than 2.9 cm. Because there are no graduations on the ruler between 2.8 and 2.9, we must estimate the pin's length between 2.8 and 2.9 cm. We do this by *imagining* that the distance between 2.8 and 2.9 is broken into 10 equal divisions [Fig. 2.5(b)] and estimating to which division the end of the pin reaches. The end of the pin appears to come

Andrew Lambert Photography/Science Source

A student performing a titration in the laboratory.

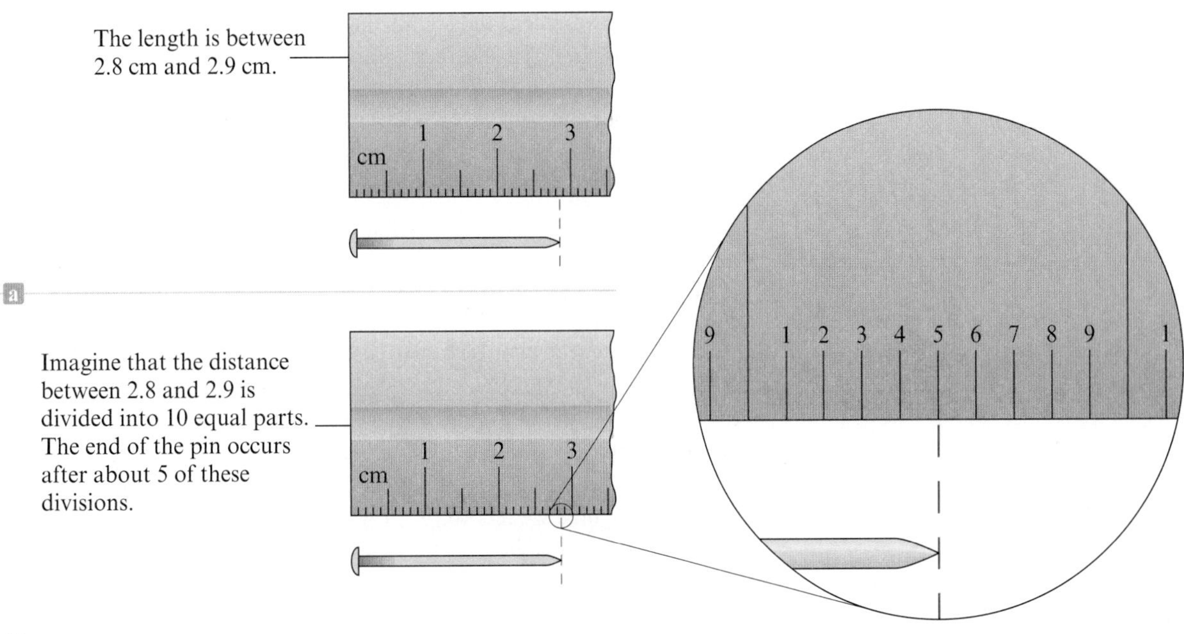

Figure 2.5 ▸ Measuring a pin.

about halfway between 2.8 and 2.9, which corresponds to 5 of our 10 imaginary divisions. So we estimate the pin's length as 2.85 cm. The result of our measurement is that the pin is approximately 2.85 cm in length, but we had to rely on a visual estimate, so it might actually be 2.84 or 2.86 cm.

Because the last number is based on a visual estimate, it may be different when another person makes the same measurement. For example, if five different people measured the pin, the results might be

Person	Result of Measurement
1	2.85 cm
2	2.84 cm
3	2.86 cm
4	2.85 cm
5	2.86 cm

Note that the first two digits in each measurement are the same regardless of who made the measurement; these are called the *certain* numbers of the measurement. However, the third digit is estimated and can vary; it is called an *uncertain* number. When one is making a measurement, the custom is to record all of the certain numbers plus the *first* uncertain number. It would not make any sense to try to measure the pin to the third decimal place (thousandths of a centimeter), because this ruler requires an estimate of even the second decimal place (hundredths of a centimeter).

It is very important to realize that *a measurement always has some degree of uncertainty.* The uncertainty of a measurement depends on the measuring device. For example, if the ruler in Fig. 2.5 had marks indicating hundredths of a centimeter, the uncertainty in the measurement of the pin would occur in the thousandths place rather than the hundredths place, but some uncertainty would still exist.

The numbers recorded in a measurement (all the certain numbers plus the first uncertain number) are called **significant figures.** The number of significant figures for a given measurement is determined by the inherent uncertainty of the measuring device. For example, the ruler used to measure the pin can give results only to hundredths of a centimeter. Thus, when we record the significant figures for a measurement, we automatically give information about the uncertainty in a measurement. The uncertainty in the last number (the estimated number) is usually assumed to be ± 1 unless otherwise indicated. For example, the measurement 1.86 kilograms can be interpreted as 1.86 ± 0.01 kilograms, where the symbol $\pm$ means plus or minus. That is, it could be $1.86 \text{ kg} - 0.01 \text{ kg} = 1.85 \text{ kg}$ or $1.86 \text{ kg} + 0.01 \text{ kg} = 1.87 \text{ kg}$.

2-5 Significant Figures

OBJECTIVE **To learn to determine the number of significant figures in a calculated result.**

We have seen that any measurement involves an estimate and thus is uncertain to some extent. We signify the degree of certainty for a particular measurement by the number of significant figures we record.

Because doing chemistry requires many types of calculations, we must consider what happens when we do arithmetic with numbers that contain uncertainties. It is important that we know the degree of uncertainty in the final result. Although we will not discuss the process here, mathematicians have studied how uncertainty accumulates and have designed a set of rules to determine how many significant figures the result of a calculation should have. You should follow these rules whenever you carry out a calculation. The first thing we need to do is learn how to count the significant figures in a given number. To do this we use the following rules:

Rules for Counting Significant Figures

1. *Nonzero integers.* Nonzero integers *always* count as significant figures. For example, the number 1457 has four nonzero integers, all of which count as significant figures.
2. *Zeros.* There are three classes of zeros:
   a. *Leading zeros* are zeros that *precede* all of the nonzero digits. They *never* count as significant figures. For example, in the number 0.0025, the three zeros simply indicate the position of the decimal point. The number has only two significant figures, the 2 and the 5.
   b. *Captive zeros* are zeros that fall *between* nonzero digits. They *always* count as significant figures. For example, the number 1.008 has four significant figures.
   c. *Trailing zeros* are zeros at the *right end* of the number. They are significant only if the number is written with a decimal point. The number one hundred written as 100 has only one significant figure, but written as 100., it has three significant figures.
3. *Exact numbers.* Often calculations involve numbers that were not obtained using measuring devices but were determined by counting: 10 experiments, 3 apples, 8 molecules. Such numbers are called *exact numbers.* They can be assumed to have an unlimited number of significant figures. Exact numbers can also arise from definitions. For example, 1 inch is defined as *exactly* 2.54 centimeters. Thus in the statement 1 in. = 2.54 cm, neither 2.54 nor 1 limits the number of significant figures when it is used in a calculation.

Math skill builder
Leading zeros are never significant figures.

Math skill builder
Captive zeros are always significant figures.

Math skill builder
Trailing zeros are sometimes significant figures.

Math skill builder
Exact numbers never limit the number of significant figures in a calculation.

Math skill builder
Significant figures are easily indicated by scientific notation.

Rules for counting significant figures also apply to numbers written in scientific notation. For example, the number 100. can also be written as 1.00×10^2, and both versions have three significant figures. Scientific notation offers two major advantages: the number of significant figures can be indicated easily, and fewer zeros are needed to write a very large or a very small number. For example, the number 0.000060 is much more conveniently represented as 6.0×10^{-5}, and the number has two significant figures, written in either form.

Interactive Example 2.3

Counting Significant Figures

Give the number of significant figures for each of the following measurements.

a. A sample of orange juice contains 0.0108 g of vitamin C.
b. A forensic chemist in a crime lab weighs a single hair and records its mass as 0.0050060 g.
c. The distance between two points was found to be 5.030×10^3 ft.
d. In yesterday's bicycle race, 110 riders started but only 60 finished.

SOLUTION

a. The number contains three significant figures. The zeros to the left of the 1 are leading zeros and are not significant, but the remaining zero (a captive zero) is significant.
b. The number contains five significant figures. The leading zeros (to the left of the 5) are not significant. The captive zeros between the 5 and the 6 are significant, and the trailing zero to the right of the 6 is significant because the number contains a decimal point.
c. This number has four significant figures. Both zeros in 5.030 are significant.
d. Both numbers are exact (they were obtained by counting the riders). Thus these numbers have an unlimited number of significant figures.

SELF CHECK

Exercise 2.2 Give the number of significant figures for each of the following measurements.

a. 0.00100 m   b. 2.0800×10^2 L   c. 480 Corvettes

See Problems 2.33 and 2.34. ■

Rounding Off Numbers

When you perform a calculation on your calculator, the number of digits displayed is usually greater than the number of significant figures that the result should possess. So you must "round off" the number (reduce it to fewer digits). The rules for **rounding off** follow.

Rules for Rounding Off

1. If the digit to be removed
   a. is less than 5, the preceding digit stays the same. For example, 1.33 rounds to 1.3.
   b. is equal to or greater than 5, the preceding digit is increased by 1. For example, 1.36 rounds to 1.4, and 3.15 rounds to 3.2.
2. In a series of calculations, carry the extra digits through to the final result and *then* round off.* This means that you should carry all of the digits that show on your calculator until you arrive at the final number (the answer) and then round off, using the procedures in Rule 1.

*This practice will not be followed in the worked-out examples in this text, because we want to show the correct number of significant figures in each step of the example.

We need to make one more point about rounding off to the correct number of significant figures. Suppose the number 4.348 needs to be rounded to two significant figures. In doing this, we look *only* at the *first number* to the right of the 3:

4.348
↑
Look at this number to round off to two significant figures.

Math skill builder

Do not round off sequentially. The number 6.8347 rounded to three significant figures is 6.83, not 6.84.

The number is rounded to 4.3 because 4 is less than 5. It is incorrect to round sequentially. For example, do *not* round the 4 to 5 to give 4.35 and then round the 3 to 4 to give 4.4.

When rounding off, *use only the first number to the right of the last significant figure.*

Determining Significant Figures in Calculations

Next we will learn how to determine the correct number of significant figures in the result of a calculation. To do this we will use the following rules.

Rules for Using Significant Figures in Calculations

1. For *multiplication* or *division,* the number of significant figures in the result is the same as that in the measurement with the *smallest number* of significant figures. We say this measurement is *limiting,* because it limits the number of significant figures in the result. For example, consider this calculation:

$$4.56 \times 1.4 = 6.384 \xrightarrow{\text{Round off}} 6.4$$

4.56: Three significant figures; 1.4: Limiting (two significant figures); 6.4: Two significant figures

Because 1.4 has only two significant figures, it limits the result to two significant figures. Thus the product is correctly written as 6.4, which has two significant figures. Consider another example. In the division $\frac{8.315}{298}$, how many significant figures should appear in the answer? Because 8.315 has four significant figures, the number 298 (with three significant figures) limits the result. The calculation is correctly represented as

$$\frac{8.315}{298} = 0.0279027 \xrightarrow{\text{Round off}} 2.79 \times 10^{-2}$$

8.315: Four significant figures; 298: Limiting (three significant figures); 0.0279027: Result shown on calculator; 2.79×10^{-2}: Three significant figures

(continued)

Math skill builder

If you need help in using your calculator, see the Appendix.

2. For *addition* or *subtraction*, the limiting term is the one with the smallest number of decimal places. For example, consider the following sum:

$$\begin{array}{r} 12.11 \\ 18.0 \\ 1.013 \\ \hline 31.123 \end{array} \quad \text{Round off} \rightarrow 31.1$$

18.0 — Limiting term (has one decimal place)

31.1 — One decimal place

Why is the answer limited by the term with the smallest number of decimal places? Recall that the last digit reported in a measurement is actually an uncertain number. Although 18, 18.0, and 18.00 are treated as the same quantities by your calculator, they are different to a scientist. The problem above can be thought of as follows:

12.11? mL
18.0?? mL
1.013 mL
31.1?? mL

Because the term 18.0 is reported only to the tenths place, our answer must be reported this way as well.

The correct result is 31.1 (it is limited to one decimal place because 18.0 has only one decimal place). Consider another example:

$$\begin{array}{r} 0.6875 \\ -0.1 \\ \hline 0.5875 \end{array} \quad \text{Round off} \rightarrow 0.6$$

−0.1 — Limiting term (one decimal place)

Note that *for multiplication and division, significant figures are counted. For addition and subtraction, the decimal places are counted.*

Now we will put together the things you have learned about significant figures by considering some mathematical operations in the following examples.

Interactive Example 2.4

Counting Significant Figures in Calculations

Without performing the calculations, tell how many significant figures each answer should contain.

a. 5.19
1.9
+0.842

b. 1081 − 7.25

c. 2.3 × 3.14

d. the total cost of 3 boxes of candy at $2.50 a box

SOLUTION

a. The answer will have one digit after the decimal place. The limiting number is 1.9, which has one decimal place, so the answer has two significant figures.

b. The answer will have no digits after the decimal point. The number 1081 has no digits to the right of the decimal point and limits the result, so the answer has four significant figures.

c. The answer will have two significant figures because the number 2.3 has only two significant figures (3.14 has three).

d. The answer will have three significant figures. The limiting factor is 2.50 because 3 (boxes of candy) is an exact number. ■

Interactive Example 2.5

Calculations Using Significant Figures

Carry out the following mathematical operations and give each result to the correct number of significant figures.

a. 5.18×0.0208

b. $(3.60 \times 10^{-3}) \times (8.123) \div 4.3$

c. $21 + 13.8 + 130.36$

d. $116.8 - 0.33$

e. $(1.33 \times 2.8) + 8.41$

SOLUTION

Limiting terms — Round to this digit.

a. $5.18 \times 0.0208 = 0.107744 \Rightarrow 0.108$

The answer should contain three significant figures because each number being multiplied has three significant figures (Rule 1). The 7 is rounded to 8 because the following digit is greater than 5.

Math skill builder

When we multiply and divide in a problem, perform all calculations before rounding the answer to the correct number of significant figures.

Round to this digit.

b. $$\frac{(3.60 \times 10^{-3})(8.123)}{4.3} = 6.8006 \times 10^{-3} \Rightarrow 6.8 \times 10^{-3}$$

Limiting term (4.3)

Because 4.3 has the least number of significant figures (two), the result should have two significant figures (Rule 1).

c.
$$\begin{array}{r} 21 \\ 13.8 \\ +130.36 \\ \hline 165.16 \end{array} \Rightarrow 165$$

In this case 21 is limiting (there are no digits after the decimal point). Thus the answer must have no digits after the decimal point, in accordance with the rule for addition (Rule 2).

Math skill builder

When we multiply (or divide) and then add (or subtract) in a problem, round the first answer from the first operation (in this case, multiplication) before performing the next operation (in this case, addition). We need to know the correct number of decimal places.

d.
$$\begin{array}{r} 116.8 \\ -\ 0.33 \\ \hline 116.47 \end{array} \Rightarrow 116.5$$

Because 116.8 has only one decimal place, the answer must have only one decimal place (Rule 2). The 4 is rounded up to 5 because the digit to the right (7) is greater than 5.

e. $1.33 \times 2.8 = 3.724 \Rightarrow 3.7$

$$\begin{array}{r} 3.7 \\ +\ 8.41 \\ \hline 12.11 \end{array} \Rightarrow 12.1$$

(3.7 ← Limiting term)

Note that in this case we multiplied and then rounded the result to the correct number of significant figures before we performed the addition so that we would know the correct number of decimal places.

SELF CHECK

Exercise 2.3 Give the answer for each calculation to the correct number of significant figures.

a. 12.6×0.53

b. $(12.6 \times 0.53) - 4.59$

c. $(25.36 - 4.15) \div 2.317$

See Problems 2.47 through 2.52. ■

Problem Solving and Dimensional Analysis

OBJECTIVE **To learn how dimensional analysis can be used to solve various types of problems.**

Suppose that the boss at the store where you work on weekends asks you to pick up 2 dozen doughnuts on the way to work. However, you find that the doughnut shop sells by the doughnut. How many doughnuts do you need?

This "problem" is an example of something you encounter all the time: converting from one unit of measurement to another. Examples of this occur in cooking (The recipe calls for 3 cups of cream, which is sold in pints. How many pints do I buy?); traveling (The purse costs 250 pesos. How much is that in dollars?); sports (A recent Tour de France bicycle race was 3215 kilometers long. How many miles is that?); and many other areas.

How do we convert from one unit of measurement to another? Let's explore this process by using the doughnut problem.

$$2 \text{ dozen doughnuts} = ? \text{ individual doughnuts}$$

where ? represents a number you don't know yet. The essential information you must have is the definition of a dozen:

$$1 \text{ dozen} = 12$$

You can use this information to make the needed conversion as follows:

$$2 \text{ dozen doughnuts} \times \frac{12}{1 \text{ dozen}} = 24 \text{ doughnuts}$$

You need to buy 24 doughnuts.

Note two important things about this process.

1. The factor $\frac{12}{1 \text{ dozen}}$ is a conversion factor based on the definition of the term *dozen.* This conversion factor is a ratio of the two parts of the definition of a dozen given above.
2. The unit "dozen" itself cancels.

Math skill builder

Since 1 dozen = 12, when we multiply by $\frac{12}{1 \text{ dozen}}$, we are multiplying by 1. The unit "dozen" cancels.

Now let's generalize a bit. To change from one unit to another, we will use a conversion factor.

$$\text{Unit}_1 \times \text{conversion factor} = \text{Unit}_2$$

The **conversion factor** is a ratio of the two parts of the statement that relates the two units. We will see this in more detail in the following discussion.

Earlier in this chapter we considered a pin that measured 2.85 cm in length. What is the length of the pin in inches? We can represent this problem as

$$2.85 \text{ cm} \rightarrow ? \text{ in.}$$

The question mark stands for the number we want to find. To solve this problem, we must know the relationship between inches and centimeters. In Table 2.7, which gives several equivalents between the English and metric systems, we find the relationship

$$2.54 \text{ cm} = 1 \text{ in.}$$

This is called an **equivalence statement.** In other words, 2.54 cm and 1 in. stand for *exactly the same distance* (Fig. 2.1). The respective numbers are different because they refer to different *scales* (*units*) of distance.

Table 2.7 ▸ English–Metric and English–English Equivalents

Length	1 m = 1.094 yd
	2.54 cm = 1 in.
	1 mi = 5280. ft
	1 mi = 1760. yd
Mass	1 kg = 2.205 lb
	453.6 g = 1 lb
Volume	1 L = 1.06 qt
	1 ft^3 = 28.32 L

The equivalence statement 2.54 cm = 1 in. can lead to either of two conversion factors:

$$\frac{2.54\text{ cm}}{1\text{ in.}} \quad \text{or} \quad \frac{1\text{ in.}}{2.54\text{ cm}}$$

Note that these *conversion factors* are *ratios of the two parts of the equivalence statement* that relates the two units. Which of the two possible conversion factors do we need? Recall our problem:

$$2.85\text{ cm} = ?\text{ in.}$$

That is, we want to convert from units of centimeters to inches:

$$2.85\text{ cm} \times \text{conversion factor} = ?\text{ in.}$$

We choose a conversion factor that cancels the units we want to discard and leaves the units we want in the result. Thus we do the conversion as follows:

Math skill builder

Units cancel just as numbers do.

$$2.85\text{ cm} \times \frac{1\text{ in.}}{2.54\text{ cm}} = \frac{2.85\text{ in.}}{2.54} = 1.12\text{ in.}$$

Note two important facts about this conversion:

1. The centimeter units cancel to give inches for the result. This is exactly what we had wanted to accomplish. Using the other conversion factor $\left(2.85\text{ cm} \times \frac{2.54\text{ cm}}{1\text{ in.}}\right)$ would not work because the units would not cancel to give inches in the result.

Math skill builder

When you finish a calculation, always check to make sure that the answer makes sense.

2. As the units changed from centimeters to inches, the number changed from 2.85 to 1.12. Thus 2.85 cm has exactly the same value (is the same length) as 1.12 in. Notice that in this conversion, the number decreased from 2.85 to 1.12. This makes sense because the inch is a larger unit of length than the centimeter is. That is, it takes fewer inches to make the same length in centimeters.

Math skill builder

When exact numbers are used in a calculation, they never limit the number of significant digits.

The result in the foregoing conversion has three significant figures as required. Caution: Noting that the term 1 appears in the conversion, you might think that because this number appears to have only one significant figure, the result should have only one significant figure. That is, the answer should be given as 1 in. rather than 1.12 in. However, in the equivalence statement 1 in. = 2.54 cm, the 1 is an exact number (by definition). In other words, exactly 1 in. equals 2.54 cm. Therefore, the 1 does not limit the number of significant digits in the result.

We have seen how to convert from centimeters to inches. What about the reverse conversion? For example, if a pencil is 7.00 in. long, what is its length in centimeters? In this case, the conversion we want to make is

$$7.00\text{ in.} \rightarrow ?\text{ cm}$$

What conversion factor do we need to make this conversion?

Remember that two conversion factors can be derived from each equivalence statement. In this case, the equivalence statement 2.54 cm = 1 in. gives

$$\frac{2.54\text{ cm}}{1\text{ in.}} \quad \text{or} \quad \frac{1\text{ in.}}{2.54\text{ cm}}$$

Again, we choose which factor to use by looking at the *direction* of the required change. For us to change from inches to centimeters, the inches must cancel. Thus the factor

$$\frac{2.54\text{ cm}}{1\text{ in.}}$$

is used, and the conversion is done as follows:

$$7.00 \text{ in.} \times \frac{2.54 \text{ cm}}{1 \text{ in.}} = (7.00)(2.54) \text{ cm} = 17.8 \text{ cm}$$

Here the inch units cancel, leaving centimeters as required.

Note that in this conversion, the number increased (from 7.00 to 17.8). This makes sense because the centimeter is a smaller unit of length than the inch. That is, it takes more centimeters to make the same length in inches. *Always take a moment to think about whether your answer makes sense.* This will help you avoid errors.

Changing from one unit to another via conversion factors (based on the equivalence statements between the units) is often called **dimensional analysis.** We will use this method throughout our study of chemistry.

We can now state some general steps for doing conversions by dimensional analysis.

Converting from One Unit to Another

Step 1 To convert from one unit to another, use the equivalence statement that relates the two units. The conversion factor needed is a ratio of the two parts of the equivalence statement.

Step 2 Choose the appropriate conversion factor by looking at the direction of the required change (make sure the unwanted units cancel).

Step 3 Multiply the quantity to be converted by the conversion factor to give the quantity with the desired units.

Step 4 Check that you have the correct number of significant figures.

Step 5 Ask whether your answer makes sense.

We will now illustrate this procedure in Example 2.6.

Interactive Example 2.6

Conversion Factors: One-Step Problems

An Italian bicycle has its frame size given as 62 cm. What is the frame size in inches?

SOLUTION

We can represent the problem as

$$62 \text{ cm} = ? \text{ in.}$$

In this problem we want to convert from centimeters to inches.

$$62 \text{ cm} \times \text{conversion factor} = ? \text{ in.}$$

Step 1 To convert from centimeters to inches, we need the equivalence statement 1 in. = 2.54 cm. This leads to two conversion factors:

$$\frac{1 \text{ in.}}{2.54 \text{ cm}} \quad \text{and} \quad \frac{2.54 \text{ cm}}{1 \text{ in.}}$$

Step 2 In this case, the direction we want is

$$\text{Centimeters} \rightarrow \text{inches}$$

so we need the conversion factor $\frac{1 \text{ in.}}{2.54 \text{ cm}}$. We know this is the one we want because using it will make the units of centimeters cancel, leaving units of inches.

Step 3 The conversion is carried out as follows:

$$62 \text{ cm} \times \frac{1 \text{ in.}}{2.54 \text{ cm}} = 24 \text{ in.}$$

Step 4 The result is limited to two significant figures by the number 62. The centimeters cancel, leaving inches as required.

Step 5 Note that the number decreased in this conversion. This makes sense; the inch is a larger unit of length than the centimeter.

SELF CHECK

Exercise 2.4 Wine is often bottled in 0.750-L containers. Using the appropriate equivalence statement from Table 2.7, calculate the volume of such a wine bottle in quarts.

See Problems 2.59 and 2.60. ■

Next we will consider a conversion that requires several steps.

Interactive Example 2.7

Conversion Factors: Multiple-Step Problems

The length of the marathon race is approximately 26.2 mi. What is this distance in kilometers?

SOLUTION

The problem before us can be represented as follows:

$$26.2 \text{ mi} = ? \text{ km}$$

We could accomplish this conversion in several different ways, but because Table 2.7 gives the equivalence statements 1 mi = 1760 yd and 1 m = 1.094 yd, we will proceed as follows:

$$\text{Miles} \longrightarrow \text{yards} \longrightarrow \text{meters} \longrightarrow \text{kilometers}$$

This process will be carried out one conversion at a time to make sure everything is clear.

MILES → YARDS We convert from miles to yards using the conversion factor $\frac{1760 \text{ yd}}{1 \text{ mi}}$.

$$26.2 \text{ mi} \times \frac{1760 \text{ yd}}{1 \text{ mi}} = 46{,}112 \text{ yd}$$

Result shown on calculator

$$46{,}112 \text{ yd} \xrightarrow{\text{Round off}} 46{,}100 \text{ yd} = 4.61 \times 10^4 \text{ yd}$$

YARDS → METERS The conversion factor used to convert yards to meters is $\frac{1 \text{ m}}{1.094 \text{ yd}}$.

$$4.61 \times 10^4 \text{ yd} \times \frac{1 \text{ m}}{1.094 \text{ yd}} = 4.213894 \times 10^4 \text{ m}$$

Result shown on calculator

$$4.213894 \times 10^4 \text{ m} \xrightarrow{\text{Round off}} 4.21 \times 10^4 \text{ m}$$

METERS → KILOMETERS Because 1000 m = 1 km, or 10^3 m = 1 km, we convert from meters to kilometers as follows:

$$4.21 \times 10^4\ \cancel{m} \times \frac{1\ km}{10^3\ \cancel{m}} = 4.21 \times 10^1\ km$$
$$= 42.1\ km$$

Thus the marathon (26.2 mi) is 42.1 km.

Once you feel comfortable with the conversion process, you can combine the steps. For the above conversion, the combined expression is

miles → yards → meters → kilometers

$$26.2\ \cancel{mi} \times \frac{1760\ \cancel{yd}}{1\ \cancel{mi}} \times \frac{1\ \cancel{m}}{1.094\ \cancel{yd}} \times \frac{1\ km}{10^3\ \cancel{m}} = 42.1\ km$$

Note that the units cancel to give the required kilometers and that the result has three significant figures.

Math skill builder

Remember that we are rounding off at the end of each step to show the correct number of significant figures. However, in doing a multistep calculation, *you* should retain the extra numbers that show on your calculator and round off only at the end of the calculation.

SELF CHECK

Exercise 2.5 Racing cars at the Indianapolis Motor Speedway now routinely travel around the track at an average speed of 225 mi/h. What is this speed in kilometers per hour?

See Problems 2.65 and 2.66. ■

Units provide a very valuable check on the validity of your solution. Always use them.

RECAP Whenever you work problems, remember the following points:

1. Always include the units (a measurement always has two parts: a number *and* a unit).
2. Cancel units as you carry out the calculations.
3. Check that your final answer has the correct units. If it doesn't, you have done something wrong.
4. Check that your final answer has the correct number of significant figures.
5. Think about whether your answer makes sense.

2-7 Temperature Conversions: An Approach to Problem Solving

OBJECTIVES

- To learn the three temperature scales.
- To learn to convert from one scale to another.
- To continue to develop problem-solving skills.

When the doctor tells you your temperature is 102 degrees and the weatherperson on TV says it will be 75 degrees tomorrow, they are using the **Fahrenheit scale.** Water boils at 212 °F and freezes at 32 °F, and normal body temperature is 98.6 °F (where °F signifies "Fahrenheit degrees"). This temperature scale is widely used in the United States and Great Britain, and it is the scale employed in most of the engineering sciences. Another temperature scale, used in Canada and Europe and in the physical and life sciences in most countries, is the **Celsius scale.** In keeping with the metric system, which is based on powers of 10, the freezing and boiling points of water on the Celsius scale are assigned as 0 °C and 100 °C, respectively. On both the Fahrenheit and the Celsius scales, the unit of temperature is called a degree, and the symbol for it is followed by the capital letter representing the scale on which the units are measured: °C or °F.

Although 373 K is often stated as 373 degrees Kelvin, it is more correct to say 373 kelvins.

Still another temperature scale used in the sciences is the **absolute** or **Kelvin scale.** On this scale water freezes at 273 K and boils at 373 K. On the Kelvin scale, the unit of temperature is called a kelvin and is symbolized by K. Thus, on the three scales, the boiling point of water is stated as 212 Fahrenheit degrees (212 °F), 100 Celsius degrees (100 °C), and 373 kelvins (373 K).

The three temperature scales are compared in Figs. 2.6 and 2.7. There are several important facts you should note.

1. The size of each temperature unit (each degree) is the same for the Celsius and Kelvin scales. This follows from the fact that the *difference* between the boiling and freezing points of water is 100 units on both of these scales.
2. The Fahrenheit degree is smaller than the Celsius and Kelvin units. Note that on the Fahrenheit scale there are 180 Fahrenheit degrees between the boiling and freezing points of water, as compared with 100 units on the other two scales.
3. The zero points are different on all three scales.

In your study of chemistry, you will sometimes need to convert from one temperature scale to another. We will consider in some detail how this is done. In addition to learning how to change temperature scales, you should also use this section as an opportunity to further develop your skills in problem solving.

Converting Between the Kelvin and Celsius Scales

It is relatively simple to convert between the Celsius and Kelvin scales because the temperature unit is the same size; only the zero points are different. Because 0 °C corresponds to 273 K, converting from Celsius to Kelvin requires that we add 273 to the Celsius temperature. We will illustrate this procedure in Example 2.8.

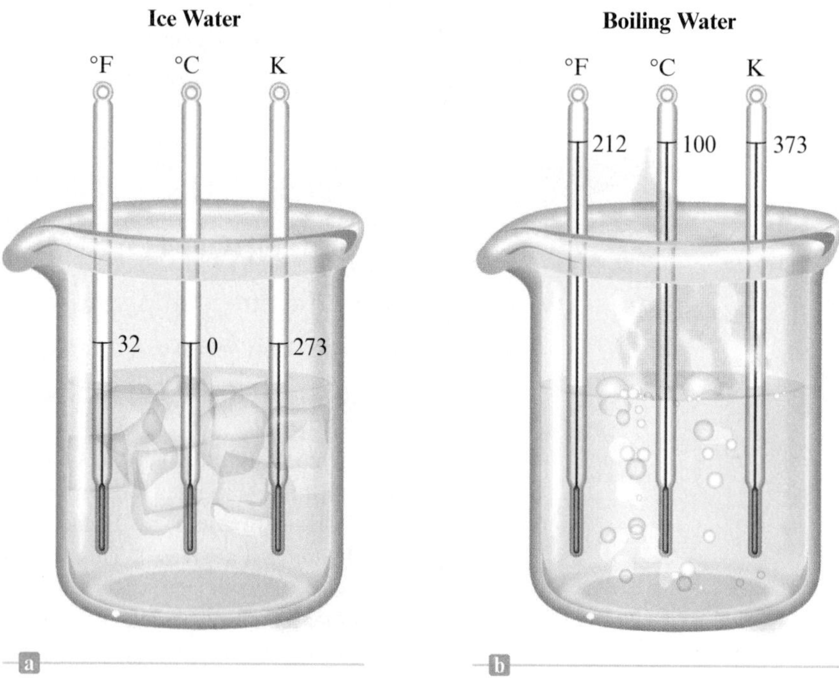

Figure 2.6 Thermometers based on the three temperature scales in (a) ice water and (b) boiling water.

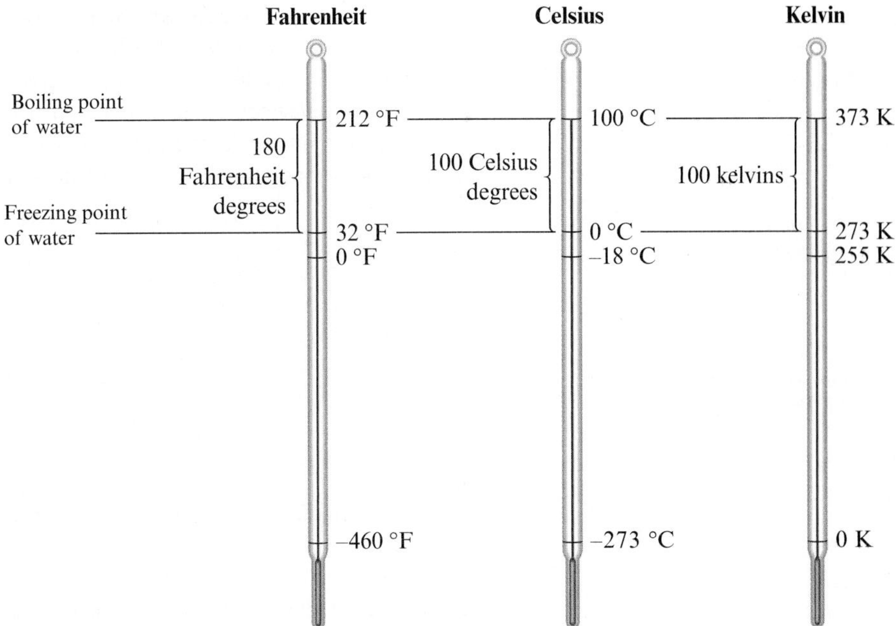

Figure 2.7 ▸ The three major temperature scales.

Interactive Example 2.8 Temperature Conversion: Celsius to Kelvin

Boiling points will be discussed further in Chapter 14.

The boiling point of water at the top of Mt. Everest is 70. °C. Convert this temperature to the Kelvin scale. (The decimal point after the temperature reading indicates that the trailing zero is significant.)

SOLUTION This problem asks us to find 70. °C in units of kelvins. We can represent this problem simply as

$$70.\ °C = ?\ K$$

In doing problems, it is often helpful to draw a diagram in which we try to represent the words in the problem with a picture. This problem can be diagramed as shown in Fig. 2.8(a).

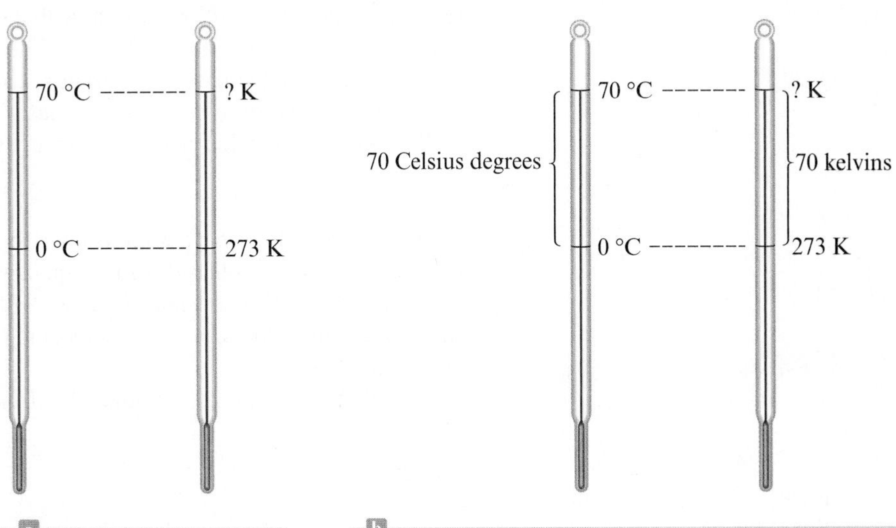

a We know 0 °C = 273 K. We want to know 70. °C = ? K.

b There are 70 degrees on the Celsius scale between 0 °C and 70. °C. Because units on these scales are the same size, there are also 70 kelvins in this same distance on the Kelvin scale.

Figure 2.8 ▸ Converting 70. °C to units measured on the Kelvin scale.

In this picture we have shown what we want to find: "What temperature (in kelvins) is the same as 70. °C?" We also know from Fig. 2.7 that 0 °C represents the same temperature as 273 K. How many degrees above 0 °C is 70. °C? The answer, of course, is 70. Thus we must add 70. to 0 °C to reach 70. °C. Because degrees are the *same size* on both the Celsius scale and the Kelvin scale [Fig. 2.8(b)], we must also add 70. to 273 K (same temperature as 0 °C) to reach ? K. That is,

$$? \text{ K} = 273 + 70. = 343 \text{ K}$$

Thus 70. °C corresponds to 343 K.

Note that to convert from the Celsius to the Kelvin scale, we simply add the temperature in °C to 273. That is,

$$\underset{\text{Temperature in Celsius degrees}}{T_{°C}} + 273 = \underset{\text{Temperature in kelvins}}{T_K}$$

Using this formula to solve the present problem gives

$$70. + 273 = 343$$

(with units of kelvins, K), which is the correct answer. ■

We can summarize what we learned in Example 2.8 as follows: to convert from the Celsius to the Kelvin scale, we can use the formula

$$\underset{\text{Temperature in Celsius degrees}}{T_{°C}} + 273 = \underset{\text{Temperature in kelvins}}{T_K}$$

Interactive Example 2.9

Temperature Conversion: Kelvin to Celsius

Liquid nitrogen boils at 77 K. What is the boiling point of nitrogen on the Celsius scale?

SOLUTION

The problem to be solved here is 77 K = ? °C. Let's explore this question by examining the picture to the left representing the two temperature scales. One key point is to recognize that 0 °C = 273 K. Also note that the difference between 273 K and 77 K is 196 kelvins (273 − 77 = 196). That is, 77 K is 196 kelvins below 273 K. The degree size is the same on these two temperature scales, so 77 K must correspond to 196 Celsius degrees below zero or −196 °C. Thus 77 K = ? °C = −196 °C.

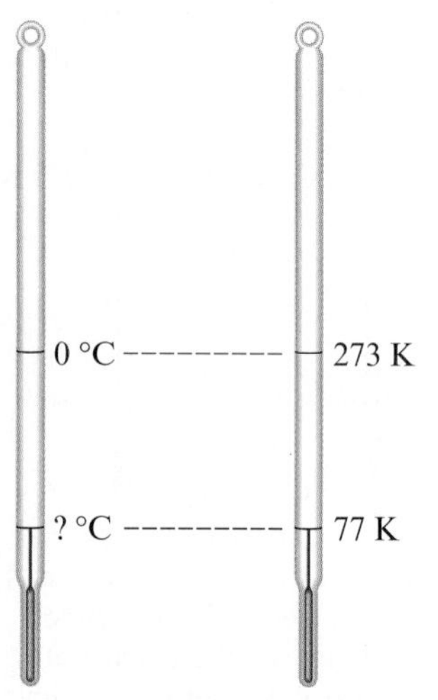

We can also solve this problem by using the formula

$$T_{°C} + 273 = T_K$$

However, in this case we want to solve for the Celsius temperature, $T_{°C}$. That is, we want to isolate $T_{°C}$ on one side of the equals sign. To do this we use an important general principle: doing *the same thing on both sides of the equals sign* preserves the equality. In other words, it's always okay to perform the same operation on both sides of the equals sign.

To isolate $T_{°C}$ we need to subtract 273 from both sides:

$$T_{°C} + \underbrace{273 - 273}_{\text{Sum is zero}} = T_K - 273$$

to give

$$T_{°C} = T_K - 273$$

CHEMISTRY IN FOCUS

Tiny Thermometers

Can you imagine a thermometer that has a diameter equal to one one-hundredth of a human hair? Such a device has actually been produced by scientists Yihua Gao and Yoshio Bando of the National Institute for Materials Science in Tsukuba, Japan. The thermometer they constructed is so tiny that it must be read using a powerful electron microscope.

It turns out that the tiny thermometers were produced by accident. The Japanese scientists were actually trying to make tiny (nanoscale) gallium nitride wires. However, when they examined the results of their experiment, they discovered tiny tubes of carbon atoms that were filled with elemental gallium. Because gallium is a liquid over an unusually large temperature range, it makes a perfect working liquid for a thermometer. Just as in mercury thermometers, which have mostly been phased out because of the toxicity of mercury, the gallium expands as the temperature increases. Therefore, gallium moves up the tube as the temperature increases.

These minuscule thermometers are not useful in the normal macroscopic world—they can't even be seen with the naked eye. However, they should be valuable for monitoring temperatures from 50 °C to 500 °C in materials in the nanoscale world.

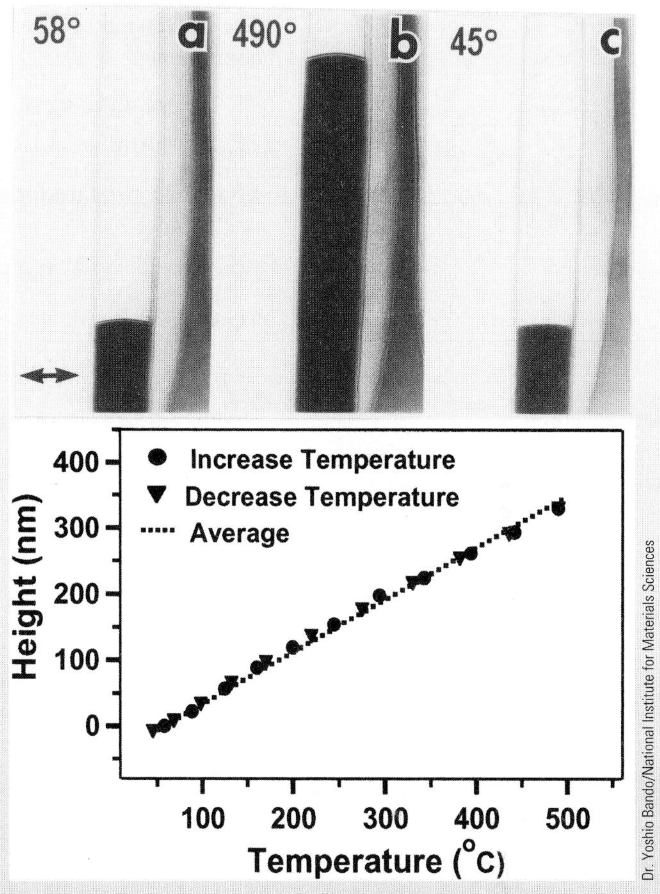

Liquid gallium expands within a carbon nanotube as the temperature increases *(left to right)*.

Using this equation to solve the problem, we have

$$T_{°C} = T_K - 273 = 77 - 273 = -196$$

So, as before, we have shown that

$$77 \text{ K} = -196 \text{ °C}$$

SELF CHECK

Exercise 2.6 Which temperature is colder, 172 K or −75 °C?

See Problems 2.73 and 2.74. ■

In summary, because the Kelvin and Celsius scales have the same size unit, to switch from one scale to the other we must simply account for the different zero points. We must add 273 to the Celsius temperature to obtain the temperature on the Kelvin scale:

$$T_K = T_{°C} + 273$$

To convert from the Kelvin scale to the Celsius scale, we must subtract 273 from the Kelvin temperature:

$$T_{°C} = T_K - 273$$

Converting Between the Fahrenheit and Celsius Scales

The conversion between the Fahrenheit and Celsius temperature scales requires two adjustments:

1. For the different size units
2. For the different zero points

To see how to adjust for the different unit sizes, consider the diagram in Fig. 2.9. Note that because 212 °F = 100 °C and 32 °F = 0 °C,

$$212 - 32 = 180 \text{ Fahrenheit degrees} = 100 - 0 = 100 \text{ Celsius degrees}$$

Thus

$$180. \text{ Fahrenheit degrees} = 100. \text{ Celsius degrees}$$

Math skill builder

Remember, it's okay to do the same thing to both sides of the equation.

Dividing both sides of this equation by 100. gives

$$\frac{180.}{100.} \text{ Fahrenheit degrees} = \frac{\cancel{100.}}{\cancel{100.}} \text{ Celsius degrees}$$

or

$$1.80 \text{ Fahrenheit degrees} = 1.00 \text{ Celsius degree}$$

The factor 1.80 is used to convert from one degree size to the other.

Next we have to account for the fact that 0 °C is *not* the same as 0 °F. In fact, 32 °F = 0 °C. Although we will not show how to derive it, the equation to convert a temperature in Celsius degrees to the Fahrenheit scale is

$$T_{°F} = 1.80(T_{°C}) + 32$$

Temperature in °F (for $T_{°F}$); Temperature in °C (for $T_{°C}$)

In this equation the term $1.80(T_{°C})$ adjusts for the difference in degree size between the two scales. The 32 in the equation accounts for the different zero points. We will now show how to use this equation.

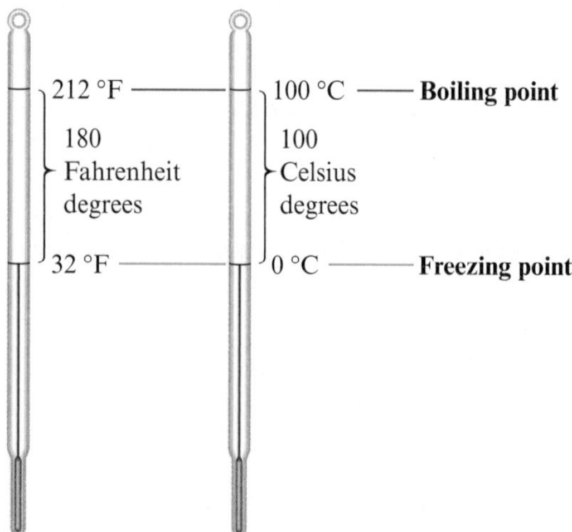

Figure 2.9 Comparison of the Celsius and Fahrenheit scales.

Interactive Example 2.10

Temperature Conversion: Celsius to Fahrenheit

On a summer day the temperature in the laboratory, as measured on a lab thermometer, is 28 °C. Express this temperature on the Fahrenheit scale.

SOLUTION This problem can be represented as 28 °C = ? °F. We will solve it using the formula

$$T_{°F} = 1.80\,(T_{°C}) + 32$$

In this case,

$$T_{°F} = ?\,°F = 1.80(\underset{T_{°C}}{28}) + 32 = \underset{\text{Rounds off to 50}}{50.4} + 32$$

$$= 50. + 32 = 82$$

Thus 28 °C = 82 °F. ■

Interactive Example 2.11

Temperature Conversion: Celsius to Fahrenheit

Express the temperature −40. °C on the Fahrenheit scale.

SOLUTION We can express this problem as −40. °C = ? °F. To solve it we will use the formula

$$T_{°F} = 1.80\,(T_{°C}) + 32$$

In this case,

$$T_{°F} = ?\,°F = 1.80(\underset{T_{°C}}{-40.}) + 32$$

$$= -72 + 32 = -40$$

So −40 °C = −40 °F. This is a very interesting result and is another useful reference point.

SELF CHECK

Exercise 2.7 Hot tubs are often maintained at 41 °C. What is this temperature in Fahrenheit degrees?

See Problems 2.75 through 2.78. ■

To convert from Celsius to Fahrenheit, we have used the equation

$$T_{°F} = 1.80\,(T_{°C}) + 32$$

To convert a Fahrenheit temperature to Celsius, we need to rearrange this equation to isolate Celsius degrees ($T_{°C}$). Remember, we can always do the same operation to both sides of the equation. First subtract 32 from each side:

$$T_{°F} - 32 = 1.80\,(T_{°C}) + \underbrace{32 - 32}_{\text{Sum is zero}}$$

to give

$$T_{°F} - 32 = 1.80(T_{°C})$$

Next divide both sides by 1.80

$$\frac{T_{°F} - 32}{1.80} = \frac{\cancel{1.80}(T_{°C})}{\cancel{1.80}}$$

to give

$$\frac{T_{°F} - 32}{1.80} = T_{°C}$$

or

Temperature in °F

$$T_{°C} = \frac{T_{°F} - 32}{1.80}$$

Temperature in °C

$$T_{°C} = \frac{T_{°F} - 32}{1.80}$$

Interactive Example 2.12

Temperature Conversion: Fahrenheit to Celsius

One of the body's responses to an infection or injury is to elevate its temperature. A certain flu victim has a body temperature of 101 °F. What is this temperature on the Celsius scale?

SOLUTION The problem is 101 °F = ? °C. Using the formula

$$T_{°C} = \frac{T_{°F} - 32}{1.80}$$

yields

$$T_{°C} = ?\,°C = \frac{\overset{T_{°F}}{101} - 32}{1.80} = \frac{69}{1.80} = 38$$

That is, 101 °F = 38 °C.

SELF CHECK

Exercise 2.8 An antifreeze solution in a car's radiator boils at 239 °F. What is this temperature on the Celsius scale?

See Problems 2.75 through 2.78. ■

In doing temperature conversions, you will need the following formulas.

Temperature Conversion Formulas

- Celsius to Kelvin $T_K = T_{°C} + 273$
- Kelvin to Celsius $T_{°C} = T_K - 273$
- Celsius to Fahrenheit $T_{°F} = 1.80(T_{°C}) + 32$
- Fahrenheit to Celsius $T_{°C} = \frac{T_{°F} - 32}{1.80}$

2-8 Density

OBJECTIVE To define density and its units.

When you were in elementary school, you may have been embarrassed by your answer to the question "Which is heavier, a pound of lead or a pound of feathers?" If you said lead, you were undoubtedly thinking about density, not mass. **Density** can be defined as the amount of matter present *in a given volume* of substance. That is, density is mass per unit volume, the ratio of the mass of an object to its volume:

$$\text{Density} = \frac{\text{mass}}{\text{volume}}$$

It takes a much bigger volume to make a pound of feathers than to make a pound of lead. This is because lead has a much greater mass per unit volume—a greater density.

The density of a liquid can be determined easily by weighing a known volume of the substance as illustrated in Example 2.13.

Interactive Example 2.13 Calculating Density

Suppose a student finds that 23.50 mL of a certain liquid weighs 35.062 g. What is the density of this liquid?

SOLUTION We can calculate the density of this liquid simply by applying the definition

$$\text{Density} = \frac{\text{mass}}{\text{volume}} = \frac{35.062 \text{ g}}{23.50 \text{ mL}} = 1.492 \text{ g/mL}$$

This result could also be expressed as 1.492 g/cm^3 because 1 mL = 1 cm^3. ■

The volume of a solid object is often determined indirectly by submerging it in water and measuring the volume of water displaced. In fact, this is the most accurate method for measuring a person's percent body fat. The person is submerged momentarily in a tank of water, and the increase in volume is measured (Fig. 2.10). It is possible to calculate the body density by using the person's weight (mass) and the volume of the person's body determined by submersion. Fat, muscle, and bone have different densities (fat is less dense than muscle tissue, for example), so the fraction of the person's

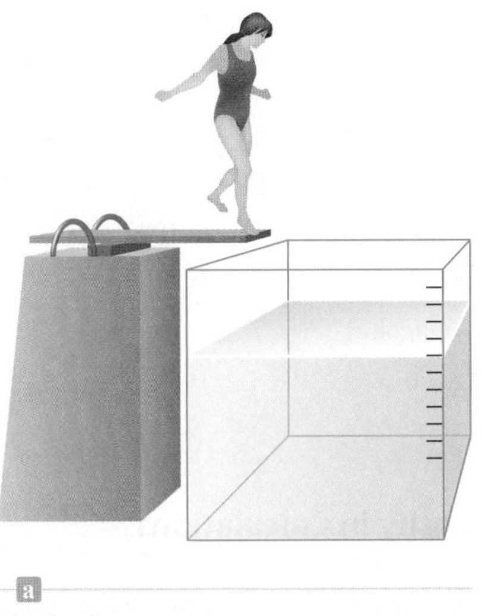

a Tank of water

b Person submerged in the tank, raising the level of the water

Figure 2.10

body that is fat can be calculated. The more muscle and the less fat a person has, the higher his or her body density. For example, a muscular person weighing 150 lb has a smaller body volume (and thus a higher density) than a fat person weighing 150 lb.

Example 2.14 Determining Density

At a local pawn shop a student finds a medallion that the shop owner insists is pure platinum. However, the student suspects that the medallion may actually be silver and thus much less valuable. The student buys the medallion only after the shop owner agrees to refund the price if the medallion is returned within two days. The student, a chemistry major, then takes the medallion to her lab and measures its density as follows. She first weighs the medallion and finds its mass to be 55.64 g. She then places some water in a graduated cylinder and reads the volume as 75.2 mL. Next she drops the medallion into the cylinder and reads the new volume as 77.8 mL. Is the medallion platinum (density = 21.4 g/cm^3) or silver (density = 10.5 g/cm^3)?

SOLUTION The densities of platinum and silver differ so much that the measured density of the medallion will show which metal is present. Because by definition

$$\text{Density} = \frac{\text{mass}}{\text{volume}}$$

to calculate the density of the medallion, we need its mass and its volume. The mass of the medallion is 55.64 g. The volume of the medallion can be obtained by taking the difference between the volume readings of the water in the graduated cylinder before and after the medallion was added.

$$\text{Volume of medallion} = 77.8\text{ mL} - 75.2\text{ mL} = 2.6\text{ mL}$$

The volume appeared to increase by 2.6 mL when the medallion was added, so 2.6 mL represents the volume of the medallion. Now we can use the measured mass and volume of the medallion to determine its density:

$$\text{Density of medallion} = \frac{\text{mass}}{\text{volume}} = \frac{55.64\text{ g}}{2.6\text{ mL}} = 21\text{ g/mL}$$

or

$$= 21\text{ g/cm}^3$$

The medallion is really platinum.

SELF CHECK

Exercise 2.9 A student wants to identify the main component in a commercial liquid cleaner. He finds that 35.8 mL of the cleaner weighs 28.1 g. Of the following possibilities, which is the main component of the cleaner?

Substance	Density, g/cm^3
chloroform	1.483
diethyl ether	0.714
isopropyl alcohol	0.785
toluene	0.867

See Problems 2.89 and 2.90. ■

Interactive Example 2.15 Using Density in Calculations

Mercury has a density of 13.6 g/mL. What volume of mercury must be taken to obtain 225 g of the metal?

SOLUTION To solve this problem, start with the definition of density,

$$\text{Density} = \frac{\text{mass}}{\text{volume}}$$

and then rearrange this equation to isolate the required quantity. In this case we want to find the volume. Remember that we maintain an equality when we do the same thing to both sides. For example, if we multiply *both sides* of the density definition by volume,

$$\text{Volume} \times \text{density} = \frac{\text{mass}}{\cancel{\text{volume}}} \times \cancel{\text{volume}}$$

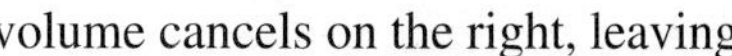

volume cancels on the right, leaving

$$\text{Volume} \times \text{density} = \text{mass}$$

We want the volume, so we now divide both sides by density,

$$\frac{\text{Volume} \times \cancel{\text{density}}}{\cancel{\text{density}}} = \frac{\text{mass}}{\text{density}}$$

to give

$$\text{Volume} = \frac{\text{mass}}{\text{density}}$$

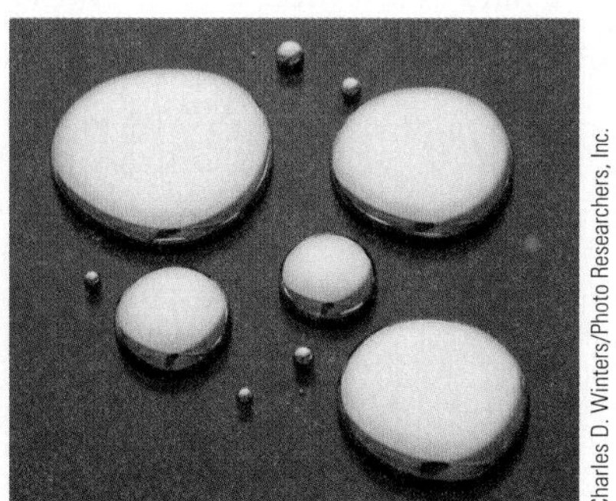

Charles D. Winters/Photo Researchers, Inc.

Drops of mercury, a very dense liquid.

Now we can solve the problem by substituting the given numbers:

$$\text{Volume} = \frac{225\text{ g}}{13.6\text{ g/mL}} = 16.5\text{ mL}$$

We must take 16.5 mL of mercury to obtain an amount that has a mass of 225 g. ■

The densities of various common substances are given in Table 2.8. Besides being a tool for the identification of substances, density has many other uses. For example, the liquid in your car's lead storage battery (a solution of sulfuric acid) changes density because the sulfuric acid is consumed as the battery discharges. In a fully charged battery, the density of the solution is about 1.30 g/cm^3. When the density falls below 1.20 g/cm^3, the battery has to be recharged. Density measurement is also used to determine the amount of antifreeze, and thus the level of protection against freezing, in the cooling system of a

Table 2.8 ▸ Densities of Various Common Substances at 20 °C

Substance	Physical State	Density (g/cm^3)
oxygen	gas	0.00133*
hydrogen	gas	0.000084*
ethanol	liquid	0.785
benzene	liquid	0.880
water	liquid	1.000
magnesium	solid	1.74
salt (sodium chloride)	solid	2.16
aluminum	solid	2.70
iron	solid	7.87
copper	solid	8.96
silver	solid	10.5
lead	solid	11.34
mercury	liquid	13.6
gold	solid	19.32

*At 1 atmosphere pressure

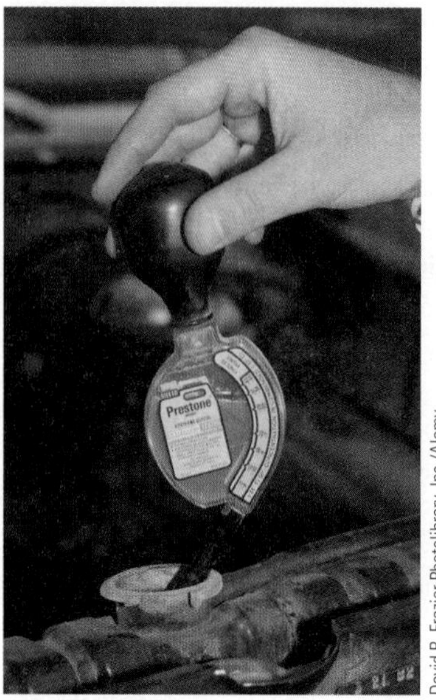

David R. Frazier Photolibrary, Inc./Alamy

Figure 2.11 ▸ A hydrometer being used to determine the density of the antifreeze solution in a car's radiator.

car. Water and antifreeze have different densities, so the measured density of the mixture tells us how much of each is present. The device used to test the density of the solution—a hydrometer—is shown in Fig. 2.11.

In certain situations, the term *specific gravity* is used to describe the density of a liquid. **Specific gravity** is defined as the ratio of the density of a given liquid to the density of water at 4 °C. Because it is a ratio of densities, specific gravity has no units.

CHAPTER 2 REVIEW

F directs you to the *Chemistry in Focus* feature in the chapter

Key Terms

scientific notation (2-1)
unit (2-2)
English system (2-2)
metric system (2-2)
SI units (2-2)
volume (2-3)
mass (2-3)
significant figures (2-4)
rounding off (2-5)
conversion factor (2-6)
equivalence statement (2-6)
dimensional analysis (2-6)
Fahrenheit scale (2-7)
Celsius scale (2-7)
Kelvin (absolute) scale (2-7)
density (2-8)
specific gravity (2-8)

For Review

- A quantitative observation is called a measurement and consists of a number and a unit.
- Very large or very small numbers are conveniently expressed by using scientific notation.
  - The number is expressed as a number between 1 and 10 multiplied by 10 and raised to a power.
- Units provide a scale on which to represent the results of a measurement. There are three commonly used unit systems.
  - English
  - Metric (uses prefixes to change the size of the unit)
  - SI (uses prefixes to change the size of the unit)
- All measurements have some uncertainty, which is reflected by the number of significant figures used to express the number.
- Rules exist for rounding off to the correct number of significant figures in a calculated result.
- We can convert from one system of units to another by a method called dimensional analysis using conversion factors.
- Conversion factors are built from an equivalence statement, which shows the relationship between the units in different systems.

English–Metric and English–English Equivalents

Length	1 m = 1.094 yd
	2.54 cm = 1 in.
	1 mi = 5280. ft
	1 mi = 1760. yd
Mass	1 kg = 2.205 lb
	453.6 g = 1 lb
Volume	1 L = 1.06 qt
	1 ft^3 = 28.32 L

- There are three commonly used temperature scales: Fahrenheit, Celsius, and Kelvin.
- We can convert among the temperature scales by adjusting the zero point and the size of the unit. Useful equations for conversions are:
  - $T_{°C} + 273 = T_K$
  - $T_{°F} = 1.80(T_{°C}) + 32$
  - Density represents the amount of matter present in a given volume:

$$\text{Density} = \frac{\text{mass}}{\text{volume}}$$

Active Learning Questions

These questions are designed to be considered by groups of students in class. Often these questions work well for introducing a particular topic in class.

1. a. There are 365 days/year, 24 hours/day, 12 months/year, and 60 minutes/hour. How many minutes are there in one month?
   b. There are 24 hours/day, 60 minutes/hour, 7 days/week, and 4 weeks/month. How many minutes are there in one month?
   c. Why are these answers different? Which (if either) is more correct and why?

2. You go to a convenience store to buy candy and find the owner to be rather odd. He allows you to buy pieces only in multiples of four, and to buy four, you need $0.23. He allows you only to use 3 pennies and 2 dimes. You have a bunch of pennies and dimes, and instead of counting them, you decide to weigh them. You have 636.3 g of pennies, and each penny weighs an average of 3.03 g. Each dime weighs an average of 2.29 g. Each piece of candy weighs an average of 10.23 g.
   a. How many pennies do you have?
   b. How many dimes do you need to buy as much candy as possible?
   c. How much would all of your dimes weigh?
   d. How many pieces of candy could you buy (based on the number of dimes from part b)?
   e. How much would this candy weigh?
   f. How many pieces of candy could you buy with twice as many dimes?

3. When a marble is dropped into a beaker of water, it sinks to the bottom. Which of the following is the best explanation?
   a. The surface area of the marble is not large enough for the marble to be held up by the surface tension of the water.
   b. The mass of the marble is greater than that of the water.
   c. The marble weighs more than an equivalent volume of the water.
   d. The force from dropping the marble breaks the surface tension of the water.
   e. The marble has greater mass and volume than the water.

   Explain each choice. That is, for choices you did not pick, explain why you feel they are wrong, and justify the choice you did pick.

4. Consider water in each graduated cylinder as shown:

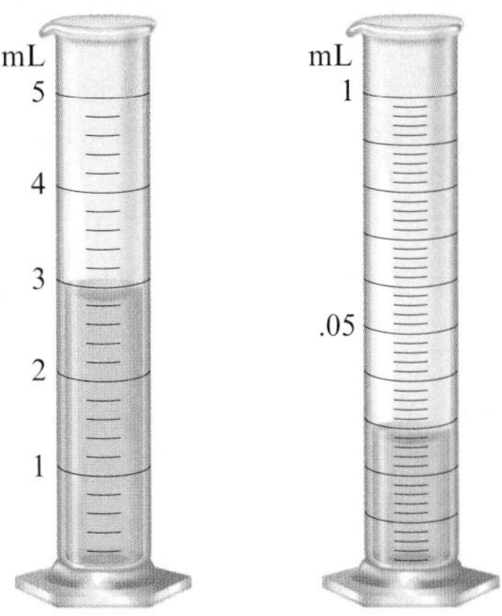

You add both samples of water to a beaker. How would you write the number describing the total volume? What limits the precision of this number?

5. What is the numerical value of a conversion factor? Why must this be true?

6. For each of the following numbers, indicate which zeros are significant and explain. Do not merely cite the rule that applies, but explain the rule.

   a. 10.020 b. 0.002050 c. 190 d. 270

7. Consider the addition of "15.4" to "28." What would a mathematician say the answer is? What would a scientist say? Justify the scientist's answer, not merely citing the rule, but explaining it.

8. Consider multiplying "26.2" by "16.43." What would a mathematician say the answer is? What would a scientist say? Justify the scientist's answer, not merely citing the rule, but explaining it.

9. In lab you report a measured volume of 128.7 mL of water. Using significant figures as a measure of the error, what range of answers does your reported volume imply? Explain.

10. Sketch two pieces of glassware: one that can measure volume to the thousandths place, and one that can measure volume only to the ones place.

11. Oil floats on water but is "thicker" than water. Why do you think this fact is true?

12. Show how converting numbers to scientific notation can help you decide which digits are significant.

13. You are driving 65 mph and take your eyes off the road "just for a second." How many feet do you travel in this time?

14. You have a 1.0-cm^3 sample of lead and a 1.0-cm^3 sample of glass. You drop each in a separate beaker of water. How do the volumes of water that are displaced by the samples compare? Explain.

15. The beakers shown below have different precisions.

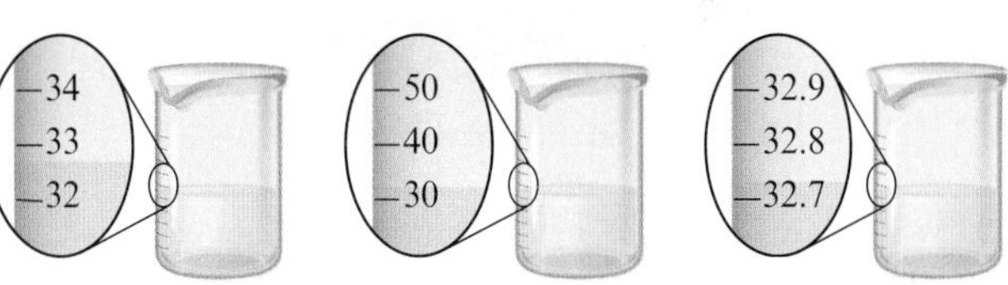

   a. Label the amount of water in each of the three beakers to the correct number of significant figures.
   b. Is it possible for each of the three beakers to contain the exact same amount of water? If no, why not? If yes, did you report the volumes as the same in part a? Explain.
   c. Suppose you pour the water from these three beakers into one container. What should be the volume in the container reported to the correct number of significant figures?

16. True or false? For any mathematical operation performed on two measurements, the number of significant figures in the answer is the same as the least number of significant figures in either of the measurements. Explain your answer.

17. Complete the following and explain each in your own words: leading zeros are (never/sometimes/always) significant; captive zeros are (never/sometimes/always) significant; and trailing zeros are (never/sometimes/always) significant.

All even-numbered Questions and Problems have answers in the back of this book and solutions in the *Student Solutions Guide*.

For any statement with an answer of "sometimes," give examples of when the zero is significant and when it is not, and explain.

18. For each of the following figures, a through d, decide which block is more dense: the orange block, the blue block, or it cannot be determined. Explain your answers.

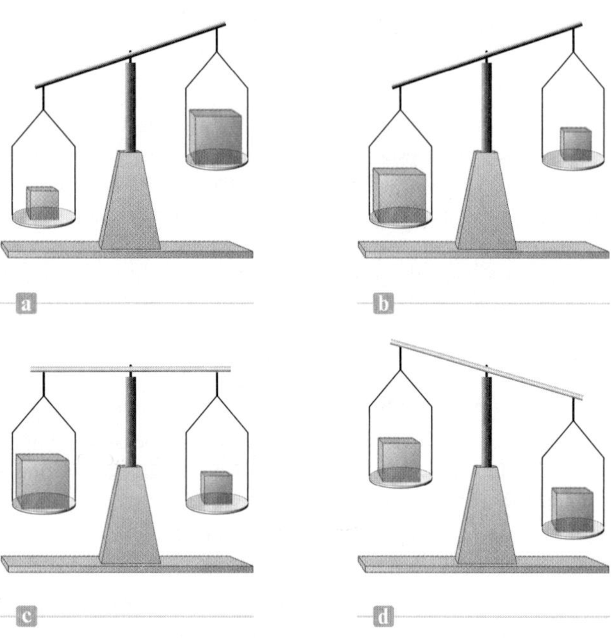

19. For the pin shown below, why is the third digit determined for the length of the pin uncertain? Considering that the third digit is uncertain, explain why the length of the pin is indicated as 2.85 cm rather than, for example, 2.83 or 2.87 cm.

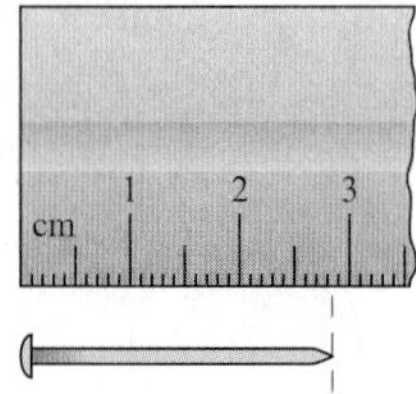

20. Why can the length of the pin shown below not be recorded as 2.850 cm?

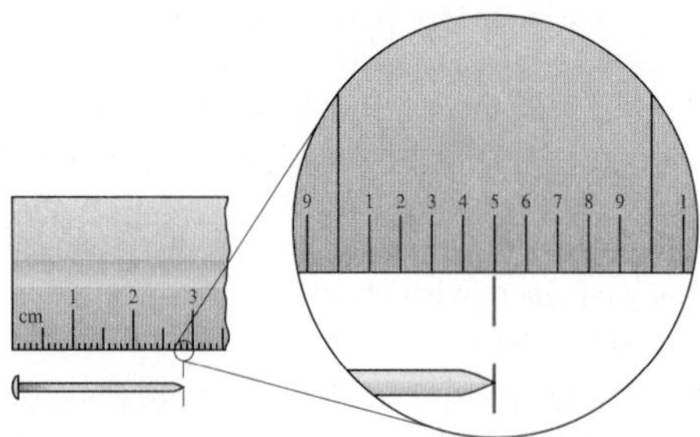

21. Use the figure below to answer the following questions.

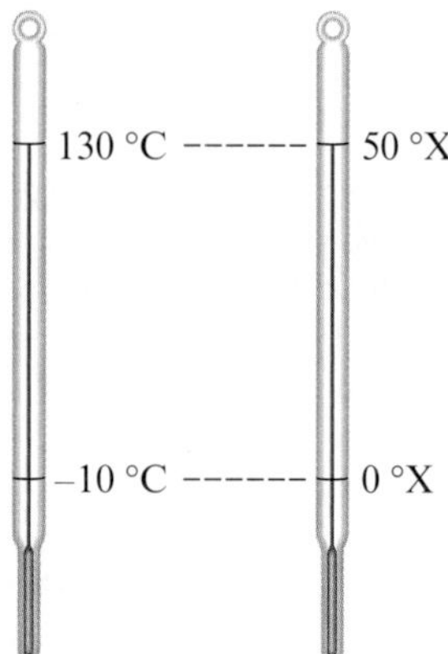

a. Derive the relationship between °C and °X.
b. If the temperature outside is 22.0 °C, what is the temperature in units of °X?
c. Convert 58.0 °X to units of °C, K, and °F.

Questions and Problems

2-1 Scientific Notation

Questions

1. A ________ represents a quantitative observation.

2. Although your textbook lists the rules for converting an ordinary number to scientific notation, oftentimes students remember such rules better if they put them into their own words. Pretend you are helping your 12-year-old niece with her math homework, and write a paragraph explaining to her how to convert the ordinary number 2421 to scientific notation.

3. When a large or small number is written in standard scientific notation, the number is expressed as the product of a *number* between 1 and 10, multiplied by the appropriate *power* of 10. For each of the following numbers, indicate what number between 1 and 10 would be appropriate when expressing the numbers in standard scientific notation.

   a. 9651   c. 93,241
   b. 0.003521   d. 0.000001002

4. When a large or small number is written in standard scientific notation, the number is expressed as the product of a *number* between 1 and 10, multiplied by the appropriate *power* of 10. For each of the following numbers, indicate what power of 10 would be appropriate when expressing the numbers in standard scientific notation.

   a. 9,367,421   c. 0.0005519
   b. 0.0624   d. 5,408,000,000

Problems

5. Will the power of 10 have a *positive* or a *negative* exponent when each of the following numbers is rewritten in standard scientific notation?

   a. 42,751   c. 0.002045
   b. 1253   d. 0.1089

All even-numbered Questions and Problems have answers in the back of this book and solutions in the *Student Solutions Guide.*

6. Will the power of 10 have a *positive, negative,* or *zero* exponent when each of the following numbers is rewritten in standard scientific notation?

a. 0.7229 c. 0.00372
b. 5.408 d. 6,319,428

7. Express each of the following numbers in *standard* scientific notation.

a. 0.5012 d. 5.012
b. 5,012,000 e. 5012
c. 0.000005012 f. 0.005012

8. Rewrite each of the following as an "ordinary" decimal number.

a. 2.789×10^3 d. 4.289×10^1
b. 2.789×10^{-3} e. 9.999×10^4
c. 9.3×10^7 f. 9.999×10^{-5}

9. By how many places must the decimal point be moved, and in which direction, to convert each of the following to "ordinary" decimal numbers?

a. 4.311×10^6 d. 4.995×10^0
b. 7.895×10^{-5} e. 2.331×10^{18}
c. 8.712×10^1 f. 1.997×10^{-16}

10. By how many places must the decimal point be moved, and in which direction, to convert each of the following to standard scientific notation?

a. 5993 d. 62.357
b. −72.14 e. 0.01014
c. 0.00008291 f. 324.9

11. Write each of the following numbers in *standard* scientific notation.

a. 97,820 d. 0.0003914
b. 42.14×10^3 e. 927.1
c. 0.08214×10^{-3} f. $4.781 \times 10^2 \times 10^{-3}$

12. Write each of the following numbers as "ordinary" decimal numbers.

a. 6.244×10^3 d. 1.771×10^{-4}
b. 9.117×10^{-2} e. 5.451×10^2
c. 8.299×10^1 f. 2.934×10^{-5}

13. Write each of the following numbers in *standard* scientific notation.

a. 1/1033 e. 1/3,093,000
b. $1/10^5$ f. $1/10^{-4}$
c. $1/10^{-7}$ g. $1/10^9$
d. 1/0.0002 h. 1/0.000015

14. Write each of the following numbers in *standard* scientific notation.

a. 1/0.00032 e. $(10^5)(10^4)(10^{-4})/(10^{-2})$
b. $10^3/10^{-3}$ f. $43.2/(4.32 \times 10^{-5})$
c. $10^3/10^3$ g. $(4.32 \times 10^{-5})/432$
d. 1/55,000 h. $1/(10^5)(10^{-6})$

2-2 Units

Questions

15. What are the fundamental units of mass, length, and temperature in the metric system?

16. Give the metric prefix that corresponds to each of the following:

a. 1000 d. 1,000,000
b. 10^{-3} e. 10^{-1}
c. 10^{-9} f. 10^{-6}

2-3 Measurements of Length, Volume, and Mass

Questions

Students often have trouble relating measurements in the metric system to the English system they have grown up with. Give the approximate English system equivalents for each of the following metric system descriptions in Exercises 17–20.

17. My new kitchen floor will require 25 square meters of linoleum.

18. My recipe for chili requires a 125-g can of tomato paste.

19. The gas tank in my new car holds 48 liters.

20. I need some 2.5-cm-long nails to hang up this picture.

21. The road sign I just passed says "New York City 100 km," which is about ________ mi.

22. The GPS in my car indicates that I have 100. mi left until I reach my destination. What is this distance in kilometers?

23. The tablecloth on my dining room table is 2 m long, which is ________ cm or about ________ in.

24. Who is taller, a man who is 1.62 m tall or a woman who is 5 ft 6 in. tall?

25. The fundamental SI unit of length is the meter. However, we often deal with larger or smaller lengths or distances for which multiples or fractions of the fundamental unit are more useful. For each of the following situations, suggest what fraction or multiple of the meter might be the most appropriate measurement.

a. the distance between Chicago and Saint Louis
b. the size of your bedroom
c. the dimensions of this textbook
d. the thickness of a hair

26. Which English unit of length or distance is most comparable in scale to each of the following metric system units for making measurements?

a. a centimeter
b. a meter
c. a kilometer

27. The unit of volume in the metric system is the liter, which consists of 1000 milliliters. How many liters or milliliters is each of the following common English system measurements approximately equivalent to?

a. a gallon of gasoline
b. a pint of milk
c. a cup of water

All even-numbered Questions and Problems have answers in the back of this book and solutions in the *Student Solutions Guide.*

28. Which metric system unit is most appropriate for measuring the length of an insect such as a beetle?

    a. meters c. megameters
    b. millimeters d. kilometers

2-4 Uncertainty in Measurement

Questions

29. When a measuring scale is used properly to the limit of precision, the last significant digit recorded for the measurement is said to be *uncertain*. Explain.
30. What does it mean to say that every measurement we make with a measuring device contains some measure of *uncertainty?*
31. For the pin shown in Fig. 2.5, why is the third figure determined for the length of the pin uncertain? Considering that the third figure is uncertain, explain why the length of the pin is indicated as 2.85 cm rather than, for example, 2.83 or 2.87 cm.
32. Why can the length of the pin shown in Fig. 2.5 not be recorded as 2.850 cm?

2-5 Significant Figures

Questions

33. Indicate the number of significant figures in each of the following:

    a. 250. b. 250 c. 2.5×10^2 d. 250.0

34. Indicate the number of significant figures implied in each of the following statements:

    a. The population of the United States is 310 million.
    b. One hour is equivalent to 60 minutes.
    c. There are 5280 feet in 1 mile.
    d. Jet airliners fly at 500 mi/h.
    e. The Daytona 500 is a 500-mile race.

Rounding Off Numbers

Questions

35. When we round off a number, if the number to the right of the digit to be rounded is greater than 5, then we should ________.
36. In a multiple-step calculation, is it better to round off the numbers to the correct number of significant figures in each step of the calculation or to round off only the final answer? Explain.
37. Round off each of the following numbers to three significant digits, and express the result in standard scientific notation.

    a. 254,931 c. 47.85×10^3
    b. 0.00025615 d. 0.08214×10^5

38. Round off each of the following numbers to three significant digits, and express the result in standard scientific notation.

    a. 1,566,311 c. 0.07759
    b. 2.7651×10^{-3} d. 0.0011672

39. Round off each of the following numbers to the indicated number of significant digits and write the answer in standard scientific notation.

    a. 4341×10^2 to three significant digits
    b. 93.441×10^3 to three significant digits
    c. 0.99155×10^2 to four significant digits
    d. 9.3265 to four significant digits

40. Round off each of the following numbers to the indicated number of significant digits and write the answer in standard scientific notation.

    a. 0.00034159 to three digits
    b. 103.351×10^2 to four digits
    c. 17.9915 to five digits
    d. 3.365×10^5 to three digits

Determining Significant Figures in Calculations

Questions

41. Consider the calculation indicated below:

$$\frac{2.21 \times 0.072333 \times 0.15}{4.995}$$

    Explain why the answer to this calculation should be reported to only two significant digits.

42. The following water measurements are made: 18 mL of water measured with a beaker, 128.7 mL of water measured with a graduated cylinder, and 23.45 mL of water measured with a buret. If all of these water samples are then poured together into one container, what total volume of water should be reported? Support your answer.
43. When the calculation $(2.31)(4.9795 \times 10^3)/(1.9971 \times 10^4)$ is performed, how many significant digits should be reported for the answer? You should *not* need to perform the calculation.
44. How many of the following measurements and/or calculations are recorded to *one* significant figure?

    a. 1.0×10^3 m
    b. 2000 in.
    c. 0.004 kg
    d. $\frac{2.8 - 2.0}{0.80} = ?$

45. When the sum 4.9965 + 2.11 + 3.887 is calculated, to how many decimal places should the answer be reported? You should *not* need to perform the calculation.
46. How many digits after the decimal point should be reported when the calculation (10,434 − 9.3344) is performed?

Problems

Note: See the Appendix for help in doing mathematical operations with numbers that contain exponents.

47. Evaluate each of the following mathematical expressions, and express the answer to the correct number of significant digits.

    a. 44.2124 + 0.81 + 7.335
    b. 9.7789 + 3.3315 − 2.21
    c. 0.8891 + 0.225 + 4.14
    d. (7.223 + 9.14 + 3.7795)/3.1

All even-numbered Questions and Problems have answers in the back of this book and solutions in the *Student Solutions Guide.*

48. Evaluate each of the following mathematical expressions, and express the answer to the correct number of significant digits.
    a. $(4.771 + 2.3)/3.1$
    b. $5.02 \times 10^2 + 4.1 \times 10^2$
    c. $1.091 \times 10^3 + 2.21 \times 10^2 + 1.14 \times 10^1$
    d. $(2.7991 \times 10^{-6})/(4.22 \times 10^6)$
49. *Without actually performing the calculations indicated,* tell to how many significant digits the answer to the calculation should be expressed.
    a. $(0.196)(0.08215)(295)/(1.1)$
    b. $(4.215 + 3.991 + 2.442)/(0.22)$
    c. $(7.881)(4.224)(0.00033)/(2.997)$
    d. $(6.219 + 2.03)/(3.1159)$
50. *Without actually performing the calculations indicated,* tell to how many significant digits the answer to the calculation should be expressed.
    a. $\dfrac{(9.7871)(2)}{(0.00182)(43.21)}$
    b. $(67.41 + 0.32 + 1.98)/(18.225)$
    c. $(2.001 \times 10^{-3})(4.7 \times 10^{-6})(68.224 \times 10^{-2})$
    d. $(72.15)(63.9)[1.98 + 4.8981]$
51. How many significant digits should be used to report the answer to each of the following calculations? Do not perform the calculations.
    a. $(2.7518 + 9.01 + 3.3349)/(2.1)$
    b. $(2.7751 \times 1.95)/(.98)$
    c. $12.0078/3.014$
    d. $(0.997 + 4.011 + 3.876)/(1.86 \times 10^{-3})$
52. Evaluate each of the following and write the answer to the appropriate number of significant figures.
    a. $(2.0944 + 0.0003233 + 12.22)/(7.001)$
    b. $(1.42 \times 10^2 + 1.021 \times 10^3)/(3.1 \times 10^{-1})$
    c. $(9.762 \times 10^{-3})/(1.43 \times 10^2 + 4.51 \times 10^1)$
    d. $(6.1982 \times 10^{-4})^2$

2-6 Problem Solving and Dimensional Analysis

Questions

53. A ________ represents a ratio based on an equivalence statement between two measurements.
54. How many significant figures are understood for the numbers in the following definition: 1 in. = 2.54 cm?
55. Given that 1 mi = 1760 yd, determine what conversion factor is appropriate to convert 1849 yd to miles; to convert 2.781 mi to yards.
56. Given that 1 mi = 5280 ft exactly, indicate what conversion factor is appropriate to convert 15.6 mi to feet; to convert 86.19 ft to miles.

For Exercises 57 and 58, apples cost $1.75 per pound.

57. What conversion factor is appropriate to express the cost of 5.3 lb of apples?
58. What conversion factor could be used to determine how many pounds of apples could be bought for $10.00?

Problems

Note: Appropriate equivalence statements for various units are found inside the back cover of this book.

59. Perform each of the following conversions, being sure to set up the appropriate conversion factor in each case.
    a. 12.5 in. to centimeters
    b. 12.5 cm to inches
    c. 2513 ft to miles
    d. 4.53 ft to meters
    e. 6.52 min to seconds
    f. 52.3 cm to meters
    g. 4.21 m to yards
    h. 8.02 oz to pounds
60. Perform each of the following conversions, being sure to set up the appropriate conversion factor in each case.
    a. 2.23 m to yards
    b. 46.2 yd to meters
    c. 292 cm to inches
    d. 881.2 in. to centimeters
    e. 1043 km to miles
    f. 445.5 mi to kilometers
    g. 36.2 m to kilometers
    h. 0.501 km to centimeters
61. Perform each of the following conversions, being sure to set up the appropriate conversion factor in each case.
    a. 1.75 mi to kilometers
    b. 2.63 gal to quarts
    c. 4.675 calories to joules
    d. 756.2 mm Hg to atmospheres
    e. 36.3 atomic mass units to kilograms
    f. 46.2 in. to centimeters
    g. 2.75 qt to fluid ounces
    h. 3.51 yd to meters
62. Perform each of the following conversions, being sure to set up the appropriate conversion factor in each case.
    a. 254.3 g to kilograms
    b. 2.75 kg to grams
    c. 2.75 kg to pounds
    d. 2.75 kg to ounces
    e. 534.1 g to pounds
    f. 1.75 lb to grams
    g. 8.7 oz to grams
    h. 45.9 g to ounces
63. 12.01 g of carbon contains 6.02×10^{23} carbon atoms. What is the mass in grams of 1.89×10^{25} carbon atoms?
64. Los Angeles and Honolulu are 2558 mi apart. What is this distance in kilometers?
65. The United States has high-speed trains running between Boston and New York capable of speeds up to 160 mi/h. Are these trains faster or slower than the fastest trains in the United Kingdom, which reach speeds of 225 km/h?
66. The radius of an atom is on the order of 10^{-10} m. What is this radius in centimeters? in inches? in nanometers?

All even-numbered Questions and Problems have answers in the back of this book and solutions in the *Student Solutions Guide.*

2-7 Temperature Conversions

Questions

67. The temperature scale used in everyday life in most of the world except the United States is the ________ scale.

68. The ________ point of water is at 32° on the Fahrenheit temperature scale.

69. The normal boiling point of water is ________ °F, or ________ °C.

70. The boiling point of water is ________ K.

71. On both the Celsius and Kelvin temperature scales, there are ________ degrees between the normal freezing and boiling points of water.

72. On which temperature scale (°F, °C, or K) does 1 degree represent the smallest change in temperature?

Problems

73. Make the following temperature conversions:

a. 44.2 °C to kelvins
b. 891 K to °C
c. −20 °C to kelvins
d. 273.1 K to °C

74. Carry out the indicated temperature conversions.

a. −201 °F to kelvins
b. −201 °C to kelvins
c. 351 °C to Fahrenheit degrees
d. −150 °F to Celsius degrees

75. Convert the following Fahrenheit temperatures to Celsius degrees.

a. a chilly morning in early autumn, 45 °F
b. a hot, dry day in the Arizona desert, 115 °F
c. the temperature in winter when my car won't start, −10 °F
d. the surface of a star, 10,000 °F

76. Convert the following Celsius temperatures to Fahrenheit degrees.

a. the boiling temperature of ethyl alcohol, 78.1 °C
b. a hot day at the beach on a Greek isle, 40. °C
c. the lowest possible temperature, −273 °C
d. the body temperature of a person with hypothermia, 32 °C

F 77. The "Chemistry in Focus" segment *Tiny Thermometers* states that the temperature range for the carbon nanotube gallium thermometers is 50 °C to 500 °C.

a. What properties of gallium make it useful in a thermometer?
b. Determine the useful temperature range for the gallium thermometer in Fahrenheit units.

78. Perform the indicated temperature conversions.

a. 275 K to °C
b. 82 °F to °C
c. −21 °C to °F
d. −40 °F to °C (Notice anything unusual about your answer?)

2-8 Density

Questions

79. What does the *density* of a substance represent?

80. The most common units for density are ________.

81. A kilogram of lead occupies a much smaller volume than a kilogram of water, because ________ has a much higher density.

82. If a solid block of glass, with a volume of exactly 100 in.^{3}, is placed in a basin of water that is full to the brim, then ________ of water will overflow from the basin.

83. Is the density of a gaseous substance likely to be larger or smaller than the density of a liquid or solid substance at the same temperature? Why?

84. What property of density makes it useful as an aid in identifying substances?

85. Referring to Table 2.8, which substance listed is most dense? Which substance is least dense? For the two substances you have identified, for which one would a 1.00-g sample occupy the larger volume?

86. Referring to Table 2.8, determine whether magnesium, ethanol, silver, or salt is the most dense.

Problems

87. For the masses and volumes indicated, calculate the density in grams per cubic centimeter.

a. mass = 452.1 g; volume = 292 cm^3
b. mass = 0.14 lb; volume = 125 mL
c. mass = 1.01 kg; volume = 1000 cm^3
d. mass = 225 mg; volume = 2.51 mL

88. For the masses and volumes indicated, calculate the density in grams per cubic centimeter.

a. mass = 4.53 kg; volume = 225 cm^3
b. mass = 26.3 g; volume = 25.0 mL
c. mass = 1.00 lb; volume = 500. cm^3
d. mass = 352 mg; volume = 0.271 cm^3

89. The element bromine at room temperature is a liquid with a density of 3.12 g/mL. Calculate the mass of 125 mL of bromine. What volume does 85.0 g of bromine occupy?

90. Sunflower oil has a density of 0.920 g/mL. What is the mass of 4.50 L of sunflower oil? What volume (in L) would 375 g of sunflower oil occupy?

91. If 1000. mL of linseed oil has a mass of 929 g, calculate the density of linseed oil.

92. A material will float on the surface of a liquid if the material has a density less than that of the liquid. Given that the density of water is approximately 1.0 g/mL under many conditions, will a block of material having a volume of 1.2×10^4 in.^{3} and weighing 3.5 lb float or sink when placed in a reservoir of water?

93. Iron has a density of 7.87 g/cm^3. If 52.4 g of iron is added to 75.0 mL of water in a graduated cylinder, to what volume reading will the water level in the cylinder rise?

94. The density of pure gold is 19.32 g/cm^3 at 20 °C. If 25.75 g of pure gold nuggets is added to a graduated cylinder containing 13.3 mL of water, to what volume level will the water in the cylinder rise?

95. Use the information in Table 2.8 to calculate the volume of 50.0 g of each of the following substances.

a. sodium chloride
b. mercury
c. benzene
d. silver

All even-numbered Questions and Problems have answers in the back of this book and solutions in the *Student Solutions Guide*.

96. Use the information in Table 2.8 to calculate the mass of 50.0 cm^3 of each of the following substances.

a. gold
b. iron
c. lead
d. aluminum

Additional Problems

97. Indicate the number of significant digits in the answer when each of the following expressions is evaluated (you do *not* have to evaluate the expression).

a. (6.25)/(74.1143)
b. (1.45)(0.08431)(6.022 × 10^{23})
c. (4.75512)(9.74441)/(3.14)

98. Express each of the following as an "ordinary" decimal number.

a. 3.011 × 10^{23}
b. 5.091 × 10^{9}
c. 7.2 × 10^{2}
d. 1.234 × 10^{5}
e. 4.32002 × 10^{-4}
f. 3.001 × 10^{-2}
g. 2.9901 × 10^{-7}
h. 4.2 × 10^{-1}

99. Write each of the following numbers in standard scientific notation, rounding off the numbers to three significant digits.

a. 424.6174
b. 0.00078145
c. 26,755
d. 0.0006535
e. 72.5654

100. If you determine that the perimeter of your textbook is 80 cm and the area is 400 cm^2, how do the numbers of significant figures between the two values compare?

a. The perimeter value has more significant figures because it is a smaller number than the area value.
b. The area value has more significant figures because it is a larger number than the perimeter value.
c. The area value has more significant figures because it contains three significant figures, whereas the perimeter value only contains two.
d. Both values have an infinite number of significant figures because you can add a decimal point after each value along with an unlimited number of zeros.
e. Both values have the same number of significant figures (they each have one significant figure). Neither value has a decimal point present.

101. Make the following conversions.

a. 1.25 in. to feet and to centimeters
b. 2.12 qt to gallons and to liters
c. 2640 ft to miles and to kilometers
d. 1.254 kg lead to its volume in cubic centimeters
e. 250. mL ethanol to its mass in grams
f. 3.5 in.^{3} of mercury to its volume in milliliters and its mass in kilograms

102. On the planet Xgnu, the most common units of length are the blim (for long distances) and the kryll (for shorter distances). Because the Xgnuese have 14 fingers, perhaps it is not surprising that 1400 kryll = 1 blim.

a. Two cities on Xgnu are 36.2 blim apart. What is this distance in kryll?
b. The average Xgnuese is 170 kryll tall. What is this height in blims?
c. This book is presently being used at Xgnu University. The area of the cover of this book is 72.5 square krylls. What is its area in square blims?

103. You pass a road sign saying "New York 110 km." If you drive at a constant speed of 100. km/h, how long should it take you to reach New York?

104. Convert 45 mi/h to m/s, showing how the units cancel appropriately.

105. Suppose your car is rated at 45 mi/gal for highway use and 38 mi/gal for city driving. If you wanted to write your friend in Spain about your car's mileage, what ratings in kilometers per liter would you report?

106. You are in Paris, and you want to buy some peaches for lunch. The sign in the fruit stand indicates that peaches are 2.76 euros per kilogram. Given that there are approximately 1.44 euros to the dollar, calculate what a pound of peaches will cost in dollars.

107. For a pharmacist dispensing pills or capsules, it is often easier to weigh the medication to be dispensed rather than to count the individual pills. If a single antibiotic capsule weighs 0.65 g, and a pharmacist weighs out 15.6 g of capsules, how many capsules have been dispensed?

108. On the planet Xgnu, the natives have 14 fingers. On the official Xgnuese temperature scale (°X), the boiling point of water (under an atmospheric pressure similar to earth's) is 140 °X, whereas water freezes at 14 °X. Derive the relationship between °X and °C.

109. For a material to float on the surface of water, the material must have a density less than that of water (1.0 g/mL) and must not react with the water or dissolve in it. A spherical ball has a radius of 0.50 cm and weighs 2.0 g. Will this ball float or sink when placed in water? (*Note:* Volume of a sphere = $\frac{4}{3}\pi r^3$.)

110. A gas cylinder having a volume of 10.5 L contains 36.8 g of gas. What is the density of the gas?

111. Using Table 2.8, calculate the volume of 25.0 g of each of the following:

a. hydrogen gas (at 1 atmosphere pressure)
b. mercury
c. lead
d. water

112. Ethanol and benzene dissolve in each other. When 100. mL of ethanol is dissolved in 1.00 L of benzene, what is the mass of the mixture? (See Table 2.8.)

All even-numbered Questions and Problems have answers in the back of this book and solutions in the *Student Solutions Guide.*

113. When 2891 is written in scientific notation, the exponent indicating the power of 10 is ________ .

114. For each of the following numbers, if the number is rewritten in scientific notation, will the exponent of the power of 10 be positive, negative, or zero?

a. $1/10^3$
b. 0.00045
c. 52,550
d. 7.21
e. 1/3

115. For each of the following numbers, if the number is rewritten in scientific notation, will the exponent of the power of 10 be positive, negative, or zero?

a. 4,915,442
b. b. 1/1000
c. 0.001
d. 3.75

116. For each of the following numbers, by how many places does the decimal point have to be moved to express the number in standard scientific notation? In each case, is the exponent positive or negative?

a. 102
b. 0.00000000003489
c. 2500
d. 0.00003489
e. 398,000
f. 1
g. 0.3489
h. 0.0000003489

117. For each of the following numbers, by how many places must the decimal point be moved to express the number in standard scientific notation? In each case, will the exponent be positive, negative, or zero?

a. 55,651
b. 0.000008991
c. 2.04
d. 883,541
e. 0.09814

118. For each of the following numbers, by how many places must the decimal point be moved to express the number in standard scientific notation? In each case, will the exponent be positive, negative, or zero?

a. 72.471
b. 0.008941
c. 9.9914
d. 6519
e. 0.000000008715

119. Express each of the following numbers in scientific (exponential) notation.

a. 529
b. 240,000,000
c. 301,000,000,000,000,000
d. 78,444
e. 0.0003442
f. 0.000000000902
g. 0.043
h. 0.0821

120. Express each of the following as an "ordinary" decimal number.

a. 2.98×10^{-5}
b. 4.358×10^{9}
c. 1.9928×10^{-6}
d. 6.02×10^{23}
e. 1.01×10^{-1}
f. 7.87×10^{-3}
g. 9.87×10^{7}
h. 3.7899×10^{2}
i. 1.093×10^{-1}
j. 2.9004×10^{0}
k. 3.9×10^{-4}
l. 1.904×10^{-8}

121. Write each of the following numbers in *standard* scientific notation.

a. 102.3×10^{-5}
b. 32.03×10^{-3}
c. 59933×10^{2}
d. 599.33×10^{4}
e. 5993.3×10^{3}
f. 2054×10^{-1}
g. $32,000,000 \times 10^{-6}$
h. 59.933×10^{5}

122. Write each of the following numbers in *standard* scientific notation. See the Appendix if you need help multiplying or dividing numbers with exponents.

a. $1/10^2$
b. $1/10^{-2}$
c. $55/10^3$
d. $(3.1 \times 10^6)/10^{-3}$
e. $(10^6)^{1/2}$
f. $(10^6)(10^4)/(10^2)$
g. $1/0.0034$
h. $3.453/10^{-4}$

123. The fundamental unit of length or distance in the metric system is the ________ .

124. Draw a piece of lab glassware that can appropriately measure the volume of a liquid as 32.87 mL.

125. Which distance is farther, 100 km or 50 mi?

126. 1 L = ________ dm^3 = ________ cm^3 = ________ mL

127. The volume 0.250 L could also be expressed as ________ mL.

128. The distance 10.5 cm could also be expressed as ________ m.

129. Would an automobile moving at a constant speed of 100 km/h violate a 65-mph speed limit?

130. Which weighs more, 0.001 g of water or 1 mg of water?

131. Which weighs more, 4.25 g of gold or 425 mg of gold?

132. The length 500 m can also be expressed as ________ nm.

133. The ratio of an object's mass to its ________ is called the *density* of the object.

134. You are working on a project where you need the volume of a box. You take the length, height, and width measurements and then multiply the values together to find the volume. You report the volume of the box as 0.310 m^3. If two of your measurements were 0.7120 m and 0.52458 m, what was the other measurement?

135. Indicate the number of significant figures in each of the following:

a. This book contains over 500 pages.
b. A mile is just over 5000 ft.
c. A liter is equivalent to 1.059 qt.
d. The population of the United States is approaching 250 million.
e. A kilogram is 1000 g.
f. The Boeing 747 cruises at around 600 mph.

136. Round off each of the following numbers to three significant digits.

a. 0.00042557
b. 4.0235×10^{-5}
c. 5,991,556
d. 399.85
e. 0.0059998

All even-numbered Questions and Problems have answers in the back of this book and solutions in the *Student Solutions Guide.*

137. Round off each of the following numbers to the indicated number of significant digits.

a. 0.75555 to four digits
b. 292.5 to three digits
c. 17.005 to four digits
d. 432.965 to five digits

138. Evaluate each of the following, and write the answer to the appropriate number of significant figures.

a. $149.2 + 0.034 + 2000.34$
b. $1.0322 \times 10^3 + 4.34 \times 10^3$
c. $4.03 \times 10^{-2} - 2.044 \times 10^{-3}$
d. $2.094 \times 10^5 - 1.073 \times 10^6$

139. Evaluate each of the following, and write the answer to the appropriate number of significant figures.

a. $(0.0432)(2.909)(4.43 \times 10^8)$
b. $(0.8922)/[(0.00932)(4.03 \times 10^2)]$
c. $(3.923 \times 10^2)(2.94)(4.093 \times 10^{-3})$
d. $(4.9211)(0.04434)/[(0.000934)(2.892 \times 10^{-7})]$

140. Evaluate each of the following, and write the answer to the appropriate number of significant figures.

a. $(2.9932 \times 10^4)[2.4443 \times 10^2 + 1.0032 \times 10^1]$
b. $[2.34 \times 10^2 + 2.443 \times 10^{-1}]/(0.0323)$
c. $(4.38 \times 10^{-3})^2$
d. $(5.9938 \times 10^{-6})^{1/2}$

141. Given that 1 L = 1000 cm^3, determine what conversion factor is appropriate to convert 350 cm^3 to liters; to convert 0.200 L to cubic centimeters.

142. Given that 12 months = 1 year, determine what conversion factor is appropriate to convert 72 months to years; to convert 3.5 years to months.

143. Perform each of the following conversions, being sure to set up clearly the appropriate conversion factor in each case.

a. 8.43 cm to millimeters
b. 2.41×10^2 cm to meters
c. 294.5 nm to centimeters
d. 404.5 m to kilometers
e. 1.445×10^4 m to kilometers
f. 42.2 mm to centimeters
g. 235.3 m to millimeters
h. 903.3 nm to micrometers

144. Perform each of the following conversions, being sure to set up clearly the appropriate conversion factor(s) in each case.

a. 908 oz to kilograms
b. 12.8 L to gallons
c. 125 mL to quarts
d. 2.89 gal to milliliters
e. 4.48 lb to grams
f. 550 mL to quarts

145. The mean distance from the earth to the sun is 9.3×10^7 mi. What is this distance in kilometers? in centimeters?

146. Given that one metric ton = 1000 kg, how many metric tons are in 5.3×10^3 lb?

147. Convert the following temperatures to kelvins.

a. 0 °C
b. 25 °C
c. 37 °C
d. 100 °C
e. −175 °C
f. 212 °C

148. Carry out the indicated temperature conversions.

a. 175 °F to kelvins
b. 255 K to Celsius degrees
c. −45 °F to Celsius degrees
d. 125 °C to Fahrenheit degrees

149. For the masses and volumes indicated, calculate the density in grams per cubic centimeter.

a. mass = 234 g; volume = 2.2 cm^3
b. mass = 2.34 kg; volume = 2.2 m^3
c. mass = 1.2 lb; volume = 2.1 ft^3
d. mass = 4.3 ton; volume = 54.2 yd^3

150. A sample of a liquid solvent has a density of 0.915 g/mL. What is the mass of 85.5 mL of the liquid?

151. An organic solvent has a density of 1.31 g/mL. What volume is occupied by 50.0 g of the liquid?

152. A solid metal sphere has a volume of 4.2 ft^3. The mass of the sphere is 155 lb. Find the density of the metal sphere in grams per cubic centimeter.

153. A sample containing 33.42 g of metal pellets is poured into a graduated cylinder initially containing 12.7 mL of water, causing the water level in the cylinder to rise to 21.6 mL. Calculate the density of the metal.

154. Convert the following temperatures to Fahrenheit degrees.

a. −5 °C
b. 273 K
c. −196 °C
d. 0 K
e. 86 °C
f. −273 °C

155. For each of the following descriptions, identify the power of 10 being indicated by the *prefix* in the measurement.

a. The sign on the interstate highway says to tune my AM radio to 540 *kilo*hertz for traffic information.
b. My new digital camera has a two-*giga*byte flash memory card.
c. The shirt I bought for my dad on my European vacation shows the sleeve length in *centi*meters.
d. My brother's camcorder records on 8-*milli*meter tape cassettes.

F **156.** The "Chemistry in Focus" segment *Critical Units!* discusses the importance of unit conversions. Read the segment and make the proper unit conversions to answer the following questions.

a. The Mars Climate Orbiter burned up because it dipped lower in the Mars atmosphere than planned. How many miles lower than planned did it dip?
b. A Canadian jetliner almost ran out of fuel because someone pumped less fuel into the aircraft than was thought. How many more pounds of fuel should have been pumped into the aircraft?

F **157.** Read the "Chemistry in Focus" segment *Measurement: Past, Present, and Future* and answer the following questions.

a. Give three examples of how developing sophisticated measuring devices is useful in our society.
b. Explain how advances in measurement abilities can be a problem.

F **158.** The "Chemistry in Focus" segment *Measurement: Past, Present, and Future* states that hormones can be detected to a level of 10^{-8} g/L. Convert this level to units of pounds per gallon.

ChemWork Problems

These multiconcept problems (and additional ones) are found interactively online with the same type of assistance a student would get from an instructor.

159. Complete the following table:

Number	Exponential Notation	Number of Significant Figures
900.0	________	________
3007	________	________
23,450	________	________
270.0	________	________
437,000	________	________

160. For each of the mathematical expressions given:

a. Tell the correct number of significant figures for the answer.
b. Evaluate the mathematical expression using correct significant figures in the result.

	Number of Significant Figures	Result
0.0394×13	________	________
$15.2 - 2.75 + 16.67$	________	________
3.984×2.16	________	________
$0.517 \div 0.2742$	________	________
$1.842 + 45.2 + 87.55$	________	________
$12.62 + 1.5 + 0.25$	________	________

161. The longest river in the world is the Nile River with a length of 4145 mi. How long is the Nile in cable lengths, meters, and nautical miles?

Use these exact conversions to help solve the problem:

$$6 \text{ ft} = 1 \text{ fathom}$$
$$100 \text{ fathoms} = 1 \text{ cable length}$$
$$10 \text{ cable lengths} = 1 \text{ nautical mile}$$
$$3 \text{ nautical miles} = 1 \text{ league}$$

162. Secretariat is known as the horse with the fastest run in the Kentucky Derby. If Secretariat's record 1.25-mi run lasted 1 minute 59.2 seconds, what was his average speed in m/s?

163. A friend tells you that it is 69.1 °F outside. What is this temperature in Celsius?

164. The hottest temperature recorded in the United States is 134 °F in Greenland Ranch, California. The melting point of phosphorus is 44 °C. At this temperature, would phosphorus be a liquid or a solid?

165. The density of osmium (the densest metal) is 22.57 g/cm^3. What is the mass of a block of osmium with dimensions 1.84 cm $\times$ 3.61 cm $\times$ 2.10 cm?

166. The radius of a neon atom is 69 pm, and its mass is 3.35×10^{-23} g. What is the density of the atom in grams per cubic centimeter (g/cm^3)? Assume the atom is a sphere with volume $= \frac{4}{3}\pi r^3$.

All even-numbered Questions and Problems have answers in the back of this book and solutions in the *Student Solutions Guide.*

Matter

3

CHAPTER 3

An iceberg in Greenland.

Frank Krahmer/Masterfile

As you look around you, you must wonder about the properties of matter. How do plants grow and why are they green? Why is the sun hot? Why does a hot dog get hot in a microwave oven? Why does wood burn whereas rocks do not? What is a flame? How does soap work? Why does soda fizz when you open the bottle? When iron rusts, what's happening? And why doesn't aluminum rust? How does a cold pack for an athletic injury, which is stored for weeks or months at room temperature, suddenly get cold when you need it? How does a hair permanent work?

The answers to these and endless other questions lie in the domain of chemistry. In this chapter we begin to explore the nature of matter: how it is organized and how and why it changes.

Why does soda fizz when you open the bottle?

3-1 Matter

OBJECTIVE **To learn about matter and its three states.**

Matter, the "stuff" of which the universe is composed, has two characteristics: it has mass and it occupies space. Matter comes in a great variety of forms: the stars, the air that you are breathing, the gasoline that you put in your car, the chair on which you are sitting, the turkey in the sandwich you may have had for lunch, the tissues in your brain that enable you to read and comprehend this sentence, and so on.

To try to understand the nature of matter, we classify it in various ways. For example, wood, bone, and steel share certain characteristics. These things are all rigid; they have definite shapes that are difficult to change. On the other hand, water and gasoline, for example, take the shape of any container into which they are poured (Fig. 3.1). Even so, 1 L of water has a volume of 1 L whether it is in a pail or a beaker. In contrast, air takes the shape of its container and fills any container uniformly.

The substances we have just described illustrate the three **states of matter: solid, liquid,** and **gas.** These are defined and illustrated in Table 3.1. The state of a given sample of matter depends on the strength of the forces among the particles contained in the matter; the stronger these forces, the more rigid the matter. We will discuss this in more detail in the next section.

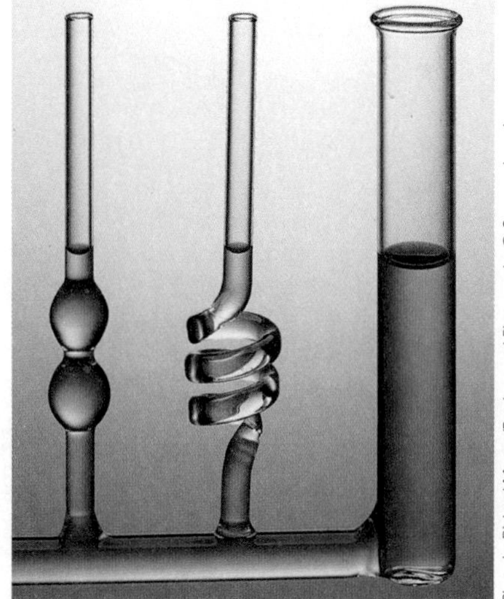

Figure 3.1 ▸ Liquid water takes the shape of its container.

Table 3.1 ▸ The Three States of Matter

State	Definition	Examples
solid	rigid; has a fixed shape and volume	ice cube, diamond, iron bar
liquid	has a definite volume but takes the shape of its container	gasoline, water, alcohol, blood
gas	has no fixed volume or shape; takes the shape and volume of its container	air, helium, oxygen

3-2 Physical and Chemical Properties and Changes

OBJECTIVES

- **To learn to distinguish between physical and chemical properties.**
- **To learn to distinguish between physical and chemical changes.**

TommL/iStockphoto.com

How does this lush vegetation grow in a tropical rain forest, and why is it green?

When you see a friend, you immediately respond and call him or her by name. We can recognize a friend because each person has unique characteristics or properties. The person may be thin and tall, may have blonde hair and blue eyes, and so on. The characteristics just mentioned are examples of **physical properties.** Substances also have physical properties. Typical physical properties of a substance include odor, color, volume, state (gas, liquid, or solid), density, melting point, and boiling point. We can also describe a pure substance in terms of its **chemical properties,** which refer to its ability to form new substances. An example of a chemical change is wood burning in a fireplace, giving off heat and gases and leaving a residue of ashes. In this process, the wood is changed to several new substances. Other examples of chemical changes include the rusting of the steel in our cars, the digestion of food in our stomachs, and the growth of grass in our yards. In a chemical change a given substance changes to a fundamentally different substance or substances.

Interactive Example 3.1

Identifying Physical and Chemical Properties

Classify each of the following as a physical or a chemical property.

a. The boiling point of a certain alcohol is 78 °C.
b. Diamond is very hard.
c. Sugar ferments to form alcohol.
d. A metal wire conducts an electric current.

SOLUTION

Items (a), (b), and (d) are physical properties; they describe inherent characteristics of each substance, and no change in composition occurs. A metal wire has the same composition before and after an electric current has passed through it. Item (c) is a chemical property of sugar. Fermentation of sugars involves the formation of a new substance (alcohol).

SELF CHECK

Exercise 3.1 Which of the following are physical properties and which are chemical properties?

a. Gallium metal melts in your hand.
b. Platinum does not react with oxygen at room temperature.
c. The page of the textbook is white.
d. The copper sheets that form the "skin" of the Statue of Liberty have acquired a greenish coating over the years.

See Problems 3.11 through 3.14. ■

© Cengage Learning

Gallium metal has such a low melting point (30 °C) that it melts from the heat of a hand.

Matter can undergo changes in both its physical and its chemical properties. To illustrate the fundamental differences between physical and chemical changes, we will consider water. As we will see in much more detail in later chapters, a sample of water contains a very large number of individual units (called molecules), each made up of two atoms of hydrogen and one atom of oxygen—the familiar H_2O. This molecule can be represented as

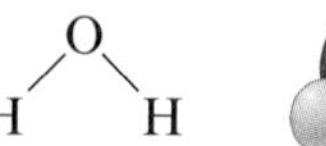

The purpose here is to give an overview. Don't worry about the precise definitions of *atom* and *molecule* now. We will explore these concepts more fully in Chapter 4.

where the letters stand for atoms and the lines show attachments (called bonds) between atoms, and the molecular model (on the right) represents water in a more three-dimensional fashion.

What is really occurring when water undergoes the following changes?

Solid (ice)	→ Melting	Liquid (water)	→ Boiling	Gas (steam)

An iron pyrite crystal.

We will describe these changes of state precisely in Chapter 14, but you already know something about these processes because you have observed them many times.

When ice melts, the rigid solid becomes a mobile liquid that takes the shape of its container. Continued heating brings the liquid to a boil, and the water becomes a gas or vapor that seems to disappear into "thin air." The changes that occur as the substance goes from solid to liquid to gas are represented in Fig. 3.2. In ice the water molecules are locked into fixed positions (although they are vibrating). In the liquid the molecules are still very close together, but some motion is occurring; the positions of the molecules are no longer fixed as they are in ice. In the gaseous state the molecules are much farther apart and move randomly, hitting each other and the walls of the container.

The most important thing about all these changes is that the water molecules are still intact. The motions of individual molecules and the distances between them change, but *H_2O molecules are still present.* These changes of state are **physical changes** because they do not affect the composition of the substance. In each state we still have water (H_2O), not some other substance.

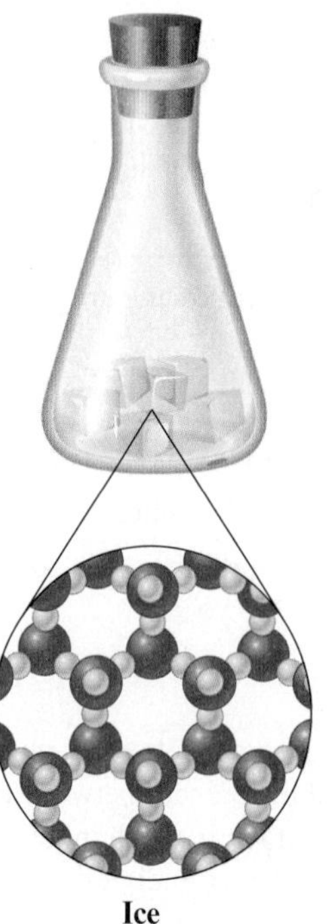

Ice

Solid: The water molecules are locked into rigid positions and are close together.

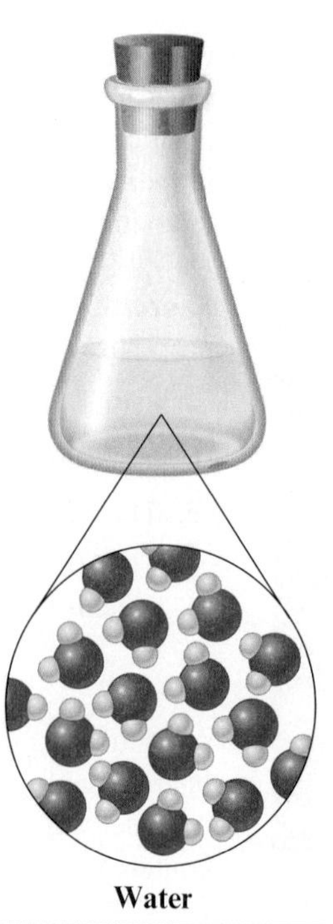

Water

Liquid: The water molecules are still close together but can move around to some extent.

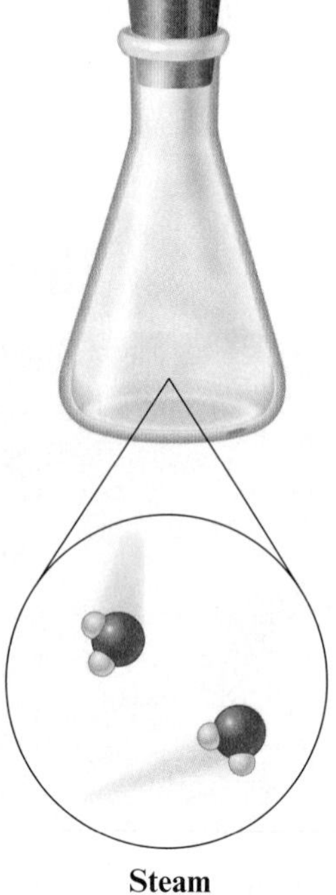

Steam

Gas: The water molecules are far apart and move randomly.

Figure 3.2 ▸ The three states of water (where red spheres represent oxygen atoms and blue spheres represent hydrogen atoms).

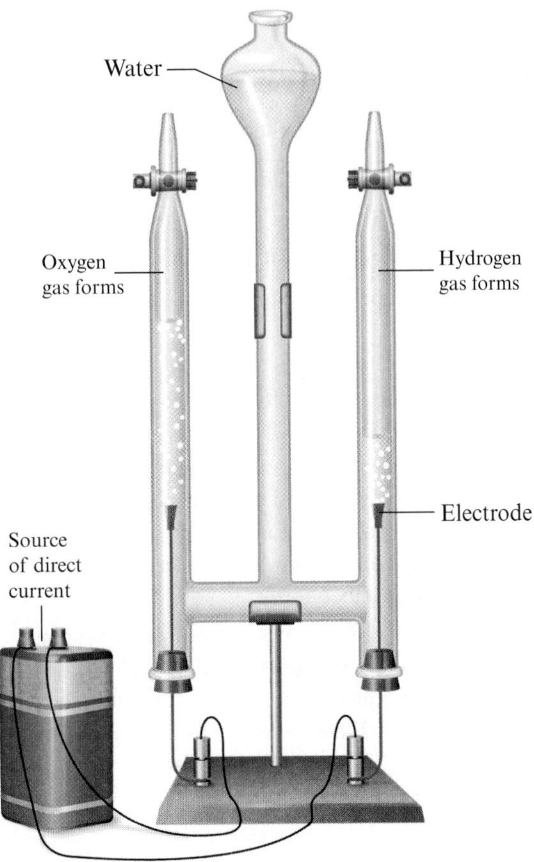

Figure 3.3 Electrolysis, the decomposition of water by an electric current, is a chemical process.

Now suppose we run an electric current through water as illustrated in Fig. 3.3. Something very different happens. The water disappears and is replaced by two new gaseous substances, hydrogen and oxygen. An electric current actually causes the water molecules to come apart—the water *decomposes* to hydrogen and oxygen. We can represent this process as follows:

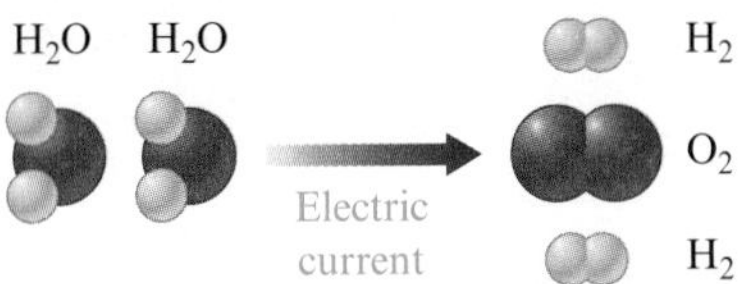

This is a **chemical change** because water (consisting of H_2O molecules) has changed into different substances: hydrogen (containing H_2 molecules) and oxygen (containing O_2 molecules). Thus in this process, the H_2O molecules have been replaced by O_2 and H_2 molecules. Let us summarize:

Physical and Chemical Changes

1. A *physical change* involves a change in one or more physical properties but no change in the fundamental components that make up the substance. The most common physical changes are changes of state: solid ⇔ liquid ⇔ gas.
2. A *chemical change* involves a change in the fundamental components of the substance; a given substance changes into a different substance or substances. Chemical changes are called **reactions:** silver tarnishes by reacting with substances in the air; a plant forms a leaf by combining various substances from the air and soil; and so on.

Interactive Example 3.2

Identifying Physical and Chemical Changes

Classify each of the following as a physical or a chemical change.

a. Iron metal is melted.
b. Iron combines with oxygen to form rust.
c. Wood burns in air.
d. A rock is broken into small pieces.

SOLUTION

a. Melted iron is just liquid iron and could cool again to the solid state. This is a physical change.
b. When iron combines with oxygen, it forms a different substance (rust) that contains iron and oxygen. This is a chemical change because a different substance forms.
c. Wood burns to form different substances (as we will see later, they include carbon dioxide and water). After the fire, the wood is no longer in its original form. This is a chemical change.
d. When the rock is broken up, all the smaller pieces have the same composition as the whole rock. Each new piece differs from the original only in size and shape. This is a physical change.

Oleksandr Prykhodko/Photos.com

Oxygen combines with the chemicals in wood to produce flames. Is a physical or chemical change taking place?

SELF CHECK

Exercise 3.2 Classify each of the following as a chemical change, a physical change, or a combination of the two.

a. Milk turns sour.
b. Wax is melted over a flame and then catches fire and burns.

See Problems 3.17 and 3.18. ■

3-3 Elements and Compounds

OBJECTIVE **To understand the definitions of elements and compounds.**

Element: a substance that cannot be broken down into other substances by chemical methods.

As we examine the chemical changes of matter, we encounter a series of fundamental substances called **elements.** Elements cannot be broken down into other substances by chemical means. Examples of elements are iron, aluminum, oxygen, and hydrogen. All of the matter in the world around us contains elements. The elements sometimes are found in an isolated state, but more often they are combined with other elements. Most substances contain several elements combined together.

The atoms of certain elements have special affinities for each other. They bind together in special ways to form **compounds,** substances that have the same composition no matter where we find them. Because compounds are made of elements, they can be broken down into elements through chemical changes:

Compounds  Elements

Chemical changes

Water is an example of a compound. Pure water always has the same composition (the same relative amounts of hydrogen and oxygen) because it consists of H_2O molecules. Water can be broken down into the elements hydrogen and oxygen by chemical means, such as by the use of an electric current (Fig. 3.3).

As we will discuss in more detail in Chapter 4, each element is made up of a particular kind of atom: a pure sample of the element aluminum contains only aluminum atoms, elemental copper contains only copper atoms, and so on. Thus an element contains only one kind of atom; a sample of iron contains many atoms, but they are all iron atoms. Samples of certain pure elements do contain molecules; for example, hydrogen gas contains H—H (usually written H_2) molecules, and oxygen gas contains O—O (O_2) molecules. However, any pure sample of an element contains only atoms of that element, *never* any atoms of any other element.

A compound *always* contains atoms of *different* elements. For example, water contains hydrogen atoms and oxygen atoms, and there are always exactly twice as many hydrogen atoms as oxygen atoms because water consists of H—O—H molecules. A different compound, carbon dioxide, consists of CO_2 molecules and so contains carbon atoms and oxygen atoms (always in the ratio 1:2).

A compound, although it contains more than one type of atom, *always has the same composition*—that is, the same combination of atoms. The properties of a compound are typically very different from those of the elements it contains. For example, the properties of water are quite different from the properties of pure hydrogen and pure oxygen.

3-4 Mixtures and Pure Substances

OBJECTIVE To learn to distinguish between mixtures and pure substances.

Virtually all of the matter around us consists of mixtures of substances. For example, if you closely observe a sample of soil, you will see that it has many types of components, including tiny grains of sand and remnants of plants. The air we breathe is a complex mixture of such gases as oxygen, nitrogen, carbon dioxide, and water vapor. Even the water from a drinking fountain contains many substances besides water.

A **mixture** can be defined as something that has variable composition. For example, wood is a mixture (its composition varies greatly depending on the tree from which it originates); wine is a mixture (it can be red or pale yellow, sweet or dry); coffee is a mixture (it can be strong, weak, or bitter); and, although it looks very pure, water pumped from deep in the earth is a mixture (it contains dissolved minerals and gases).

A **pure substance,** on the other hand, will always have the same composition. Pure substances are either elements or compounds. For example, pure water is a compound containing individual H_2O molecules. However, as we find it in nature, liquid water always contains other substances in addition to pure water—it is a mixture. This is obvious from the different tastes, smells, and colors of water samples obtained from various locations. However, if we take great pains to purify samples of water from various sources (such as oceans, lakes, rivers, and the earth's interior), we always end up with the same pure substance—water, which is made up only of H_2O molecules. Pure water always has the same physical and chemical properties and is always made of molecules containing hydrogen and oxygen in exactly the same proportions, regardless of the original source of the water. The properties of a pure substance make it possible to identify that substance conclusively.

Although we say we can separate mixtures into pure substances, it is virtually impossible to separate mixtures into totally pure substances. No matter how hard we try, some impurities (components of the original mixture) remain in each of the "pure substances."

Mixtures can be separated into pure substances: elements and/or compounds.

Mixtures  Two or more pure substances

Concrete—An Ancient Material Made New

Concrete, which was invented more than 2000 years ago by the ancient Romans, is being transformed into a high-tech building material through the use of our knowledge of chemistry. There is little doubt that concrete is the world's most important material. It is used to construct highways, bridges, buildings, floors, countertops, and countless other objects. In its simplest form concrete consists of about 70% sand and gravel, 15% water, and 15% cement (a mixture prepared by heating and grinding limestone, clay, shale, and gypsum). Because concrete forms the skeleton of much of our society, improvements to make it last longer and perform better are crucial.

One new type of concrete is Ductal, which was developed by the French company Lafarge. Unlike traditional concrete, which is brittle and can rupture suddenly under a heavy load, Ductal can bend. Even better, Ductal is five times stronger than traditional concrete. The secret behind Ductal's near-magical properties lies in the addition of small steel or polymeric fibers, which are dispersed throughout the structure. The fibers eliminate the need for steel reinforcing bars (rebar) for structures such as bridges. Bridges built of Ductal are lighter, thinner, and much more corrosion resistant than bridges built with traditional concrete containing rebar.

In another innovation, the Hungarian company Litracon has developed a translucent concrete material by incorporating optical fibers of various diameters into the concrete. With this light-transmitting concrete, architects can design buildings with translucent concrete walls and concrete floors that can be lighted from below.

Another type of concrete being developed by the Italian company Italcementi Group has a self-cleaning surface. This new material is made by mixing titanium oxide particles into the concrete. Titanium oxide can absorb ultraviolet light and promote the decomposition of pollutants that would otherwise darken the surface of the building. This material has already been used for several buildings in Italy. One additional bonus of using this material for buildings and roads in cities is that it may actually act to reduce air pollution very significantly.

Concrete is an ancient material, but one that is showing the flexibility to be a high-tech material. Its adaptability will ensure that it finds valuable uses far into the future.

An object made of translucent concrete.

For example, the mixture known as air can be separated into oxygen (element), nitrogen (element), water (compound), carbon dioxide (compound), argon (element), and other pure substances.

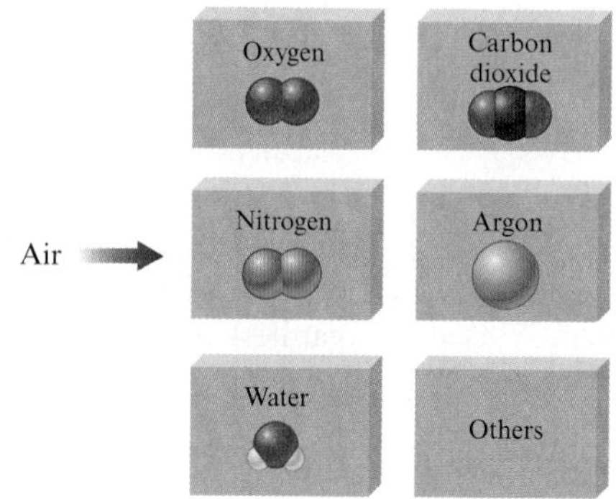

Mixtures can be classified as either homogeneous or heterogeneous. A **homogeneous mixture** is *the same throughout.* For example, when we dissolve some salt in water and stir well, all regions of the resulting mixture have the same properties. A homogeneous

Photo by Richard Megna/Fundamental Photographs © Cengage Learning

Figure 3.4 When table salt is stirred into water (*right*), a homogeneous mixture called a solution forms (*left*).

Photo by Richard Megna/Fundamental Photographs © Cengage Learning

Figure 3.5 Sand and water do not mix to form a uniform mixture. After the mixture is stirred, the sand settles back to the bottom (*left*).

mixture is also called a **solution.** Of course, different amounts of salt and water can be mixed to form various solutions, but a homogeneous mixture (a solution) does not vary in composition from one region to another (Fig. 3.4).

The air around you is a solution—it is a homogeneous mixture of gases. Solid solutions also exist. Brass is a homogeneous mixture of the metals copper and zinc.

A **heterogeneous mixture** contains regions that have different properties from those of other regions. For example, when we pour sand into water, the resulting mixture has one region containing water and another, very different region containing mostly sand (Fig. 3.5).

Interactive Example 3.3

Distinguishing Between Mixtures and Pure Substances

Identify each of the following as a pure substance, a homogeneous mixture, or a heterogeneous mixture.

a. gasoline
b. a stream with gravel at the bottom
c. air
d. brass
e. copper metal

SOLUTION

a. Gasoline is a homogeneous mixture containing many compounds.
b. A stream with gravel on the bottom is a heterogeneous mixture.
c. Air is a homogeneous mixture of elements and compounds.
d. Brass is a homogeneous mixture containing the elements copper and zinc. Brass is not a pure substance because the relative amounts of copper and zinc are different in different brass samples.
e. Copper metal is a pure substance (an element).

SELF CHECK

Exercise 3.3 Classify each of the following as a pure substance, a homogeneous mixture, or a heterogeneous mixture.

a. maple syrup
b. the oxygen and helium in a scuba tank
c. oil and vinegar salad dressing
d. common salt (sodium chloride)

See Problems 3.29 through 3.31. ■

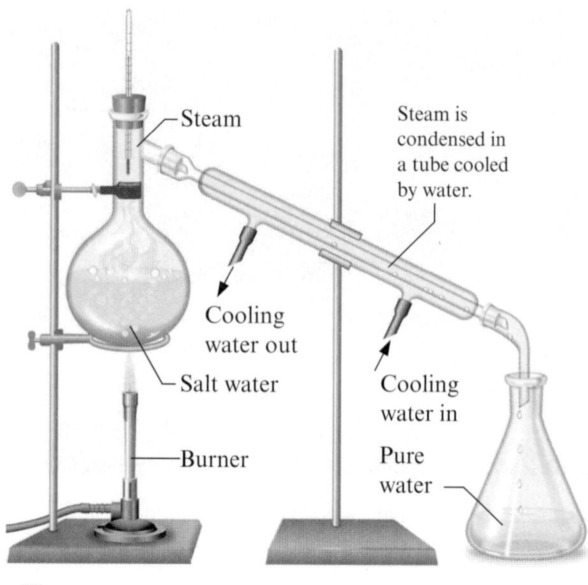

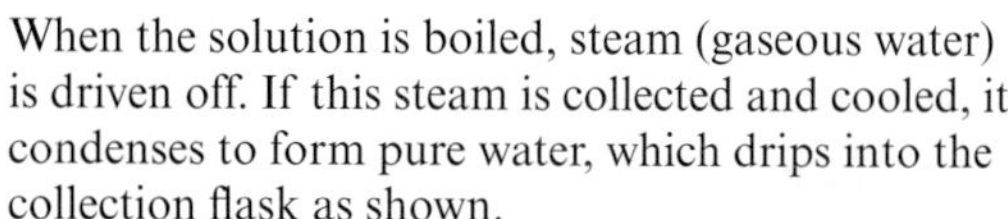

a
When the solution is boiled, steam (gaseous water) is driven off. If this steam is collected and cooled, it condenses to form pure water, which drips into the collection flask as shown.

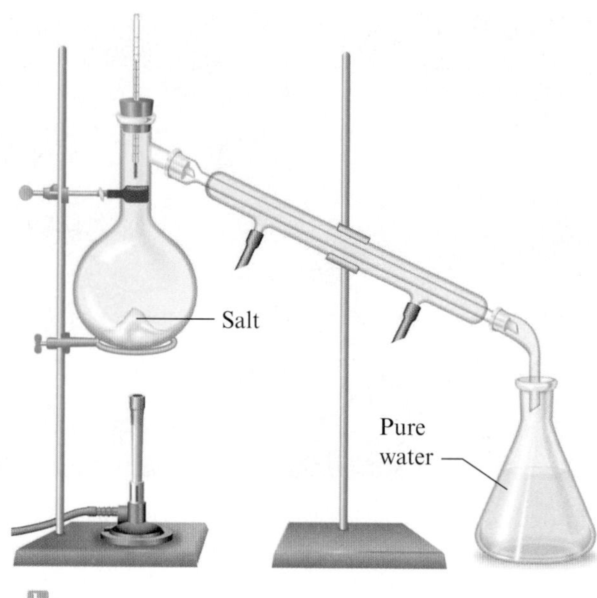

b
After all of the water has been boiled off, the salt remains in the original flask and the water is in the collection flask.

Figure 3.6 ▸ Distillation of a solution consisting of salt dissolved in water.

3-5 Separation of Mixtures

OBJECTIVE To learn two methods of separating mixtures.

We have seen that the matter found in nature is typically a mixture of pure substances. For example, seawater is water containing dissolved minerals. We can separate the water from the minerals by boiling, which changes the water to steam (gaseous water) and leaves the minerals behind as solids. If we collect and cool the steam, it condenses to pure water. This separation process, called **distillation,** is shown in Fig. 3.6.

When we carry out the distillation of salt water, water is changed from the liquid state to the gaseous state and then back to the liquid state. These changes of state are examples of physical changes. We are separating a mixture of substances, but we are not changing the composition of the individual substances. We can represent this as shown in Fig. 3.7.

Suppose we scooped up some sand with our sample of seawater. This sample is a heterogeneous mixture, because it contains an undissolved solid as well as the saltwater solution. We can separate out the sand by simple **filtration.** We pour the mixture

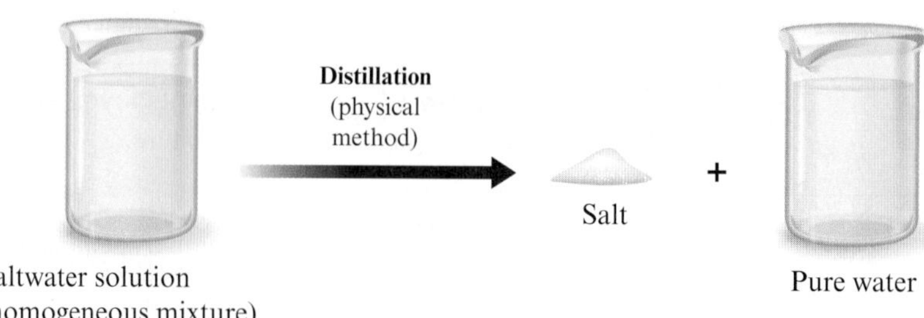

Figure 3.7 ▸ No chemical change occurs when salt water is distilled.

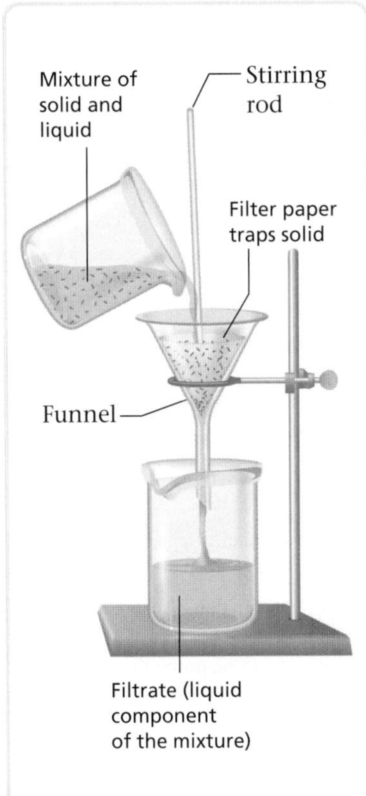

Figure 3.8 ▸ Filtration separates a liquid from a solid. The liquid passes through the filter paper, but the solid particles are trapped.

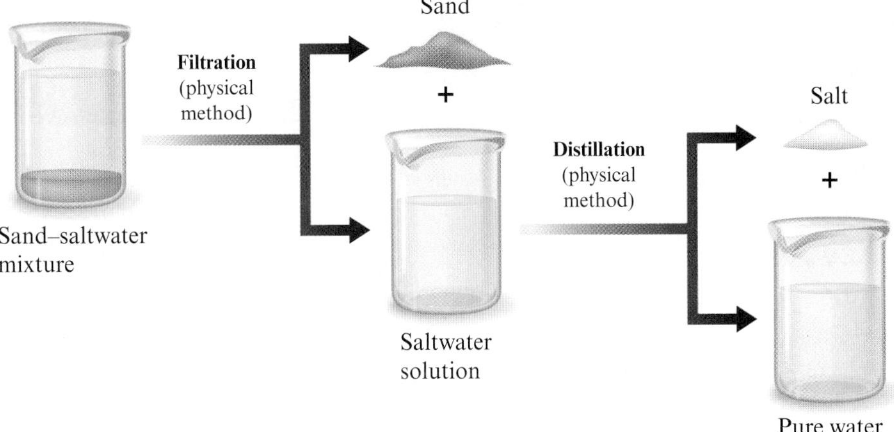

Figure 3.9 ▸ Separation of a sand–saltwater mixture.

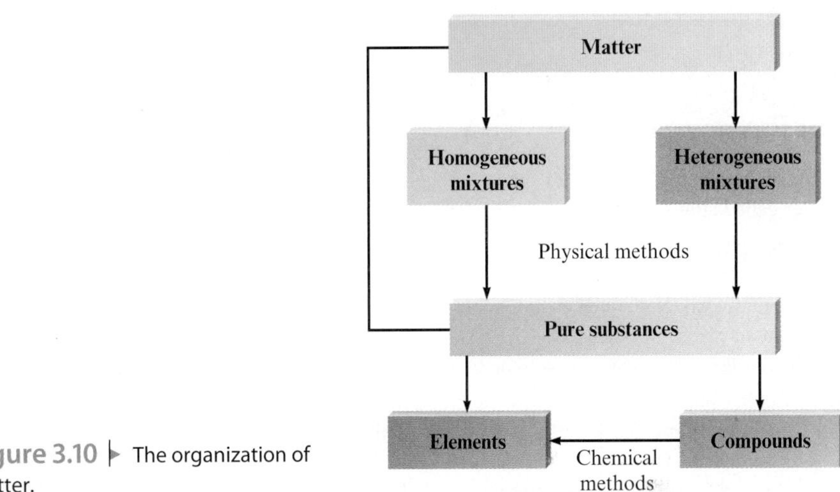

Figure 3.10 ▸ The organization of matter.

onto a mesh, such as a filter paper, which allows the liquid to pass through and leaves the solid behind (Fig. 3.8). The salt can then be separated from the water by distillation. The total separation process is represented in Fig. 3.9. All the changes involved are physical changes.

Critical Thinking

The scanning tunneling microscope allows us to "see" atoms. What if you were sent back in time before the invention of the scanning tunneling microscope? What evidence could you give to support the theory that all matter is made of atoms and molecules?

We can summarize the description of matter given in this chapter with the diagram shown in Fig. 3.10. Note that a given sample of matter can be a pure substance (either an element or a compound) or, more commonly, a mixture (homogeneous or heterogeneous). We have seen that all matter exists as elements or can be broken down into elements, the most fundamental substances we have encountered up to this point. We will have more to say about the nature of elements in the next chapter.

CHAPTER 3 REVIEW

F directs you to the *Chemistry in Focus* feature in the chapter

Key Terms

matter (3-1)
states of matter (3-1)
solid (3-1)
liquid (3-1)
gas (3-1)
physical properties (3-2)
chemical properties (3-2)
physical change (3-2)
chemical change (3-2)
reaction (3-2)
element (3-3)
compound (3-3)
mixture (3-4)
pure substance (3-4)
homogeneous mixture (3-4)
solution (3-4)
heterogeneous mixture (3-4)
distillation (3-5)
filtration (3-5)

For Review

- Matter has mass and occupies space. It is composed of tiny particles called atoms.
- Matter exists in three states:
  - Solid—is a rigid substance with a definite shape
  - Liquid—has a definite volume but takes the shape of its container
  - Gas—takes the shape and volume of its container
- Matter has both physical and chemical properties.
  - Chemical properties describe a substance's ability to change to a different substance.
  - Physical properties are the characteristics of a substance that do not involve changing to another substance.
    - Examples are shape, size, and color.
- Matter undergoes physical and chemical changes.
  - A physical change involves a change in one or more physical properties but no change in composition.
  - A chemical change transforms a substance into one or more new substances.
- Elements contain only one kind of atom—elemental copper contains only copper atoms, and elemental gold contains only gold atoms.
- Compounds are substances that contain two or more kinds of atoms.
- Compounds often contain discrete molecules.
- A molecule contains atoms bound together in a particular way—an example is , the water molecule, which is written H_2O.
- Matter can be classified as a mixture or a pure substance.
  - A mixture has variable composition.
    - A homogeneous mixture has the same properties throughout.
    - A heterogeneous mixture has different properties in different parts of the mixture.
  - A pure substance always has the same composition.
- Mixtures can be separated into pure substances by various means including distillation and filtration.
- Pure substances are of two types:
  - Elements, which cannot be broken down chemically into simpler substances
  - Compounds, which can be broken down chemically into elements

Active Learning Questions

These questions are designed to be considered by groups of students in class. Often these questions work well for introducing a particular topic in class.

1. When water boils, you can see bubbles rising to the surface of the water. Of what are these bubbles made?
   a. air
   b. hydrogen and oxygen gas
   c. oxygen gas
   d. water vapor
   e. carbon dioxide gas
2. If you place a glass rod over a burning candle, the glass turns black. What is happening to each of the following (physical change, chemical change, both, or neither) as the candle burns? Explain.
   a. the wax
   b. the wick
   c. the glass rod
3. The boiling of water is a
   a. physical change because the water disappears.
   b. physical change because the gaseous water is chemically the same as the liquid.
   c. chemical change because heat is needed for the process to occur.
   d. chemical change because hydrogen and oxygen gases are formed from water.
   e. chemical and physical change.

   Explain your answer.
4. Is there a difference between a homogeneous mixture of hydrogen and oxygen in a 2:1 ratio and a sample of water vapor? Explain.
5. Sketch a magnified view (showing atoms and/or molecules) of each of the following and explain why the specified type of mixture is
   a. a heterogeneous mixture of two different compounds
   b. a homogeneous mixture of an element and a compound

All even-numbered Questions and Problems have answers in the back of this book and solutions in the *Student Solutions Guide*.

6. Are all physical changes accompanied by chemical changes? Are all chemical changes accompanied by physical changes? Explain.
7. Why would a chemist find fault with the phrase "pure orange juice"?
8. Are separations of mixtures physical or chemical changes? Explain.
9. Explain the terms *element, atom,* and *compound.* Provide an example and microscopic drawing of each.
10. Mixtures can be classified as either homogeneous or heterogeneous. Compounds cannot be classified in this way. Why not? In your answer, explain what is meant by heterogeneous and homogeneous.
11. Provide microscopic drawings down to the atoms for Fig. 3.10 in your text.
12. Look at Table 2.8 in your text. How do the densities of gases, liquids, and solids compare to each other? Use microscopic pictures to explain why this is true.
13. Label each of the following as an atomic element, a molecular element, or a compound.

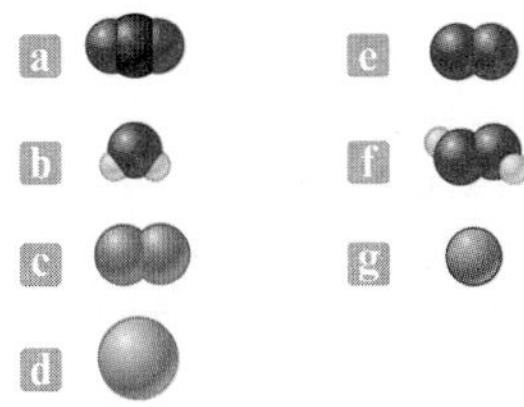

14. Match each description below with the following microscopic pictures. More than one picture may fit each description. A picture may be used more than once or not used at all.

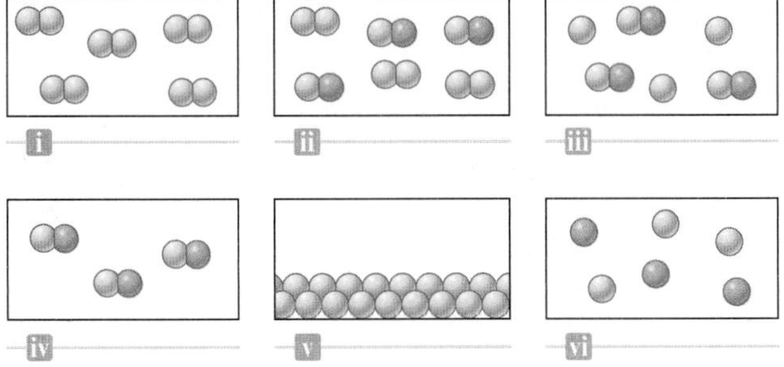

   a. a gaseous compound
   b. a mixture of two gaseous elements
   c. a solid element
   d. a mixture of gaseous element and a gaseous compound

Questions and Problems

3-1 Matter

Questions

1. What are the two characteristic properties of *matter?*
2. What is the chief factor that determines the *physical state* of a sample of matter?
3. Of the three states of matter, ________ and ________ are not very compressible.
4. ________ have a fixed shape and volume.
5. Compare and contrast the ease with which molecules are able to move relative to each other in the three states of matter.
6. Matter in the ________ state has no shape and fills completely whatever container holds it.
7. What similarities are there between the liquid and gaseous states of matter? What differences are there between these two states?
8. A sample of matter that is "rigid" has (stronger/weaker) forces among the particles in the sample than does a sample that is not rigid.
9. Consider three 10-g samples of water: one as ice, one as liquid, and one as vapor. How do the volumes of these three samples compare with one another? How is this difference in volume related to the physical state involved?
10. In a sample of a gaseous substance, more than 99% of the overall volume of the sample is empty space. How is this fact reflected in the properties of a gaseous substance, compared with the properties of a liquid or solid substance?

3-2 Physical and Chemical Properties and Changes

Questions

11. Elemental bromine is a dense, dark-red, pungent-smelling liquid. Are these characteristics of elemental bromine physical or chemical properties?
12. Is the process represented below a physical or chemical change?

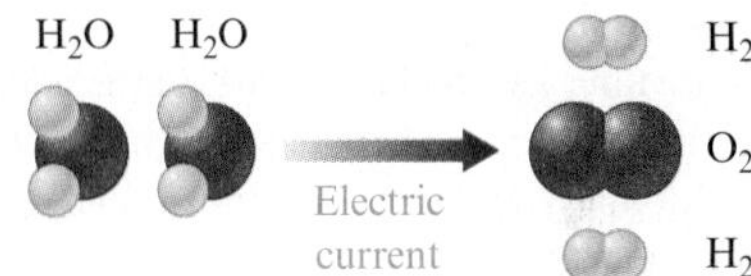

(For Exercises 13–14) Magnesium metal is very malleable, and is able to be pounded and stretched into long, thin, narrow "ribbons" that are often used in the introductory chemistry lab as a source of the metal. If a strip of magnesium ribbon is ignited in a Bunsen burner flame, the magnesium burns brightly and produces a quantity of white magnesium oxide powder.

13. From the information given above, indicate one *chemical* property of magnesium metal.
14. From the information given above, indicate one *physical* property of magnesium metal.
15. Choose a chemical substance with which you are familiar, and give an example of a *chemical change* that might take place to the substance.
16. Which of the following is/are examples of a chemical change?
   a. carving wood
   b. snow melting
   c. dry ice subliming (solid CO_2 vaporizing into a gas, passing the liquid state)
   d. burning cookies in the oven
17. Classify each of the following as a *physical* or *chemical* change or property.

a. Oven cleaners contain sodium hydroxide, which converts the grease/oil spatters inside the oven to water-soluble materials, which can be washed away.
b. A rubber band stretches when you pull on it.
c. A cast-iron frying pan will rust if it is not dried after washing.
d. Concentrated hydrochloric acid has a choking, pungent odor.
e. Concentrated hydrochloric acid will burn a hole in cotton jeans because the acid breaks down the cellulose fibers in cotton.
f. Copper compounds often form beautiful blue crystals when a solution of a given copper compound is evaporated slowly.
g. Copper metal combines with substances in the air to form a green "patina" that protects the copper from further reaction.
h. Bread turns brown when you heat it in a toaster.
i. When you use the perfume your boyfriend gave you for your birthday, the liquid of the perfume evaporates quickly from your skin.
j. If you leave your steak on the gas grill too long, the steak will turn black and char.
k. Hydrogen peroxide fizzes when it is applied to a cut or scrape.

18. Classify each of the following as a *physical* or *chemical* change or property.

a. A fireplace poker glows red when you heat it in the fire.
b. A marshmallow turns black when toasted too long in a campfire.
c. Hydrogen peroxide dental strips will make your teeth whiter.
d. If you wash your jeans with chlorine bleach, they will fade.
e. If you spill some nail polish remover on your skin, it will evaporate quickly.
f. When making ice cream at home, salt is added to lower the temperature of the ice being used to freeze the mixture.
g. A hair clog in your bathroom sink drain can be cleared with drain cleaner.
h. The perfume your boyfriend gave you for your birthday smells like flowers.
i. Mothballs pass directly into the gaseous state in your closet without first melting.
j. A log of wood is chopped up with an axe into smaller pieces of wood.
k. A log of wood is burned in a fireplace.

3-3 Elements and Compounds

Questions

19. Although some elements are found in an isolated state, most elements are found combined as ________ with other elements.

20. What is a *compound?* What are compounds composed of? What is true about the composition of a compound, no matter where we happen to find the compound?

21. Certain elements have special affinities for other elements. This causes them to bind together in special ways to form ________.

22. ________ can be broken down into the component elements by chemical changes.

23. The composition of a given pure compound is always ________ no matter what the source of the compound.

24. Which of the following are considered elements (as opposed to compounds)?

$$He, F_2, HCl, S_8$$

3-4 Mixtures and Pure Substances

Questions

25. If iron filings are placed with excess powdered sulfur in a beaker, the iron filings are still attracted by a magnet and could be separated from the sulfur with the magnet. Would this combination of iron and sulfur represent a *mixture* or a *pure substance?*

26. If the combination of iron filings and sulfur in Question 25 is heated strongly, the iron reacts with the sulfur to form a solid that is no longer attracted by the magnet. Would this still represent a "mixture?" Why or why not?

27. What does it mean to say that a solution is a *homogeneous mixture?*

28. Give three examples of heterogeneous *mixtures* and three examples of *solutions* that you might use in everyday life.

29. Classify the following as *mixtures* or *pure substances.*

a. the vegetable soup you had for lunch
b. the fertilizer your dad spreads on the front lawn in the spring
c. the salt you sprinkle on your French fries
d. the hydrogen peroxide you cleaned a cut finger with

30. Classify the following as *mixtures* or *pure substances.*

a. a multivitamin tablet
b. the blue liquid in your car's windshield reservoir
c. a ham and cheese omelet
d. a diamond

31. Classify the following mixtures as *heterogeneous* or *homogeneous.*

a. soil
b. mayonnaise
c. Italian salad dressing
d. the wood from which the desk you are studying on is made
e. sand at the beach

F **32.** Read the "Chemistry in Focus" segment *Concrete—An Ancient Material Made New* and classify concrete as an element, a mixture, or a compound. Defend your answer.

3-5 Separation of Mixtures

Questions

33. Describe how the process of *distillation* could be used to separate a solution into its component substances. Give an example.

34. Describe how the process of *filtration* could be used to separate a mixture into its components. Give an example.

35. In a common laboratory experiment in general chemistry, students are asked to determine the relative amounts of benzoic acid and charcoal in a solid mixture. Benzoic acid is relatively soluble in hot water, but charcoal is not. Devise a method for separating the two components of this mixture.

All even-numbered Questions and Problems have answers in the back of this book and solutions in the *Student Solutions Guide.*

36. During a filtration or distillation experiment, we separate a mixture into its individual components. Do the chemical identities of the components of the mixture change during such a process? Explain.

Additional Problems

37. If powdered elemental zinc and powdered elemental sulfur are poured into a metal beaker and then heated strongly, a very vigorous chemical reaction takes place, and the ________ zinc sulfide is formed.

38. Classify each of the following as a: element, compound, pure substance, homogeneous mixture, and/or heterogeneous mixture. More than one classification is possible and not all of them may be used.

a. calcium carbonate ($CaCO_3$)
b. iron
c. water you regularly drink (from your faucet or a bottle)

39. If a piece of hard white blackboard chalk is heated strongly in a flame, the mass of the piece of chalk will decrease, and eventually the chalk will crumble into a fine white dust. Does this change suggest that the chalk is composed of an element or a compound?

40. During a very cold winter, the temperature may remain below freezing for extended periods. However, fallen snow can still disappear, even though it cannot melt. This is possible because a solid can vaporize directly, without passing through the liquid state. Is this process (sublimation) a physical or a chemical change?

41. Discuss the similarities and differences between a liquid and a gas.

42. True or false? Salad dressing (such as oil and vinegar dressing) separating into layers after standing is an example of a chemical change because the end result looks different from how it started. Explain your answer.

43. The fact that solutions of potassium chromate are bright yellow is an example of a ________ property.

44. Which of the following are true?

a. P_4 is considered a compound.
b. Metal rusting on a car is a chemical change.
c. Dissolving sugar in water is a chemical change.
d. Sodium chloride (NaCl) is a homogeneous mixture.

(For Exercises 45–46) Solutions containing nickel(II) ion are usually bright green in color. When potassium hydroxide is added to such a nickel(II) solution, a pale-green fluffy solid forms and settles out of the solution.

45. The fact that a reaction takes place when potassium hydroxide is added to a solution of nickel(II) ions is an example of a ________ property.

46. The fact that a solution of nickel(II) ion is bright green is an example of a ________ property.

47. The processes of melting and evaporation involve changes in the ________ of a substance.

48. A/n ________ always has the same composition.

49. Classify each of the following as a *physical* or *chemical* change or property.

a. Milk curdles if a few drops of lemon juice are added to it.
b. Butter turns rancid if it is left exposed at room temperature.
c. Salad dressing separates into layers after standing.
d. Milk of magnesia neutralizes stomach acid.
e. The steel in a car has rust spots.
f. A person is asphyxiated by breathing carbon monoxide.
g. Sulfuric acid spilled on a laboratory notebook page causes the paper to char and disintegrate.
h. Sweat cools the body as the sweat evaporates from the skin.
i. Aspirin reduces fever.
j. Oil feels slippery.
k. Alcohol burns, forming carbon dioxide and water.

50. Classify the following mixtures as *homogeneous* or *heterogeneous.*

a. dirt from a corn field
b. flavored sports drink
c. nut party mix
d. pepperoni and sausage pizza
e. air

51. Classify the following mixtures as *homogeneous* or *heterogeneous.*

a. potting soil
b. white wine
c. your sock drawer
d. window glass
e. granite

52. Mixtures can be heterogeneous or homogeneous. Give two examples of each type. Explain why you classified each example as you did.

53. Give three examples each of *heterogeneous* mixtures and *homogeneous* mixtures.

54. True or false? Mixtures always result in a chemical reaction because they consist of two or more substances and thus combine to create a new product.

55. Choose an element or compound with which you are familiar in everyday life. Give two *physical* properties and two *chemical* properties of your choice of element or compound.

56. Oxygen forms molecules in which there are two oxygen atoms, O_2. Phosphorus forms molecules in which there are four phosphorus atoms, P_4. Does this mean that O_2 and P_4 are "compounds" because they contain multiple atoms? O_2 and P_4 react with each other to form diphosphorus pentoxide, P_2O_5. Is P_2O_5 a "compound"? Why (or why not)?

57. Give an example of each of the following:

a. a heterogeneous mixture
b. a homogeneous mixture
c. an element
d. a compound
e. a physical property or change
f. a chemical property or change
g. a solution

58. Distillation and filtration are important methods for separating the components of mixtures. Suppose we had a mixture of sand, salt, and water. Describe how filtration and distillation could be used sequentially to separate this mixture into the three separate components.

59. Sketch the apparatus commonly used for simple distillation in the laboratory and identify each component.

60. The properties of a compound are often very different from the properties of the elements making up the compound. Water is an excellent example of this idea. Discuss.

61. Which of the following best describes the substance XeF_4?

a. element
b. compound
c. heterogeneous mixture
d. homogeneous mixture

ChemWork Problems

These multiconcept problems (and additional ones) are found interactively online with the same type of assistance a student would get from an instructor.

62. Which of the following statements is(are) *true*?

a. A spoonful of sugar is a mixture.
b. Only elements are pure substances.
c. Air is a mixture of gases.
d. Gasoline is a pure substance.
e. Compounds can be broken down only by chemical means.

63. Which of the following describes a chemical property?

a. The density of iron is 7.87 g/cm^3.
b. A platinum wire glows red when heated.
c. An iron bar rusts.
d. Aluminum is a silver-colored metal.

64. Which of the following describes a physical change?

a. Paper is torn into several smaller pieces.
b. Two clear solutions are mixed together to produce a yellow solid.
c. A match burns in the air.
d. Sugar is dissolved in water.

All even-numbered Questions and Problems have answers in the back of this book and solutions in the *Student Solutions Guide.*

CUMULATIVE REVIEW CHAPTERS 1-3

Questions

1. In the exercises for Chapter 1 of this text, you were asked to give your *own* definition of what chemistry represents. After having completed a few more chapters in this book, has your definition changed? Do you have a better appreciation for what chemists do? Explain.
2. Early on in this text, some aspects of the best way to go about learning chemistry were presented. In *beginning* your study of chemistry, you may initially have approached studying chemistry as you would any of your other academic subjects (taking notes in class, reading the text, memorizing facts, and so on). Discuss why the ability to sort through and analyze facts and the ability to propose and solve problems are so much more important in learning chemistry.
3. You have learned the basic way in which scientists analyze problems, propose models to explain the systems under consideration, and then experiment to test their models. Suppose you have a sample of a liquid material. You are not sure whether the liquid is a pure *compound* (for example, water or alcohol) or a *solution.* How could you apply the scientific method to study the liquid and to determine which type of material the liquid is?
4. Many college students would not choose to take a chemistry course if it were not required for their major. Do you have a better appreciation of *why* chemistry is a required course for your own particular major or career choice? Discuss.
5. In Chapter 2 of this text, you were introduced to the International System (SI) of measurements. What are the basic units of this system for mass, distance, time, and temperature? What are some of the prefixes used to indicate common multiples and subdivisions of these basic units? Give three examples of the *use* of such prefixes, and explain why the prefix is appropriate to the quantity or measurement being indicated.
6. Most people think of science as being a specific, exact discipline, with a "correct" answer for every problem. Yet you were introduced to the concept of *uncertainty* in scientific measurements. What is meant by "uncertainty"? How does uncertainty creep into measurements? How is uncertainty *indicated* in scientific measurements? Can uncertainty ever be completely eliminated in experiments? Explain.
7. After studying a few chapters of this text, and perhaps having done a few lab experiments and taken a few quizzes in chemistry, you are probably sick of hearing the term *significant figures.* Most chemistry teachers make a big deal about significant figures. Why is reporting the correct number of significant figures so important in science? Summarize the rules for deciding whether a figure in a calculation is "significant." Summarize the rules for rounding off numbers. Summarize the rules for doing arithmetic with the correct number of significant figures.
8. This chemistry course may have been the first time you have encountered the method of *dimensional analysis* in problem solving. Explain what are meant by a *conversion factor* and an *equivalence statement.* Give an everyday example of how you might use dimensional analysis to solve a simple problem.
9. You have learned about several temperature scales so far in this text. Describe the Fahrenheit, Celsius, and Kelvin temperature scales. How are these scales defined? Why were they defined this way? Which of these temperature scales is the most fundamental? Why?
10. What is *matter?* What is matter composed of? What are some of the different types of matter? How do these types of matter differ and how are they the same?
11. It is important to be able to distinguish between the *physical* and the *chemical* properties of chemical substances. Choose a chemical substance you are familiar with, then use the Internet or a handbook of chemical information to list three physical properties and three chemical properties of the substance.
12. What is an *element* and what is a *compound?* Give examples of each. What does it mean to say that a compound has a *constant composition?* Would samples of a particular compound here and in another part of the world have the same composition and properties?
13. What is a mixture? What is a solution? How do mixtures differ from pure substances? What are some of the techniques by which mixtures can be resolved into their components?

Problems

14. For each of the following, make the indicated conversion.
    a. 0.0008917 to standard scientific notation
    b. 2.795×10^{-4} to ordinary decimal notation
    c. 4.913×10^{3} to ordinary decimal notation
    d. 85,100,000 to standard scientific notation
    e. $5.751 \times 10^{5} \times 2.119 \times 10^{-4}$ to standard scientific notation.
    f. $\dfrac{2.791 \times 10^{-5}}{8.219 \times 10^{3}}$
15. For each of the following, make the indicated conversion, showing explicitly the conversion factor(s) you used.
    a. 493.2 g to kilograms
    b. 493.2 g to pounds
    c. 9.312 mi to kilometers
    d. 9.312 mi to feet
    e. 4.219 m to feet
    f. 4.219 m to centimeters
    g. 429.2 mL to liters
    h. 2.934 L to quarts
16. Without performing the actual calculations, determine to how many significant figures the results of the following calculations should be reported.
    a. $\dfrac{(2.991)(4.3785)(1.97)}{(2.1)}$
    b. $\dfrac{(5.2)}{(1.9311 + 0.4297)}$
    c. $1.782 + 0.00035 + 2.11$
    d. $(6.521)(5.338 + 2.11)$
    e. $9 - 0.000017$

f. $(4.2005 + 2.7)(7.99118)$
g. $(5.12941 \times 10^4)(4.91 \times 10^{-3})(0.15)$
h. $97.215 + 42.1 - 56.3498$

17. Chapter 2 introduced the Kelvin and Celsius temperature scales and related them to the Fahrenheit temperature scale commonly used in the United States.

a. How is the size of the temperature unit (degree) related between the Kelvin and Celsius scale?
b. How does the size of the temperature unit (degree) on the Fahrenheit scale compare to the temperature unit on the Celsius scale?
c. What is the normal freezing point of water on each of the three temperature scales?
d. Convert 27.5 °C to kelvins and to Fahrenheit degrees.
e. Convert 298.1 K to Celsius degrees and to Fahrenheit degrees.
f. Convert 98.6 °F to kelvins and to Celsius degrees.

18. a. Given that 100. mL of ethyl alcohol weighs 78.5 g, calculate the density of ethyl alcohol.
b. What volume would 1.59 kg of ethyl alcohol occupy?
c. What is the mass of 1.35 L of ethyl alcohol?
d. Pure aluminum metal has a density of 2.70 g/cm^3. Calculate the volume of 25.2 g of pure aluminum.
e. What will a rectangular block of pure aluminum having dimensions of 12.0 cm × 2.5 cm × 2.5 cm weigh?

19. Which of the following represent physical properties or changes, and which represent chemical properties or changes?

a. You curl your hair with a curling iron.
b. You curl your hair by getting a "permanent wave" at the hair salon.
c. Ice on your sidewalk melts when you put salt on it.
d. A glass of water evaporates overnight when it is left on the bedside table.
e. Your steak chars if the skillet is too hot.
f. Alcohol feels cool when it is spilled on the skin.
g. Alcohol ignites when a flame is brought near it.
h. Baking powder causes biscuits to rise.

All even-numbered Questions and Problems have answers in the back of this book and solutions in the *Student Solutions Guide*.

Chemical Foundations: Elements, Atoms, and Ions

4

CHAPTER 4

Foyer at the National Theatre in San Jose, Costa Rica showing ornate gold decorations.

Richard Cummins/Lonely Planet Images/Getty Images

The chemical elements are very important to each of us in our daily lives. Although certain elements are present in our bodies in tiny amounts, they can have a profound impact on our health and behavior. As we will see in this chapter, lithium can be a miracle treatment for someone with bipolar disorder, and our cobalt levels can have a remarkable impact on whether we behave violently.

Since ancient times, humans have used chemical changes to their advantage. The processing of ores to produce metals for ornaments and tools and the use of embalming fluids are two applications of chemistry that were used before 1000 B.C.

The Greeks were the first to try to explain why chemical changes occur. By about 400 B.C. they had proposed that all matter was composed of four fundamental substances: fire, earth, water, and air.

The next 2000 years of chemical history were dominated by alchemy. Some alchemists were mystics and fakes who were obsessed with the idea of turning cheap metals into gold. However, many alchemists were sincere scientists, and this period saw important events: the elements mercury, sulfur, and antimony were discovered, and alchemists learned how to prepare acids.

The first scientist to recognize the importance of careful measurements was the Irishman Robert Boyle (1627–1691). Boyle is best known for his pioneering work on the properties of gases, but his most important contribution to science was probably his insistence that science should be firmly grounded in experiments. For example, Boyle held no preconceived notions about how many elements there might be. His definition of the term *element* was based on experiments: a substance was an element unless it could be broken down into two or more simpler substances. For example, air could not be an element as the Greeks believed, because it could be broken down into many pure substances.

Lithium is administered in the form of lithium carbonate pills. © Cengage Learning

As Boyle's experimental definition of an element became generally accepted, the list of known elements grew, and the Greek system of four elements died. But although Boyle was an excellent scientist, he was not always right. For some reason he ignored his own definition of an element and clung to the alchemists' views that metals were not true elements and that a way would be found eventually to change one metal into another.

Robert Boyle at 62 years of age. The Granger Collection, New York

4-1 The Elements

OBJECTIVES

- **To learn about the relative abundances of the elements.**
- **To learn the names of some elements.**

In studying the materials of the earth (and other parts of the universe), scientists have found that all matter can be broken down chemically into about 100 different elements. At first it might seem amazing that the millions of known substances are composed of so few fundamental elements. Fortunately for those trying to understand and systematize it, nature often uses a relatively small number of fundamental units to assemble even extremely complex materials. For example, proteins, a group of substances that serve the human body in almost uncountable ways, are all made by linking together a few fundamental units to form huge molecules. A nonchemical example is the English language, where hundreds of thousands of words are constructed from only 26 letters. If you take apart the thousands of words in an English dictionary, you will find only these 26 fundamental components. In much the same way, when we take apart all of the substances in the world around us, we find only about 100 fundamental building blocks—the elements. Compounds are made by combining atoms of the various elements, just as words are constructed from the 26 letters of the alphabet. And just as you had to learn the letters of the alphabet before you learned to read and write, you need to learn the names and symbols of the chemical elements before you can read and write chemistry.

Presently about 118 different elements are known,* 88 of which occur naturally. (The rest have been made in laboratories.) The elements vary tremendously in abundance. In fact, only 9 elements account for most of the compounds found in the earth's crust. In Table 4.1, the elements are listed in order of their abundance (mass percent) in the earth's crust, oceans, and atmosphere. Note that nearly half of the mass is accounted for by oxygen alone. Also note that the 9 most abundant elements account for over 98% of the total mass.

Jeremy Woodhouse/PhotoDisc/Getty Images

Footprints in the sand of the Namib Desert in Namibia.

Oxygen, in addition to accounting for about 20% of the earth's atmosphere (where it occurs as O_2 molecules), is found in virtually all the rocks, sand, and soil on the earth's crust. In these latter materials, oxygen is not present as O_2 molecules but exists in compounds that usually contain silicon and aluminum atoms. The familiar substances of the geological world, such as rocks and sand, contain large groups of silicon and oxygen atoms bound together to form huge clusters.

The list of elements found in living matter is very different from the list of elements found in the earth's crust. Table 4.2 shows the distribution of elements in the human body. Oxygen, carbon, hydrogen, and nitrogen form the basis for all biologically important molecules. Some elements found in the body (called trace elements) are crucial for life, even though they are present in relatively small amounts. For example, chromium helps the body use sugars to provide energy.

One more general comment is important at this point. As we have seen, elements are fundamental to understanding chemistry. However, students are often confused by the many different ways that chemists use the term *element.* Sometimes when we say *element,* we mean a single atom of that element. We might call this the microscopic form of an element. Other times when we use the term *element,* we mean a sample of the element large enough to weigh on a balance. Such a sample contains many, many

Table 4.1 Distribution (Mass Percent) of the 18 Most Abundant Elements in the Earth's Crust, Oceans, and Atmosphere

Element	Mass Percent	Element	Mass Percent
oxygen	49.2	titanium	0.58
silicon	25.7	chlorine	0.19
aluminum	7.50	phosphorus	0.11
iron	4.71	manganese	0.09
calcium	3.39	carbon	0.08
sodium	2.63	sulfur	0.06
potassium	2.40	barium	0.04
magnesium	1.93	nitrogen	0.03
hydrogen	0.87	fluorine	0.03
		all others	0.49

*This number changes as new elements are made in particle accelerators.

Table 4.2 ▸ Abundance of Elements in the Human Body

Major Elements	Mass Percent	Trace Elements (in alphabetical order)
oxygen	65.0	arsenic
carbon	18.0	chromium
hydrogen	10.0	cobalt
nitrogen	3.0	copper
calcium	1.4	fluorine
phosphorus	1.0	iodine
magnesium	0.50	manganese
potassium	0.34	molybdenum
sulfur	0.26	nickel
sodium	0.14	selenium
chlorine	0.14	silicon
iron	0.004	vanadium
zinc	0.003	

atoms of the element, and we might call this the macroscopic form of the element. There is yet a further complication. As we will see in more detail in Section 4-9 the macroscopic forms of several elements contain molecules rather than individual atoms as the fundamental components. For example, chemists know that oxygen gas consists of molecules with two oxygen atoms connected together (represented as O—O or more commonly as O_2). Thus when we refer to the element oxygen we might mean a single atom of oxygen, a single O_2 molecule, or a macroscopic sample containing many O_2 molecules. Finally, we often use the term *element* in a generic fashion. When we say the human body contains the element sodium or lithium, we do not mean that free elemental sodium or lithium is present. Rather, we mean that atoms of these elements are present in some form. In this text we will try to make clear what we mean when we use the term *element* in a particular case.

4-2 Symbols for the Elements

OBJECTIVE **To learn the symbols of some elements.**

The names of the chemical elements have come from many sources. Often an element's name is derived from a Greek, Latin, or German word that describes some property of the element. For example, gold was originally called *aurum,* a Latin word meaning "shining dawn," and lead was known as *plumbum,* which means "heavy." The names for chlorine and iodine come from Greek words describing their colors, and the name for bromine comes from a Greek word meaning "stench." In addition, it is very common for an element to be named for the place where it was discovered. You can guess where the elements francium, germanium, californium,* and americium* were first found. Some of the heaviest elements are named after famous scientists—for example, einsteinium* and nobelium.*

We often use abbreviations to simplify the written word. For example, it is much easier to put MA on an envelope than to write out Massachusetts, and we often write USA instead of United States of America. Likewise, chemists have invented a set of

*These elements are made artificially. They do not occur naturally.

CHEMISTRY IN FOCUS

Trace Elements: Small but Crucial

We all know that certain chemical elements, such as calcium, carbon, nitrogen, phosphorus, and iron, are essential for humans to live. However, many other elements that are present in tiny amounts in the human body are also essential to life. Examples are chromium, cobalt, iodine, manganese, and copper. Chromium assists in the metabolism of sugars, cobalt is present in vitamin B_{12}, iodine is necessary for the proper functioning of the thyroid gland, manganese appears to play a role in maintaining the proper calcium levels in bones, and copper is involved in the production of red blood cells.

It is becoming clear that certain trace elements are very important in determining human behavior. For example, lithium (administered as lithium carbonate) has been a miracle drug for some people afflicted with bipolar disorder, a disease that produces oscillatory behavior between inappropriate "highs" and the blackest of depressions. Although its exact function remains unknown, lithium seems to moderate the levels of neurotransmitters (compounds that are essential to nerve function), thus relieving some of the extreme emotions in sufferers of bipolar disorder.

In addition, a chemist named William Walsh has done some very interesting studies on the inmates of Stateville Prison in Illinois. By analyzing the trace elements in the hair of prisoners, he has found intriguing relationships between the behavior of the inmates and their trace element profile. For example, Walsh found an inverse relationship between the level of cobalt in the prisoner's body and the degree of violence in his behavior.

Besides the levels of trace elements in our bodies, the various substances in the water, the food we consume, and the air we breathe also are of great importance to our health. For example, many scientists are concerned about our exposure to aluminum, through aluminum compounds used in water purification, baked goods and cheese (sodium aluminum phosphate acts as a leavening agent and also is added to cheese to make it softer and easier to melt), and the aluminum that dissolves from our cookware and utensils. The effects of exposure to low levels of aluminum on humans are not presently clear, but there are some indications that we should limit our intake of this element.

Another example of low-level exposure to an element is the fluoride placed in many water supplies and toothpastes to control tooth decay by making tooth enamel more resistant to dissolving. However, the exposure of large numbers of people to fluoride is quite controversial—many people think it is harmful.

The chemistry of trace elements is fascinating and important. Keep your eye on the news for further developments.

abbreviations or **element symbols** for the chemical elements. These symbols usually consist of the first letter or the first two letters of the element names. The first letter is always capitalized, and the second is not. Examples include

fluorine	F	neon	Ne
oxygen	O	silicon	Si
carbon	C		

Sometimes, however, the two letters used are not the first two letters in the name. For example,

zinc	Zn	cadmium	Cd
chlorine	Cl	platinum	Pt

The symbols for some other elements are based on the original Latin or Greek name.

Current Name	Original Name	Symbol
gold	aurum	Au
lead	plumbum	Pb
sodium	natrium	Na
iron	ferrum	Fe

Jesus Ayala/Getty Images

Various forms of the element gold.

Table 4.3 ▸ The Names and Symbols of the Most Common Elements

Element	Symbol	Element	Symbol
aluminum	Al	lithium	Li
antimony (stibium)*	Sb	magnesium	Mg
argon	Ar	manganese	Mn
arsenic	As	mercury (hydrargyrum)	Hg
barium	Ba	neon	Ne
bismuth	Bi	nickel	Ni
boron	B	nitrogen	N
bromine	Br	oxygen	O
cadmium	Cd	phosphorus	P
calcium	Ca	platinum	Pt
carbon	C	potassium (kalium)	K
chlorine	Cl	radium	Ra
chromium	Cr	silicon	Si
cobalt	Co	silver (argentium)	Ag
copper (cuprum)	Cu	sodium (natrium)	Na
fluorine	F	strontium	Sr
gold (aurum)	Au	sulfur	S
helium	He	tin (stannum)	Sn
hydrogen	H	titanium	Ti
iodine	I	tungsten (wolfram)	W
iron (ferrum)	Fe	uranium	U
lead (plumbum)	Pb	zinc	Zn

*Where appropriate, the original name is shown in parentheses so that you can see where some of the symbols came from.

A list of the most common elements and their symbols is given in Table 4.3. You can also see the elements represented on a table in Fig. 4.9. We will explain the form of this table (which is called the periodic table) in later chapters.

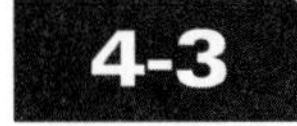

4-3 Dalton's Atomic Theory

OBJECTIVES

- **To learn about Dalton's theory of atoms.**
- **To understand and illustrate the law of constant composition.**

As scientists of the eighteenth century studied the nature of materials, several things became clear:

1. Most natural materials are mixtures of pure substances.
2. Pure substances are either elements or combinations of elements called compounds.
3. A given compound always contains the same proportions (by mass) of the elements. For example, water *always* contains 8 g of oxygen for every 1 g of hydrogen, and carbon dioxide *always* contains 2.7 g of oxygen for every 1 g of carbon. This principle became known as the **law of constant composition.** It means that a given compound always has the same composition, regardless of where it comes from.

John Dalton (Fig. 4.1), an English scientist and teacher, was aware of these observations, and in about 1808 he offered an explanation for them that became known as

Dalton's atomic theory. The main ideas of this theory (model) can be stated as follows:

Dalton's Atomic Theory

1. Elements are made of tiny particles called **atoms.**
2. All atoms of a given element are identical.
3. The atoms of a given element are different from those of any other element.
4. Atoms of one element can combine with atoms of other elements to form compounds. A given compound always has the same relative numbers and types of atoms.
5. Atoms are indivisible in chemical processes. That is, atoms are not created or destroyed in chemical reactions. A chemical reaction simply changes the way the atoms are grouped together.

Figure 4.1 ▸ John Dalton (1766–1844) was an English scientist who made his living as a teacher in Manchester. Although Dalton is best known for his atomic theory, he made contributions in many other areas, including meteorology (he recorded daily weather conditions for 46 years, producing a total of 200,000 data entries). A rather shy man, Dalton was colorblind to red (a special handicap for a chemist) and suffered from lead poisoning contracted from drinking stout (strong beer or ale) that had been drawn through lead pipes.

Dalton's model successfully explained important observations such as the law of constant composition. This law makes sense because if a compound always contains the same relative numbers of atoms, it will always contain the same proportions by mass of the various elements.

Like most new ideas, Dalton's model was not accepted immediately. However, Dalton was convinced he was right and *used his model to predict* how a given pair of elements might combine to form more than one compound. For example, nitrogen and oxygen might form a compound containing one atom of nitrogen and one atom of oxygen (written NO), a compound containing two atoms of nitrogen and one atom of oxygen (written N_2O), a compound containing one atom of nitrogen and two atoms of oxygen (written NO_2), and so on (Fig. 4.2). When the existence of these substances was verified, it was a triumph for Dalton's model. Because Dalton was able to predict correctly the formation of multiple compounds between two elements, his atomic theory became widely accepted.

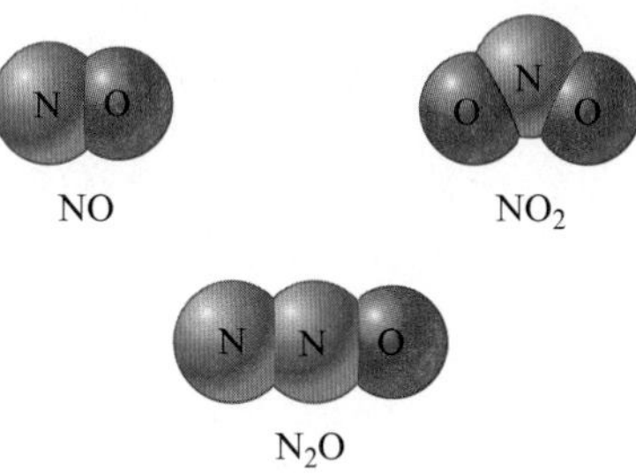

Figure 4.2 ▸ Dalton pictured compounds as collections of atoms. Here NO, NO_2, and N_2O are represented. Note that the number of atoms of each type in a molecule is given by a subscript, except that the number 1 is always assumed and never written.

4-4 Formulas of Compounds

OBJECTIVE To learn how a formula describes a compound's composition.

A **compound** is a distinct substance that is composed of the atoms of two or more elements and always contains exactly the same relative masses of those elements. In light of Dalton's atomic theory, this simply means that a compound always contains the same relative *numbers* of atoms of each element. For example, water always contains two hydrogen atoms for each oxygen atom. In this context, the term *relative* refers to ratios.

The types of atoms and the number of each type in each unit (molecule) of a given compound are conveniently expressed by a **chemical formula.** In a chemical formula the atoms are indicated by the element symbols, and the number of each type of atom is indicated by a subscript, a number that appears to the right of and below the symbol for the element. The formula for water is written H_2O, indicating that each molecule of water contains two atoms of hydrogen and one atom of oxygen (the subscript 1 is always understood and not written). Following are some general rules for writing formulas:

Rules for Writing Formulas

1. Each atom present is represented by its element symbol.
2. The number of each type of atom is indicated by a subscript written to the right of the element symbol.
3. When only one atom of a given type is present, the subscript 1 is not written.

Interactive Example 4.1

Writing Formulas of Compounds

Write the formula for each of the following compounds, listing the elements in the order given.

a. Each molecule of a compound that has been implicated in the formation of acid rain contains one atom of sulfur and three atoms of oxygen.

b. Each molecule of a certain compound contains two atoms of nitrogen and five atoms of oxygen.

c. Each molecule of glucose, a type of sugar, contains six atoms of carbon, twelve atoms of hydrogen, and six atoms of oxygen.

SOLUTION

a. 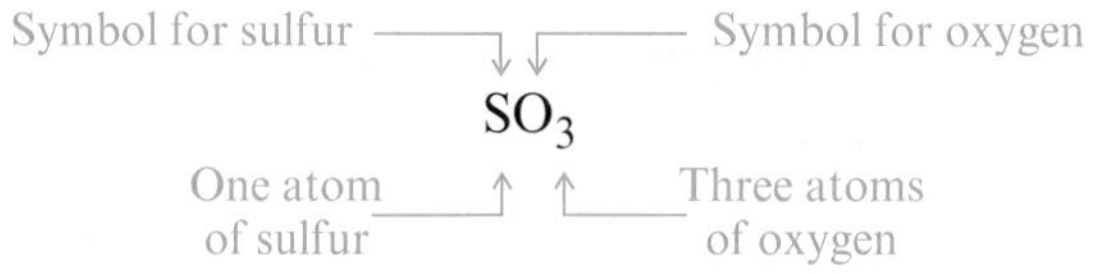 Symbol for sulfur → S; Symbol for oxygen → O

SO_3

One atom of sulfur; Three atoms of oxygen

b. 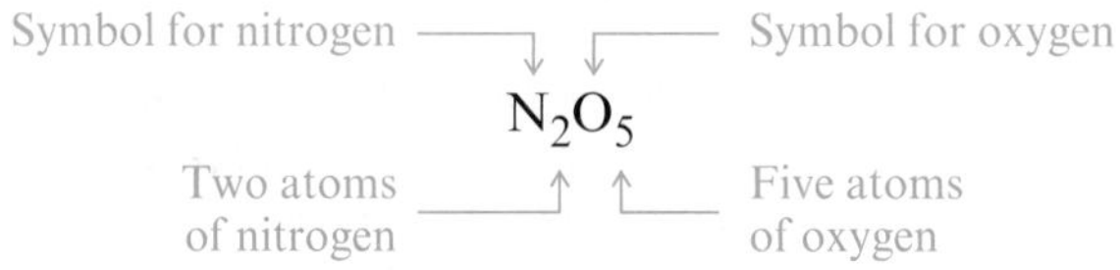 Symbol for nitrogen → N; Symbol for oxygen → O

N_2O_5

Two atoms of nitrogen; Five atoms of oxygen

c. 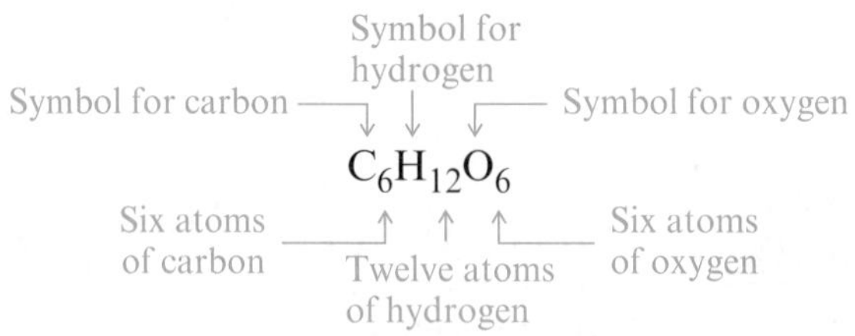 Symbol for carbon → C; Symbol for hydrogen → H; Symbol for oxygen → O

$C_6H_{12}O_6$

Six atoms of carbon; Twelve atoms of hydrogen; Six atoms of oxygen

SELF CHECK

Exercise 4.1 Write the formula for each of the following compounds, listing the elements in the order given.

a. A molecule contains four phosphorus atoms and ten oxygen atoms.

b. A molecule contains one uranium atom and six fluorine atoms.

c. A molecule contains one aluminum atom and three chlorine atoms.

See Problems 4.19 and 4.20. ■

CHEMISTRY IN FOCUS

A Four-Wheel-Drive Nanocar

A special kind of "microscope" called a scanning tunneling microscope (STM) has been developed that allows scientists to "see" individual atoms and to manipulate individual atoms and molecules on various surfaces. One very interesting application of this technique is the construction of tiny "machines" made of atoms. A recent example of this activity was performed by a group of scientists from the University of Groningen in the Netherlands. They used carbon

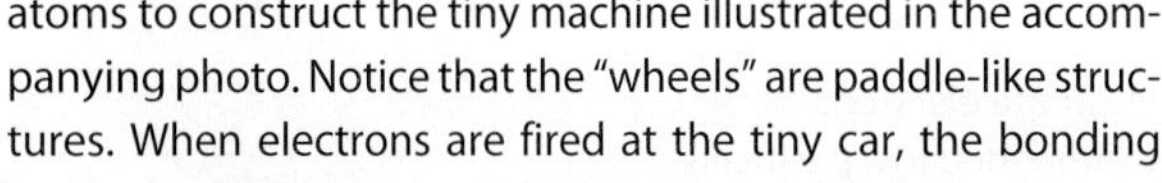

atoms to construct the tiny machine illustrated in the accompanying photo. Notice that the "wheels" are paddle-like structures. When electrons are fired at the tiny car, the bonding between the carbon atoms changes in such a way that the paddles twist and propel the car forward. The scientists have been able to move the car forward as much as ten car lengths on the copper surface.

Although at this point it is not quite clear how the little car could perform useful tasks, the research is yielding useful results on how the carbon-based structure interacts with the copper atoms comprising the surface.

Measuring approximately 4×2 nm, the molecular car is forging ahead on a copper surface on four electrically driven wheels.

Courtesy EMPA

4-5 The Structure of the Atom

OBJECTIVES

- **To learn about the internal parts of an atom.**
- **To understand Rutherford's experiment to characterize the atom's structure.**

Dalton's atomic theory, proposed in about 1808, provided such a convincing explanation for the composition of compounds that it became generally accepted. Scientists came to believe that *elements consist of atoms* and that *compounds are a specific collection of atoms* bound together in some way. But what is an atom like? It might be a tiny ball of matter that is the same throughout with no internal structure—like a ball bearing. Or the atom might be composed of parts—it might be made up of a number of subatomic particles. But if the atom contains parts, there should be some way to break up the atom into its components.

Many scientists pondered the nature of the atom during the 1800s, but it was not until almost 1900 that convincing evidence became available that the atom has a number of different parts.

A physicist in England named J. J. Thomson showed in the late 1890s that the atoms of any element can be made to emit tiny negative particles. (He knew the particles had a negative charge because he could show that they were repelled by the negative part of an electric field.) Thus he concluded that all types of atoms must contain these negative particles, which are now called **electrons.**

On the basis of his results, Thomson wondered what an atom must be like. Although he knew that atoms contain these tiny negative particles, he also knew that whole atoms are not negatively *or* positively charged. Thus he concluded that the atom must also contain positive particles that balance exactly the negative charge carried by the electrons, giving the atom a zero overall charge.

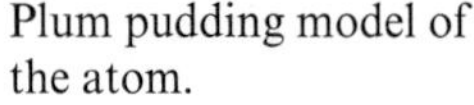

Plum pudding model of the atom.

English plum pudding.

Figure 4.3 One of the early models of the atom was the plum pudding model, in which the electrons were pictured as embedded in a positively charged spherical cloud, much as raisins are distributed in an old-fashioned plum pudding.

Another scientist pondering the structure of the atom was William Thomson (better known as Lord Kelvin and no relation to J. J. Thomson). Lord Kelvin got the idea (which might have occurred to him during dinner) that the atom might be something like plum pudding (a pudding with raisins randomly distributed throughout). Kelvin reasoned that the atom might be thought of as a uniform "pudding" of positive charge with enough negative electrons scattered within to counterbalance that positive charge (Fig. 4.3). Thus the plum pudding model of the atom came into being.

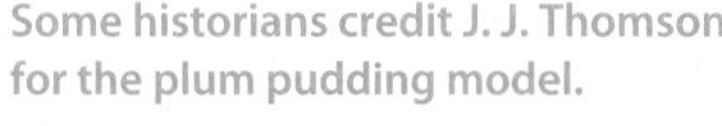

Some historians credit J. J. Thomson for the plum pudding model.

If you had taken this course in 1910, the plum pudding model would have been the only picture of the atom described. However, our ideas about the atom were changed dramatically in 1911 by a physicist named Ernest Rutherford (Fig. 4.4), who learned physics in J. J. Thomson's laboratory in the late 1890s. By 1911 Rutherford had become a distinguished scientist with many important discoveries to his credit. One of his main areas of interest involved alpha particles (α particles), positively charged particles with a mass approximately 7500 times that of an electron. In studying the flight of these particles through air, Rutherford found that some of the α particles were deflected by something in the air. Puzzled by this, he designed an experiment that involved directing α particles toward a thin metal foil. Surrounding the foil was a detector coated with a substance that produced tiny flashes wherever it was hit by an α particle (Fig. 4.5). The results of the experiment were very different from those Rutherford anticipated. Although most of the α particles passed straight through the foil, some of the particles were deflected at large angles, as shown in Fig. 4.5, and some were reflected backward.

This outcome was a great surprise to Rutherford. (He described this result as comparable to shooting a gun at a piece of paper and having the bullet bounce back.) Rutherford knew that if the plum pudding model of the atom was correct, the massive α particles would crash through the thin foil like cannonballs through paper [as shown in Fig. 4.6(a)]. So he expected the α particles to travel through the foil experiencing, at most, very minor deflections of their paths.

Figure 4.4 Ernest Rutherford (1871–1937) was born on a farm in New Zealand. In 1895 he placed second in a scholarship competition to attend Cambridge University but was awarded the scholarship when the winner decided to stay home and get married. Rutherford was an intense, hard-driving person who became a master at designing just the right experiment to test a given idea. He was awarded the Nobel Prize in chemistry in 1908.

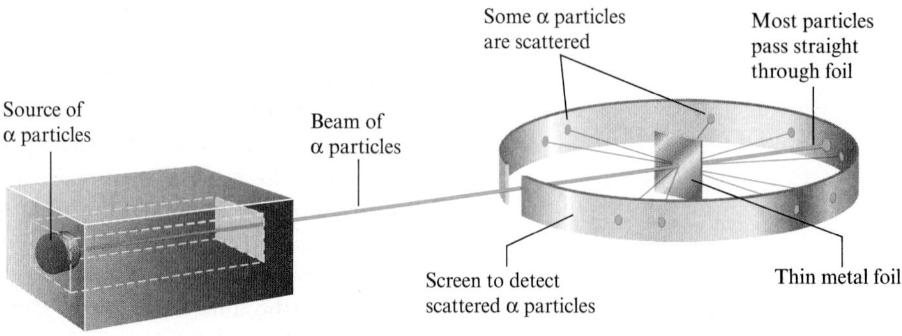

Figure 4.5 Rutherford's experiment on α-particle bombardment of metal foil.

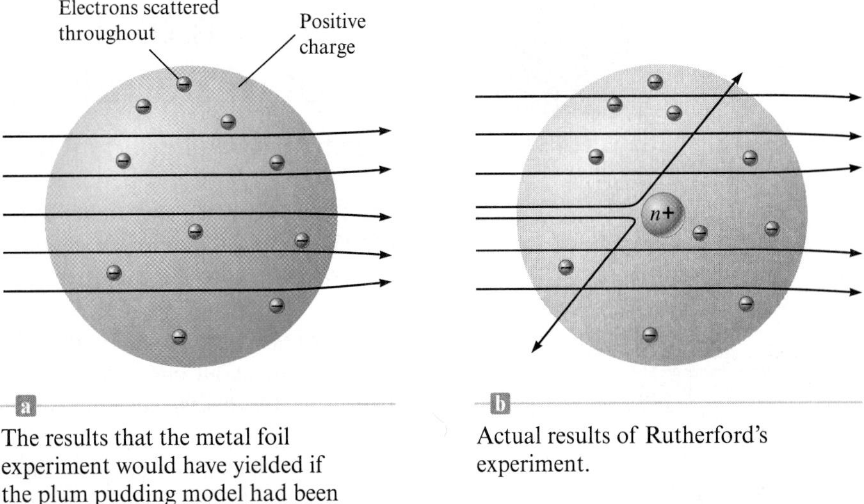

Figure 4.6 ▸ a The results that the metal foil experiment would have yielded if the plum pudding model had been correct. b Actual results of Rutherford's experiment.

One of Rutherford's coworkers in this experiment was an undergraduate named Ernest Marsden who, like Rutherford, was from New Zealand.

Rutherford concluded from these results that the plum pudding model for the atom could not be correct. The large deflections of the α particles could be caused only by a center of concentrated positive charge that would repel the positively charged α particles, as illustrated in Fig. 4.6(b). Most of the α particles passed directly through the foil because the atom is mostly open space. The deflected α particles were those that had a "close encounter" with the positive center of the atom, and the few reflected α particles were those that scored a "direct hit" on the positive center. In Rutherford's mind these results could be explained only in terms of a **nuclear atom**—an atom with a dense center of positive charge (the **nucleus**) around which tiny electrons moved in a space that was otherwise empty.

Critical Thinking

The average diameter of an atom is 1.3×10^{-10} m. What if the average diameter of an atom were 1 cm? How tall would you be?

If the atom were expanded to the size of a huge stadium, the nucleus would be only about as big as a fly at the center.

He concluded that the nucleus must have a positive charge to balance the negative charge of the electrons and that it must be small and dense. What was it made of? By 1919 Rutherford concluded that the nucleus of an atom contained what he called protons. A **proton** has the same magnitude (size) of charge as the electron, but its charge is *positive*. We say that the proton has a charge of 1+ and the electron a charge of 1−.

Rutherford reasoned that the hydrogen atom has a single proton at its center and one electron moving through space at a relatively large distance from the proton (the hydrogen nucleus). He also reasoned that other atoms must have nuclei (the plural of *nucleus*) composed of many protons bound together somehow. In addition, Rutherford and a coworker, James Chadwick, were able to show in 1932 that most nuclei also contain a neutral particle that they named the **neutron.** A neutron is slightly more massive than a proton but has no charge.

Critical Thinking

You have learned about three different models of the atom: Dalton's model, Thomson's model, and Rutherford's model. What if Dalton was correct? What would Rutherford have expected from his experiments with gold foil? What if Thomson was correct? What would Rutherford have expected from his experiments with gold foil?

4-6 Introduction to the Modern Concept of Atomic Structure

OBJECTIVE **To understand some important features of subatomic particles.**

In the years since Thomson and Rutherford, a great deal has been learned about atomic structure. The simplest view of the atom is that it consists of a tiny nucleus (about 10^{-13} cm in diameter) and electrons that move about the nucleus at an average distance of about 10^{-8} cm from it (Fig. 4.7). To visualize how small the nucleus is compared with the size of the atom, consider that if the nucleus were the size of a grape, the electrons would be about one *mile* away on average. The nucleus contains protons, which have a positive charge equal in magnitude to the electron's negative charge, and neutrons, which have almost the same mass as a proton but no charge. The neutrons' function in the nucleus is not obvious. They may help hold the protons (which repel each other) together to form the nucleus, but we will not be concerned with that here. The relative masses and charges of the electron, proton, and neutron are shown in Table 4.4.

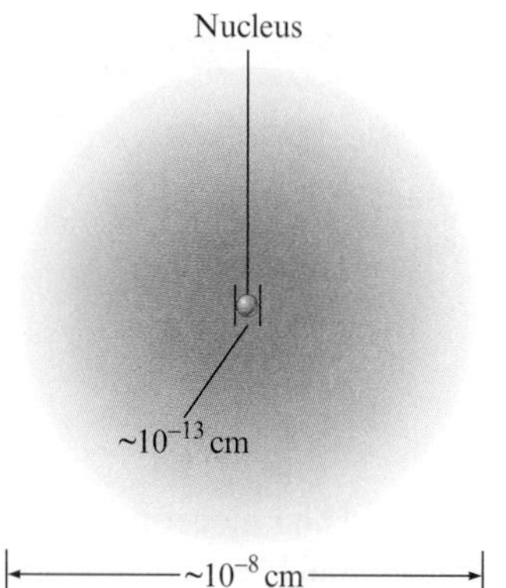

Figure 4.7 ▸ A nuclear atom viewed in cross section. (The symbol ~ means approximately.) This drawing does not show the actual scale. The nucleus is actually *much* smaller compared with the size of an atom.

An important question arises at this point: *"If all atoms are composed of these same components, why do different atoms have different chemical properties?"* The answer lies in the number and arrangement of the electrons. The space in which the electrons move accounts for most of the atomic volume. The electrons are the parts of atoms that "intermingle" when atoms combine to form molecules. Therefore, the number of electrons a given atom possesses greatly affects the way it can interact with other atoms. As a result, atoms of different elements, which have different numbers of electrons, show different chemical behavior. Although the atoms of different elements also differ in their numbers of protons, it is the number of electrons that really determines chemical behavior. We will discuss how this happens in later chapters.

Table 4.4 ▸ The Mass and Charge of the Electron, Proton, and Neutron

Particle	Relative Mass*	Relative Charge
electron	1	1−
proton	1836	1+
neutron	1839	none

*The electron is arbitrarily assigned a mass of 1 for comparison.

4-7 Isotopes

OBJECTIVES

- **To learn about the terms *isotope, atomic number,* and *mass number.***
- **To understand the use of the symbol $^{A}_{Z}X$ to describe a given atom.**

We have seen that an atom has a nucleus with a positive charge due to its protons and has electrons in the space surrounding the nucleus at relatively large distances from it.

As an example, consider a sodium atom, which has 11 protons in its nucleus. Because an atom has no overall charge, the number of electrons must equal the number of protons. Therefore, a sodium atom has 11 electrons in the space around its nucleus. It is *always* true that a sodium atom has 11 protons and 11 electrons. However, each sodium atom also has neutrons in its nucleus, and different types of sodium atoms exist that have different numbers of neutrons.

When Dalton stated his atomic theory in the early 1800s, he assumed all of the atoms of a given element were identical. This idea persisted for over a hundred years, until James Chadwick discovered that the nuclei of most atoms contain neutrons as well as protons. (This is a good example of how a theory changes as new observations are made.) After the discovery of the neutron, Dalton's statement that all atoms of a given element are identical had to be changed to "All atoms of the same element contain the same number of protons and electrons, but atoms of a given element may have different numbers of neutrons."

To illustrate this idea, consider the sodium atoms represented in Fig. 4.8. These atoms are **isotopes,** or *atoms with the same number of protons but different numbers of neutrons.* The number of protons in a nucleus is called the atom's **atomic number (Z).** The *sum* of the number of neutrons and the number of protons in a given nucleus is called the atom's **mass number (*A*).** To specify which of the isotopes of an element we are talking about, we use the symbol

$${}^{A}_{Z}\mathrm{X}$$

where

X = the symbol of the element
A = the mass number (number of protons and neutrons)
Z = the atomic number (number of protons)

For example, the symbol for one particular type of sodium atom is written

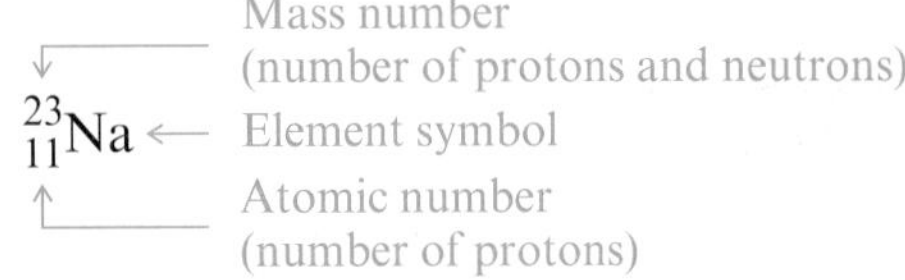

The particular atom represented here is called sodium-23, because it has a mass number of 23. Let's specify the number of each type of subatomic particle. From the atomic number 11 we know that the nucleus contains 11 protons. And because the number of electrons is equal to the number of protons, we know that this atom contains 11 electrons. How many neutrons are present? We can calculate the number of neutrons from the definition of the mass number

$$\text{Mass number} = \text{number of protons} + \text{number of neutrons}$$

or, in symbols,

$$A = Z + \text{number of neutrons}$$

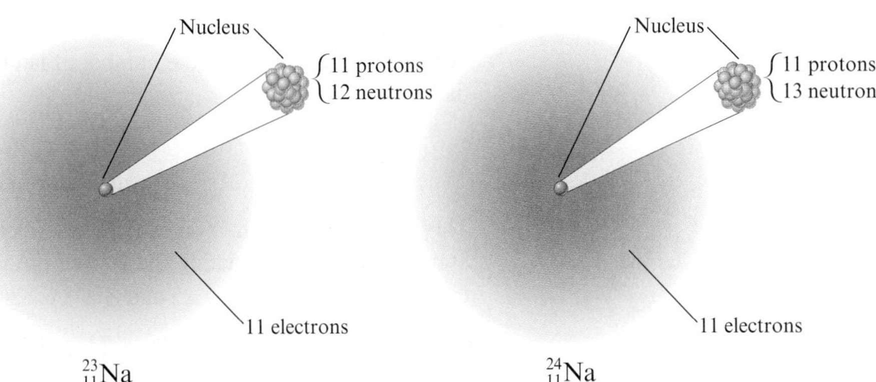

Figure 4.8 Two isotopes of sodium. Both have 11 protons and 11 electrons, but they differ in the number of neutrons in their nuclei.

"Whair" Do You Live?

Picture a person who has been the victim of a crime in a large city in the eastern United States. The person has been hit in the head and, as a result, has total amnesia. The person's ID has been stolen, but the authorities suspect he may not be from the local area. Is there any way to find out where the person might be from? The answer is yes. Recent research indicates that the relative amounts of the isotopes of hydrogen and oxygen in a person's hair indicate in which part of the United States a person lives.

Support for this idea has come from a recent study by James Ehleringer, a chemist at the University of Utah in Salt Lake City. Noting that the concentrations of hydrogen-2 (deuterium) and oxygen-18 in drinking water vary significantly from region to region in the United States (see accompanying illustration), Ehleringer and his colleagues collected hair samples from barbershops in 65 cities and 18 states. Their analyses showed that 86% of the variations in the hydrogen and oxygen isotopes in the hair samples result from the isotopic composition of the local water. Based on their results, the group was able to develop estimates of the isotopic signature of peoples' hair from various regions of the country. Although this method cannot be used to pinpoint a person's place of residence, it can give a general region. This method might be helpful for the amnesia victim described above by showing where to look for his family. His picture could be shown on TV in the region indicated by analysis of his hair. Another possible use of this technique is identifying the country of origin of victims of a natural disaster in a tourist region with visitors from all over the world. In fact, a similar technique was used to specify the countries of origin of the victims of the tsunami that devastated southern Asia in December 2004.

An interesting verification of this technique occurred when the researchers examined a strand of hair from a person who had recently moved from Beijing, China, to Salt Lake City. Analysis of various parts of the hair showed a distinct change in isotopic distribution corresponding to his change of residence. Thus the isotopes of elements can provide useful information in unexpected ways.

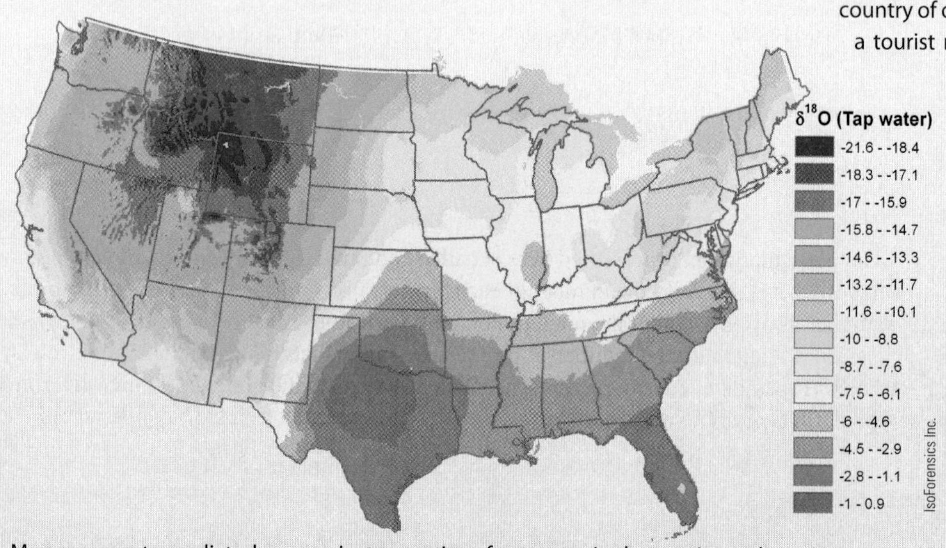

Map represents predicted oxygen isotope ratios of tap water in the continental United States, 2013.

We can isolate (solve for) the number of neutrons by subtracting Z from both sides of the equation

$$A - Z = Z - Z + \text{number of neutrons}$$
$$A - Z = \text{number of neutrons}$$

This is a general result. You can always determine the number of neutrons present in a given atom by subtracting the atomic number from the mass number. In this case $({}^{23}_{11}\text{Na})$, we know that $A = 23$ and $Z = 11$. Thus

$$A - Z = 23 - 11 = 12 = \text{number of neutrons}$$

In summary, sodium-23 has 11 electrons, 11 protons, and 12 neutrons.

Interactive Example 4.2

Interpreting Symbols for Isotopes

In nature, elements are usually found as a mixture of isotopes. Three isotopes of elemental carbon are $^{12}_{6}C$ (carbon-12), $^{13}_{6}C$ (carbon-13), and $^{14}_{6}C$ (carbon-14). Determine the number of each of the three types of subatomic particles in each of these carbon atoms.

SOLUTION The number of protons and electrons is the same in each of the isotopes and is given by the atomic number of carbon, 6. The number of neutrons can be determined by subtracting the atomic number (Z) from the mass number (A):

$$A - Z = \text{number of neutrons}$$

The numbers of neutrons in the three isotopes of carbon are

$$^{12}_{6}C\text{: number of neutrons} = A - Z = 12 - 6 = 6$$

$$^{13}_{6}C\text{: number of neutrons} = 13 - 6 = 7$$

$$^{14}_{6}C\text{: number of neutrons} = 14 - 6 = 8$$

In summary,

Symbol	Number of Protons	Number of Electrons	Number of Neutrons
$^{12}_{6}C$	6	6	6
$^{13}_{6}C$	6	6	7
$^{14}_{6}C$	6	6	8

SELF CHECK

Exercise 4.2 Give the number of protons, neutrons, and electrons in the atom symbolized by $^{90}_{38}Sr$. Strontium-90 occurs in fallout from nuclear testing. It can accumulate in bone marrow and may cause leukemia and bone cancer.

See Problems 4.39 and 4.42. ■

SELF CHECK

Exercise 4.3 Give the number of protons, neutrons, and electrons in the atom symbolized by $^{201}_{80}Hg$.

See Problems 4.39 and 4.42. ■

Interactive Example 4.3

Writing Symbols for Isotopes

Write the symbol for the magnesium atom (atomic number 12) with a mass number of 24. How many electrons and how many neutrons does this atom have?

SOLUTION The atomic number 12 means the atom has 12 protons. The element magnesium is symbolized by Mg. The atom is represented as

$$^{24}_{12}Mg$$

and is called magnesium-24. Because the atom has 12 protons, it must also have 12 electrons. The mass number gives the total number of protons and neutrons, which means that this atom has 12 neutrons (24 − 12 = 12). ■

Magnesium burns in air to give a bright white flame.

Isotope Tales

The atoms of a given element typically consist of several isotopes—atoms with the same number of protons but different numbers of neutrons. It turns out that the ratio of isotopes found in nature can be very useful in natural detective work. One reason is that the ratio of isotopes of elements found in living animals and humans reflects their diets. For example, African elephants that feed on grasses have a different $^{13}C/^{12}C$ ratio in their tissues than elephants that primarily eat tree leaves. This difference arises because grasses have a different growth pattern than leaves do, resulting in different amounts of ^{13}C and ^{12}C being incorporated from the CO_2 in the air. Because leaf-eating and grass-eating elephants live in different areas of Africa, the observed differences in the $^{13}C/^{12}C$ isotope ratios in elephant ivory samples have enabled authorities to identify the sources of illegal samples of ivory.

Paul Chesley/National Geographic/Getty Images

Ancient Anasazi Indian cliff dwellings.

Another case of isotope detective work involves the tomb of King Midas, who ruled the kingdom Phyrgia in the eighth century B.C. Analysis of nitrogen isotopes in the king's decayed casket has revealed details about the king's diet. Scientists have learned that the $^{15}N/^{14}N$ ratios of carnivores are higher than those of herbivores, which in turn are higher than those of plants. It turns out that the organism responsible for decay of the king's wooden casket has an unusually large requirement for nitrogen. The source of this nitrogen was the body of the dead king. Because the decayed wood under his now-decomposed body showed a high $^{15}N/^{14}N$ ratio, researchers feel sure that the king's diet was rich in meat.

A third case of historical isotope detective work concerns the Pueblo ancestor people (commonly called the Anasazi), who lived in what is now northwestern New Mexico between A.D. 900 and 1150. The center of their civilization, Chaco Canyon, was a thriving cultural center boasting dwellings made of hand-hewn sandstone and more than 200,000 logs. The sources of the logs have always been controversial. Many theories have been advanced concerning the distances over which the logs were hauled. Recent research by Nathan B. English, a geochemist at the University of Arizona in Tucson, has used the distribution of strontium isotopes in the wood to identify the probable sources of the logs. This effort has enabled scientists to understand more clearly the Anasazi building practices.

These stories illustrate how isotopes can serve as valuable sources of biologic and historical information.

Interactive Example 4.4

Calculating Mass Number

Write the symbol for the silver atom ($Z = 47$) that has 61 neutrons.

SOLUTION The element symbol is $^{A}_{Z}\text{Ag}$, where we know that $Z = 47$. We can find A from its definition, $A = Z +$ number of neutrons. In this case,

$$A = 47 + 61 = 108$$

The complete symbol for this atom is $^{108}_{47}\text{Ag}$.

SELF CHECK **Exercise 4.4** Give the symbol for the phosphorus atom ($Z = 15$) that contains 17 neutrons.

See Problem 4.42. ■

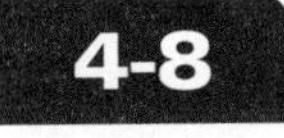

4-8 Introduction to the Periodic Table

OBJECTIVES

- **To learn about various features of the periodic table.**
- **To learn some of the properties of metals, nonmetals, and metalloids.**

In any room where chemistry is taught or practiced, you are almost certain to find a chart called the **periodic table** hanging on the wall. This chart shows all of the known elements and gives a good deal of information about each. As our study of chemistry progresses, the usefulness of the periodic table will become more obvious. This section will simply introduce it.

A simple version of the periodic table is shown in Fig. 4.9. Note that each box of this table contains a number written over one, two, or three letters. The letters are the symbols for the elements. The number shown above each symbol is the atomic number (the number of protons and also the number of electrons) for that element. For example, carbon (C) has atomic number 6:

6 C

1 1A (Alkali metals)	2 2A (Alkaline earth metals)	3	4	5	6	7	8	9	10	11	12	13 3A	14 4A	15 5A	16 6A	17 7A (Halogens)	18 8A (Noble gases)
1 H																	2 He
3 Li	4 Be											5 B	6 C	7 N	8 O	9 F	10 Ne
11 Na	12 Mg	Transition metals										13 Al	14 Si	15 P	16 S	17 Cl	18 Ar
19 K	20 Ca	21 Sc	22 Ti	23 V	24 Cr	25 Mn	26 Fe	27 Co	28 Ni	29 Cu	30 Zn	31 Ga	32 Ge	33 As	34 Se	35 Br	36 Kr
37 Rb	38 Sr	39 Y	40 Zr	41 Nb	42 Mo	43 Tc	44 Ru	45 Rh	46 Pd	47 Ag	48 Cd	49 In	50 Sn	51 Sb	52 Te	53 I	54 Xe
55 Cs	56 Ba	57 La*	72 Hf	73 Ta	74 W	75 Re	76 Os	77 Ir	78 Pt	79 Au	80 Hg	81 Tl	82 Pb	83 Bi	84 Po	85 At	86 Rn
87 Fr	88 Ra	89 Ac†	104 Rf	105 Db	106 Sg	107 Bh	108 Hs	109 Mt	110 Ds	111 Rg	112 Cn	113 Uut	114 Fl	115 Uup	116 Lv	117 Uus	118 Uuo

*Lanthanides	58 Ce	59 Pr	60 Nd	61 Pm	62 Sm	63 Eu	64 Gd	65 Tb	66 Dy	67 Ho	68 Er	69 Tm	70 Yb	71 Lu
†Actinides	90 Th	91 Pa	92 U	93 Np	94 Pu	95 Am	96 Cm	97 Bk	98 Cf	99 Es	100 Fm	101 Md	102 No	103 Lr

Figure 4.9 The periodic table.

Lead (Pb) has atomic number 82:

82 Pb

Notice that elements 112 through 115 and 118 have unusual three-letter designations beginning with U. These are abbreviations for the systematic names of the atomic numbers of these elements. "Regular" names for these elements will be chosen eventually by the scientific community.

Note that the elements are listed on the periodic table in order of increasing atomic number. They are also arranged in specific horizontal rows and vertical columns. The elements were first arranged in this way in 1869 by Dmitri Mendeleev, a Russian scientist. Mendeleev arranged the elements in this way because of similarities in the chemical properties of various "families" of elements. For example, fluorine and chlorine are reactive gases that form similar compounds. It was also known that sodium and potassium behave very similarly. Thus the name *periodic table* refers to the fact that as we increase the atomic numbers, every so often an element occurs with properties similar to those of an earlier (lower-atomic-number) element. For example, the elements

Mendeleev actually arranged the elements in order of increasing atomic mass rather than atomic number.

9 F
17 Cl
35 Br
53 I
85 At

Throughout the text, we will highlight the location of various elements by presenting a small version of the periodic table.

all show similar chemical behavior and so are listed vertically, as a "family" of elements.

A family of elements with similar chemical properties that lie in the same vertical column on the periodic table is called a **group.** Groups are often referred to by the number over the column (see Fig. 4.9). Note that the group numbers are accompanied by the letter A on the periodic table in Fig. 4.9. For simplicity we will delete the A's when we refer to groups in the text. Many of the groups have special names. For example, the first column of elements (Group 1) has the name **alkali metals.** The Group 2 elements are called the **alkaline earth metals,** the Group 7 elements are the **halogens,** and the elements in Group 8 are called the **noble gases.** A large collection of elements that spans many vertical columns consists of the **transition metals.**

There's another convention recommended by the International Union of Pure and Applied Chemistry for group designations that uses numbers 1 through 18 and includes the transition metals (see Fig. 4.9). Do not confuse that system with the one used in this text, where only the representative elements have group numbers (1 through 8).

Most of the elements are **metals.** Metals have the following characteristic physical properties:

Physical Properties of Metals

1. Efficient conduction of heat and electricity
2. Malleability (they can be hammered into thin sheets)
3. Ductility (they can be pulled into wires)
4. A lustrous (shiny) appearance

Nonmetals sometimes have one or more metallic properties. For example, solid iodine is lustrous, and graphite (a form of pure carbon) conducts electricity.

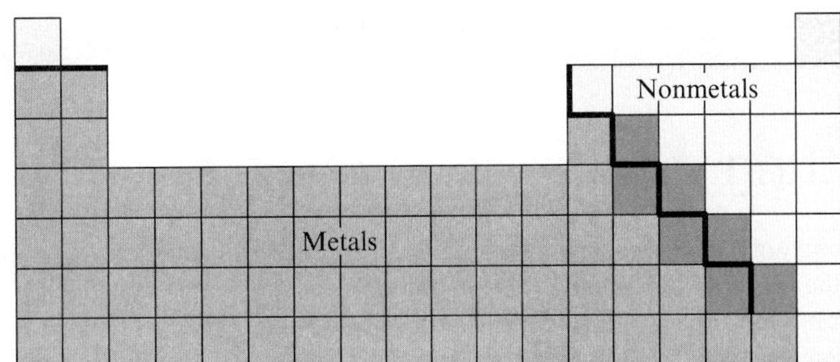

Figure 4.10 ▸ The elements classified as metals and as nonmetals.

For example, copper is a typical metal. It is lustrous (although it tarnishes readily); it is an excellent conductor of electricity (it is widely used in electrical wires); and it is readily formed into various shapes, such as pipes for water systems. Copper is one of the transition metals—the metals shown in the center of the periodic table. Iron, aluminum, and gold are other familiar elements that have metallic properties. All of the elements shown to the left of and below the heavy "stair-step" black line in Fig. 4.9 are classified as metals, except for hydrogen (Fig. 4.10).

The relatively small number of elements that appear in the upper-right corner of the periodic table (to the right of the heavy line in Figs. 4.9 and 4.10) are called **nonmetals.** Nonmetals generally lack those properties that characterize metals and show much more variation in their properties than metals do. Whereas almost all metals are solids at normal temperatures, many nonmetals (such as nitrogen, oxygen, chlorine, and neon) are gaseous and one (bromine) is a liquid. Several nonmetals (such as carbon, phosphorus, and sulfur) are also solids.

Daniel Osterkamp/Flickr/Getty Images

Sulfur miner carrying sulfur in Java.

The elements that lie close to the "stair-step" line as shown in blue in Fig. 4.10 often show a mixture of metallic and nonmetallic properties. These elements, which are called **metalloids** or **semimetals,** include silicon, germanium, arsenic, antimony, and tellurium.

As we continue our study of chemistry, we will see that the periodic table is a valuable tool for organizing accumulated knowledge and that it helps us predict the properties we expect a given element to exhibit. We will also develop a model for atomic structure that will explain why there are groups of elements with similar chemical properties.

Interactive Example 4.5

Interpreting the Periodic Table

For each of the following elements, use the periodic table in the front of the book to give the symbol and atomic number and to specify whether the element is a metal or a nonmetal. Also give the named family to which the element belongs (if any).

a. iodine b. magnesium c. gold d. lithium

SOLUTION

a. Iodine (symbol I) is element 53 (its atomic number is 53). Iodine lies to the right of the stair-step line in Fig. 4.10 and thus is a nonmetal. Iodine is a member of Group 7, the family of halogens.

b. Magnesium (symbol Mg) is element 12 (atomic number 12). Magnesium is a metal and is a member of the alkaline earth metal family (Group 2).

c. Gold (symbol Au) is element 79 (atomic number 79). Gold is a metal and is not a member of a named vertical family. It is classed as a transition metal.

d. Lithium (symbol Li) is element 3 (atomic number 3). Lithium is a metal in the alkali metal family (Group 1).

CHEMISTRY IN FOCUS

Putting the Brakes on Arsenic

The toxicity of arsenic is well known. Indeed, arsenic has often been the poison of choice in classic plays and films—watch *Arsenic and Old Lace* sometime. Contrary to its treatment in the aforementioned movie, arsenic poisoning is a serious, contemporary problem. For example, the World Health Organization estimates that 77 million people in Bangladesh are at risk from drinking water that contains large amounts of naturally occurring arsenic. Recently, the Environmental Protection Agency announced more stringent standards for arsenic in U.S. public drinking water supplies. Studies show that prolonged exposure to arsenic can lead to a higher risk of bladder, lung, and skin cancers as well as other ailments, although the levels of arsenic that induce these symptoms remain in dispute in the scientific community.

Tara Piasio/IFAS/University of Florida

Lena Q. Ma and *Pteris vittata*—called the Chinese brake fern.

Cleaning up arsenic-contaminated soil and water poses a significant problem. One approach is to find plants that will remove arsenic from the soil. Such a plant, the Chinese brake fern, recently has been shown to have a voracious appetite for arsenic. Research led by Lena Q. Ma, a chemist at the University of Florida in Gainesville, has shown that the brake fern accumulates arsenic at a rate 200 times that of the average plant. The arsenic, which becomes concentrated in fronds that grow up to 5 feet long, can be easily harvested and hauled away. Researchers are now investigating the best way to dispose of the plants so the arsenic can be isolated. The fern (*Pteris vittata*) looks promising for putting the brakes on arsenic pollution.

Exercise 4.5 Give the symbol and atomic number for each of the following elements. Also indicate whether each element is a metal or a nonmetal and whether it is a member of a named family.

a. argon b. chlorine c. barium d. cesium

See Problems 4.53 and 4.54. ■

4-9 Natural States of the Elements

OBJECTIVE **To learn the natures of the common elements.**

As we have noted, the matter around us consists mainly of mixtures. Most often these mixtures contain compounds, in which atoms from different elements are bound together. Most elements are quite reactive: their atoms tend to combine with those of other elements to form compounds. Thus we do not often find elements in nature in pure form—uncombined with other elements. However, there are notable exceptions. The gold nuggets found at Sutter's Mill in California that launched the Gold Rush in 1849 are virtually pure elemental gold. And platinum and silver are often found in nearly pure form.

Gold, silver, and platinum are members of a class of metals called *noble metals* because they are relatively unreactive. (The term *noble* implies a class set apart.)

Other elements that appear in nature in the uncombined state are the elements in Group 8: helium, neon, argon, krypton, xenon, and radon. Because the atoms of these elements do not combine readily with those of other elements, we call them the *noble gases*. For example, helium gas is found in uncombined form in underground deposits with natural gas.

Recall that a molecule is a collection of atoms that behaves as a unit. Molecules are always electrically neutral (zero charge).

When we take a sample of air (the mixture of gases that constitute the earth's atmosphere) and separate it into its components, we find several pure elements present. One of these is argon. Argon gas consists of a collection of separate argon atoms, as shown in Fig. 4.11.

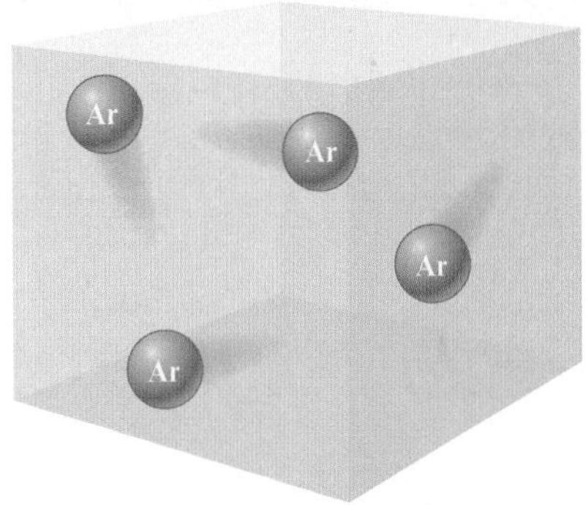

Figure 4.11 ▸ Argon gas consists of a collection of separate argon atoms.

Air also contains nitrogen gas and oxygen gas. When we examine these two gases, however, we find that they do not contain single atoms, as argon does, but instead contain **diatomic molecules:** molecules made up of *two atoms,* as represented in Fig. 4.12. In fact, any sample of elemental oxygen gas at normal temperatures contains O_2 molecules. Likewise, nitrogen gas contains N_2 molecules.

Hydrogen is another element that forms diatomic molecules. Though virtually all of the hydrogen found on earth is present in compounds with other elements (such as with oxygen in water), when hydrogen is prepared as a free element it contains diatomic H_2 molecules. For example, an electric current can be used to decompose water (Fig. 4.13 and Fig. 3.3) into elemental hydrogen and oxygen containing H_2 and O_2 molecules, respectively.

Several other elements, in addition to hydrogen, nitrogen, and oxygen, exist as diatomic molecules. For example, when sodium chloride is melted and subjected to an electric current, chlorine gas is produced (along with sodium metal). This chemical

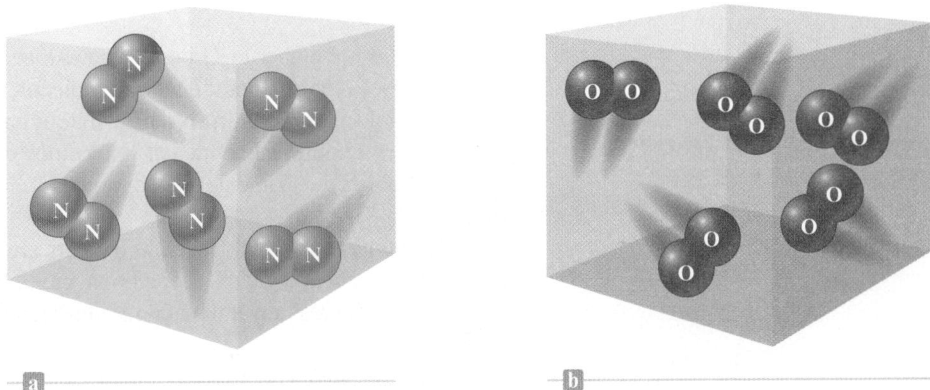

Figure 4.12 ▸ Gaseous nitrogen and oxygen contain diatomic (two-atom) molecules.

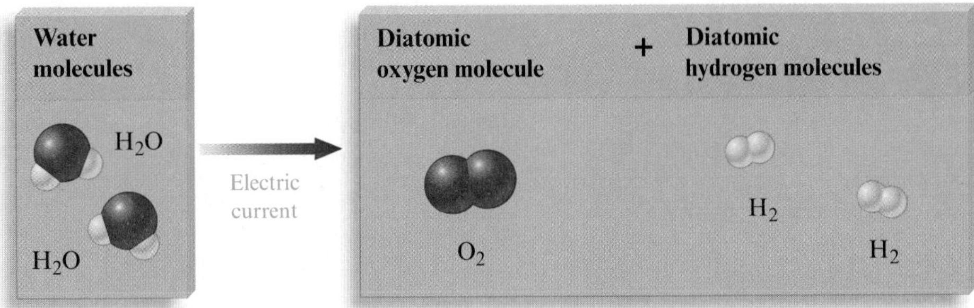

Figure 4.13 ▸ The decomposition of two water molecules (H_2O) to form two hydrogen molecules (H_2) and one oxygen molecule (O_2). Note that only the grouping of the atoms changes in this process; no atoms are created or destroyed. There must be the same number of H atoms and O atoms before and after the process. Thus the decomposition of two H_2O molecules (containing four H atoms and two O atoms) yields one O_2 molecule (containing two O atoms) and two H_2 molecules (containing a total of four H atoms).

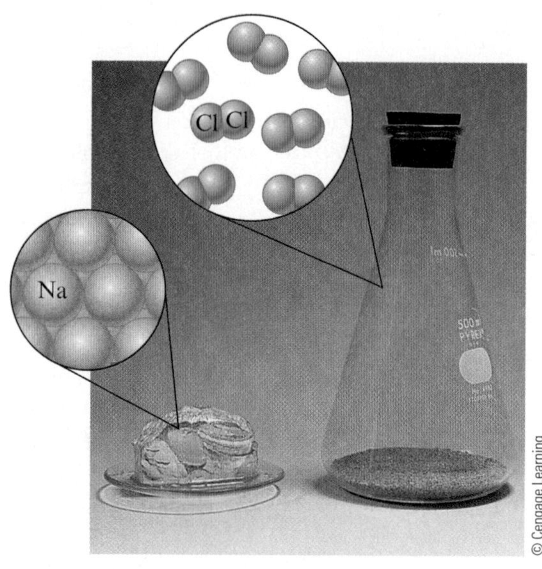

Figure 4.14 ▸ a Sodium chloride (common table salt) can be decomposed to its elements. b Sodium metal (on the left) and chlorine gas.

change is represented in Fig. 4.14. Chlorine gas is a pale green gas that contains Cl_2 molecules.

Chlorine is a member of Group 7, the halogen family. All the elemental forms of the Group 7 elements contain diatomic molecules. Fluorine is a pale yellow gas containing F_2 molecules. Bromine is a brown liquid made up of Br_2 molecules. Iodine is a lustrous, purple solid that contains I_2 molecules.

Table 4.5 lists the elements that contain diatomic molecules in their pure, elemental forms.

So far we have seen that several elements are gaseous in their elemental forms at normal temperatures (~25 °C). The noble gases (the Group 8 elements) contain individual atoms, whereas several other gaseous elements contain diatomic molecules (H_2, N_2, O_2, F_2, and Cl_2).

~ means "approximately."

Only two elements are liquids in their elemental forms at 25 °C: the nonmetal bromine (containing Br_2 molecules) and the metal mercury. The metals gallium and cesium almost qualify in this category; they are solids at 25 °C, but both melt at ~30 °C.

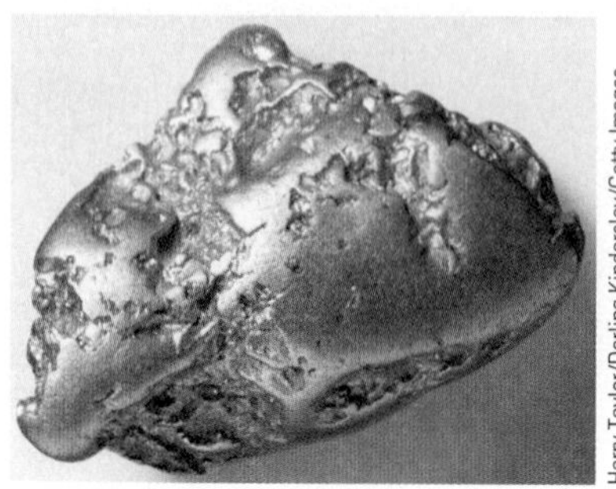

Platinum is a noble metal used in jewelry and in many industrial processes.

Table 4.5 ▸ Elements That Exist as Diatomic Molecules in Their Elemental Forms

Element Present	Elemental State at 25 °C	Molecule
hydrogen	colorless gas	H_2
nitrogen	colorless gas	N_2
oxygen	pale blue gas	O_2
fluorine	pale yellow gas	F_2
chlorine	pale green gas	Cl_2
bromine	reddish brown liquid	Br_2
iodine	lustrous, dark purple solid	I_2

Liquid bromine in a flask with bromine vapor.

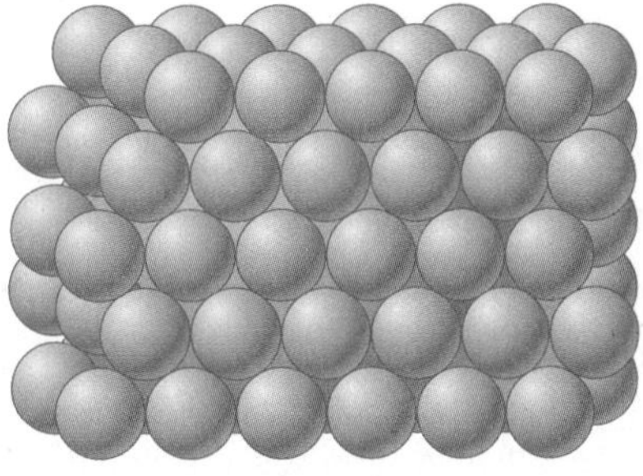

Figure 4.15 ▶ In solid metals, the spherical atoms are packed closely together.

The other elements are solids in their elemental forms at 25 °C. For metals these solids contain large numbers of atoms packed together much like marbles in a jar (Fig. 4.15).

The structures of solid nonmetallic elements are more varied than those of metals. In fact, different forms of the same element often occur. For example, solid carbon occurs in three forms. Different forms of a given element are called *allotropes.* The three allotropes of carbon are the familiar diamond and graphite forms plus a form that has only recently been discovered called *buckminsterfullerene.* These elemental forms have very different properties because of their different structures (Fig. 4.16). Diamond is the hardest natural substance known and is often used for industrial cutting tools. Diamonds are also valued as gemstones. Graphite, by contrast, is a rather soft material useful for writing (pencil "lead" is really graphite) and (in the form of a powder) for lubricating locks. The rather odd name given to buckminsterfullerene comes from the structure of the C_{60} molecules that form this allotrope. The soccer-ball-like structure contains five- and six-member rings reminiscent of the structure of geodesic domes suggested by the late industrial designer Buckminster Fuller. Other "fullerenes" containing molecules with more than 60 carbon atoms have also been discovered, leading to a new area of chemistry.

Cut diamond, held over coal.

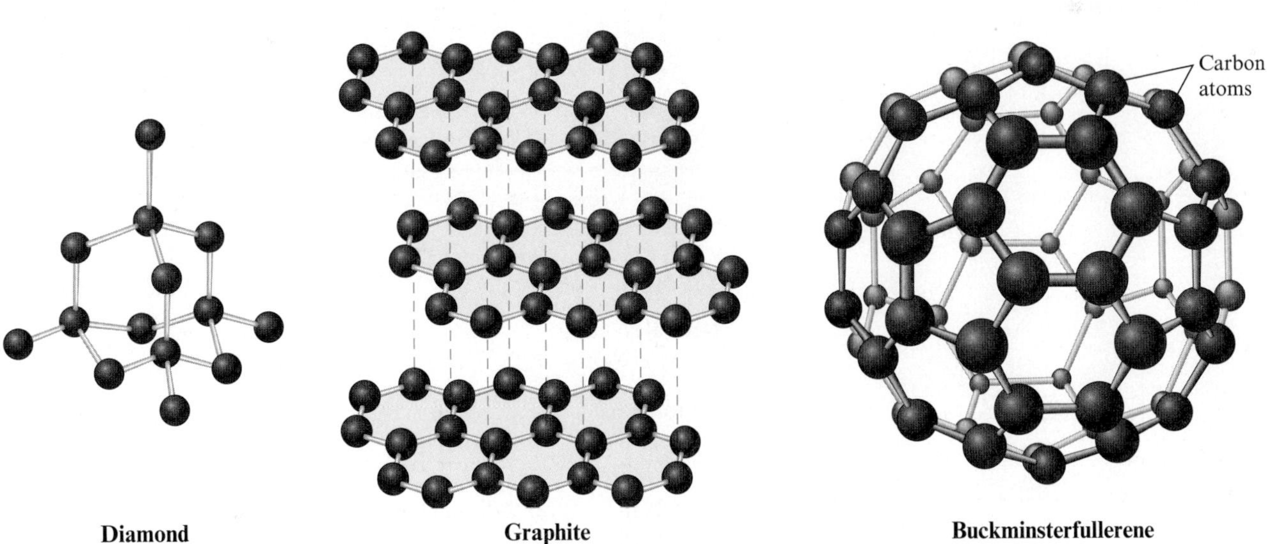

Figure 4.16 ▶ The three solid elemental (allotropes) forms of carbon. The representations of diamond and graphite are just fragments of much larger structures that extend in all directions from the parts shown here. Buckminsterfullerene contains C_{60} molecules, one of which is shown.

4-10 Ions

OBJECTIVES

- **To understand the formation of ions from their parent atoms, and learn to name them.**
- **To learn how the periodic table can help predict which ion a given element forms.**

We have seen that an atom has a certain number of protons in its nucleus and an equal number of electrons in the space around the nucleus. This results in an exact balance of positive and negative charges. We say that an atom is a neutral entity—it has *zero net charge.*

We can produce a charged entity, called an **ion,** by taking a neutral atom and adding or removing one or more electrons. For example, a sodium atom ($Z = 11$) has eleven protons in its nucleus and eleven electrons outside its nucleus.

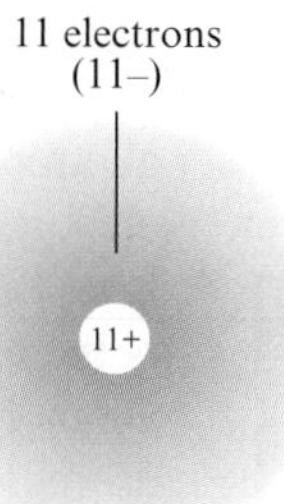

If one of the electrons is lost, there will be eleven positive charges but only ten negative charges. This gives an ion with a net positive one (1+) charge: $(11+) + (10-) = 1+$. We can represent this process as follows:

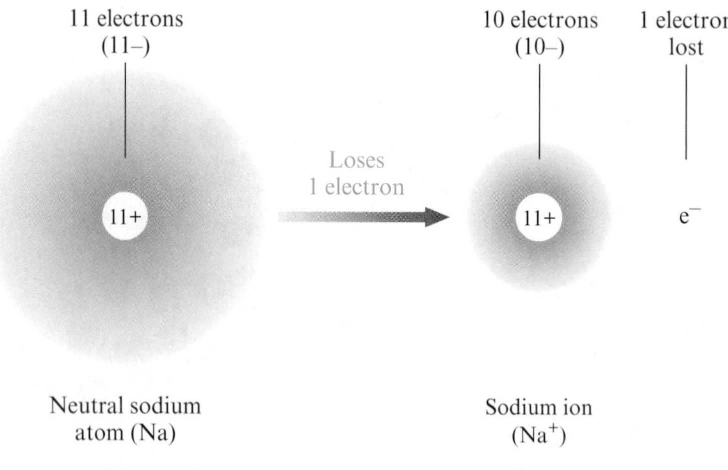

or, in shorthand form, as

$$Na \rightarrow Na^+ + e^-$$

where Na represents the neutral sodium atom, Na^+ represents the 1+ ion formed, and e^- represents an electron.

A positive ion, called a **cation** (pronounced *cat' eye on*), is produced when one or more electrons are *lost* from a neutral atom. We have seen that sodium loses one electron to become a 1+ cation. Some atoms lose more than one electron. For example, a magnesium atom typically loses two electrons to form a 2+ cation:

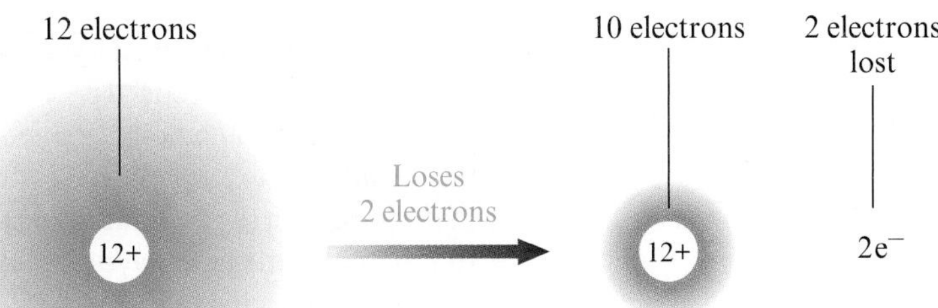

We usually represent this process as follows:

$$Mg \rightarrow Mg^{2+} + 2e^-$$

Aluminum forms a 3+ cation by losing three electrons:

Note the size decreases dramatically when an atom loses one or more electrons to form a positive ion.

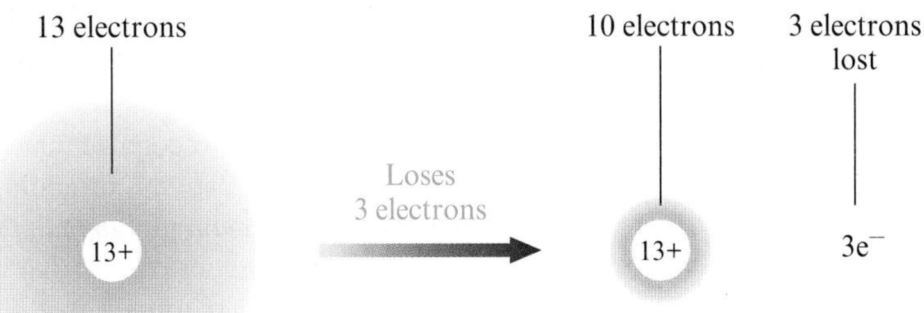

or

$$Al \rightarrow Al^{3+} + 3e^-$$

A cation is named using the name of the parent atom. Thus Na^+ is called the sodium ion (or sodium cation), Mg^{2+} is called the magnesium ion (or magnesium cation), and Al^{3+} is called the aluminum ion (or aluminum cation).

When electrons are *gained* by a neutral atom, an ion with a negative charge is formed. A negatively charged ion is called an **anion** (pronounced *an' ion*). An atom that gains one extra electron forms an anion with a 1− charge. An example of an atom that forms a 1− anion is the chlorine atom, which has seventeen protons and seventeen electrons:

Note the size increases dramatically when an atom gains one or more electrons to form a negative ion.

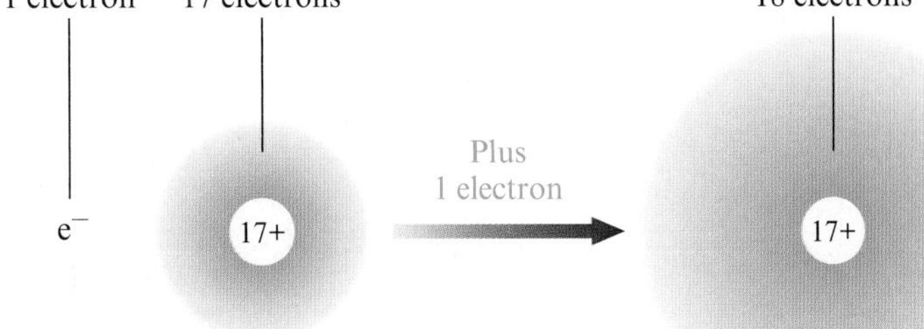

This process is usually represented as

$$Cl + e^- \rightarrow Cl^-$$

Note that the anion formed by chlorine has eighteen electrons but only seventeen protons, so the net charge is (18−) + (17+) = 1−. Unlike a cation, which is named for the parent atom, an anion is named by taking the root name of the atom and changing the ending. For example, the Cl^- anion produced from the Cl (chlorine) atom is called the *chloride* ion (or chloride anion). Notice that the word *chloride* is obtained from the root of the atom name (*chlor-*) plus the suffix *-ide*. Other atoms that add one electron to form 1− ions include

fluorine	$F + e^- \rightarrow F^-$	(*fluor*ide ion)
bromine	$Br + e^- \rightarrow Br^-$	(*brom*ide ion)
iodine	$I + e^- \rightarrow I^-$	(*iod*ide ion)

Note that the name of each of these anions is obtained by adding *-ide* to the root of the atom name.

Some atoms can add two electrons to form 2− anions. Examples include

oxygen	$O + 2e^- \rightarrow O^{2-}$	(*ox*ide ion)
sulfur	$S + 2e^- \rightarrow S^{2-}$	(*sulf*ide ion)

Note that the names for these anions are derived in the same way as those for the 1− anions.

It is important to recognize that ions are always formed by removing electrons from an atom (to form cations) or adding electrons to an atom (to form anions). *Ions are never formed by changing the number of protons* in an atom's nucleus.

It is essential to understand that isolated atoms do not form ions on their own. Most commonly, ions are formed when metallic elements combine with nonmetallic elements. As we will discuss in detail in Chapter 7, when metals and nonmetals react, the metal atoms tend to lose one or more electrons, which are in turn gained by the atoms of the nonmetal. Thus reactions between metals and nonmetals tend to form compounds that contain metal cations and nonmetal anions. We will have more to say about these compounds in Section 4-11.

Ion Charges and the Periodic Table

We find the periodic table very useful when we want to know what type of ion is formed by a given atom. Fig. 4.17 shows the types of ions formed by atoms in several of the groups on the periodic table. Note that the Group 1 metals all form 1+ ions (M^+), the Group 2 metals all form 2+ ions (M^{2+}), and the Group 3 metals form 3+

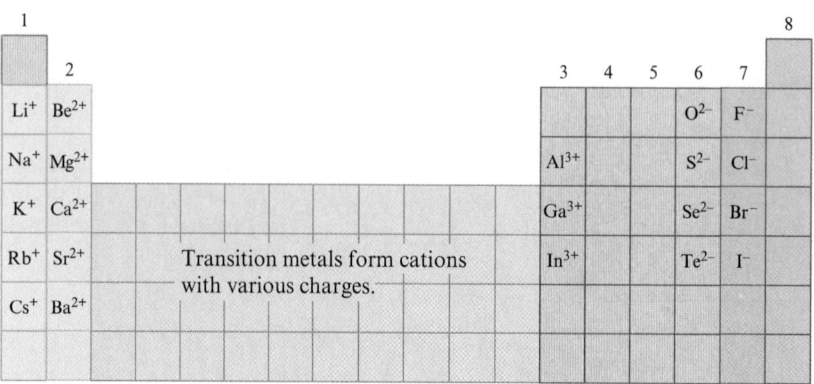

Figure 4.17 ▸ The ions formed by selected members of Groups 1, 2, 3, 6, and 7.

ions (M^{3+}). Thus for Groups 1 through 3 the charges of the cations formed are identical to the group numbers.

In contrast to the Group 1, 2, and 3 metals, most of the many *transition metals* form cations with various positive charges. For these elements there is no easy way to predict the charge of the cation that will be formed.

Note that metals always form positive ions. This tendency to lose electrons is a fundamental characteristic of metals. Nonmetals, on the other hand, form negative ions by gaining electrons. Note that the Group 7 atoms all gain one electron to form 1− ions and that all the nonmetals in Group 6 gain two electrons to form 2− ions.

At this point you should memorize the relationships between the group number and the type of ion formed, as shown in Fig. 4.17. You will understand why these relationships exist after we further discuss the theory of the atom in Chapter 11.

4-11 Compounds That Contain Ions

OBJECTIVE **To learn how ions combine to form neutral compounds.**

Chemists have good reasons to believe that many chemical compounds contain ions. For instance, consider some of the properties of common table salt, sodium chloride (NaCl). It must be heated to about 800 °C to melt and to almost 1500 °C to boil (compare to water, which boils at 100 °C). As a solid, salt will not conduct an electric current, but when melted it is a very good conductor. Pure water does not conduct electricity (does not allow an electric current to flow), but when salt is dissolved in water, the resulting solution readily conducts electricity (Fig. 4.18).

Melting means that the solid, where the ions are locked into place, is changed to a liquid, where the ions can move.

Chemists have come to realize that we can best explain these properties of sodium chloride (NaCl) by picturing it as containing Na^+ ions and Cl^- ions packed together as shown in Fig. 4.19. Because the positive and negative charges attract each other very strongly, it must be heated to a very high temperature (800 °C) before it melts.

To explore further the significance of the electrical conductivity results, we need to discuss briefly the nature of electric currents. An electric current can travel along a metal wire because *electrons are free to move* through the wire; the moving electrons carry the current. In ionic substances the ions carry the current. Thus substances that contain ions can conduct an electric current *only if the ions can move*—the current travels by the movement of the charged ions. In solid NaCl the ions are tightly held and

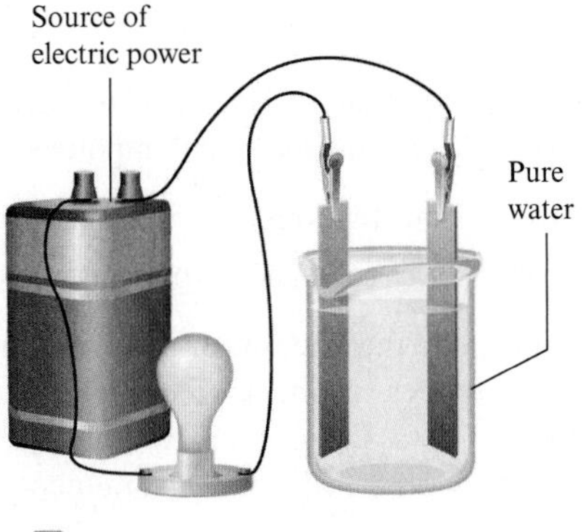

a
Pure water does not conduct a current, so the circuit is not complete and the bulb does not light.

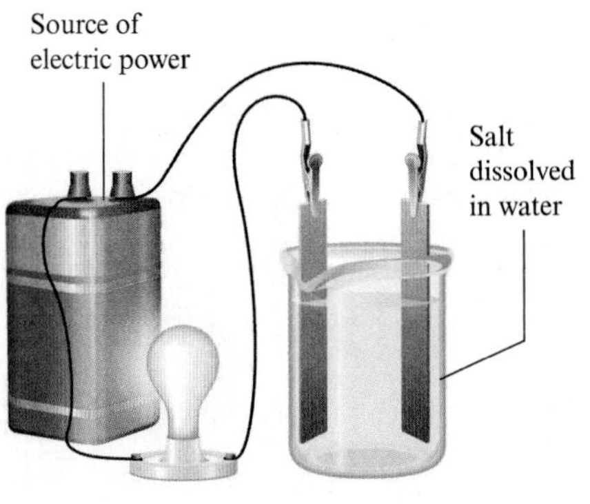

b
Water containing dissolved salt conducts electricity and the bulb lights.

Figure 4.18

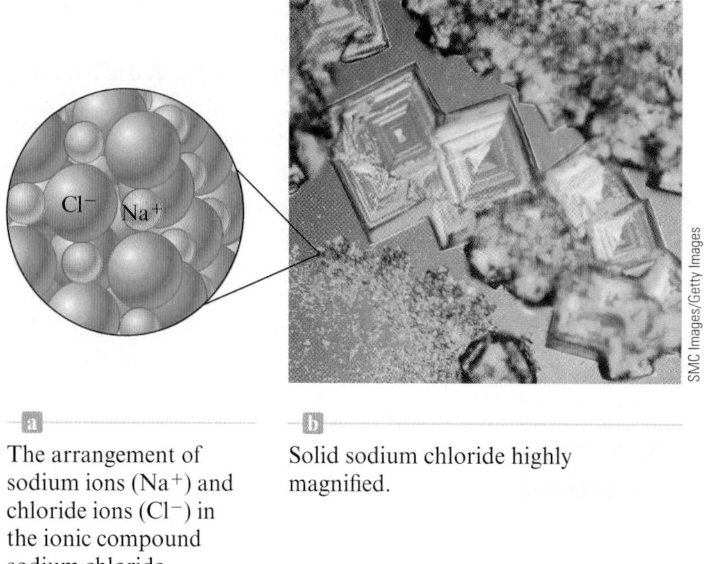

Figure 4.19 a The arrangement of sodium ions (Na^+) and chloride ions (Cl^-) in the ionic compound sodium chloride. b Solid sodium chloride highly magnified.

cannot move, but when the solid is melted and changed to a liquid, the structure is disrupted and the ions can move. As a result, an electric current can travel through the melted salt.

The same reasoning applies to NaCl dissolved in water. When the solid dissolves, the ions are dispersed throughout the water and can move around in the water, allowing it to conduct a current.

Thus, we recognize substances that contain ions by their characteristic properties. They often have very high melting points, and they conduct an electric current when melted or when dissolved in water.

Many substances contain ions. In fact, whenever a compound forms between a metal and a nonmetal, it can be expected to contain ions. We call these substances **ionic compounds.**

Critical Thinking

Thomson and Rutherford helped to show that atoms consist of three types of subatomic particles, two of which are charged. What if subatomic particles had no charge? How would this affect what you have learned?

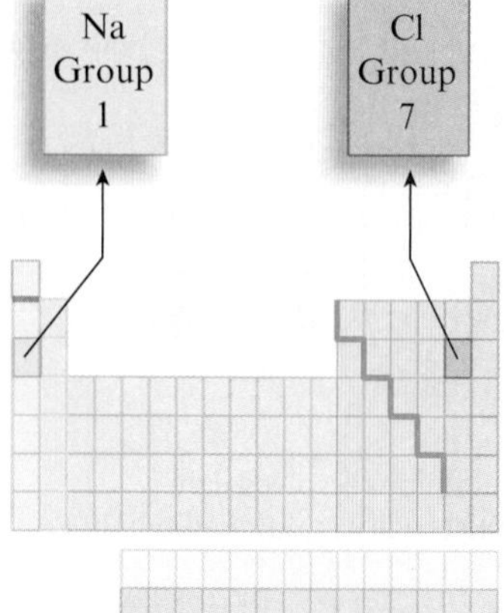

One fact very important to remember is that *a chemical compound must have a net charge of zero.* This means that if a compound contains ions, then

1. Both positive ions (cations) and negative ions (anions) must be present.
2. The numbers of cations and anions must be such that the net charge is zero.

For example, note that the formula for sodium chloride is written NaCl, indicating one of each type of these elements. This makes sense because sodium chloride contains Na^+ ions and Cl^- ions. Each sodium ion has a 1+ charge and each chloride ion has a 1− charge, so they must occur in equal numbers to give a net charge of zero.

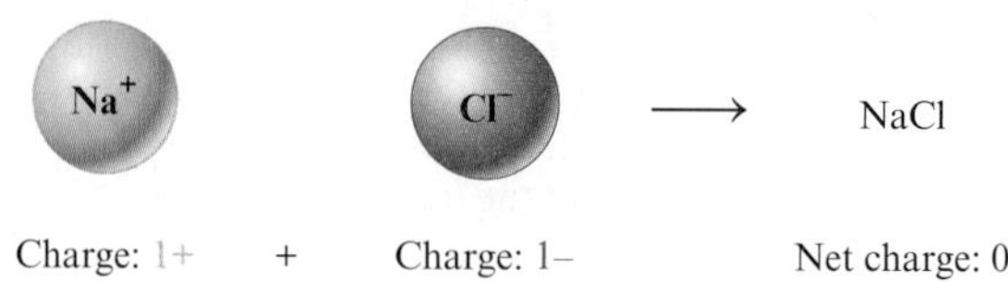

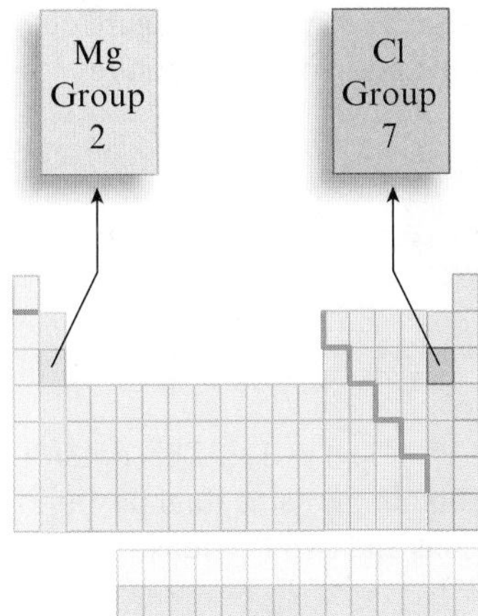

And for *any* ionic compound,

$$\text{Total charge of cations} + \text{Total charge of anions} = \text{Zero net charge}$$

Consider an ionic compound that contains the ions Mg^{2+} and Cl^-. What combination of these ions will give a net charge of zero? To balance the 2+ charge on Mg^{2+}, we will need two Cl^- ions to give a net charge of zero.

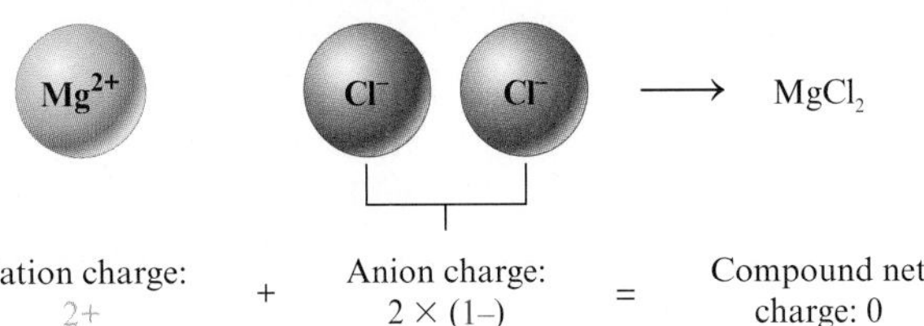

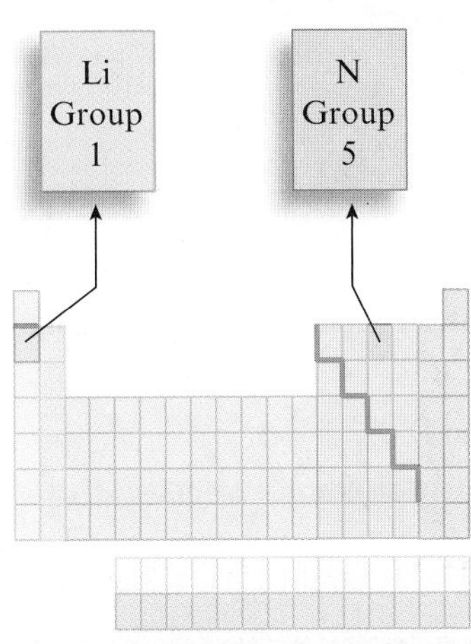

This means that the formula of the compound must be $MgCl_2$. Remember that subscripts are used to give the relative numbers of atoms (or ions).

Now consider an ionic compound that contains the ions Ba^{2+} and O^{2-}. What is the correct formula? These ions have charges of the same size (but opposite sign), so they must occur in equal numbers to give a net charge of zero. The formula of the compound is BaO, because $(2+) + (2-) = 0$.

Similarly, the formula of a compound that contains the ions Li^+ and N^{3-} is Li_3N, because three Li^+ cations are needed to balance the charge of the N^{3-} anion.

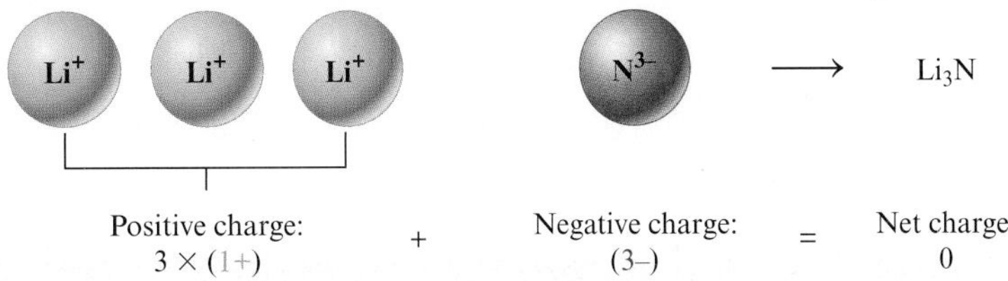

Interactive Example 4.6

Writing Formulas for Ionic Compounds

The pairs of ions contained in several ionic compounds are listed below. Give the formula for each compound.

a. Ca^{2+} and Cl^- b. Na^+ and S^{2-} c. Ca^{2+} and P^{3-}

SOLUTION

a. Ca^{2+} has a 2+ charge, so two Cl^- ions (each with the charge 1−) will be needed.

The subscript 1 in a formula is not written.

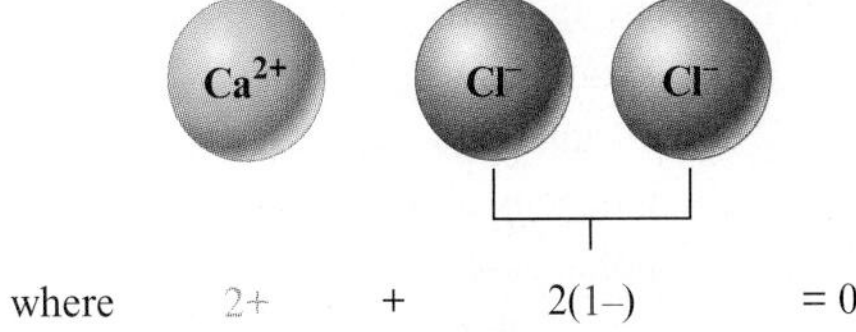

The formula is $CaCl_2$.

b. In this case S^{2-}, with its 2− charge, requires two Na^+ ions to produce a zero net charge.

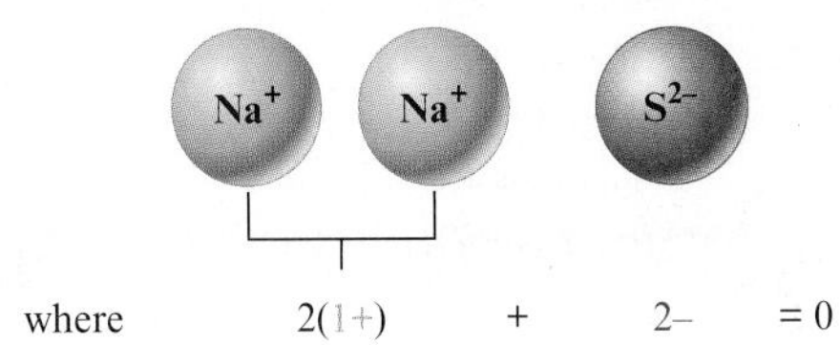

The formula is Na_2S.

c. We have the ions Ca^{2+} (charge 2+) and P^{3-} (charge 3−). We must figure out how many of each are needed to balance exactly the positive and negative charges. Let's try two Ca^{2+} and one P^{3-}.

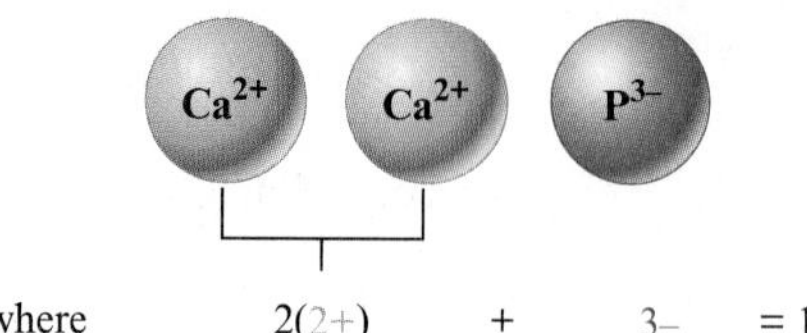

where 2(2+) + 3− = 1+

The resulting net charge is 2(2+) + (3−) = (4+) + (3−) = 1−. This doesn't work because the net charge is not zero. We can obtain the same total positive and total negative charges by having three Ca^{2+} ions and two P^{3-} ions.

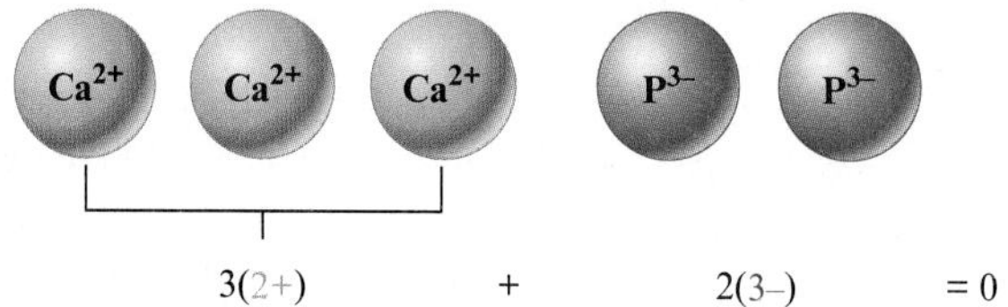

where 3(2+) + 2(3−) = 0

Thus the formula must be Ca_3P_2.

SELF CHECK

Exercise 4.6 Give the formulas for the compounds that contain the following pairs of ions.

a. K^+ and I^- b. Mg^{2+} and N^{3-} c. Al^{3+} and O^{2-}

See Problems 4.83 and 4.84. ■

CHAPTER 4 REVIEW

Ⓕ directs you to the *Chemistry in Focus* feature in the chapter

Key Terms

element symbols (4-2)
law of constant composition (4-3)
Dalton's atomic theory (4-3)
atom (4-3)
compound (4-4)
chemical formula (4-4)
electron (4-5)
nuclear atom (4-5)
nucleus (4-5)
proton (4-5)
neutron (4-5)
isotopes (4-7)
atomic number (*Z*) (4-7)
mass number (*A*) (4-7)
periodic table (4-8)
group (4-8)
alkali metals (4-8)
alkaline earth metals (4-8)
halogens (4-8)
noble gases (4-8)
transition metals (4-8)
metals (4-8)
nonmetals (4-8)
metalloids (4-8)
diatomic molecule (4-9)
ion (4-10)
cation (4-10)
anion (4-10)
ionic compound (4-11)

For Review

- All of the materials in the universe can be chemically broken down into about 100 different elements.
- Nine elements account for about 98% of the earth's crust, oceans, and atmosphere.
- In the human body, oxygen, carbon, hydrogen, and nitrogen are the most abundant elements.
- Each element has a name and a symbol.
  - The symbol usually consists of the first one or two letters of the element's name.
  - Sometimes the symbol is taken from the element's original Latin or Greek name.
- The law of constant composition states that a given compound always contains the same proportion by mass of the elements of which it is composed.
- Dalton's atomic theory states:
  - All elements are composed of atoms.
  - All atoms of a given element are identical.
  - Atoms of different elements are different.
  - Compounds consist of the atoms of different elements.
  - Atoms are not created or destroyed in a chemical reaction.

- A compound is represented by a chemical formula in which the number and type of atoms present are shown by using the element symbols and subscripts.
- Experiments by J. J. Thomson and Ernest Rutherford showed that atoms have internal structure.
  - The nucleus, which is at the center of the atom, contains protons (positively charged) and neutrons (uncharged).
  - Electrons move around the nucleus.
    - Electrons have a small mass (1/1836 of the proton mass).
    - Electrons have a negative charge equal and opposite to that of the proton.
- Isotopes are atoms with the same number of protons but a different number of neutrons.
- A particular isotope is represented by the symbol $^{A}_{Z}X$, in which Z represents the number of protons (atomic number) and A represents the total number of protons and neutrons (mass number) in the nucleus.
- The periodic table shows all of the known elements in order of increasing atomic number; the table is organized to group elements with similar properties in vertical columns.
- Most elements have metallic properties (the metals) and appear on the left side of the periodic table.
- Nonmetals appear on the right side of the periodic table.
- Metalloids are elements that have some metallic and some nonmetallic properties.
- Atoms can form ions (species with a charge) by gaining or losing electrons.
  - Metals tend to lose one or more electrons to form positive ions called cations; these are generally named by using the name of the parent atom.
  - Nonmetals tend to gain one or more electrons to form negative ions called anions; these are named by using the root of the atom name followed by the suffix *-ide*.
- The ion that a particular atom will form can be predicted from the atom's position on the periodic table.
  - Elements in Group 1 and 2 form 1+ and 2+ ions, respectively.
  - Group 7 atoms form anions with 1− charges.
  - Group 6 atoms form anions with 2− charges.
- Ions combine to form compounds. Compounds are electrically neutral, so the sum of the charges on the anions and cations in the compound must equal zero.

Alkaline earth metals (Group 2) · Alkali metals (Group 1, Li–Fr) · Transition metals (Groups 3–12) · Halogens (Group 17) · Noble gases (Group 18)

1 1A	2 2A	3	4	5	6	7	8	9	10	11	12	13 3A	14 4A	15 5A	16 6A	17 7A	18 8A
1 H																	2 He
3 Li	4 Be											5 B	6 C	7 N	8 O	9 F	10 Ne
11 Na	12 Mg											13 Al	14 Si	15 P	16 S	17 Cl	18 Ar
19 K	20 Ca	21 Sc	22 Ti	23 V	24 Cr	25 Mn	26 Fe	27 Co	28 Ni	29 Cu	30 Zn	31 Ga	32 Ge	33 As	34 Se	35 Br	36 Kr
37 Rb	38 Sr	39 Y	40 Zr	41 Nb	42 Mo	43 Tc	44 Ru	45 Rh	46 Pd	47 Ag	48 Cd	49 In	50 Sn	51 Sb	52 Te	53 I	54 Xe
55 Cs	56 Ba	57 La*	72 Hf	73 Ta	74 W	75 Re	76 Os	77 Ir	78 Pt	79 Au	80 Hg	81 Tl	82 Pb	83 Bi	84 Po	85 At	86 Rn
87 Fr	88 Ra	89 Ac†	104 Rf	105 Db	106 Sg	107 Bh	108 Hs	109 Mt	110 Ds	111 Rg	112 Cn	113 Uut	114 Fl	115 Uup	116 Lv	117 Uus	118 Uuo

*Lanthanides	58 Ce	59 Pr	60 Nd	61 Pm	62 Sm	63 Eu	64 Gd	65 Tb	66 Dy	67 Ho	68 Er	69 Tm	70 Yb	71 Lu
†Actinides	90 Th	91 Pa	92 U	93 Np	94 Pu	95 Am	96 Cm	97 Bk	98 Cf	99 Es	100 Fm	101 Md	102 No	103 Lr

All even-numbered Questions and Problems have answers in the back of this book and solutions in the *Student Solutions Guide*.

Active Learning Questions

These questions are designed to be considered by groups of students in class. Often these questions work well for introducing a particular topic in class.

1. Knowing the number of protons in the atom of a neutral element enables you to determine which of the following?
   a. the number of neutrons in the atom of the neutral element
   b. the number of electrons in the atom of the neutral element
   c. the name of the element
   d. two of the above
   e. none of the above

   Explain.
2. The average mass of a carbon atom is 12.011. Assuming you could pick up one carbon atom, what is the chance that you would randomly get one with a mass of 12.011?
   a. 0%
   b. 0.011%
   c. about 12%
   d. 12.011%
   e. greater than 50%
   f. none of the above

   Explain.

3. How is an ion formed?
   a. by either adding or subtracting protons from the atom
   b. by either adding or subtracting neutrons from the atom
   c. by either adding or subtracting electrons from the atom
   d. all of the above
   e. two of the above

   Explain.
4. The formula of water, H_2O, suggests which of the following?
   a. There is twice as much mass of hydrogen as oxygen in each molecule.
   b. There are two hydrogen atoms and one oxygen atom per water molecule.
   c. There is twice as much mass of oxygen as hydrogen in each molecule.
   d. There are two oxygen atoms and one hydrogen atom per water molecule.
   e. Two of the above.

   Explain.
5. The vitamin niacin (nicotinic acid, $C_6H_5NO_2$) can be isolated from a variety of natural sources, such as liver, yeast, milk, and whole grain. It also can be synthesized from commercially available materials. Which source of nicotinic acid, from a nutritional view, is best for use in a multivitamin tablet? Why?
6. One of the best indications of a useful theory is that it raises more questions for further experimentation than it originally answered. How does this apply to Dalton's atomic theory? Give examples.
7. Dalton assumed that all atoms of the same element are identical in all their properties. Explain why this assumption is not valid.
8. How does Dalton's atomic theory account for the law of constant composition?
9. Which of the following is true about the state of an individual atom?
   a. An individual atom should be considered to be a solid.
   b. An individual atom should be considered to be a liquid.
   c. An individual atom should be considered to be a gas.
   d. The state of the atom depends on which element it is.
   e. An individual atom cannot be considered to be a solid, liquid, or gas.

   For choices you did not pick, explain what you feel is wrong with them, and justify the choice you did pick.
10. These questions concern the work of J. J. Thomson:
    a. From Thomson's work, which particles do you think he would feel are most important in the formation of compounds (chemical changes) and why?
    b. Of the remaining two subatomic particles, which do you place second in importance for forming compounds and why?
    c. Come up with three models that explain Thomson's findings and evaluate them. To be complete you should include Thomson's findings.
11. Heat is applied to an ice cube until only steam is present. Draw a sketch of this process, assuming you can see it at an extremely high level of magnification. What happens to the size of the molecules? What happens to the total mass of the sample?
12. What makes a carbon atom different from a nitrogen atom? How are they alike?
13. Hundreds of years ago, alchemists tried to turn lead into gold. Is this possible? If not, why not? If yes, how would you do it?
14. Chlorine has two prominent isotopes, ^{37}Cl and ^{35}Cl. Which is more abundant? How do you know?
15. Differentiate between an atomic element and a molecular element. Provide an example and microscopic drawing of each.
16. Science often develops by using the known theories and expanding, refining, and perhaps changing these theories. As discussed in Section 4-5, Rutherford used Thomson's ideas when thinking about his model of the atom. What if Rutherford had not known about Thomson's work? How might Rutherford's model of the atom have been different?
17. Rutherford was surprised when some of the α-particles bounced back. He was surprised because he was thinking of Thomson's model of the atom. What if Rutherford believed atoms were as Dalton envisioned them? What do you suppose Rutherford would have expected, and what would have surprised him?
18. It is good practice to actively read the textbook and to try to verify claims that are made when you can. The following claim is made in your textbook: ". . . if the nucleus were the size of a grape, the electrons would be about one mile away on average."

    Provide mathematical support for this statement.
19. Why is the term "sodium chloride molecule" incorrect but the term "carbon dioxide molecule" is correct?
20. Both atomic elements and molecular elements exist. Are there such entities as atomic compounds and molecular compounds? If so, provide an example and microscopic drawing. If not, explain why not.
21. Now that you have gone through Chapter 4, go back to Section 4-3 and review Dalton's Atomic Theory. Which of the premises are no longer accepted? Explain your answer.
22. Write the formula for each of the following substances, listing the elements in the order given.
    a. 

    List the phosphorus atom first.
    b. a molecule containing two boron atoms and six hydrogen atoms
    c. a compound containing one calcium atom for every two chlorine atoms
    d. 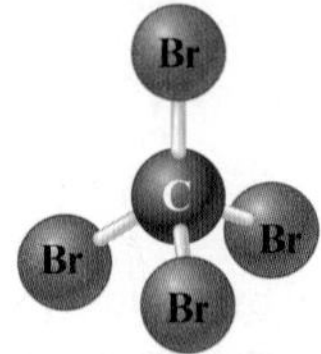

    List the carbon atom first.
    e. a compound containing two iron atoms for every three oxygen atoms
    f. a molecule containing three hydrogen atoms, one phosphorus atom, and four oxygen atoms

All even-numbered Questions and Problems have answers in the back of this book and solutions in the *Student Solutions Guide.*

23. Use the following figures to identify the element or ion. Write the symbol for each, using the $^{A}_{Z}X$ format.

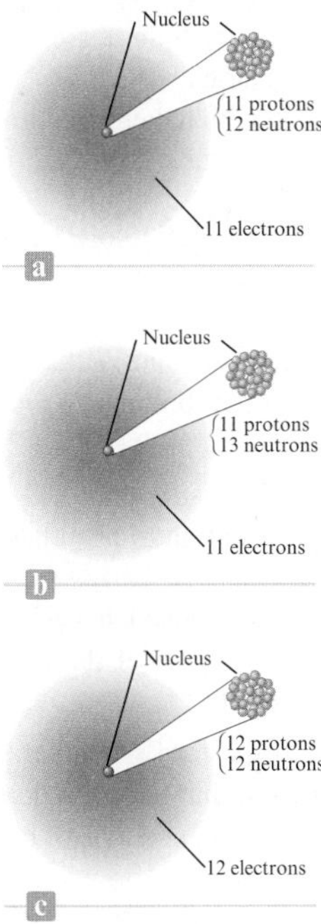

Questions and Problems

4-1 The Elements

Questions

1. What were the four fundamental substances postulated by the Greeks?
2. ________ was the first scientist to recognize the importance of careful measurements.
3. In addition to his important work on the properties of gases, what other valuable contributions did Robert Boyle make to the development of the study of chemistry?
4. What are the three most abundant elements (by mass) in the human body?
5. What are the five most abundant elements (by mass) in the earth's crust, oceans, and atmosphere?
6. (F) Read the "Chemistry in Focus" segment *Trace Elements: Small but Crucial,* and answer the following questions.
   a. What is meant by the term *trace element?*
   b. Name two essential trace elements in the body and list their function(s).

4-2 Symbols for the Elements

Note: Refer to Fig. 4.9 when appropriate.

Questions

7. Give the symbols and names for the elements whose chemical symbols consist of only one letter.
8. The symbols for most elements are based on the first few letters of the respective element's common English name. In some cases, however, the symbol seems to have nothing to do with the element's common name. Give three examples of elements whose symbols are not directly derived from the element's common English name.
9. Find the symbol in Column 2 for each name in Column 1.

Column 1	Column 2
a. helium	1. Si
b. sodium	2. So
c. silver	3. S
d. sulfur	4. He
e. bromine	5. C
f. potassium	6. Co
g. neon	7. Ba
h. barium	8. Br
i. cobalt	9. K
j. carbon	10. Po
	11. Na
	12. Ag
	13. Ne
	14. Ca

10. Several elements have chemical symbols beginning with the letter C. For each of the following chemical symbols, give the name of the corresponding element.

a. Cu	e. Cr
b. Co	f. Cs
c. Ca	g. Cl
d. C	h. Cd

11. Use the periodic table shown in Fig. 4.9 to find the symbol or name for each of the following elements.

Symbol	Name
Co	________
________	rubidium
Rn	________
________	radium
U	________

12. Use the periodic table shown in Fig. 4.9 to find the symbol or name for each of the following elements.

Symbol	Name
Si	________
________	nickel
Ag	________
________	potassium
Ca	________

13. For each of the following chemical symbols, give the name of the corresponding element.

a. K	e. N
b. Ge	f. Na
c. P	g. Ne
d. C	h. I

All even-numbered Questions and Problems have answers in the back of this book and solutions in the *Student Solutions Guide.*

14. Several chemical elements have English names beginning with the letters B, N, P, or S. For each letter, list the English *names* for two elements whose names begin with that letter, and give the symbols for the elements you choose (the symbols do not necessarily need to begin with the same letters).

4-3 Dalton's Atomic Theory

Questions

15. A given compound always contains the same proportion (by mass) of the elements. This principle became known as ________.

16. Correct each of the following misstatements from Dalton's atomic theory.
    a. Elements are made of tiny particles called molecules.
    b. All atoms of a given element are very similar.
    c. The atoms of a given element may be the same as those of another element.
    d. A given compound may vary in the relative number and types of atoms depending on the source of the compound.
    e. A chemical reaction may involve the gain or loss of atoms as it takes place.

4-4 Formulas of Compounds

Questions

17. What is a compound?

18. A given compound always contains the same relative masses of its constituent elements. How is this related to the relative numbers of each kind of atom present?

19. Based on the following word descriptions, write the formula for each of the indicated substances.
    a. a compound whose molecules each contain six carbon atoms and six hydrogen atoms
    b. an aluminum compound in which there are three chlorine atoms for each aluminum atom
    c. a compound in which there are two sodium atoms for every sulfur atom
    d. a compound whose molecules each contain two nitrogen atoms and four oxygen atoms
    e. a compound in which there is an equal number of sodium, hydrogen, and carbon atoms but there are three times as many oxygen atoms as atoms of the other three elements
    f. a compound that has equal numbers of potassium and iodide atoms

20. Based on the following word descriptions, write the formula for each of the indicated substances.
    a. a compound whose molecules contain twice as many oxygen atoms as carbon atoms
    b. a compound whose molecules contain an equal number of carbon and oxygen atoms
    c. a compound in which there is an equal number of calcium and carbon atoms but there are three times as many atoms of oxygen as of the other two elements
    d. a compound whose molecules contain twice as many hydrogen atoms as sulfur atoms and four times as many oxygen atoms as sulfur atoms
    e. a compound in which there are twice as many chlorine atoms as barium atoms
    f. a compound in which there are three sulfur atoms for every two aluminum atoms

4-5 The Structure of the Atom

Questions

21. Scientists J. J. Thomson and William Thomson (Lord Kelvin) made numerous contributions to our understanding of the atom's structure.
    a. Which subatomic particle did J. J. Thomson discover, and what did this lead him to postulate about the nature of the atom?
    b. William Thomson postulated what became known as the "plum pudding" model of the atom's structure. What did this model suggest?

22. True or false? Rutherford's bombardment experiments with metal foil suggested that the α particles were being deflected by coming near a large, positively charged atomic nucleus.

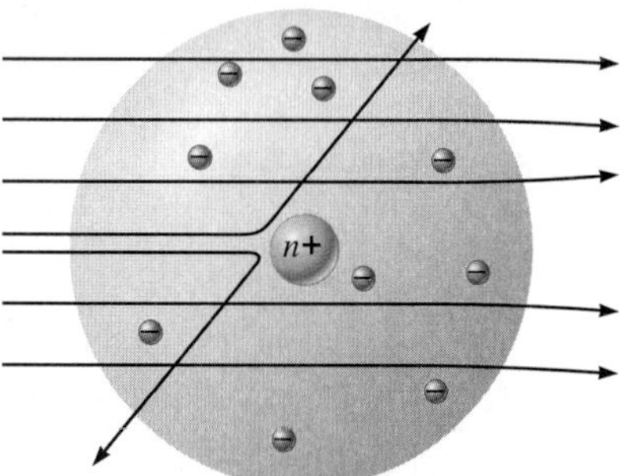

4-6 Introduction to the Modern Concept of Atomic Structure

Questions

23. Where are neutrons found in an atom? Are neutrons positively charged, negatively charged, or electrically uncharged?

24. What are the positively charged particles found in the nuclei of atoms called?

25. Do the proton and the neutron have exactly the same mass? How do the masses of the proton and the neutron compare to the mass of the electron? Which particles make the greatest contribution to the mass of an atom? Which particles make the greatest contribution to the chemical properties of an atom?

26. The proton and the (electron/neutron) have almost equal masses. The proton and the (electron/neutron) have charges that are equal in magnitude but opposite in nature.

27. An average atomic nucleus has a diameter of about ________ m.

28. Although the nucleus of an atom is very important, it is the ________ of the atom that determine its chemical properties.

All even-numbered Questions and Problems have answers in the back of this book and solutions in the *Student Solutions Guide*.

4-7 Isotopes

Questions

29. Explain what we mean when we say that a particular element consists of several *isotopes*.
30. True or false? The mass number of a nucleus represents the number of protons in the nucleus.
31. For an isolated atom, why do we expect the number of electrons present in the atom to be the *same* as the number of protons in the nucleus of the atom?
32. Why do we not necessarily expect the number of neutrons in the nucleus of an atom to be the same as the number of protons?
33. Dalton's original atomic theory proposed that all atoms of a given element are *identical*. Did this turn out to be true after further experimentation was carried out? Explain.
34. Are all atoms of the same element identical? If not, how can they differ?
35. For each of the following elements, use the periodic table shown in Fig. 4.9 to write the element's atomic number, symbol, or name.

Atomic Number	Symbol	Name
8	______	______
______	Cu	______
78	______	______
______	______	phosphorus
17	______	______
______	Sn	______
______	______	zinc

36. For each of the following elements, use the periodic table shown in Fig 4.9 to write the element's atomic number, symbol, or name.

Atomic Number	Symbol	Name
32	Ge	______
______	Zn	zinc
24	______	chromium
74	______	tungsten
______	Sr	strontium
______	Co	cobalt
4	______	beryllium

37. Write the atomic symbol ($^{A}_{Z}X$) for each of the isotopes described below.
    a. the isotope of carbon with 7 neutrons
    b. the isotope of carbon with 6 neutrons
    c. $Z = 6$, number of neutrons $= 8$
    d. atomic number 5, mass number 11
    e. number of protons $= 5$, number of neutrons $= 5$
    f. the isotope of boron with mass number 10
38. Write the atomic symbol ($^{A}_{Z}X$) for each of the isotopes described below.
    a. $Z = 26, A = 54$
    b. the isotope of iron with 30 neutrons
    c. number of protons-26, number of neutrons-31
    d. the isotope of nitrogen with 7 neutrons
    e. $Z = 7, A = 15$
    f. atomic number 7, number of neutrons-8
39. How many protons and neutrons are contained in the nucleus of each of the following atoms? Assuming each atom is uncharged, how many electrons are present?
    a. $^{130}_{56}Ba$ c. $^{46}_{22}Ti$ e. $^{6}_{3}Li$
    b. $^{136}_{56}Ba$ d. $^{48}_{22}Ti$ f. $^{7}_{3}Li$
40. **F** Read the "Chemistry in Focus" segment *"Whair" Do You Live?* How can isotopes be used to identify the general region of a person's place of residence?
41. **F** Read the "Chemistry in Focus" segment *Isotope Tales.* Define the term *isotope,* and explain how isotopes can be used to answer scientific and historical questions.
42. Complete the following table.

Name	Symbol	Atomic Number	Mass Number	Number of Neutrons
______	$^{17}_{8}O$	______	______	______
______	______	8	______	9
______	______	10	20	______
iron	______	______	56	______
______	$^{244}_{94}Pu$	______	______	______
______	$^{202}_{80}Hg$	______	______	______
cobalt	______	______	59	______
______	______	28	56	______
______	$^{19}_{9}F$	______	______	______
chromium	______	______	______	26

4-8 Introduction to the Periodic Table

Questions

43. True or false? The elements are arranged in the periodic table in order of increasing mass.
44. In which direction on the periodic table, horizontal or vertical, are elements with similar chemical properties aligned? What are families of elements with similar chemical properties called?
45. List the characteristic physical properties that distinguish the metallic elements from the nonmetallic elements.
46. True or false? Nitrogen and phosphorus are both nonmetals.
47. Most, but not all, metallic elements are solids under ordinary laboratory conditions. Which metallic elements are *not* solids?
48. List five nonmetallic elements that exist as gaseous substances under ordinary conditions. Do any metallic elements ordinarily occur as gases?
49. Under ordinary conditions, only a few pure elements occur as liquids. Give an example of a metallic and a nonmetallic element that ordinarily occur as liquids.

All even-numbered Questions and Problems have answers in the back of this book and solutions in the *Student Solutions Guide*.

50. The elements that lie close to the "stair-step" line as shown below in blue are called ________.

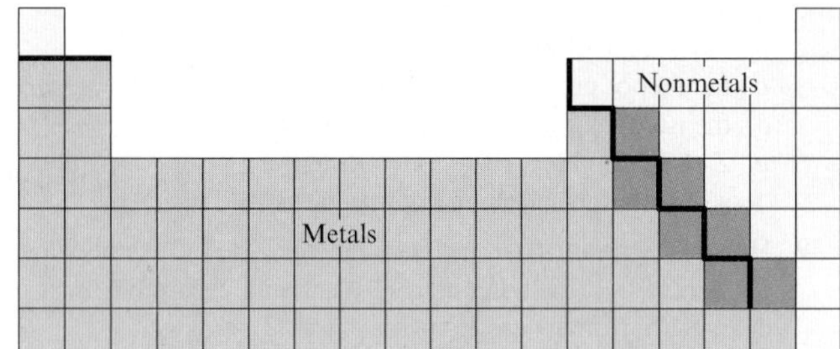

51. Write the number and name (if any) of the group (family) to which each of the following elements belongs.

a. cesium
b. Ra
c. Rn
d. chlorine
e. strontium
f. Xe
g. Rb

52. Without looking at your textbook or the periodic table, name three elements in each of the following groups (families).

a. halogens
b. alkali metals
c. alkaline earth metals
d. noble/inert gases

53. For each of the following elements, use the periodic table shown in Fig. 4.9 to give the chemical symbol, atomic number, and group number, and to specify whether each element is a metal, nonmetal, or metalloid.

a. strontium
b. iodine
c. silicon
d. cesium
e. sulfur

F **54.** The "Chemistry in Focus" segment *Putting the Brakes on Arsenic* discusses the dangers of arsenic and a possible help against arsenic pollution. Is arsenic a metal, a nonmetal, or a metalloid? What other elements are in the same group on the periodic table as arsenic?

4-9 Natural States of the Elements

Questions

55. Most substances are composed of ________ rather than elemental substances.

56. Are most of the chemical elements found in nature in the elemental form or combined in compounds? Why?

57. The noble gas present in relatively large concentrations in the atmosphere is ________.

58. Why are the elements of Group 8 referred to as the noble or inert gas elements?

59. Molecules of nitrogen gas and oxygen gas are said to be ________, which means they consist of pairs of atoms.

60. Give three examples of gaseous elements that exist as diatomic molecules. Give three examples of gaseous elements that exist as monatomic species.

61. A simple way to generate elemental hydrogen gas is to pass ________ through water.

62. If sodium chloride (table salt) is melted and then subjected to an electric current, elemental ________ gas is produced, along with sodium metal.

63. Most of the elements are solids at room temperature. Give three examples of elements that are *liquids* at room temperature, and three examples of elements that are *gases* at room temperature.

64. Graphite and diamond are two of the most common elemental forms of ________.

4-10 Ions

Questions

65. An isolated atom has a net charge of ________.

66. Ions are produced when an atom gains or loses ________.

67. A simple ion with a 3+ charge (for example, Al^{3+}) results when an atom (gains/loses) ________ electrons.

68. An ion that has two more electrons outside the nucleus than there are protons in the nucleus will have a charge of ________.

69. Positive ions are called ________, whereas negative ions are called ________.

70. Simple negative ions formed from single atoms are given names that end in ________.

71. Based on their location in the periodic table, give the symbols for three elements that would be expected to form positive ions in their reactions.

72. True or false? N^{3-} and P^{3-} contain a different number of protons but the same number of electrons. Justify your answer.

73. How many electrons are present in each of the following ions?

a. Ba^{2+}
b. P^{3}
c. Mn^{2+}
d. Mg^{2+}
e. Cs^{+}
f. Pb^{2+}

74. State the number of protons, electrons, and neutrons for $^{56}_{26}Fe^{3+}$.

75. For the following processes that show the formation of ions, use the periodic table to indicate the number of electrons and protons present in both the *ion* and the *neutral atom* from which the ion is made.

a. $Ca \rightarrow Ca^{2+} + 2e^-$
b. $P + 3e^- \rightarrow P^{3-}$
c. $Br + e^- \rightarrow Br^-$
d. $Fe \rightarrow Fe^{3+} + 3e^-$
e. $Al \rightarrow Al^{3+} + 3e^-$
f. $N + 3e^- \rightarrow N^{3-}$

76. For the following ions, indicate whether electrons must be *gained* or *lost* from the parent neutral atom, and *how many* electrons must be gained or lost.

a. O^{2-}
b. P^{3-}
c. Cr^{3+}
d. Sn^{2+}
e. Rb^{+}
f. Pb^{2+}

77. For each of the following atomic numbers, use the periodic table to write the formula (including the charge) for the simple *ion* that the element is most likely to form.

a. 53
b. 38
c. 55
d. 88
e. 9
f. 13

All even-numbered Questions and Problems have answers in the back of this book and solutions in the *Student Solutions Guide*.

78. On the basis of the element's location in the periodic table, indicate what simple ion each of the following elements is most likely to form.

a. P
b. Ra
c. At
d. Rn
e. Cs
f. Se

4-11 Compounds That Contain Ions

Questions

79. List some properties of a substance that would lead you to believe it consists of ions. How do these properties differ from those of nonionic compounds?

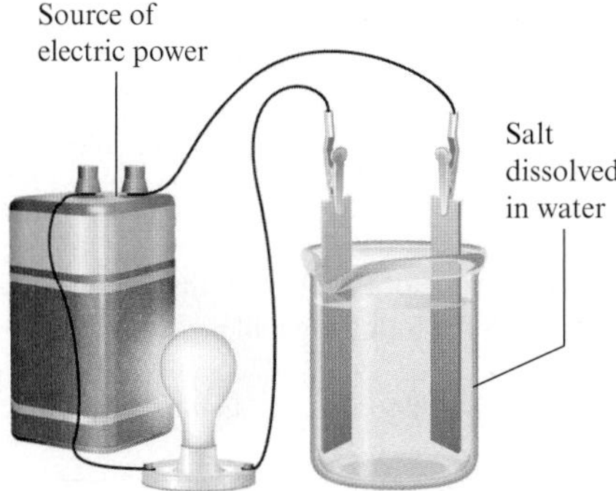

80. Why does a solution of sodium chloride in water conduct an electric current?

81. Why does an ionic compound conduct an electric current when the compound is melted but not when it is in the solid state?

82. Why must the total number of positive charges in an ionic compound equal the total number of negative charges?

83. For each of the following positive ions, use the concept that a chemical compound must have a net charge of zero to predict the formula of the simple compounds that the positive ions would form with the Cl^-, S^{2-}, and N^{3-} ions.

a. K^+
b. Mg^{2+}
c. Al^{3+}
d. Ca^{2+}
e. Li^+

84. For each of the following negative ions, use the concept that a chemical compound must have a net charge of zero to predict the formula of the simple compounds that the negative ions would form with the Cs^+, Ba^{2+}, and Al^{3+} ions.

a. I^-
b. O^{2-}
c. P^{3-}
d. Se^{2-}
e. H^-

Additional Problems

85. For each of the following elements, give the chemical symbol and atomic number.

a. astatine
b. xenon
c. radium
d. strontium
e. lead
f. selenium
g. argon
h. cesium

86. Give the group number (if any) in the periodic table for the elements listed in Problem 85. If the group has a family name, give that name.

87. List the names, symbols, and atomic numbers of the top four elements in Groups 1, 2, 6, and 7.

88. Which of the following is/are true regarding the number of neutrons in an atom?

a. Identifies the element.
b. Gives the number of protons in a neutral atom.
c. Gives the number of electrons in a neutral atom.
d. Contributes to the mass number.

89. What is the difference between the atomic number and the mass number of an element? Can atoms of two different elements have the same atomic number? Could they have the same mass number? Why or why not?

90. Which subatomic particles contribute most to the atom's mass? Which subatomic particles determine the atom's chemical properties?

91. Is it possible for the same two elements to form more than one compound? Is this consistent with Dalton's atomic theory? Give an example.

92. Carbohydrates, a class of compounds containing the elements carbon, hydrogen, and oxygen, were originally thought to contain one water molecule (H_2O) for each carbon atom present. The carbohydrate glucose contains six carbon atoms. Write a general formula showing the relative numbers of each type of atom present in glucose.

93. When iron rusts in moist air, the product is typically a mixture of two iron–oxygen compounds. In one compound, there is an equal number of iron and oxygen atoms. In the other compound, there are three oxygen atoms for every two iron atoms. Write the formulas for the two iron oxides.

94. How many protons and neutrons are contained in the nucleus of each of the following atoms? For an atom of the element, how many electrons are present?

a. $^{63}_{29}Cu$
b. $^{80}_{35}Br$
c. $^{24}_{12}Mg$

95. Though the common isotope of aluminum has a mass number of 27, isotopes of aluminum have been isolated (or prepared in nuclear reactors) with mass numbers of 24, 25, 26, 28, 29, and 30. How many neutrons are present in each of these isotopes? Why are they all considered aluminum atoms, even though they differ greatly in mass? Write the atomic symbol for each isotope.

96. The principal goal of alchemists was to convert cheaper, more common metals into gold. Considering that gold had no particular practical uses (for example, it was too soft to be used for weapons), why do you think early civilizations placed such emphasis on the value of gold?

97. How did Robert Boyle define an element?

98. What is the symbol for an ion with a 1− charge, 36 electrons, and 46 neutrons?

99. Give the chemical symbol for each of the following elements.

a. barium
b. potassium
c. cesium
d. lead
e. platinum
f. gold

100. Which of the following is/are true regarding the ion $^{133}Cs^+$?

a. The number of neutrons is 78.
b. The ion contains more protons than electrons.
c. The mass number is 133.
d. The ion is an alkali metal.

101. Give the chemical symbol for each of the following elements.

a. silver
b. aluminum
c. cadmium
d. antimony
e. tin
f. arsenic

102. A metal ion with a 2+ charge contains 34 neutrons and 27 electrons. Identify the metal ion and determine its mass number.

103. For each of the following chemical symbols, give the name of the corresponding element.

a. Te
b. Pd
c. Zn
d. Si
e. Cs
f. Bi
g. F
h. Ti

104. Write the simplest formula for each of the following substances, listing the elements in the order given.

a. a molecule containing one carbon atom and two oxygen atoms
b. a compound containing one aluminum atom for every three chlorine atoms
c. perchloric acid, which contains one hydrogen atom, one chlorine atom, and four oxygen atoms
d. a molecule containing one sulfur atom and six chlorine atoms

105. For each of the following atomic numbers, write the name and chemical symbol of the corresponding element. (Refer to Figure 4.9.)

a. 7
b. 10
c. 11
d. 28
e. 22
f. 18
g. 36
h. 54

106. Write the atomic symbol ($^A_Z X$) for each of the isotopes described below.

a. $Z = 6$, number of neutrons $= 7$
b. the isotope of carbon with a mass number of 13
c. $Z = 6, A = 13$
d. $Z = 19, A = 44$
e. the isotope of calcium with a mass number of 41
f. the isotope with 19 protons and 16 neutrons

107. How many protons and neutrons are contained in the nucleus of each of the following atoms? In an atom of each element, how many electrons are present?

a. $^{41}_{22}Ti$
b. $^{64}_{30}Zn$
c. $^{76}_{32}Ge$
d. $^{86}_{36}Kr$
e. $^{75}_{33}As$
f. $^{41}_{19}K$

108. Complete the following table.

Symbol	Protons	Neutrons	Mass Number
$^{41}_{20}Ca$	______	______	______
______	25	30	______
______	47	______	109
$^{45}_{21}Sc$	______	______	______

109. For each of the following elements, use the table shown in Fig. 4.9 to give the chemical symbol and atomic number and to specify whether the element is a metal or a nonmetal. Also give the named family to which the element belongs (if any).

a. carbon
b. selenium
c. radon
d. beryllium

F 110. Read the "Chemistry in Focus" segment *A Four-Wheel-Drive Nanocar*. It discusses carbon and copper atoms. Both atoms have stable isotopes. Example 4.2 had you consider the isotopes of carbon. Copper exists as copper-63 and copper-65. Determine the number of each of the three types of subatomic particles in each of the copper atoms and write their symbols.

ChemWork Problems

These multiconcept problems (and additional ones) are found interactively online with the same type of assistance a student would get from an instructor.

111. Provide the name of the element that corresponds to the symbols given in the following table.

Symbol	Element Name
Au	______
Kr	______
He	______
C	______
Li	______
Si	______

112. Provide the symbols for the elements given in the following table.

Element Name	Symbol
tin	______
beryllium	______
hydrogen	______
chlorine	______
radium	______
xenon	______
zinc	______
oxygen	______

113. Complete the following table.

Number of Protons	Number of Neutrons	Symbol
34	45	______
19	20	______
53	74	______
4	5	______
24	32	______

All even-numbered Questions and Problems have answers in the back of this book and solutions in the *Student Solutions Guide.*

114. Complete the following table to predict whether the given atom will gain or lose electrons in forming an ion.

Atom	Gain (G) or Lose (L) Electrons	Ion Formed
O	______	______
Mg	______	______
Rb	______	______
Br	______	______
Cl	______	______

115. Using the periodic table, complete the following table.

Atoms	Number of Protons	Number of Neutrons
${}^{55}_{25}Mn$	______	______
${}^{18}_{8}O$	______	______
${}^{59}_{28}Ni$	______	______
${}^{238}_{92}U$	______	______
${}^{201}_{80}Hg$	______	______

116. Complete the following table.

Atom/Ion	Protons	Neutrons	Electrons
${}^{120}_{50}Sn$	______	______	______
${}^{25}_{12}Mg^{2+}$	______	______	______
${}^{56}_{26}Fe^{2+}$	______	______	______
${}^{79}_{34}Se$	______	______	______
${}^{35}_{17}Cl$	______	______	______
${}^{63}_{29}Cu$	______	______	______

117. Which of the following is(are) correct?

a. ${}^{40}Ca^{2+}$ contains 20 protons and 18 electrons.
b. Rutherford created the cathode-ray tube and was the founder of the charge-to-mass ratio of an electron.
c. An electron is heavier than a proton.
d. The nucleus contains protons, neutrons, and electrons.

All even-numbered Questions and Problems have answers in the back of this book and solutions in the *Student Solutions Guide.*

5 Nomenclature

CHAPTER 5

Clouds over tufa towers in Mono Lake, California. Provided by jp2pix.com/Getty Images

When chemistry was an infant science, there was no system for naming compounds. Names such as sugar of lead, blue vitriol, quicklime, Epsom salts, milk of magnesia, gypsum, and laughing gas were coined by early chemists. Such names are called *common names.* As our knowledge of chemistry grew, it became clear that using common names for compounds was not practical. More than four million chemical compounds are currently known. Memorizing common names for all these compounds would be impossible.

The solution, of course, is a *system* for naming compounds in which the name tells something about the composition of the compound. After learning the system, you should be able to name a compound when you are given its formula. And, conversely, you should be able to construct a compound's formula, given its name. In the next few sections we will specify the most important rules for naming compounds other than organic compounds (those based on chains of carbon atoms).

An artist using plaster of Paris, a gypsum plaster. Bob Daemmrich/The Image Works

5-1 Naming Compounds

OBJECTIVE **To learn two broad classes of binary compounds.**

We will begin by discussing the system for naming **binary compounds**—compounds composed of two elements. We can divide binary compounds into two broad classes:

1. Compounds that contain a metal and a nonmetal
2. Compounds that contain two nonmetals

We will describe how to name compounds in each of these classes in the next several sections. Then, in succeeding sections, we will describe the systems used for naming more complex compounds.

5-2 Naming Binary Compounds That Contain a Metal and a Nonmetal (Types I and II)

OBJECTIVE **To learn to name binary compounds of a metal and a nonmetal.**

As we saw in Section 4-11, when a metal such as sodium combines with a nonmetal such as chlorine, the resulting compound contains ions. The metal loses one or more electrons to become a cation, and the nonmetal gains one or more electrons to form an anion. The resulting substance is called a **binary ionic compound.** Binary ionic compounds contain a positive ion (cation), which is always written first in the formula, and a negative ion (anion). *To name these compounds we simply name the ions.*

Sugar of Lead

In ancient Roman society it was common to boil wine in a lead-lined vessel, driving off much of the water to produce a very sweet, viscous syrup called *sapa.* This syrup was commonly used as a sweetener for many types of food and drink.

We now realize that a major component of this syrup was lead acetate, $Pb(C_2H_3O_2)_2$. This compound has a very sweet taste—hence its original name, sugar of lead.

Many historians believe that the fall of the Roman Empire was due at least in part to lead poisoning, which causes lethargy and mental malfunctions. One major source of this lead was the sapa syrup. In addition, the Romans' highly advanced plumbing system employed lead water pipes, which allowed lead to be leached into their drinking water.

Sadly, this story is more relevant to today's society than you might think. Lead-based solder was widely used for many years to connect the copper pipes in water systems in homes and commercial buildings. There is evidence that dangerous amounts of lead can be leached from these soldered joints into drinking water. In fact, large quantities of lead have been found in the water that some drinking fountains and water coolers dispense. In response to these problems, the U.S. Congress has passed a law banning lead from the solder used in plumbing systems for drinking water.

Art Resource, NY

An ancient painting showing Romans drinking wine.

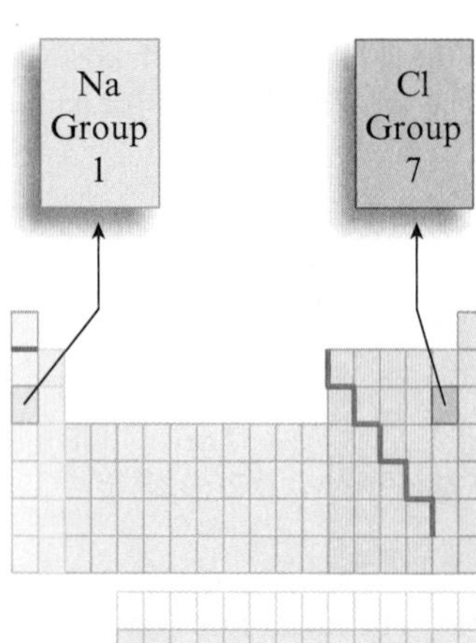

In this section we will consider binary ionic compounds of two types based on the cations they contain. Certain metal atoms form only one cation. For example, the Na atom always forms Na^+, *never* Na^{2+} or Na^{3+}. Likewise, Cs always forms Cs^+, Ca always forms Ca^{2+}, and Al always forms Al^{3+}. Wc will call compounds that contain this type of metal atom Type I binary compounds and the cations they contain Type I cations. Examples of Type I cations are Na^+, Ca^{2+}, Cs^+, and Al^{3+}.

Other metal atoms can form two or more cations. For example, Cr can form Cr^{2+} and Cr^{3+} and Cu can form Cu^+ and Cu^{2+}. We will call such ions Type II cations and their compounds Type II binary compounds.

In summary:

Type I compounds: The metal present forms only one type of cation.

Type II compounds: The metal present can form two (or more) cations that have different charges.

Some common cations and anions and their names are listed in Table 5.1. You should memorize these. They are an essential part of your chemical vocabulary.

Table 5.1 ▸ Common Simple Cations and Anions

Cation	Name	Anion	Name*
H^+	hydrogen	H^-	hydride
Li^+	lithium	F^-	fluoride
Na^+	sodium	Cl^-	chloride
K^+	potassium	Br^-	bromide
Cs^+	cesium	I^-	iodide
Be^{2+}	beryllium	O^{2-}	oxide
Mg^{2+}	magnesium	S^{2-}	sulfide
Ca^{2+}	calcium		
Ba^{2+}	barium		
Al^{3+}	aluminum		
Ag^+	silver		
Zn^{2+}	zinc		

*The root is given in color.

Type I Binary Ionic Compounds

The following rules apply for Type I ionic compounds:

Rules for Naming Type I Ionic Compounds

1. The cation is always named first and the anion second.
2. A simple cation (obtained from a single atom) takes its name from the name of the element. For example, Na^+ is called sodium in the names of compounds containing this ion.
3. A simple anion (obtained from a single atom) is named by taking the first part of the element name (the root) and adding *-ide*. Thus the Cl^- ion is called chloride.

We will illustrate these rules by naming a few compounds. For example, the compound NaI is called sodium iodide. It contains Na^+ (the sodium cation, named for the parent metal) and I^- (iodide: the root of *iodine* plus *-ide*). Similarly, the compound CaO is called calcium oxide because it contains Ca^{2+} (the calcium cation) and O^{2-} (the oxide anion).

The rules for naming binary compounds are also illustrated by the following examples:

Compound	Ions Present	Name
NaCl	Na^+, Cl^-	sodium chloride
KI	K^+, I^-	potassium iodide
CaS	Ca^{2+}, S^{2-}	calcium sulfide
CsBr	Cs^+, Br^-	cesium bromide
MgO	Mg^{2+}, O^{2-}	magnesium oxide

It is important to note that in the *formulas* of ionic compounds, simple ions are represented by the element symbol: Cl means Cl^-, Na means Na^+, and so on. However, when *individual ions* are shown, the charge is always included. Thus the formula of potassium bromide is written KBr, but when the potassium and bromide ions are shown individually, they are written K^+ and Br^-.

Interactive Example 5.1

Naming Type I Binary Compounds

Name each binary compound.

a. CsF b. $AlCl_3$ c. MgI_2

SOLUTION We will name these compounds by systematically following the rules given above.

a. CsF

Step 1 Identify the cation and anion. Cs is in Group 1, so we know it will form the 1+ ion Cs^+. Because F is in Group 7, it forms the 1− ion F^-.

Step 2 Name the cation. Cs^+ is simply called cesium, the same as the element name.

Step 3 Name the anion. F^- is called fluoride: we use the root name of the element plus *-ide*.

Step 4 Name the compound by combining the names of the individual ions. The name for CsF is cesium fluoride. (Remember that the name of the cation is always given first.)

b.

Compound		Ions Present	Ion Names	Comments
$AlCl_3$	Cation →	Al^{3+}	aluminum	Al (Group 3) always forms Al^{3+}.
	Anion →	Cl^-	chloride	Cl (Group 7) always forms Cl^-.

The name of $AlCl_3$ is aluminum chloride.

c.

Compound		Ions Present	Ion Names	Comments
MgI_2	Cation →	Mg^{2+}	magnesium	Mg (Group 2) always forms Mg^{2+}.
	Anion →	I^-	iodide	I (Group 7) gains one electron to form I^-.

The name of MgI_2 is magnesium iodide.

SELF CHECK

Exercise 5.1 Name the following compounds.

a. Rb_2O b. SrI_2 c. K_2S

See Problems 5.9 and 5.10. ■

Example 5.1 reminds us of three things:

1. Compounds formed from metals and nonmetals are ionic.
2. In an ionic compound the cation is always named first.
3. The *net* charge on an ionic compound is always zero. Thus, in CsF, one of each type of ion (Cs^+ and F^-) is required: (1+) + (1−) = 0 charge. In $AlCl_3$, however, three Cl^- ions are needed to balance the charge of Al^{3+}: (3+) + 3(1−) = 0 charge. In MgI_2, two I^- ions are needed for each Mg^{2+} ion: (2+) + 2(1−) = 0 charge.

Type II Binary Ionic Compounds

So far we have considered binary ionic compounds (Type I) containing metals that always give the same cation. For example, sodium always forms the Na^+ ion, calcium always forms the Ca^{2+} ion, and aluminum always forms the Al^{3+} ion. As we said in the previous section, we can predict with certainty that each Group 1 metal will give a 1+ cation and each Group 2 metal will give a 2+ cation. Aluminum always forms Al^{3+}.

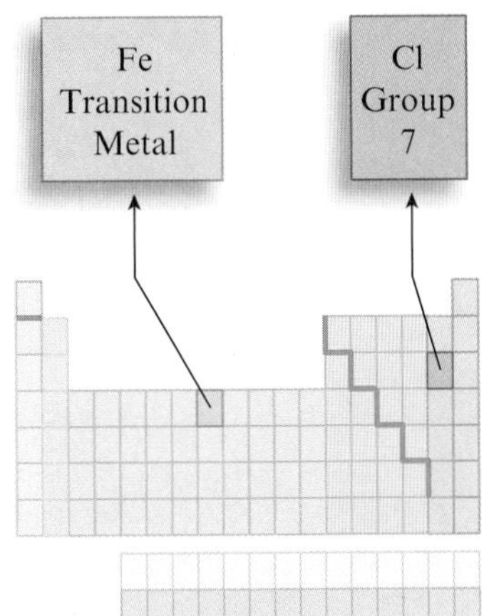

However, there are many metals that can form more than one type of cation. For example, lead (Pb) can form Pb^{2+} or Pb^{4+} in ionic compounds. Also, iron (Fe) can produce Fe^{2+} or Fe^{3+}, chromium (Cr) can produce Cr^{2+} or Cr^{3+}, gold (Au) can produce Au^+ or Au^{3+}, and so on. This means that if we saw the name gold chloride, we wouldn't know whether it referred to the compound AuCl (containing Au^+ and Cl^-) or the compound $AuCl_3$ (containing Au^{3+} and three Cl^- ions). Therefore, we need a way of specifying which cation is present in compounds containing metals that can form more than one type of cation.

Chemists have decided to deal with this situation by using a Roman numeral to specify the charge on the cation. To see how this works, consider the compound $FeCl_2$. Iron can form Fe^{2+} or Fe^{3+}, so we must first decide which of these cations is present. We can determine the charge on the iron cation, because we know it must just balance the charge on the two 1− anions (the chloride ions). Thus if we represent the charges as

$$(?+) \; + \; 2\,(1-) \; = \; 0$$

$(?+)$	$+\,2\,(1-)$	$=\,0$
↑ Charge on iron cation	↑ Charge on Cl^-	↑ Net charge

we know that ? must represent 2 because

$$(2+) + 2(1-) = 0$$

FeCl₃ must contain Fe³⁺ to balance the charge of three Cl⁻ ions.

The compound $FeCl_2$, then, contains one Fe^{2+} ion and two Cl^- ions. We call this compound iron(II) chloride, where the II tells the charge of the iron cation. That is, Fe^{2+} is called iron(II). Likewise, Fe^{3+} is called iron(III). And $FeCl_3$, which contains one Fe^{3+} ion and three Cl^- ions, is called iron(III) chloride. Remember that the Roman numeral tells the *charge* on the ion, not the number of ions present in the compound.

Note that in the preceding examples the Roman numeral for the cation turned out to be the same as the subscript needed for the anion (to balance the charge). This is often not the case. For example, consider the compound PbO_2. Since the oxide ion is O^{2-}, for PbO_2 we have

$$(?+) \; + \; 2\,(2-) \; = \; 0$$

$(?+)$	$+\,2\,(2-)$	$=\,0$
↑ Charge on lead ion	↑ (4−) Charge on two O^{2-} ions	↑ Net charge

Thus the charge on the lead ion must be 4+ to balance the 4− charge of the two oxide ions. The name of PbO_2 is therefore lead(IV) oxide, where the IV indicates the presence of the Pb^{4+} cation.

There is another system for naming ionic compounds containing metals that form two cations. *The ion with the higher charge has a name ending in* -ic, *and the one with the lower charge has a name ending in* -ous. In this system, for example, Fe^{3+} is called the ferric ion, and Fe^{2+} is called the ferrous ion. The names for $FeCl_3$ and $FeCl_2$, in this

Copper(II) sulfate crystals.

Table 5.2 Common Type II Cations

Ion	Systematic Name	Older Name
Fe^{3+}	iron(III)	ferric
Fe^{2+}	iron(II)	ferrous
Cu^{2+}	copper(II)	cupric
Cu^{+}	copper(I)	cuprous
Co^{3+}	cobalt(III)	cobaltic
Co^{2+}	cobalt(II)	cobaltous
Sn^{4+}	tin(IV)	stannic
Sn^{2+}	tin(II)	stannous
Pb^{4+}	lead(IV)	plumbic
Pb^{2+}	lead(II)	plumbous
Hg^{2+}	mercury(II)	mercuric
Hg_2^{2+}*	mercury(I)	mercurous

*Mercury(I) ions always occur bound together in pairs to form Hg_2^{2+}.

system, are ferric chloride and ferrous chloride, respectively. Table 5.2 gives both names for many Type II cations. We will use the system of Roman numerals exclusively in this text; the other system is falling into disuse.

To help distinguish between Type I and Type II cations, remember that Group 1 and 2 metals are always Type I. On the other hand, transition metals are almost always Type II.

Rules for Naming Type II Ionic Compounds

1. The cation is always named first and the anion second.
2. Because the cation can assume more than one charge, the charge is specified by a Roman numeral in parentheses.

Interactive Example 5.2

Naming Type II Binary Compounds

Give the systematic name of each of the following compounds.

a. CuCl b. HgO c. Fe_2O_3 d. MnO_2 e. $PbCl_4$

SOLUTION All these compounds include a metal that can form more than one type of cation; thus we must first determine the charge on each cation. We do this by recognizing that a compound must be electrically neutral; that is, the positive and negative charges must balance exactly. We will use the known charge on the anion to determine the charge of the cation.

a. In CuCl we recognize the anion as Cl^-. To determine the charge on the copper cation, we invoke the principle of charge balance.

$$(?+) + (1-) = 0$$

Charge on copper ion ↑ (?+); Charge on Cl^- ↑ (1−); Net charge (must be zero) ↑ 0

In this case, ?+ must be 1+ because (1+) + (1−) = 0. Thus the copper cation must be Cu^+. Now we can name the compound by using the regular steps.

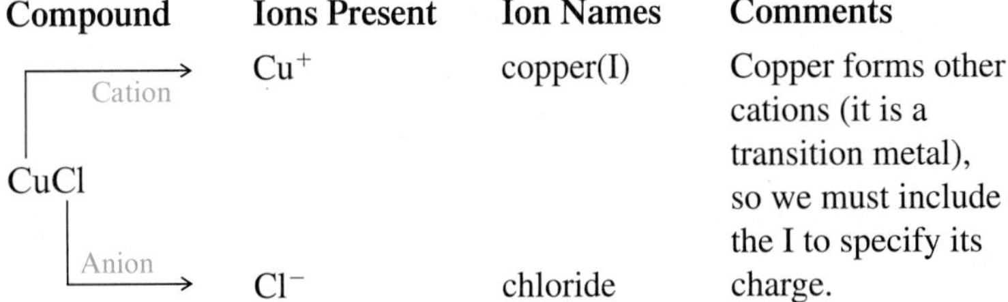

Compound	Ions Present	Ion Names	Comments
CuCl (Cation →)	Cu^+	copper(I)	Copper forms other cations (it is a transition metal), so we must include the I to specify its charge.
(Anion →)	Cl^-	chloride	

The name of CuCl is copper(I) chloride.

b. In HgO we recognize the O^{2-} anion. To yield zero net charge, the cation must be Hg^{2+}.

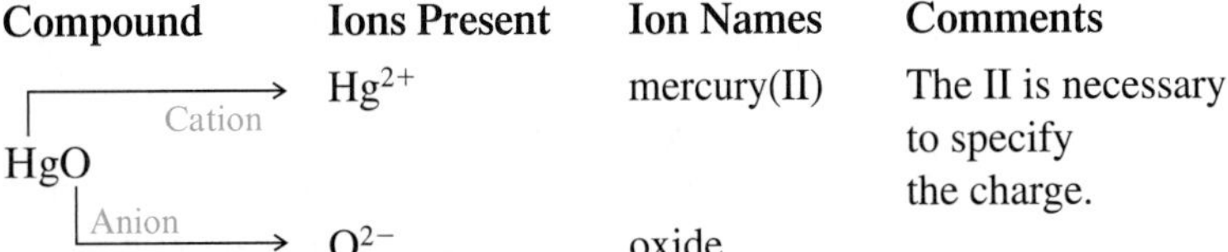

Compound	Ions Present	Ion Names	Comments
HgO (Cation →)	Hg^{2+}	mercury(II)	The II is necessary to specify the charge.
(Anion →)	O^{2-}	oxide	

The name of HgO is mercury(II) oxide.

c. Because Fe_2O_3 contains three O^{2-} anions, the charge on the iron cation must be 3+.

$$\underset{Fe^{3+}}{2(3+)} + \underset{O^{2-}}{3(2-)} = \underset{\text{Net charge}}{0}$$

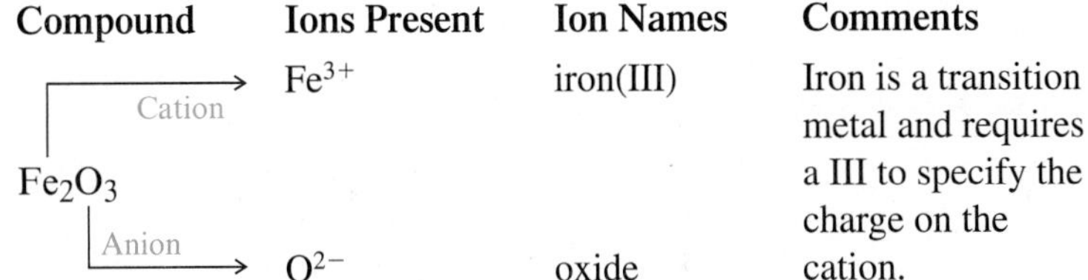

Compound	Ions Present	Ion Names	Comments
Fe_2O_3 (Cation →)	Fe^{3+}	iron(III)	Iron is a transition metal and requires a III to specify the charge on the cation.
(Anion →)	O^{2-}	oxide	

The name of Fe_2O_3 is iron(III) oxide.

d. MnO_2 contains two O^{2-} anions, so the charge on the manganese cation is 4+.

$$\underset{Mn^{4+}}{(4+)} + \underset{O^{2-}}{2(2-)} = \underset{\text{Net charge}}{0}$$

Compound	Ions Present	Ion Names	Comments
MnO_2 (Cation →)	Mn^{4+}	manganese(IV)	Manganese is a transition metal and requires a IV to specify the charge on the cation.
(Anion →)	O^{2-}	oxide	

The name of MnO_2 is manganese(IV) oxide.

Critical Thinking

We can use the periodic table to tell us something about the stable ions formed by many atoms. For example, the atoms in column 1 always form +1 ions. The transition metals, however, can form more than one type of stable ion. What if each transition metal ion had only one possible charge? How would the naming of compounds be different?

e. Because $PbCl_4$ contains four Cl^- anions, the charge on the lead cation is 4+.

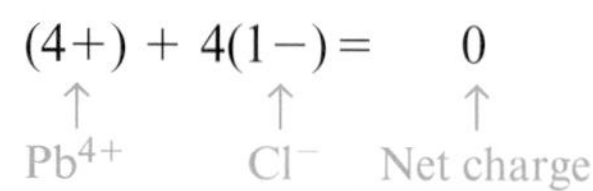

$$(4+) + 4(1-) = 0$$

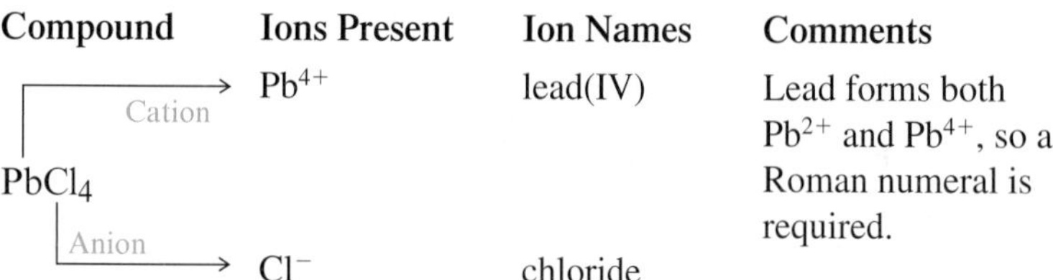

Compound	Ions Present	Ion Names	Comments
$PbCl_4$ (Cation)	Pb^{4+}	lead(IV)	Lead forms both Pb^{2+} and Pb^{4+}, so a Roman numeral is required.
(Anion)	Cl^-	chloride	

The name for $PbCl_4$ is lead(IV) chloride. ■

Sometimes transition metals form only one ion, such as silver, which commonly forms Ag^+; zinc, which forms Zn^{2+}; and cadmium, which forms Cd^{2+}. In these cases, chemists do not use a Roman numeral, although it is not "wrong" to do so.

The use of a Roman numeral in a systematic name for a compound is required only in cases where more than one ionic compound forms between a given pair of elements. This occurs most often for compounds that contain transition metals, which frequently form more than one cation. *Metals that form only one cation do not need to be identified by a Roman numeral.* Common metals that do not require Roman numerals are the Group 1 elements, which form only 1+ ions; the Group 2 elements, which form only 2+ ions; and such Group 3 metals as aluminum and gallium, which form only 3+ ions.

As shown in Example 5.2, when a metal ion that forms more than one type of cation is present, the charge on the metal ion must be determined by balancing the positive and negative charges of the compound. To do this, you must be able to recognize the common anions and you must know their charges (see Table 5.1).

Interactive Example 5.3

Naming Binary Ionic Compounds: Summary

Give the systematic name of each of the following compounds.

a. $CoBr_2$
b. $CaCl_2$
c. Al_2O_3
d. $CrCl_3$

SOLUTION

	Compound	Ions and Names	Compound Name	Comments
a.	$CoBr_2$	Co^{2+} cobalt(II)	cobalt(II) bromide	Cobalt is a transition metal; the name of the compound must have a Roman numeral. The two Br^- ions must be balanced by a Co^{2+} cation.
		Br^- bromide		
b.	$CaCl_2$	Ca^{2+} calcium	calcium chloride	Calcium, a Group 2 metal, forms only the Ca^{2+} ion. A Roman numeral is not necessary.
		Cl^- chloride		
c.	Al_2O_3	Al^{3+} aluminum	aluminum oxide	Aluminum forms only Al^{3+}. A Roman numeral is not necessary.
		O^{2-} oxide		

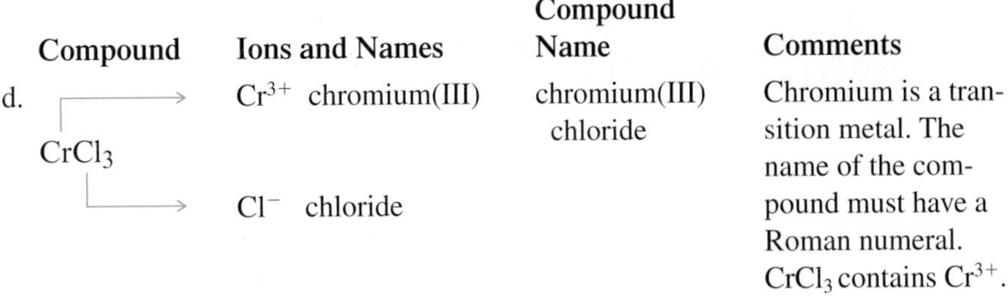

	Compound	Ions and Names	Compound Name	Comments
d.	$CrCl_3$	Cr^{3+} chromium(III)	chromium(III) chloride	Chromium is a transition metal. The name of the compound must have a Roman numeral. $CrCl_3$ contains Cr^{3+}.
		Cl^- chloride		

SELF CHECK

Exercise 5.2 Give the names of the following compounds.

a. $PbBr_2$ and $PbBr_4$ b. FeS and Fe_2S_3 c. $AlBr_3$ d. Na_2S e. $CoCl_3$

See Problems 5.9, 5.10, and 5.13 through 5.16. ■

The following flow chart is useful when you are naming binary ionic compounds:

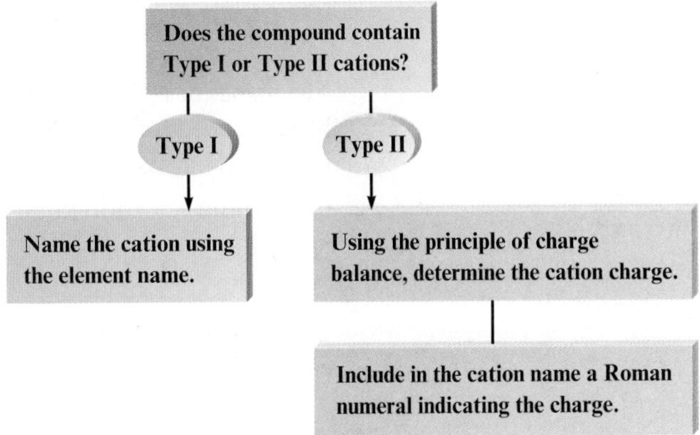

Naming Binary Compounds That Contain Only Nonmetals (Type III)

OBJECTIVE

To learn how to name binary compounds containing only nonmetals.

Binary compounds that contain only nonmetals are named in accordance with a system similar in some ways to the rules for naming binary ionic compounds, but there are important differences. *Type III binary compounds contain only nonmetals.* The following rules cover the naming of these compounds.

Table 5.3 ▸ Prefixes Used to Indicate Numbers in Chemical Names

Prefix	Number Indicated
mono-	1
di-	2
tri-	3
tetra-	4
penta-	5
hexa-	6
hepta-	7
octa-	8

Rules for Naming Type III Binary Compounds

1. The first element in the formula is named first, and the full element name is used.
2. The second element is named as though it were an anion.
3. Prefixes are used to denote the numbers of atoms present. These prefixes are given in Table 5.3.
4. The prefix *mono-* is never used for naming the first element. For example, CO is called carbon monoxide, *not* monocarbon monoxide.

We will illustrate the application of these rules in Example 5.4.

Interactive Example 5.4

Naming Type III Binary Compounds

Name the following binary compounds, which contain two nonmetals (Type III).

a. BF_3 b. NO c. N_2O_5

SOLUTION a. BF_3

Rule 1 Name the first element, using the full element name: boron.

Rule 2 Name the second element as though it were an anion: fluoride.

Rules 3 and 4 Use prefixes to denote numbers of atoms. One boron atom: do not use *mono-* in first position. Three fluorine atoms: use the prefix *tri-*.

The name of BF_3 is boron trifluoride.

b.

Compound	Individual Names	Prefixes	Comments
NO	nitrogen	none	*Mono-* is not used for the
	oxide	*mono-*	first element.

The name for NO is nitrogen monoxide. Note that the second *o* in *mono-* has been dropped for easier pronunciation. The *common* name for NO, which is often used by chemists, is nitric oxide.

c.

Compound	Individual Names	Prefixes	Comments
N_2O_5	nitrogen	*di-*	two N atoms
	oxide	*penta-*	five O atoms

The name for N_2O_5 is dinitrogen pentoxide. The *a* in *penta-* has been dropped for easier pronunciation.

A piece of copper metal about to be placed in nitric acid (*left*). Copper reacts with nitric acid to produce colorless NO (which immediately reacts with the oxygen in the air to form reddish-brown NO_2 gas) and Cu^{2+} ions in solution (which produce the green color) (*right*).

SELF CHECK

Exercise 5.3 Name the following compounds.

a. CCl_4 b. NO_2 c. IF_5

See Problems 5.17 and 5.18. ■

The previous examples illustrate that, to avoid awkward pronunciation, we often drop the final *o* or *a* of the prefix when the second element is oxygen. For example, N_2O_4 is called dinitrogen tetroxide, *not* dinitrogen tetr*a*oxide, and CO is called carbon monoxide, *not* carbon mono*o*xide.

Some compounds are always referred to by their common names. The two best examples are water and ammonia. The systematic names for H_2O and NH_3 are never used.

To make sure you understand the procedures for naming binary nonmetallic compounds (Type III), study Example 5.5 and then do Self Check Exercise 5.4.

Naming Type III Binary Compounds: Summary

Name each of the following compounds.

a. PCl_5 c. SF_6 e. SO_2
b. P_4O_6 d. SO_3 f. N_2O_3

SOLUTION

Compound	Name
a. PCl_5	phosphorus pentachloride
b. P_4O_6	tetraphosphorus hexoxide
c. SF_6	sulfur hexafluoride
d. SO_3	sulfur trioxide
e. SO_2	sulfur dioxide
f. N_2O_3	dinitrogen trioxide

SELF CHECK

Exercise 5.4 Name the following compounds.

a. SiO_2 b. O_2F_2 c. XeF_6

See Problems 5.17 and 5.18. ■

5-4 Naming Binary Compounds: A Review

OBJECTIVE **To review the naming of Type I, Type II, and Type III binary compounds.**

Because different rules apply for naming various types of binary compounds, we will now consider an overall strategy to use for these compounds. We have considered three types of binary compounds, and naming each of them requires different procedures.

Type I: Ionic compounds with metals that always form a cation with the same charge

Type II: Ionic compounds with metals (usually transition metals) that form cations with various charges

Type III: Compounds that contain only nonmetals

In trying to determine which type of compound you are naming, use the periodic table to help identify metals and nonmetals and to determine which elements are transition metals.

The flow chart given in Fig. 5.1 should help you as you name binary compounds of the various types.

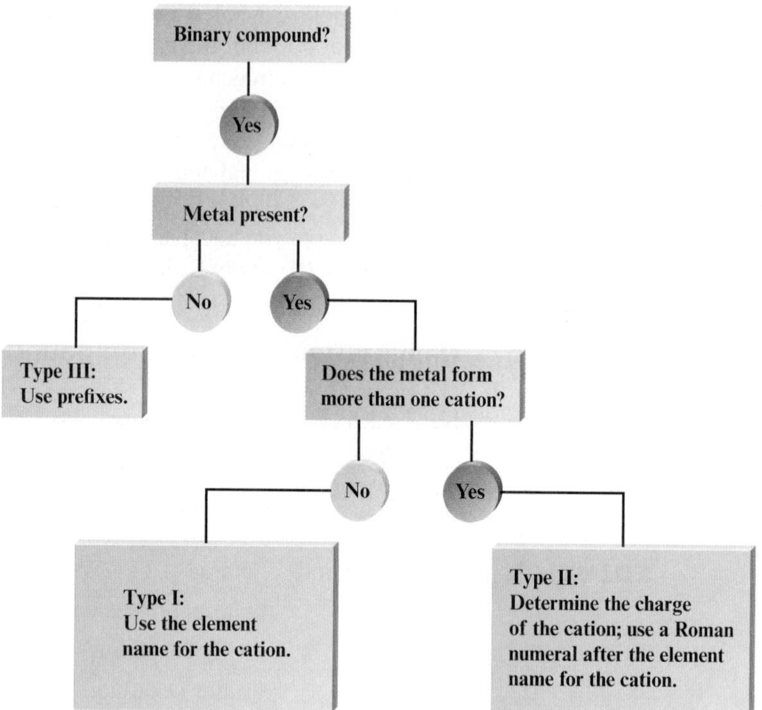

Figure 5.1 A flow chart for naming binary compounds.

Interactive Example 5.6

Naming Binary Compounds: Summary

Name the following binary compounds.

a. CuO c. B_2O_3 e. K_2S g. NH_3

b. SrO d. $TiCl_4$ f. OF_2

SOLUTION

a.

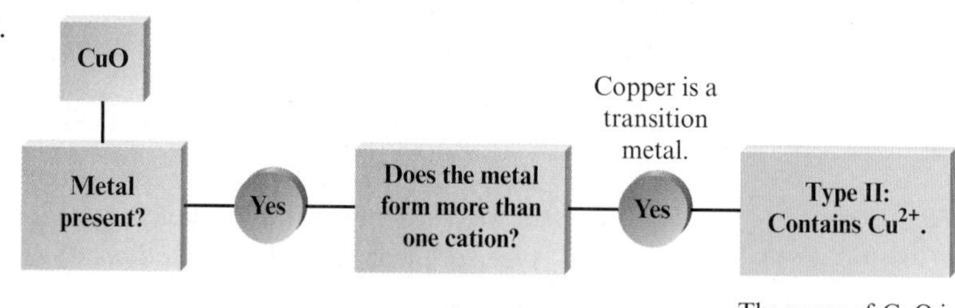

The name of CuO is copper(II) oxide.

b.

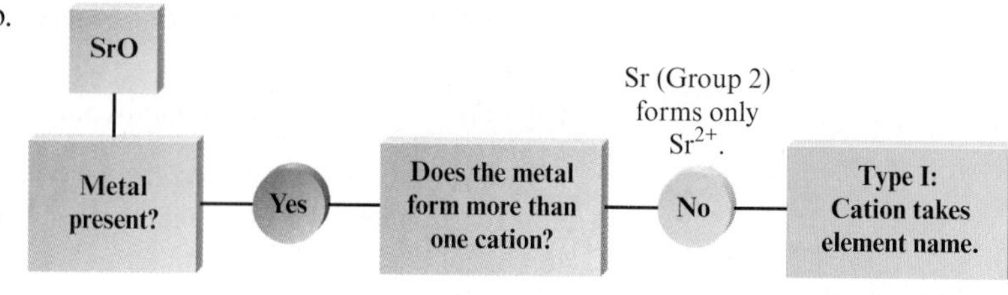

The name of SrO is strontium oxide.

c.

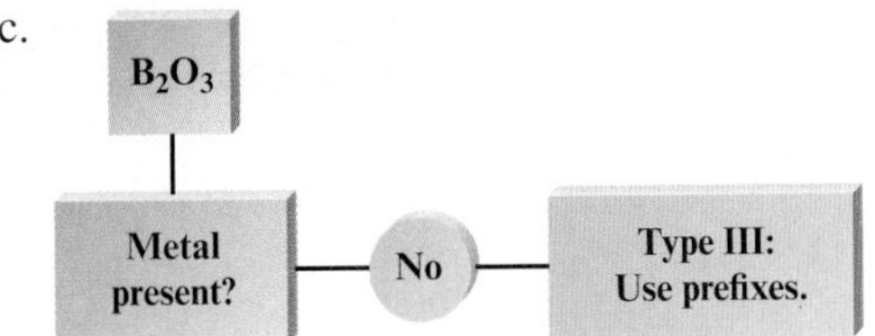

The name of B_2O_3 is diboron trioxide.

d.

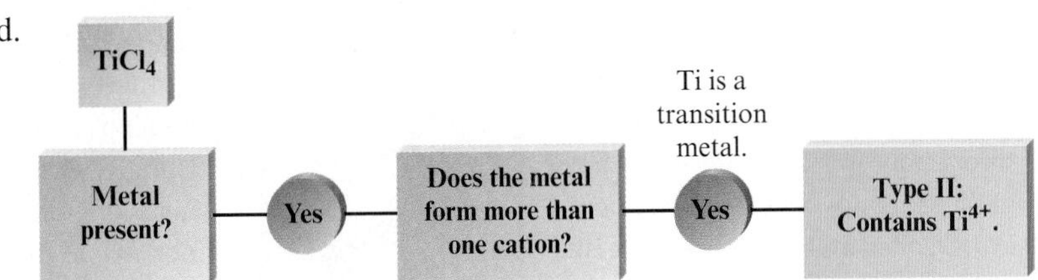

Ti is a transition metal.

The name of $TiCl_4$ is titanium(IV) chloride.

e.

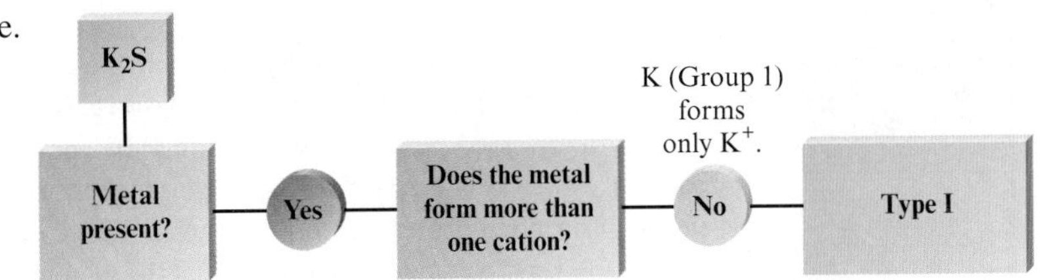

K (Group 1) forms only K^+.

The name of K_2S is potassium sulfide.

f.

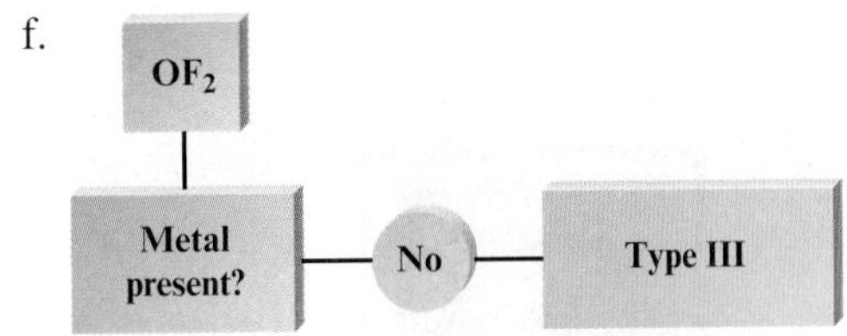

The name of OF_2 is oxygen difluoride.

g.

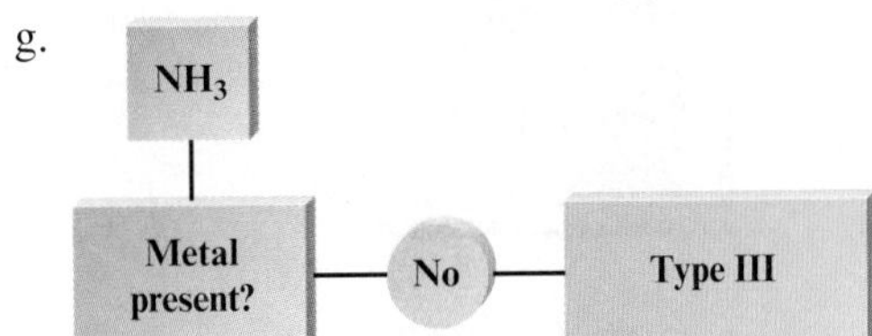

The name of NH_3 is ammonia. The systematic name is never used.

SELF CHECK

Exercise 5.5 Name the following binary compounds.

a. ClF_3
b. VF_5
c. $CuCl$
d. MnO_2
e. MgO
f. H_2O

See Problems 5.19 through 5.22. ■

Chemophilately

Philately is the study of postage stamps. Chemophilately, a term coined by the Israeli chemist Zvi Rappoport, refers to the study of stamps that have some sort of chemical connection. Collectors estimate that more than 2000 chemical-related stamps have been printed throughout the world. Relatively few of these stamps have been produced in the United States. One example is a 29¢ stamp honoring minerals that shows a copper nugget.

Courtesy, Daniel Rabinovich/University of North Carolina at Charlotte (UNCC)

Courtesy, Daniel Rabinovich/University of North Carolina at Charlotte (UNCC)

Chemists also have been honored on U.S. postage stamps. One example is a 29¢ stamp printed in 1993 honoring Percy L. Julian, an African-American chemist who was the grandson of slaves. Julian is noted for his synthesis of steroids used to treat glaucoma and rheumatoid arthritis. As a holder of more than 100 patents, he was inducted into the National Inventors Hall of Fame in 1990.

In 1983 the United States issued a stamp honoring Joseph Priestley, whose experiments led to the discovery of oxygen.

Courtesy, Whitman Publishing LLC

Olga Popova/Shutterstock.com

A Russian stamp from 2009 pictures Dmitri Mendeleev, who, in 1869, arranged the 63 known elements in the present form of the periodic table. Mendeleev's arrangement allowed for the prediction of yet unknown elements and their properties.

In 2008, a 41¢ stamp was issued that honors Linus C. Pauling, who pioneered the concept of the chemical bond. Pauling received two Nobel Prizes: one for his work on chemical bonds and the other for his work championing world peace. His stamp includes drawings of red blood cells to commemorate his work on the study of hemoglobin, which led to the classification of sickle cell anemia as a molecular disease.

Courtesy, Daniel Rabinovich/University of North Carolina at Charlotte (UNCC)

Marie Curie also received two Nobel Prizes and was the first person to be so honored. She shared her Nobel Prize in Physics in 1903 with her husband, Pierre Curie, and Henri Becquerel for their research on radiation. She was the sole winner of the Nobel Prize in Chemistry in 1911 for the discovery and study of the elements radium and polonium. In 2011 (The International Year of Chemistry), many countries issued stamps honoring the 100th anniversary of Marie Curie winning the Nobel Prize in Chemistry.

Courtesy, Daniel Rabinovich/University of North Carolina at Charlotte (UNCC)

Postal chemistry also shows up in postmarks from places in the United States with chemical names. Examples include Radium, KS; Neon, KY; Boron, CA; Bromide, OK; and Telluride, CO.

Courtesy, Daniel Rabinovich/University of North Carolina at Charlotte (UNCC)

Chemophilately—further proof that chemistry is everywhere!

5-5 Naming Compounds That Contain Polyatomic Ions

OBJECTIVE **To learn the names of common polyatomic ions and how to use them in naming compounds.**

A type of ionic compound that we have not yet considered is exemplified by ammonium nitrate, NH_4NO_3, which contains the **polyatomic ions** NH_4^+ and NO_3^-. As their name suggests, polyatomic ions are charged entities composed of several atoms bound together. Polyatomic ions are assigned special names that you *must memorize* to name the compounds containing them. The most important polyatomic ions and their names are listed in Table 5.4.

Note in Table 5.4 that several series of polyatomic anions exist that contain an atom of a given element and different numbers of oxygen atoms. These anions are called **oxyanions.** When there are two members in such a series, the name of the one with the smaller number of oxygen atoms ends in *-ite,* and the name of the one with the larger number ends in *-ate.* For example, SO_3^{2-} is sulfite and SO_4^{2-} is sulfate. When more than two oxyanions make up a series, *hypo-* (less than) and *per-* (more than) are used as prefixes to name the members of the series with the fewest and the most oxygen atoms, respectively. The best example involves the oxyanions containing chlorine:

ClO^-	*hypo*chlor*ite*
ClO_2^-	chlor*ite*
ClO_3^-	chlor*ate*
ClO_4^-	*per*chlor*ate*

Naming ionic compounds that contain polyatomic ions is very similar to naming binary ionic compounds. For example, the compound NaOH is called sodium hydroxide, because it contains the Na^+ (sodium) cation and the OH^- (hydroxide) anion. To name these compounds, *you must learn to recognize the common polyatomic ions.* That is, you must learn the *composition* and *charge* of each of the ions in Table 5.4.

Table 5.4 ▸ Names of Common Polyatomic Ions

Ion	Name	Ion	Name
NH_4^+	ammonium	CO_3^{2-}	carbonate
NO_2^-	nitrite	HCO_3^-	hydrogen carbonate (bicarbonate is a widely used common name)
NO_3^-	nitrate		
SO_3^{2-}	sulfite	ClO^-	hypochlorite
SO_4^{2-}	sulfate	ClO_2^-	chlorite
HSO_4^-	hydrogen sulfate (bisulfate is a widely used common name)	ClO_3^-	chlorate
		ClO_4^-	perchlorate
OH^-	hydroxide	$C_2H_3O_2^-$	acetate
CN^-	cyanide	MnO_4^-	permanganate
PO_4^{3-}	phosphate	$Cr_2O_7^{2-}$	dichromate
HPO_4^{2-}	hydrogen phosphate	CrO_4^{2-}	chromate
$H_2PO_4^-$	dihydrogen phosphate	O_2^{2-}	peroxide

Then when you see the formula $NH_4C_2H_3O_2$, you should immediately recognize its two "parts":

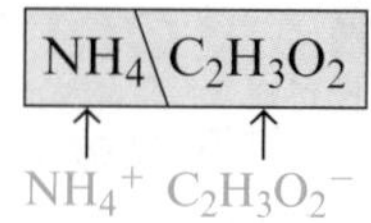

The correct name is ammonium acetate.

Remember that when a metal is present that forms more than one cation, a Roman numeral is required to specify the cation charge, just as in naming Type II binary ionic compounds. For example, the compound $FeSO_4$ is called iron(II) sulfate, because it contains Fe^{2+} (to balance the 2− charge on SO_4^{2-}). Note that to determine the charge on the iron cation, you must know that sulfate has a 2− charge.

Interactive Example 5.7

Naming Compounds That Contain Polyatomic Ions

Give the systematic name of each of the following compounds.

a. Na_2SO_4　　c. $Fe(NO_3)_3$　　e. Na_2SO_3
b. KH_2PO_4　　d. $Mn(OH)_2$　　f. NH_4ClO_3

SOLUTION

Compound	Ions Present	Ion Names	Compound Name
a. Na_2SO_4	two Na^+ SO_4^{2-}	sodium sulfate	sodium sulfate
b. KH_2PO_4	K^+ $H_2PO_4^-$	potassium dihydrogen phosphate	potassium dihydrogen phosphate
c. $Fe(NO_3)_3$	Fe^{3+} three NO_3^-	iron(III) nitrate	iron(III) nitrate
d. $Mn(OH)_2$	Mn^{2+} two OH^-	manganese(II) hydroxide	manganese(II) hydroxide
e. Na_2SO_3	two Na^+ SO_3^{2-}	sodium sulfite	sodium sulfite
f. NH_4ClO_3	NH_4^+ ClO_3^-	ammonium chlorate	ammonium chlorate

SELF CHECK

Exercise 5.6 Name each of the following compounds.

a. $Ca(OH)_2$　　d. $(NH_4)_2Cr_2O_7$　　g. $Cu(NO_2)_2$
b. Na_3PO_4　　e. $Co(ClO_4)_2$
c. $KMnO_4$　　f. $KClO_3$

See Problems 5.35 and 5.36. ■

Example 5.7 illustrates that when more than one polyatomic ion appears in a chemical formula, parentheses are used to enclose the ion and a subscript is written after the closing parenthesis. Other examples are $(NH_4)_2SO_4$ and $Fe_3(PO_4)_2$.

In naming chemical compounds, use the strategy summarized in Fig. 5.2. If the compound being considered is binary, use the procedure summarized in Fig. 5.1. If the compound has more than two elements, ask yourself whether it has any polyatomic ions. Use Table 5.4 to help you recognize these ions until you have committed them to memory. If a polyatomic ion is present, name the compound using procedures very similar to those for naming binary ionic compounds.

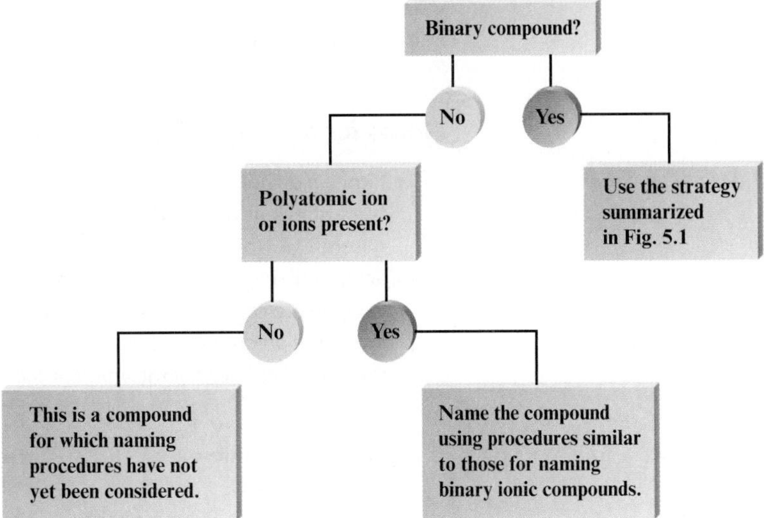

Figure 5.2 Overall strategy for naming chemical compounds.

Interactive Example 5.8

Summary of Naming Binary Compounds and Compounds That Contain Polyatomic Ions

Name the following compounds.

a. Na_2CO_3
b. $FeBr_3$
c. $CsClO_4$
d. PCl_3
e. $CuSO_4$

SOLUTION

Compound	Name	Comments
a. Na_2CO_3	sodium carbonate	Contains $2Na^+$ and CO_3^{2-}.
b. $FeBr_3$	iron(III) bromide	Contains Fe^{3+} and $3Br^-$.
c. $CsClO_4$	cesium perchlorate	Contains Cs^+ and ClO_4^-.
d. PCl_3	phosphorus trichloride	Type III binary compound (both P and Cl are nonmetals).
e. $CuSO_4$	copper(II) sulfate	Contains Cu^{2+} and SO_4^{2-}.

SELF CHECK

Exercise 5.7 Name the following compounds.

a. $NaHCO_3$
b. $BaSO_4$
c. $CsClO_4$
d. BrF_5
e. NaBr
f. KOCl
g. $Zn_3(PO_4)_2$

See Problems 5.29 through 5.36. ■

5-6 Naming Acids

OBJECTIVES

- To learn how the anion composition determines the acid's name.
- To learn names for common acids.

When dissolved in water, certain molecules produce H^+ ions (protons). These substances, which are called **acids,** were first recognized by the sour taste of their solutions. For example, citric acid is responsible for the tartness of lemons and limes. Acids will be discussed in detail later. Here we simply present the rules for naming acids.

An acid can be viewed as a molecule with one or more H^+ ions attached to an anion. The rules for naming acids depend on whether the anion contains oxygen.

Rules for Naming Acids

1. If the *anion does not contain oxygen,* the acid is named with the prefix *hydro-* and the suffix *-ic* attached to the root name for the element. For example, when gaseous HCl (hydrogen chloride) is dissolved in water, it forms hydrochloric acid. Similarly, hydrogen cyanide (HCN) and dihydrogen sulfide (H_2S) dissolved in water are called hydrocyanic acid and hydrosulfuric acid, respectively.
2. When the *anion contains oxygen,* the acid name is formed from the root name of the central element of the anion or the anion name, with a suffix of *-ic* or *-ous.* When the anion name ends in *-ate,* the suffix *-ic* is used. For example,

Acid	Anion	Name
H_2SO_4	SO_4^{2-} (sulfate)	sulfuric acid
H_3PO_4	PO_4^{3-} (phosphate)	phosphoric acid
$HC_2H_3O_2$	$C_2H_3O_2^-$ (acetate)	acetic acid

When the anion name ends in *-ite,* the suffix *-ous* is used in the acid name. For example,

Acid	Anion	Name
H_2SO_3	SO_3^{2-} (sulfite)	sulfurous acid
HNO_2	NO_2^- (nitrite)	nitrous acid

Table 5.5 ▸ Names of Acids That Do Not Contain Oxygen

Acid	Name
HF	hydrofluoric acid
HCl	hydrochloric acid
HBr	hydrobromic acid
HI	hydroiodic acid
HCN	hydrocyanic acid
H_2S	hydrosulfuric acid

The application of Rule 2 can be seen in the names of the acids of the oxyanions of chlorine below.

Acid	Anion	Name
$HClO_4$	perchlor*ate*	perchlor*ic* acid
$HClO_3$	chlor*ate*	chlor*ic* acid
$HClO_2$	chlor*ite*	chlor*ous* acid
HClO	hypochlor*ite*	hypochlor*ous* acid

Table 5.6 ▸ Names of Some Oxygen-Containing Acids

Acid	Name
HNO_3	nitric acid
HNO_2	nitrous acid
H_2SO_4	sulfuric acid
H_2SO_3	sulfurous acid
H_3PO_4	phosphoric acid
$HC_2H_3O_2$	acetic acid

The rules for naming acids are given in schematic form in Fig. 5.3. The names of the most important acids are given in Tables 5.5 and 5.6. These should be memorized.

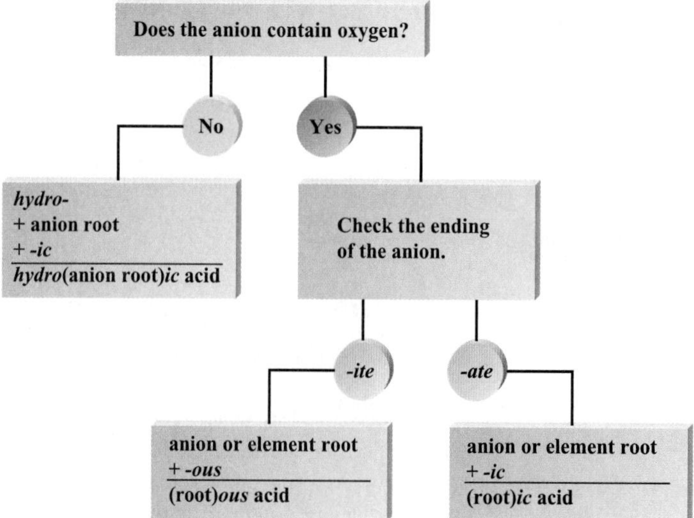

Figure 5.3 ▸ A flowchart for naming acids. An acid is best considered as one or more H^+ ions attached to an anion.

5-7 Writing Formulas from Names

OBJECTIVE **To learn to write the formula of a compound, given its name.**

So far we have started with the chemical formula of a compound and decided on its systematic name. Being able to reverse the process is also important. Often a laboratory procedure describes a compound by name, but the label on the bottle in the lab shows only the formula of the chemical it contains. It is essential that you are able to get the formula of a compound from its name. In fact, you already know enough about compounds to do this. For example, given the name calcium hydroxide, you can write the formula as $Ca(OH)_2$ because you know that calcium forms only Ca^{2+} ions and that, since hydroxide is OH^-, two of these anions are required to give a neutral compound. Similarly, the name iron(II) oxide implies the formula FeO, because the Roman numeral II indicates the presence of the cation Fe^{2+} and the oxide ion is O^{2-}.

We emphasize at this point that it is essential to learn the name, composition, and charge of each of the common polyatomic anions (and the NH_4^+ cation). If you do not recognize these ions by formula and by name, you will not be able to write the compound's name given its formula or the compound's formula given its name. You must also learn the names of the common acids.

Critical Thinking

In this chapter, you have learned a systematic way to name chemical compounds. What if all compounds had only common names? What problems would this cause?

Interactive Example 5.9

Writing Formulas from Names

Give the formula for each of the following compounds.

a. potassium hydroxide
b. sodium carbonate
c. nitric acid
d. cobalt(III) nitrate
e. calcium chloride
f. lead(IV) oxide
g. dinitrogen pentoxide
h. ammonium perchlorate

SOLUTION

Name	Formula	Comments
a. potassium hydroxide	KOH	Contains K^+ and OH^-.
b. sodium carbonate	Na_2CO_3	We need two Na^+ to balance CO_3^{2-}.
c. nitric acid	HNO_3	Common strong acid; memorize.
d. cobalt(III) nitrate	$Co(NO_3)_3$	Cobalt(III) means Co^{3+}; we need three NO_3^- to balance Co^{3+}.
e. calcium chloride	$CaCl_2$	We need two Cl^- to balance Ca^{2+}; Ca (Group 2) always forms Ca^{2+}.
f. lead(IV) oxide	PbO_2	Lead(IV) means Pb^{4+}; we need two O^{2-} to balance Pb^{4+}.
g. dinitrogen pentoxide	N_2O_5	*Di-* means two; *pent(a)-* means five.
h. ammonium perchlorate	NH_4ClO_4	Contains NH_4^+ and ClO_4^-.

SELF CHECK

Exercise 5.8 Write the formula for each of the following compounds.

a. ammonium sulfate
b. vanadium(V) fluoride
c. disulfur dichloride
d. rubidium peroxide
e. aluminum oxide

See Problems 5.41 through 5.46. ■

CHAPTER 5 REVIEW

F directs you to the *Chemistry in Focus* feature in the chapter

Key Terms

binary compound (5-1)
binary ionic compound (5-2)
polyatomic ion (5-5)
oxyanion (5-5)
acid (5-6)

For Review

- Binary compounds are named by following a set of rules.
  - For compounds containing both a metal and a nonmetal, the metal is always named first. The nonmetal is named from the root element name.
  - If the metal ion can have more than one charge (Type II), a Roman numeral is used to specify the charge.
  - For binary compounds containing only nonmetals (Type III), prefixes are used to specify the numbers of atoms present.

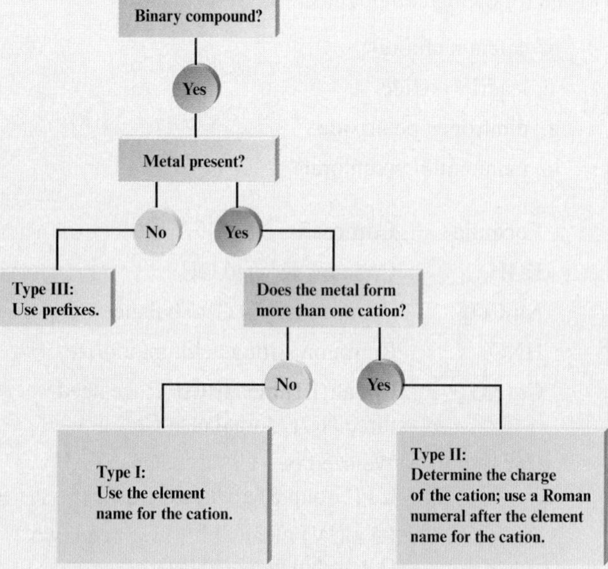

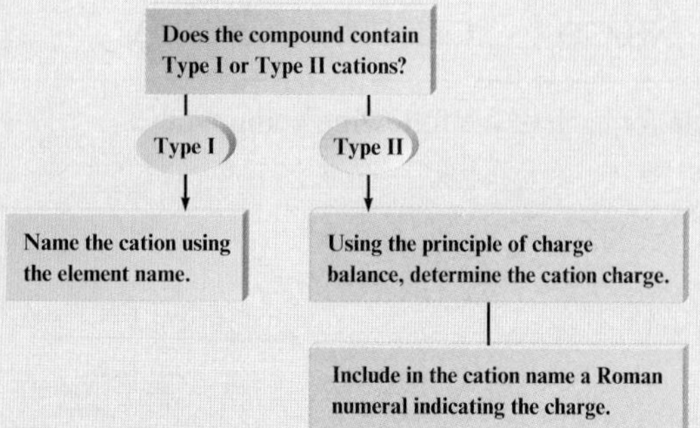

- Polyatomic ions are charged entities composed of several atoms bound together. They have special names and must be memorized.
- Naming ionic compounds containing polyatomic ions follows rules similar to those for naming binary compounds.
- The names of acids (molecules with one or more H^+ ions attached to an anion) depend on whether the acid contains oxygen.

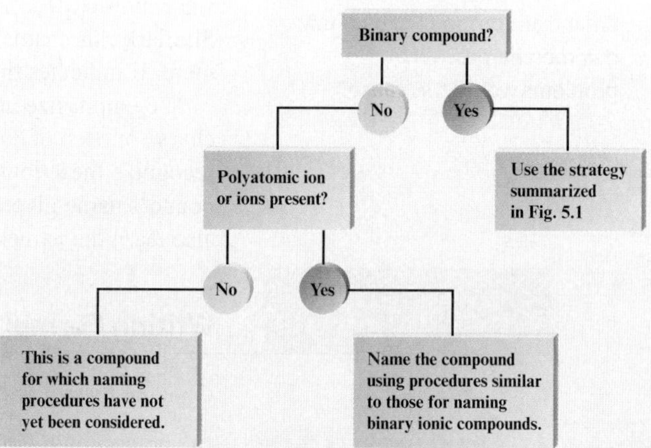

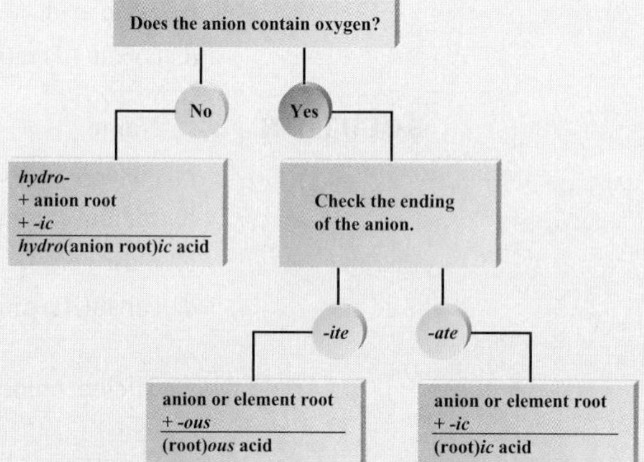

Active Learning Questions

These questions are designed to be considered by groups of students in class. Often these questions work well for introducing a particular topic in class.

1. In some cases the Roman numeral in a name is the same as a subscript in the formula, and in some cases it is not. Provide an example (formula and name) for each of these cases. Explain why the Roman numeral is not necessarily the same as the subscript.
2. The formulas $CaCl_2$ and $CoCl_2$ look very similar. What is the name for each compound? Why do we name them differently?
3. The formulas MgO and CO look very similar. What is the name for each compound? Why do we name them differently?
4. Explain how to use the periodic table to determine that there are two chloride ions for every magnesium ion in magnesium chloride and one chloride ion for every sodium ion in sodium chloride. Then write the formulas for calcium oxide and potassium oxide and explain how you got them.
5. What is the general formula for an ionic compound formed by elements in the following groups? Explain your reasoning and provide an example for each (name and formula).
   a. Group 1 with group 7
   b. Group 2 with group 7
   c. Group 1 with group 6
   d. Group 2 with group 6
6. An element forms an ionic compound with chlorine, leading to a compound having the formula XCl_2. The ion of element X has mass number 89 and 36 electrons. Identify the element X, tell how many neutrons it has, and name the compound.
7. Name each of the following compounds.

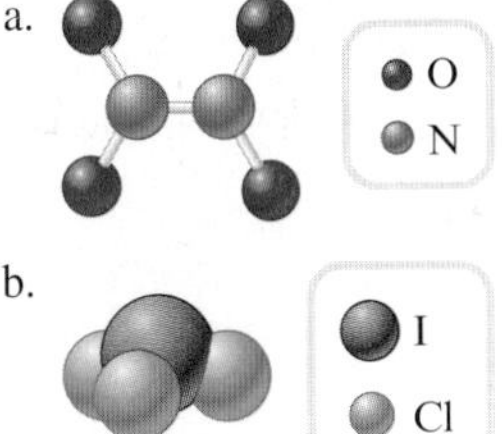

   c. SO_5
   d. P_2S_5
8. Why do we call $Ba(NO_3)_2$ barium nitrate but call $Fe(NO_3)_2$ iron(II) nitrate?
9. What is the difference between sulfuric acid and hydrosulfuric acid?

Questions and Problems

5-1 Naming Compounds

Questions

F 1. The "Chemistry in Focus" segment *Sugar of Lead* discusses $Pb(C_2H_3O_2)_2$, which originally was known as sugar of lead.
   a. Why was it called sugar of lead?
   b. What is the systematic name for $Pb(C_2H_3O_2)_2$?
   c. Why is it necessary to have a *system* for the naming of chemical compounds?

2. What is a *binary* chemical compound? What are the two major *types* of binary chemical compounds? Give three examples of each type of binary compound.

5-2 Naming Binary Compounds That Contain a Metal and a Nonmetal (Types I and II)

Questions

3. Cations are ______ ions, and anions are ______ ions.
4. In naming ionic compounds, we always name the ______ first.
5. In a simple binary ionic compound, which ion (cation/anion) has the same name as its parent element?
6. Although we write the formula of sodium chloride as NaCl, we realize that NaCl is an ionic compound and contains no molecules. Explain.
7. For a metallic element that forms two stable cations, the ending ______ is used to indicate the cation of lower charge and the ending ______ is used to indicate the cation of higher charge.
8. We indicate the charge of a metallic element that forms more than one cation by adding a ______ after the name of the cation.
9. Give the name of each of the following simple binary ionic compounds.
   a. NaBr
   b. $MgCl_2$
   c. AlP
   d. $SrBr_2$
   e. AgI
   f. K_2S
10. Give the name of each of the following simple binary ionic compounds.
   a. LiI
   b. MgF_2
   c. SrO
   d. $AlBr_3$
   e. CaS
   f. Na_2O
11. In each of the following, identify which names are incorrect for the given formulas, and give the correct name.
   a. CaH_2, calcium hydride
   b. $PbCl_2$, lead(IV) chloride
   c. CrI_3, chromium(III) iodide
   d. Na_2S, disodium sulfide
   e. $CuBr_2$, cupric bromide
12. In each of the following, identify which names are incorrect for the given formulas, and give the correct name.
   a. $CuCl_2$, copper(II) chloride
   b. Ag_2O, silver(II) oxide
   c. Li_2O, lithium(I) oxide
   d. CaS, calcium sulfide
   e. Cs_2S, cesium(II) sulfide
13. Write the name of each of the following ionic substances, using the system that includes a Roman numeral to specify the charge of the cation.
   a. $SnCl_4$
   b. Fe_2S_3
   c. PbO_2
   d. Cr_2S_3
   e. CuO
   f. Cu_2O

All even-numbered Questions and Problems have answers in the back of this book and solutions in the *Student Solutions Guide*.

14. Write the name of each of the following ionic substances, using the system that includes a Roman numeral to specify the charge of the cation.

a. FeI_3
b. $MnCl_2$
c. HgO
d. Cu_2S
e. CoO
f. $SnBr_4$

15. Write the name of each of the following ionic substances, using *-ous* or *-ic* endings to indicate the charge of the cation.

a. $CuCl$
b. Fe_2O_3
c. Hg_2Cl_2
d. $MnCl_2$
e. TiO_2
f. PbO

16. Write the name of each of the following ionic substances, using *-ous* or *-ic* endings to indicate the charge of the cation.

a. $CoCl_2$
b. $CrBr_3$
c. PbO
d. SnO_2
e. Co_2O_3
f. $FeCl_3$

5-3 Naming Binary Compounds That Contain Only Nonmetals (Type III)

Questions

17. Write the name of each of the following binary compounds of nonmetallic elements.

a. KrF_2
b. Se_2S_6
c. AsH_3
d. XeO_4
e. BrF_3
f. P_2S_5

18. Write the name for each of the following binary compounds of nonmetallic elements.

a. GeH_4
b. N_2Br_4
c. P_2O_5
d. CO_2
e. NH_3
f. SiO_2

5-4 Naming Binary Compounds: A Review

Questions

19. Name each of the following binary compounds, using the periodic table to determine whether the compound is likely to be ionic (containing a metal and a nonmetal) or nonionic (containing only nonmetals).

a. Fe_3P_2
b. $CaBr_2$
c. N_2O_5
d. $PbCl_4$
e. S_2F_{10}
f. Cu_2O

20. The formulas Na_2O and N_2O look very similar. What is the name for each compound? Why do we use a different naming convention between the two compounds?

21. Name each of the following binary compounds, using the periodic table to determine whether the compound is likely to be ionic (containing a metal and a nonmetal) or nonionic (containing only nonmetals).

a. MgS
b. $AlCl_3$
c. PH_3
d. $ClBr$
e. Li_2O
f. P_4O_{10}

22. Name each of the following binary compounds, using the periodic table to determine whether the compound is likely to be ionic (containing a metal or a nonmetal) or nonionic (containing only nonmetals).

a. $RaCl_2$
b. $SeCl_2$
c. PCl_3
d. Na_3P
e. MnF_2
f. ZnO

5-5 Naming Compounds That Contain Polyatomic Ions

Questions

23. What is a *polyatomic* ion? Give examples of five common polyatomic ions.

24. What is an *oxyanion?* List the series of oxyanions that chlorine and bromine form and give their names.

25. For the oxyanions of sulfur, the ending *-ite* is used for SO_3^{2-} to indicate that it contains ________ than does SO_4^{2-}.

26. In naming oxyanions, when there are more than two members in the series for a given element, what prefixes are used to indicate the oxyanions in the series with the *fewest* and the *most* oxygen atoms?

27. Complete the following list by filling in the missing names or formulas of the oxyanions of chlorine.

ClO_4^-	________
________	hypochlorite
ClO_3^-	________
________	chlorite

28. A series of oxyanions of iodine, comparable to the series for chlorine discussed in the text, also exists. Write the formulas and names for the oxyanions of iodine.

29. Write the formula for each of the following phosphorus-containing ions, including the overall charge of the ion.

a. phosphide
b. phosphate
c. phosphite
d. hydrogen phosphate

30. Write the formula for each of the following chlorine-containing ions, including the overall charge of the ion.

a. chloride
b. hypochlorite
c. chlorate
d. perchlorate

31. Write the formulas of the compounds below (refer to your answers to Problem 30).

a. magnesium chloride
b. calcium hypochlorite
c. potassium chlorate
d. barium perchlorate

32. Carbon occurs in several common polyatomic anions. List the formulas of as many such anions as you can, along with the names of the anions.

All even-numbered Questions and Problems have answers in the back of this book and solutions in the *Student Solutions Guide*.

33. Give the name of each of the following polyatomic ions.

a. HCO_3^-
b. $C_2H_3O_2^-$
c. CN^-
d. OH^-
e. NO_2^-
f. HPO_4^{2-}

34. Give the name of each of the following polyatomic ions.

a. NH_4^+
b. $H_2PO_4^-$
c. SO_4^{2-}
d. HSO_3^-
e. ClO_4^-
f. IO_3^-

35. Name each of the following compounds, which contain polyatomic ions.

a. NH_4NO_3
b. $Ca(HCO_3)_2$
c. $MgSO_4$
d. Na_2HPO_4
e. $KClO_4$
f. $Ba(C_2H_3O_2)_2$

36. Name each of the following compounds, which contain polyatomic ions.

a. $NaMnO_4$
b. $AlPO_4$
c. $CrCO_3$
d. $Ca(ClO)_2$
e. $BaCO_3$
f. $CaCrO_4$

5-6 Naming Acids

Questions

37. Give a simple definition of an *acid.*

38. Many acids contain the element ________ in addition to hydrogen.

39. Name each of the following acids.

a. HCl
b. H_2SO_4
c. HNO_3
d. HI
e. HNO_2
f. $HClO_3$
g. HBr
h. HF
i. $HC_2H_3O_2$

40. Name each of the following acids.

a. HOCl
b. H_2SO_3
c. $HBrO_3$
d. HOI
e. $HBrO_4$
f. H_2S
g. H_2Se
h. H_3PO_3

5-7 Writing Formulas from Names

Problems

41. Write the formula for each of the following simple binary ionic compounds.

a. cobalt(II) chloride
b. cobaltic chloride
c. sodium phosphide
d. iron(II) oxide
e. calcium hydride
f. manganese(IV) oxide
g. magnesium iodide
h. copper(I) sulfide

42. Write the formula for each of the following simple binary ionic compounds.

a. magnesium fluoride
b. ferric iodide
c. mercuric sulfide
d. barium nitride
e. plumbous chloride
f. stannic fluoride
g. silver oxide
h. potassium selenide

43. Write the formula for each of the following binary compounds of nonmetallic elements.

a. carbon disulfide
b. water
c. dinitrogen trioxide
d. dichlorine heptoxide
e. carbon dioxide
f. ammonia
g. xenon tetrafluoride

44. Write the formula for each of the following binary compounds of nonmetallic elements.

a. dinitrogen oxide
b. nitrogen dioxide
c. dinitrogen tetraoxide (tetroxide)
d. sulfur hexafluoride
e. phosphorus tribromide
f. carbon tetraiodide
g. oxygen dichloride

45. Write the formula for each of the following compounds that contain polyatomic ions. Be sure to enclose the polyatomic ion in parentheses if more than one such ion is needed to balance the oppositely charged ion(s).

a. ammonium nitrate
b. magnesium acetate
c. calcium peroxide
d. potassium hydrogen sulfate
e. iron(II) sulfate
f. potassium hydrogen carbonate
g. cobalt(II) sulfate
h. lithium perchlorate

46. Write the formula for each of the following compounds that contain polyatomic ions. Be sure to enclose the polyatomic ion in parentheses if more than one such ion is needed to balance the oppositely charged ions.

a. ammonium acetate
b. ferrous hydroxide
c. cobalt(III) carbonate
d. barium dichromate
e. lead(II) sulfate
f. potassium dihydrogen phosphate
g. lithium peroxide
h. zinc chlorate

All even-numbered Questions and Problems have answers in the back of this book and solutions in the *Student Solutions Guide.*

47. Write the formula for each of the following acids.
   a. hydrosulfuric acid
   b. perbromic acid
   c. acetic acid
   d. hydrobromic acid
   e. chlorous acid
   f. hydroselenic acid
   g. sulfurous acid
   h. perchloric acid
48. Write the formula for each of the following acids.
   a. hydrocyanic acid
   b. nitric acid
   c. sulfuric acid
   d. phosphoric acid
   e. hypochlorous acid
   f. hydrobromic acid
   g. bromous acid
   h. hydrofluoric acid
49. Write the formula for each of the following substances.
   a. sodium peroxide
   b. calcium chlorate
   c. rubidium hydroxide
   d. zinc nitrate
   e. ammonium dichromate
   f. hydrosulfuric acid
   g. calcium bromide
   h. hypochlorous acid
   i. potassium sulfate
   j. nitric acid
   k. barium acetate
   l. lithium sulfite
50. Write the formula for each of the following substances.
   a. calcium hydrogen sulfate
   b. zinc phosphate
   c. iron(III) perchlorate
   d. cobaltic hydroxide
   e. potassium chromate
   f. aluminum dihydrogen phosphate
   g. lithium bicarbonate
   h. manganese(II) acetate
   i. magnesium hydrogen phosphate
   j. cesium chlorite
   k. barium peroxide
   l. nickelous carbonate

Additional Problems

51. Iron forms both 2+ and 3+ cations. Write formulas for the oxide, sulfide, and chloride compound of each iron cation, and give the name of each compound in both the nomenclature method that uses Roman numerals to specify the charge of the cation and the *-ous*/*-ic* notation.
52. Before an electrocardiogram (ECG) is recorded for a cardiac patient, the ECG leads are usually coated with a moist paste containing sodium chloride. What property of an ionic substance such as NaCl is being made use of here?
53. Nitrogen and oxygen form numerous binary compounds, including NO, NO_2, N_2O_4, N_2O_5, and N_2O. Give the name of each of these oxides of nitrogen.
54. On some periodic tables, hydrogen is listed both as a member of Group 1 and as a member of Group 7. Write an equation showing the formation of H^+ ion and an equation showing the formation of H^- ion.
55. Examine the following table of formulas and names. Which of the compounds is/are named correctly?

	Formula	Name
a.	$Fe_3(PO_4)_2$	iron(III) phosphate
b.	K_3N	potassium nitride
c.	MnO_2	manganese(II) oxide
d.	SiO_2	monosilicon dioxide

56. Complete the following list by filling in the missing oxyanion or oxyacid for each pair.

ClO_4^-	________
________	HIO_3
ClO^-	________
BrO_2^-	________
________	$HClO_2$

57. Name the following compounds.
   a. $Ca(C_2H_3O_2)_2$
   b. PCl_3
   c. $Cu(MnO_4)_2$
   d. $Fe_2(CO_3)_3$
   e. $LiHCO_3$
   f. Cr_2S_3
   g. $Ca(CN)_2$
58. Name the following compounds.
   a. $AuBr_3$
   b. $Co(CN)_3$
   c. $MgHPO_4$
   d. B_2H_6
   e. NH_3
   f. Ag_2SO_4
   g. $Be(OH)_2$
59. Name the following compounds.
   a. $HClO_3$
   b. $CoCl_3$
   c. B_2O_3
   d. H_2O
   e. $HC_2H_3O_2$
   f. $Fe(NO_3)_3$
   g. $CuSO_4$
60. A compound has the general formula X_2O, with X representing an unknown element or ion and O representing oxygen. Which of the following could not be a name for this compound?
   a. sodium oxide
   b. iron(II) oxide
   c. copper(I) oxide
   d. dinitrogen monoxide
   e. water
61. Most metallic elements form *oxides*, and often the oxide is the most common compound of the element that is found in the earth's crust. Write the formulas for the oxides of the following metallic elements.
   a. potassium
   b. magnesium
   c. iron(II)
   d. iron(III)
   e. zinc(II)
   f. lead(II)
   g. aluminum
62. Consider a hypothetical simple ion M^{2+}. Determine the formula of the compound this ion would form with each of the following anions.
   a. acetate
   b. permanganate
   c. oxide
   d. hydrogen phosphate
   e. hydroxide
   f. nitrite

All even-numbered Questions and Problems have answers in the back of this book and solutions in the *Student Solutions Guide*.

63. Consider a hypothetical element M, which is capable of forming stable simple cations that have charges of 1+, 2+, and 3+, respectively. Write the formulas of the compounds formed by the various M cations with each of the following anions.

a. chromate
b. dichromate
c. sulfide
d. bromide
e. bicarbonate
f. hydrogen phosphate

64. A metal ion with a 2+ charge has 23 electrons and forms a compound with a halogen ion that contains 17 protons.

a. What is the identity of the metal ion?
b. What is the identity of the halogen ion and how many electrons does it contain?
c. Determine the compound that it forms and name it.

65. Complete Table 5.A (below) by writing the names and formulas for the ionic compounds formed when the cations listed across the top combine with the anions shown in the left-hand column.

66. Complete Table 5.B (below) by writing the formulas for the ionic compounds formed when the anions listed across the top combine with the cations shown in the left-hand column.

67. The noble metals gold, silver, and platinum are often used in fashioning jewelry because they are relatively ________.

68. The formula for ammonium phosphate is ________.

69. The elements of Group 7 (fluorine, chlorine, bromine, and iodine) consist of molecules containing ________ atom(s).

70. Under what physical state at room temperature do each of the halogen elements exist?

71. When an atom gains two electrons, the ion formed has a charge of ________.

72. An ion with one less electron than it has protons has a ________ charge.

73. An atom that has lost three electrons will have a charge of ________.

74. An ion with two more electrons than it has protons has a ________ charge.

75. For each of the negative ions listed in column 1, use the periodic table to find in column 2 the total number of electrons the ion contains. A given answer may be used more than once.

Column 1	Column 2
[1] Se^{2-}	[a] 18
[2] S^{2-}	[b] 35
[3] P^{3-}	[c] 52
[4] O^{2-}	[d] 34
[5] N^{3-}	[e] 36
[6] I^-	[f] 54
[7] F^-	[g] 10
[8] Cl^-	[h] 9
[9] Br^-	[i] 53
[10] At^-	[j] 86

Table 5.A

Ions	Fe^{2+}	Al^{3+}	Na^+	Ca^{2+}	NH_4^+	Fe^{3+}	Ni^{2+}	Hg_2^{2+}	Hg^{2+}
CO_3^{2-}	____	____	____	____	____	____	____	____	____
BrO_3^-	____	____	____	____	____	____	____	____	____
$C_2H_3O_2^-$	____	____	____	____	____	____	____	____	____
OH^-	____	____	____	____	____	____	____	____	____
HCO_3^-	____	____	____	____	____	____	____	____	____
PO_4^{3-}	____	____	____	____	____	____	____	____	____
SO_3^{2-}	____	____	____	____	____	____	____	____	____
ClO_4^-	____	____	____	____	____	____	____	____	____
SO_4^{2-}	____	____	____	____	____	____	____	____	____
O^{2-}	____	____	____	____	____	____	____	____	____
Cl^-	____	____	____	____	____	____	____	____	____

Table 5.B

Ions	nitrate	sulfate	hydrogen sulfate	dihydrogen phosphate	oxide	chloride
calcium	____	____	____	____	____	____
strontium	____	____	____	____	____	____
ammonium	____	____	____	____	____	____
aluminum	____	____	____	____	____	____
iron(III)	____	____	____	____	____	____
nickel(II)	____	____	____	____	____	____
silver(I)	____	____	____	____	____	____
gold(III)	____	____	____	____	____	____
potassium	____	____	____	____	____	____
mercury(II)	____	____	____	____	____	____
barium	____	____	____	____	____	____

All even-numbered Questions and Problems have answers in the back of this book and solutions in the *Student Solutions Guide*.

76. For each of the following processes that show the formation of ions, complete the process by indicating the number of electrons that must be gained or lost to form the ion. Indicate the total number of electrons in the ion, and in the atom from which it was made.

a. $Al \rightarrow Al^{3+}$
b. $S \rightarrow S^{2-}$
c. $Cu \rightarrow Cu^{+}$
d. $F \rightarrow F^{-}$
e. $Zn \rightarrow Zn^{2+}$
f. $P \rightarrow P^{3-}$

77. For each of the following atomic numbers, use the periodic table to write the formula (including the charge) for the simple *ion* that the element is most likely to form.

a. 36 d. 81
b. 31 e. 35
c. 52 f. 87

78. For the following pairs of ions, use the principle of electrical neutrality to predict the formula of the binary compound that the ions are most likely to form.

a. Na^{+} and S^{2-} e. Cu^{2+} and Br^{-}
b. K^{+} and Cl^{-} f. Al^{3+} and I^{-}
c. Ba^{2+} and O^{2-} g. Al^{3+} and O^{2-}
d. Mg^{2+} and Se^{2-} h. Ca^{2+} and N^{3-}

79. Give the name of each of the following simple binary ionic compounds.

a. BeO e. HCl
b. MgI_2 f. LiF
c. Na_2S g. Ag_2S
d. Al_2O_3 h. CaH_2

80. In which of the following pairs is the name incorrect? Give the correct name for the formulas indicated.

a. Ag_2O, disilver monoxide
b. N_2O, dinitrogen monoxide
c. Fe_2O_3, iron(II) oxide
d. PbO_2, plumbous oxide
e. $Cr_2(SO_4)_3$, chromium(III) sulfate

81. Write the name of each of the following ionic substances, using the system that includes a Roman numeral to specify the charge of the cation.

a. $FeBr_2$ d. SnO_2
b. CoS e. Hg_2Cl_2
c. Co_2S_3 f. $HgCl_2$

82. Write the name of each of the following ionic substances, using *-ous* or *-ic* endings to indicate the charge of the cation.

a. $SnCl_2$ d. PbS
b. FeO e. Co_2S_3
c. SnO_2 f. $CrCl_2$

83. Name each of the following binary compounds.

a. XeF_6 d. N_2O_4
b. OF_2 e. Cl_2O
c. AsI_3 f. SF_6

84. Name each of the following compounds.

a. $Fe(C_2H_3O_2)_3$ d. $SiBr_4$
b. BrF e. $Cu(MnO_4)_2$
c. K_2O_2 f. $CaCrO_4$

85. Which oxyanion of nitrogen contains a larger number of oxygen atoms, the nitr*ate* ion or the nitr*ite* ion?

86. Examine the following table of formulas and names. Which of the compounds is/are named correctly?

	Formula	Name
a.	P_2O_5	diphosphorus pentoxide
b.	ClO_2	chlorine oxide
c.	PbI_4	lead iodide
d.	$CuSO_4$	copper(I) sulfate

87. Write the formula for each of the following chromium-containing ions, including the overall charge of the ion.

a. chromous c. chromic
b. chromate d. dichromate

88. Give the name of each of the following polyatomic anions.

a. CO_3^{2-} d. PO_4^{3-}
b. ClO_3^{-} e. ClO_4^{-}
c. SO_4^{2-} f. MnO_4^{-}

89. Name each of the following compounds, which contain polyatomic ions.

a. LiH_2PO_4 d. Na_2HPO_4
b. $Cu(CN)_2$ e. $NaClO_2$
c. $Pb(NO_3)_2$ f. $Co_2(SO_4)_3$

90. A member of the alkaline earth metal family whose most stable ion contains 36 electrons forms a compound with bromine. What is the correct formula for this compound?

91. Write the formula for each of the following binary compounds of nonmetallic elements.

a. sulfur dioxide
b. dinitrogen monoxide
c. xenon tetrafluoride
d. tetraphosphorus decoxide
e. phosphorus pentachloride
f. sulfur hexafluoride
g. nitrogen dioxide

92. Write the formula of each of the following ionic substances.

a. sodium dihydrogen phosphate
b. lithium perchlorate
c. copper(II) hydrogen carbonate
d. potassium acetate
e. barium peroxide
f. cesium sulfite

93. Write the formula for each of the following compounds, which contain polyatomic ions. Be sure to enclose the polyatomic ion in parentheses if more than one such ion is needed to balance the oppositely charged ion(s).

a. silver(I) perchlorate (usually called silver perchlorate)
b. cobalt(III) hydroxide
c. sodium hypochlorite
d. potassium dichromate
e. ammonium nitrite
f. ferric hydroxide
g. ammonium hydrogen carbonate
h. potassium perbromate

All even-numbered Questions and Problems have answers in the back of this book and solutions in the *Student Solutions Guide*.

ChemWork Problems

These multiconcept problems (and additional ones) are found interactively online with the same type of assistance a student would get from an instructor.

94. Complete the following table to predict whether the given atom will gain or lose electrons in forming the ion most likely to form when in ionic compounds.

Atom	Gain (G) or Lose (L) Electrons	Ion Formed
K	______	______
Cs	______	______
Br	______	______
S	______	______
Se	______	______

95. What are the formulas of the compounds that correspond to the names given in the following table?

Compound Name	Formula
Carbon tetrabromide	______
Cobalt(II) phosphate	______
Magnesium chloride	______
Nickel(II) acetate	______
Calcium nitrate	______

96. What are the names of the compounds that correspond to the formulas given in the following table?

Formula	Compound Name
$Co(NO_2)_2$	______
AsF_5	______
LiCN	______
K_2SO_3	______
Li_3N	______
$PbCrO_4$	______

97. Provide the name of the acid that corresponds to the formula given in the following table.

Formula	Acid Name
H_2SO_3	______
$HC_2H_3O_2$	______
$HClO_4$	______
HOCl	______
HCN	______

98. Which of the following statements is(are) correct?

a. The symbols for the elements magnesium, aluminum, and xenon are Mn, Al, and Xe, respectively.
b. The elements P, As, and Bi are in the same family on the periodic table.
c. All of the following elements are expected to gain electrons to form ions in ionic compounds: Ga, Se, and Br.
d. The elements Co, Ni, and Hg are all transition elements.
e. The correct name for TiO_2 is titanium dioxide.

All even-numbered Questions and Problems have answers in the back of this book and solutions in the *Student Solutions Guide.*

Questions

1. What is an element? Which elements are most abundant on the earth? Which elements are most abundant in the human body?
2. Without consulting any reference, write the name and symbol for as many elements as you can. How many could you name? How many symbols did you write correctly?
3. The symbols for the elements silver (Ag), gold (Au), and tungsten (W) seem to bear no relation to their English names. Explain and give three additional examples.
4. Without consulting your textbook or notes, state as many points as you can of Dalton's atomic theory. Explain in your own words each point of the theory.
5. What is a compound? What is meant by the *law of constant composition* for compounds and why is this law so important to our study of chemistry?
6. What is meant by a *nuclear atom?* Describe the points of Rutherford's model for the nuclear atom and how he tested this model. Based on his experiments, how did Rutherford envision the structure of the atom? How did Rutherford's model of the atom's structure differ from Kelvin's "plum pudding" model?
7. Consider the neutron, the proton, and the electron.
   a. Which is(are) found in the nucleus?
   b. Which has the largest relative mass?
   c. Which has the smallest relative mass?
   d. Which is negatively charged?
   e. Which is electrically neutral?
8. What are *isotopes?* To what do the *atomic number* and the *mass number* of an isotope refer? How are specific isotopes indicated symbolically (give an example and explain)? Do the isotopes of a given element have the same chemical and physical properties? Explain.
9. Complete the following table by giving the symbol, name, atomic number, and/or group(family) number as required.

Symbol	Name	Atomic Number	Group Number
Ca	______	______	______
I	______	______	______
______	cesium	______	______
______	______	16	______
______	arsenic	______	______
Sr	______	______	______
______	______	14	______
Rn	______	______	______
______	radium	______	______
Se	______	______	______

10. Are most elements found in nature in the elemental or the combined form? Why? Name several elements that are usually found in the elemental form.
11. What are *ions?* How are ions formed from atoms? Do isolated atoms form ions spontaneously? To what do the terms *cation* and *anion* refer? In terms of subatomic particles, how is an ion related to the atom from which it is formed? Does the nucleus of an atom change when the atom is converted into an ion? How can the periodic table be used to predict what ion an element's atoms will form?
12. What are some general physical properties of ionic compounds such as sodium chloride? How do we know that substances such as sodium chloride consist of positively and negatively charged particles? Since ionic compounds are made up of electrically charged particles, why doesn't such a compound have an overall electric charge? Can an ionic compound consist only of cations or anions (but not both)? Why not?
13. What principle do we use in writing the formula of an ionic compound such as NaCl or MgI_2? How do we know that *two* iodide ions are needed for each magnesium ion, whereas only one chloride ion is needed per sodium ion?
14. When writing the name of an ionic compound, which is named first, the anion or the cation? Give an example. What ending is added to the root name of an element to show that it is a simple anion in a Type I ionic compound? Give an example. What *two* systems are used to show the charge of the cation in a Type II ionic compound? Give examples of each system for the same compound. What general type of element is involved in Type II compounds?
15. Describe the system used to name Type III binary compounds (compounds of nonmetallic elements). Give several examples illustrating the method. How does this system differ from that used for ionic compounds? How is the system for Type III compounds similar to those for ionic compounds?
16. What is a *polyatomic* ion? Without consulting a reference, list the formulas and names of at least ten polyatomic ions. When writing the overall formula of an ionic compound involving polyatomic ions, why are parentheses used around the formula of a polyatomic ion when more than one such ion is present? Give an example.
17. What is an *oxyanion?* What special system is used in a series of related oxyanions that indicates the relative number of oxygen atoms in each ion? Give examples.
18. What is an *acid?* How are acids that do *not* contain oxygen named? Give several examples. Describe the naming system for the oxyacids. Give examples of a series of oxyacids illustrating this system.

All even-numbered Questions and Problems have answers in the back of this book and solutions in the *Student Solutions Guide.*

Problems

19. Complete the following table by giving the symbol, name, atomic number, and/or group (family) number as required.

Symbol	Name	Atomic Number	Group Number
Al	_______	_______	_______
_______	radon	_______	_______
_______	sulfur	_______	_______
_______	_______	38	_______
Br	_______	_______	_______
_______	carbon	_______	_______
Ba	_______	_______	_______
_______	_______	88	_______
_______	_______	11	_______
K	_______	_______	_______
_______	germanium	_______	_______
_______	_______	17	_______

20. Your text indicates that the Group 1, Group 2, Group 7, and Group 8 elements all have "family" names (alkali metals, alkaline earth metals, halogens, and noble gases, respectively). Without looking at your textbook, name as many elements in each family as you can. What similarities are there among the members of a family? Why?

21. Using the periodic table shown in Fig. 4.9, for each of the following symbols, write the name of the element and its atomic number.

a. Mg
b. Ga
c. Sn
d. Sb
e. Sr
f. Si
g. Cs
h. Ca
i. Cr
j. Co
k. Cu
l. Ag
m. U
n. As
o. At
p. Ar
q. Zn
r. Mn
s. Se
t. W
u. Ra
v. Rn
w. Ce
x. Zr
y. Al
z. Pd

22. How many electrons, protons, and neutrons are found in isolated atoms having the following atomic symbols?

a. $^{17}_{8}O$
b. $^{235}_{92}U$
c. $^{37}_{17}Cl$
d. $^{3}_{1}H$
e. $^{4}_{2}He$
f. $^{119}_{50}Sn$
g. $^{124}_{54}Xe$
h. $^{64}_{30}Zn$

23. What simple ion does each of the following elements most commonly form?

a. Mg
b. F
c. Ag
d. Al
e. O
f. Ba
g. Na
h. Br
i. K
j. Ca
k. S
l. Li
m. Cl

24. For each of the following simple ions, indicate the number of protons and electrons the ion contains.

a. Mg^{2+}
b. Fe^{2+}
c. Fe^{3+}
d. F^{-}
e. Ni^{2+}
f. Zn^{2+}
g. Co^{3+}
h. N^{3-}
i. S^{2-}
j. Rb^{+}
k. Se^{2-}
l. K^{+}

25. Using the ions indicated in Problem 24, write the formulas and give the names for all possible simple ionic compounds involving these ions.

26. Write the formula for each of the following binary ionic compounds.

a. copper(I) iodide
b. cobaltous chloride
c. silver sulfide
d. mercurous bromide
e. mercuric oxide
f. chromium(III) sulfide
g. plumbic oxide
h. potassium nitride
i. stannous fluoride
j. ferric oxide

27. Which of the following formula–name pairs are incorrect? Explain why for each case.

a. $Ag(NO_3)_2$ silver nitrate
b. Fe_2Cl ferrous chloride
c. NaH_2PO_4 sodium hydrogen phosphate
d. NH_4S ammonium sulfide
e. $KC_2H_3O_2$ potassium acetate
f. $Ca(ClO_4)_2$ calcium perchlorate
g. $K_2Cr_2O_7$ potassium dichromate
h. BaOH barium hydroxide
i. Na_2O_2 sodium peroxide
j. $Ca(CO_3)_2$ calcium carbonate

28. Give the name of each of the following polyatomic ions.

a. NH_4^{+}
b. SO_3^{2-}
c. NO_3^{-}
d. SO_4^{2-}
e. NO_2^{-}
f. CN^{-}
g. OH^{-}
h. ClO_4^{-}
i. ClO^{-}
j. PO_4^{3-}

29. Using the negative polyatomic ions listed in Table 5.4, write formulas for each of their sodium and calcium compounds.

30. Give the name of each of the following compounds.

a. XeO_2
b. ICl_5
c. PCl_3
d. CO
e. OF_2
f. P_2O_5
g. AsI_3
h. SO_3

31. Write formulas for each of the following compounds.

a. mercuric chloride
b. iron(III) oxide
c. sulfurous acid
d. calcium hydride
e. potassium nitrate
f. aluminum fluoride
g. dinitrogen monoxide
h. sulfuric acid
i. potassium nitride
j. nitrogen dioxide
k. silver acetate
l. acetic acid
m. platinum(IV) chloride
n. ammonium sulfide
o. cobalt(III) bromide
p. hydrofluoric acid

All even-numbered Questions and Problems have answers in the back of this book and solutions in the *Student Solutions Guide*.

Chemical Reactions: An Introduction

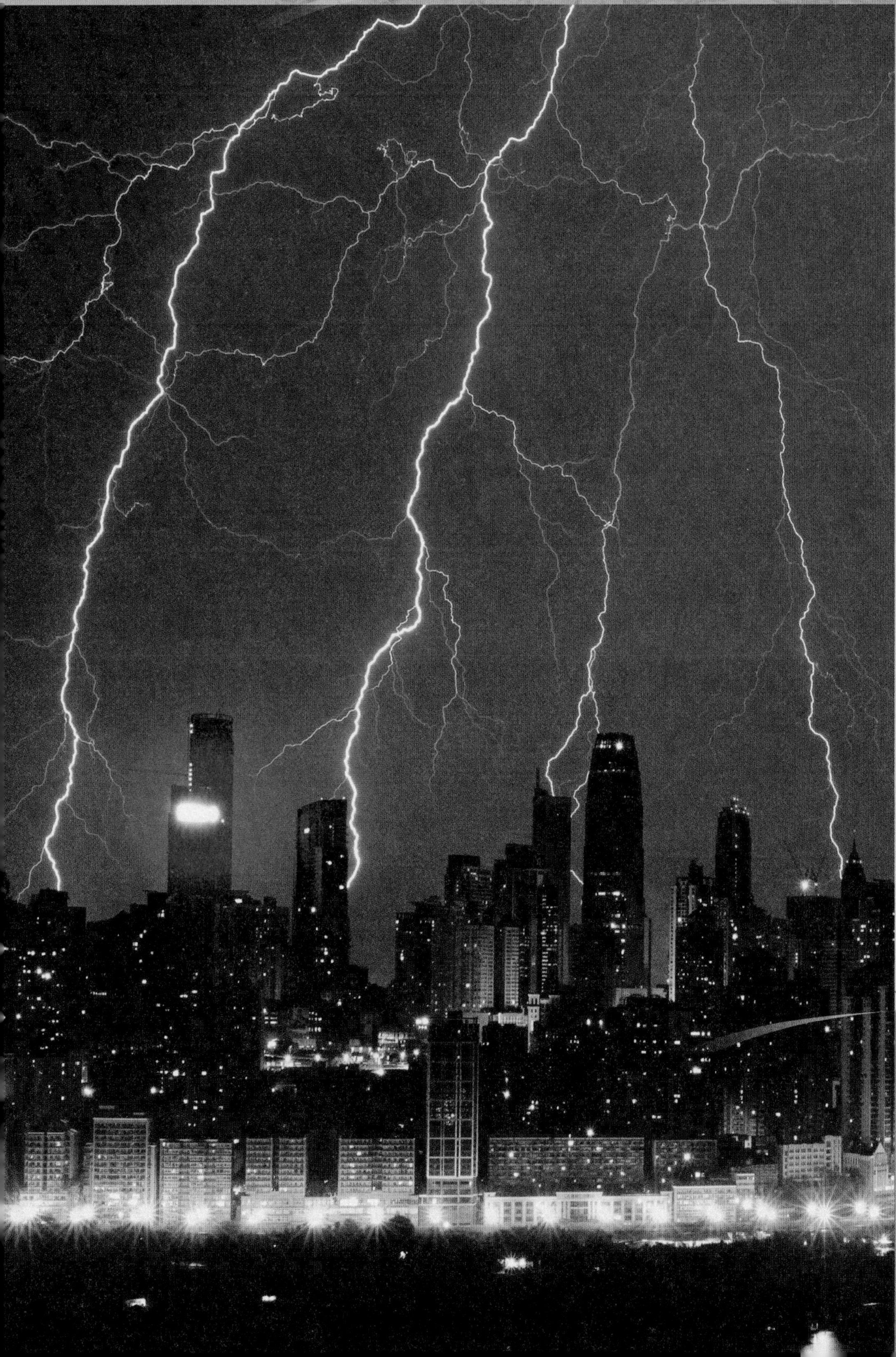

CHAPTER 6

Lightning in Shancheng, Chongqing, China. View Stock/Getty Images

Chemistry is about change. Grass grows. Steel rusts. Hair is bleached, dyed, "permed," or straightened. Natural gas burns to heat houses. Nylon is produced for jackets, swimsuits, and pantyhose. Water is decomposed to hydrogen and oxygen gas by an electric current. Grape juice ferments in the production of wine. The bombardier beetle concocts a toxic spray to shoot at its enemies (see the "Chemistry in Focus" segment *The Beetle That Shoots Straight*).

These are just a few examples of chemical changes that affect each of us. Chemical reactions are the heart and soul of chemistry, and in this chapter we will discuss the fundamental ideas about chemical reactions.

Nylon is a strong material that makes sturdy parasails. © iStockphoto.com/Jose Ignacio Soto

Kim Steele/Photodisc/Getty Images

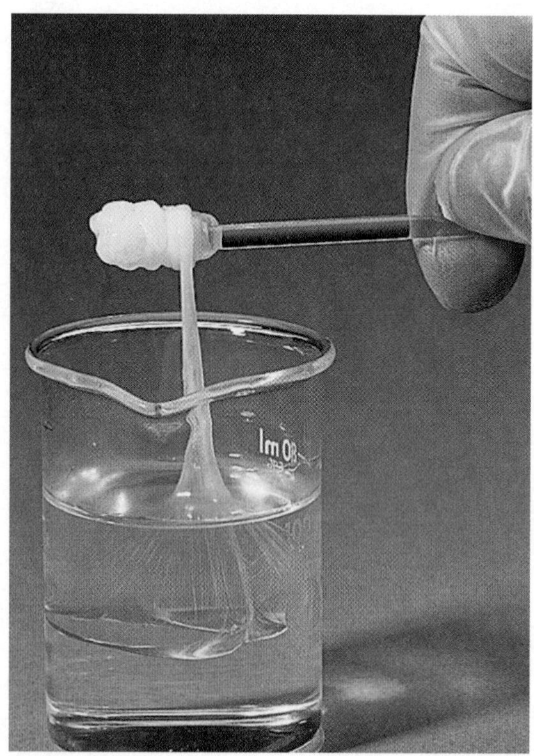

© Cengage Learning

Production of plastic film for use in containers such as soft drink bottles (*left*). Nylon being drawn from the boundary between two solutions containing different reactants (*right*).

6-1 Evidence for a Chemical Reaction

OBJECTIVE **To learn the signals that show a chemical reaction has occurred.**

Energy and chemical reactions will be discussed in more detail in Chapters 7 and 10.

How do we know when a chemical reaction has occurred? That is, what are the clues that a chemical change has taken place? A glance back at the processes in the introduction suggests that *chemical reactions often give a visual signal.* Steel changes from a smooth, shiny material to a reddish-brown, flaky substance when it rusts. Hair changes color when it is bleached. Solid nylon is formed when two particular liquid solutions are brought into contact. A blue flame appears when natural gas reacts with oxygen. Chemical reactions, then, often give *visual* clues: a color changes, a solid forms, bubbles are produced (Fig. 6.1), a flame occurs, and so on. However, reactions are not always visible. Sometimes the only signal that a reaction is occurring is a change in temperature as heat is produced or absorbed (Fig. 6.2).

Table 6.1 summarizes common clues to the occurrence of a chemical reaction, and Fig. 6.3 gives some examples of reactions that show these clues.

Table 6.1 Some Clues That a Chemical Reaction Has Occurred

1. The color changes.
2. A solid forms.
3. Bubbles form.
4. Heat and/or a flame is produced, or heat is absorbed.

Figure 6.1 Bubbles of hydrogen and oxygen gas form when an electric current is used to decompose water.

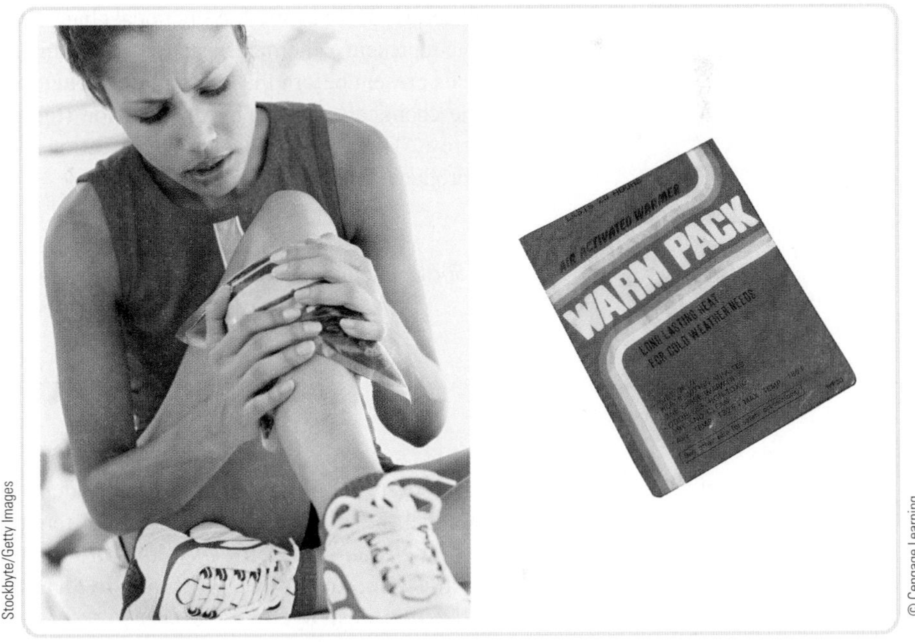

Figure 6.2

a An injured girl using an ice pack to prevent swelling. The pack is activated by breaking an ampule; this initiates a chemical reaction that absorbs heat rapidly, lowering the temperature of the area to which the pack is applied.

b A hot pack used to warm hands and feet in winter. When the package is opened, oxygen from the air penetrates a bag containing solid chemicals. The resulting reaction produces heat for several hours.

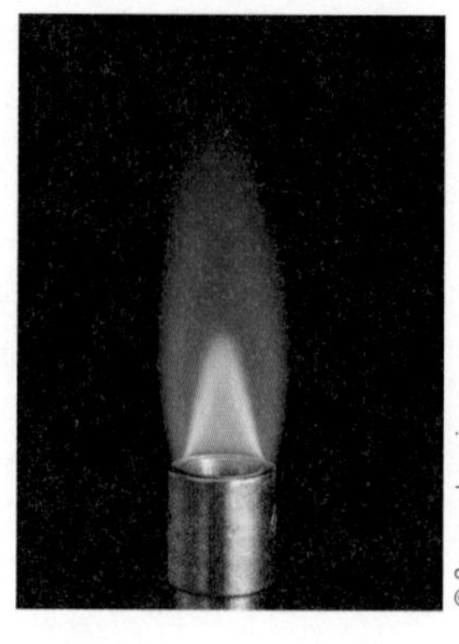

© Cengage Learning

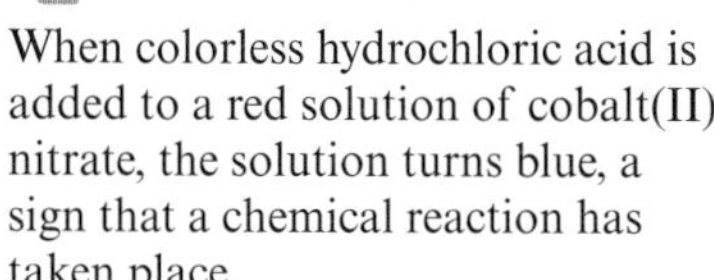

a When colorless hydrochloric acid is added to a red solution of cobalt(II) nitrate, the solution turns blue, a sign that a chemical reaction has taken place.

b A solid forms when a solution of sodium dichromate is added to a solution of lead nitrate.

c Bubbles of hydrogen gas form when calcium metal reacts with water.

d Methane gas reacts with oxygen to produce a flame in a Bunsen burner.

Figure 6.3

6-2 Chemical Equations

OBJECTIVE **To learn to identify the characteristics of a chemical reaction and the information given by a chemical equation.**

Chemists have learned that a chemical change always involves a rearrangement of the ways in which the atoms are grouped. For example, when the methane, CH_4, in natural gas combines with oxygen, O_2, in the air and burns, carbon dioxide, CO_2, and water, H_2O, are formed. A chemical change such as this is called a **chemical reaction.** We represent a chemical reaction by writing a **chemical equation** in which the chemicals present before the reaction (the **reactants**) are shown to the left of an arrow and the chemicals formed by the reaction (the **products**) are shown to the right of an arrow. The arrow indicates the direction of the change and is read as "yields" or "produces":

$$\text{Reactants} \rightarrow \text{Products}$$

For the reaction of methane with oxygen, we have

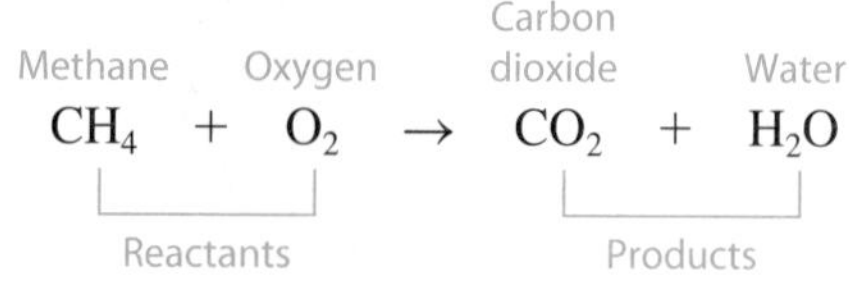

$$CH_4 + O_2 \rightarrow CO_2 + H_2O$$

Note from this equation that the products contain the same atoms as the reactants but that the atoms are associated in different ways. That is, a *chemical reaction involves changing the ways the atoms are grouped.*

It is important to recognize that **in a chemical reaction, atoms are neither created nor destroyed.** *All atoms present in the reactants must be accounted for among the products.* In other words, there must be the same number of each type of atom on the product side as on the reactant side of the arrow. Making sure that the equation for a reaction obeys this rule is called **balancing the chemical equation** for a reaction.

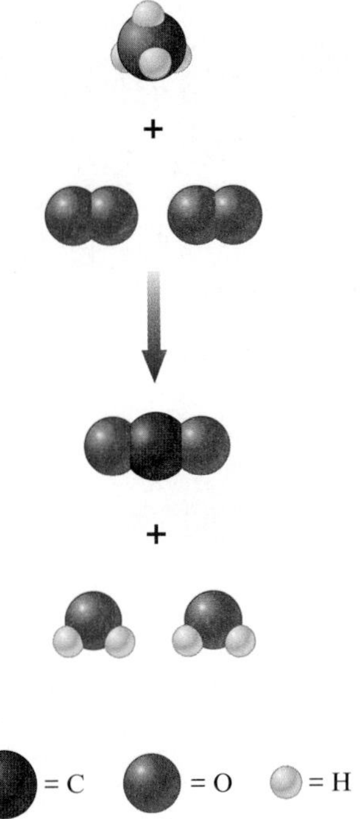

Figure 6.4 ▸ The reaction between methane and oxygen to give water and carbon dioxide. Note that there are four oxygen atoms in the products *and* in the reactants; none has been gained or lost in the reaction. Similarly, there are four hydrogen atoms and one carbon atom in the reactants *and* in the products. The reaction simply changes the way the atoms are grouped.

The equation that we have shown for the reaction between CH_4 and O_2 is not balanced. We can see that it is not balanced by taking the reactants and products apart.

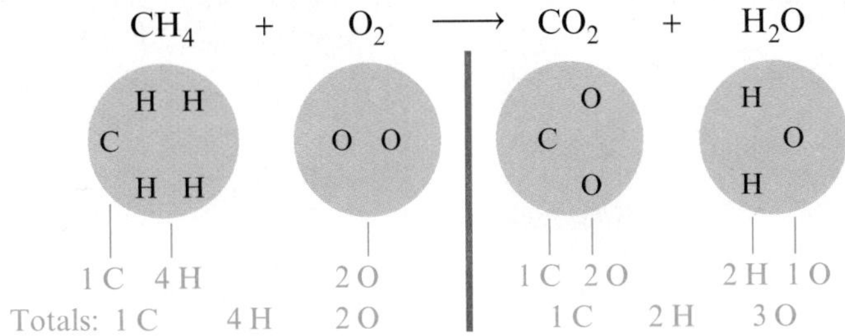

The reaction cannot happen this way because, as it stands, this equation states that one oxygen atom is created and two hydrogen atoms are destroyed. A reaction is only a rearrangement of the way the atoms are grouped; atoms are not created or destroyed. The total number of each type of atom must be the same on both sides of the arrow. We can fix the imbalance in this equation by involving one more O_2 molecule on the left and by showing the production of one more H_2O molecule on the right.

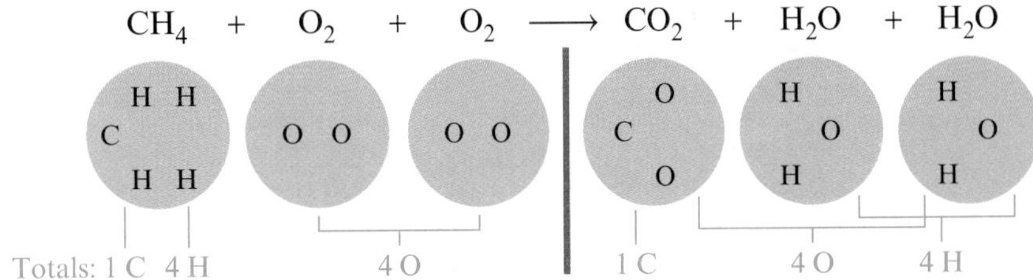

This *balanced chemical equation* shows the actual numbers of molecules involved in this reaction (Fig. 6.4).

When we write the balanced equation for a reaction, we group like molecules together. Thus

$$CH_4 + O_2 + O_2 \longrightarrow CO_2 + H_2O + H_2O$$

is written

$$CH_4 + 2O_2 \longrightarrow CO_2 + 2H_2O$$

The chemical equation for a reaction provides us with two important types of information:

1. The identities of the reactants and products
2. The relative numbers of each

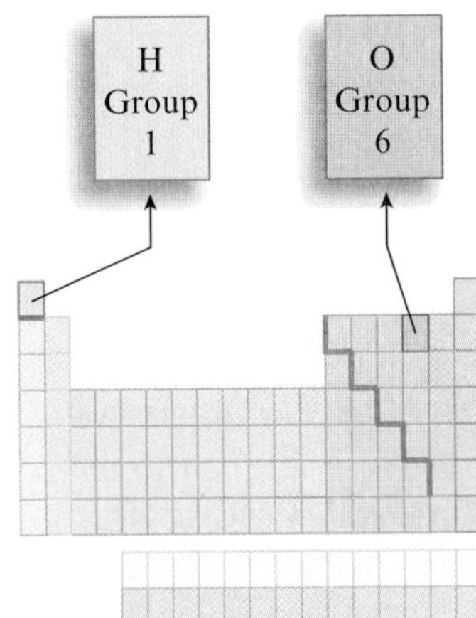

Physical States

Besides specifying the compounds involved in the reaction, we often indicate in the equation the *physical states* of the reactants and products by using the following symbols:

Symbol	State
(*s*)	solid
(*l*)	liquid
(*g*)	gas
(*aq*)	dissolved in water (in aqueous solution)

a The reactant potassium metal (stored in mineral oil to prevent oxidation).

b The reactant water.

c The reaction of potassium with water. The flame occurs because hydrogen gas, $H_2(g)$, produced by the reaction burns in air [reacts with $O_2(g)$] at the high temperatures caused by the reaction.

Figure 6.5

For example, when solid potassium reacts with liquid water, the products are hydrogen gas and potassium hydroxide; the latter remains dissolved in the water. From this information about the reactants and products, we can write the equation for the reaction. Solid potassium is represented by K(*s*); liquid water is written as $H_2O(l)$; hydrogen gas contains diatomic molecules and is represented as $H_2(g)$; potassium hydroxide dissolved in water is written as KOH(*aq*). So the *unbalanced* equation for the reaction is

Solid potassium		Water		Hydrogen gas		Potassium hydroxide dissolved in water
$K(s)$	$+$	$H_2O(l)$	$\rightarrow$	$H_2(g)$	$+$	$KOH(aq)$

This reaction is shown in Fig. 6.5.

The hydrogen gas produced in this reaction then reacts with the oxygen gas in the air, producing gaseous water and a flame. The *unbalanced* equation for this second reaction is

$$H_2(g) + O_2(g) \rightarrow H_2O(g)$$

Both of these reactions produce a great deal of heat. In Example 6.1 we will practice writing the unbalanced equations for reactions. Then, in the next section, we will discuss systematic procedures for balancing equations.

Example 6.1

Chemical Equations: Recognizing Reactants and Products

Write the *unbalanced* chemical equation for each of the following reactions.

a. Solid mercury(II) oxide decomposes to produce liquid mercury metal and gaseous oxygen.

b. Solid carbon reacts with gaseous oxygen to form gaseous carbon dioxide.

c. Solid zinc is added to an aqueous solution containing dissolved hydrogen chloride to produce gaseous hydrogen that bubbles out of the solution and zinc chloride that remains dissolved in the water.

Zinc metal reacts with hydrochloric acid to produce bubbles of hydrogen gas.

SOLUTION

a. In this case we have only one reactant, mercury(II) oxide. The name mercury(II) oxide means that the Hg^{2+} cation is present, so one O^{2-} ion is required for a zero net charge. Thus the formula is HgO, which is written HgO(*s*) in this case because it is given as a solid. The products are liquid mercury, written Hg(*l*), and gaseous oxygen, written $O_2(g)$. (Remember that oxygen exists as a diatomic molecule under normal conditions.) The unbalanced equation is

$$\underset{\text{Reactant}}{HgO(s)} \rightarrow \underset{\text{Products}}{Hg(l) + O_2(g)}$$

b. In this case, solid carbon, written C(*s*), reacts with oxygen gas, $O_2(g)$, to form gaseous carbon dioxide, which is written $CO_2(g)$. The equation (which happens to be balanced) is

$$\underset{\text{Reactants}}{C(s) + O_2(g)} \rightarrow \underset{\text{Product}}{CO_2(g)}$$

c. In this reaction solid zinc, Zn(*s*), is added to an aqueous solution of hydrogen chloride, which is written HCl(*aq*) and called hydrochloric acid. These are the reactants. The products of the reaction are gaseous hydrogen, $H_2(g)$, and aqueous zinc chloride. The name zinc chloride means that the Zn^{2+} ion is present, so two Cl^- ions are needed to achieve a zero net charge. Thus zinc chloride dissolved in water is written $ZnCl_2(aq)$. The unbalanced equation for the reaction is

Because Zn forms only the Zn^{2+} ion, a Roman numeral is usually not used. Thus $ZnCl_2$ is commonly called zinc chloride.

$$\underset{\text{Reactants}}{Zn(s) + HCl(aq)} \rightarrow \underset{\text{Products}}{H_2(g) + ZnCl_2(aq)}$$

SELF CHECK

Exercise 6.1 Identify the reactants and products and write the *unbalanced* equation (including symbols for states) for each of the following chemical reactions.

a. Solid magnesium metal reacts with liquid water to form solid magnesium hydroxide and hydrogen gas.

b. Solid ammonium dichromate (review Table 5.4 if this compound is unfamiliar) decomposes to solid chromium(III) oxide, gaseous nitrogen, and gaseous water.

c. Gaseous ammonia reacts with gaseous oxygen to form gaseous nitrogen monoxide and gaseous water.

See Problems 6.13 through 6.34. ■

6-3 Balancing Chemical Equations

OBJECTIVE **To learn how to write a balanced equation for a chemical reaction.**

As we saw in the previous section, an unbalanced chemical equation is not an accurate representation of the reaction that occurs. Whenever you see an equation for a reaction, you should ask yourself whether it is balanced. The principle that lies at the heart of the balancing process is that **atoms are conserved in a chemical reaction.** That is, atoms are neither created nor destroyed. They are just grouped differently. The same number of each type of atom is found among the reactants and among the products.

Chemists determine the identity of the reactants and products of a reaction by experimental observation. For example, when methane (natural gas) is burned in the presence of sufficient oxygen gas, the products are always carbon dioxide and water. **The identities (formulas) of the compounds must never be changed in balancing a chemical equation.** In other words, the subscripts in a formula cannot be changed, nor can atoms be added to or subtracted from a formula.

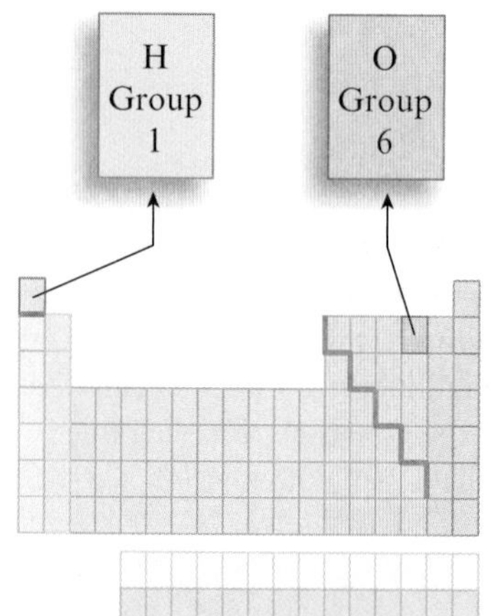

Most chemical equations can be balanced by trial and error—that is, by inspection. Keep trying until you find the numbers of reactants and products that give the same number of each type of atom on both sides of the arrow. For example, consider the reaction of hydrogen gas and oxygen gas to form liquid water. First, we write the unbalanced equation from the description of the reaction.

$$H_2(g) + O_2(g) \rightarrow H_2O(l)$$

We can see that this equation is unbalanced by counting the atoms on both sides of the arrow.

Reactants	Products
2 H	2 H
2 O	1 O

$$H_2(g) \quad + \quad O_2(g) \longrightarrow H_2O(l)$$

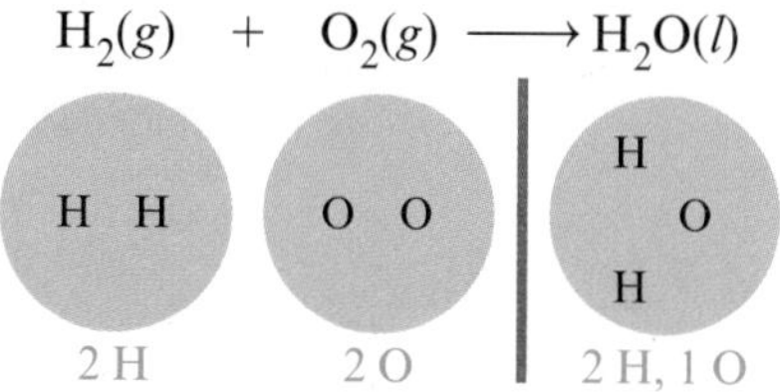

We have one more oxygen atom in the reactants than in the product. Because we cannot create or destroy atoms and because we *cannot change the formulas* of the reactants or products, we must balance the equation by adding more molecules of reactants and/or products. In this case we need one more oxygen atom on the right, so we add another water molecule (which contains one O atom). Then we count all of the atoms again.

Reactants	Products
2 H	4 H
2 O	2 O

$$H_2(g) \quad + \quad O_2(g) \longrightarrow H_2O(l) \quad + \quad H_2O(l)$$

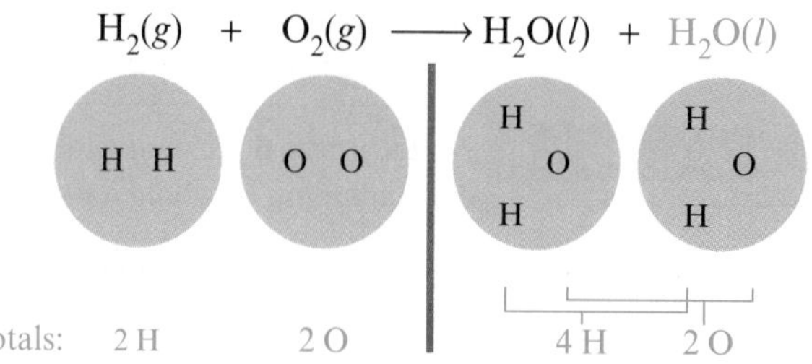

We have balanced the oxygen atoms, but now the hydrogen atoms have become unbalanced. There are more hydrogen atoms on the right than on the left. We can solve this problem by adding another hydrogen molecule (H_2) to the reactant side.

$$H_2(g) + H_2(g) + O_2(g) \longrightarrow H_2O(l) + H_2O(l)$$

H H | H H | O O | H O H | H O H

Totals: 4 H 2 O | 4 H 2 O

Reactants	Products
4 H	4 H
2 O	2 O

The equation is now balanced. We have the same numbers of hydrogen and oxygen atoms represented on both sides of the arrow. Collecting like molecules, we write the balanced equation as

$$2H_2(g) + O_2(g) \rightarrow 2H_2O(l)$$

Consider next what happens if we multiply every part of this balanced equation by 2:

$$2 \times [2H_2(g) + O_2(g) \rightarrow 2H_2O(l)]$$

to give

$$4H_2(g) + 2O_2(g) \rightarrow 4H_2O(l)$$

This equation is balanced (count the atoms to verify this). In fact, we can multiply or divide *all parts* of the original balanced equation by any number to give a new balanced equation. Thus each chemical reaction has many possible balanced equations. Is one of the many possibilities preferred over the others? Yes.

The accepted convention is that the "best" balanced equation is the one with the *smallest integers (whole numbers).* These integers are called the **coefficients** for the balanced equation. Therefore, for the reaction of hydrogen and oxygen to form water, the "correct" balanced equation is

$$2H_2(g) + O_2(g) \rightarrow 2H_2O(l)$$

The coefficients 2, 1 (never written), and 2, respectively, are the smallest *integers* that give a balanced equation for this reaction.

Critical Thinking

What if a friend was balancing chemical equations by changing the values of the subscripts instead of using the coefficients? How would you explain to your friend that this tactic is the wrong approach?

Next we will balance the equation for the reaction of liquid ethanol, C_2H_5OH, with oxygen gas to form gaseous carbon dioxide and water. This reaction, among many others, occurs in engines that burn a gasoline–ethanol mixture called gasohol.

The first step in obtaining the balanced equation for a reaction is always to identify the reactants and products from the description given for the reaction. In this case we are told that liquid ethanol, $C_2H_5OH(l)$, reacts with gaseous oxygen, $O_2(g)$, to produce gaseous carbon dioxide, $CO_2(g)$, and gaseous water, $H_2O(g)$. Therefore, the unbalanced equation is

$$C_2H_5OH(l) + O_2(g) \rightarrow CO_2(g) + H_2O(g)$$

Liquid ethanol | Gaseous oxygen | Gaseous carbon dioxide | Gaseous water

CHEMISTRY IN FOCUS

The Beetle That Shoots Straight

If someone said to you, "Name something that protects itself by spraying its enemies," your answer would almost certainly be "a skunk." Of course, you would be correct, but there is another correct answer—the bombardier beetle. When threatened, this beetle shoots a boiling stream of toxic chemicals at its enemy. How does this clever beetle accomplish this? Obviously, the boiling mixture cannot be stored inside the beetle's body all the time. Instead, when endangered, the beetle mixes chemicals that produce the hot spray. The chemicals involved are stored in two compartments. One compartment contains the chemicals hydrogen peroxide (H_2O_2) and methylhydroquinone ($C_7H_8O_2$). The key reaction is the decomposition of hydrogen peroxide to form oxygen gas and water:

A bombardier beetle defending itself.

Reuters Pictures

$$2H_2O_2(aq) \rightarrow 2H_2O(l) + O_2(g)$$

Hydrogen peroxide also reacts with the hydroquinones to produce other compounds that become part of the toxic spray.

However, none of these reactions occurs very fast unless certain enzymes are present. (Enzymes are natural substances that speed up biological reactions by means we will not discuss here.) When the beetle mixes the hydrogen peroxide and hydroquinones with the enzyme, the decomposition of H_2O_2 occurs rapidly, producing a hot mixture pressurized by the formation of oxygen gas. When the gas pressure becomes high enough, the hot spray is ejected in one long stream or in short bursts. The beetle has a highly accurate aim and can shoot several attackers with one batch of spray.

When one molecule in an equation is more complicated (contains more elements) than the others, it is best to start with that molecule. The most complicated molecule here is C_2H_5OH, so we begin by considering the products that contain the atoms in C_2H_5OH. We start with carbon. The only product that contains carbon is CO_2. Because C_2H_5OH contains two carbon atoms, we place a 2 before the CO_2 to balance the carbon atoms.

C_2H_5OH

H
CH
HOH
CH
H

2 C, 6 H, 1 O

$$C_2H_5OH(l) + O_2(g) \longrightarrow 2CO_2(g) + H_2O(g)$$

2 C atoms   2 C atoms

Remember, we cannot change the formula of any reactant or product when we balance an equation. We can only place coefficients in front of the formulas.

Next we consider hydrogen. The only product containing hydrogen is H_2O. C_2H_5OH contains six hydrogen atoms, so we need six hydrogen atoms on the right. Because each H_2O contains two hydrogen atoms, we need three H_2O molecules to yield six hydrogen atoms. So we place a 3 before the H_2O.

$$C_2H_5OH(l) + O_2(g) \longrightarrow 2CO_2(g) + 3H_2O(g)$$

(5 + 1) H   (3 × 2) H
6 H   6 H

O—C—O	H—O—H
O—C—O	H—O—H
	H—O—H
4 O atoms	3 O atoms

Finally, we count the oxygen atoms. On the left we have three oxygen atoms (one in C_2H_5OH and two in O_2), and on the right we have seven oxygen atoms (four in $2CO_2$

and three in $3H_2O$). We can correct this imbalance if we have three O_2 molecules on the left. That is, we place a coefficient of 3 before the O_2 to produce the balanced equation.

$$\underbrace{\underset{\uparrow \atop 1\,O}{C_2H_5OH(l)} + \underset{\uparrow\ \uparrow \atop (3 \times 2)\,O}{3O_2(g)}}_{7\,O} \longrightarrow \underbrace{\underset{\uparrow\ \uparrow \atop (2 \times 2)\,O}{2CO_2(g)} + \underset{\uparrow \atop 3\,O}{3H_2O(g)}}_{7\,O}$$

At this point you may have a question: why did we choose O_2 on the left when we balanced the oxygen atoms? Why not use C_2H_5OH, which has an oxygen atom? The answer is that if we had changed the coefficient in front of C_2H_5OH, we would have unbalanced the hydrogen and carbon atoms. Now we count all of the atoms as a check to make sure the equation is balanced.

$$C_2H_5OH(l) + 3O_2(g) \longrightarrow 2CO_2(g) + 3H_2O(g)$$

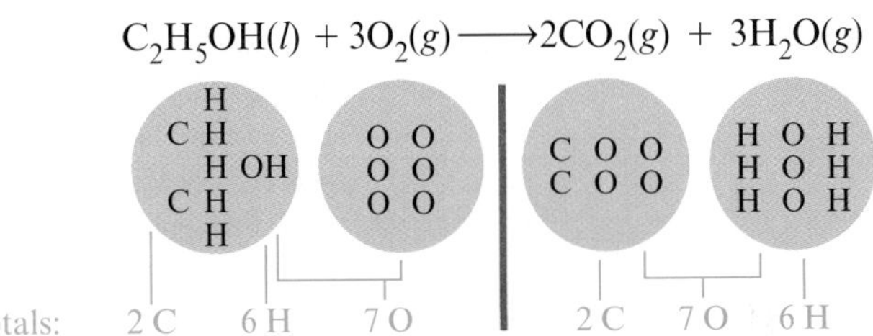

Reactants	Products
2 C	2 C
6 H	6 H
7 O	7 O

The equation is now balanced. We have the same numbers of all types of atoms on both sides of the arrow. Notice that these coefficients are the smallest integers that give a balanced equation.

The process of writing and balancing the equation for a chemical reaction consists of several steps:

How to Write and Balance Equations

Step 1 Read the description of the chemical reaction. What are the reactants, the products, and their states? Write the appropriate formulas.

Step 2 Write the *unbalanced* equation that summarizes the information from step 1.

Step 3 Balance the equation by inspection, starting with the most complicated molecule. Proceed element by element to determine what coefficients are necessary so that the same number of each type of atom appears on both the reactant side and the product side. Do not change the identities (formulas) of any of the reactants or products.

Step 4 Check to see that the coefficients used give the same number of each type of atom on both sides of the arrow. (Note that an "atom" may be present in an element, a compound, or an ion.) Also check to see that the coefficients used are the smallest integers that give the balanced equation. This can be done by determining whether all coefficients can be divided by the same integer to give a set of smaller *integer* coefficients.

Critical Thinking

One part of the problem-solving strategy for balancing chemical equations is "starting with the most complicated molecule." What if you started with a different molecule? Could you still eventually balance the chemical equation? How would this approach be different from the suggested technique?

Interactive Example 6.2

Balancing Chemical Equations I

For the following reaction, write the unbalanced equation and then balance the equation: solid potassium reacts with liquid water to form gaseous hydrogen and potassium hydroxide that dissolves in the water.

SOLUTION

Step 1 From the description given for the reaction, we know that the reactants are solid potassium, $K(s)$, and liquid water, $H_2O(l)$. The products are gaseous hydrogen, $H_2(g)$, and dissolved potassium hydroxide, $KOH(aq)$.

Step 2 The unbalanced equation for the reaction is

$$K(s) + H_2O(l) \rightarrow H_2(g) + KOH(aq)$$

Step 3 Although none of the reactants or products is very complicated, we will start with KOH because it contains the most elements (three). We will arbitrarily consider hydrogen first. Note that on the reactant side of the equation in step 2, there are two hydrogen atoms but on the product side there are three. If we place a coefficient of 2 in front of both H_2O and KOH, we now have four H atoms on each side.

$$K(s) + \underbrace{2H_2O(l)}_{\text{4 H atoms}} \rightarrow \underbrace{H_2(g)}_{\text{2 H atoms}} + \underbrace{2KOH(aq)}_{\text{2 H atoms}}$$

Also note that the oxygen atoms balance.

$$K(s) + \underbrace{2H_2O(l)}_{\text{2 O atoms}} \rightarrow H_2(g) + \underbrace{2KOH(aq)}_{\text{2 O atoms}}$$

However, the K atoms do not balance; we have one on the left and two on the right. We can fix this easily by placing a coefficient of 2 in front of $K(s)$ to give the balanced equation:

$$2K(s) + 2H_2O(l) \rightarrow H_2(g) + 2KOH(aq)$$

Step 4

Reactants	Products
2 K	2 K
4 H	4 H
2 O	2 O

CHECK There are 2 K, 4 H, and 2 O on both sides of the arrow, and the coefficients are the smallest integers that give a balanced equation. We know this because we cannot divide through by a given integer to give a set of smaller *integer* (whole-number) coefficients. For example, if we divide all of the coefficients by 2, we get

$$K(s) + H_2O(l) \rightarrow \tfrac{1}{2}H_2(g) + KOH(aq)$$

This is not acceptable because the coefficient for H_2 is not an integer. ■

Interactive Example 6.3

Balancing Chemical Equations II

Under appropriate conditions at 1000 °C, ammonia gas reacts with oxygen gas to produce gaseous nitrogen monoxide (common name, nitric oxide) and gaseous water. Write the unbalanced and balanced equations for this reaction.

SOLUTION

Step 1 The reactants are gaseous ammonia, $NH_3(g)$, and gaseous oxygen, $O_2(g)$. The products are gaseous nitrogen monoxide, $NO(g)$, and gaseous water, $H_2O(g)$.

Reactants	Products
1 N	1 N
3 H	2 H
2 O	2 O

Step 2 The unbalanced equation for the reaction is

$$NH_3(g) + O_2(g) \rightarrow NO(g) + H_2O(g)$$

Step 3 In this equation there is no molecule that is obviously the most complicated. Three molecules contain two elements, so we arbitrarily start with NH_3. We arbitrarily begin by looking at hydrogen. A coefficient of 2 for NH_3 and a coefficient of 3 for H_2O give six atoms of hydrogen on both sides.

$$\underbrace{2NH_3(g)}_{6\text{ H}} + O_2(g) \rightarrow NO(g) + \underbrace{3H_2O(g)}_{6\text{ H}}$$

We can balance the nitrogen by giving NO a coefficient of 2.

$$\underbrace{2NH_3(g)}_{2\text{ N}} + O_2(g) \rightarrow \underbrace{2NO(g)}_{2\text{ N}} + 3H_2O(g)$$

$\frac{5}{2} = 2\frac{1}{2}$

O—O

O—O

O+O

$2\frac{1}{2}\ O_2$ contains 5 O atoms

Finally, we note that there are two atoms of oxygen on the left and five on the right. The oxygen can be balanced with a coefficient of $\frac{5}{2}$ for O_2, because $\frac{5}{2} \times O_2$ gives five oxygen atoms.

$$2NH_3(g) + \underbrace{\tfrac{5}{2}O_2(g)}_{5\text{ O}} \rightarrow \underbrace{2NO(g)}_{2\text{ O}} + \underbrace{3H_2O(g)}_{3\text{ O}}$$

However, the convention is to have integer (whole-number) coefficients, so we multiply the entire equation by 2.

$$2 \times [2NH_3(g) + \tfrac{5}{2}O_2(g) \rightarrow 2NO(g) + 3H_2O(g)]$$

or

$$2 \times 2NH_3(g) + 2 \times \tfrac{5}{2}O_2(g) \rightarrow 2 \times 2NO(g) + 2 \times 3H_2O(g)$$
$$4NH_3(g) + 5O_2(g) \rightarrow 4NO(g) + 6H_2O(g)$$

Reactants	Products
4 N	4 N
12 H	12 H
10 O	10 O

Step 4

CHECK There are 4 N, 12 H, and 10 O atoms on both sides, so the equation is balanced. These coefficients are the smallest integers that give a balanced equation. That is, we cannot divide all coefficients by the same integer and obtain a smaller set of *integers*.

SELF CHECK

Exercise 6.2 Propane, C_3H_8, a liquid at 25 °C under high pressure, is often used for gas grills and as a fuel in rural areas where there is no natural gas pipeline. When liquid propane is released from its storage tank, it changes to propane gas that reacts with oxygen gas (it "burns") to give gaseous carbon dioxide and gaseous water. Write and balance the equation for this reaction.

HINT This description of a chemical process contains many words, some of which are crucial to solving the problem and some of which are not. First sort out the important information and use symbols to represent it.

See Problems 6.37 through 6.44. ■

Interactive Example 6.4

Balancing Chemical Equations III

Glass is sometimes decorated by etching patterns on its surface. Etching occurs when hydrofluoric acid (an aqueous solution of HF) reacts with the silicon dioxide in the glass to form gaseous silicon tetrafluoride and liquid water. Write and balance the equation for this reaction.

Richard Megna/Fundamental Photographs © Cengage Learning

Decorations on glass are produced by etching with hydrofluoric acid.

SOLUTION

Step 1 From the description of the reaction we can identify the reactants:

hydrofluoric acid	$HF(aq)$
solid silicon dioxide	$SiO_2(s)$

and the products:

gaseous silicon tetrafluoride	$SiF_4(g)$
liquid water	$H_2O(l)$

Step 2 The unbalanced equation is

$$SiO_2(s) + HF(aq) \rightarrow SiF_4(g) + H_2O(l)$$

Reactants	Products
1 Si	1 Si
1 H	2 H
1 F	4 F
2 O	1 O

Step 3 There is no clear choice here for the most complicated molecule. We arbitrarily start with the elements in SiF_4. The silicon is balanced (one atom on each side), but the fluorine is not. To balance the fluorine, we need a coefficient of 4 before the HF.

$$SiO_2(s) + 4HF(aq) \rightarrow SiF_4(g) + H_2O(l)$$

Reactants	Products
1 Si	1 Si
4 H	2 H
4 F	4 F
2 O	1 O

Hydrogen and oxygen are not balanced. Because we have four hydrogen atoms on the left and two on the right, we place a 2 before the H_2O:

$$SiO_2(s) + 4HF(aq) \rightarrow SiF_4(g) + 2H_2O(l)$$

This balances the hydrogen *and* the oxygen (two atoms on each side).

Reactants	Products
1 Si	1 Si
4 H	4 H
4 F	4 F
2 O	2 O

Step 4

CHECK $SiO_2(s) + 4HF(aq) \rightarrow SiF_4(g) + 2H_2O(l)$

Totals: 1 Si, 2 O, 4 H, 4 F → 1 Si, 4 F, 4 H, 2 O

All atoms check, so the equation is balanced.

SELF CHECK

Exercise 6.3 Give the balanced equation for each of the following reactions.

If you are having trouble writing formulas from names, review the appropriate sections of Chapter 5. It is very important that you are able to do this.

a. When solid ammonium nitrite is heated, it produces nitrogen gas and water vapor.

b. Gaseous nitrogen monoxide (common name, nitric oxide) decomposes to produce dinitrogen monoxide gas (common name, nitrous oxide) and nitrogen dioxide gas.

c. Liquid nitric acid decomposes to reddish-brown nitrogen dioxide gas, liquid water, and oxygen gas. (This is why bottles of nitric acid become yellow upon standing.)

See Problems 6.37 through 6.44. ■

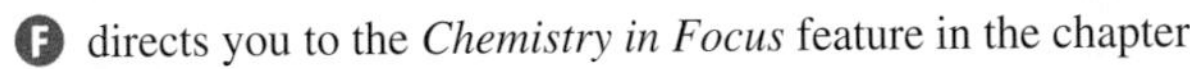

CHAPTER 6 REVIEW

F directs you to the *Chemistry in Focus* feature in the chapter

Key Terms

chemical reaction (6-2)
chemical equation (6-2)
reactants (6-2)
products (6-2)
balancing the chemical equation (6-2)
coefficients (6-3)

For Review

- A chemical reaction produces a signal that it has occurred. These signals include
  - Color change
  - Solid formation
  - Bubble formation
  - Heat
  - Flame
- The physical states of reactants and products in a reaction are indicated by the following symbols.

Physical States	
Symbol	**State**
(*s*)	solid
(*l*)	liquid
(*g*)	gas
(*aq*)	dissolved in water (in aqueous solution)

- Chemical reactions involve a rearrangement of the ways atoms are grouped together.
- A chemical equation represents a chemical reaction.
  - Reactants are shown to the left of an arrow.
  - Products are shown to the right of the arrow.
- In a chemical reaction, atoms are neither created nor destroyed. A balanced chemical equation must have the same number of each type of atom on the reactant and product sides.
- A balanced chemical equation uses numbers (coefficients) in front of the reactant and product formulas to show the relative numbers of each.
- A chemical reaction is balanced by using a systematic approach.
  - Write the formulas of the reactants and products to give the unbalanced chemical equation.
  - Balance by trial and error, starting with the most complicated molecule(s).
  - Check to be sure the equation is balanced (same numbers of all types of atoms on the reactant and product sides).

Active Learning Questions

These questions are designed to be considered by groups of students in class. Often these questions work well for introducing a particular topic in class.

1. The following are actual student responses to the question: Why is it necessary to balance chemical equations?
   a. The chemicals will not react until you have added the correct ratios.
   b. The correct products will not form unless the right amounts of reactants have been added.
   c. A certain number of products cannot form without a certain number of reactants.
   d. The balanced equation tells you how much reactant you need, and allows you to predict how much product you will make.
   e. A ratio must be established for the reaction to occur as written.

   Justify the best choice, and, for choices you did not pick, explain what is wrong with them.
2. What information do we get from a formula? From an equation?
3. Given the equation for the reaction: $N_2 + H_2 \rightarrow NH_3$, draw a molecular diagram that represents the reaction (make sure it is balanced).
4. What do the subscripts in a chemical formula represent? What do the coefficients in a balanced chemical equation represent?
5. Can the subscripts in a chemical formula be fractions? Explain.
6. Can the coefficients in a balanced chemical equation be fractions? Explain.
7. Changing the subscripts of chemicals can mathematically balance the equations. Why is this unacceptable?
8. Table 6.1 lists some clues that a chemical reaction has occurred. However, these events do not necessarily prove the existence of a chemical change. Give an example for each of the clues that is not a chemical reaction but a physical change.
9. Use molecular-level drawings to show the difference between physical and chemical changes.
10. It is stated in Section 6-3 of the text that to balance equations by inspection you start "with the most complicated molecule." What does this mean? Why is it best to do this?
11. Which of the following statements concerning balanced chemical equations are true? There may be more than one true statement.
    a. Atoms are neither created nor destroyed.
    b. The coefficients indicate the mass ratios of the substances used.
    c. The sum of the coefficients on the reactant side always equals the sum of the coefficients on the product side.
12. Consider the generic chemical equation $aA + bB \rightarrow cC + dD$ (where a, b, c, and d represent coefficients for the chemicals A, B, C, and D, respectively).
    a. How many possible values are there for "c"? Explain your answer.
    b. How many possible values are there for "c/d"? Explain your answer.
13. How is the balancing of chemical equations related to the law of conservation of mass?
14. Which of the following correctly describes the balanced chemical equation given below? There may be more than one true statement. If a statement is incorrect, explain why it is incorrect.

    $$4Al + 3O_2 \rightarrow 2Al_2O_3$$

    a. For every 4 atoms of aluminum that reacts with 6 atoms of oxygen, 2 molecules of aluminum oxide are produced.
    b. For every 4 moles of aluminum that reacts with 3 moles of oxygen, 2 moles of aluminum(III) oxide are produced.
    c. For every 4 g of aluminum that reacts with 3 g of oxygen, 2 g of aluminum oxide are produced.
15. Which of the following correctly balances the chemical equation given below? There may be more than one correct balanced equation. If a balanced equation is incorrect, explain why it is incorrect.

    $$CaO + C \rightarrow CaC_2 + CO_2$$

    a. $CaO_2 + 3C \rightarrow CaC_2 + CO_2$
    b. $2CaO + 5C \rightarrow 2CaC_2 + CO_2$
    c. $CaO + 2\frac{1}{2}C \rightarrow CaC_2 + \frac{1}{2}CO_2$
    d. $4CaO + 10C \rightarrow 4CaC_2 + 2CO_2$
16. The reaction of an element X (△) with element Y (○) is represented in the following diagram. Which of the elements best describes this reaction?

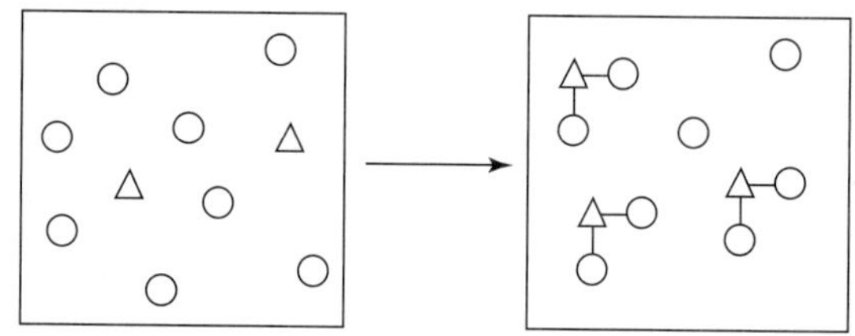

    a. $3X + 8Y \rightarrow X_3Y_8$
    b. $3X + 6Y \rightarrow X_3Y_6$
    c. $X + 2Y \rightarrow XY_2$
    d. $3X + 8Y \rightarrow 3XY_2 + 2Y$

Questions and Problems

6-1 Evidence for a Chemical Reaction

Questions

1. How do we *know* when a chemical reaction is taking place? Can you think of an example of how each of the five senses (sight, hearing, taste, touch, smell) might be used in detecting when a chemical reaction has taken place?
2. These days many products are available to whiten teeth at home. Many of these products contain a peroxide that bleaches stains from the teeth. What evidence is there that the bleaching process is a chemical reaction?
3. Although these days many people have "self-cleaning" ovens, if your oven gets *really* dirty you may have to resort to one of the spray-on oven cleaner preparations sold in supermarkets. What evidence is there that such oven cleaners work by a chemical reaction?
4. Small cuts and abrasions on the skin are frequently cleaned using hydrogen peroxide solution. What evidence is there that treating a wound with hydrogen peroxide causes a chemical reaction to take place?

All even-numbered Questions and Problems have answers in the back of this book and solutions in the *Student Solutions Guide.*

5. You have probably had the unpleasant experience of discovering that a flashlight battery has gotten old and begun to leak. Is there evidence that this change is due to a chemical reaction?

6. If you've ever left bread in a toaster too long, you know that the bread eventually burns and turns black. What evidence is there that this represents a chemical process?

6-2 Chemical Equations

Questions

7. What are the substances to the *left* of the arrow in a chemical equation called? To the *right* of the arrow? What does the arrow itself mean?

8. For the unbalanced chemical equation $N_2(g) + H_2(g) \rightarrow NH_3(g)$
   a. list the reactant(s).
   b. list the product(s).

9. In a chemical reaction, the total number of atoms present after the reaction is complete is (larger than/smaller than/the same as) the total number of atoms present before the reaction began.

10. What does "balancing" an equation accomplish?

11. Why are the *physical states* of the reactants and products often indicated when writing a chemical equation?

12. The notation "(g)" after a substance's formula indicates it exists in the ________ state.

Problems

Note: In some of the following problems you will need to write a chemical formula from the name of the compound. Review Chapter 5 if you are having trouble.

13. A common experiment to determine the relative reactivity of metallic elements is to place a pure sample of one metal into an aqueous solution of a compound of another metallic element. If the pure metal you are adding is more reactive than the metallic element in the compound, then the pure metal will *replace* the metallic element in the compound. For example, if you place a piece of pure zinc metal into a solution of copper(II) sulfate, the zinc will slowly dissolve to produce zinc sulfate solution, and the copper(II) ion of the copper(II) sulfate will be converted to metallic copper. Write the unbalanced equation for this process.

14. If calcium carbonate is heated strongly, carbon dioxide gas is driven off, leaving a residue of calcium oxide. Write the unbalanced chemical equation for this process.

15. If a sample of pure hydrogen gas is ignited very carefully, the hydrogen burns gently, combining with the oxygen gas of the air to form water vapor. Write the unbalanced chemical equation for this reaction.

16. Liquid hydrazine, N_2H_4, has been used as a fuel for rockets. When the rocket is to be launched, a catalyst causes the liquid hydrazine to decompose quickly into elemental nitrogen and hydrogen gases. The rapid expansion of the product gases and the heat released by the reaction provide the thrust for the rocket. Write the unbalanced equation for the reaction of hydrazine to produce nitrogen and hydrogen gases.

17. If electricity of sufficient voltage is passed into a solution of potassium iodide in water, a reaction takes place in which elemental hydrogen gas and elemental iodine are produced, leaving a solution of potassium hydroxide. Write the unbalanced equation for this process.

18. Silver oxide may be decomposed by strong heating into silver metal and oxygen gas. Write the unbalanced chemical equation for this process.

19. Elemental boron is produced in one industrial process by heating diboron trioxide with magnesium metal, also producing magnesium oxide as a by-product. Write the unbalanced chemical equation for this process.

20. Many over-the-counter antacid tablets are now formulated using calcium carbonate as the active ingredient, which enables such tablets to also be used as dietary calcium supplements. As an antacid for gastric hyperacidity, calcium carbonate reacts by combining with hydrochloric acid found in the stomach, producing a solution of calcium chloride, converting the stomach acid to water, and releasing carbon dioxide gas (which the person suffering from stomach problems may feel as a "burp"). Write the unbalanced chemical equation for this process.

21. Phosphorus trichloride is used in the manufacture of certain pesticides, and may be synthesized by direct combination of its constituent elements. Write the unbalanced chemical equation for this process.

22. Pure silicon, which is needed in the manufacturing of electronic components, may be prepared by heating silicon dioxide (sand) with carbon at high temperatures, releasing carbon monoxide gas. Write the unbalanced chemical equation for this process.

23. Nitrous oxide gas (systematic name: dinitrogen monoxide) is used by some dental practitioners as an anesthetic. Nitrous oxide (and water vapor as by-product) can be produced in small quantities in the laboratory by careful heating of ammonium nitrate. Write the unbalanced chemical equation for this reaction.

24. Solid zinc is added to an aqueous solution containing dissolved hydrogen chloride to produce gaseous hydrogen that bubbles out of the solution and zinc chloride that remains dissolved in the water. Write the unbalanced chemical equation for this process.

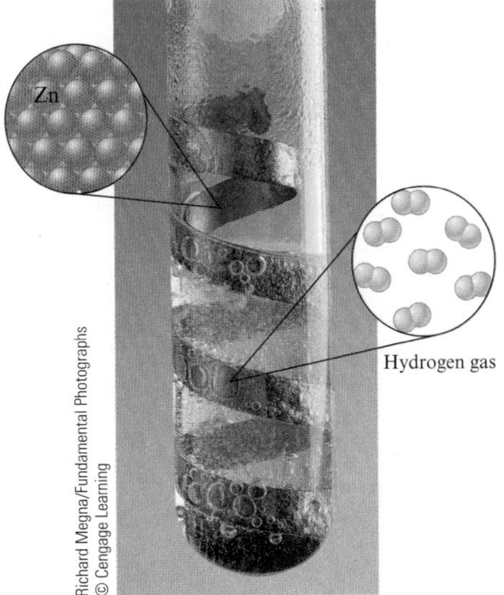

25. Acetylene gas (C_2H_2) is often used by plumbers, welders, and glass blowers because it burns in oxygen with an intensely hot flame. The products of the combustion of acetylene are carbon dioxide and water vapor. Write the unbalanced chemical equation for this process.

26. The burning of high-sulfur fuels has been shown to cause the phenomenon of "acid rain." When a high-sulfur fuel is burned, the sulfur is converted to sulfur dioxide (SO_2) and sulfur trioxide (SO_3). When sulfur dioxide and sulfur trioxide gas dissolve in water in the atmosphere, sulfurous acid and sulfuric acid are produced, respectively. Write the unbalanced chemical equations for the reactions of sulfur dioxide and sulfur trioxide with water.

27. The Group 2 metals (Ba, Ca, Sr) can be produced in the elemental state by the reaction of their oxides with aluminum metal at high temperatures, also producing solid aluminum oxide as a by-product. Write the unbalanced chemical equations for the reactions of barium oxide, calcium oxide, and strontium oxide with aluminum.

28. There are fears that the protective ozone layer around the earth is being depleted. Ozone, O_3, is produced by the interaction of ordinary oxygen gas in the atmosphere with ultraviolet light and lightning discharges. The oxides of nitrogen (which are common in automobile exhaust gases), in particular, are known to decompose ozone. For example, gaseous nitric oxide (NO) reacts with ozone gas to produce nitrogen dioxide gas and oxygen gas. Write the unbalanced chemical equation for this process.

29. Carbon tetrachloride was widely used for many years as a solvent until its harmful properties became well established. Carbon tetrachloride may be prepared by the reaction of natural gas (methane, CH_4) and elemental chlorine gas in the presence of ultraviolet light. Write the unbalanced chemical equation for this process.

30. When elemental phosphorus, P_4, burns in oxygen gas, it produces an intensely bright light, a great deal of heat, and massive clouds of white solid phosphorus(V) oxide (P_2O_5) product. Given these properties, it is not surprising that phosphorus has been used to manufacture incendiary bombs for warfare. Write the unbalanced equation for the reaction of phosphorus with oxygen gas to produce phosphorus(V) oxide.

31. Calcium oxide is sometimes very challenging to store in the chemistry laboratory. This compound reacts with moisture in the air and is converted to calcium hydroxide. If a bottle of calcium oxide is left on the shelf too long, it gradually absorbs moisture from the humidity in the laboratory. Eventually the bottle cracks and spills the calcium hydroxide that has been produced. Write the unbalanced chemical equation for this process.

32. Although they were formerly called the inert gases, the heavier elements of Group 8 do form relatively stable compounds. For example, at high temperatures in the presence of an appropriate catalyst, xenon gas will combine directly with fluorine gas to produce solid xenon tetrafluoride. Write the unbalanced chemical equation for this process.

33. The element tin often occurs in nature as the oxide, SnO_2. To produce pure tin metal from this sort of tin ore, the ore usually is heated with coal (carbon). This produces pure molten tin, with the carbon being removed from the reaction system as the gaseous byproduct carbon monoxide. Write the unbalanced equation for this process.

34. Nitric acid, HNO_3, can be produced by reacting high-pressure ammonia gas with oxygen gas at around 750 °C in the presence of a platinum catalyst. Water is a by-product of the reaction. Write the unbalanced chemical equation for this process.

6-3 Balancing Chemical Equations

Questions

35. When balancing chemical equations, beginning students are often tempted to change the numbers *within* a formula (the subscripts) to balance the equation. Why is this never permitted? What effect does changing a subscript have?

F 36. The "Chemistry in Focus" segment *The Beetle That Shoots Straight* discusses the bombardier beetle and the chemical reaction of the decomposition of hydrogen peroxide.

$$H_2O_2(aq) \rightarrow H_2O(l) + O_2(g)$$

The balanced equation given in the segment is

$$2H_2O_2(aq) \rightarrow 2H_2O(l) + O_2(g)$$

Why can't we balance the equation in the following way?

$$H_2O_2(aq) \rightarrow H_2(g) + O_2(g)$$

Use molecular level pictures like those in Section 6-3 to support your answer.

Problems

37. Balance each of the following chemical equations.
   a. $FeCl_3(aq) + KOH(aq) \rightarrow Fe(OH)_3(s) + KCl(aq)$
   b. $Pb(C_2H_3O_2)_2(aq) + KI(aq) \rightarrow PbI_2(s) + KC_2H_3O_2(aq)$
   c. $P_4O_{10}(s) + H_2O(l) \rightarrow H_3PO_4(aq)$
   d. $Li_2O(s) + H_2O(l) \rightarrow LiOH(aq)$
   e. $MnO_2(s) + C(s) \rightarrow Mn(s) + CO_2(g)$
   f. $Sb(s) + Cl_2(g) \rightarrow SbCl_3(s)$
   g. $CH_4(g) + H_2O(g) \rightarrow CO(g) + H_2(g)$
   h. $FeS(s) + HCl(aq) \rightarrow FeCl_2(aq) + H_2S(g)$

38. Balance the equation for the reaction of potassium with water.

$$K(s) + H_2O(l) \rightarrow H_2(g) + KOH(aq)$$

Richard Megna/Fundamental Photographs © Cengage Learning

All even-numbered Questions and Problems have answers in the back of this book and solutions in the *Student Solutions Guide*.

39. Balance each of the following chemical equations.
   a. $K_2SO_4(aq) + BaCl_2(aq) \rightarrow BaSO_4(s) + KCl(aq)$
   b. $Fe(s) + H_2O(g) \rightarrow FeO(s) + H_2(g)$
   c. $NaOH(aq) + HClO_4(aq) \rightarrow NaClO_4(aq) + H_2O(l)$
   d. $Mg(s) + Mn_2O_3(s) \rightarrow MgO(s) + Mn(s)$
   e. $KOH(s) + KH_2PO_4(aq) \rightarrow K_3PO_4(aq) + H_2O(l)$
   f. $NO_2(g) + H_2O(l) + O_2(g) \rightarrow HNO_3(aq)$
   g. $BaO_2(s) + H_2O(l) \rightarrow Ba(OH)_2(aq) + O_2(g)$
   h. $NH_3(g) + O_2(g) \rightarrow NO(g) + H_2O(l)$

40. Balance each of the following chemical equations.
   a. $Na_2SO_4(aq) + CaCl_2(aq) \rightarrow CaSO_4(s) + NaCl(aq)$
   b. $Fe(s) + H_2O(g) \rightarrow Fe_3O_4(s) + H_2(g)$
   c. $Ca(OH)_2(aq) + HCl(aq) \rightarrow CaCl_2(aq) + H_2O(l)$
   d. $Br_2(g) + H_2O(l) + SO_2(g) \rightarrow HBr(aq) + H_2SO_4(aq)$
   e. $NaOH(s) + H_3PO_4(aq) \rightarrow Na_3PO_4(aq) + H_2O(l)$
   f. $NaNO_3(s) \rightarrow NaNO_2(s) + O_2(g)$
   g. $Na_2O_2(s) + H_2O(l) \rightarrow NaOH(aq) + O_2(g)$
   h. $Si(s) + S_8(s) \rightarrow Si_2S_4(s)$

41. Balance each of the following chemical equations.
   a. $Fe_3O_4(s) + H_2(g) \rightarrow Fe(l) + H_2O(g)$
   b. $K_2SO_4(aq) + BaCl_2(aq) \rightarrow BaSO_4(s) + KCl(aq)$
   c. $HCl(aq) + FeS(s) \rightarrow FeCl_2(aq) + H_2S(g)$
   d. $Br_2(g) + H_2O(l) + SO_2(g) \rightarrow HBr(aq) + H_2SO_4(aq)$
   e. $CS_2(l) + Cl_2(g) \rightarrow CCl_4(l) + S_2Cl_2(g)$
   f. $Cl_2O_7(g) + Ca(OH)_2(aq) \rightarrow Ca(ClO_4)_2(aq) + H_2O(l)$
   g. $PBr_3(l) + H_2O(l) \rightarrow H_3PO_3(aq) + HBr(g)$
   h. $Ba(ClO_3)_2(s) \rightarrow BaCl_2(s) + O_2(s)$

42. Balance each of the following chemical equations.
   a. $NaCl(s) + SO_2(g) + H_2O(g) + O_2(g) \rightarrow Na_2SO_4(s) + HCl(g)$
   b. $Br_2(l) + I_2(s) \rightarrow IBr_3(s)$
   c. $Ca_3N_2(s) + H_2O(l) \rightarrow Ca(OH)_2(aq) + PH_3(g)$
   d. $BF_3(g) + H_2O(g) \rightarrow B_2O_3(s) + HF(g)$
   e. $SO_2(g) + Cl_2(g) \rightarrow SOCl_2(l) + Cl_2O(g)$
   f. $Li_2O(s) + H_2O(l) \rightarrow LiOH(aq)$
   g. $Mg(s) + CuO(s) \rightarrow MgO(s) + Cu(l)$
   h. $Fe_3O_4(s) + H_2(g) \rightarrow Fe(l) + H_2O(g)$

43. Balance each of the following chemical equations.
   a. $KO_2(s) + H_2O(l) \rightarrow KOH(aq) + O_2(g) + H_2O_2(aq)$
   b. $Fe_2O_3(s) + HNO_3(aq) \rightarrow Fe(NO_3)_3(aq) + H_2O(l)$
   c. $NH_3(g) + O_2(g) \rightarrow NO(g) + H_2O(g)$
   d. $PCl_5(l) + H_2O(l) \rightarrow H_3PO_4(aq) + HCl(g)$
   e. $C_2H_5OH(l) + O_2(g) \rightarrow CO_2(g) + H_2O(l)$
   f. $CaO(s) + C(s) \rightarrow CaC_2(s) + CO_2(g)$
   g. $MoS_2(s) + O_2(g) \rightarrow MoO_3(s) + SO_2(g)$
   h. $FeCO_3(s) + H_2CO_3(aq) \rightarrow Fe(HCO_3)_2(aq)$

44. Balance each of the following chemical equations.
   a. $Ba(NO_3)_2(aq) + Na_2CrO_4(aq) \rightarrow BaCrO_4(s) + NaNO_3(aq)$
   b. $PbCl_2(aq) + K_2SO_4(aq) \rightarrow PbSO_4(s) + KCl(aq)$
   c. $C_2H_5OH(l) + O_2(g) \rightarrow CO_2(g) + H_2O(l)$
   d. $CaC_2(s) + H_2O(l) \rightarrow Ca(OH)_2(s) + C_2H_2(g)$
   e. $Sr(s) + HNO_3(aq) \rightarrow Sr(NO_3)_2(aq) + H_2(g)$
   f. $BaO_2(s) + H_2SO_4(aq) \rightarrow BaSO_4(s) + H_2O_2(aq)$
   g. $AsI_3(s) \rightarrow As(s) + I_2(s)$
   h. $CuSO_4(aq) + KI(s) \rightarrow CuI(s) + I_2(s) + K_2SO_4(aq)$

All even-numbered Questions and Problems have answers in the back of this book and solutions in the *Student Solutions Guide*.

Additional Problems

45. Acetylene gas, C_2H_2, is used in welding because it generates an extremely hot flame when it is combusted with oxygen. The heat generated is sufficient to melt the metals being welded together. Carbon dioxide gas and water vapor are the chemical products of this reaction. Write the unbalanced chemical equation for the reaction of acetylene with oxygen.

46. When balancing a chemical equation, which of the following statements is *false*?
   a. Subscripts in the reactants must be conserved in the products.
   b. Coefficients are used to balance the atoms on both sides.
   c. The Law of Conservation of Matter must be followed.
   d. Phases are often shown for each compound but are not critical to balancing an equation.

47. Crude gunpowders often contain a mixture of potassium nitrate and charcoal (carbon). When such a mixture is heated until reaction occurs, a solid residue of potassium carbonate is produced. The explosive force of the gunpowder comes from the fact that two gases are also produced (carbon monoxide and nitrogen), which increase in volume with great force and speed. Write the unbalanced chemical equation for the process.

48. After balancing a chemical equation, we ordinarily make sure that the coefficients are the smallest ________ possible.

49. Methanol (methyl alcohol), CH_3OH, is a very important industrial chemical. Formerly, methanol was prepared by heating wood to high temperatures in the absence of air. The complex compounds present in wood are degraded by this process into a charcoal residue and a volatile portion that is rich in methanol. Today, methanol is instead synthesized from carbon monoxide and elemental hydrogen. Write the balanced chemical equation for this latter process.

50. The Hall process is an important method by which pure aluminum is prepared from its oxide (alumina, Al_2O_3) by indirect reaction with graphite (carbon). Balance the following equation, which is a simplified representation of this process.

$$Al_2O_3(s) + C(s) \rightarrow Al(s) + CO_2(g)$$

51. Iron oxide ores, commonly a mixture of FeO and Fe_2O_3, are given the general formula Fe_3O_4. They yield elemental iron when heated to a very high temperature with either carbon monoxide or elemental hydrogen. Balance the following equations for these processes.

$$Fe_3O_4(s) + H_2(g) \rightarrow Fe(s) + H_2O(g)$$
$$Fe_3O_4(s) + CO(g) \rightarrow Fe(s) + CO_2(g)$$

52. True or false? Coefficients can be fractions when balancing a chemical equation. Whether true or false, explain why this can or cannot occur.

53. When steel wool (iron) is heated in pure oxygen gas, the steel wool bursts into flame and a fine powder consisting of a mixture of iron oxides (FeO and Fe_2O_3) forms. Write *separate* unbalanced equations for the reaction of iron with oxygen to give each of these products.

54. One method of producing hydrogen peroxide is to add barium peroxide to water. A precipitate of barium oxide forms, which may then be filtered off to leave a solution of hydrogen peroxide. Write the balanced chemical equation for this process.

55. When elemental boron, B, is burned in oxygen gas, the product is diboron trioxide. If the diboron trioxide is then reacted with a measured quantity of water, it reacts with the water to form what is commonly known as boric acid, $B(OH)_3$. Write a balanced chemical equation for each of these processes.

56. A common experiment in introductory chemistry courses involves heating a weighed mixture of potassium chlorate, $KClO_3$, and potassium chloride. Potassium chlorate decomposes when heated, producing potassium chloride and evolving oxygen gas. By measuring the volume of oxygen gas produced in this experiment, students can calculate the relative percentage of $KClO_3$ and KCl in the original mixture. Write the balanced chemical equation for this process.

57. A common demonstration in chemistry courses involves adding a tiny speck of manganese(IV) oxide to a concentrated hydrogen peroxide, H_2O_2, solution. Hydrogen peroxide is unstable, and it decomposes quite spectacularly under these conditions to produce oxygen gas and steam (water vapor). Manganese(IV) oxide is a catalyst for the decomposition of hydrogen peroxide and is not consumed in the reaction. Write the balanced equation for the decomposition reaction of hydrogen peroxide.

58. Balance the following chemical equation.

$$FeO(s) + O_2(g) \rightarrow Fe_2O_3(s)$$

59. Glass is a mixture of several compounds, but a major constituent of most glass is calcium silicate, $CaSiO_3$. Glass can be etched by treatment with hydrogen fluoride: HF attacks the calcium silicate of the glass, producing gaseous and water-soluble products (which can be removed by washing the glass). Balance the following equation for the reaction of hydrogen fluoride with calcium silicate.

$$CaSiO_3(s) + HF(g) \rightarrow CaF_2(aq) + SiF_4(g) + H_2O(l)$$

Richard Megna/Fundamental Photographs © Cengage Learning

60. Balance the following chemical equation.

$$LiAlH_4(s) + AlCl_3(s) \rightarrow AlH_3(s) + LiCl(s)$$

61. If you had a "sour stomach," you might try an over-the-counter antacid tablet to relieve the problem. Can you think of evidence that the action of such an antacid is a chemical reaction?

62. When iron wire is heated in the presence of sulfur, the iron soon begins to glow, and a chunky, blue-black mass of iron(II) sulfide is formed. Write the unbalanced chemical equation for this reaction.

63. When finely divided solid sodium is dropped into a flask containing chlorine gas, an explosion occurs and a fine powder of sodium chloride is deposited on the walls of the flask. Write the unbalanced chemical equation for this process.

64. If aqueous solutions of potassium chromate and barium chloride are mixed, a bright yellow solid (barium chromate) forms and settles out of the mixture, leaving potassium chloride in solution. Write a balanced chemical equation for this process.

65. When hydrogen sulfide, H_2S, gas is bubbled through a solution of lead(II) nitrate, $Pb(NO_3)_2$, a black precipitate of lead(II) sulfide, PbS, forms, and nitric acid, HNO_3, is produced. Write the unbalanced chemical equation for this reaction.

66. If an electric current is passed through aqueous solutions of sodium chloride, sodium bromide, and sodium iodide, the elemental halogens are produced at one electrode in each case, with hydrogen gas being evolved at the other electrode. If the liquid is then evaporated from the mixture, a residue of sodium hydroxide remains. Write balanced chemical equations for these electrolysis reactions.

67. When a strip of magnesium metal is heated in oxygen, it bursts into an intensely white flame and produces a finely powdered dust of magnesium oxide. Write the unbalanced chemical equation for this process.

68. Which of the following statements is *false* for the reaction of hydrogen gas with oxygen gas to produce water? (a, b, and c represent coefficients)

$$a\,H_2(g) + b\,O_2(g) \rightarrow c\,H_2O(g)$$

a. The ratio of "a/c" must always equal one.
b. The sum of $a + b + c$ equals 5 when balanced using the lowest whole-number coefficients.
c. Coefficient b can equal ½ because coefficients can be fractions.
d. The number of atoms on the reactant side must equal the number of atoms on the product side.
e. Subscripts can be changed to balance this equation, just as they can be changed to balance the charges when writing the formula for an ionic compound.

69. When solid red phosphorus, P_4, is burned in air, the phosphorus combines with oxygen, producing a choking cloud of tetraphosphorus decoxide. Write the unbalanced chemical equation for this reaction.

70. When copper(II) oxide is boiled in an aqueous solution of sulfuric acid, a strikingly blue solution of copper(II) sulfate forms along with additional water. Write the unbalanced chemical equation for this reaction.

71. When lead(II) sulfide is heated to high temperatures in a stream of pure oxygen gas, solid lead(II) oxide forms with the release of gaseous sulfur dioxide. Write the unbalanced chemical equation for this reaction.

All even-numbered Questions and Problems have answers in the back of this book and solutions in the *Student Solutions Guide.*

72. Which of the following statements about chemical reactions is *false*?

a. When balancing a chemical equation, all subscripts must be conserved.
b. When one coefficient is doubled, the rest of the coefficients in the balanced equation must also be doubled.
c. The subscripts in a balanced equation tell us the number of atoms in a molecule.
d. The phases in a chemical reaction tell us the nature of the reactants and products.

73. Balance each of the following chemical equations.

a. $Cl_2(g) + KBr(aq) \rightarrow Br_2(l) + KCl(aq)$
b. $Cr(s) + O_2(g) \rightarrow Cr_2O_3(s)$
c. $P_4(s) + H_2(g) \rightarrow PH_3(g)$
d. $Al(s) + H_2SO_4(aq) \rightarrow Al_2(SO_4)_3(aq) + H_2(g)$
e. $PCl_3(l) + H_2O(l) \rightarrow H_3PO_3(aq) + HCl(aq)$
f. $SO_2(g) + O_2(g) \rightarrow SO_3(g)$
g. $C_7H_{16}(l) + O_2(g) \rightarrow CO_2(g) + H_2O(g)$
h. $C_2H_6(g) + O_2(g) \rightarrow CO_2(g) + H_2O(g)$

74. Balance the following chemical equation.

$$Na_2S_2O_3(aq) + I_2(aq) \rightarrow Na_2S_4O_6(aq) + NaI(aq)$$

75. Balance each of the following chemical equations.

a. $SiCl_4(l) + Mg(s) \rightarrow Si(s) + MgCl_2(s)$
b. $NO(g) + Cl_2(g) \rightarrow NOCl(g)$
c. $MnO_2(s) + Al(s) \rightarrow Mn(s) + Al_2O_3(s)$
d. $Cr(s) + S_8(s) \rightarrow Cr_2S_3(s)$
e. $NH_3(g) + F_2(g) \rightarrow NH_4F(s) + NF_3(g)$
f. $Ag_2S(s) + H_2(g) \rightarrow Ag(s) + H_2S(g)$
g. $O_2(g) \rightarrow O_3(g)$
h. $Na_2SO_3(aq) + S_8(s) \rightarrow Na_2S_2O_3(aq)$

76. Using different shapes to distinguish between different elements, draw a balanced equation for the following reaction on the "microscopic" level.

$$NH_3(g) + O_2(g) \rightarrow N_2(g) + H_2O(g)$$

ChemWork Problems

These multiconcept problems (and additional ones) are found interactively online with the same type of assistance a student would get from an instructor.

77. Which of the following statements about chemical equations is (are) *true*?

a. When balancing a chemical equation, you can never change the coefficient in front of any chemical formula.
b. The coefficients in a balanced chemical equation refer to the number of grams of reactants and products.
c. In a chemical equation, the reactants are on the right and the products are on the left.
d. When balancing a chemical equation, you can never change the subscripts of any chemical formula.
e. In chemical reactions, matter is neither created nor destroyed, so a chemical equation must have the same number of atoms on both sides of the equation.

78. Balance the following chemical equations.

$$Fe(s) + O_2(g) \rightarrow Fe_2O_3(s)$$
$$PbO_2(s) \rightarrow PbO(s) + O_2(g)$$
$$H_2O_2(l) \rightarrow O_2(g) + H_2O(l)$$

79. Balance the following chemical equations.

$$MnO_2(s) + CO(g) \rightarrow Mn_2O_3(aq) + CO_2(g)$$
$$Al(s) + H_2SO_4(aq) \rightarrow Al_2(SO_4)_3(aq) + H_2(g)$$
$$C_4H_{10}(g) + O_2(g) \rightarrow CO_2(g) + H_2O(l)$$
$$NH_4I(aq) + Cl_2(g) \rightarrow NH_4Cl(aq) + I_2(g)$$
$$KOH(aq) + H_2SO_4(aq) \rightarrow K_2SO_4(aq) + H_2O(l)$$

7 Reactions in Aqueous Solutions

CHAPTER 7

A yellow precipitate of lead(II) iodide is formed by the reaction of lead(II) nitrate solution and potassium iodide solution.
David Taylor/Science Source

The chemical reactions that are most important to us occur in water—in aqueous solutions. Virtually all of the chemical reactions that keep each of us alive and well take place in the aqueous medium present in our bodies. For example, the oxygen you breathe dissolves in your blood, where it associates with the hemoglobin in the red blood cells. While attached to the hemoglobin it is transported to your cells, where it reacts with fuel (from the food you eat) to provide energy for living. However, the reaction between oxygen and fuel is not direct—the cells are not tiny furnaces. Instead, electrons are transferred from the fuel to a series of molecules that pass them along (this is called the respiratory chain) until they eventually reach oxygen. Many other reactions are also crucial to our health and well-being. You will see numerous examples of these as you continue your study of chemistry.

In this chapter we will study some common types of reactions that take place in water, and we will become familiar with some of the driving forces that make these reactions occur. We will also learn how to predict the products for these reactions and how to write various equations to describe them.

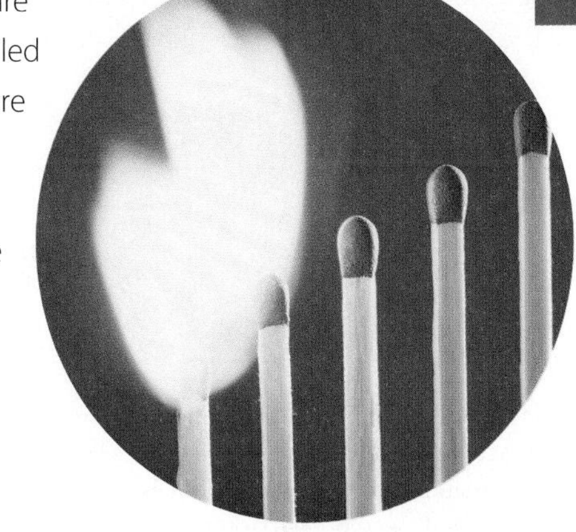

A burning match involves several chemical reactions. Royalty-Free Corbis

7-1 Predicting Whether a Reaction Will Occur

OBJECTIVE To learn about some of the factors that cause reactions to occur.

In this text we have already seen many chemical reactions. Now let's consider an important question: Why does a chemical reaction occur? What causes reactants to "want" to form products? As chemists have studied reactions, they have recognized several "tendencies" in reactants that drive them to form products. That is, there are several "driving forces" that pull reactants toward products—changes that tend to make reactions go in the direction of the arrow. The most common of these driving forces are

1. Formation of a solid
2. Formation of water
3. Transfer of electrons
4. Formation of a gas

When two or more chemicals are brought together, if any of these things can occur, a chemical change (a reaction) is likely to take place. Accordingly, when we are confronted with a set of reactants and want to predict whether a reaction will occur and what products might form, we will consider these driving forces. They will help us organize our thoughts as we encounter new reactions.

7-2 Reactions in Which a Solid Forms

OBJECTIVE To learn to identify the solid that forms in a precipitation reaction.

One driving force for a chemical reaction is the formation of a solid, a process called **precipitation.** The solid that forms is called a **precipitate,** and the reaction is known as a **precipitation reaction.** For example, when an aqueous (water) solution of potas-

Figure 7.1 The precipitation reaction that occurs when yellow potassium chromate, $K_2CrO_4(aq)$, is mixed with a colorless barium nitrate solution, $Ba(NO_3)_2(aq)$.

sium chromate, $K_2CrO_4(aq)$, which is yellow, is added to a colorless aqueous solution containing barium nitrate, $Ba(NO_3)_2(aq)$, a yellow solid forms (Fig. 7.1). The fact that a solid forms tells us that a reaction—a chemical change—has occurred. That is, we have a situation where

$$\text{Reactants} \rightarrow \text{Products}$$

What is the equation that describes this chemical change? To write the equation, we must decipher the identities of the reactants and products. The reactants have already been described: $K_2CrO_4(aq)$ and $Ba(NO_3)_2(aq)$. Is there some way in which we can predict the identities of the products? What is the yellow solid? The best way to predict the identity of this solid is to first *consider what products are possible.* To do this we need to know what chemical species are present in the solution that results when the reactant solutions are mixed. First, let's think about the nature of each reactant in an aqueous solution.

What Happens When an Ionic Compound Dissolves in Water?

The designation $Ba(NO_3)_2(aq)$ means that barium nitrate (a white solid) has been dissolved in water. Note from its formula that barium nitrate contains the Ba^{2+} and NO_3^- ions. *In virtually every case when a solid containing ions dissolves in water, the ions separate* and move around independently. That is, $Ba(NO_3)_2(aq)$ does not contain $Ba(NO_3)_2$ units. Rather, it contains separated Ba^{2+} and NO_3^- ions. In the solution there are two NO_3^- ions for every Ba^{2+} ion. Chemists know that separated ions are present in this solution because it is an excellent conductor of electricity (Fig. 7.2). Pure water does not conduct an electric current. Ions must be present in water for a current to flow.

When each unit of a substance that dissolves in water produces separated ions, the substance is called a **strong electrolyte.** Barium nitrate is a strong electrolyte in water, because each $Ba(NO_3)_2$ unit produces the separated ions (Ba^{2+}, NO_3^-, NO_3^-).

Similarly, aqueous K_2CrO_4 also behaves as a strong electrolyte. Potassium chromate contains the K^+ and CrO_4^{2-} ions, so an aqueous solution of potassium chromate (which is prepared by dissolving solid K_2CrO_4 in water) contains these separated ions. That is, $K_2CrO_4(aq)$ does not contain K_2CrO_4 units but instead contains K^+ cations and CrO_4^{2-} anions, which move around independently. (There are two K^+ ions for each CrO_4^{2-} ion.)

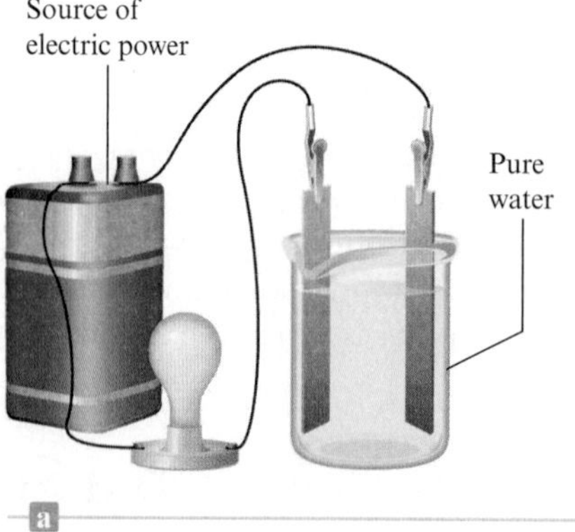

a

Pure water does not conduct an electric current. The lamp does not light.

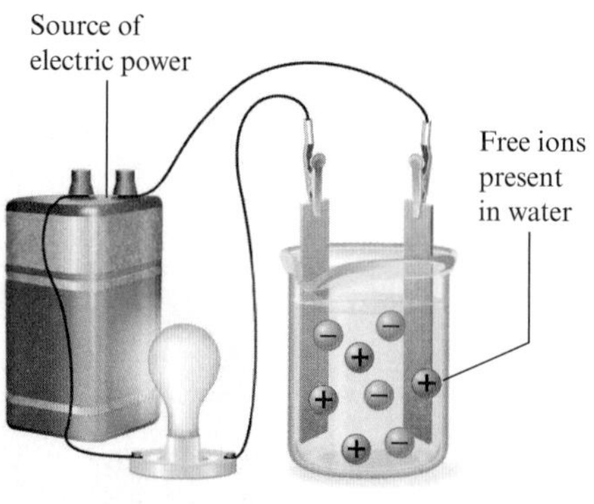

b

When an ionic compound is dissolved in water, current flows and the lamp lights. The result of this experiment is strong evidence that ionic compounds dissolved in water exist in the form of separated ions.

Figure 7.2 Electrical conductivity of aqueous solutions.

The idea introduced here is very important: when ionic compounds dissolve, the *resulting solution contains the separated ions*. Therefore, we can represent the mixing of $K_2CrO_4(aq)$ and $Ba(NO_3)_2(aq)$ in two ways. We usually write these reactants as

$$K_2CrO_4(aq) + Ba(NO_3)_2(aq) \rightarrow \text{Products}$$

However, a more accurate representation of the situation is

We can express this information in equation form as follows:

$$2K^+(aq) + CrO_4^{2-}(aq) + Ba^{2+}(aq) + 2NO_3^-(aq) \rightarrow \text{Products}$$

The ions in $K_2CrO_4(aq)$ The ions in $Ba(NO_3)_2(aq)$

Thus the *mixed solution* contains four types of ions: K^+, CrO_4^{2-}, Ba^{2+}, and NO_3^-. Now that we know what the reactants are, we can make some educated guesses about the possible products.

How to Decide What Products Form

Which of these ions combine to form the yellow solid observed when the original solutions are mixed? This is not an easy question to answer. Even an experienced chemist is not sure what will happen in a new reaction. The chemist tries to think of the various possibilities, considers the likelihood of each possibility, and then makes a prediction (an educated guess). Only after identifying each product experimentally can the chemist be sure what reaction actually has taken place. However, an educated guess is very useful because it indicates what kinds of products are most likely. It gives us a place to start. So the best way to proceed is first to think of the various possibilities and then to decide which of them is most likely.

What are the possible products of the reaction between $K_2CrO_4(aq)$ and $Ba(NO_3)_2(aq)$ or, more accurately, what reaction can occur among the ions K^+, CrO_4^{2-}, Ba^{2+}, and NO_3^-? We already know some things that will help us decide. We know that a *solid compound must have a zero net charge.* This means that the product of our reaction must contain *both anions and cations* (negative and positive ions). For example, K^+ and Ba^{2+} could not combine to form the solid because such a solid would have a positive charge. Similarly, CrO_4^{2-} and NO_3^- could not combine to form a solid because that solid would have a negative charge.

Something else that will help us is an observation that chemists have made by examining many compounds: *most ionic materials contain only two types of ions*—one type of cation and one type of anion. This idea is illustrated by the following compounds (among many others):

Compound	Cation	Anion
$NaCl$	Na^+	Cl^-
KOH	K^+	OH^-
Na_2SO_4	Na^+	SO_4^{2-}
NH_4Cl	NH_4^+	Cl^-
Na_2CO_3	Na^+	CO_3^{2-}

All the possible combinations of a cation and an anion to form uncharged compounds from among the ions K^+, CrO_4^{2-}, Ba^{2+}, and NO_3^- are shown below:

	NO_3^-	CrO_4^{2-}
K^+	KNO_3	K_2CrO_4
Ba^{2+}	$Ba(NO_3)_2$	$BaCrO_4$

So the compounds that *might* make up the solid are

K_2CrO_4	$BaCrO_4$
KNO_3	$Ba(NO_3)_2$

Which of these possibilities is most likely to represent the yellow solid? We know it's not K_2CrO_4 or $Ba(NO_3)_2$; these are the reactants. They were present (dissolved) in the separate solutions that were mixed initially. The only real possibilities are KNO_3 and $BaCrO_4$. To decide which of these is more likely to represent the yellow solid, we need more facts. An experienced chemist, for example, knows that KNO_3 is a white solid. On the other hand, the CrO_4^{2-} ion is yellow. Therefore, the yellow solid most likely is $BaCrO_4$.

We have determined that one product of the reaction between $K_2CrO_4(aq)$ and $Ba(NO_3)_2(aq)$ is $BaCrO_4(s)$, but what happened to the K^+ and NO_3^- ions? The answer is that these ions are left dissolved in the solution. That is, KNO_3 does not form a solid when the K^+ and NO_3^- ions are present in water. In other words, if we took the white solid $KNO_3(s)$ and put it in water, it would totally dissolve (the white solid would "disappear," yielding a colorless solution). So when we mix $K_2CrO_4(aq)$ and $Ba(NO_3)_2(aq)$, $BaCrO_4(s)$ forms but KNO_3 is left behind in solution [we write it as $KNO_3(aq)$]. (If we poured the mixture through a filter to remove the solid $BaCrO_4$ and then evaporated all of the water, we would obtain the white solid KNO_3.)

After all this thinking, we can finally write the unbalanced equation for the precipitation reaction:

$$K_2CrO_4(aq) + Ba(NO_3)_2(aq) \rightarrow BaCrO_4(s) + KNO_3(aq)$$

We can represent this reaction in pictures as follows:

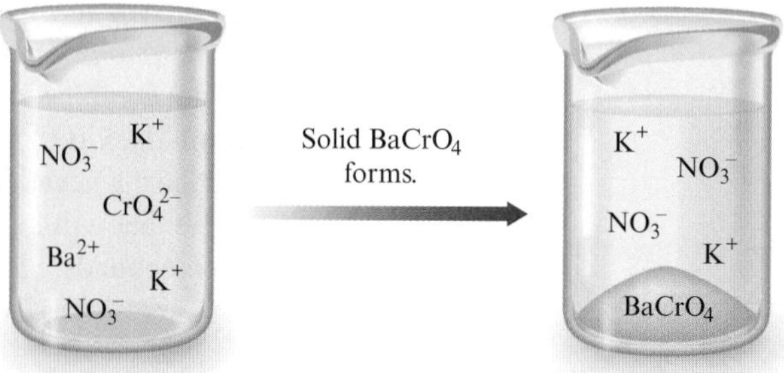

Note that the K^+ and NO_3^- ions are not involved in the chemical change. They remain dispersed in the water before and after the reaction.

Using Solubility Rules

In the example considered above we were finally able to identify the products of the reaction by using two types of chemical knowledge:

1. Knowledge of facts
2. Knowledge of concepts

For example, knowing the colors of the various compounds proved very helpful. This represents factual knowledge. Awareness of the concept that solids always have a net charge of zero was also essential. These two kinds of knowledge allowed us to make a good guess about the identity of the solid that formed. As you continue to study chemistry, you will see that a balance of factual and conceptual knowledge is always required. You must both *memorize* important facts and *understand* crucial concepts to succeed.

In the present case we are dealing with a reaction in which an ionic solid forms—that is, a process in which ions that are dissolved in water combine to give a solid. We know that for a solid to form, both positive and negative ions must be present in relative numbers that give zero net charge. However, oppositely charged ions in water do not always react to form a solid, as we have seen for K^+ and NO_3^-. In addition, Na^+ and Cl^- can coexist in water in very large numbers with no formation of solid NaCl. In other words, when solid NaCl (common salt) is placed in water, it dissolves—the white solid "disappears" as the Na^+ and Cl^- ions are dispersed throughout the water. (You probably have observed this phenomenon in preparing salt water to cook food.) The following two statements, then, are really saying the same thing.

1. Solid NaCl is very soluble in water.
2. Solid NaCl does not form when one solution containing Na^+ is mixed with another solution containing Cl^-.

To predict whether a given pair of dissolved ions will form a solid when mixed, we must know some facts about the solubilities of various types of ionic compounds. In this text we will use the term **soluble solid** to mean a solid that readily dissolves in water; the solid "disappears" as the ions are dispersed in the water. The terms **insoluble solid** and **slightly soluble** solid are taken to mean the same thing: a solid where such a tiny amount dissolves in water that it is undetectable with the naked eye. The solubility information about common solids that is summarized in Table 7.1 is based on observations of the behavior of many compounds. This is factual knowledge that you will need to predict what will happen in chemical reactions where a solid might form. This information is summarized in Fig. 7.3.

Notice that in Table 7.1 and Fig. 7.3 the term *salt* is used to mean *ionic compound.* Many chemists use the terms *salt* and *ionic compound* interchangeably. In Example 7.1, we will illustrate how to use the solubility rules to predict the products of reactions among ions.

Table 7.1 General Rules for Solubility of Ionic Compounds (Salts) in Water at 25 °C

1. Most nitrate (NO_3^-) salts are soluble.
2. Most salts of Na^+, K^+, and NH_4^+ are soluble.
3. Most chloride salts are soluble. Notable exceptions are AgCl, $PbCl_2$, and Hg_2Cl_2.
4. Most sulfate salts are soluble. Notable exceptions are $BaSO_4$, $PbSO_4$, and $CaSO_4$.
5. Most hydroxide compounds are only slightly soluble.* The important exceptions are NaOH and KOH. $Ba(OH)_2$ and $Ca(OH)_2$ are only moderately soluble.
6. Most sulfide (S^{2-}), carbonate (CO_3^{2-}), and phosphate (PO_4^{3-}) salts are only slightly soluble.*

*The terms *insoluble* and *slightly soluble* really mean the same thing: such a tiny amount dissolves that it is not possible to detect it with the naked eye.

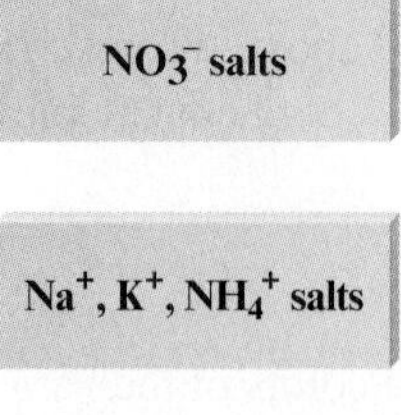

Cl^-, Br^-, I^- salts — Except for those containing Ag^+, Hg_2^{2+}, Pb^{2+}

SO_4^{2-} salts — Except for those containing Ba^{2+}, Pb^{2+}, Ca^{2+}

Insoluble compounds

S^{2-}, CO_3^{2-}, PO_4^{3-} salts

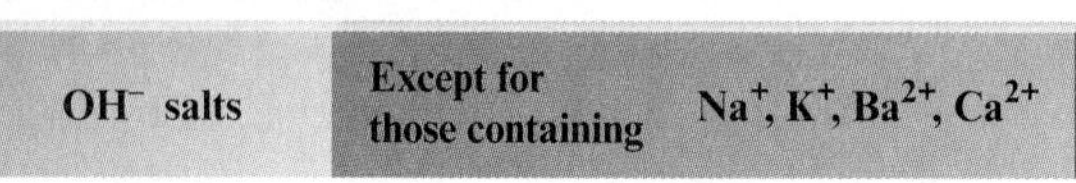

Figure 7.3 ▸ Solubilities of common compounds.

Critical Thinking

What if no ionic solids were soluble in water? Could reactions occur in aqueous solutions?

Interactive Example 7.1

Identifying Precipitates in Reactions Where a Solid Forms

$AgNO_3$ is usually called silver nitrate rather than silver(I) nitrate because silver forms only Ag^+.

When an aqueous solution of silver nitrate is added to an aqueous solution of potassium chloride, a white solid forms. Identify the white solid and write the balanced equation for the reaction that occurs.

SOLUTION First let's use the description of the reaction to represent what we know:

$$AgNO_3(aq) + KCl(aq) \rightarrow \text{White solid}$$

Remember, try to determine the essential facts from the words and represent these facts by symbols or diagrams. To answer the main question (What is the white solid?), we must establish what ions are present in the mixed solution. That is, we must know what the reactants are really like. Remember that *when ionic substances dissolve in water, the ions separate.* So we can write the equation

$$\underbrace{Ag^+(aq) + NO_3^-(aq)}_{\text{Ions in } AgNO_3(aq)} + \underbrace{K^+(aq) + Cl^-(aq)}_{\text{Ions in } KCl(aq)} \rightarrow \text{Products}$$

or using pictures

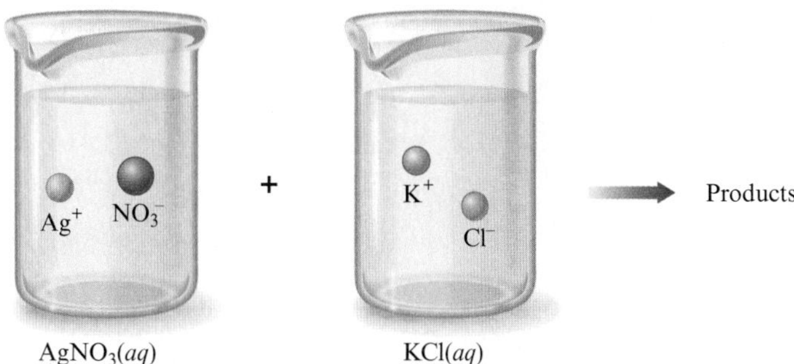

to represent the ions present in the mixed solution before any reaction occurs. In summary:

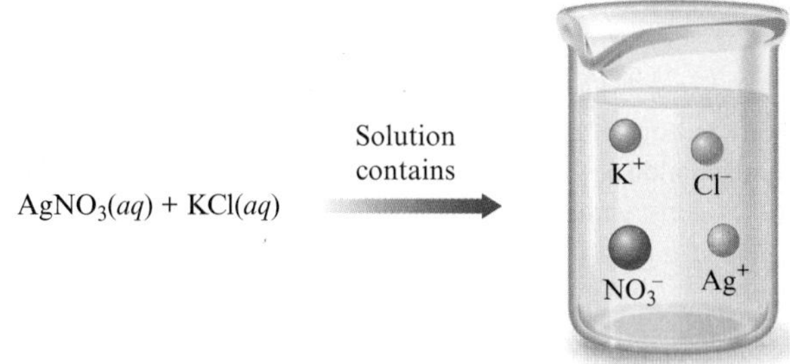

	NO_3^-	Cl^-
Ag^+	$AgNO_3$	$AgCl$
K^+	KNO_3	KCl

Now we will consider what solid *might* form from this collection of ions. Because the solid must contain both positive and negative ions, the possible compounds that can be assembled from this collection of ions are

$AgNO_3$	$AgCl$
KNO_3	KCl

$AgNO_3$ and KCl are the substances already dissolved in the reactant solutions, so we know that they do not represent the white solid product. We are left with two possibilities:

AgCl

KNO_3

Another way to obtain these two possibilities is by *ion interchange*. This means that in the reaction of $AgNO_3(aq)$ and $KCl(aq)$, we take the cation from one reactant and combine it with the anion of the other reactant.

$$Ag^+ + NO_3^- + K^+ + Cl^- \rightarrow \text{Products}$$

Possible solid products

Ion interchange also leads to the following possible solids:

AgCl or KNO_3

To decide whether AgCl or KNO_3 is the white solid, we need the solubility rules (see Table 7.1). Rule 2 states that most salts containing K^+ are soluble in water. Rule 1 says that most nitrate salts (those containing NO_3^-) are soluble. So the salt KNO_3 is water-soluble. That is, when K^+ and NO_3^- are mixed in water, a solid (KNO_3) does *not* form.

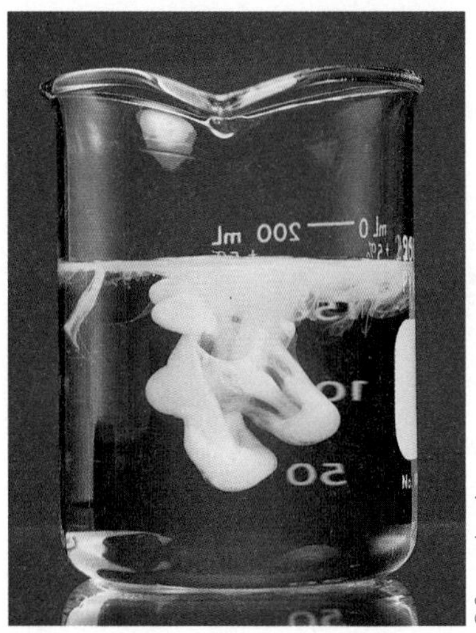

Figure 7.4 Precipitation of silver chloride occurs when solutions of silver nitrate and potassium chloride are mixed. The K^+ and NO_3^- ions remain in solution.

On the other hand, Rule 3 states that although most chloride salts (salts that contain Cl^-) are soluble, AgCl is an exception. That is, AgCl(*s*) is insoluble in water. Thus the white solid must be AgCl. Now we can write

$$AgNO_3(aq) + KCl(aq) \rightarrow AgCl(s) + ?$$

What is the other product?

To form AgCl(*s*), we have used the Ag^+ and Cl^- ions:

$$Ag^+(aq) + NO_3^-(aq) + K^+(aq) + Cl^-(aq) \rightarrow AgCl(s)$$

This leaves the K^+ and NO_3^- ions. What do they do? Nothing. Because KNO_3 is very soluble in water (Rules 1 and 2), the K^+ and NO_3^- ions remain separate in the water; the KNO_3 remains dissolved and we represent it as $KNO_3(aq)$. We can now write the full equation:

$$AgNO_3(aq) + KCl(aq) \rightarrow AgCl(s) + KNO_3(aq)$$

Figure 7.4 shows the precipitation of AgCl(*s*) that occurs when this reaction takes place. In graphic form, the reaction is

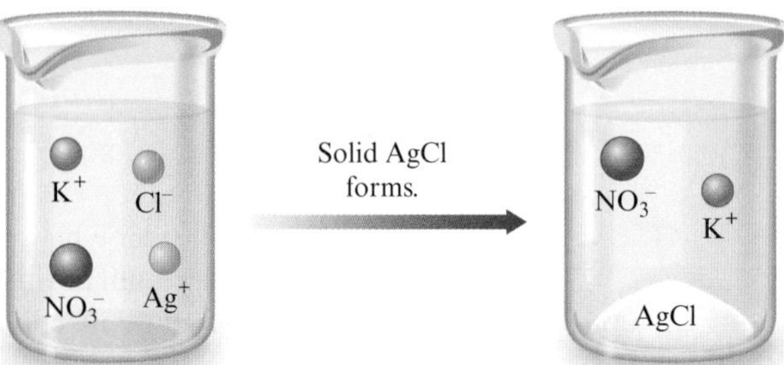

■

The following strategy is useful for predicting what will occur when two solutions containing dissolved salts are mixed.

How to Predict Precipitates When Solutions of Two Ionic Compounds Are Mixed

Step 1 Write the reactants as they actually exist before any reaction occurs. Remember that when a salt dissolves, its ions separate.

Step 2 Consider the various solids that could form. To do this, simply *exchange the anions* of the added salts.

Step 3 Use the solubility rules (Table 7.1) to decide whether a solid forms and, if so, to predict the identity of the solid.

Interactive Example 7.2

Using Solubility Rules to Predict the Products of Reactions

Using the solubility rules in Table 7.1, predict what will happen when the following solutions are mixed. Write the balanced equation for any reaction that occurs.

a. $KNO_3(aq)$ and $BaCl_2(aq)$

b. $Na_2SO_4(aq)$ and $Pb(NO_3)_2(aq)$

c. $KOH(aq)$ and $Fe(NO_3)_3(aq)$

SOLUTION (a)

Step 1 $KNO_3(aq)$ represents an aqueous solution obtained by dissolving solid KNO_3 in water to give the ions $K^+(aq)$ and $NO_3^-(aq)$. Likewise, $BaCl_2(aq)$ is a solution formed by dissolving solid $BaCl_2$ in water to produce $Ba^{2+}(aq)$ and $Cl^-(aq)$. When these two solutions are mixed, the following ions will be present:

$$\underbrace{K^+, \quad NO_3^-,}_{\text{From } KNO_3(aq)} \quad \underbrace{Ba^{2+}, \quad Cl^-}_{\text{From } BaCl_2(aq)}$$

Step 2 To get the possible products, we exchange the anions.

$$K^+ \quad NO_3^- \quad Ba^{2+} \quad Cl^-$$

This yields the possibilities KCl and $Ba(NO_3)_2$. These are the solids that *might* form. Notice that two NO_3^- ions are needed to balance the 2+ charge on Ba^{2+}.

Step 3 The rules listed in Table 7.1 indicate that both KCl and $Ba(NO_3)_2$ are soluble in water. So no precipitate forms when $KNO_3(aq)$ and $BaCl_2(aq)$ are mixed. All of the ions remain dissolved in the solution. This means that no reaction takes place. That is, no chemical change occurs.

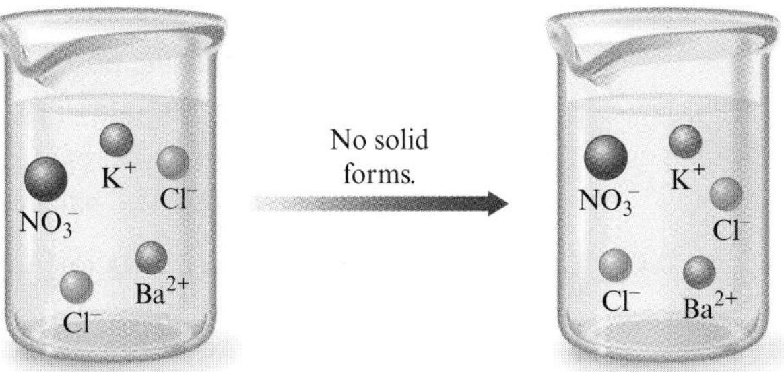

SOLUTION (b)

Step 1 The following ions are present in the mixed solution before any reaction occurs:

$$\underbrace{Na^+, \quad SO_4^{2-},}_{\text{From } Na_2SO_4(aq)} \quad \underbrace{Pb^{2+}, \quad NO_3^-}_{\text{From } Pb(NO_3)_2(aq)}$$

Step 2 Exchanging anions

$$Na^+ \quad SO_4^{2-} \quad Pb^{2+} \quad NO_3^-$$

yields the *possible* solid products $PbSO_4$ and $NaNO_3$.

Step 3 Using Table 7.1, we see that $NaNO_3$ is soluble in water (Rules 1 and 2) but that $PbSO_4$ is only slightly soluble (Rule 4). Thus, when these solutions are mixed, solid $PbSO_4$ forms. The balanced reaction is

$$Na_2SO_4(aq) + Pb(NO_3)_2(aq) \rightarrow PbSO_4(s) + \underbrace{2NaNO_3(aq)}_{\text{Remains dissolved}}$$

which can be represented as

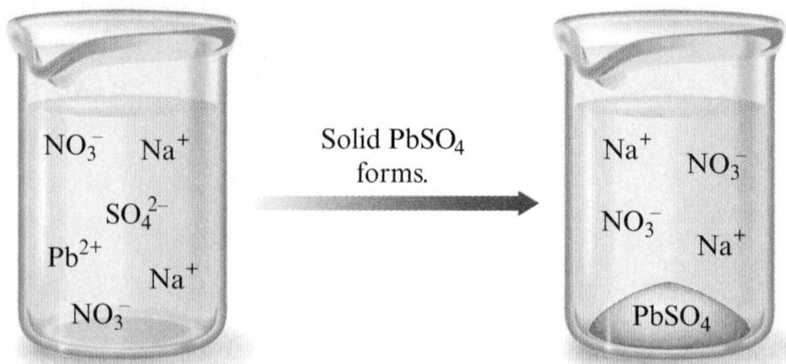

SOLUTION (c)

Step 1 The ions present in the mixed solution before any reaction occurs are

$$K^+, \quad OH^-, \quad Fe^{3+}, \quad NO_3^-$$

From KOH(*aq*) From $Fe(NO_3)_3(aq)$

Step 2 Exchanging anions

$$K^+ \quad OH^- \quad Fe^{3+} \quad NO_3^-$$

yields the possible solid products KNO_3 and $Fe(OH)_3$.

Step 3 Rules 1 and 2 (Table 7.1) state that KNO_3 is soluble, whereas $Fe(OH)_3$ is only slightly soluble (Rule 5). Thus, when these solutions are mixed, solid $Fe(OH)_3$ forms. The balanced equation for the reaction is

$$3KOH(aq) + Fe(NO_3)_3(aq) \rightarrow Fe(OH)_3(s) + 3KNO_3(aq)$$

which can be represented as

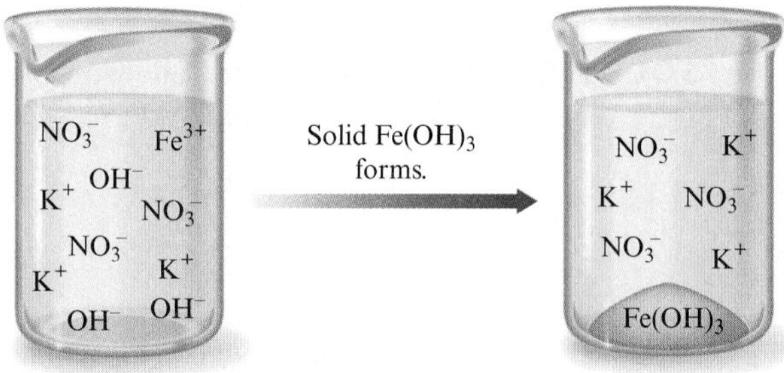

SELF CHECK

Exercise 7.1 Predict whether a solid will form when the following pairs of solutions are mixed. If so, identify the solid and write the balanced equation for the reaction.

a. $Ba(NO_3)_2(aq)$ and $NaCl(aq)$

b. $Na_2S(aq)$ and $Cu(NO_3)_2(aq)$

c. $NH_4Cl(aq)$ and $Pb(NO_3)_2(aq)$

See Problems 7.17 and 7.18. ■

Describing Reactions in Aqueous Solutions

OBJECTIVE **To learn to describe reactions in solutions by writing molecular, complete ionic, and net ionic equations.**

Much important chemistry, including virtually all of the reactions that make life possible, occurs in aqueous solutions. We will now consider the types of equations used to represent reactions that occur in water. For example, as we saw earlier, when we mix aqueous potassium chromate with aqueous barium nitrate, a reaction occurs to form solid barium chromate and dissolved potassium nitrate. One way to represent this reaction is by the equation

$$K_2CrO_4(aq) + Ba(NO_3)_2(aq) \rightarrow BaCrO_4(s) + 2KNO_3(aq)$$

This is called the **molecular equation** for the reaction; it shows the complete formulas of all reactants and products. However, although this equation shows the reactants and products of the reaction, it does not give a very clear picture of what actually occurs in solution. As we have seen, aqueous solutions of potassium chromate, barium nitrate, and potassium nitrate contain the individual ions, not molecules as is implied by the molecular equation. Thus the **complete ionic equation,**

Ions from K_2CrO_4 — Ions from $Ba(NO_3)_2$

$$\underbrace{2K^+(aq) + CrO_4^{2-}(aq)}_{\text{Ions from } K_2CrO_4} + \underbrace{Ba^{2+}(aq) + 2NO_3^-(aq)}_{\text{Ions from } Ba(NO_3)_2} \rightarrow BaCrO_4(s) + 2K^+(aq) + 2NO_3^-(aq)$$

better represents the actual forms of the reactants and products in solution. *In a complete ionic equation, all substances that are strong electrolytes are represented as ions.* Notice that $BaCrO_4$ is not written as the separate ions, because it is present as a solid; it is not dissolved.

The complete ionic equation reveals that only some of the ions participate in the reaction. Notice that the K^+ and NO_3^- ions are present in solution both before and after the reaction. Ions such as these, which do not participate directly in a reaction in solution, are called **spectator ions.** The ions that participate in this reaction are the Ba^{2+} and CrO_4^{2-} ions, which combine to form solid $BaCrO_4$:

$$Ba^{2+}(aq) + CrO_4^{2-}(aq) \rightarrow BaCrO_4(s)$$

This equation, called the **net ionic equation,** includes only those components that are directly involved in the reaction. Chemists usually write the net ionic equation for a reaction in solution, because it gives the actual forms of the reactants and products and includes only the species that undergo a change.

Types of Equations for Reactions in Aqueous Solutions

Three types of equations are used to describe reactions in solutions.

1. The *molecular equation* shows the overall reaction but not necessarily the actual forms of the reactants and products in solution.
2. The *complete ionic equation* represents all reactants and products that are strong electrolytes as ions. All reactants and products are included.
3. The *net ionic equation* includes only those components that undergo a change. Spectator ions are not included.

To make sure these ideas are clear, we will do another example. In Example 7.2 we considered the reaction between aqueous solutions of lead nitrate and sodium sulfate. The molecular equation for this reaction is

$$Pb(NO_3)_2(aq) + Na_2SO_4(aq) \rightarrow PbSO_4(s) + 2NaNO_3(aq)$$

Because any ionic compound that is dissolved in water is present as the separated ions, we can write the complete ionic equation as follows:

$$Pb^{2+}(aq) + 2NO_3^-(aq) + 2Na^+(aq) + SO_4^{2-}(aq) \rightarrow PbSO_4(s) + 2Na^+(aq) + 2NO_3^-(aq)$$

The $PbSO_4$ is not written as separate ions because it is present as a solid. The ions that take part in the chemical change are the Pb^{2+} and the SO_4^{2-} ions, which combine to form solid $PbSO_4$. Thus the net ionic equation is

$$Pb^{2+}(aq) + SO_4^{2-}(aq) \rightarrow PbSO_4(s)$$

The Na^+ and NO_3^- ions do not undergo any chemical change; they are spectator ions.

Interactive Example 7.3

Writing Equations for Reactions

For each of the following reactions, write the molecular equation, the complete ionic equation, and the net ionic equation.

a. Aqueous sodium chloride is added to aqueous silver nitrate to form solid silver chloride plus aqueous sodium nitrate.

b. Aqueous potassium hydroxide is mixed with aqueous iron(III) nitrate to form solid iron(III) hydroxide and aqueous potassium nitrate.

SOLUTION

a. *Molecular equation:*

$$NaCl(aq) + AgNO_3(aq) \rightarrow AgCl(s) + NaNO_3(aq)$$

Complete ionic equation:

$$Na^+(aq) + Cl^-(aq) + Ag^+(aq) + NO_3^-(aq) \rightarrow AgCl(s) + Na^+(aq) + NO_3^-(aq)$$

Net ionic equation:

$$Cl^-(aq) + Ag^+(aq) \rightarrow AgCl(s)$$

b. *Molecular equation:*

$$3KOH(aq) + Fe(NO_3)_3(aq) \rightarrow Fe(OH)_3(s) + 3KNO_3(aq)$$

Complete ionic equation:

$$3K^+(aq) + 3OH^-(aq) + Fe^{3+}(aq) + 3NO_3^-(aq) \rightarrow Fe(OH)_3(s) + 3K^+(aq) + 3NO_3^-(aq)$$

Net ionic equation:

$$3OH^-(aq) + Fe^{3+}(aq) \rightarrow Fe(OH)_3(s)$$

SELF CHECK

Exercise 7.2 For each of the following reactions, write the molecular equation, the complete ionic equation, and the net ionic equation.

a. Aqueous sodium sulfide is mixed with aqueous copper(II) nitrate to produce solid copper(II) sulfide and aqueous sodium nitrate.

b. Aqueous ammonium chloride and aqueous lead(II) nitrate react to form solid lead(II) chloride and aqueous ammonium nitrate.

See Problems 7.25 through 7.30. ■

7-4 Reactions That Form Water: Acids and Bases

OBJECTIVE To learn the key characteristics of the reactions between strong acids and strong bases.

Don't taste chemicals!

In this section we encounter two very important classes of compounds: acids and bases. Acids were first associated with the sour taste of citrus fruits. In fact, the word *acid* comes from the Latin word *acidus,* which means "sour." Vinegar tastes sour because it is a dilute solution of acetic acid; citric acid is responsible for the sour taste of a lemon. Bases, sometimes called *alkalis,* are characterized by their bitter taste and slippery feel, like wet soap. Most commercial preparations for unclogging drains are highly basic.

Acids have been known for hundreds of years. For example, the *mineral acids* sulfuric acid, H_2SO_4, and nitric acid, HNO_3, so named because they were originally obtained by the treatment of minerals, were discovered around 1300. However, it was not until the late 1800s that the essential nature of acids was discovered by Svante Arrhenius, then a Swedish graduate student in physics.

Arrhenius, who was trying to discover why only certain solutions could conduct an electric current, found that conductivity arose from the presence of ions. In his studies of solutions, Arrhenius observed that when the substances HCl, HNO_3, and H_2SO_4 were dissolved in water, they behaved as strong electrolytes. He suggested that this was the result of ionization reactions in water.

$$HCl \xrightarrow{H_2O} H^+(aq) + Cl^-(aq)$$

$$HNO_3 \xrightarrow{H_2O} H^+(aq) + NO_3^-(aq)$$

$$H_2SO_4 \xrightarrow{H_2O} H^+(aq) + HSO_4^-(aq)$$

Arrhenius proposed that an **acid** *is a substance that produces* H^+ *ions (protons) when it is dissolved in water.*

Studies show that when HCl, HNO_3, and H_2SO_4 are placed in water, *virtually every molecule* dissociates to give ions. This means that when 100 molecules of HCl are dissolved in water, 100 H^+ ions and 100 Cl^- ions are produced. Virtually no HCl molecules exist in aqueous solution (Fig. 7.5). Because these substances are strong electrolytes that produce H^+ ions, they are called **strong acids.**

Arrhenius also found that *aqueous solutions that exhibit basic behavior* always contain hydroxide ions. He defined a **base** as a *substance that produces hydroxide ions* (OH^-) *in water.* The base most commonly used in the chemical laboratory is sodium hydroxide, NaOH, which contains Na^+ and OH^- ions and is very soluble in water.

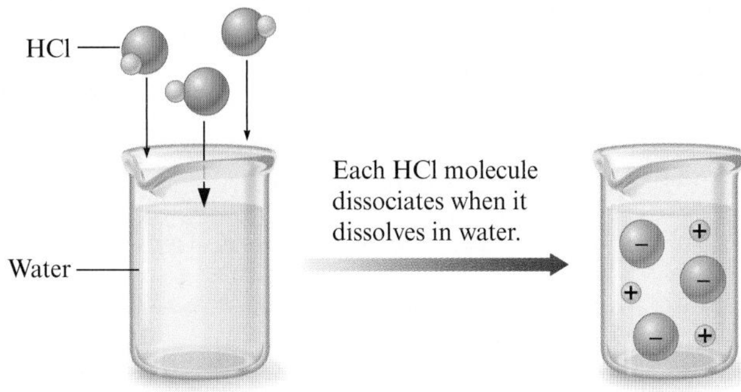

Figure 7.5 When gaseous HCl is dissolved in water, each molecule dissociates to produce H^+ and Cl^- ions. That is, HCl behaves as a strong electrolyte.

Kuttelvaserova Stuchelova/Shutterstock

The marsh marigold is a beautiful but poisonous plant. Its toxicity results partly from the presence of erucic acid.

Sodium hydroxide, like all ionic substances, produces separated cations and anions when it is dissolved in water.

$$NaOH(s) \xrightarrow{H_2O} Na^+(aq) + OH^-(aq)$$

Although dissolved sodium hydroxide is usually represented as NaOH(*aq*), you should remember that the solution really contains separated Na^+ and OH^- ions. In fact, for every 100 units of NaOH dissolved in water, 100 Na^+ and 100 OH^- ions are produced.

Potassium hydroxide (KOH) has properties markedly similar to those of sodium hydroxide. It is very soluble in water and produces separated ions.

$$KOH(s) \xrightarrow{H_2O} K^+(aq) + OH^-(aq)$$

Because these hydroxide compounds are strong electrolytes that contain OH^- ions, they are called **strong bases.**

When strong acids and strong bases (hydroxides) are mixed, the fundamental chemical change that always occurs is that *H^+ ions react with OH^- ions to form water.*

$$H^+(aq) + OH^-(aq) \rightarrow H_2O(l)$$

Water is a very stable compound, as evidenced by the abundance of it on the earth's surface. Therefore, when substances that can form water are mixed, there is a strong tendency for the reaction to occur. In particular, the hydroxide ion OH^- has a high affinity for H^+ ions, because water is produced in the reaction between these ions.

Hydrochloric acid is an aqueous solution that contains dissolved hydrogen chloride. It is a strong electrolyte.

The tendency to form water is the second of the driving forces for reactions that we mentioned in Section 7-1. Any compound that produces OH^- ions in water reacts vigorously with any compound that can furnish H^+ ions to form H_2O. For example, the reaction between hydrochloric acid and aqueous sodium hydroxide is represented by the following molecular equation:

$$HCl(aq) + NaOH(aq) \rightarrow H_2O(l) + NaCl(aq)$$

Because HCl, NaOH, and NaCl exist as completely separated ions in water, the complete ionic equation for this reaction is

$$H^+(aq) + Cl^-(aq) + Na^+(aq) + OH^-(aq) \rightarrow H_2O(l) + Na^+(aq) + Cl^-(aq)$$

Notice that the Cl^- and Na^+ are spectator ions (they undergo no changes), so the net ionic equation is

$$H^+(aq) + OH^-(aq) \rightarrow H_2O(l)$$

Thus the only chemical change that occurs when these solutions are mixed is that water is formed from H^+ and OH^- ions.

Interactive Example 7.4

Writing Equations for Acid–Base Reactions

Nitric acid is a strong acid. Write the molecular, complete ionic, and net ionic equations for the reaction of aqueous nitric acid and aqueous potassium hydroxide.

SOLUTION *Molecular equation:*

$$HNO_3(aq) + KOH(aq) \rightarrow H_2O(l) + KNO_3(aq)$$

Complete ionic equation:

$$H^+(aq) + NO_3^-(aq) + K^+(aq) + OH^-(aq) \rightarrow H_2O(l) + K^+(aq) + NO_3^-(aq)$$

Net ionic equation:

$$H^+(aq) + OH^-(aq) \rightarrow H_2O(l)$$

Note that K^+ and NO_3^- are spectator ions and that the formation of water is the driving force for this reaction. ■

There are two important things to note as we examine the reaction of hydrochloric acid with aqueous sodium hydroxide and the reaction of nitric acid with aqueous potassium hydroxide.

1. The net ionic equation is the same in both cases; water is formed.

$$H^+(aq) + OH^-(aq) \rightarrow H_2O(l)$$

2. Besides water, which is *always a product* of the reaction of an acid with OH^-, the second product is an ionic compound, which might precipitate or remain dissolved, depending on its solubility.

$$HCl(aq) + NaOH(aq) \rightarrow H_2O(l) + NaCl(aq)$$

$$HNO_3(aq) + KOH(aq) \rightarrow H_2O(l) + KNO_3(aq)$$

Dissolved ionic compounds

This ionic compound is called a **salt.** In the first case the salt is sodium chloride, and in the second case the salt is potassium nitrate. We can obtain these soluble salts in solid form (both are white solids) by evaporating the water.

Drano contains a strong base.

Summary of Strong Acids and Strong Bases

The following points about strong acids and strong bases are particularly important.

1. The common strong acids are aqueous solutions of HCl, HNO_3, and H_2SO_4.
2. A strong acid is a substance that completely dissociates (ionizes) in water. (Each molecule breaks up into an H^+ ion plus an anion.)
3. A strong base is a metal hydroxide compound that is very soluble in water. The most common strong bases are NaOH and KOH, which completely break up into separated ions (Na^+ and OH^- or K^+ and OH^-) when they are dissolved in water.
4. The net ionic equation for the reaction of a strong acid and a strong base (contains OH^-) is always the same: it shows the production of water.

$$H^+(aq) + OH^-(aq) \rightarrow H_2O(l)$$

5. In the reaction of a strong acid and a strong base, one product is always water and the other is always an ionic compound called a salt, which remains dissolved in the water. This salt can be obtained as a solid by evaporating the water.
6. The reaction of H^+ and OH^- is often called an acid–base reaction, where H^+ is the acidic ion and OH^- is the basic ion.

7-5 Reactions of Metals with Nonmetals (Oxidation–Reduction)

OBJECTIVES

- **To learn the general characteristics of a reaction between a metal and a nonmetal.**
- **To understand electron transfer as a driving force for a chemical reaction.**

In Chapter 4 we spent considerable time discussing ionic compounds—compounds formed in the reaction of a metal and a nonmetal. A typical example is sodium chloride, formed by the reaction of sodium metal and chlorine gas:

$$2Na(s) + Cl_2(g) \rightarrow 2NaCl(s)$$

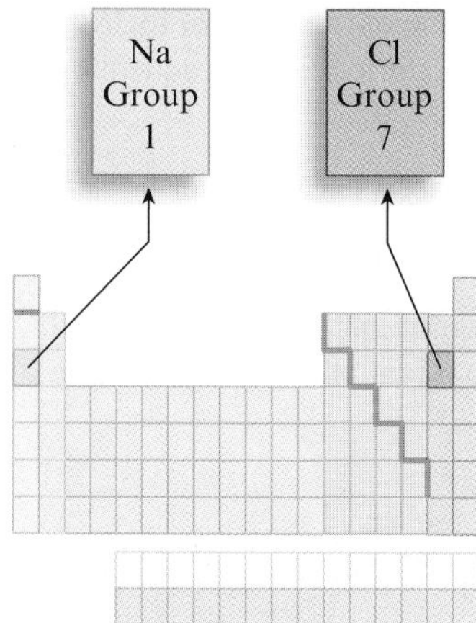

Let's examine what happens in this reaction. Sodium metal is composed of sodium atoms, each of which has a net charge of zero. (The positive charges of the 11 protons in its nucleus are exactly balanced by the negative charges on the 11 electrons.) Similarly, the chlorine molecule consists of two uncharged chlorine atoms (each has 17 protons and 17 electrons). However, in the product (sodium chloride), the sodium is present as Na^+ and the chlorine as Cl^-. By what process do the neutral atoms become ions? The answer is that one electron is transferred from each sodium atom to each chlorine atom.

$$Na + Cl \rightarrow Na^+ + Cl^-$$

(e^- transferred from Na to Cl)

After the electron transfer, each sodium has ten electrons and eleven protons (a net charge of 1+), and each chlorine has eighteen electrons and seventeen protons (a net charge of 1−).

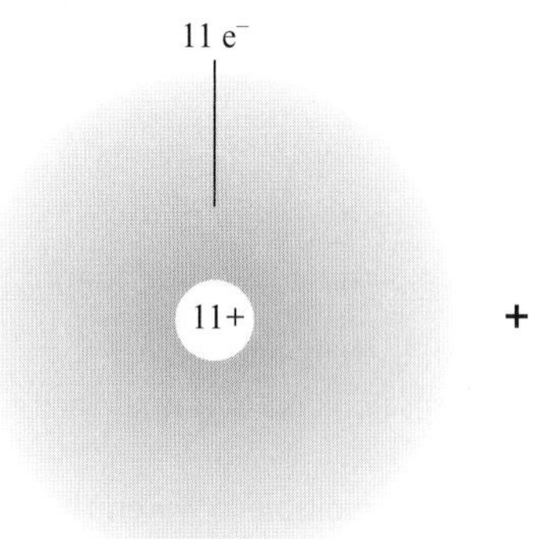

Thus the reaction of a metal with a nonmetal to form an ionic compound involves the transfer of one or more electrons from the metal (which forms a cation) to the nonmetal (which forms an anion). This tendency to transfer electrons from metals to nonmetals is the third driving force for reactions that we listed in Section 7-1. A reaction that *involves a transfer of electrons* is called an **oxidation–reduction reaction.**

There are many examples of oxidation–reduction reactions in which a metal reacts with a nonmetal to form an ionic compound. Consider the reaction of magnesium metal with oxygen,

$$2Mg(s) + O_2(g) \rightarrow 2MgO(s)$$

which produces a bright, white light that was once useful in camera flash units. Note that the reactants contain uncharged atoms, but the product contains ions:

MgO

Contains Mg^{2+}, O^{2-}

Therefore, in this reaction, each magnesium atom loses two electrons ($Mg \rightarrow Mg^{2+} + 2e^-$) and each oxygen atom gains two electrons ($O + 2e^- \rightarrow O^{2-}$). We might represent this reaction as follows:

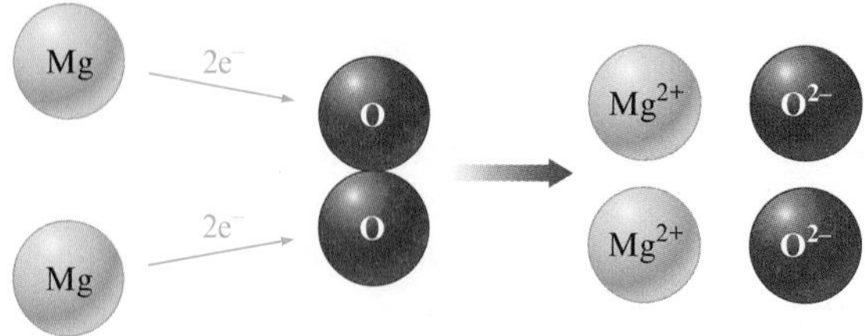

Figure 7.6 The thermite reaction gives off so much heat that the iron formed is molten.

Another example is

$$2Al(s) + Fe_2O_3(s) \rightarrow 2Fe(s) + Al_2O_3(s)$$

which is a reaction (called the thermite reaction) that produces so much energy (heat) that the iron is initially formed as a liquid (Fig. 7.6). In this case the aluminum is originally present as the elemental metal (which contains uncharged Al atoms) and ends up in Al_2O_3, where it is present as Al^{3+} cations (the $2Al^{3+}$ ions just balance the charge of the $3O^{2-}$ ions). Therefore, in the reaction each aluminum atom loses three electrons.

This equation is read, "An aluminum atom yields an aluminum ion with a 3+ charge and three electrons."

$$Al \rightarrow Al^{3+} + 3e^-$$

The opposite process occurs with the iron, which is initially present as Fe^{3+} ions in Fe_2O_3 and ends up as uncharged atoms in the elemental iron. Thus each iron cation gains three electrons to form an uncharged atom:

$$Fe^{3+} + 3e^- \rightarrow Fe$$

We can represent this reaction in schematic form as follows:

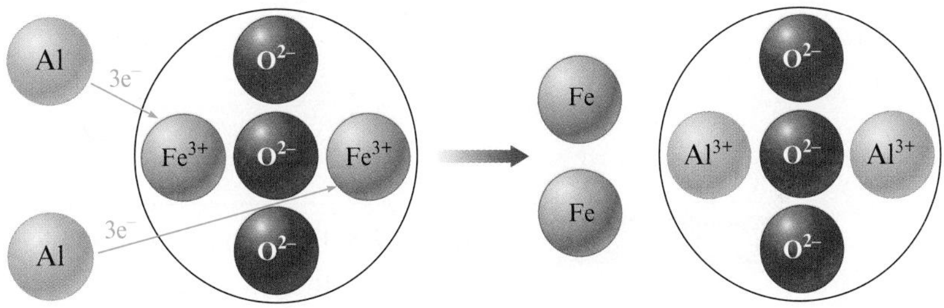

Example 7.5 Identifying Electron Transfer in Oxidation–Reduction Reactions

For each of the following reactions, show how electrons are gained and lost.

a. $2Al(s) + 3I_2(s) \rightarrow 2AlI_3(s)$ (This reaction is shown in Fig. 7.7. Note the purple "smoke," which is excess I_2 being driven off by the heat.)

b. $2Cs(s) + F_2(g) \rightarrow 2CsF(s)$

SOLUTION

a. In AlI_3 the ions are Al^{3+} and I^- (aluminum always forms Al^{3+}, and iodine always forms I^-). In $Al(s)$ the aluminum is present as uncharged atoms. Thus aluminum goes from Al to Al^{3+} by losing three electrons ($Al \rightarrow Al^{3+} + 3e^-$). In I_2 each iodine atom is uncharged. Thus each iodine atom goes from I to I^- by gaining one electron ($I + e^- \rightarrow I^-$). A schematic for this reaction is

Figure 7.7 When powdered aluminum and iodine (shown in the foreground) are mixed (and a little water added), they react vigorously.

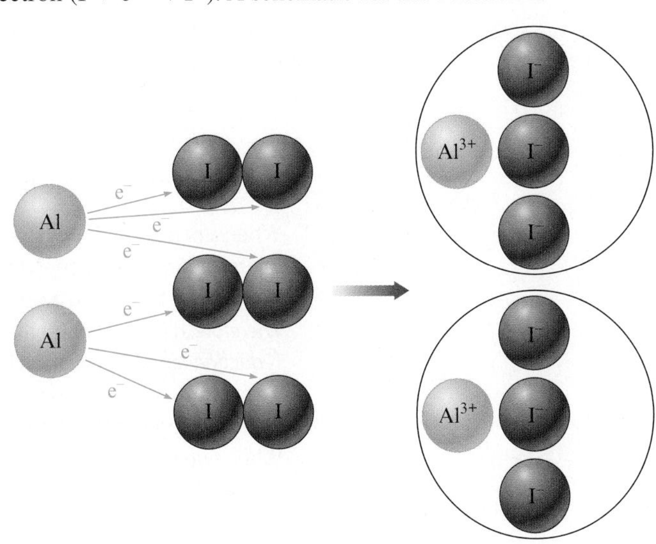

b. In CsF the ions present are Cs^+ and F^-. Cesium metal, Cs(*s*), contains uncharged cesium atoms, and fluorine gas, $F_2(g)$, contains uncharged fluorine atoms. Thus in the reaction each cesium atom loses one electron ($Cs \rightarrow Cs^+ + e^-$) and each fluorine atom gains one electron ($F + e^- \rightarrow F^-$). The schematic for this reaction is

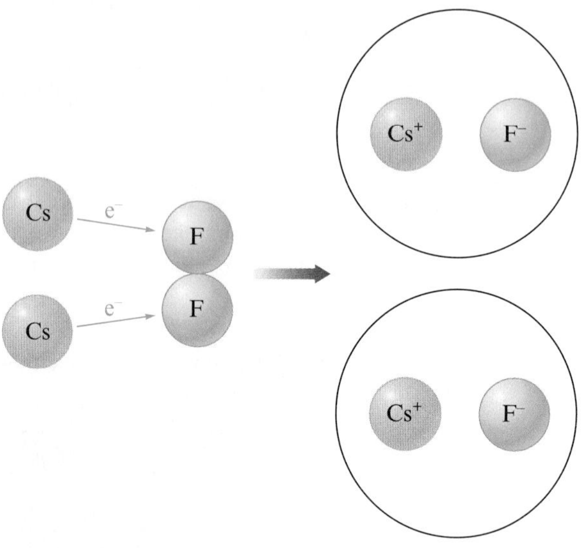

SELF CHECK

Exercise 7.3 For each reaction, show how electrons are gained and lost.

a. $2Na(s) + Br_2(l) \rightarrow 2NaBr(s)$

b. $2Ca(s) + O_2(g) \rightarrow 2CaO(s)$

See Problems 7.47 and 7.48. ■

So far we have emphasized electron transfer (oxidation–reduction) reactions that involve a metal and a nonmetal. Electron transfer reactions can also take place between two nonmetals. We will not discuss these reactions in detail here. All we will say at this point is that one sure sign of an oxidation–reduction reaction between nonmetals is the presence of oxygen, $O_2(g)$, as a reactant or product. In fact, oxidation got its name from oxygen. Thus the reactions

$$CH_4(g) + 2O_2(g) \rightarrow CO_2(g) + 2H_2O(g)$$

and

$$2SO_2(g) + O_2(g) \rightarrow 2SO_3(g)$$

are electron transfer reactions, even though it is not obvious at this point.

We can summarize what we have learned about oxidation–reduction reactions as follows:

Characteristics of Oxidation–Reduction Reactions

1. When a metal reacts with a nonmetal, an ionic compound is formed. The ions are formed when the metal transfers one or more electrons to the nonmetal, the metal atom becoming a cation and the nonmetal atom becoming an anion. *Therefore, a metal–nonmetal reaction can always be assumed to be an oxidation–reduction reaction, which involves electron transfer.*
2. Two nonmetals can also undergo an oxidation–reduction reaction. At this point we can recognize these cases only by looking for O_2 as a reactant or product. When two nonmetals react, the compound formed is not ionic.

7-6 Ways to Classify Reactions

OBJECTIVE **To learn various classification schemes for reactions.**

So far in our study of chemistry we have seen many, many chemical reactions—and this is just Chapter 7. In the world around us and in our bodies, literally millions of chemical reactions are taking place. Obviously, we need a system for putting reactions into meaningful classes that will make them easier to remember and easier to understand.

In Chapter 7 we have so far considered the following "driving forces" for chemical reactions:

- Formation of a solid
- Formation of water
- Transfer of electrons

We will now discuss how to classify reactions involving these processes. For example, in the reaction

$$\underset{\text{Solution}}{K_2CrO_4(aq)} + \underset{\text{Solution}}{Ba(NO_3)_2(aq)} \rightarrow \underset{\text{Solid formed}}{BaCrO_4(s)} + \underset{\text{Solution}}{2KNO_3(aq)}$$

solid $BaCrO_4$ (a precipitate) is formed. Because the *formation of a solid when two solutions are mixed* is called *precipitation,* we call this a **precipitation reaction.**

Notice in this reaction that two anions (NO_3^- and CrO_4^{2-}) are simply exchanged. Note that CrO_4^{2-} was originally associated with K^+ in K_2CrO_4 and that NO_3^- was associated with Ba^{2+} in $Ba(NO_3)_2$. In the products these associations are reversed. Because of this double exchange, we sometimes call this reaction a double-exchange reaction or **double-displacement reaction.** We might represent such a reaction as

$$AB + CD \rightarrow AD + CB$$

So we can classify a reaction such as this one as a precipitation reaction or as a double-displacement reaction. Either name is correct, but the former is more commonly used by chemists.

In this chapter we have also considered reactions in which water is formed when a strong acid is mixed with a strong base. All of these reactions had the same net ionic equation:

$$H^+(aq) + OH^-(aq) \rightarrow H_2O(l)$$

The H^+ ion comes from a strong acid, such as $HCl(aq)$ or $HNO_3(aq)$, and the origin of the OH^- ion is a strong base, such as $NaOH(aq)$ or $KOH(aq)$. An example is

$$HCl(aq) + KOH(aq) \rightarrow H_2O(l) + KCl(aq)$$

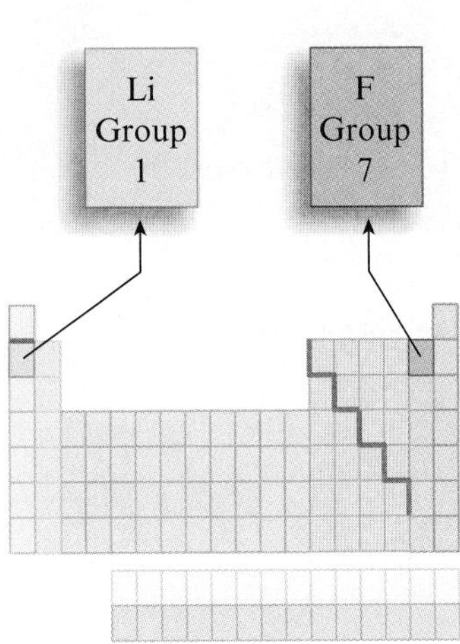

We classify these reactions as **acid–base reactions.** You can recognize this as an acid–base reaction because it *involves an* H^+ *ion that ends up in the product water.*

The third driving force is electron transfer. We see evidence of this driving force particularly in the "desire" of a metal to donate electrons to nonmetals. An example is

$$2Li(s) + F_2(g) \rightarrow 2LiF(s)$$

where each lithium atom loses one electron to form Li^+, and each fluorine atom gains one electron to form the F^- ion. The process of electron transfer is also called oxidation–reduction. Thus we classify the preceding reaction as an **oxidation–reduction reaction.**

CHEMISTRY IN FOCUS

Oxidation–Reduction Reactions Launch the Space Shuttle

Launching into space a vehicle that weighs millions of pounds requires unimaginable quantities of energy—all furnished by oxidation–reduction reactions.

Notice from Fig. 7.8 that three cylindrical objects are attached to the shuttle orbiter. In the center is a tank about 28 feet in diameter and 154 feet long that contains liquid oxygen and liquid hydrogen (in separate compartments). These fuels are fed to the orbiter's rocket engines, where they react to form water and release a huge quantity of energy.

$$2H_2 + O_2 \rightarrow 2H_2O + \text{energy}$$

Note that we can recognize this reaction as an oxidation–reduction reaction because O_2 is a reactant.

Two solid-fuel rockets 12 feet in diameter and 150 feet long are also attached to the orbiter. Each rocket contains 1.1 million pounds of fuel: ammonium perchlorate (NH_4ClO_4) and powdered aluminum mixed with a binder ("glue"). Because the rockets are so large, they are built in segments and assembled at the launch site as shown in Fig. 7.9. Each segment is filled with the syrupy propellant (Fig. 7.10), which then solidifies to a consistency much like that of a hard rubber eraser.

The oxidation–reduction reaction between the ammonium perchlorate and the aluminum is represented as follows:

$$3NH_4ClO_4(s) + 3Al(s) \rightarrow Al_2O_3(s) + AlCl_3(s) + 3NO(g) + 6H_2O(g) + \text{energy}$$

It produces temperatures of about 5700 °F and 3.3 million pounds of thrust in each rocket.

Thus we can see that oxidation–reduction reactions furnish the energy to launch the space shuttle.

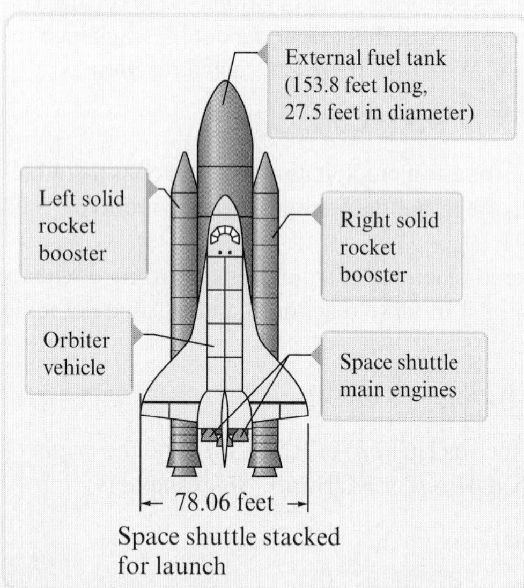

Figure 7.8 For launch, the Space Shuttle Orbiter is attached to two solid-fuel rockets (left and right) and a fuel tank (center) that supplies hydrogen and oxygen to the orbiter's engines. *(Reprinted with permission from* Chemical and Engineering News, *September 19, 1988. Copyright © 1988 American Chemical Society.)*

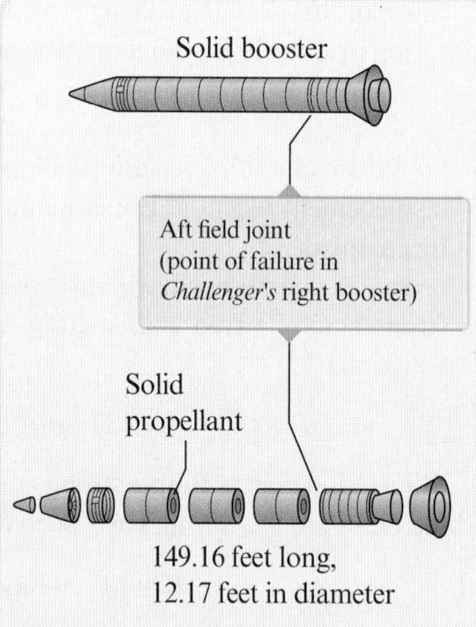

Figure 7.9 The solid-fuel rockets are assembled from segments to make loading the fuel more convenient. *(Reprinted with permission from* Chemical and Engineering News, *September 19, 1988. Copyright © 1988 American Chemical Society.)*

ATK Launch Systems Group

Figure 7.10 A mix bowl mixing propellant for a rocket motor.

An additional driving force for chemical reactions that we have not yet discussed is *formation of a gas.* A reaction in aqueous solution that forms a gas (which escapes as bubbles) is pulled toward the products by this event. An example is the reaction

$$2HCl(aq) + Na_2CO_3(aq) \rightarrow CO_2(g) + H_2O(l) + NaCl(aq)$$

for which the net ionic equation is

$$2H^+(aq) + CO_3^{2-}(aq) \rightarrow CO_2(g) + H_2O(l)$$

Note that this reaction forms carbon dioxide gas as well as water, so it illustrates two of the driving forces that we have considered. Because this reaction involves H^+ that ends up in the product water, we classify it as an acid–base reaction.

Consider another reaction that forms a gas:

$$Zn(s) + 2HCl(aq) \rightarrow H_2(g) + ZnCl_2(aq)$$

How might we classify this reaction? A careful look at the reactants and products shows the following:

Note that in the reactant zinc metal, Zn exists as uncharged atoms, whereas in the product it exists as Zn^{2+}. Thus each Zn atom loses two electrons. Where have these electrons gone? They have been transferred to two H^+ ions to form H_2. The schematic for this reaction is

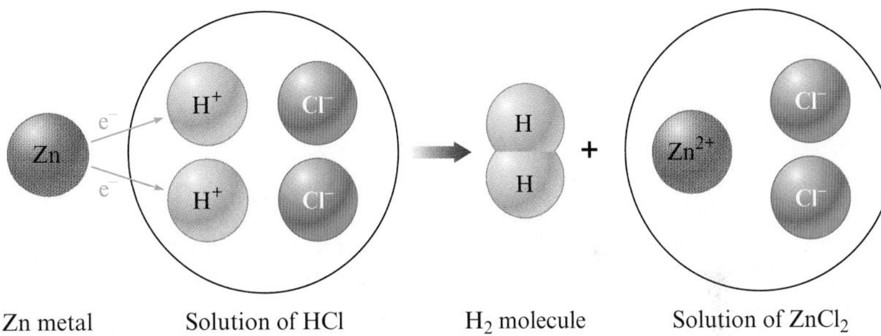

This is an electron transfer process, so the reaction can be classified as an oxidation–reduction reaction.

Another way this reaction is sometimes classified is based on the fact that a *single* type of anion (Cl^-) has been exchanged between H^+ and Zn^{2+}. That is, Cl^- is originally associated with H^+ in HCl and ends up associated with Zn^{2+} in the product $ZnCl_2$. We can call this a *single-replacement reaction* in contrast to double-displacement reactions, in which two types of anions are exchanged. We can represent a single replacement as

$$A + BC \rightarrow B + AC$$

7-7 Other Ways to Classify Reactions

OBJECTIVE **To consider additional classes of chemical reactions.**

So far in this chapter we have classified chemical reactions in several ways. The most commonly used of these classifications are

- Precipitation reactions
- Acid–base reactions
- Oxidation–reduction reactions

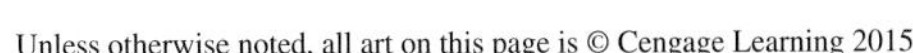

However, there are still other ways to classify reactions that you may encounter in your future studies of chemistry. We will consider several of these in this section.

Combustion Reactions

Many chemical reactions that involve oxygen produce energy (heat) so rapidly that a flame results. Such reactions are called **combustion reactions.** We have considered some of these reactions previously. For example, the methane in natural gas reacts with oxygen according to the following balanced equation:

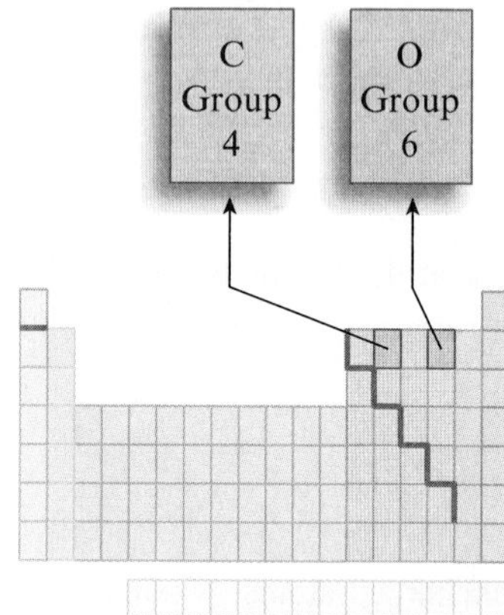

$$CH_4(g) + 2O_2(g) \rightarrow CO_2(g) + 2H_2O(g)$$

This reaction produces the flame of the common laboratory burner and is used to heat most homes in the United States. Recall that we originally classified this reaction as an oxidation–reduction reaction in Section 7-5. Thus we can say that the reaction of methane with oxygen is both an oxidation–reduction reaction and a combustion reaction. Combustion reactions, in fact, are a special class of oxidation–reduction reactions (Fig. 7.11).

There are many combustion reactions, most of which are used to provide heat or electricity for homes or businesses or energy for transportation. Some examples are:

- Combustion of propane (used to heat some rural homes)

$$C_3H_8(g) + 5O_2(g) \rightarrow 3CO_2(g) + 4H_2O(g)$$

- Combustion of gasoline* (used to power cars and trucks)

$$2C_8H_{18}(l) + 25O_2(g) \rightarrow 16CO_2(g) + 18H_2O(g)$$

- Combustion of coal* (used to generate electricity)

$$C(s) + O_2(g) \rightarrow CO_2(g)$$

Synthesis (Combination) Reactions

One of the most important activities in chemistry is the synthesis of new compounds. Each of our lives has been greatly affected by synthetic compounds such as plastic, polyester, and aspirin. When a given compound is formed from simpler materials, we call this a **synthesis** (or **combination**) **reaction.**

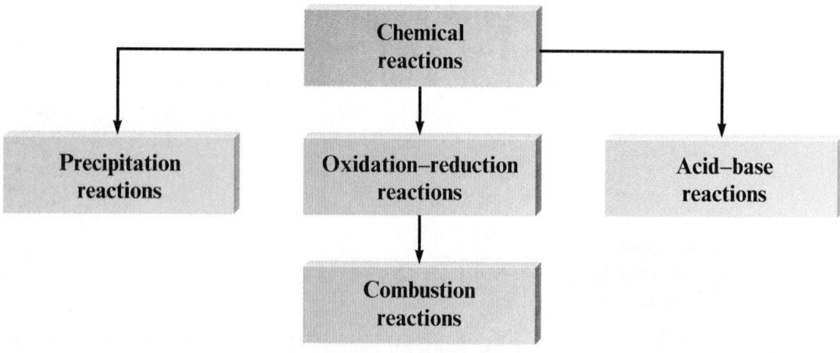

Figure 7.11 Classes of reactions. Combustion reactions are a special type of oxidation–reduction reaction.

*This substance is really a complex mixture of compounds, but the reaction shown is representative of what takes place.

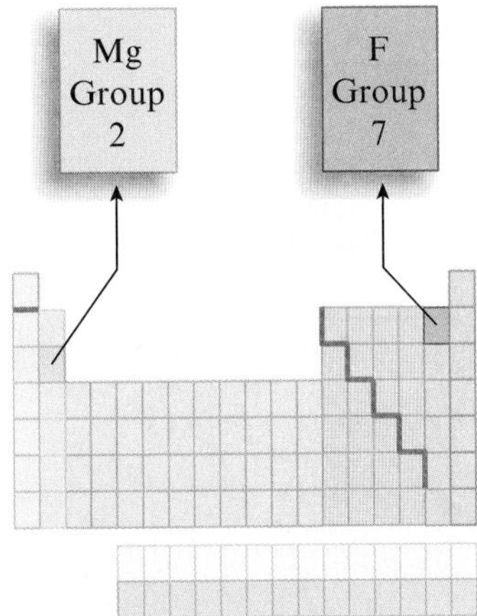

In many cases synthesis reactions start with elements, as shown by the following examples:

- Synthesis of water $2H_2(g) + O_2(g) \rightarrow 2H_2O(l)$
- Synthesis of carbon dioxide $C(s) + O_2(g) \rightarrow CO_2(g)$
- Synthesis of nitrogen monoxide $N_2(g) + O_2(g) \rightarrow 2NO(g)$

Notice that each of these reactions involves oxygen, so each can be classified as an oxidation–reduction reaction. The first two reactions are also commonly called combustion reactions because they produce flames. The reaction of hydrogen with oxygen to produce water, then, can be classified three ways: as an oxidation–reduction reaction, as a combustion reaction, and as a synthesis reaction.

There are also many synthesis reactions that do not involve oxygen:

- Synthesis of sodium chloride $2Na(s) + Cl_2(g) \rightarrow 2NaCl(s)$
- Synthesis of magnesium fluoride $Mg(s) + F_2(g) \rightarrow MgF_2(s)$

We have discussed the formation of sodium chloride before and have noted that it is an oxidation–reduction reaction; uncharged sodium atoms lose electrons to form Na^+ ions, and uncharged chlorine atoms gain electrons to form Cl^- ions. The synthesis of magnesium fluoride is also an oxidation–reduction reaction because Mg^{2+} and F^- ions are produced from the uncharged atoms.

We have seen that synthesis reactions in which the reactants are elements are oxidation–reduction reactions as well. In fact, we can think of these synthesis reactions as another subclass of the oxidation–reduction class of reactions.

Cristian Lucaci/Photos.com

Formation of the colorful plastics used in these zippers is an example of a synthetic reaction.

Decomposition Reactions

In many cases a compound can be broken down into simpler compounds or all the way to the component elements. This is usually accomplished by heating or by the application of an electric current. Such reactions are called **decomposition reactions.** We have discussed decomposition reactions before, including

- Decomposition of water

$$2H_2O(l) \xrightarrow[\text{Electric current}]{} 2H_2(g) + O_2(g)$$

- Decomposition of mercury(II) oxide

$$2HgO(s) \xrightarrow[\text{Heat}]{} 2Hg(l) + O_2(g)$$

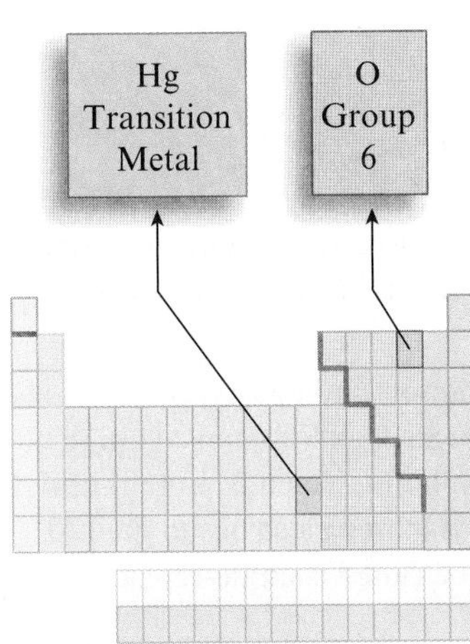

Because O_2 is involved in the first reaction, we recognize it as an oxidation–reduction reaction. In the second reaction, HgO, which contains Hg^{2+} and O^{2-} ions, is decomposed to the elements, which contain uncharged atoms. In this process each Hg^{2+} gains two electrons and each O^{2-} loses two electrons, so this is both a decomposition reaction and an oxidation–reduction reaction.

A decomposition reaction, in which a compound is broken down into its elements, is just the opposite of the synthesis (combination) reaction, in which elements combine to form the compound. For example, we have just discussed the synthesis of sodium chloride from its elements. Sodium chloride can be decomposed into its elements by melting it and passing an electric current through it:

$$2NaCl(l) \xrightarrow[\text{Electric current}]{} 2Na(l) + Cl_2(g)$$

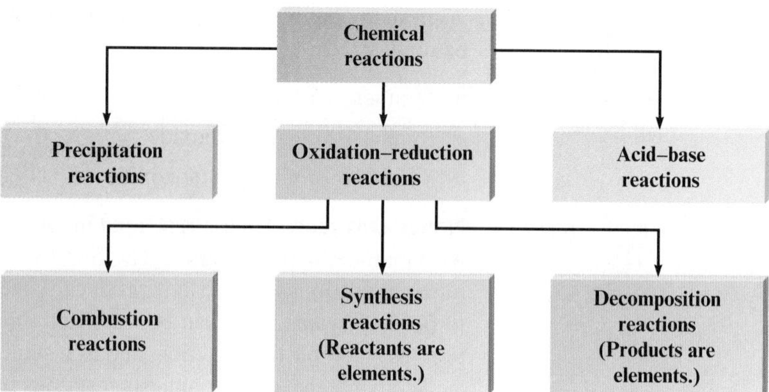

Figure 7.12 Summary of classes of reactions.

There are other schemes for classifying reactions that we have not considered. However, we have covered many of the classifications that are commonly used by chemists as they pursue their science in laboratories and industrial plants.

It should be apparent that many important reactions can be classified as oxidation–reduction reactions. As shown in Fig. 7.12, various types of reactions can be viewed as subclasses of the overall oxidation–reduction category.

Critical Thinking

Dalton believed that atoms were indivisible. Thomson and Rutherford helped to show that this was not true. What if atoms were indivisible? How would this affect the types of reactions you have learned about in this chapter?

Interactive Example 7.6

Classifying Reactions

Classify each of the following reactions in as many ways as possible.

a. $2K(s) + Cl_2(g) \rightarrow 2KCl(s)$
b. $Fe_2O_3(s) + 2Al(s) \rightarrow Al_2O_3(s) + 2Fe(s)$
c. $2Mg(s) + O_2(g) \rightarrow 2MgO(s)$
d. $HNO_3(aq) + NaOH(aq) \rightarrow H_2O(l) + NaNO_3(aq)$
e. $KBr(aq) + AgNO_3(aq) \rightarrow AgBr(s) + KNO_3(aq)$
f. $PbO_2(s) \rightarrow Pb(s) + O_2(g)$

SOLUTION

a. This is both a synthesis reaction (elements combine to form a compound) and an oxidation–reduction reaction (uncharged potassium and chlorine atoms are changed to K^+ and Cl^- ions in KCl).

b. This is an oxidation–reduction reaction. Iron is present in $Fe_2O_3(s)$ as Fe^{3+} ions and in elemental iron, $Fe(s)$, as uncharged atoms. So each Fe^{3+} must gain three electrons to form Fe. The reverse happens to aluminum, which is present initially as uncharged aluminum atoms, each of which loses three electrons to give Al^{3+} ions in Al_2O_3. Note that this reaction might also be called a single-replacement reaction because O is switched from Fe to Al.

c. This is both a synthesis reaction (elements combine to form a compound) and an oxidation–reduction reaction (each magnesium atom loses two electrons to give Mg^{2+} ions in MgO, and each oxygen atom gains two electrons to give O^{2-} in MgO).

d. This is an acid–base reaction. It might also be called a double-displacement reaction because NO_3^- and OH^- "switch partners."

e. This is a precipitation reaction that might also be called a double-displacement reaction in which the anions Br^- and NO_3^- are exchanged.

f. This is a decomposition reaction (a compound breaks down into elements). It also is an oxidation–reduction reaction, because the ions in PbO_2 (Pb^{4+} and O^{2-}) are changed to uncharged atoms in the elements $Pb(s)$ and $O_2(g)$. That is, electrons are transferred from O^{2-} to Pb^{4+} in the reaction.

SELF CHECK

Exercise 7.4 Classify each of the following reactions in as many ways as possible.

a. $4NH_3(g) + 5O_2(g) \rightarrow 4NO(g) + 6H_2O(g)$

b. $S_8(s) + 8O_2(g) \rightarrow 8SO_2(g)$

c. $2Al(s) + 3Cl_2(g) \rightarrow 2AlCl_3(s)$

d. $2AlN(s) \rightarrow 2Al(s) + N_2(g)$

e. $BaCl_2(aq) + Na_2SO_4(aq) \rightarrow BaSO_4(s) + 2NaCl(aq)$

f. $2Cs(s) + Br_2(l) \rightarrow 2CsBr(s)$

g. $KOH(aq) + HCl(aq) \rightarrow H_2O(l) + KCl(aq)$

h. $2C_2H_2(g) + 5O_2(g) \rightarrow 4CO_2(g) + 2H_2O(l)$

See Problems 7.53 and 7.54. ■

CHAPTER 7 REVIEW

F directs you to the *Chemistry in Focus* feature in the chapter

Key Terms

precipitation (7-2)
precipitate (7-2)
precipitation reaction (7-2, 7-6)
strong electrolyte (7-2)
soluble solid (7-2)
insoluble solid (7-2)
slightly soluble (7-2)
molecular equation (7-3)
complete ionic equation (7-3)
spectator ions (7-3)
net ionic equation (7-3)
acid (7-4)
strong acid (7-4)
base (7-4)
strong base (7-4)
salt (7-4)
oxidation–reduction reaction (7-5, 7-6)
precipitation reaction (7-6)
double-displacement reaction (7-6)
acid–base reaction (7-6)
combustion reaction (7-7)
synthesis (combination) reaction (7-7)
decomposition reaction (7-7)

For Review

- Four driving forces favor chemical change.
  - Formation of a solid
  - Formation of water
  - Transfer of electrons
  - Formation of a gas
- A reaction in which a solid forms is called a precipitation reaction.
- Solubility rules help predict what solid (if any) will form when solutions are mixed.
- Three types of equations are used to describe reactions in solution.
  - Molecular (formula) equation, which shows the complete formulas of all reactants and products
  - Complete ionic equation in which all strong electrolytes are shown as ions
  - Net ionic equation which includes only those components of the solution that undergo a change
    - Spectator ions (those that remain unchanged) are not shown in the net ionic equation.
- A strong acid is one in which virtually every molecule dissociates (ionizes) in water to an H^+ ion and an anion.
- A strong base is a metal hydroxide that is completely soluble in water, giving separate OH^- ions and cations.
- The products of the reaction of a strong acid and a strong base are water and a salt.
- Reactions between metals and nonmetals involve a transfer of electrons from the metal to the nonmetal, which is called an oxidation–reduction reaction.
- Reactions can be classified in various ways.
  - A synthesis reaction is one in which a compound forms from simpler substances, such as elements.
  - A decomposition reaction occurs when a compound is broken down into simpler substances.
  - A combustion reaction is an oxidation–reduction reaction that involves O_2.

Active Learning Questions

These questions are designed to be considered by groups of students in class. Often these questions work well for introducing a particular topic in class.

1. Consider the mixing of aqueous solutions of lead(II) nitrate and sodium iodide to form a solid.
   a. Name the possible products, and determine the formulas of these possible products.
   b. What is the precipitate? How do you know?
   c. Must the subscript for an ion in a reactant stay the same as the subscript of that ion in a product? Explain your answer.
2. Assume a highly magnified view of a solution of HCl that allows you to "see" the HCl. Draw this magnified view. If you dropped in a piece of magnesium, the magnesium would disappear and hydrogen gas would be released. Represent this change using symbols for the elements, and write the balanced equation.
3. Why is the formation of a solid evidence of a chemical reaction? Use a molecular-level drawing in your explanation.
4. Sketch molecular-level drawings to differentiate between two soluble compounds: one that is a strong electrolyte, and one that is not an electrolyte.
5. Mixing an aqueous solution of potassium nitrate with an aqueous solution of sodium chloride does not result in a chemical reaction. Why?
6. Why is the formation of water evidence of a chemical reaction? Use a molecular-level drawing in your explanation.
7. Use the Arrhenius definition of acids and bases to write the net ionic equation for the reaction of an acid with a base.
8. Why is the transfer of electrons evidence of a chemical reaction? Use a molecular-level drawing in your explanation.
9. Why is the formation of a gas evidence of a chemical reaction? Use a molecular-level drawing in your explanation.
10. Label each of the following statements as true or false. Explain your answers, and provide an example for each that supports your answer.
    a. All nonelectrolytes are insoluble.
    b. All insoluble substances are nonelectrolytes.
    c. All strong electrolytes are soluble.
    d. All soluble substances are strong electrolytes.
11. Look at Fig. 7.2 in the text. It is possible for a weak electrolyte solution to cause the bulb to glow brighter than a strong electrolyte. Explain how this is possible.
12. What is the purpose of spectator ions? If they are not present as part of the reaction, why are they present at all?
13. Which of the following **must** be an oxidation–reduction reaction? Explain your answer, and include an example oxidation–reduction reaction for all that apply.
    a. A metal reacts with a nonmetal.
    b. A precipitation reaction.
    c. An acid–base reaction.
14. If an element is a reactant or product in a chemical reaction, the reaction must be an oxidation–reduction reaction. Why is this true?
15. Match each name below with the following microscopic pictures of that compound in aqueous solution.

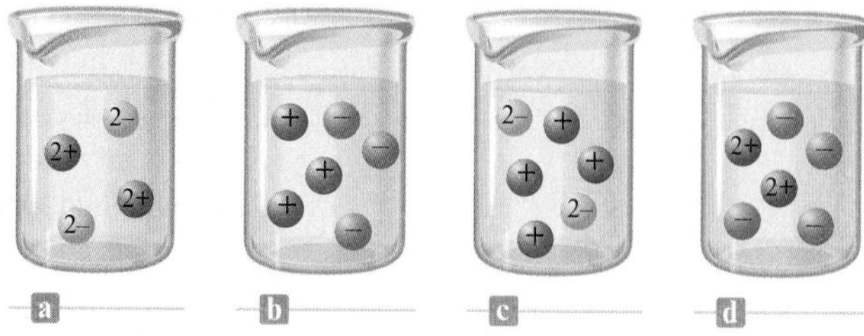

a. barium nitrate
b. sodium chloride
c. potassium carbonate
d. magnesium sulfate

Which picture best represents $HNO_3(aq)$? Why aren't any of the pictures a good representation of $HC_2H_3O_2(aq)$?

16. On the basis of the general solubility rules given in Table 7.1, predict the identity of the precipitate that forms when aqueous solutions of the following substances are mixed. If no precipitate is likely, indicate which rules apply.

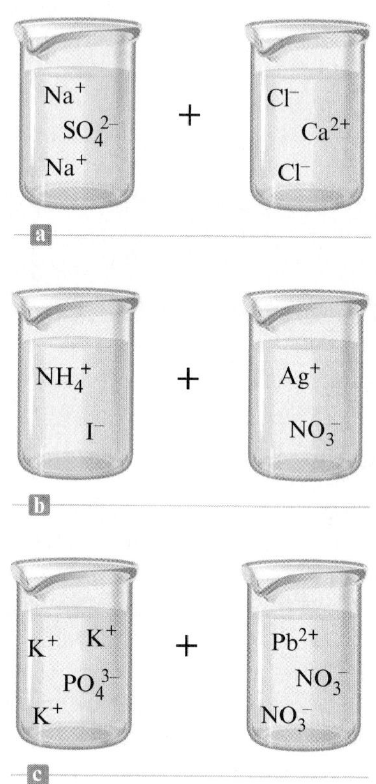

17. Write the balanced formula and net ionic equation for the reaction that occurs when the contents of the two beakers are added together. What colors represent the spectator ions in each reaction?

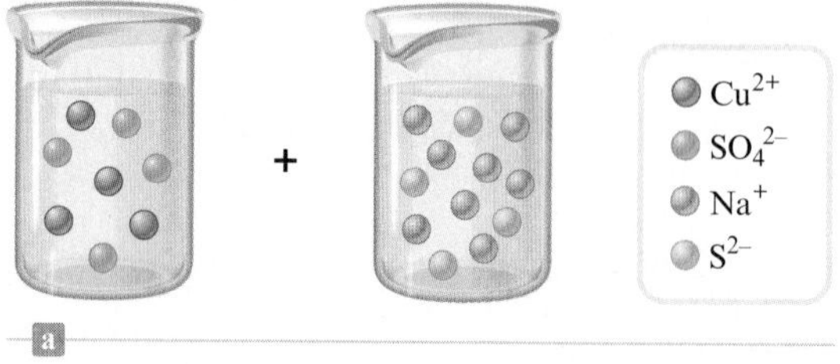

All even-numbered Questions and Problems have answers in the back of this book and solutions in the *Student Solutions Guide*.

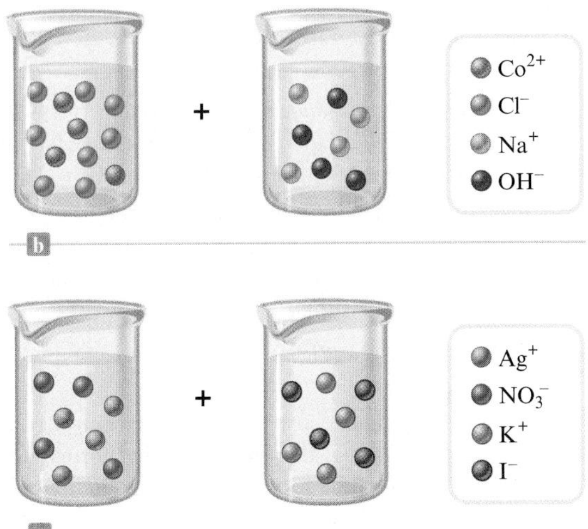

All even-numbered Questions and Problems have answers in the back of this book and solutions in the *Student Solutions Guide.*

Questions and Problems

7-1 Predicting Whether a Reaction Will Occur

Questions

1. Why is water an important solvent? Although you have not yet studied water in detail, can you think of some properties of water that make it so important?
2. What is a "driving force"? What are some of the driving forces discussed in this section that tend to make reactions likely to occur? Can you think of any other possible driving forces?

7-2 Reactions in Which a Solid Forms

Questions

3. A reaction in aqueous solution that results in the formation of a *solid* is called a ________ reaction.

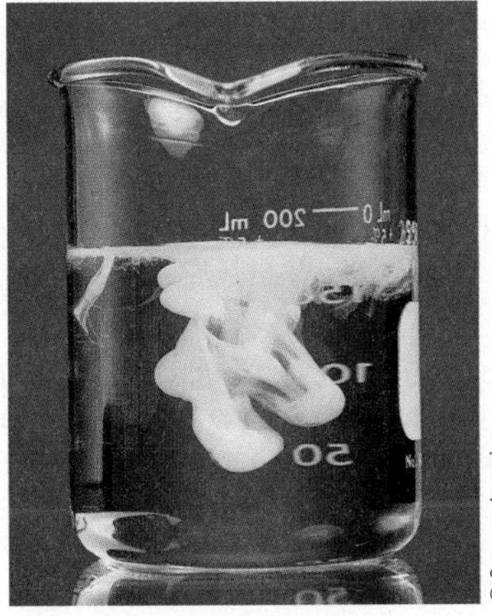
© Cengage Learning

4. When writing the chemical equation for a reaction, how do you indicate that a given reactant is dissolved in water? How do you indicate that a precipitate has formed as a result of the reaction?
5. Describe briefly what happens when an ionic substance is dissolved in water.
6. When the ionic solute $MgCl_2$ is dissolved in water, what can you say about the number of chloride ions present in the solution compared to the number of magnesium ions in the solution?
7. What is meant by a *strong electrolyte?* Give two examples of substances that behave in solution as strong electrolytes.
8. How do chemists know that the ions behave independently of one another when an ionic solid is dissolved in water?
9. Suppose you are trying to help your friend understand the general solubility rules for ionic substances in water. Explain in general terms to your friend what the solubility rules mean, and give an example of how the rules could be applied in determining the identity of the precipitate in a reaction between solutions of two ionic compounds.
10. Using the general solubility rules given in Table 7.1, which of the following ions will form a precipitate with SO_4^{2-}?
    a. Ba^{2+}
    b. Na^+
    c. NH_4^+
    d. At least two of the above ions will form a precipitate with SO_4^{2-}.
    e. All of the above ions will form a precipitate with SO_4^{2-}.
11. On the basis of the general solubility rules given in Table 7.1, predict which of the following substances are *not* likely to be soluble in water. Indicate which specific rule(s) led to your conclusion.
    a. PbS
    b. $Mg(OH)_2$
    c. Na_2SO_4
    d. $(NH_4)_2S$
    e. $BaCO_3$
    f. $AlPO_4$
    g. $PbCl_2$
    h. $CaSO_4$
12. On the basis of the general solubility rules given in Table 7.1, predict which of the following substances are likely to be appreciably soluble in water. Indicate which specific rule(s) led to your conclusion.
    a. $Ba(NO_3)_2$
    b. K_2SO_4
    c. $PbSO_4$
    d. $Cu(OH)_2$
    e. KCl
    f. Hg_2Cl_2
    g. $(NH_4)_2CO_3$
    h. Cr_2S_3
13. On the basis of the general solubility rules given in Table 7.1, for each of the following compounds, explain why the compound would be expected to be appreciably soluble in water. Indicate which of the solubility rules covers each substance's particular situation.
    a. potassium sulfide
    b. cobalt(III) nitrate
    c. ammonium phosphate
    d. cesium sulfate
    e. strontium chloride
14. On the basis of the general solubility rules given in Table 7.1, for each of the following compounds, explain why the compound would *not* be expected to be appreciably soluble in water. Indicate which of the solubility rules covers each substance's particular situation.
    a. iron(III) hydroxide
    b. calcium carbonate
    c. cobalt(III) phosphate
    d. silver chloride
    e. barium sulfate

15. On the basis of the general solubility rules given in Table 7.1, predict the identity of the precipitate that forms when aqueous solutions of the following substances are mixed. If no precipitate is likely, indicate which rules apply.
    a. copper(II) chloride, $CuCl_2$, and ammonium sulfide, $(NH_4)_2S$
    b. barium nitrate, $Ba(NO_3)_2$, and potassium phosphate, K_3PO_4
    c. silver acetate, $AgC_2H_3O_2$, and calcium chloride, $CaCl_2$
    d. potassium carbonate, K_2CO_3, and cobalt(II) chloride, $CoCl_2$
    e. sulfuric acid, H_2SO_4, and calcium nitrate, $Ca(NO_3)_2$
    f. mercurous acetate, $Hg_2(C_2H_3O_2)_2$, and hydrochloric acid, HCl
16. On the basis of the general solubility rules given in Table 7.1, predict the identity of the precipitate that forms when aqueous solutions of the following substances are mixed. If no precipitate is likely, indicate which rules apply.
    a. sodium carbonate, Na_2CO_3, and manganese(II) chloride, $MnCl_2$
    b. potassium sulfate, K_2SO_4, and calcium acetate, $Ca(C_2H_3O_2)_2$
    c. hydrochloric acid, HCl, and mercurous acetate, $Hg_2(C_2H_3O_2)_2$
    d. sodium nitrate, $NaNO_3$, and lithium sulfate, Li_2SO_4
    e. potassium hydroxide, KOH, and nickel(II) chloride, $NiCl_2$
    f. sulfuric acid, H_2SO_4, and barium chloride, $BaCl_2$

Problems

17. On the basis of the general solubility rules given in Table 7.1, write a balanced molecular equation for the precipitation reactions that take place when the following aqueous solutions are mixed. Underline the formula of the precipitate (solid) that forms. If no precipitation reaction is likely for the reactants given, explain why.
    a. ammonium chloride, NH_4Cl, and sulfuric acid, H_2SO_4
    b. potassium carbonate, K_2CO_3, and tin(IV) chloride, $SnCl_4$
    c. ammonium chloride, NH_4Cl, and lead(II) nitrate, $Pb(NO_3)_2$
    d. copper(II) sulfate, $CuSO_4$, and potassium hydroxide, KOH
    e. sodium phosphate, Na_3PO_4, and chromium(III) chloride, $CrCl_3$
    f. ammonium sulfide, $(NH_4)_2S$, and iron(III) chloride, $FeCl_3$
18. On the basis of the general solubility rules given in Table 7.1, write a balanced molecular equation for the precipitation reactions that take place when the following aqueous solutions are mixed. Underline the formula of the precipitate (solid) that forms. If no precipitation reaction is likely for the solutes given, so indicate.
    a. sodium carbonate, Na_2CO_3, and copper(II) sulfate, $CuSO_4$
    b. hydrochloric acid, HCl, and silver acetate, $AgC_2H_3O_2$
    c. barium chloride, $BaCl_2$, and calcium nitrate, $Ca(NO_3)_2$
    d. ammonium sulfide, $(NH_4)_2S$, and iron(III) chloride, $FeCl_3$
    e. sulfuric acid, H_2SO_4, and lead(II) nitrate, $Pb(NO_3)_2$
    f. potassium phosphate, K_3PO_4, and calcium chloride, $CaCl_2$
19. Balance each of the following equations that describe precipitation reactions.
    a. $Na_2SO_4(aq) + CaCl_2(aq) \rightarrow CaSO_4(s) + NaCl(aq)$
    b. $Co(C_2H_3O_2)_2(aq) + Na_2S(aq) \rightarrow CoS(s) + NaC_2H_3O_2(aq)$
    c. $KOH(aq) + NiCl_2(aq) \rightarrow Ni(OH)_2(s) + KCl(aq)$
20. Balance each of the following equations that describe precipitation reactions.
    a. $CaCl_2(aq) + AgNO_3(aq) \rightarrow Ca(NO_3)_2(aq) + AgCl(s)$
    b. $AgNO_3(aq) + K_2CrO_4(aq) \rightarrow Ag_2CrO_4(s) + KNO_3(aq)$
    c. $BaCl_2(aq) + K_2SO_4(aq) \rightarrow BaSO_4(s) + KCl(aq)$
21. For each of the following precipitation reactions, complete and balance the equation, indicating clearly which product is the precipitate. If no reaction would be expected, so indicate.
    a. $(NH_4)_2SO_4(aq) + Ba(NO_3)_2(aq) \rightarrow$
    b. $H_2S(aq) + NiSO_4(aq) \rightarrow$
    c. $FeCl_3(aq) + NaOH(aq) \rightarrow$
22. A solution of sodium phosphate is mixed with a solution of lead(II) nitrate. What is the precipitate that forms? Complete and balance the equation for this reaction, including the phases of each reactant and product.

7-3 Describing Reactions in Aqueous Solutions

Questions

23. What is a net ionic equation? What species are shown in such an equation, and which species are not shown?
24. Which of the following most accurately describes a spectator ion?
    a. An ion that is used up in a chemical reaction; it is limiting.
    b. An ion that participates in a chemical reaction but is always present in excess
    c. An ion that becomes part of the precipitate in a chemical reaction
    d. An ion that does not have a charge but can dissolve in solution and thus does not conduct electricity
    e. An ion that is present in solution but does not participate directly in the chemical reaction

Problems

25. Based on the general solubility rules given in Table 7.1, propose five combinations of aqueous ionic reagents that likely would form a precipitate when they are mixed. Write the balanced full molecular equation and the balanced net ionic equation for each of your choices.
26. Write the balanced molecular, complete ionic, and net ionic equations for the reaction between potassium sulfate and lead(II) nitrate.
27. Many chromate (CrO_4^{2-}) salts are insoluble, and most have brilliant colors that have led to their being used as pigments. Write balanced net ionic equations for the reactions of Cu^{2+}, Co^{3+}, Ba^{2+}, and Fe^{3+} with chromate ion.
28. The procedures and principles of qualitative analysis are covered in many introductory chemistry laboratory courses. In qualitative analysis, students learn to analyze mixtures of the common positive and negative ions, separating and confirming the presence of the particular ions in the mixture. One of the first steps in such an analysis is to treat the mixture with hydrochloric acid, which precipitates and removes silver ion, lead(II) ion, and mercury(I) ion from the aqueous mixture as the insoluble chloride salts. Write balanced net ionic equations for the precipitation reactions of these three cations with chloride ion.

All even-numbered Questions and Problems have answers in the back of this book and solutions in the *Student Solutions Guide*.

29. Many plants are poisonous because their stems and leaves contain oxalic acid, $H_2C_2O_4$, or sodium oxalate, $Na_2C_2O_4$; when ingested, these substances cause swelling of the respiratory tract and suffocation. A standard analysis for determining the amount of oxalate ion, $C_2O_4^{2-}$, in a sample is to precipitate this species as calcium oxalate, which is insoluble in water. Write the net ionic equation for the reaction between sodium oxalate and calcium chloride, $CaCl_2$, in aqueous solution.

30. Another step in the qualitative analysis of cations (see Exercise 28) involves precipitating some of the metal ions as the insoluble sulfides (followed by subsequent treatment of the mixed sulfide precipitate to separate the individual ions). Write balanced net ionic equations for the reactions of Co(II), Co(III), Fe(II), and Fe(III) ions with sulfide ion, S^{2-}.

7-4 Reactions That Form Water: Acids and Bases

Questions

31. What is meant by a *strong acid?* Are the strong acids also strong *electrolytes?* Explain.

32. What is meant by a *strong base?* Are the strong bases also strong *electrolytes?* Explain.

33. The same net ionic process takes place when any strong acid reacts with any strong base. Write the equation for that process.

34. Write the formulas and names of three common strong acids and strong bases.

35. If 1000 NaOH units were dissolved in a sample of water, the NaOH would produce ________ Na^+ ions and ________ OH^- ions.

36. What is a *salt?* Give two balanced chemical equations showing how a salt is formed when an acid reacts with a base.

Problems

37. Write balanced equations showing how three of the common strong acids ionize to produce hydrogen ion.

38. Along with the three strong acids emphasized in the chapter (HCl, HNO_3, and H_2SO_4), hydrobromic acid, HBr, and perchloric acid, $HClO_4$, are also strong acids. Write equations for the dissociation of each of these additional strong acids in water.

39. What salt would form when each of the following strong acid/strong base reactions takes place?
    a. $HCl(aq) + KOH(aq) \rightarrow$
    b. $RbOH(aq) + HNO_3(aq) \rightarrow$
    c. $HClO_4(aq) + NaOH(aq) \rightarrow$
    d. $HBr(aq) + CsOH(aq) \rightarrow$

40. Complete the following acid–base reactions by indicating the acid and base that must have reacted in each case to produce the indicated salt.
    a. ________ + ________ $\rightarrow K_2SO_4(aq) + 2H_2O(l)$
    b. ________ + ________ $\rightarrow NaNO_3(aq) + H_2O(l)$
    c. ________ + ________ $\rightarrow CaCl_2(aq) + 2H_2O(l)$
    d. ________ + ________ $\rightarrow Ba(ClO_4)_2(aq) + 2H_2O(l)$

7-5 Reactions of Metals with Nonmetals (Oxidation–Reduction)

Questions

41. What is an oxidation–reduction reaction? What is transferred during such a reaction?

42. Give an example of a simple chemical reaction that involves the *transfer of electrons* from a metallic element to a nonmetallic element.

43. What do we mean when we say that the transfer of electrons can be the "driving force" for a reaction? Give an example of a reaction where this happens.

44. The thermite reaction produces so much energy (heat) that the iron is initially formed as a liquid.

© Cengage Learning

$$2Al(s) + Fe_2O_3(s) \rightarrow 2Fe(s) + Al_2O_3(s)$$

Describe the transfer of electrons for both the aluminum and iron.

45. If atoms of the metal calcium were to react with molecules of the nonmetal fluorine, F_2, how many electrons would each calcium atom lose? How many electrons would each fluorine atom gain? How many calcium atoms would be needed to react with one fluorine molecule? What charges would the resulting calcium and fluoride ions have?

46. If oxygen molecules, O_2, were to react with magnesium atoms, how many electrons would each magnesium atom lose? How many electrons would each oxygen atom gain? How many magnesium atoms would be needed to react with each oxygen molecule? What charges would the resulting magnesium and oxide ions have?

Problems

47. For the reaction $Mg(s) + Cl_2(g) \rightarrow MgCl_2(s)$, illustrate how electrons are gained and lost during the reaction.

48. For the reaction $2Al(s) + 3Br_2(l) \rightarrow 2AlBr_3(s)$, show how electrons are gained and lost by the atoms.

All even-numbered Questions and Problems have answers in the back of this book and solutions in the *Student Solutions Guide.*

49. Balance each of the following oxidation–reduction reactions.

a. $Co(s) + Br_2(l) \rightarrow CoBr_3(s)$
b. $Al(s) + H_2SO_4(aq) \rightarrow Al_2(SO_4)_3(aq) + H_2(g)$
c. $Na(s) + H_2O(l) \rightarrow NaOH(aq) + H_2(g)$
d. $Cu(s) + O_2(g) \rightarrow Cu_2O(s)$

50. Balance each of the following oxidation–reduction chemical reactions.

a. $P_4(s) + O_2(g) \rightarrow P_4O_{10}(s)$
b. $MgO(s) + C(s) \rightarrow Mg(s) + CO(g)$
c. $Sr(s) + H_2O(l) \rightarrow Sr(OH)_2(aq) + H_2(g)$
d. $Co(s) + HCl(aq) \rightarrow CoCl_2(aq) + H_2(g)$

7-6 Ways to Classify Reactions

Questions

51. a. Give two examples each of a single-displacement reaction and of a double-replacement reaction. How are the two reaction types similar, and how are they different?
b. Give two examples each of a reaction in which formation of water is the driving force and in which formation of a gas is the driving force.

F 52. The reaction between ammonium perchlorate and aluminum is discussed in the "Chemistry in Focus" segment *Oxidation–Reduction Reactions Launch the Space Shuttle*. The reaction is labeled as an oxidation–reduction reaction. Explain why this is an oxidation–reduction reaction and defend your answer.

53. Identify each of the following unbalanced reaction equations as belonging to one or more of the following categories: precipitation, acid–base, or oxidation–reduction.

a. $K_2SO_4(aq) + Ba(NO_3)_2(aq) \rightarrow BaSO_4(s) + KNO_3(aq)$
b. $HCl(aq) + Zn(s) \rightarrow H_2(g) + ZnCl_2(aq)$
c. $HCl(aq) + AgNO_3(aq) \rightarrow HNO_3(aq) + AgCl(s)$
d. $HCl(aq) + KOH(aq) \rightarrow H_2O(l) + KCl(aq)$
e. $Zn(s) + CuSO_4(aq) \rightarrow ZnSO_4(aq) + Cu(s)$
f. $NaH_2PO_4(aq) + NaOH(aq) \rightarrow Na_3PO_4(aq) + H_2O(l)$
g. $Ca(OH)_2(aq) + H_2SO_4(aq) \rightarrow CaSO_4(s) + H_2O(l)$
h. $ZnCl_2(aq) + Mg(s) \rightarrow Zn(s) + MgCl_2(aq)$
i. $BaCl_2(aq) + H_2SO_4(aq) \rightarrow BaSO_4(s) + HCl(aq)$

54. Identify each of the following unbalanced reaction equations as belonging to one or more of the following categories: precipitation, acid–base, or oxidation–reduction.

a. $H_2O_2(aq) \rightarrow H_2O(l) + O_2(g)$
b. $H_2SO_4(aq) + Zn(s) \rightarrow ZnSO_4(aq) + H_2(g)$
c. $H_2SO_4(aq) + NaOH(aq) \rightarrow Na_2SO_4(aq) + H_2O(l)$
d. $H_2SO_4(aq) + Ba(OH)_2(aq) \rightarrow BaSO_4(s) + H_2O(l)$
e. $AgNO_3(aq) + CuCl_2(aq) \rightarrow Cu(NO_3)_2(aq) + AgCl(s)$
f. $KOH(aq) + CuSO_4(aq) \rightarrow Cu(OH)_2(s) + K_2SO_4(aq)$
g. $Cl_2(g) + F_2(g) \rightarrow ClF(g)$
h. $NO(g) + O_2(g) \rightarrow NO_2(g)$
i. $Ca(OH)_2(s) + HNO_3(aq) \rightarrow Ca(NO_3)_2(aq) + H_2O(l)$

7-7 Other Ways to Classify Reactions

Questions

55. How do we define a *combustion* reaction? In addition to the chemical products, what other products do combustion reactions produce? Give two examples of balanced chemical equations for combustion reactions.

56. Reactions involving the combustion of fuel substances make up a subclass of ________ reactions.

57. What is a *synthesis* or *combination* reaction? Give an example. Can such reactions also be classified in other ways? Give an example of a synthesis reaction that is also a *combustion* reaction. Give an example of a synthesis reaction that is also an *oxidation–reduction* reaction, but that does not involve combustion.

58. What is a *decomposition* reaction? Give an example. Can such reactions also be classified in other ways?

Problems

59. Complete and balance each of the following combustion reactions.

a. $C_6H_6(l) + O_2(g) \rightarrow$
b. $C_5H_{12}(l) + O_2(g) \rightarrow$
c. $C_2H_6O(l) + O_2(g) \rightarrow$

60. Complete and balance each of the following combustion reactions.

a. $C_3H_8(g) + O_2(g) \rightarrow$
b. $C_2H_4(g) + O_2(g) \rightarrow$
c. $C_8H_{18}(l) + O_2(g) + H_2O(g) \rightarrow$

61. By now, you are familiar with enough chemical compounds to begin to write your own chemical reaction equations. Write two examples of what we mean by a *combustion* reaction.

62. By now, you are familiar with enough chemical compounds to begin to write your own chemical reaction equations. Write two examples each of what we mean by a *synthesis* reaction and by a *decomposition* reaction.

63. Balance each of the following equations that describe synthesis reactions.

a. $CaO(s) + H_2O(l) \rightarrow Ca(OH)_2(s)$
b. $Fe(s) + O_2(g) \rightarrow Fe_2O_3(s)$
c. $P_2O_5(s) + H_2O(l) \rightarrow H_3PO_4(aq)$

64. Balance each of the following equations that describe synthesis reactions.

a. $Fe(s) + S_8(s) \rightarrow FeS(s)$
b. $Co(s) + O_2(g) \rightarrow Co_2O_3(s)$
c. $Cl_2O_7(g) + H_2O(l) \rightarrow HClO_4(aq)$

65. Balance each of the following equations that describe decomposition reactions.

a. $CaSO_4(s) \rightarrow CaO(s) + SO_3(g)$
b. $Li_2CO_3(s) \rightarrow Li_2O(s) + CO_2(g)$
c. $LiHCO_3(s) \rightarrow Li_2CO_3(s) + H_2O(g) + CO_2(g)$
d. $C_6H_6(l) \rightarrow C(s) + H_2(g)$
e. $PBr_3(l) \rightarrow P_4(s) + Br_2(l)$

All even-numbered Questions and Problems have answers in the back of this book and solutions in the *Student Solutions Guide*.

66. Balance each of the following equations that describe oxidation–reduction reactions.

    a. $Al(s) + Br_2(l) \rightarrow AlBr_3(s)$
    b. $Zn(s) + HClO_4(aq) \rightarrow Zn(ClO_4)_2(aq) + H_2(g)$
    c. $Na(s) + P(s) \rightarrow Na_3P(s)$
    d. $CH_4(g) + Cl_2(g) \rightarrow CCl_4(l) + HCl(g)$
    e. $Cu(s) + AgNO_3(aq) \rightarrow Cu(NO_3)_2(aq) + Ag(s)$

Additional Problems

All even-numbered Questions and Problems have answers in the back of this book and solutions in the Student Solutions Guide.

67. Distinguish between the *molecular* equation, the *complete ionic* equation, and the *net ionic* equation for a reaction in solution. Which type of equation most clearly shows the species that actually react with one another?

68. Which of the following ions form compounds with Pb^{2+} that are generally soluble in water?

    a. S^{2-}
    b. Cl^-
    c. NO_3^-
    d. SO_4^{2-}
    e. Na^+

69. Without first writing a full molecular or ionic equation, write the net ionic equations for any precipitation reactions that occur when aqueous solutions of the following compounds are mixed. If no reaction occurs, so indicate.

    a. iron(III) nitrate and sodium carbonate
    b. mercurous nitrate and sodium chloride
    c. sodium nitrate and ruthenium nitrate
    d. copper(II) sulfate and sodium sulfide
    e. lithium chloride and lead(II) nitrate
    f. calcium nitrate and lithium carbonate
    g. gold(III) chloride and sodium hydroxide

70. Complete and balance each of the following molecular equations for strong acid/strong base reactions. Underline the formula of the *salt* produced in each reaction.

    a. $HNO_3(aq) + KOH(aq) \rightarrow$
    b. $H_2SO_4(aq) + Ba(OH)_2(aq) \rightarrow$
    c. $HClO_4(aq) + NaOH(aq) \rightarrow$
    d. $HCl(aq) + Ca(OH)_2(aq) \rightarrow$

71. For the cations listed in the left-hand column, give the formulas of the precipitates that would form with each of the anions in the right-hand column. If no precipitate is expected for a particular combination, so indicate.

Cations	**Anions**
Ag^+	$C_2H_3O_2^-$
Ba^{2+}	Cl^-
Ca^{2+}	CO_3^{2-}
Fe^{3+}	NO_3^-
Hg_2^{2+}	OH^-
Na^+	PO_4^{3-}
Ni^{2+}	S^{2-}
Pb^{2+}	SO_4^{2-}

72. Balance each of the following equations that describe precipitation reactions.

    a. $AgNO_3(aq) + H_2SO_4(aq) \rightarrow Ag_2SO_4(s) + HNO_3(aq)$
    b. $Ca(NO_3)_2(aq) + H_2SO_4(aq) \rightarrow CaSO_4(s) + HNO_3(aq)$
    c. $Pb(NO_3)_2(aq) + H_2SO_4(aq) \rightarrow PbSO_4(s) + HNO_3(aq)$

73. On the basis of the general solubility rules given in Table 7.1, predict the identity of the precipitate that forms when aqueous solutions of the following substances are mixed. If no precipitate is likely, indicate why (which rules apply).

    a. iron(III) chloride and sodium hydroxide
    b. nickel(II) nitrate and ammonium sulfide
    c. silver nitrate and potassium chloride
    d. sodium carbonate and barium nitrate
    e. potassium chloride and mercury(I) nitrate
    f. barium nitrate and sulfuric acid

74. Below are indicated the formulas of some salts. Such salts could be formed by the reaction of the appropriate strong acid and strong base (with the other product of the reaction being, of course, water). For each salt, write an equation showing the formation of the salt from reaction of the appropriate strong acid and strong base.

    a. Na_2SO_4
    b. $RbNO_3$
    c. $KClO_4$
    d. KCl

75. For each of the following *un*balanced molecular equations, write the corresponding *balanced net ionic equation* for the reaction.

    a. $HCl(aq) + AgNO_3(aq) \rightarrow AgCl(s) + HNO_3(aq)$
    b. $CaCl_2(aq) + Na_3PO_4(aq) \rightarrow Ca_3(PO_4)_2(s) + NaCl(aq)$
    c. $Pb(NO_3)_2(aq) + BaCl_2(aq) \rightarrow PbCl_2(s) + Ba(NO_3)_2(aq)$
    d. $FeCl_3(aq) + NaOH(aq) \rightarrow Fe(OH)_3(s) + NaCl(aq)$

76. Write the balanced molecular, complete ionic, and net ionic equations for the reaction of sodium sulfate with calcium chloride.

77. What strong acid and what strong base would react in aqueous solution to produce the following salts?

    a. potassium perchlorate, $KClO_4$
    b. cesium nitrate, $CsNO_3$
    c. potassium chloride, KCl
    d. sodium sulfate, Na_2SO_4

78. For the reaction $2Al(s) + 3I_2(s) \rightarrow 2AlI_3(s)$, show how electrons are gained and lost by the atoms.

© Cengage Learning

79. For the reaction $16Fe(s) + 3S_8(s) \rightarrow 8Fe_2S_3(s)$, show how electrons are gained and lost by the atoms.

80. Balance the equation for each of the following oxidation–reduction chemical reactions.

    a. $Na(s) + O_2(g) \rightarrow Na_2O_2(s)$
    b. $Fe(s) + H_2SO_4(aq) \rightarrow FeSO_4(aq) + H_2(g)$
    c. $Al_2O_3(s) \rightarrow Al(s) + O_2(g)$
    d. $Fe(s) + Br_2(l) \rightarrow FeBr_3(s)$
    e. $Zn(s) + HNO_3(aq) \rightarrow Zn(NO_3)_2(aq) + H_2(g)$

81. Identify each of the following unbalanced reaction equations as belonging to one or more of the following categories: precipitation, acid–base, or oxidation–reduction.

a. $Fe(s) + H_2SO_4(aq) \rightarrow Fe_3(SO_4)_2(aq) + H_2(g)$
b. $HClO_4(aq) + RbOH(aq) \rightarrow RbClO_4(aq) + H_2O(l)$
c. $Ca(s) + O_2(g) \rightarrow CaO(s)$
d. $H_2SO_4(aq) + NaOH(aq) \rightarrow Na_2SO_4(aq) + H_2O(l)$
e. $Pb(NO_3)_2(aq) + Na_2CO_3(aq) \rightarrow PbCO_3(s) + NaNO_3(aq)$
f. $K_2SO_4(aq) + CaCl_2(aq) \rightarrow KCl(aq) + CaSO_4(s)$
g. $HNO_3(aq) + KOH(aq) \rightarrow KNO_3(aq) + H_2O(l)$
h. $Ni(C_2H_3O_2)_2(aq) + Na_2S(aq) \rightarrow NiS(s) + NaC_2H_3O_2(aq)$
i. $Ni(s) + Cl_2(g) \rightarrow NiCl_2(s)$

82. Which of the following statements is/are true regarding solutions?

a. If a solute is dissolved in water, then the resulting solution is considered aqueous.
b. If two solutions are mixed and no chemical reaction occurs, then a net ionic equation cannot be written.
c. If two clear solutions are mixed and then cloudiness results, this indicates that a precipitate formed.

83. Balance each of the following equations that describe synthesis reactions.

a. $FeO(s) + O_2(g) \rightarrow Fe_2O_3(s)$
b. $CO(g) + O_2(g) \rightarrow CO_2(g)$
c. $H_2(g) + Cl_2(g) \rightarrow HCl(g)$
d. $K(s) + S_8(s) \rightarrow K_2S(s)$
e. $Na(s) + N_2(g) \rightarrow Na_3N(s)$

84. Balance each of the following equations that describe decomposition reactions.

a. $NaHCO_3(s) \rightarrow Na_2CO_3(s) + H_2O(g) + CO_2(g)$
b. $NaClO_3(s) \rightarrow NaCl(s) + O_2(g)$
c. $HgO(s) \rightarrow Hg(l) + O_2(g)$
d. $C_{12}H_{22}O_{11}(s) \rightarrow C(s) + H_2O(g)$
e. $H_2O_2(l) \rightarrow H_2O(l) + O_2(g)$

85. Write a balanced oxidation–reduction equation for the reaction of each of the metals in the left-hand column with each of the nonmetals in the right-hand column.

Ba	O_2
K	S
Mg	Cl_2
Rb	N_2
Ca	Br_2
Li	

86. Write the balanced oxidation–reduction equation for the reaction between sodium metal and nitrogen gas to produce sodium nitride (do not worry about including the phases).

87. Although the metals of Group 2 of the periodic table are not nearly as reactive as those of Group 1, many of the Group 2 metals will combine with common nonmetals, especially at elevated temperatures. Write balanced chemical equations for the reactions of Mg, Ca, Sr, and Ba with Cl_2, Br_2, and O_2.

88. For each of the following metals, how many electrons will the metal atoms lose when the metal reacts with a nonmetal?

a. sodium
b. potassium
c. magnesium
d. barium
e. aluminum

89. For each of the following nonmetals, how many electrons will each atom of the nonmetal gain in reacting with a metal?

a. oxygen
b. fluorine
c. nitrogen
d. chlorine
e. sulfur

90. True or false? When solutions of barium hydroxide and sulfuric acid are mixed, the net ionic equation is: $Ba^{2+}(aq) + SO_4^{2-}(aq) \rightarrow BaSO_4(s)$ because only the species involved in making the precipitate are included. Whether true or false, include a balanced molecular equation and complete ionic equation for the reaction between barium hydroxide and sulfuric acid to support your answer.

91. Classify the reactions represented by the following unbalanced equations by as many methods as possible. Balance the equations.

a. $I_4O_9(s) \rightarrow I_2O_6(s) + I_2(s) + O_2(g)$
b. $Mg(s) + AgNO_3(aq) \rightarrow Mg(NO_3)_2(aq) + Ag(s)$
c. $SiCl_4(l) + Mg(s) \rightarrow MgCl_2(s) + Si(s)$
d. $CuCl_2(aq) + AgNO_3(aq) \rightarrow Cu(NO_3)_2(aq) + AgCl(s)$
e. $Al(s) + Br_2(l) \rightarrow AlBr_3(s)$

92. When a sodium chromate solution and aluminum bromide solution are mixed, a precipitate forms. Complete and balance the equation for this reaction, including the phases of each reactant and product.

93. Corrosion of metals costs us billions of dollars annually, slowly destroying cars, bridges, and buildings. Corrosion of a metal involves the oxidation of the metal by the oxygen in the air, typically in the presence of moisture. Write a balanced equation for the reaction of each of the following metals with O_2: Zn, Al, Fe, Cr, and Ni.

94. Consider a solution with the following ions present:

$$NO_3^-, Pb^{2+}, K^+, Ag^+, Cl^-, SO_4^{2-}, PO_4^{3-}$$

All are allowed to react and there are plenty available of each. List all of the solids that will form using the correct formulas in your explanation.

95. Give a balanced molecular chemical equation to illustrate each of the following types of reactions.

a. a synthesis (combination) reaction
b. a precipitation reaction
c. a double-displacement reaction
d. an acid–base reaction
e. an oxidation–reduction reaction
f. a combustion reaction

All even-numbered Questions and Problems have answers in the back of this book and solutions in the *Student Solutions Guide*.

ChemWork Problems

These multiconcept problems (and additional ones) are found interactively online with the same type of assistance a student would get from an instructor.

96. For the following chemical reactions, determine the precipitate produced when the two reactants listed below are mixed together. Indicate "none" if no precipitate will form.

	Formula of Precipitate
$Na_2SO_4(aq) + Pb(NO_3)_2(aq) \rightarrow$	______________ (s)
$AgNO_3(aq) + KCl(aq) \rightarrow$	______________ (s)
$KCl(aq) + NaNO_3(aq) \rightarrow$	______________ (s)

97. For the following chemical reactions, determine the precipitate produced when the two reactants listed below are mixed together. Indicate "none" if no precipitate will form.

	Formula of Precipitate
$Sr(NO_3)_2(aq) + K_3PO_4(aq) \rightarrow$	______________ (s)
$K_2CO_3(aq) + AgNO_3(aq) \rightarrow$	______________ (s)
$NaCl(aq) + KNO_3(aq) \rightarrow$	______________ (s)
$KCl(aq) + AgNO_3(aq) \rightarrow$	______________ (s)
$FeCl_3(aq) + Pb(NO_3)_2(aq) \rightarrow$	______________ (s)

All even-numbered Questions and Problems have answers in the back of this book and solutions in the *Student Solutions Guide*.

Questions

1. What kind of *visual* evidence indicates that a chemical reaction has occurred? Give an example of each type of evidence you have mentioned. Do *all* reactions produce visual evidence that they have taken place?
2. What, in general terms, does a chemical equation indicate? What are the substances indicated to the left of the arrow called in a chemical equation? To the right of the arrow?
3. What does it mean to "balance" an equation? Why is it so important that equations be balanced? What does it mean to say that atoms must be *conserved* in a balanced chemical equation? How are the physical states of reactants and products indicated when writing chemical equations?
4. When balancing a chemical equation, why is it *not* permissible to adjust the subscripts in the formulas of the reactants and products? What would changing the subscripts within a formula do? What do the *coefficients* in a balanced chemical equation represent? Why is it acceptable to adjust a substance's coefficient but not permissible to adjust the subscripts within the substance's formula?
5. What is meant by the *driving force* for a reaction? Give some examples of driving forces that make reactants tend to form products. Write a balanced chemical equation illustrating each type of driving force you have named.
6. Explain to your friend what chemists mean by a *precipitation* reaction. What is the driving force in a precipitation reaction? Using the information provided about solubility in these chapters, write balanced molecular and net ionic equations for five examples of precipitation reactions.
7. Define the term *strong electrolyte.* What types of substances tend to be strong electrolytes? What does a solution of a strong electrolyte contain? Give a way to determine if a substance is a strong electrolyte.
8. Summarize the simple solubility rules for ionic compounds. How do we use these rules in determining the identity of the solid formed in a precipitation reaction? Give examples including balanced complete and net ionic equations.
9. In general terms, what are the *spectator ions* in a precipitation reaction? Why are the spectator ions not included in writing the net ionic equation for a precipitation reaction? Does this mean that the spectator ions do not have to be present in the solution?
10. Describe some physical and chemical properties of *acids* and *bases.* What is meant by a *strong* acid or base? Are strong acids and bases also strong electrolytes? Give several examples of strong acids and strong bases.
11. What is a *salt?* How are salts formed by acid–base reactions? Write chemical equations showing the formation of three different salts. What other product is formed when an aqueous acid reacts with an aqueous base? Write the net ionic equation for the formation of this substance.
12. What do we call reactions in which electrons are transferred between atoms or ions? What do we call a *loss* of electrons by an atom or ion? What is it called when an atom or ion *gains* electrons? Can we have a process in which electrons are lost by one species without there also being a process in which the electrons are gained by another species? Why? Give three examples of equations in which there is a transfer of electrons between a metallic element and a nonmetallic element. In your examples, identify which species loses electrons and which species gains electrons.
13. What is a *combustion* reaction? Are combustion reactions a unique type of reaction, or are they a special case of a more general type of reaction? Write an equation that illustrates a combustion reaction.
14. Give an example of a *synthesis* reaction and of a *decomposition* reaction. Are synthesis and decomposition reactions always also oxidation–reduction reactions? Explain.
15. List and define all the ways of classifying chemical reactions that have been discussed in the text. Give a balanced chemical equation as an example of each type of reaction, and show clearly how your example fits the definition you have given.

Problems

16. The element carbon undergoes many inorganic reactions, as well as being the basis for the field of organic chemistry. Write balanced chemical equations for the reactions of carbon described below.
    a. Carbon burns in an excess of oxygen (for example, in the air) to produce carbon dioxide.
    b. If the supply of oxygen is limited, carbon will still burn, but will produce carbon monoxide rather than carbon dioxide.
    c. If molten lithium metal is treated with carbon, lithium carbide, Li_2C_2, is produced.
    d. Iron(II) oxide reacts with carbon above temperatures of about 700 °C to produce carbon monoxide gas and molten elemental iron.
    e. Carbon reacts with fluorine gas at high temperatures to make carbon tetrafluoride.
17. Balance each of the following chemical equations.
    a. $Na_2SO_4(aq) + BaCl_2(aq) \rightarrow BaSO_4(s) + NaCl(aq)$
    b. $Zn(s) + H_2O(g) \rightarrow ZnO(s) + H_2(g)$
    c. $NaOH(aq) + H_3PO_4(aq) \rightarrow Na_3PO_4(aq) + H_2O(l)$
    d. $Al(s) + Mn_2O_3(s) \rightarrow Al_2O_3(s) + Mn(s)$
    e. $C_7H_6O_2(s) + O_2(g) \rightarrow CO_2(g) + H_2O(g)$
    f. $C_6H_{14}(l) + O_2(g) \rightarrow CO_2(g) + H_2O(g)$
    g. $C_3H_8O(l) + O_2(g) \rightarrow CO_2(g) + H_2O(g)$
    h. $Mg(s) + HClO_4(aq) \rightarrow Mg(ClO_4)_2(aq) + H_2(g)$

All even-numbered Questions and Problems have answers in the back of this book and solutions in the *Student Solutions Guide.*

18. The reagent shelf in a general chemistry lab contains aqueous solutions of the following substances: silver nitrate, sodium chloride, acetic acid, nitric acid, sulfuric acid, potassium chromate, barium nitrate, phosphoric acid, hydrochloric acid, lead nitrate, sodium hydroxide, and sodium carbonate. Suggest how you might prepare the following pure substances using these reagents and any normal laboratory equipment. If it is *not* possible to prepare a substance using these reagents, indicate why.

a. $BaCrO_4(s)$
b. $NaC_2H_3O_2(s)$
c. $AgCl(s)$
d. $PbSO_4(s)$
e. $Na_2SO_4(s)$
f. $BaCO_3(s)$

19. The common strong acids are HCl, HNO_3, and H_2SO_4, whereas NaOH and KOH are the common strong bases. Write the neutralization reaction equations for each of these strong acids with each of these strong bases in aqueous solution.

20. Classify each of the following chemical equations in as *many* ways as possible based on what you have learned. Balance each equation.

a. $FeO(s) + HNO_3(aq) \rightarrow Fe(NO_3)_2(aq) + H_2O(l)$
b. $Mg(s) + CO_2(g) + O_2(g) \rightarrow MgCO_3(s)$
c. $NaOH(s) + CuSO_4(aq) \rightarrow Cu(OH)_2(s) + Na_2SO_4(aq)$
d. $HI(aq) + KOH(aq) \rightarrow KI(aq) + H_2O(l)$
e. $C_3H_8(g) + O_2(g) \rightarrow CO_2(g) + H_2O(g)$
f. $Co(NH_3)_6Cl_2(s) \rightarrow CoCl_2(s) + NH_3(g)$
g. $HCl(aq) + Pb(C_2H_3O_2)_2(aq) \rightarrow HC_2H_3O_2(aq) + PbCl_2(s)$
h. $C_{12}H_{22}O_{11}(s) \rightarrow C(s) + H_2O(g)$
i. $Al(s) + HNO_3(aq) \rightarrow Al(NO_3)_3(aq) + H_2(g)$
j. $B(s) + O_2(g) \rightarrow B_2O_3(s)$

21. In Column 1 are listed some reactive metals; in Column 2 are listed some nonmetals. Write a balanced chemical equation for the combination/synthesis reaction of each element in Column 1 with each element in Column 2.

Column 1	**Column 2**
sodium, Na	fluorine gas, F_2
calcium, Ca	oxygen gas, O_2
aluminum, Al	sulfur, S
magnesium, Mg	chlorine gas, Cl_2

22. Give balanced equations for two examples of each of the following types of reactions.

a. precipitation
b. single-displacement
c. combustion
d. synthesis
e. oxidation–reduction
f. decomposition
g. acid–base neutralization

23. Using the general solubility rules discussed in Chapter 7, give the formulas of five substances that would be expected to be readily soluble in water and five substances that would be expected to *not* be very soluble in water. For each of the substances you choose, indicate the specific solubility rule you applied to make your prediction.

24. Write the balanced net ionic equation for the reaction that takes place when aqueous solutions of the following solutes are mixed. If no reaction is likely, explain why no reaction would be expected for that combination of solutes.

a. potassium nitrate and sodium chloride
b. calcium nitrate and sulfuric acid
c. ammonium sulfide and lead(II) nitrate
d. sodium carbonate and iron(III) chloride
e. mercurous nitrate and calcium chloride
f. silver acetate and potassium chloride
g. phosphoric acid (H_3PO_4) and calcium nitrate
h. sulfuric acid and nickel(II) sulfate

25. Complete and balance the following equations.

a. $Pb(NO_3)_2(aq) + Na_2S(aq) \rightarrow$
b. $AgNO_3(aq) + HCl(aq) \rightarrow$
c. $Mg(s) + O_2(g) \rightarrow$
d. $H_2SO_4(aq) + KOH(aq) \rightarrow$
e. $BaCl_2(aq) + H_2SO_4(aq) \rightarrow$
f. $Mg(s) + H_2SO_4(aq) \rightarrow$
g. $Na_3PO_3(aq) + CaCl_2(aq) \rightarrow$
h. $C_4H_{10}(l) + O_2(g) \rightarrow$

All even-numbered Questions and Problems have answers in the back of this book and solutions in the *Student Solutions Guide*.

8 Chemical Composition

CHAPTER 8

These glass bottles contain silicon dioxide. James L. Amos/Photo Researchers/Getty Images

One very important chemical activity is the synthesis of new substances. Nylon, the artificial sweetener aspartame (NutraSweet®), Kevlar used in bulletproof vests and the body parts of exotic cars, polyvinyl chloride (PVC) for plastic water pipes, Teflon, Nitinol (the alloy that remembers its shape even after being severely distorted), and so many other materials that make our lives easier—all originated in some chemist's laboratory. Some of the new materials have truly amazing properties such as the plastic that listens and talks, described in the "Chemistry in Focus" segment *Plastic That Talks and Listens!* When a chemist makes a new substance, the first order of business is to identify it. What is its composition? What is its chemical formula?

In this chapter we will learn to determine a compound's formula. Before we can do that, however, we need to think about counting atoms. How do we determine the number of each type of atom in a substance so that we can write its formula? Of course, atoms are too small to count individually. As we will see in this chapter, we typically count atoms by weighing them. So let us first consider the general principle of counting by weighing.

The Enzo Ferrari has a body made of carbon fiber composite materials.
oksana.perkins/Shutterstock.com

8-1 Counting by Weighing

OBJECTIVE **To understand the concept of average mass and explore how counting can be done by weighing.**

Suppose you work in a candy store that sells gourmet jelly beans by the bean. People come in and ask for 50 beans, 100 beans, 1000 beans, and so on, and you have to count them out—a tedious process at best. As a good problem solver, you try to come up with a better system. It occurs to you that it might be far more efficient to buy a scale and count the jelly beans by weighing them. How can you count jelly beans by weighing them? What information about the individual beans do you need to know?

Assume that all of the jelly beans are identical and that each has a mass of 5 g. If a customer asks for 1000 jelly beans, what mass of jelly beans would be required? Each bean has a mass of 5 g, so you would need 1000 beans $\times$ 5 g/bean, or 5000 g (5 kg). It takes just a few seconds to weigh out 5 kg of jelly beans. It would take much longer to count out 1000 of them.

In reality, jelly beans are not identical. For example, let's assume that you weigh 10 beans individually and get the following results:

Bean	Mass
1	5.1 g
2	5.2 g
3	5.0 g
4	4.8 g
5	4.9 g
6	5.0 g
7	5.0 g
8	5.1 g
9	4.9 g
10	5.0 g

CHEMISTRY IN FOCUS

Plastic That Talks and Listens!

Imagine a plastic so "smart" that it can be used to sense a baby's breath, measure the force of a karate punch, sense the presence of a person 100 ft away, or make a balloon that sings. There is a plastic film capable of doing all these things. It's called **polyvinylidene difluoride (PVDF),** which has the structure

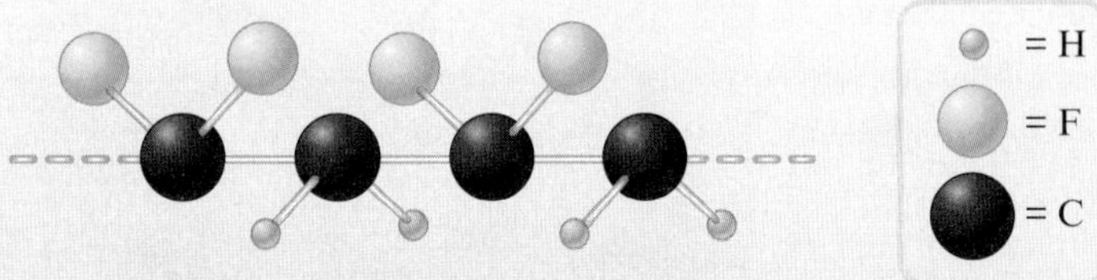

When this polymer is processed in a particular way, it becomes piezoelectric and pyroelectric. A *piezoelectric* substance produces an electric current when it is physically deformed or, alternatively, undergoes a deformation when a current is applied. A *pyroelectric* material is one that develops an electrical potential in response to a change in its temperature.

Because PVDF is piezoelectric, it can be used to construct a paper-thin microphone; it responds to sound by producing a current proportional to the deformation caused by the sound waves. A ribbon of PVDF plastic one-quarter of an inch wide could be strung along a hallway and used to listen to all the conversations going on as people walk through. On the other hand, electric pulses can be applied to the PVDF film to produce a speaker. A strip of PVDF film glued to the inside of a balloon can play any song stored on a microchip attached to the film—hence a balloon that can sing happy birthday at a party. The PVDF film also can be used to construct a sleep apnea monitor, which, when placed beside the mouth of a sleeping infant, will set off an alarm if the breathing stops, thus helping to prevent sudden infant death syndrome (SIDS). The same type of film is used by the U.S. Olympic karate team to measure the force of kicks and punches as the team trains. Also, gluing two strips of film together gives a material that curls in response to a current, creating an artificial muscle. In addition, because the PVDF film is pyroelectric, it responds to the infrared (heat) radiation emitted by a human as far away as 100 ft, making it useful for burglar alarm systems. Making the PVDF polymer piezoelectric and pyroelectric requires some very special processing, which makes it costly ($10 per square foot), but this seems a small price to pay for its near-magical properties.

Can we count these nonidentical beans by weighing? Yes. The key piece of information we need is the *average mass* of the jelly beans. Let's compute the average mass for our 10-bean sample.

$$\text{Average mass} = \frac{\text{total mass of beans}}{\text{number of beans}}$$

$$= \frac{5.1\text{ g} + 5.2\text{ g} + 5.0\text{ g} + 4.8\text{ g} + 4.9\text{ g} + 5.0\text{ g} + 5.0\text{ g} + 5.1\text{ g} + 4.9\text{ g} + 5.0\text{ g}}{10}$$

$$= \frac{50.0}{10} = 5.0\text{ g}$$

The average mass of a jelly bean is 5.0 g. Thus, to count out 1000 beans, we need to weigh out 5000 g of beans. This sample of beans, in which the beans have an average mass of 5.0 g, can be treated exactly like a sample where all of the beans are identical. Objects do not need to have identical masses to be counted by weighing. We simply need to know the average mass of the objects. For purposes of counting, the objects *behave as though they were all identical,* as though they each actually had the average mass.

Suppose a customer comes into the store and says, "I want to buy a bag of candy for each of my kids. One of them likes jelly beans and the other one likes mints. Please put a scoopful of jelly beans in a bag and a scoopful of mints in another bag." Then the customer recognizes a problem. "Wait! My kids will fight unless I bring home exactly the same number of candies for each one. Both bags must have the same number of pieces because they'll definitely count them and compare. But I'm really in a hurry, so we don't have time to count them here. Is there a simple way you can be sure the bags will contain the same number of candies?"

You need to solve this problem quickly. Suppose you know the average masses of the two kinds of candy:

$$\text{Jelly beans: average mass} = 5\text{ g}$$
$$\text{Mints: average mass} = 15\text{ g}$$

You fill the scoop with jelly beans and dump them onto the scale, which reads 500 g. Now the key question: What mass of mints do you need to give the same number of mints as there are jelly beans in 500 g of jelly beans? Comparing the average masses of the jelly beans (5 g) and mints (15 g), you realize that each mint has three times the mass of each jelly bean:

$$\frac{15\text{ g}}{5\text{ g}} = 3$$

This means that you must weigh out an amount of mints that is three times the mass of the jelly beans:

$$3 \times 500\text{ g} = 1500\text{ g}$$

You weigh out 1500 g of mints and put them in a bag. The customer leaves with your assurance that both the bag containing 500 g of jelly beans and the bag containing 1500 g of mints contain the same number of candies.

In solving this problem, you have discovered a principle that is very important in chemistry: two samples containing different types of components, A and B, both *contain the same number of components if the ratio of the sample masses is the same as the ratio of the masses of the individual components* of A and B.

Let's illustrate this rather intimidating statement by using the example we just discussed. The individual components have the masses 5 g (jelly beans) and 15 g (mints). Consider several cases.

- Each sample contains 1 component:

$$\text{Mass of mint} = 15\text{ g}$$
$$\text{Mass of jelly bean} = 5\text{ g}$$

- Each sample contains 10 components:

$$10\ \cancel{\text{mints}} \times \frac{15\text{ g}}{\cancel{\text{mint}}} = 150\text{ g of mints}$$
$$10\ \cancel{\text{jelly beans}} \times \frac{5\text{ g}}{\cancel{\text{jelly bean}}} = 50\text{ g of jelly beans}$$

- Each sample contains 100 components:

$$100\ \cancel{\text{mints}} \times \frac{15\text{ g}}{\cancel{\text{mint}}} = 1500\text{ g of mints}$$
$$100\ \cancel{\text{jelly beans}} \times \frac{5\text{ g}}{\cancel{\text{jelly bean}}} = 500\text{ g of jelly beans}$$

Note in each case that the ratio of the masses is always 3 to 1:

$$\frac{1500}{500} = \frac{150}{50} = \frac{15}{5} = \frac{3}{1}$$

This is the ratio of the masses of the individual components:

$$\frac{\text{Mass of mint}}{\text{Mass of jelly bean}} = \frac{15}{5} = \frac{3}{1}$$

Any two samples, one of mints and one of jelly beans, that have a *mass ratio* of 15/5 = 3/1 will contain the same number of components. And these same ideas apply also to atoms, as we will see in the next section.

8-2 Atomic Masses: Counting Atoms by Weighing

OBJECTIVE To understand atomic mass and its experimental determination.

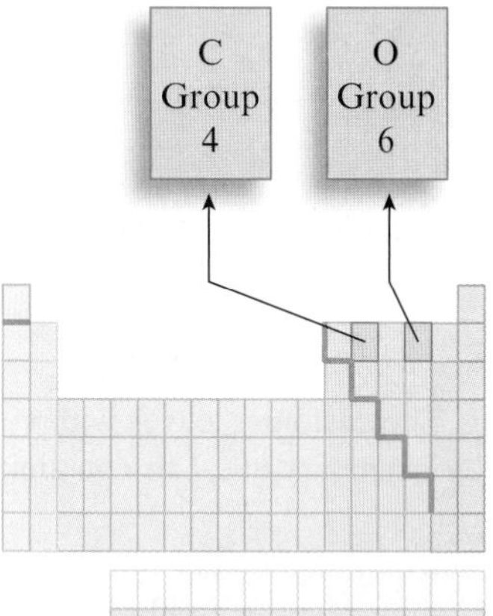

In Chapter 6 we considered the balanced equation for the reaction of solid carbon and gaseous oxygen to form gaseous carbon dioxide:

$$C(s) + O_2(g) \rightarrow CO_2(g)$$

Now suppose you have a small pile of solid carbon and want to know how many oxygen molecules are required to convert all of this carbon into carbon dioxide. The balanced equation tells us that one oxygen molecule is required for each carbon atom.

$$C(s) \quad + \quad O_2(g) \quad \rightarrow \quad CO_2(g)$$

1 atom reacts with 1 molecule to yield 1 molecule

To determine the number of oxygen molecules required, we must know how many carbon atoms are present in the pile of carbon. But individual atoms are far too small to see. We must learn to count atoms by weighing samples containing large numbers of them.

In the last section we saw that we can easily count things like jelly beans and mints by weighing. Exactly the same principles can be applied to counting atoms.

The symbol **u** is sometimes used for **amu.**

Because atoms are so tiny, the normal units of mass—the gram and the kilogram—are much too large to be convenient. For example, the mass of a single carbon atom is 1.99×10^{-23} g. To avoid using terms like 10^{-23} when describing the mass of an atom, scientists have defined a much smaller unit of mass called the **atomic mass unit (amu).** In terms of grams,

$$1 \text{ amu} = 1.66 \times 10^{-24} \text{ g}$$

Now let's return to our problem of counting carbon atoms. To count carbon atoms by weighing, we need to know the mass of individual atoms, just as we needed to know the mass of the individual jelly beans. Recall from Chapter 4 that the atoms of a given element exist as isotopes. The isotopes of carbon are $^{12}_{6}C$, $^{13}_{6}C$, and $^{14}_{6}C$. Any sample of carbon contains a mixture of these isotopes, always in the same proportions. Each of these isotopes has a slightly different mass. Therefore, just as with the nonidentical jelly beans, we need to use an average mass for the carbon atoms. The **average atomic mass** for carbon atoms is 12.01 amu. This means that any sample of carbon from nature *can be treated as though it were composed of identical carbon atoms,* each with a mass of 12.01 amu. Now that we know the average mass of the carbon atom, we can count carbon atoms by weighing samples of natural carbon. For example, what mass of natural carbon must we take to have 1000 carbon atoms present? Because 12.01 amu is the average mass,

Math skill builder
Remember that 1000 is an exact number here.

$$\text{Mass of 1000 natural carbon atoms} = (1000 \text{ atoms})\left(12.01 \frac{\text{amu}}{\text{atom}}\right)$$
$$= 12{,}010 \text{ amu} = 12.01 \times 10^3 \text{ amu}$$

Now let's assume that when we weigh the pile of natural carbon mentioned earlier, the result is 3.00×10^{20} amu. How many carbon atoms are present in this sample? We know that an average carbon atom has the mass 12.01 amu, so we can compute the number of carbon atoms by using the equivalence statement

$$1 \text{ carbon atom} = 12.01 \text{ amu}$$

to construct the appropriate conversion factor,

$$\frac{1 \text{ carbon atom}}{12.01 \text{ amu}}$$

The calculation is carried out as follows:

$$3.00 \times 10^{20} \cancel{\text{amu}} \times \frac{1 \text{ carbon atom}}{12.01 \cancel{\text{amu}}} = 2.50 \times 10^{19} \text{ carbon atoms}$$

The principles we have just discussed for carbon apply to all the other elements as well. All the elements as found in nature typically consist of a mixture of various isotopes. So to count the atoms in a sample of a given element by weighing, we must know the mass of the sample and the average mass for that element. Some average masses for common elements are listed in Table 8.1.

Table 8.1 ▸ Average Atomic Mass Values for Some Common Elements

Element	Average Atomic Mass (amu)
Hydrogen	1.008
Carbon	12.01
Nitrogen	14.01
Oxygen	16.00
Sodium	22.99
Aluminum	26.98

Interactive Example 8.1

Calculating Mass Using Atomic Mass Units (amu)

Calculate the mass, in amu, of a sample of aluminum that contains 75 atoms.

SOLUTION To solve this problem we use the average mass for an aluminum atom: 26.98 amu. We set up the equivalence statement:

$$1 \text{ Al atom} = 26.98 \text{ amu}$$

It gives the conversion factor we need:

$$75 \cancel{\text{Al atoms}} \times \frac{26.98 \text{ amu}}{1 \cancel{\text{Al atom}}} = 2024 \text{ amu}$$

Math skill builder

The 75 in this problem is an exact number—the number of atoms.

SELF CHECK

Exercise 8.1 Calculate the mass of a sample that contains 23 nitrogen atoms.

See Problems 8.5 and 8.8. ■

The opposite calculation can also be carried out. That is, if we know the mass of a sample, we can determine the number of atoms present. This procedure is illustrated in Example 8.2.

Interactive Example 8.2

Calculating the Number of Atoms from the Mass

Calculate the number of sodium atoms present in a sample that has a mass of 1172.49 amu.

SOLUTION We can solve this problem by using the average atomic mass for sodium (see Table 8.1) of 22.99 amu. The appropriate equivalence statement is

$$1 \text{ Na atom} = 22.99 \text{ amu}$$

which gives the conversion factor we need:

$$1172.49 \cancel{\text{amu}} \times \frac{1 \text{ Na atom}}{22.99 \cancel{\text{amu}}} = 51.00 \text{ Na atoms}$$

SELF CHECK

Exercise 8.2 Calculate the number of oxygen atoms in a sample that has a mass of 288 amu.

See Problems 8.6 and 8.7. ■

To summarize, we have seen that we can count atoms by weighing if we know the average atomic mass for that type of atom. This is one of the fundamental operations in chemistry, as we will see in the next section.

The average atomic mass for each element is listed in tables found inside the front cover of this book. Chemists often call these values the *atomic weights* for the elements, although this terminology is passing out of use.

8-3 The Mole

OBJECTIVES

- **To understand the mole concept and Avogadro's number.**
- **To learn to convert among moles, mass, and number of atoms in a given sample.**

In the previous section we used atomic mass units for mass, but these are extremely small units. In the laboratory a much larger unit, the gram, is the convenient unit for mass. In this section we will learn to count atoms in samples with masses given in grams.

Let's assume we have a sample of aluminum that has a mass of 26.98 g. What mass of copper contains exactly the same number of atoms as this sample of aluminum?

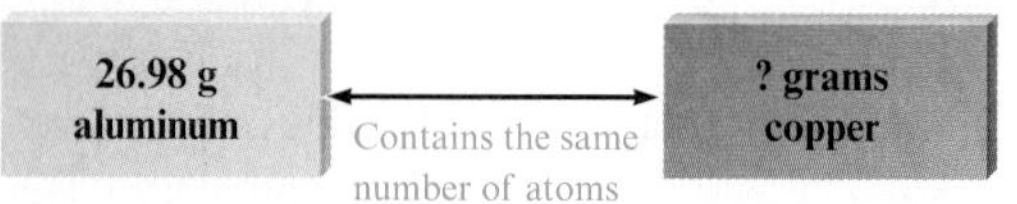

To answer this question, we need to know the average atomic masses for aluminum (26.98 amu) and copper (63.55 amu). Which atom has the greater atomic mass, aluminum or copper? The answer is copper. If we have 26.98 g of aluminum, do we need more or less than 26.98 g of copper to have the same number of copper atoms as aluminum atoms? We need more than 26.98 g of copper because each copper atom has a greater mass than each aluminum atom. Therefore, a given number of copper atoms will weigh more than an equal number of aluminum atoms. How much copper do we need? Because the average masses of aluminum and copper atoms are 26.98 amu and 63.55 amu, respectively, 26.98 g of aluminum and 63.55 g of copper contain exactly the same number of atoms. So we need 63.55 g of copper. As we saw in the first section when we were discussing candy, *samples in which the ratio of the masses is the same as the ratio of the masses of the individual atoms always contain the same number of atoms.* In the case just considered, the ratios are

$$\underbrace{\frac{26.98\text{ g}}{63.55\text{ g}}}_{\text{Ratio of sample masses}} = \underbrace{\frac{26.98\text{ amu}}{63.55\text{ amu}}}_{\text{Ratio of atomic masses}}$$

Therefore, 26.98 g of aluminum contains the same number of aluminum atoms as 63.55 g of copper contains copper atoms.

Now compare carbon (average atomic mass, 12.01 amu) and helium (average atomic mass, 4.003 amu). A sample of 12.01 g of carbon contains the same number of atoms as 4.003 g of helium. In fact, if we weigh out samples of all the elements such that each sample has a mass equal to that element's average atomic mass in grams, these samples all contain the same number of atoms (Fig. 8.1). This number (the number of atoms present in all of these samples) assumes special importance in chemistry. It is called the mole, the unit all chemists use in describing numbers of atoms. The **mole (mol)** can be defined as *the number equal to the number of carbon atoms in 12.01 grams of carbon.* Techniques for counting atoms very precisely have been used

This definition of the mole is slightly different from the SI definition but is used because it is easier to understand at this point.

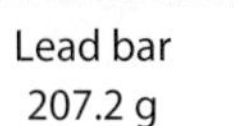
Lead bar
207.2 g

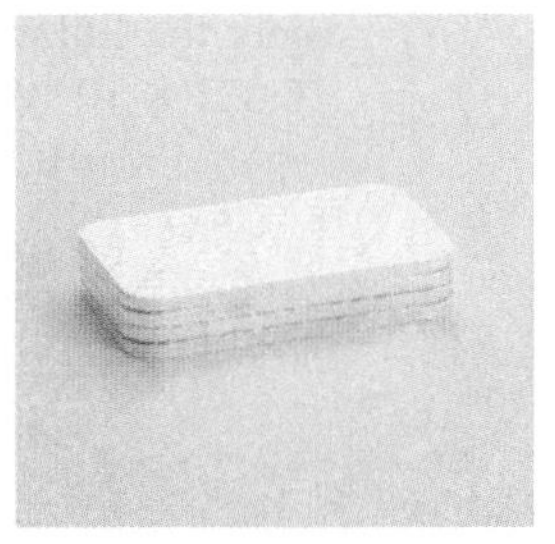
Silver bars
107.9 g

Pile of copper
63.55 g

Ken O'Donoghue © Cengage Learning

Figure 8.1 All these samples of pure elements contain the *same number* (a mole) of atoms: 6.022×10^{23} atoms.

Avogadro's number (to four significant figures) is 6.022×10^{23}. One mole of *anything* is 6.022×10^{23} units of that substance.

to determine this number to be 6.022×10^{23}. This number is called **Avogadro's number.** *One mole of something consists of 6.022×10^{23} units of that substance.* Just as a dozen eggs is 12 eggs, a mole of eggs is 6.022×10^{23} eggs. And a mole of water contains 6.022×10^{23} H_2O molecules.

The magnitude of the number 6.022×10^{23} is very difficult to imagine. To give you some idea, 1 mole of seconds represents a span of time 4 million times as long as the earth has already existed! One mole of marbles is enough to cover the entire earth to a depth of 50 miles! However, because atoms are so tiny, a mole of atoms or molecules is a perfectly manageable quantity to use in a reaction (Fig. 8.2).

Ken O'Donoghue © Cengage Learning

Figure 8.2 One-mole samples of iron (nails), iodine crystals, liquid mercury, and powdered sulfur.

Critical Thinking

What if you were offered \$1 million to count from 1 to 6×10^{23} at a rate of one number each second? Determine your hourly wage. Would you do it? Could you do it?

How do we use the mole in chemical calculations? Recall that Avogadro's number is defined such that a 12.01-g sample of carbon contains 6.022×10^{23} atoms. By the same token, because the average atomic mass of hydrogen is 1.008 amu (see Table 8.1), 1.008 g of hydrogen contains 6.022×10^{23} hydrogen atoms. Similarly, 26.98 g of aluminum contains 6.022×10^{23} aluminum atoms. The point is that a sample of *any* element that weighs a number of grams equal to the average atomic mass of that element contains 6.022×10^{23} atoms (1 mole) of that element.

Table 8.2 shows the masses of several elements that contain 1 mole of atoms.

In summary, *a sample of an element with a mass equal to that element's average atomic mass expressed in grams contains 1 mole of atoms.*

To do chemical calculations, you *must* understand what the mole means and how to determine the number of moles in a given mass of a substance. However, before we do any calculations, let's be sure that the process of counting by weighing is clear. Con-

Table 8.2 Comparison of 1-Mole Samples of Various Elements

Element	Number of Atoms Present	Mass of Sample (g)
Aluminum	6.022×10^{23}	26.98
Gold	6.022×10^{23}	196.97
Iron	6.022×10^{23}	55.85
Sulfur	6.022×10^{23}	32.07
Boron	6.022×10^{23}	10.81
Xenon	6.022×10^{23}	131.3

sider the following "bag" of H atoms (symbolized by dots), which contains 1 mole (6.022×10^{23}) of H atoms and has a mass of 1.008 g. Assume the bag itself has no mass.

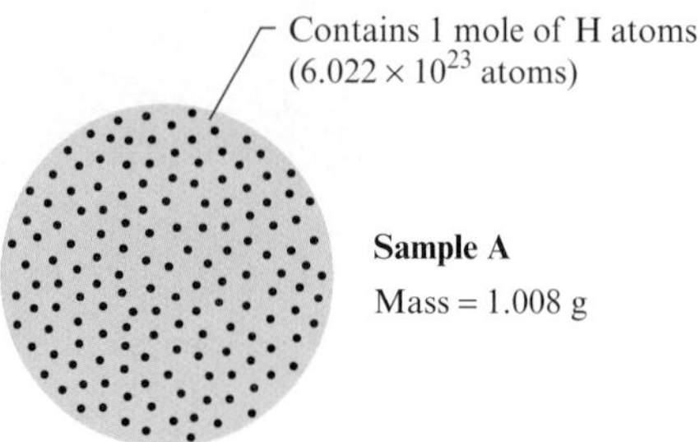

Now consider another "bag" of hydrogen atoms in which the number of hydrogen atoms is unknown.

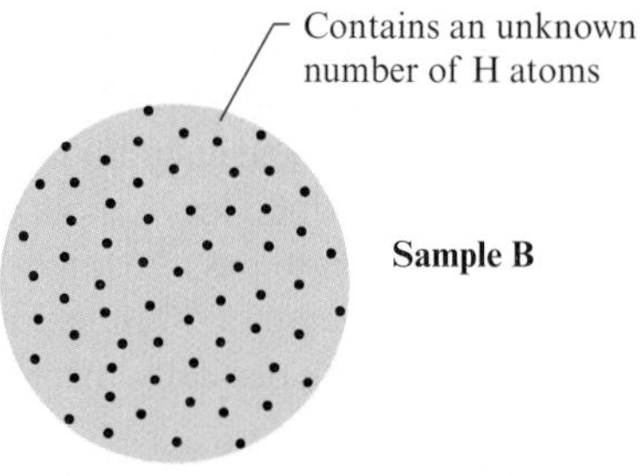

A 1-mole sample of graphite (a form of carbon) weighs 12.01 g.

We want to find out how many H atoms are present in sample ("bag") B. How can we do that? We can do it by weighing the sample. We find the mass of sample B to be 0.500 g.

How does this measured mass help us determine the number of atoms in sample B? We know that 1 mole of H atoms has a mass of 1.008 g. Sample B has a mass of 0.500 g, which is approximately half the mass of a mole of H atoms.

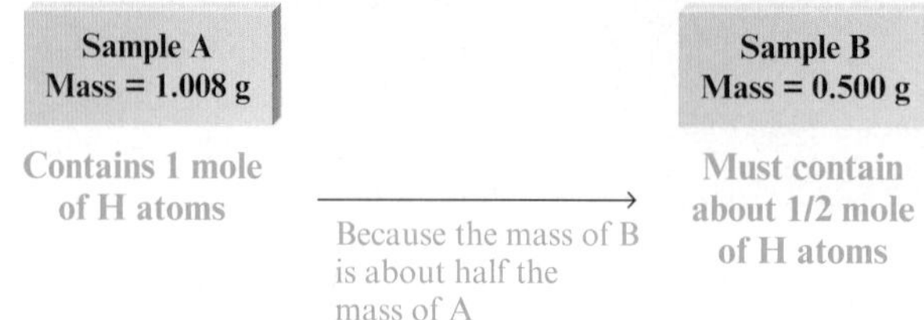

We carry out the actual calculation by using the equivalence statement

$$1 \text{ mol H atoms} = 1.008 \text{ g H}$$

to construct the conversion factor we need:

$$0.500 \cancel{\text{g H}} \times \frac{1 \text{ mol H}}{1.008 \cancel{\text{g H}}} = 0.496 \text{ mol H in sample B}$$

Math skill builder

In demonstrating how to solve problems requiring more than one step, we will often break the problem into smaller steps and report the answer to each step in the correct number of significant figures. While it may not always affect the final answer, it is a better idea to wait until the final step to round your answer to the correct number of significant figures.

Let's summarize. We know the mass of 1 mole of H atoms, so we can determine the number of moles of H atoms in any other sample of pure hydrogen by weighing the sample and *comparing* its mass to 1.008 g (the mass of 1 mole of H atoms). We can follow this same process for any element, because we know the mass of 1 mole for each of the elements.

Also, because we know that 1 mole is 6.022×10^{23} units, once we know the *moles* of atoms present, we can easily determine the *number* of atoms present. In the case considered above, we have approximately 0.5 mole of H atoms in sample B. This means that about 1/2 of 6×10^{23}, or 3×10^{23}, H atoms is present. We carry out the actual calculation by using the equivalence statement

$$1 \text{ mol} = 6.022 \times 10^{23}$$

to determine the conversion factor we need:

$$0.496 \text{ mol H atoms} \times \frac{6.022 \times 10^{23} \text{ H atoms}}{1 \text{ mol H atoms}} = 2.99 \times 10^{23} \text{ H atoms in sample B}$$

These procedures are illustrated in Example 8.3.

Critical Thinking

What if you discovered Avogadro's number was not 6.02×10^{23} but 3.01×10^{23}? Would this affect the relative masses given on the periodic table? If so, how? If not, why not?

Interactive Example 8.3

Calculating Moles and Number of Atoms

Aluminum (Al), a metal with a high strength-to-weight ratio and a high resistance to corrosion, is often used for structures such as high-quality bicycle frames. Compute both the number of moles of atoms and the number of atoms in a 10.0-g sample of aluminum.

SOLUTION In this case we want to change from mass to moles of atoms:

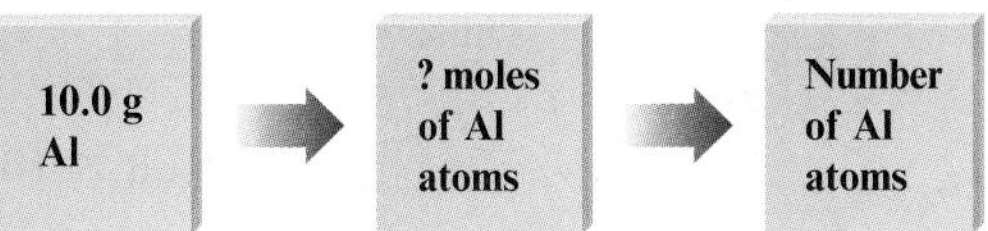

© Cengage Learning

A bicycle with an aluminum frame.

The mass of 1 mole (6.022×10^{23} atoms) of aluminum is 26.98 g. The sample we are considering has a mass of 10.0 g. Its mass is less than 26.98 g, so this sample contains less than 1 mole of aluminum atoms. We calculate the number of moles of aluminum atoms in 10.0 g by using the equivalence statement

$$1 \text{ mol Al} = 26.98 \text{ g Al}$$

to construct the appropriate conversion factor:

$$10.0 \text{ g Al} \times \frac{1 \text{ mol Al}}{26.98 \text{ g Al}} = 0.371 \text{ mol Al}$$

Next we convert from moles of atoms to the number of atoms, using the equivalence statement

$$6.022 \times 10^{23} \text{ Al atoms} = 1 \text{ mol Al atoms}$$

We have

$$0.371 \text{ mol Al} \times \frac{6.022 \times 10^{23} \text{ Al atoms}}{1 \text{ mol Al}} = 2.23 \times 10^{23} \text{ Al atoms}$$

We can summarize this calculation as follows:

10.0 g Al $\times \frac{1 \text{ mol Al}}{26.98 \text{ g Al}}$ → **0.371 mol Al**

0.371 mol Al $\times \frac{6.022 \times 10^{23} \text{ Al atoms}}{\text{mol Al}}$ → **2.23×10^{23} Al atoms** ■

Interactive Example 8.4 Calculating the Number of Atoms

A silicon chip used in an integrated circuit of a microcomputer has a mass of 5.68 mg. How many silicon (Si) atoms are present in this chip? The average atomic mass for silicon is 28.09 amu.

SOLUTION

Stockbyte/Getty Images

A silicon chip of the type used in electronic equipment.

Our strategy for doing this problem is to convert from milligrams of silicon to grams of silicon, then to moles of silicon, and finally to atoms of silicon:

where each arrow in the schematic represents a conversion factor. Because 1 g = 1000 mg, we have

$$5.68\ \cancel{\text{mg Si}} \times \frac{1\text{ g Si}}{1000\ \cancel{\text{mg Si}}} = 5.68 \times 10^{-3}\text{ g Si}$$

Next, because the average mass of silicon is 28.09 amu, we know that 1 mole of Si atoms weighs 28.09 g. This leads to the equivalence statement

$$1\text{ mol Si atoms} = 28.09\text{ g Si}$$

Thus,

$$5.68 \times 10^{-3}\ \cancel{\text{g Si}} \times \frac{1\text{ mol Si}}{28.09\ \cancel{\text{g Si}}} = 2.02 \times 10^{-4}\text{ mol Si}$$

Using the definition of a mole (1 mol = 6.022×10^{23}), we have

$$2.02 \times 10^{-4}\ \cancel{\text{mol Si}} \times \frac{6.022 \times 10^{23}\text{ atoms}}{1\ \cancel{\text{mol Si}}} = 1.22 \times 10^{20}\text{ Si atoms}$$

We can summarize this calculation as follows:

$$\boxed{5.68\text{ mg Si}} \times \frac{1\text{ g}}{1000\text{ mg}} \rightarrow \boxed{5.68 \times 10^{-3}\text{ g Si}}$$

$$\boxed{5.68 \times 10^{-3}\text{ g Si}} \times \frac{1\text{ mol}}{28.09\text{ g}} \rightarrow \boxed{2.02 \times 10^{-4}\text{ mol Si}}$$

$$\boxed{2.02 \times 10^{-4}\text{ mol Si}} \times \frac{6.022 \times 10^{23}\text{ Si atoms}}{\text{mol}} \rightarrow \boxed{1.22 \times 10^{20}\text{ Si atoms}}$$

PROBLEM SOLVING: DOES THE ANSWER MAKE SENSE?

When you finish a problem, always think about the "reasonableness" of your answers. In Example 8.4, 5.68 mg of silicon is clearly much less than 1 mole of silicon (which has a mass of 28.09 g), so the final answer of 1.22×10^{20} atoms (compared to 6.022×10^{23} atoms in a mole) at least lies in the right direction. That is, 1.22×10^{20} atoms is a smaller number than 6.022×10^{23}. Also, always include the units as you perform calculations and make sure the correct units are obtained at the end. Paying careful attention to units and making this type of general check can help you detect errors such as an inverted conversion factor or a number that was incorrectly entered into your calculator.

As you can see, the problems are getting more complicated to solve. In the next section we will discuss strategies that will help you become a better problem solver.

SELF CHECK

The values for the average masses of the atoms of the elements are listed inside the front cover of this book.

Exercise 8.3 Chromium (Cr) is a metal that is added to steel to improve its resistance to corrosion (for example, to make stainless steel). Calculate both the number of moles in a sample of chromium containing 5.00×10^{20} atoms and the mass of the sample.

See Problems 8.19 through 8.24. ■

8-4 Learning to Solve Problems

OBJECTIVE To understand how to solve problems by asking and answering a series of questions.

Imagine today is the first day of your new job. The problem is that you don't know how to get there. However, as luck would have it, a friend does know the way and offers to drive you. What should you do while you sit in the passenger seat? If your goal is simply to get to work today, you might not pay attention to how to get there. However, you will need to get there on your own tomorrow, so you should pay attention to distances, signs, and turns. The difference between these two approaches is the difference between taking a passive role (going along for the ride) and an active role (learning how to do it yourself). In this section, we will emphasize that you should take an active role in reading the text, especially the solutions to the practice problems.

One of the great rewards of studying chemistry is that you become a good problem solver. Being able to solve complex problems is a talent that will serve you well in all walks of life. It is our purpose in this text to help you learn to solve problems in a flexible, creative way based on understanding the fundamental ideas of chemistry. We call this approach **conceptual problem solving.** The ultimate goal is to be able to solve new problems (that is, problems you have not seen before) on your own. In this text, we will provide problems, but instead of giving solutions for you to memorize, we will explain how to think about the solutions to the problems. Although the answers to these problems are important, it is even more important that you understand the process—the thinking necessary to get to the answer. At first we will be solving the problem for you (we will be "driving"). However, it is important that you do not take a passive role. While studying the solution, it is crucial that you interact—think through the problem with us, that is, take an active role so that eventually you can "drive" by yourself. Do not skip the discussion and jump to the answer. Usually, the solution involves asking a series of questions. Make sure that you understand each step in the process.

Although actively studying our solutions to the problems is helpful, at some point you will need to know how to think about these problems on your own. If we help you too much as you solve the problems, you won't really learn effectively. If we always "drive," you won't interact as meaningfully with the material. Eventually you need to learn to drive by yourself. Because of this, we will provide more help on the earlier problems and less as we proceed in later chapters. The goal is for you to learn how to solve a problem because you understand the main concepts and ideas in the problem.

Consider, for example, that you now know how to get from home to work. Does this mean that you can drive from work to home? Not necessarily, as you probably know from experience. If you have only memorized the directions from home to work and do not understand fundamental principles such as "I traveled north to get to the workplace, so my house is south of the workplace," you may find yourself stranded. Part of conceptual problem solving is understanding these fundamental principles.

Of course, there are many more places to go than from home to work and back. In a more complicated example, suppose you know how to get from your house to work (and back) and from your house to the library (and back). Can you get from work to the library without having to go back home? Probably not, if you have only memorized directions and you do not have a "big picture" of where your house, your workplace, and the library are relative to one another. Getting this big picture—a real understanding of the situation—is the other part of conceptual problem solving.

In conceptual problem solving, we let the problem guide us as we solve it. We ask a series of questions as we proceed and use our knowledge of fundamental principles to answer these questions. Learning this approach requires some patience, but the reward is that you become an effective solver of any new problem that confronts you in daily life or in your work in any field.

To help us as we proceed to solve a problem, the following organizing principles will be useful to us.

1. First, we need to read the problem and decide on the final goal. Then we sort through the facts given, focusing on keywords and often drawing a diagram of the problem. In this part of the analysis, we need to state the problem as simply and as visually as possible. We can summarize this process as **"Where Are We Going?"**
2. We need to work backward from the final goal in order to decide where to start. For example, in a stoichiometry problem we always start with the chemical reaction. Then as we proceed, we ask a series of questions, such as "What are the reactants and products?," "What is the balanced equation?," and "What are the amounts of the reactants?" Our understanding of the fundamental principles of chemistry will enable us to answer each of these simple questions and eventually will lead us to the final solution. We can summarize this process as **"How Do We Get There?"**
3. Once we get the solution of the problem, then we ask ourselves: "Does it make sense?" That is, does our answer seem reasonable? We call this the **Reality Check.** It always pays to check your answer.

Using a conceptual approach to problem solving will enable you to develop real confidence as a problem solver. You will no longer panic when you see a problem that is different in some ways from those you have solved in the past. Although you might be frustrated at times as you learn this method, we guarantee that it will pay dividends later and should make your experience with chemistry a positive one that will prepare you for any career you choose.

To summarize, a creative problem solver has an understanding of fundamental principles and a big picture of the situation. One of our major goals in this text is to help you become a creative problem solver. We will do this first by giving you lots of guidance on how to solve problems. We will "drive," but we hope you will be paying attention instead of just "going along for the ride." As we move forward, we will gradually shift more of the responsibility to you. As you gain confidence in letting the problem guide you, you will be amazed at how effective you can be at solving some really complex problems, just like the ones you will confront in real life.

An Example of Conceptual Problem Solving

Let's look at how conceptual problem solving works in practice. Because we used a driving analogy before, let's consider a problem about driving.

> Estimate the amount of money you would spend on gasoline to drive from New York, New York, to Los Angeles, California.

Where Are We Going?

The first thing we need to do is state the problem in words or as a diagram so that we understand the problem.

In this case, we are trying to estimate how much money we will spend on gasoline. How are we going to do this? We need to understand what factors cause us to spend more or less money. This requires us to ask, **"What Information Do We Need?,"** and **"What Do We Know?"**

Consider two people traveling in separate cars. Why might one person spend more money on gasoline than does the other person? In other words, if you were told that the two people spent different amounts of money on gasoline for a trip, what are some reasons you could give? Consider this, and write down some ideas before you continue reading.

Three factors that are important in this case are

- The price of a gallon of gasoline
- The distance of the trip between New York and Los Angeles
- The average gas mileage of the car we are driving

What do we know, or what are we given in the problem? In this problem, we are not given any of these values but are asked to estimate the cost of gasoline. So we need to estimate the required information. For example, the distance between New York and Los Angeles is about 3000 miles. The cost of gasoline varies over time and location, but a reasonable estimate is \$4.00 a gallon. Gas mileage also varies, but we will assume it is about 30 miles per gallon.

Now that we have the necessary information, we will solve the problem.

How Do We Get There?

To set up the solution, we need to understand how the information affects our answer. Let's consider the relationship between the three factors we identified and our final answer.

- Price of gasoline: directly related. The more a gallon of gasoline costs, the more we will spend in total.
- Distance: directly related. The farther we travel, the more we will spend on gasoline.
- Gas mileage: inversely related. The better our gas mileage (the higher the number), the less we will spend on gasoline.

It should make sense, then, that we multiply the distance and price (because they are directly related) and then divide by the gas mileage (because it is inversely related). We will use dimensional analysis as discussed in Chapter 2. First let's determine how much gasoline we will need for our trip.

$$3000\ \cancel{\text{miles}} \times \frac{1\ \text{gal}}{30\ \cancel{\text{miles}}} = 100 \text{ gallons of gasoline}$$

Notice how the distance is in the numerator and the gas mileage is in the denominator, just as we determined they each should be. So, we will need about 100 gallons of gasoline. How much will this much gasoline cost?

$$100\ \cancel{\text{gallons}} \times \frac{\$4.00}{1\ \cancel{\text{gallon}}} = \$400$$

Notice that the price of a gallon of gas is in the numerator, just as predicted. So, given our information, we estimate the total cost of gasoline to be \$400. The final step is to consider if this answer is reasonable.

REALITY CHECK Does our answer make sense? This is always a good question to consider, and our answer will depend on our familiarity with the situation. Sometimes we may not have a good feel for what the answer should be, especially when we are learning a new concept. Other times we may have only a rough idea and may be able to claim that the answer seems reasonable, although we cannot say it is exactly right. This will usually be the case, and it is the case here if you are familiar with how much you spend on gasoline. For example, the price to fill up the tank for an average car (at \$4.00 per gallon) is around \$40 to \$80. So, if our answer is under \$100, we should be suspicious. An answer in the thousands of dollars is way too high. So, an answer in the hundreds of dollars seems reasonable.

8-5 Molar Mass

OBJECTIVES

- **To understand the definition of molar mass.**
- **To learn to convert between moles and mass of a given sample of a chemical compound.**

Note that when we say 1 mole of methane, we mean 1 mole of methane *molecules*.

A chemical compound is, fundamentally, a collection of atoms. For example, methane (the major component of natural gas) consists of molecules each containing one carbon atom and four hydrogen atoms (CH_4). How can we calculate the mass of 1 mole of methane? That is, what is the mass of 6.022×10^{23} CH_4 molecules? Because each CH_4 molecule contains one carbon atom and four hydrogen atoms, 1 mole of CH_4 molecules consists of 1 mole of carbon atoms and 4 moles of hydrogen atoms (Fig. 8.3).

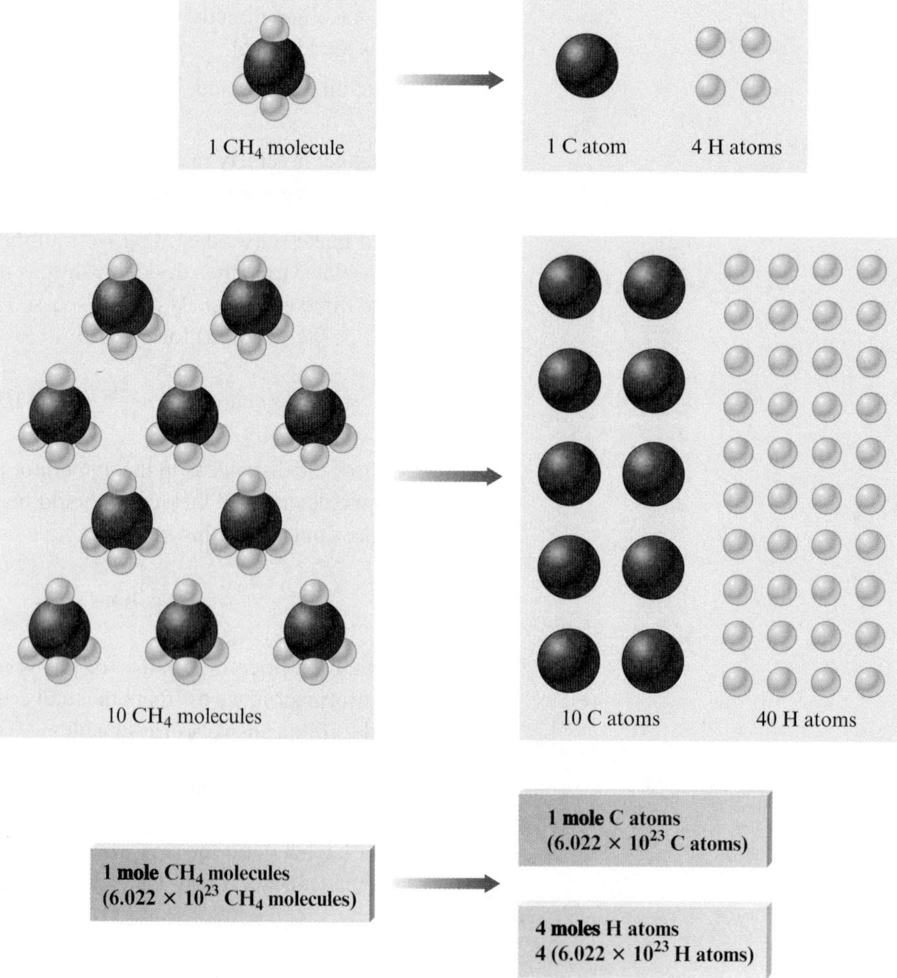

Figure 8.3 Various numbers of methane molecules showing their constituent atoms.

Math skill builder

Remember that the least number of decimal places limits the number of significant figures in addition.

The mass of 1 mole of methane can be found by summing the masses of carbon and hydrogen present:

$$\begin{aligned} \text{Mass of 1 mol C} &= 1 \times 12.01\text{ g} = 12.01\text{ g} \\ \text{Mass of 4 mol H} &= 4 \times 1.008\text{ g} = \underline{4.032\text{ g}} \\ \text{Mass of 1 mol } CH_4 &= 16.04\text{ g} \end{aligned}$$

The quantity 16.04 g is called the molar mass for methane: the mass of 1 mole of CH_4 molecules. The **molar mass*** of any substance is the *mass (in grams) of 1 mole of the substance*. The molar mass is obtained by summing the masses of the component atoms.

Interactive Example 8.5

Calculating Molar Mass

Calculate the molar mass of sulfur dioxide, a gas produced when sulfur-containing fuels are burned. Unless "scrubbed" from the exhaust, sulfur dioxide can react with moisture in the atmosphere to produce acid rain.

SOLUTION

Where Are We Going?

We want to determine the molar mass of sulfur dioxide in units of g/mol.

What Do We Know?

- The formula for sulfur dioxide is SO_2, which means that 1 mole of SO_2 molecules contains 1 mole of sulfur atoms and 2 moles of oxygen atoms.

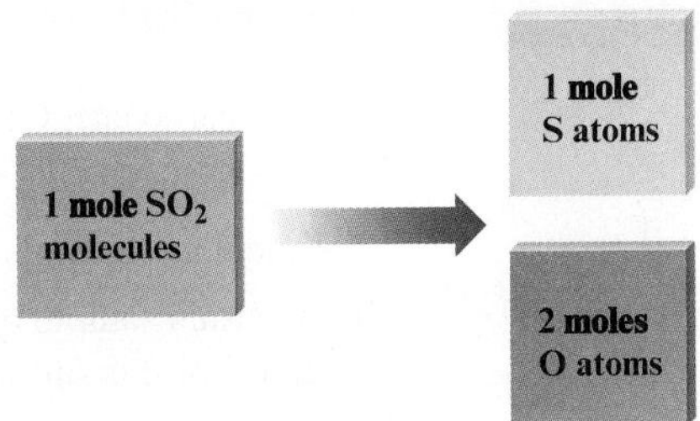

- We know the atomic masses of sulfur (32.07 g/mol) and oxygen (16.00 g/mol).

How Do We Get There?

We need to find the mass of 1 mole of SO_2 molecules, which is the molar mass of SO_2.

$$\begin{aligned} \text{Mass of 1 mol S} &= 1 \times 32.07\text{ g} = 32.07\text{ g} \\ \text{Mass of 2 mol O} &= 2 \times 16.00\text{ g} = \underline{32.00\text{ g}} \\ \text{Mass of 1 mol } SO_2 &= 64.07\text{ g} = \text{molar mass} \end{aligned}$$

The molar mass of SO_2 is 64.07 g. It represents the mass of 1 mole of SO_2 molecules.

REALITY CHECK The answer is greater than the atomic masses of sulfur and oxygen. The units (g/mol) are correct, and the answer is reported to the correct number of significant figures (to two decimal places).

SELF CHECK

Exercise 8.4 Polyvinyl chloride (called PVC), which is widely used for floor coverings ("vinyl") and for plastic pipes in plumbing systems, is made from a molecule with the formula C_2H_3Cl. Calculate the molar mass of this substance.

See Problems 8.27 through 8.30. ■

*The term *molecular weight* was traditionally used instead of *molar mass*. The terms *molecular weight* and *molar mass* mean exactly the same thing. Because the term *molar mass* more accurately describes the concept, it will be used in this text.

Some substances exist as a collection of ions rather than as separate molecules. For example, ordinary table salt, sodium chloride (NaCl), is composed of an array of Na^+ and Cl^- ions. There are no NaCl molecules present. In some books the term *formula weight* is used instead of molar mass for ionic compounds. However, in this book we will apply the term *molar mass* to both ionic and molecular substances.

To calculate the molar mass for sodium chloride, we must realize that 1 mole of NaCl contains 1 mole of Na^+ ions and 1 mole of Cl^- ions.

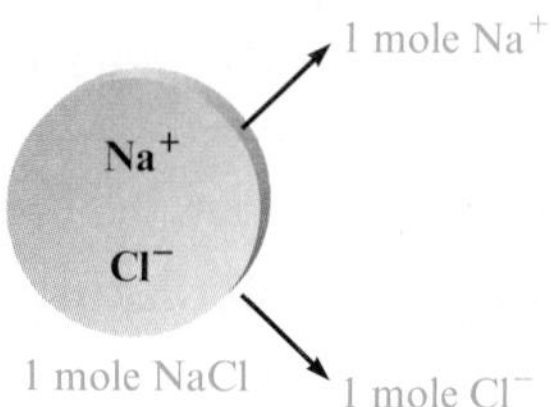

The mass of the electron is so small that Na^+ and Na have the same mass for our purposes, even though Na^+ has one electron fewer than Na. Also, the mass of Cl virtually equals the mass of Cl^- even though it has one more electron than Cl.

Therefore, the molar mass (in grams) for sodium chloride represents the sum of the mass of 1 mole of sodium ions and the mass of 1 mole of chloride ions.

$$\begin{aligned}
\text{Mass of 1 mol } Na^+ &= 22.99 \text{ g} \\
\text{Mass of 1 mol } Cl^- &= \underline{35.45 \text{ g}} \\
\text{Mass of 1 mol NaCl} &= 58.44 \text{ g} = \text{molar mass}
\end{aligned}$$

The molar mass of NaCl is 58.44 g. It represents the mass of 1 mole of sodium chloride.

Interactive Example 8.6

Calculating Mass from Moles

Calcium carbonate, $CaCO_3$ (also called calcite), is the principal mineral found in limestone, marble, chalk, pearls, and the shells of marine animals such as clams.

a. Calculate the molar mass of calcium carbonate.

b. A certain sample of calcium carbonate contains 4.86 moles. What is the mass in grams of this sample?

SOLUTION

a. Where Are We Going?

We want to determine the molar mass of calcium carbonate in units of g/mol.

What Do We Know?

- The formula for calcium carbonate is $CaCO_3$. One mole of $CaCO_3$ contains 1 mole of Ca, 1 mole of C, and 3 moles of O.
- We know the atomic masses of calcium (40.08 g/mol), carbon (12.01 g/mol), and oxygen (16.00 g/mol).

How Do We Get There?

Calcium carbonate is an ionic compound composed of Ca^{2+} and CO_3^{2-} ions. One mole of calcium carbonate contains 1 mole of Ca^{2+} and 1 mole of CO_3^{2-} ions. We calculate the molar mass by summing the masses of the components.

$$\begin{aligned}
&\text{Mass of 1 mol } Ca^{2+} = 1 \times 40.08 \text{ g} = 40.08 \text{ g} \\
&\text{Mass of 1 mol } CO_3^{2-} \text{ (contains 1 mol C and 3 mol O):} \\
&\quad \text{1 mol C} = 1 \times 12.01 \text{ g} = 12.01 \text{ g} \\
&\quad \text{3 mol O} = 3 \times 16.00 \text{ g} = \underline{48.00 \text{ g}} \\
&\quad\quad \text{Mass of 1 mol } CaCO_3 = 100.09 \text{ g} = \text{molar mass}
\end{aligned}$$

REALITY CHECK The answer is greater than the atomic masses of calcium, carbon, and oxygen. The units (g/mol) are correct, and the answer is reported to the correct number of significant figures (to two decimal places).

b. Where Are We Going?

We want to determine the mass of 4.86 moles of $CaCO_3$.

What Do We Know?

- From part a, we know that the molar mass of $CaCO_3$ is 100.09 g/mol.
- We have 4.86 moles of $CaCO_3$.

How Do We Get There?

We determine the mass of 4.86 moles of $CaCO_3$ by using the molar mass.

$$4.86 \text{ } \cancel{\text{mol CaCO}_3} \times \frac{100.09 \text{ g CaCO}_3}{1 \text{ } \cancel{\text{mol CaCO}_3}} = 486 \text{ g CaCO}_3$$

This can be diagrammed as follows:

$$\boxed{4.86 \text{ mol CaCO}_3} \times \frac{100.09 \text{ g}}{\text{mol}} \rightarrow \boxed{486 \text{ g CaCO}_3}$$

REALITY CHECK We have a bit less than 5 moles of $CaCO_3$, which has a molar mass of about 100 g/mol. We should expect an answer a bit less than 500 g, so our answer makes sense. The number of significant figures in our answer (486 g) is three, as required by the initial number of moles (4.86 moles).

SELF CHECK

For average atomic masses, look inside the front cover of this book.

Exercise 8.5 Calculate the molar mass for sodium sulfate, Na_2SO_4. A sample of sodium sulfate with a mass of 300.0 g represents what number of moles of sodium sulfate?

See Problems 8.35 through 8.38. ■

In summary, the molar mass of a substance can be obtained by summing the masses of the component atoms. The molar mass (in grams) represents the mass of 1 mole of the substance. Once we know the molar mass of a compound, we can compute the number of moles present in a sample of known mass. The reverse, of course, is also true as illustrated in Example 8.7.

Interactive Example 8.7

Calculating Moles from Mass

Juglone, a dye known for centuries, is produced from the husks of black walnuts. It is also a natural herbicide (weed killer) that kills off competitive plants around the black walnut tree but does not affect grass and other noncompetitive plants. The formula for juglone is $C_{10}H_6O_3$.

a. Calculate the molar mass of juglone.

b. A sample of 1.56 g of pure juglone was extracted from black walnut husks. How many moles of juglone does this sample represent?

Black walnuts encased in green husks growing on a tree.

SOLUTION

a. Where Are We Going?

We want to determine the molar mass of juglone in units of g/mol.

What Do We Know?

- The formula for juglone is $C_{10}H_6O_3$. One mole of juglone contains 10 moles of C, 6 moles of H, and 3 moles of O.
- We know the atomic masses of carbon (12.01 g/mol), hydrogen (1.008 g/mol), and oxygen (16.00 g/mol).

How Do We Get There?

The molar mass is obtained by summing the masses of the component atoms. In 1 mole of juglone there are 10 moles of carbon atoms, 6 moles of hydrogen atoms, and 3 moles of oxygen atoms.

Math skill builder

The 120.1 limits the sum to one decimal place.

$$\begin{aligned}\text{Mass of 10 mol C} &= 10 \times 12.01\text{ g} = 120.1\text{ g}\\ \text{Mass of 6 mol H} &= 6 \times 1.008\text{ g} = 6.048\text{ g}\\ \text{Mass of 3 mol O} &= 3 \times 16.00\text{ g} = \underline{48.00\text{ g}}\\ \text{Mass of 1 mol } C_{10}H_6O_3 &= 174.1\text{ g} = \text{molar mass}\end{aligned}$$

REALITY CHECK Ten moles of carbon would have a mass of about 120 g, and our answer is higher than this. The units (g/mol) are correct, and the answer is reported to the correct number of significant figures (to one decimal place).

b. Where Are We Going?

We want to determine the number of moles of juglone in a sample with mass of 1.56 g.

What Do We Know?

- From part a, we know that the molar mass of juglone is 174.1 g/mol.
- We have 1.56 g of juglone.

How Do We Get There?

The mass of 1 mole of this compound is 174.1 g, so 1.56 g is much less than a mole. We can determine the exact fraction of a mole by using the equivalence statement

$$1\text{ mol} = 174.1\text{ g juglone}$$

to derive the appropriate conversion factor:

$$1.56\text{ }\cancel{\text{g juglone}} \times \frac{1\text{ mol juglone}}{174.1\text{ }\cancel{\text{g juglone}}} = 0.00896\text{ mol juglone}$$

$$= 8.96 \times 10^{-3}\text{ mol juglone}$$

1.56 g juglone $\times \frac{1\text{ mol}}{174.1\text{ g}}$ → 8.96×10^{-3} mol juglone

REALITY CHECK The mass of 1 mole of juglone is 174.1 g, so 1.56 g is much less than 1 mole. Our answer has units of moles, and the number of significant figures in our answer is three, as required by the initial mass of 1.56 g. ■

Interactive Example 8.8

Calculating Number of Molecules

Isopentyl acetate, $C_7H_{14}O_2$, the compound responsible for the scent of bananas, can be produced commercially. Interestingly, bees release about 1 μg (1×10^{-6} g) of this compound when they sting. This attracts other bees, which then join the attack. How many moles and how many molecules of isopentyl acetate are released in a typical bee sting?

SOLUTION

Where Are We Going?

We want to determine the number of moles and the number of molecules of isopentyl acetate in a sample with mass of 1×10^{-6} g.

What Do We Know?

- The formula for isopentyl acetate is $C_7H_{14}O_2$.
- We know the atomic masses of carbon (12.01 g/mol), hydrogen (1.008 g/mol), and oxygen (16.00 g/mol).
- The mass of isopentyl acetate is 1×10^{-6} g.
- There are 6.022×10^{23} molecules in 1 mole.

How Do We Get There?

We are given a mass of isopentyl acetate and want the number of molecules, so we must first compute the molar mass.

$$7 \cancel{\text{mol C}} \times 12.01 \frac{\text{g}}{\cancel{\text{mol}}} = 84.07 \text{ g C}$$

$$14 \cancel{\text{mol H}} \times 1.008 \frac{\text{g}}{\cancel{\text{mol}}} = 14.11 \text{ g H}$$

$$2 \cancel{\text{mol O}} \times 16.00 \frac{\text{g}}{\cancel{\text{mol}}} = 32.00 \text{ g O}$$

$$\text{Molar mass} = 130.18 \text{ g}$$

This means that 1 mole of isopentyl acetate (6.022×10^{23} molecules) has a mass of 130.18 g.

Next we determine the number of moles of isopentyl acetate in 1 μg, which is 1×10^{-6} g. To do this, we use the equivalence statement

$$1 \text{ mol isopentyl acetate} = 130.18 \text{ g isopentyl acetate}$$

which yields the conversion factor we need:

$$1 \times 10^{-6} \cancel{\text{g } C_7H_{14}O_2} \times \frac{1 \text{ mol } C_7H_{14}O_2}{130.18 \cancel{\text{g } C_7H_{14}O_2}} = 8 \times 10^{-9} \text{ mol } C_7H_{14}O_2$$

Using the equivalence statement 1 mol $= 6.022 \times 10^{23}$ units, we can determine the number of molecules:

$$8 \times 10^{-9} \cancel{\text{mol } C_7H_{14}O_2} \times \frac{6.022 \times 10^{23} \text{ molecules}}{1 \cancel{\text{mol } C_7H_{14}O_2}} = 5 \times 10^{15} \text{ molecules}$$

This very large number of molecules is released in each bee sting.

REALITY CHECK The mass of isopentyl acetate released in each sting (1×10^{-6} g) is much less than the mass of 1 mole of $C_7H_{14}O_2$, so the number of moles should be less than 1 mole, and it is (8×10^{-9} mol). The number of molecules should be much less than 6.022×10^{23}, and it is (5×10^{15} molecules).

Our answers have the proper units, and the number of significant figures in our answer is one, as required by the initial mass.

SELF CHECK

Exercise 8.6 The substance Teflon, the slippery coating on many frying pans, is made from the C_2F_4 molecule. Calculate the number of C_2F_4 units present in 135 g of Teflon.

See Problems 8.39 and 8.40. ■

8-6 Percent Composition of Compounds

OBJECTIVE **To learn to find the mass percent of an element in a given compound.**

So far we have discussed the composition of compounds in terms of the numbers of constituent atoms. It is often useful to know a compound's composition in terms of the *masses* of its elements. We can obtain this information from the formula of the compound by comparing the mass of each element present in 1 mole of the compound to the total mass of 1 mole of the compound. The mass fraction for each element is calculated as follows:

Math skill builder

$$\text{Percent} = \frac{\text{Part}}{\text{Whole}} \times 100\%$$

$$\text{Mass fraction for a given element} = \frac{\text{mass of the element present in 1 mole of compound}}{\text{mass of 1 mole of compound}}$$

The mass fraction is converted to *mass percent* by multiplying by 100%.

We will illustrate this concept using the compound ethanol, an alcohol obtained by fermenting the sugar in grapes, corn, and other fruits and grains. Ethanol is often added to gasoline as an octane enhancer to form a fuel called gasohol. The added ethanol has the effect of increasing the octane of the gasoline and also lowering the carbon monoxide in automobile exhaust.

The formula for ethanol is written C_2H_5OH, although you might expect it to be written simply as C_2H_6O.

Note from its formula that each molecule of ethanol contains two carbon atoms, six hydrogen atoms, and one oxygen atom. This means that each mole of ethanol contains 2 moles of carbon atoms, 6 moles of hydrogen atoms, and 1 mole of oxygen atoms. We calculate the mass of each element present and the molar mass for ethanol as follows:

$$\text{Mass of C} = 2\ \cancel{\text{mol}} \times 12.01\ \frac{\text{g}}{\cancel{\text{mol}}} = 24.02\ \text{g}$$

$$\text{Mass of H} = 6\ \cancel{\text{mol}} \times 1.008\ \frac{\text{g}}{\cancel{\text{mol}}} = 6.048\ \text{g}$$

$$\text{Mass of O} = 1\ \cancel{\text{mol}} \times 16.00\ \frac{\text{g}}{\cancel{\text{mol}}} = 16.00\ \text{g}$$

$$\text{Mass of 1 mol } C_2H_5OH = 46.07\ \text{g} = \text{molar mass}$$

The **mass percent** (sometimes called the weight percent) of carbon in ethanol can be computed by comparing the mass of carbon in 1 mole of ethanol with the total mass of 1 mole of ethanol and multiplying the result by 100%.

$$\text{Mass percent of C} = \frac{\text{mass of C in 1 mol } C_2H_5OH}{\text{mass of 1 mol } C_2H_5OH} \times 100\%$$

$$= \frac{24.02\ \text{g}}{46.07\ \text{g}} \times 100\% = 52.14\%$$

That is, ethanol contains 52.14% by mass of carbon. The mass percents of hydrogen and oxygen in ethanol are obtained in a similar manner.

$$\text{Mass percent of H} = \frac{\text{mass of H in 1 mol } C_2H_5OH}{\text{mass of 1 mol } C_2H_5OH} \times 100\%$$

$$= \frac{6.048\ \text{g}}{46.07\ \text{g}} \times 100\% = 13.13\%$$

$$\text{Mass percent of O} = \frac{\text{mass of O in 1 mol } C_2H_5OH}{\text{mass of 1 mol } C_2H_5OH} \times 100\%$$

$$= \frac{16.00\ \text{g}}{46.07\ \text{g}} \times 100\% = 34.73\%$$

Math skill builder

Sometimes, because of rounding-off effects, the sum of the mass percents in a compound is not exactly 100%.

The mass percents of all the elements in a compound add up to 100%, although rounding-off effects may produce a small deviation. Adding up the percentages is a good way to check the calculations. In this case, the sum of the mass percents is 52.14% + 13.13% + 34.73% = 100.00%.

Interactive Example 8.9

Calculating Mass Percent

Carvone is a substance that occurs in two forms, both of which have the same molecular formula ($C_{10}H_{14}O$) and molar mass. One type of carvone gives caraway seeds their characteristic smell; the other is responsible for the smell of spearmint oil. Compute the mass percent of each element in carvone.

SOLUTION

Where Are We Going?

We want to determine the mass percent of each element in carvone.

What Do We Know?

- The formula for carvone is $C_{10}H_{14}O$.
- We know the atomic masses of carbon (12.01 g/mol), hydrogen (1.008 g/mol), and oxygen (16.00 g/mol).
- The mass of isopentyl acetate is 1×10^{-6} g.
- There are 6.022×10^{23} molecules in 1 mole.

What Do We Need To Know?

- The mass of each element (we'll use 1 mole of carvone)
- Molar mass of carvone

How Do We Get There?

Because the formula for carvone is $C_{10}H_{14}O$, the masses of the various elements in 1 mole of carvone are

$$\text{Mass of C in 1 mol} = 10\ \cancel{\text{mol}} \times 12.01\ \frac{\text{g}}{\cancel{\text{mol}}} = 120.1\ \text{g}$$

$$\text{Mass of H in 1 mol} = 14\ \cancel{\text{mol}} \times 1.008\ \frac{\text{g}}{\cancel{\text{mol}}} = 14.11\ \text{g}$$

$$\text{Mass of O in 1 mol} = 1\ \cancel{\text{mol}} \times 16.00\ \frac{\text{g}}{\cancel{\text{mol}}} = 16.00\ \text{g}$$

$$\text{Mass of 1 mol } C_{10}H_{14}O = 150.21\ \text{g}$$

$$\text{Molar mass} = 150.2\ \text{g}$$

(rounding to the correct number of significant figures)

Math skill builder

The 120.1 limits the sum to one decimal place.

Next we find the fraction of the total mass contributed by each element and convert it to a percentage.

$$\text{Mass percent of C} = \frac{120.1\ \text{g C}}{150.2\ \text{g } C_{10}H_{14}O} \times 100\% = 79.96\%$$

$$\text{Mass percent of H} = \frac{14.11\ \text{g H}}{150.2\ \text{g } C_{10}H_{14}O} \times 100\% = 9.394\%$$

$$\text{Mass percent of O} = \frac{16.00\ \text{g O}}{150.2\ \text{g } C_{10}H_{14}O} \times 100\% = 10.65\%$$

REALITY CHECK Add the individual mass percent values—they should total 100% within a small range due to rounding off. In this case, the percentages add up to 100.00%.

SELF CHECK

Exercise 8.7 Penicillin, an important antibiotic (antibacterial agent), was discovered accidentally by the Scottish bacteriologist Alexander Fleming in 1928, although he was never able to isolate it as a pure compound. This and similar antibiotics have saved millions of lives that would otherwise have been lost to infections. Penicillin, like many of the molecules produced by living systems, is a large molecule containing many atoms. One type of penicillin, penicillin F, has the formula $C_{14}H_{20}N_2SO_4$. Compute the mass percent of each element in this compound.

See Problems 8.45 through 8.50. ■

8-7 Formulas of Compounds

OBJECTIVE **To understand the meaning of empirical formulas of compounds.**

Assume that you have mixed two solutions, and a solid product (a precipitate) forms. How can you find out what the solid is? What is its formula? There are several possible approaches you can take to answering these questions. For example, we saw in Chapter 7 that we can usually predict the identity of a precipitate formed when two solutions are mixed in a reaction of this type if we know some facts about the solubilities of ionic compounds.

However, although an experienced chemist can often predict the product expected in a chemical reaction, the only sure way to identify the product is to perform experiments. Usually we compare the physical properties of the product to the properties of known compounds.

Sometimes a chemical reaction gives a product that has never been obtained before. In such a case, a chemist determines what compound has been formed by determining which elements are present and how much of each. These data can be used to obtain the formula of the compound. In Section 8-6 we used the formula of the compound to determine the mass of each element present in a mole of the compound. To obtain the formula of an unknown compound, we do the opposite. That is, we use the measured masses of the elements present to determine the formula.

Recall that the formula of a compound represents the relative numbers of the various types of atoms present. For example, the molecular formula CO_2 tells us that for each carbon atom there are two oxygen atoms in each molecule of carbon dioxide. So to determine the formula of a substance we need to count the atoms. As we have seen in this chapter, we can do this by weighing. Suppose we know that a compound contains only the elements carbon, hydrogen, and oxygen, and we weigh out a 0.2015-g sample for analysis. Using methods we will not discuss here, we find that this 0.2015-g sample of compound contains 0.0806 g of carbon, 0.01353 g of hydrogen, and 0.1074 g of oxygen. We have just learned how to convert these masses to numbers of atoms by using the atomic mass of each element. We begin by converting to moles.

Carbon

$$0.0806\ \text{g}\,\cancel{\text{C}} \times \frac{1\ \text{mol C atoms}}{12.01\ \text{g}\,\cancel{\text{C}}} = 0.00671\ \text{mol C atoms}$$

Hydrogen

$$0.01353\ \text{g}\,\cancel{\text{H}} \times \frac{1\ \text{mol H atoms}}{1.008\ \text{g}\,\cancel{\text{H}}} = 0.01342\ \text{mol H atoms}$$

Oxygen

$$0.1074\ \cancel{g\ O} \times \frac{1\ \text{mol O atoms}}{16.00\ \cancel{g\ O}} = 0.006713\ \text{mol O atoms}$$

Let's review what we have established. We now know that 0.2015 g of the compound contains 0.00671 mole of C atoms, 0.01342 mole of H atoms, and 0.006713 mole of O atoms. Because 1 mole is 6.022×10^{23}, these quantities can be converted to actual numbers of atoms.

Carbon

$$0.00671\ \cancel{\text{mol C atoms}}\ \frac{6.022 \times 10^{23}\ \text{C atoms}}{1\ \cancel{\text{mol C atoms}}} = 4.04 \times 10^{21}\ \text{C atoms}$$

Hydrogen

$$0.01342\ \cancel{\text{mol H atoms}}\ \frac{6.022 \times 10^{23}\ \text{H atoms}}{1\ \cancel{\text{mol H atoms}}} = 8.08 \times 10^{21}\ \text{H atoms}$$

Oxygen

$$0.006713\ \cancel{\text{mol O atoms}}\ \frac{6.022 \times 10^{23}\ \text{O atoms}}{1\ \cancel{\text{mol O atoms}}} = 4.043 \times 10^{21}\ \text{O atoms}$$

These are the numbers of the various types of atoms *in 0.2015 g of compound.* What do these numbers tell us about the formula of the compound? Note the following:

1. The compound contains the same number of C and O atoms.
2. There are twice as many H atoms as C atoms or O atoms.

We can represent this information by the formula CH_2O, which expresses the *relative* numbers of C, H, and O atoms present. Is this the true formula for the compound? In other words, is the compound made up of CH_2O molecules? It may be. However, it might also be made up of $C_2H_4O_2$ molecules, $C_3H_6O_3$ molecules, $C_4H_8O_4$ molecules, $C_5H_{10}O_5$ molecules, $C_6H_{12}O_6$ molecules, and so on. Note that each of these molecules has the required 1:2:1 ratio of carbon to hydrogen to oxygen atoms (the ratio shown by experiment to be present in the compound).

When we break a compound down into its separate elements and "count" the atoms present, we learn only the ratio of atoms—we get only the *relative* numbers of atoms. The formula of a compound that expresses the smallest whole-number ratio of the atoms present is called the **empirical formula** or *simplest formula.* A compound that contains the molecules $C_4H_8O_4$ has the same empirical formula as a compound that contains $C_6H_{12}O_6$ molecules. The empirical formula for both is CH_2O. The actual formula of a compound—the one that gives the composition of the molecules that are present—is called the **molecular formula.** The sugar called glucose is made of molecules with the molecular formula $C_6H_{12}O_6$ (Fig. 8.4). Note from the molecular formula for glucose that the empirical formula is CH_2O. We can represent the molecular formula as a multiple (by 6) of the empirical formula:

$$C_6H_{12}O_6 = (CH_2O)_6$$

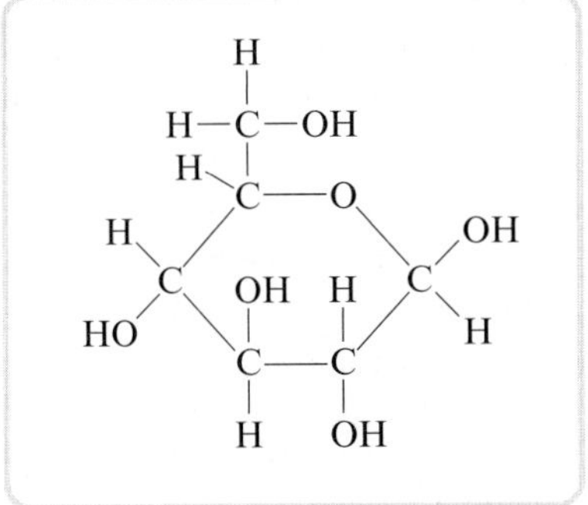

Figure 8.4 ▸ The glucose molecule. The molecular formula is $C_6H_{12}O_6$, as can be verified by counting the atoms. The empirical formula for glucose is CH_2O.

In the next section, we will explore in more detail how to calculate the empirical formula for a compound from the relative masses of the elements present. As we will see in Sections 8-8 and 8-9, we must know the molar mass of a compound to determine its molecular formula.

Interactive Example 8.10

Determining Empirical Formulas

In each case below, the molecular formula for a compound is given. Determine the empirical formula for each compound.

a. C_6H_6. This is the molecular formula for benzene, a liquid commonly used in industry as a starting material for many important products.

b. $C_{12}H_4Cl_4O_2$. This is the molecular formula for a substance commonly called dioxin, a powerful poison that sometimes occurs as a by-product in the production of other chemicals.

c. $C_6H_{16}N_2$. This is the molecular formula for one of the reactants used to produce nylon.

SOLUTION

a. $C_6H_6 = (CH)_6$; CH is the empirical formula. Each subscript in the empirical formula is multiplied by 6 to obtain the molecular formula.

b. $C_{12}H_4Cl_4O_2$; $C_{12}H_4Cl_4O_2 = (C_6H_2Cl_2O)_2$; $C_6H_2Cl_2O$ is the empirical formula. Each subscript in the empirical formula is multiplied by 2 to obtain the molecular formula.

c. $C_6H_{16}N_2 = (C_3H_8N)_2$; C_3H_8N is the empirical formula. Each subscript in the empirical formula is multiplied by 2 to obtain the molecular formula. ■

8-8 Calculation of Empirical Formulas

OBJECTIVE To learn to calculate empirical formulas.

As we said in the previous section, one of the most important things we can learn about a new compound is its chemical formula. To calculate the empirical formula of a compound, we first determine the relative masses of the various elements that are present.

One way to do this is to measure the masses of elements that react to form the compound. For example, suppose we weigh out 0.2636 g of pure nickel metal into a crucible and heat this metal in the air so that the nickel can react with oxygen to form a nickel oxide compound. After the sample has cooled, we weigh it again and find its mass to be 0.3354 g. The gain in mass is due to the oxygen that reacts with the nickel to form the oxide. Therefore, the mass of oxygen present in the compound is the total mass of the product minus the mass of the nickel:

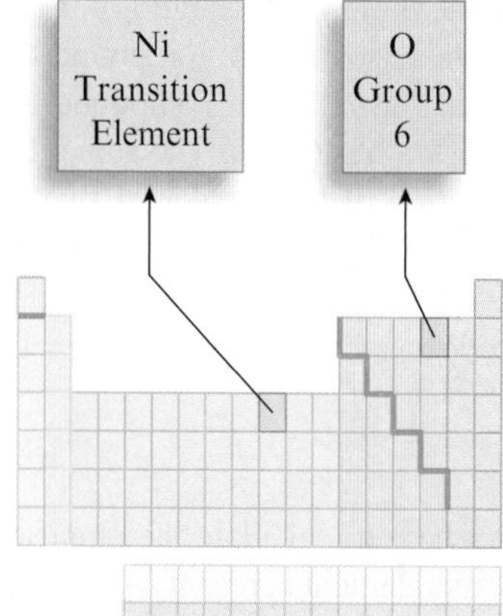

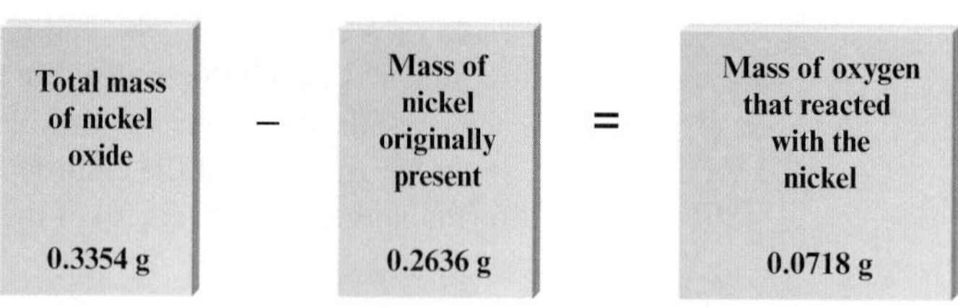

Note that the mass of nickel present in the compound is the nickel metal originally weighed out. So we know that the nickel oxide contains 0.2636 g of nickel and 0.0718 g of oxygen. What is the empirical formula of this compound?

To answer this question we must convert the masses to numbers of atoms, using atomic masses:

Four significant figures allowed.

$$0.2636 \text{ g Ni} \times \frac{1 \text{ mol Ni atoms}}{58.69 \text{ g Ni}} = 0.004491 \text{ mol Ni atoms}$$

Three significant figures allowed.

$$0.0718 \text{ g O} \times \frac{1 \text{ mol O atoms}}{16.00 \text{ g O}} = 0.00449 \text{ mol O atoms}$$

These mole quantities represent numbers of atoms (remember that a mole of atoms is 6.022×10^{23} atoms). It is clear from the moles of atoms that the compound contains an equal number of Ni and O atoms, so the formula is NiO. This is the *empirical formula;* it expresses the smallest whole-number (integer) ratio of atoms:

$$\frac{0.004491 \text{ mol Ni atoms}}{0.00449 \text{ mol O atoms}} = \frac{1 \text{ Ni}}{1 \text{ O}}$$

That is, this compound contains equal numbers of nickel atoms and oxygen atoms. We say the ratio of nickel atoms to oxygen atoms is 1:1 (1 to 1).

Interactive Example 8.11

Calculating Empirical Formulas

An oxide of aluminum is formed by the reaction of 4.151 g of aluminum with 3.692 g of oxygen. Calculate the empirical formula for this compound.

SOLUTION

Where Are We Going?

We want to determine the empirical formula for the aluminum oxide, Al_xO_y. That is, we want to solve for *x* and *y*.

What Do We Know?

- The compound contains 4.151 g of aluminum and 3.692 g of oxygen.
- We know the atomic masses of aluminum (26.98 g/mol), and oxygen (16.00 g/mol).

What Do We Need To Know?

- *x* and *y* represent moles of atoms in 1 mole of the compound, so we need to determine the relative number of moles of Al and O.

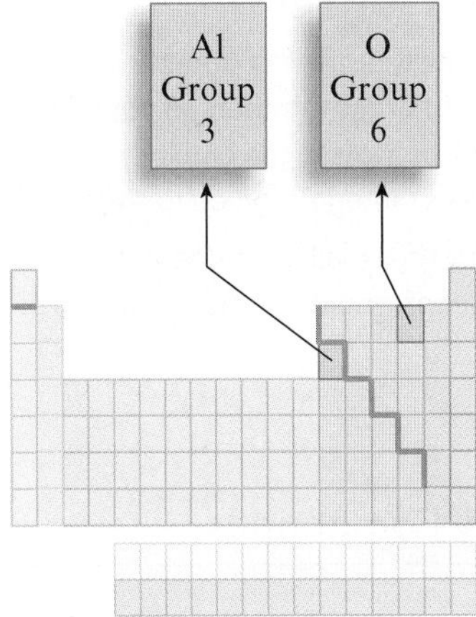

How Do We Get There?

We need to know the relative numbers of each type of atom to write the formula, so we must convert these masses to moles of atoms to get the empirical formula. We carry out the conversion by using the atomic masses of the elements.

$$4.151 \cancel{\text{g Al}} \times \frac{1 \text{ mol Al}}{26.98 \cancel{\text{g Al}}} = 0.1539 \text{ mol Al atoms}$$

$$3.692 \cancel{\text{g O}} \times \frac{1 \text{ mol O}}{16.00 \cancel{\text{g O}}} = 0.2308 \text{ mol O atoms}$$

Because chemical formulas use only whole numbers, we next find the integer (whole-number) ratio of the atoms. To do this we start by dividing both numbers by the smallest of the two. This converts the smallest number to 1.

$$\frac{0.1539 \text{ mol Al}}{0.1539} = 1.000 \text{ mol Al atoms}$$

$$\frac{0.2308 \text{ mol O}}{0.1539} = 1.500 \text{ mol O atoms}$$

Note that dividing both numbers of moles of atoms by the *same* number does not change the *relative* numbers of oxygen and aluminum atoms. That is,

$$\frac{0.2308 \text{ mol O}}{0.1539 \text{ mol Al}} = \frac{1.500 \text{ mol O}}{1.000 \text{ mol Al}}$$

Thus we know that the compound contains 1.500 moles of O atoms for every 1.000 mole of Al atoms, or, in terms of individual atoms, we could say that the compound contains

We might express these data as:

$Al_{1.000\ mol}O_{1.500\ mol}$

or

$Al_{2.000\ mol}O_{3.000\ mol}$

or

Al_2O_3

1.500 O atoms for every 1.000 Al atom. However, because only *whole* atoms combine to form compounds, we must find a set of *whole numbers* to express the empirical formula. When we multiply both 1.000 and 1.500 by 2, we get the integers we need.

$$1.500\ \text{O} \times 2 = 3.000 = 3\ \text{O atoms}$$
$$1.000\ \text{Al} \times 2 = 2.000 = 2\ \text{Al atoms}$$

Therefore, this compound contains two Al atoms for every three O atoms, and the empirical formula is Al_2O_3. Note that the *ratio* of atoms in this compound is given by each of the following fractions:

$$\frac{0.2308\ \text{O}}{0.1539\ \text{Al}} = \frac{1.500\ \text{O}}{1.000\ \text{Al}} = \frac{\frac{3}{2}\text{O}}{1\ \text{Al}} = \frac{3\ \text{O}}{2\ \text{Al}}$$

The smallest whole-number ratio corresponds to the subscripts of the empirical formula, Al_2O_3.

REALITY CHECK The values for x and y are whole numbers. ■

Sometimes the relative numbers of moles you get when you calculate an empirical formula will turn out to be nonintegers, as was the case in Example 8.11. When this happens, you must convert to the appropriate whole numbers. This is done by multiplying all the numbers by the same small integer, which can be found by trial and error. The multiplier needed is almost always between 1 and 6. We will now summarize what we have learned about calculating empirical formulas.

Steps for Determining the Empirical Formula of a Compound

Step 1 Obtain the mass of each element present (in grams).

Step 2 Determine the number of moles of each type of atom present.

Step 3 Divide the number of moles of each element by the smallest number of moles to convert the smallest number to 1. If all of the numbers so obtained are integers (whole numbers), these are the subscripts in the empirical formula. If one or more of these numbers are not integers, go on to step 4.

Step 4 Multiply the numbers you derived in step 3 by the smallest integer that will convert all of them to whole numbers. This set of whole numbers represents the subscripts in the empirical formula.

Interactive Example 8.12

Calculating Empirical Formulas for Binary Compounds

When a 0.3546-g sample of vanadium metal is heated in air, it reacts with oxygen to achieve a final mass of 0.6330 g. Calculate the empirical formula of this vanadium oxide.

SOLUTION

Where Are We Going?

We want to determine the empirical formula for the vanadium oxide, V_xO_y. That is, we want to solve for x and y.

What Do We Know?

- The compound contains 0.3546 g of vanadium and has a total mass of 0.6330 g.
- We know the atomic masses of vanadium (50.94 g/mol) and oxygen (16.00 g/mol).

What Do We Need To Know?

- We need to know the mass of oxygen in the sample.
- x and y represent moles of atoms in 1 mole of the compound, so we need to determine the relative number of moles of V and O.

How Do We Get There?

Step 1 All the vanadium that was originally present will be found in the final compound, so we can calculate the mass of oxygen that reacted by taking the following difference:

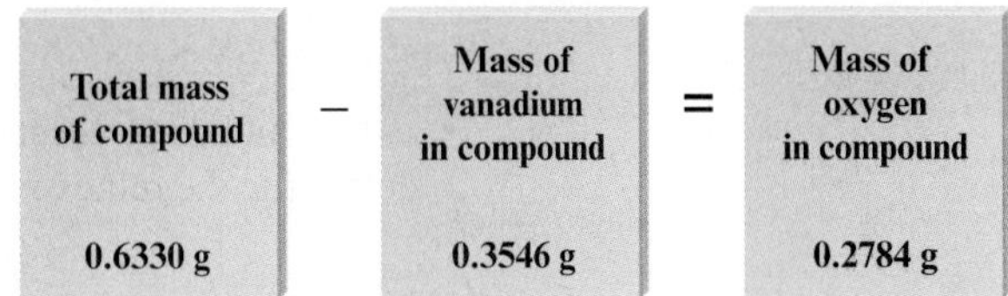

Step 2 Using the atomic masses (50.94 for V and 16.00 for O), we obtain

$$0.3546 \text{ g V} \times \frac{1 \text{ mol V atoms}}{50.94 \text{ g V}} = 0.006961 \text{ mol V atoms}$$

$$0.2784 \text{ g O} \times \frac{1 \text{ mol O atoms}}{16.00 \text{ g O}} = 0.01740 \text{ mol O atoms}$$

Step 3 Then we divide both numbers of moles by the smaller, 0.006961.

$$\frac{0.006961 \text{ mol V atoms}}{0.006961} = 1.000 \text{ mol V atoms}$$

$$\frac{0.01740 \text{ mol O atoms}}{0.006961} = 2.500 \text{ mol O atoms}$$

Because one of these numbers (2.500) is not an integer, we go on to step 4.

Math skill builder

$V_{1.000}O_{2.500}$ becomes V_2O_5.

Step 4 We note that $2 \times 2.500 = 5.000$ and $2 \times 1.000 = 2.000$, so we multiply both numbers by 2 to get integers.

$$2 \times 1.000 \text{ V} = 2.000 \text{ V} = 2 \text{ V}$$

$$2 \times 2.500 \text{ O} = 5.000 \text{ O} = 5 \text{ O}$$

This compound contains 2 V atoms for every 5 O atoms, and the empirical formula is V_2O_5.

REALITY CHECK The values for x and y are whole numbers.

SELF CHECK

Exercise 8.8 In a lab experiment it was observed that 0.6884 g of lead combines with 0.2356 g of chlorine to form a binary compound. Calculate the empirical formula of this compound.

See Problems 8.61, 8.63, 8.65, and 8.66. ■

The same procedures we have used for binary compounds also apply to compounds containing three or more elements, as Example 8.13 illustrates.

Interactive Example 8.13 Calculating Empirical Formulas for Compounds Containing Three or More Elements

A sample of lead arsenate, an insecticide used against the potato beetle, contains 1.3813 g of lead, 0.00672 g of hydrogen, 0.4995 g of arsenic, and 0.4267 g of oxygen. Calculate the empirical formula for lead arsenate.

SOLUTION

Where Are We Going?

We want to determine the empirical formula for lead arsenate, $Pb_aH_bAs_cO_d$. That is, we want to solve for *a, b, c,* and *d.*

What Do We Know?

- The compound contains 1.3813 g of Pb, 0.00672 g of H, 0.4995 g of As, and 0.4267 g of O.
- We know the atomic masses of lead (207.2 g/mol), hydrogen (1.008 g/mol), arsenic (74.92 g/mol), and oxygen (16.00 g/mol).

What Do We Need To Know?

- *a, b, c,* and *d* represent moles of atoms in 1 mole of the compound, so we need to determine the relative number of moles of Pb, H, As, and O.

How Do We Get There?

Step 1 The compound contains 1.3813 g Pb, 0.00672 g H, 0.4995 g As, and 0.4267 g O.

Step 2 We use the atomic masses of the elements present to calculate the moles of each.

$$1.3813\ \cancel{\text{g Pb}} \times \frac{1\text{ mol Pb}}{207.2\ \cancel{\text{g Pb}}} = 0.006667\text{ mol Pb}$$

$$0.00672\ \cancel{\text{g H}} \times \frac{1\text{ mol H}}{1.008\ \cancel{\text{g H}}} = 0.00667\text{ mol H}$$

$$0.4995\ \cancel{\text{g As}} \times \frac{1\text{ mol As}}{74.92\ \cancel{\text{g As}}} = 0.006667\text{ mol As}$$

$$0.4267\ \cancel{\text{g O}} \times \frac{1\text{ mol O}}{16.00\ \cancel{\text{g O}}} = 0.02667\text{ mol O}$$

Step 3 Now we divide by the smallest number of moles.

$$\frac{0.006667\text{ mol Pb}}{0.006667} = 1.000\text{ mol Pb}$$

$$\frac{0.00667\text{ mol H}}{0.006667} = 1.00\text{ mol H}$$

$$\frac{0.006667\text{ mol As}}{0.006667} = 1.000\text{ mol As}$$

$$\frac{0.02667\text{ mol O}}{0.006667} = 4.000\text{ mol O}$$

The numbers of moles are all whole numbers, so the empirical formula is $PbHAsO_4$.

REALITY CHECK The values for *a, b, c,* and *d* are whole numbers.

SELF CHECK

Exercise 8.9 Sevin, the commercial name for an insecticide used to protect crops such as cotton, vegetables, and fruit, is made from carbamic acid. A chemist analyzing a sample of carbamic acid finds 0.8007 g of carbon, 0.9333 g of nitrogen, 0.2016 g of hydrogen, and 2.133 g of oxygen. Determine the empirical formula for carbamic acid.

See Problems 8.57 and 8.59. ■

When a compound is analyzed to determine the relative amounts of the elements present, the results are usually given in terms of percentages by masses of the various elements. In Section 8-6 we learned to calculate the percent composition of a com-

Math skill builder

Percent by mass for a given element means the grams of that element in 100 g of the compound.

pound from its formula. Now we will do the opposite. Given the percent composition, we will calculate the empirical formula.

To understand this procedure, you must understand the meaning of *percent.* Remember that percent means parts of a given component per 100 parts of the total mixture. For example, if a given compound is 15% carbon (by mass), the compound contains 15 g of carbon per 100 g of compound.

Calculation of the empirical formula of a compound when one is given its percent composition is illustrated in Example 8.14.

Interactive Example 8.14

Calculating Empirical Formulas from Percent Composition

Cisplatin, the common name for a platinum compound that is used to treat cancerous tumors, has the composition (mass percent) 65.02% platinum, 9.34% nitrogen, 2.02% hydrogen, and 23.63% chlorine. Calculate the empirical formula for cisplatin.

SOLUTION

Where Are We Going?

We want to determine the empirical formula for cisplatin, $Pt_aN_bH_cCl_d$. That is, we want to solve for *a, b, c,* and *d.*

What Do We Know?

- The compound has the composition (mass percent) 65.02% Pt, 9.34% N, 2.02% H, and 23.63% Cl.
- We know the atomic masses of platinum (195.1 g/mol), nitrogen (14.01 g/mol), hydrogen (1.008 g/mol), and chlorine (35.45 g/mol).

What Do We Need To Know?

- *a, b, c,* and *d* represent moles of atoms in 1 mole of the compound, so we need to determine the relative number of moles of Pt, N, H, and Cl.
- We have mass percent data, and to get to the number of moles we need to know the mass of each element (g) in the sample.

How Do We Get There?

Step 1 Determine how many grams of each element are present in 100 g of compound. Cisplatin is 65.02% platinum (by mass), which means there is 65.02 g of platinum (Pt) per 100.00 g of compound. Similarly, a 100.00-g sample of cisplatin contains 9.34 g of nitrogen (N), 2.02 g of hydrogen (H), and 26.63 g of chlorine (Cl).

If we have a 100.00-g sample of cisplatin, we have 65.02 g Pt, 9.34 g N, 2.02 g H, and 23.63 g Cl.

Step 2 Determine the number of moles of each type of atom. We use the atomic masses to calculate moles.

$$65.02\ \text{g}\ \cancel{\text{Pt}} \times \frac{1\ \text{mol Pt}}{195.1\ \text{g}\ \cancel{\text{Pt}}} = 0.3333\ \text{mol Pt}$$

$$9.34\ \text{g}\ \cancel{\text{N}} \times \frac{1\ \text{mol N}}{14.01\ \text{g}\ \cancel{\text{N}}} = 0.667\ \text{mol N}$$

$$2.02\ \text{g}\ \cancel{\text{H}} \times \frac{1\ \text{mol H}}{1.008\ \text{g}\ \cancel{\text{H}}} = 2.00\ \text{mol H}$$

$$23.63\ \text{g}\ \cancel{\text{Cl}} \times \frac{1\ \text{mol Cl}}{35.45\ \text{g}\ \cancel{\text{Cl}}} = 0.6666\ \text{mol Cl}$$

Step 3 Divide through by the smallest number of moles.

$$\frac{0.3333 \text{ mol Pt}}{0.3333} = 1.000 \text{ mol Pt}$$

$$\frac{0.667 \text{ mol N}}{0.3333} = 2.00 \text{ mol N}$$

$$\frac{2.00 \text{ mol H}}{0.3333} = 6.01 \text{ mol H}$$

$$\frac{0.6666 \text{ mol Cl}}{0.3333} = 2.000 \text{ mol Cl}$$

The empirical formula for cisplatin is $PtN_2H_6Cl_2$. Note that the number for hydrogen is slightly greater than 6 because of rounding-off effects.

REALITY CHECK The values for *a, b, c,* and *d* are whole numbers.

Critical Thinking

One part of the problem-solving strategy for empirical formula determination is to base the calculation on 100 g of compound. What if you chose a mass other than 100 g? Would this work? What if you chose to base the calculation on 100 moles of compound? Would this work?

SELF CHECK

Exercise 8.10 The most common form of nylon (Nylon-6) is 63.68% carbon, 12.38% nitrogen, 9.80% hydrogen, and 14.14% oxygen. Calculate the empirical formula for Nylon-6.

See Problems 8.67 through 8.74. ■

Note from Example 8.14 that once the percentages are converted to masses, this example is the same as earlier examples in which the masses were given directly.

8-9 Calculation of Molecular Formulas

OBJECTIVE **To learn to calculate the molecular formula of a compound, given its empirical formula and molar mass.**

If we know the composition of a compound in terms of the masses (or mass percentages) of the elements present, we can calculate the empirical formula but not the molecular formula. For reasons that will become clear as we consider Example 8.15, to obtain the molecular formula we must know the molar mass. In this section we will consider compounds where both the percent composition and the molar mass are known.

Interactive Example 8.15 Calculating Molecular Formulas

A white powder is analyzed and found to have an empirical formula of P_2O_5. The compound has a molar mass of 283.88 g. What is the compound's molecular formula?

SOLUTION

Where Are We Going?

We want to determine the molecular formula (P_xO_y) for a compound. That is, we want to solve for x and y.

What Do We Know?

- The empirical formula of the compound is P_2O_5.
- The molar mass of the compound is 283.88 g/mol.
- We know the atomic masses of phosphorus (30.97 g/mol) and oxygen (16.00 g/mol).
- The molecular formula contains a whole number of empirical formula units. So, the molecular formula will be $P_{2x}O_{5y}$.

What Do We Need To Know?

- We need to know the empirical formula mass.

How Do We Get There?

To obtain the molecular formula, we must compare the empirical formula mass to the molar mass. The empirical formula mass for P_2O_5 is the mass of 1 mole of P_2O_5 units.

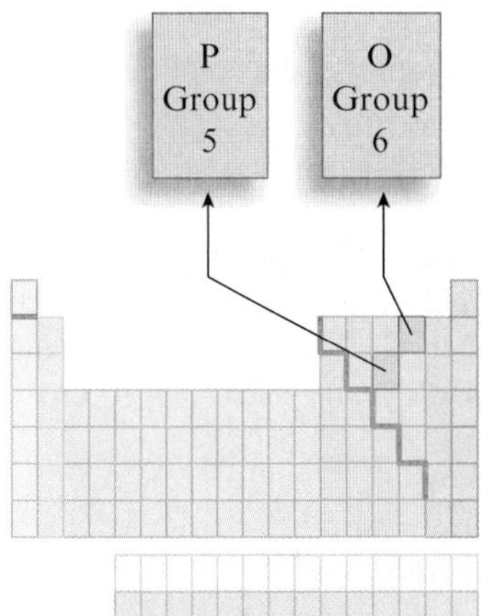

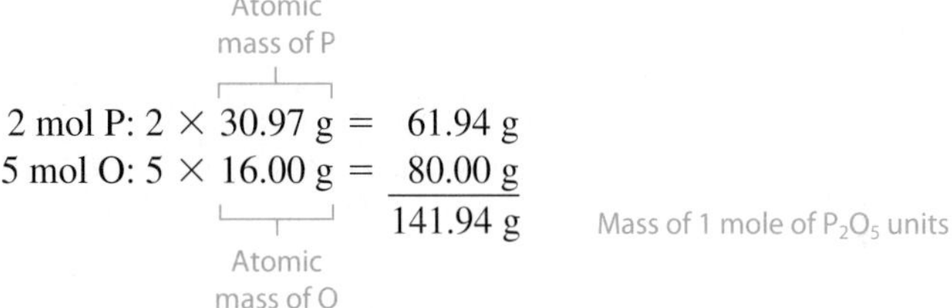

$$\begin{aligned} 2 \text{ mol P: } 2 \times 30.97 \text{ g} &= 61.94 \text{ g} \\ 5 \text{ mol O: } 5 \times 16.00 \text{ g} &= \underline{80.00 \text{ g}} \\ & 141.94 \text{ g} \end{aligned}$$

Recall that the molecular formula contains a whole number of empirical formula units. That is,

$$\text{Molecular formula} = (\text{empirical formula})_n$$

where n is a small whole number. Now, because

$$\text{Molecular formula} = n \times \text{empirical formula}$$

then

$$\text{Molar mass} = n \times \text{empirical formula mass}$$

Solving for n gives

$$n = \frac{\text{molar mass}}{\text{empirical formula mass}}$$

Thus, to determine the molecular formula, we first divide the molar mass by the empirical formula mass. This tells us how many empirical formula masses there are in one molar mass.

$$\frac{\text{Molar mass}}{\text{Empirical formula mass}} = \frac{283.88 \text{ g}}{141.94 \text{ g}} = 2$$

This result means that $n = 2$ for this compound, so the molecular formula consists of two empirical formula units, and the molecular formula is $(P_2O_5)_2$, or P_4O_{10}. The structure of this interesting compound is shown in Fig. 8.5.

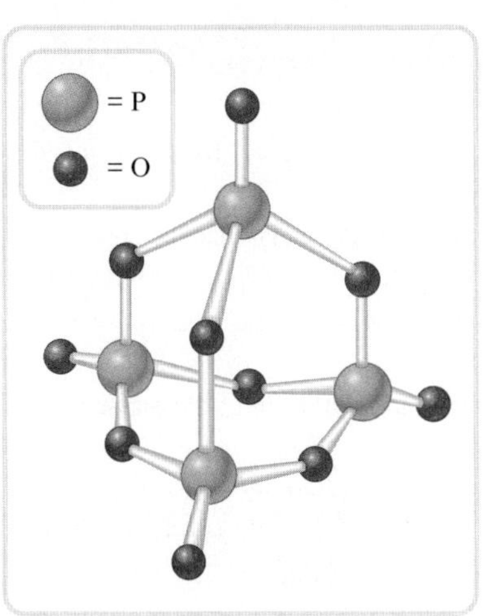

Figure 8.5 ▸ The structure of P_4O_{10} as a "ball-and-stick" model. This compound has a great affinity for water and is often used as a desiccant, or drying agent.

REALITY CHECK The values for x and y are whole numbers. Also, the ratio of P:O in the molecular formula (4:10) is 2:5.

SELF CHECK

Exercise 8.11 A compound used as an additive for gasoline to help prevent engine knock shows the following percent composition:

71.65% Cl 24.27% C 4.07% H

The molar mass is known to be 98.96 g. Determine the empirical formula and the molecular formula for this compound.

See Problems 8.81 and 8.82. ■

It is important to realize that the molecular formula is always an integer multiple of the empirical formula. For example, the sugar glucose (Fig. 8.4) has the empirical formula CH_2O and the molecular formula $C_6H_{12}O_6$. In this case there are six empirical formula units in each glucose molecule:

$$(CH_2O)_6 = C_6H_{12}O_6$$

In general, we can represent the molecular formula in terms of the empirical formula as follows:

$$(\text{Empirical formula})_n = \text{molecular formula}$$

where n is an integer. If $n = 1$, the molecular formula is the same as the empirical formula. For example, for carbon dioxide the empirical formula (CO_2) and the molecular formula (CO_2) are the same, so $n = 1$. On the other hand, for tetraphosphorus decoxide the empirical formula is P_2O_5 and the molecular formula is $P_4O_{10} = (P_2O_5)_2$. In this case $n = 2$.

CHAPTER 8 REVIEW

F directs you to the *Chemistry in Focus* feature in the chapter

Key Terms

atomic mass unit (amu) (8-2)
average atomic mass (8-2)
mole (mol) (8-3)
Avogadro's number (8-3)
conceptual problem solving (8-4)
molar mass (8-5)
mass percent (8-6)
empirical formula (8-7)
molecular formula (8-7)

For Review

- Objects do not need to have identical masses to be counted by weighing. All we need to know is the average mass of the objects.
- To count the atoms in a sample of a given element by weighing, we must know the mass of the sample and the average mass for that element.
- Samples in which the ratio of the masses is the same as the ratio of the masses of the individual atoms always contain the same number of atoms.
- One mole of anything contains 6.022×10^{23} units of that substance.
- A sample of an element with a mass equal to that element's average atomic mass (expressed in grams) contains 1 mole of atoms.
- The molar mass of any compound is the mass in grams of 1 mole of the compound.
- The molar mass of a compound is the sum of the masses of the component atoms.
- $\text{Moles of a compound} = \dfrac{\text{mass of the sample (g)}}{\text{molar mass of the compound (g/mol)}}$
- $\text{Mass of a sample (g)} = (\text{moles of sample}) \times (\text{molar mass of compound})$
- Percent composition consists of the mass percent of each element in a compound:

  Mass percent =

  $$\frac{\text{mass of a given element in 1 mole of a compound}}{\text{mass of 1 mole of compound}} \times 100\%$$

- The empirical formula of a compound is the simplest whole-number ratio of the atoms present in the compound.
- The empirical formula can be found from the percent composition of the compound.
- The molecular formula is the exact formula of the molecules present in a substance.
- The molecular formula is always a whole-number multiple of the empirical formula.
- The following diagram shows these different ways of expressing the same information.

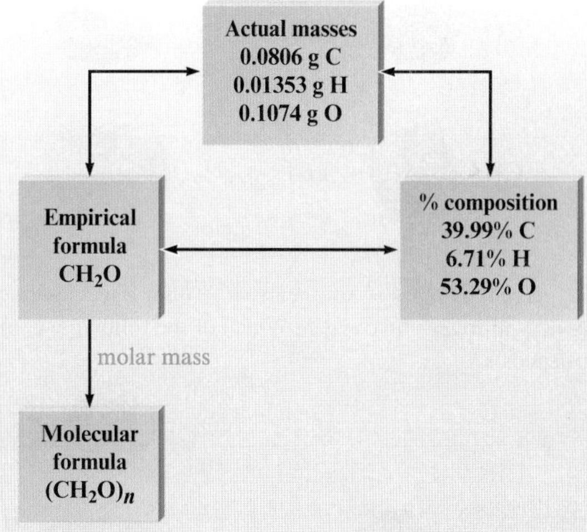

Active Learning Questions

These questions are designed to be considered by groups of students in class. Often these questions work well for introducing a particular topic in class.

1. In chemistry, what is meant by the term *mole?* What is the importance of the mole concept?
2. What is the difference between the empirical and molecular formulas of a compound? Can they ever be the same? Explain.
3. A substance A_2B is 60% A by mass. Calculate the percent B (by mass) for AB_2.
4. Give the formula for calcium phosphate and then answer the following questions:
   a. Calculate the percent composition of each of the elements in this compound.
   b. If you knew that there was 50.0 g of phosphorus in your sample, how many grams of calcium phosphate would you have? How many moles of calcium phosphate would this be? How many formula units of calcium phosphate?
5. How would you find the number of "chalk molecules" it takes to write your name on the board? Explain what you would need to do, and provide a sample calculation.
6. A 0.821-mol sample of a substance composed of diatomic molecules has a mass of 131.3 g. Identify this molecule.
7. How many molecules of water are there in a 10.0-g sample of water? How many hydrogen atoms are there in this sample?
8. What is the mass (in grams) of one molecule of ammonia?
9. Consider separate 100.0-g samples of each of the following: NH_3, N_2O, N_2H_4, HCN, HNO_3. Arrange these samples from largest mass of nitrogen to smallest mass of nitrogen and prove/explain your order.
10. A molecule has a mass of 4.65×10^{-23} g. Provide two possible chemical formulas for such a molecule.
11. Differentiate between the terms *atomic mass* and *molar mass.*
12. Consider Figure 4.19 in the text. Why is it that the formulas for ionic compounds are always empirical formulas?
13. Why do we need to count atoms by weighing them?
14. The following claim is made in your text: 1 mole of marbles is enough to cover the entire earth to a depth of 50 miles. Provide mathematical support for this claim. Is it reasonably accurate?
15. Estimate the length of time it would take you to count to Avogadro's number. Provide mathematical support.
16. Suppose Avogadro's number was 1000 instead of 6.022×10^{23}. How, if at all, would this affect the relative masses on the periodic table? How, if at all, would this affect the absolute masses of the elements?
17. Estimate the number of atoms in your body and provide mathematical support. Because it is an estimate, it need not be exact, although you should choose your number wisely.
18. Consider separate equal mass samples of magnesium, zinc, and silver. Rank them from greatest to least number of atoms and support your answer.
19. You have a 20.0-g sample of silver metal. You are given 10.0 g of another metal and told that this sample contains twice the number of atoms as the sample of silver metal. Identify this metal.
20. How would you find the number of "ink molecules" it takes to write your name on a piece of paper with your pen? Explain what you would need to do, and provide a sample calculation.
21. True or false? The atom with the largest subscript in a formula is the atom with the largest percent by mass in the compound.

    If true, explain why with an example. If false, explain why and provide a counterexample. In either case, provide mathematical support.
22. Which of the following compounds have the same empirical formulas?

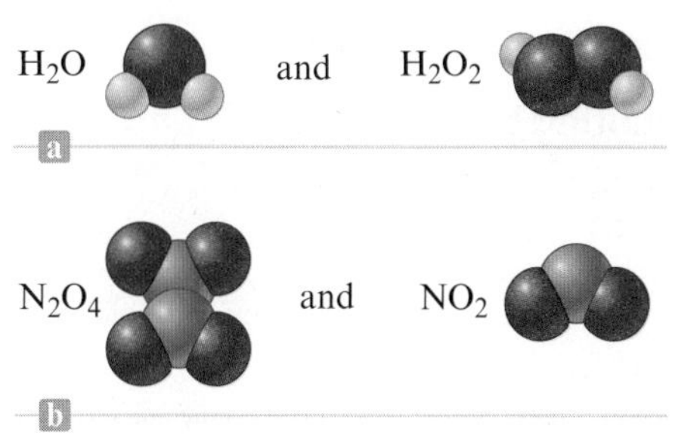

(continued)

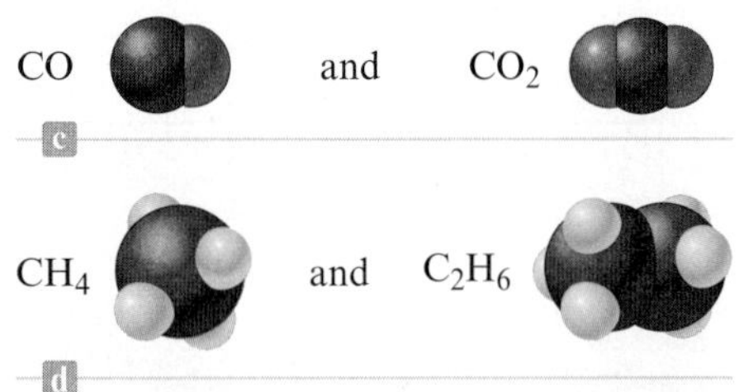

23. The percent by mass of nitrogen is 46.7% for a species containing only nitrogen and oxygen. Which of the following could be this species?

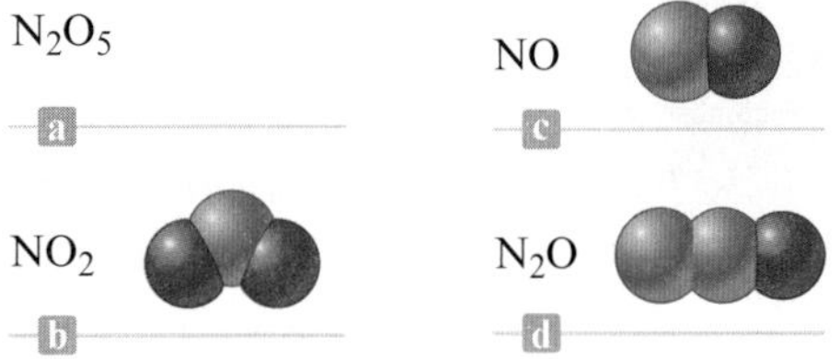

24. Calculate the molar mass of the following substances.

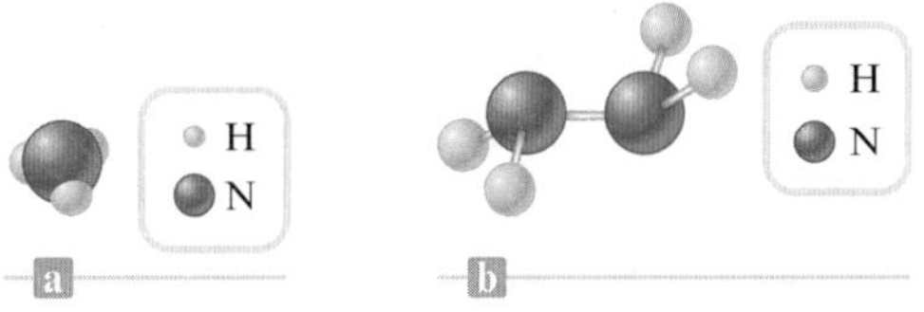

25. Give the empirical formula for each of the compounds represented below.

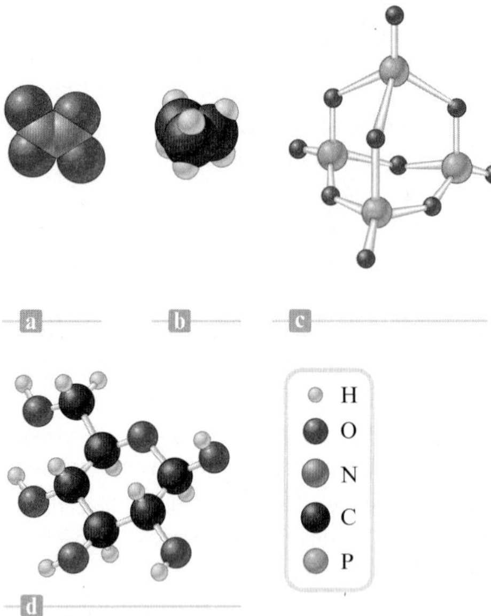

Questions and Problems*

8-1 Counting by Weighing

Problems

1. Merchants usually sell small nuts, washers, and bolts by weight (like jelly beans!) rather than by individually counting the items. Suppose a particular type of washer weighs 0.110 g on the average. What would 100 such washers weigh? How many washers would there be in 100. g of washers?

F 2. The "Chemistry in Focus" segment *Plastic That Talks and Listens!* discusses polyvinylidene difluoride (PVDF). What is the empirical formula of PVDF? Note: An empirical formula is the simplest whole-number ratio of atoms in a compound. This is discussed more fully in Sections 8-7 and 8-8 of your text.

8-2 Atomic Masses: Counting Atoms by Weighing

Questions

3. Define the *amu.* What is one amu equivalent to in grams?
4. What do we mean by the *average* atomic mass of an element? What is "averaged" to arrive at this number?

Problems

5. Using the average atomic masses for each of the following elements (see the table inside the cover of this book), calculate the mass (in amu) of each of the following samples.
   a. 125 carbon atoms
   b. 5 million potassium atoms
   c. 1.04×10^{22} lithium atoms
   d. 1 atom of magnesium
   e. 3.011×10^{23} iodine atoms
6. Using the average atomic masses for each of the following elements (see the table inside the front cover of this book), calculate the number of atoms present in each of the following samples.
   a. 40.08 amu of calcium
   b. 919.5 amu of tungsten
   c. 549.4 amu of manganese
   d. 6345 amu of iodine
   e. 2072 amu of lead
7. What is the average atomic mass (in amu) of iron atoms? What would 299 iron atoms weigh? How many iron atoms are present in a sample of iron that has a mass of 5529.2 amu?
8. The atomic mass of bromine is 79.90 amu. What would be the mass of 54 bromine atoms? How many bromine atoms are contained in a sample of bromine that has a mass of 5672.9 amu?

8-3 The Mole

Questions

9. There are ________ iron atoms present in 55.85 g of iron.
10. There are 6.022×10^{23} tin atoms present in ________ g of tin.

Problems

11. Suppose you have a sample of sodium weighing 11.50 g. How many atoms of sodium are present in the sample? What mass of potassium would you need to have the same number of potassium atoms as there are sodium atoms in the sample of sodium?
12. Consider a sample of silver weighing 300.0 g. How many atoms of silver are present in the sample? What mass of copper would you need for the copper sample to contain the same number of atoms as the silver sample?

*The element symbols and formulas are given in some problems but not in others to help you learn this necessary "vocabulary."

All even-numbered Questions and Problems have answers in the back of this book and solutions in the *Student Solutions Guide.*

13. What mass of hydrogen contains the same number of atoms as 7.00 g of nitrogen?
14. What mass of cobalt contains the same number of atoms as 57.0 g of fluorine?
15. If an average sodium atom has a mass of 3.82×10^{-23} g, what is the mass of a magnesium atom in grams?
16. If an average fluorine atom has a mass of 3.16×10^{-23} g, what is the average mass of a chlorine atom in grams?
17. Which has the smaller mass, 1 mole of He atoms or 4 moles of H atoms?
18. Which weighs less, 0.25 mole of xenon atoms or 2.0 moles of carbon atoms?
19. Use the average atomic masses given inside the front cover of this book to calculate the number of *moles* of the element present in each of the following samples.
    a. 4.95 g of neon
    b. 72.5 g of nickel
    c. 115 mg of silver
    d. 6.22 μg of uranium (μ is a standard abbreviation meaning "micro")
    e. 135 g of iodine
20. Use the average atomic masses given inside the front cover of this book to calculate the number of *moles* of the element present in each of the following samples.
    a. 49.2 g of sulfur
    b. 7.44×10^4 kg of lead
    c. 3.27 mg of chlorine
    d. 4.01 g of lithium
    e. 100.0 g of copper
    f. 82.6 mg of strontium
21. Use the average atomic masses given inside the front cover of this book to calculate the mass in grams of each of the following samples.
    a. 0.251 mole of lithium
    b. 1.51 moles of aluminum
    c. 8.75×10^{-2} moles of lead
    d. 125 moles of chromium
    e. 4.25×10^3 moles of iron
    f. 0.000105 mole of magnesium
22. Use the average atomic masses given inside the front cover of this book to calculate the mass in grams of each of the following samples.
    a. 0.00552 mole of calcium
    b. 6.25 mmol of boron (1 mmol = 1/1000 mole)
    c. 135 moles of aluminum
    d. 1.34×10^{-7} moles of barium
    e. 2.79 moles of phosphorus
    f. 0.0000997 mole of arsenic
23. Using the average atomic masses given inside the front cover of this book, calculate the number of *atoms* present in each of the following samples.
    a. 1.50 g of silver, Ag
    b. 0.0015 mole of copper, Cu
    c. 0.0015 g of copper, Cu
    d. 2.00 kg of magnesium, Mg
    e. 2.34 oz of calcium, Ca
    f. 2.34 g of calcium, Ca
    g. 2.34 moles of calcium, Ca
24. Using the average atomic masses given inside the front cover of this book, calculate the indicated quantities.
    a. the mass in grams of 125 iron atoms
    b. the mass in amu of 125 iron atoms
    c. the number of moles of iron atoms in 125 g of iron
    d. the mass in grams of 125 moles of iron
    e. the number of iron atoms in 125 g of iron
    f. the number of iron atoms in 125 moles of iron

8-5 Molar Mass

Questions

25. The ________ of a substance is the mass (in grams) of 1 mole of the substance.
26. Describe in your own words how the molar mass of the compound below may be calculated.

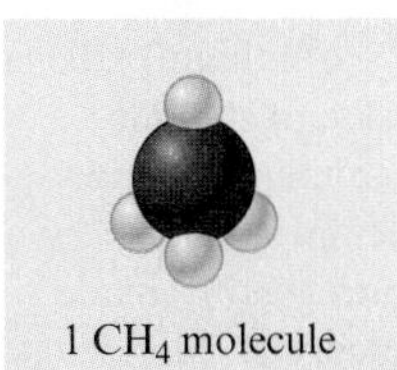
1 CH_4 molecule

Problems

27. Give the name and calculate the molar mass for each of the following substances.
    a. H_3PO_4
    b. Fe_2O_3
    c. $NaClO_4$
    d. $PbCl_2$
    e. HBr
    f. $Al(OH)_3$
28. Give the name and calculate the molar mass for each of the following substances.
    a. $KHCO_3$
    b. Hg_2Cl_2
    c. H_2O_2
    d. $BeCl_2$
    e. $Al_2(SO_4)_3$
    f. $KClO_3$
29. Write the formula and calculate the molar mass for each of the following substances.
    a. barium chloride
    b. aluminum nitrate
    c. iron(II) chloride
    d. sulfur dioxide
    e. calcium acetate
30. Write the formula and calculate the molar mass for each of the following substances.
    a. barium perchlorate
    b. magnesium sulfate
    c. lead(II) chloride
    d. copper(II) nitrate
    e. tin(IV) chloride

31. Calculate the number of *moles* of the indicated substance present in each of the following samples.
    a. 21.4 mg of nitrogen dioxide
    b. 1.56 g of copper(II) nitrate
    c. 2.47 g of carbon disulfide
    d. 5.04 g of aluminum sulfate
    e. 2.99 g of lead(II) chloride
    f. 62.4 g of calcium carbonate
32. Calculate the number of *moles* of the indicated substance present in each of the following samples.
    a. 47.2 g of aluminum oxide
    b. 1.34 kg of potassium bromide
    c. 521 mg of germanium
    d. 56.2 μg of uranium
    e. 29.7 g of sodium acetate
    f. 1.03 g of sulfur trioxide
33. Calculate the number of *moles* of the indicated substance in each of the following samples.
    a. 41.5 g of $MgCl_2$
    b. 135 mg of Li_2O
    c. 1.21 kg of Cr
    d. 62.5 g of H_2SO_4
    e. 42.7 g of C_6H_6
    f. 135 g of H_2O_2
34. Calculate the number of moles of the indicated substance present in each of the following samples.
    a. 1.95×10^{-3} g of lithium carbonate
    b. 4.23 kg of calcium chloride
    c. 1.23 mg of strontium chloride
    d. 4.75 g of calcium sulfate
    e. 96.2 mg of nitrogen(IV) oxide
    f. 12.7 g of mercury(I) chloride
35. Calculate the mass in grams of each of the following samples.
    a. 1.25 moles of aluminum chloride
    b. 3.35 moles of sodium hydrogen carbonate
    c. 4.25 millimoles of hydrogen bromide (1 millimole = $\frac{1}{1000}$ mole)
    d. 1.31×10^{-3} moles of uranium
    e. 0.00104 mole of carbon dioxide
    f. 1.49×10^{2} moles of iron
36. Calculate the mass in grams of each of the following samples.
    a. 6.14×10^{-4} moles of sulfur trioxide
    b. 3.11×10^{5} moles of lead(IV) oxide
    c. 0.495 mole of chloroform, $CHCl_3$
    d. 2.45×10^{-8} moles of trichloroethane, $C_2H_3Cl_3$
    e. 0.167 mole of lithium hydroxide
    f. 5.26 moles of copper(I) chloride
37. Calculate the mass in grams of each of the following samples.
    a. 0.251 mole of ethyl alcohol, C_2H_6O
    b. 1.26 moles of carbon dioxide
    c. 9.31×10^{-4} moles of gold(III) chloride
    d. 7.74 moles of sodium nitrate
    e. 0.000357 mole of iron
38. Calculate the mass in grams of each of the following samples.
    a. 0.994 mole of benzene, C_6H_6
    b. 4.21 moles of calcium hydride
    c. 1.79×10^{-4} moles of hydrogen peroxide, H_2O_2
    d. 1.22 mmol of glucose, $C_6H_{12}O_6$ (1 mmol = $\frac{1}{1000}$ mole).
    e. 10.6 moles of tin
    f. 0.000301 mole of strontium fluoride
39. Calculate the number of *molecules* present in each of the following samples.
    a. 4.75 mmol of phosphine, PH_3
    b. 4.75 g of phosphine, PH_3
    c. 1.25×10^{-2} g of lead(II) acetate, $Pb(CH_3CO_2)_2$
    d. 1.25×10^{-2} moles of lead(II) acetate, $Pb(CH_3CO_2)_2$
    e. a sample of benzene, C_6H_6, which contains a total of 5.40 moles of carbon
40. Calculate the number of *molecules* present in each of the following samples.
    a. 3.54 moles of sulfur dioxide, SO_2
    b. 3.54 g of sulfur dioxide, SO_2
    c. 4.46×10^{-5} g of ammonia, NH_3
    d. 4.46×10^{-5} moles of ammonia, NH_3
    e. 1.96 mg of ethane, C_2H_6
41. Calculate the number of *moles* of carbon atoms present in each of the following samples.
    a. 1.271 g of ethanol, C_2H_5OH
    b. 3.982 g of 1,4-dichlorobenzene, $C_6H_4Cl_2$
    c. 0.4438 g of carbon suboxide, C_3O_2
    d. 2.910 g of methylene chloride, CH_2Cl_2
42. Calculate the number of *moles* of sulfur atoms present in each of the following samples.
    a. 2.01 g of sodium sulfate
    b. 2.01 g of sodium sulfite
    c. 2.01 g of sodium sulfide
    d. 2.01 g of sodium thiosulfate, $Na_2S_2O_3$

8-6 Percent Composition of Compounds

Questions

43. The mass fraction of an element present in a compound can be obtained by comparing the mass of the particular element present in 1 mole of the compound to the ________ mass of the compound.
44. If the amount of a sample doubles, what happens to the percent composition of each element in the compound?

Problems

45. Calculate the percent by mass of each element in the following compounds.
    a. $HClO_3$
    b. UF_4
    c. CaH_2
    d. Ag_2S
    e. $NaHSO_3$
    f. MnO_2

All even-numbered Questions and Problems have answers in the back of this book and solutions in the *Student Solutions Guide*.

46. Calculate the percent by mass of each element in the following compounds.

a. ZnO
b. Na_2S
c. $Mg(OH)_2$
d. H_2O_2
e. CaH_2
f. K_2O

47. Calculate the percent by mass of the element listed *first* in the formulas for each of the following compounds.

a. methane, CH_4
b. sodium nitrate, $NaNO_3$
c. carbon monoxide, CO
d. nitrogen dioxide, NO_2
e. 1-octanol, $C_8H_{18}O$
f. calcium phosphate, $Ca_3(PO_4)_2$
g. 3-phenylphenol, $C_{12}H_{10}O$
h. aluminum acetate, $Al(C_2H_3O_2)_3$

48. Calculate the percent by mass of the element listed *first* in the formulas for each of the following compounds.

a. copper(II) bromide, $CuBr_2$
b. copper(I) bromide, $CuBr$
c. iron(II) chloride, $FeCl_2$
d. iron(III) chloride, $FeCl_3$
e. cobalt(II) iodide, CoI_2
f. cobalt(III) iodide, CoI_3
g. tin(II) oxide, SnO
h. tin(IV) oxide, SnO_2

49. Calculate the percent by mass of the element listed *first* in the formulas for each of the following compounds.

a. adipic acid, $C_6H_{10}O_4$
b. ammonium nitrate, NH_4NO_3
c. caffeine, $C_8H_{10}N_4O_2$
d. chlorine dioxide, ClO_2
e. cyclohexanol, $C_6H_{11}OH$
f. dextrose, $C_6H_{12}O_6$
g. eicosane, $C_{20}H_{42}$
h. ethanol, C_2H_5OH

50. What is the mass percent of oxygen in each of the following compounds?

a. carbon monoxide
b. manganese(IV) oxide
c. potassium chlorate
d. iron(II) oxide
e. calcium hydroxide

51. For each of the following samples of ionic substances, calculate the number of moles and mass of the positive ions present in each sample.

a. 4.25 g of ammonium iodide, NH_4I
b. 6.31 moles of ammonium sulfide, $(NH_4)_2S$
c. 9.71 g of barium phosphide, Ba_3P_2
d. 7.63 moles of calcium phosphate, $Ca_3(PO_4)_2$

52. For each of the following ionic substances, calculate the percentage of the overall molar mass of the compound that is represented by the *negative* ions in the substance.

a. ammonium sulfide
b. calcium chloride
c. barium oxide
d. nickel(II) sulfate

8-7 Formulas of Compounds

Questions

53. What experimental evidence about a new compound must be known before its formula can be determined?

54. Explain to a friend who has not yet taken a chemistry course what is meant by the *empirical formula* of a compound.

55. Give the empirical formula that corresponds to each of the following molecular formulas.

a. sodium peroxide, Na_2O_2
b. terephthalic acid, $C_8H_6O_4$
c. phenobarbital, $C_{12}H_{12}N_2O_3$
d. 1,4-dichloro-2-butene, $C_4H_6Cl_2$

56. Which of the following pairs of compounds have the same *empirical* formula?

a. acetylene, C_2H_2, and benzene, C_6H_6
b. ethane, C_2H_6, and butane, C_4H_{10}
c. nitrogen dioxide, NO_2, and dinitrogen tetroxide, N_2O_4
d. diphenyl ether, $C_{12}H_{10}O$, and phenol, C_6H_5OH

8-8 Calculation of Empirical Formulas

Problems

57. A compound was analyzed and was found to contain the following percentages of the elements by mass: barium, 89.56%; oxygen, 10.44%. Determine the empirical formula of the compound.

58. A compound was analyzed and was found to contain the following percentages of the elements by mass: nitrogen, 11.64%; chlorine, 88.36%. Determine the empirical formula of the compound.

59. A 0.5998-g sample of a new compound has been analyzed and found to contain the following masses of elements: carbon, 0.2322 g; hydrogen, 0.05848 g; oxygen, 0.3091 g. Calculate the empirical formula of the compound.

60. A compound was analyzed and was found to contain the following percentages of the elements by mass: boron, 78.14%; hydrogen, 21.86%. Determine the empirical formula of the compound.

61. If a 1.271-g sample of aluminum metal is heated in a chlorine gas atmosphere, the mass of aluminum chloride produced is 6.280 g. Calculate the empirical formula of aluminum chloride.

62. A compound was analyzed and was found to contain the following percentages of the elements by mass: tin, 45.56%; chlorine, 54.43%. Determine the empirical formula of the compound.

63. When 3.269 g of zinc is heated in pure oxygen, the sample gains 0.800 g of oxygen in forming the oxide. Calculate the empirical formula of zinc oxide.

64. If cobalt metal is mixed with excess sulfur and heated strongly, a sulfide is produced that contains 55.06% cobalt by mass. Calculate the empirical formula of the sulfide.

65. If 1.25 g of aluminum metal is heated in an atmosphere of fluorine gas, 3.89 g of aluminum fluoride results. Determine the empirical formula of aluminum fluoride.

All even-numbered Questions and Problems have answers in the back of this book and solutions in the *Student Solutions Guide.*

66. If 2.50 g of aluminum metal is heated in a stream of fluorine gas, it is found that 5.28 g of fluorine will combine with the aluminum. Determine the empirical formula of the compound that results.

67. A compound used in the nuclear industry has the following composition: uranium, 67.61%; fluorine, 32.39%. Determine the empirical formula of the compound.

68. A compound was analyzed and was found to contain the following percentages of the elements by mass: lithium, 46.46%; oxygen, 53.54%. Determine the empirical formula of the compound.

69. A compound has the following percentage composition by mass: copper, 33.88%; nitrogen, 14.94%; oxygen, 51.18%. Determine the empirical formula of the compound.

70. When lithium metal is heated strongly in an atmosphere of pure nitrogen, the product contains 59.78% Li and 40.22% N on a mass basis. Determine the empirical formula of the compound.

71. A compound has been analyzed and has been found to have the following composition: copper, 66.75%; phosphorus, 10.84%; oxygen, 22.41%. Determine the empirical formula of the compound.

72. A compound that contains only carbon, hydrogen, and oxygen is 48.64% C and 8.16% H by mass. What is the empirical formula of this substance?

73. When 1.00 mg of lithium metal is reacted with fluorine gas (F_2), the resulting fluoride salt has a mass of 3.73 mg. Calculate the empirical formula of lithium fluoride.

74. Phosphorus and chlorine form two binary compounds, in which the percentages of phosphorus are 22.55% and 14.87%, respectively. Calculate the empirical formulas of the two binary phosphorus–chlorine compounds.

8-9 Calculation of Molecular Formulas

Questions

75. How does the *molecular* formula of a compound differ from the *empirical* formula? Can a compound's empirical and molecular formulas be the same? Explain.

76. What are the molecular and empirical formulas for the following molecule? Explain your reasoning.

```
         H
         |
       H-C-OH
       H |
        \C---O
   H   /      \   OH
    \ /        \ /
     C  OH  H   C
    /    |  |    \
  HO     C--C     H
         |  |
         H  OH
```

Problems

77. A binary compound of boron and hydrogen has the following percentage composition: 78.14% boron, 21.86% hydrogen. If the molar mass of the compound is determined by experiment to be between 27 and 28 g, what are the empirical and molecular formulas of the compound?

78. A compound with empirical formula CH was found by experiment to have a molar mass of approximately 78 g. What is the molecular formula of the compound?

79. A compound with the empirical formula CH_2 was found to have a molar mass of approximately 84 g. What is the molecular formula of the compound?

80. A compound with empirical formula C_2H_5O was found in a separate experiment to have a molar mass of approximately 90 g. What is the molecular formula of the compound?

81. A compound having an approximate molar mass of 165–170 g has the following percentage composition by mass: carbon, 42.87%; hydrogen, 3.598%; oxygen, 28.55%; nitrogen, 25.00%. Determine the empirical and molecular formulas of the compound.

82. A compound consists of carbon and hydrogen. The molar mass of this compound is 44.1 g/mol. The percent by mass of carbon is 81.71%. Determine the empirical and molecular formulas of the compound.

Additional Problems

83. Use the periodic table shown in Fig. 4.9 to determine the atomic mass (per mole) or molar mass of each of the substances in column 1, and find that mass in column 2.

Column 1	Column 2
(1) molybdenum	(a) 33.99 g
(2) lanthanum	(b) 79.9 g
(3) carbon tetrabromide	(c) 95.94 g
(4) mercury(II) oxide	(d) 125.84 g
(5) titanium(IV) oxide	(e) 138.9 g
(6) manganese(II) chloride	(f) 143.1 g
(7) phosphine, PH_3	(g) 156.7 g
(8) tin(II) fluoride	(h) 216.6 g
(9) lead(II) sulfide	(i) 239.3 g
(10) copper(I) oxide	(j) 331.6 g

84. Complete the following table.

Mass of Sample	Moles of Sample	Atoms in Sample
5.00 g Al	________	________
________	0.00250 mol Fe	________
________	________	2.6×10^{24} atoms Cu
0.00250 g Mg	________	________
________	2.7×10^{-3} mol Na	________
________	________	1.00×10^{4} atoms U

All even-numbered Questions and Problems have answers in the back of this book and solutions in the *Student Solutions Guide*.

85. Complete the following table.

Mass of Sample	Moles of Sample	Molecules in Sample	Atoms in Sample
4.24 g C_6H_6	______	______	______
______	0.224 mol H_2O	______	______
______	______	2.71×10^{22} molecules CO_2	______
______	1.26 mol HCl	______	______
______	______	4.21×10^{24} molecules H_2O	______
0.297 g CH_3OH	______	______	______

86. Consider a hypothetical compound composed of elements X, Y, and Z with the empirical formula X_2YZ_3. Given that the atomic masses of X, Y, and Z are 41.2, 57.7, and 63.9, respectively, calculate the percentage composition by mass of the compound. If the molecular formula of the compound is found by molar mass determination to be actually $X_4Y_2Z_6$, what is the percentage of each element present? Explain your results.

87. A binary compound of magnesium and nitrogen is analyzed, and 1.2791 g of the compound is found to contain 0.9240 g of magnesium. When a second sample of this compound is treated with water and heated, the nitrogen is driven off as ammonia, leaving a compound that contains 60.31% magnesium and 39.69% oxygen by mass. Calculate the empirical formulas of the two magnesium compounds.

88. When a 2.118-g sample of copper is heated in an atmosphere in which the amount of oxygen present is restricted, the sample gains 0.2666 g of oxygen in forming a reddish-brown oxide. However, when 2.118 g of copper is heated in a stream of pure oxygen, the sample gains 0.5332 g of oxygen. Calculate the empirical formulas of the two oxides of copper.

89. Hydrogen gas reacts with each of the halogen elements to form the hydrogen halides (HF, HCl, HBr, HI). Calculate the percent by mass of hydrogen in each of these compounds.

90. Calculate the number of atoms of each element present in each of the following samples.

a. 4.21 g of water
b. 6.81 g of carbon dioxide
c. 0.000221 g of benzene, C_6H_6
d. 2.26 moles of $C_{12}H_{22}O_{11}$

91. Calculate the mass in grams of each of the following samples.

a. 10,000,000,000 nitrogen molecules
b. 2.49×10^{20} carbon dioxide molecules
c. 7.0983 moles of sodium chloride
d. 9.012×10^{-6} moles of 1,2-dichloroethane, $C_2H_4Cl_2$

92. Calculate the mass of carbon in grams, the percent carbon by mass, and the number of individual carbon atoms present in each of the following samples.

a. 7.819 g of carbon suboxide, C_3O_2
b. 1.53×10^{21} molecules of carbon monoxide
c. 0.200 mole of phenol, C_6H_6O

93. Find the item in column 2 that best explains or completes the statement or question in column 1.

Column 1

(1) 1 amu
(2) 1008 amu
(3) mass of the "average" atom of an element
(4) number of carbon atoms in 12.01 g of carbon
(5) 6.022×10^{23} molecules
(6) total mass of all atoms in 1 mole of a compound
(7) smallest whole-number ratio of atoms present in a molecule
(8) formula showing actual number of atoms present in a molecule
(9) product formed when any carbon-containing compound is burned in O_2
(10) have the same empirical formulas, but different molecular formulas

Column 2

(a) 6.022×10^{23}
(b) atomic mass
(c) mass of 1000 hydrogen atoms
(d) benzene, C_6H_6, and acetylene, C_2H_2
(e) carbon dioxide
(f) empirical formula
(g) 1.66×10^{-24} g
(h) molecular formula
(i) molar mass
(j) 1 mole

94. If you have 6.022×10^{23} water molecules present in a sample, which of the following must be *true*?

a. 1 mole of water is present in the sample
b. 18.016 g of water is present in the sample
c. 1.807×10^{24} atoms are present in the sample
d. 2.016 g of hydrogen are present in the sample
e. 2 moles of hydrogen are present in the sample

95. Calculate the number of grams of cobalt that contain the same number of atoms as 2.24 g of iron.

96. Which of the following samples contains the greatest number of atoms?

a. 10.0 g Na
b. 10.0 g Fe
c. 10.0 g Ag
d. 10.0 g Cu
e. 10.0 g Ba

97. Calculate the number of grams of lithium that contain the same number of atoms as 1.00 kg of zirconium.

98. Given that the molar mass of carbon tetrachloride, CCl_4, is 153.8 g, calculate the mass in grams of 1 molecule of CCl_4.

99. Calculate the mass in grams of hydrogen present in 2.500 g of each of the following compounds.

a. benzene, C_6H_6
b. calcium hydride, CaH_2
c. ethyl alcohol, C_2H_5OH
d. serine, $C_3H_7O_3N$

All even-numbered Questions and Problems have answers in the back of this book and solutions in the *Student Solutions Guide*.

100. If you have equal mole samples of NO_2 and F_2, which of the following must be *true*?

a. The number of molecules in each sample is the same.
b. The number of atoms in each sample is the same.
c. The masses of the samples are the same.
d. 6.022×10^{23} molecules are present in each sample.

101. A strikingly beautiful copper compound with the common name "blue vitriol" has the following elemental composition: 25.45% Cu, 12.84% S, 4.036% H, 57.67% O. Determine the empirical formula of the compound.

102. A compound sample by mass contains 60.87% C, 4.38% H, and the rest oxygen. What is the empirical formula of the compound?

103. The mass 1.66×10^{-24} g is equivalent to 1 ________.

104. Although exact isotopic masses are known with great precision for most elements, we use the *average* mass of an element's atoms in most chemical calculations. Explain.

105. Using the average atomic masses given in Table 8.1, calculate the number of atoms present in each of the following samples.

a. 160,000 amu of oxygen
b. 8139.81 amu of nitrogen
c. 13,490 amu of aluminum
d. 5040 amu of hydrogen
e. 367,495.15 amu of sodium

106. If an average sodium atom weighs 22.99 amu, how many sodium atoms are contained in 1.98×10^{13} amu of sodium? What will 3.01×10^{23} sodium atoms weigh?

107. Using the average atomic masses given inside the front cover of this text, calculate how many *moles* of each element the following *masses* represent.

a. 1.5 mg of chromium
b. 2.0×10^{-3} g of strontium
c. 4.84×10^{4} g of boron
d. 3.6×10^{-6} μg of californium
e. 1.0 ton (2000 lb) of iron
f. 20.4 g of barium
g. 62.8 g of cobalt

108. Using the average atomic masses given inside the front cover of this text, calculate the *mass in grams* of each of the following samples.

a. 5.0 moles of potassium
b. 0.000305 mole of mercury
c. 2.31×10^{-5} moles of manganese
d. 10.5 moles of phosphorus
e. 4.9×10^{4} moles of iron
f. 125 moles of lithium
g. 0.01205 mole of fluorine

109. Using the average atomic masses given inside the front cover of this text, calculate the number of *atoms* present in each of the following samples.

a. 2.89 g of gold
b. 0.000259 mole of platinum
c. 0.000259 g of platinum
d. 2.0 lb of magnesium
e. 1.90 mL of liquid mercury (density = 13.6 g/mL)
f. 4.30 moles of tungsten
g. 4.30 g of tungsten

110. Calculate the molar mass for each of the following substances.

a. ferrous sulfate
b. mercuric iodide
c. stannic oxide
d. cobaltous chloride
e. cupric nitrate

111. Calculate the molar mass for each of the following substances.

a. adipic acid, $C_6H_{10}O_4$
b. caffeine, $C_8H_{10}N_4O_2$
c. eicosane, $C_{20}H_{42}$
d. cyclohexanol, $C_6H_{11}OH$
e. vinyl acetate, $C_4H_6O_2$
f. dextrose, $C_6H_{12}O_6$

112. Calculate the number of *moles* of the indicated substance present in each of the following samples.

a. 21.2 g of ammonium sulfide
b. 44.3 g of calcium nitrate
c. 4.35 g of dichlorine monoxide
d. 1.0 lb of ferric chloride
e. 1.0 kg of ferric chloride

113. Calculate the number of *moles* of the indicated substance present in each of the following samples.

a. 1.28 g of iron(II) sulfate
b. 5.14 mg of mercury(II) iodide
c. 9.21 μg of tin(IV) oxide
d. 1.26 lb of cobalt(II) chloride
e. 4.25 g of copper(II) nitrate

114. Consider equal mole samples of tetraphosphorus decoxide, copper(II) carbonate, and sodium phosphate. Rank these from lowest to highest mass in grams for each sample.

115. Calculate the mass in grams of each of the following samples.

a. 3.09 moles of ammonium carbonate
b. 4.01×10^{-6} moles of sodium hydrogen carbonate
c. 88.02 moles of carbon dioxide
d. 1.29 mmol of silver nitrate
e. 0.0024 mole of chromium(III) chloride

116. Calculate the number of *molecules* present in each of the following samples.

a. 3.45 g of $C_6H_{12}O_6$
b. 3.45 moles of $C_6H_{12}O_6$
c. 25.0 g of ICl_5
d. 1.00 g of B_2H_6
e. 1.05 mmol of $Al(NO_3)_3$

117. Calculate the number of moles of hydrogen atoms present in each of the following samples.

a. 2.71 g of ammonia
b. 0.824 mole of water
c. 6.25 mg of sulfuric acid
d. 451 g of ammonium carbonate

118. How many nitrogen atoms are present in 5.00 g of magnesium nitride?

All even-numbered Questions and Problems have answers in the back of this book and solutions in the *Student Solutions Guide*.

119. Calculate the percent by mass of the element mentioned *first* in the formulas for each of the following compounds.

a. sodium azide, NaN_3
b. copper(II) sulfate, $CuSO_4$
c. gold(III) chloride, $AuCl_3$
d. silver nitrate, $AgNO_3$
e. rubidium sulfate, Rb_2SO_4
f. sodium chlorate, $NaClO_3$
g. nitrogen triiodide, NI_3
h. cesium bromide, CsBr

120. Which of the following samples contains the greatest percent by mass of nitrogen? Explain.

a. 1.00 mole of $Ca(NO_3)_2$
b. 2.00 moles of $Ca(NO_3)_2$
c. 3.00 moles of $Ca(NO_3)_2$
d. All three samples contain the same percent by mass of nitrogen.
e. Cannot be determined without more information given.

121. A 1.2569-g sample of a new compound has been analyzed and found to contain the following masses of elements: carbon, 0.7238 g; hydrogen, 0.07088 g; nitrogen, 0.1407 g; oxygen, 0.3214 g. Calculate the empirical formula of the compound.

122. What mass of sodium hydroxide has the same number of oxygen atoms as 100.0 g of ammonium carbonate?

123. When 2.004 g of calcium is heated in pure nitrogen gas, the sample gains 0.4670 g of nitrogen. Calculate the empirical formula of the calcium nitride formed.

124. You find a compound composed only of element "X" and chlorine, and you know that it is 13.102% element X by mass. Each molecule has 6 times as many chlorine atoms as X atoms. What is element X?

125. When 1.00 g of metallic chromium is heated with elemental chlorine gas, 3.045 g of a chromium chloride salt results. Calculate the empirical formula of the compound.

126. When barium metal is heated in chlorine gas, a binary compound forms that consists of 65.95% Ba and 34.05% Cl by mass. Calculate the empirical formula of the compound.

ChemWork Problems

These multiconcept problems (and additional ones) are found interactively online with the same type of assistance a student would get from an instructor.

127. Determine the molar mass for the following compounds to four significant figures.

Compound	Molar Mass (g/mol)
water	______
iron(III) chloride	______
potassium bromide	______
ammonium nitrate	______
sodium hydroxide	______

128. Vitamin B_{12}, cyanocobalamin, is essential for human nutrition. Its molecular formula is $C_{63}H_{88}CoN_{14}O_{14}P$. A lack of this vitamin in the diet can lead to anemia. Cyanocobalamin is the form of the vitamin found in vitamin supplements.

a. What is the molar mass of cyanocobalamin to two decimal places?
b. How many moles of cyanocobalamin molecules are present in 250 mg of cyanocobalamin?
c. What is the mass of 0.60 mole of cyanocobalamin?
d. How many atoms of hydrogen are in 1.0 mole of cyanocobalamin?
e. What is the mass of 1.0×10^7 molecules of cyanocobalamin?
f. What is the mass (in grams) of one molecule of cyanocobalamin?

129. Calculate the number of moles for each compound in the following table.

Compound	Mass	Moles
Magnesium phosphate	326.4 g	______
Calcium nitrate	303.0 g	______
Potassium chromate	141.6 g	______
Dinitrogen pentoxide	406.3 g	______

130. a. How many atoms of carbon are present in 1.0 g of CH_4O?
b. How many atoms of carbon are present in 1.0 g of CH_3CH_2OH?
c. How many atoms of nitrogen are present in 25.0 g of $CO(NH_2)_2$?

131. Consider samples of phosphine (PH_3), water (H_2O), hydrogen sulfide (H_2S), and hydrogen fluoride (HF), each with a mass of 119 g. Rank the compounds from the least to the greatest number of hydrogen atoms contained in the samples.

132. The chemical formula for aspirin is $C_9H_8O_4$. What is the mass percent for each element in 1 mole of aspirin? (Give your answer to four significant figures.)

carbon ______ %
hydrogen ______ %
oxygen ______ %

133. Arrange the following substances in order of increasing mass percent of nitrogen.

a. NO
b. N_2O
c. NH_3
d. SNH

134. A compound with molar mass 180.1 g/mol has the following composition by mass:

C	40.0%
H	6.70%
O	53.3%

Determine the empirical and molecular formulas of the compound.

All even-numbered Questions and Problems have answers in the back of this book and solutions in the *Student Solutions Guide*.

Chemical Quantities

CHAPTER 9

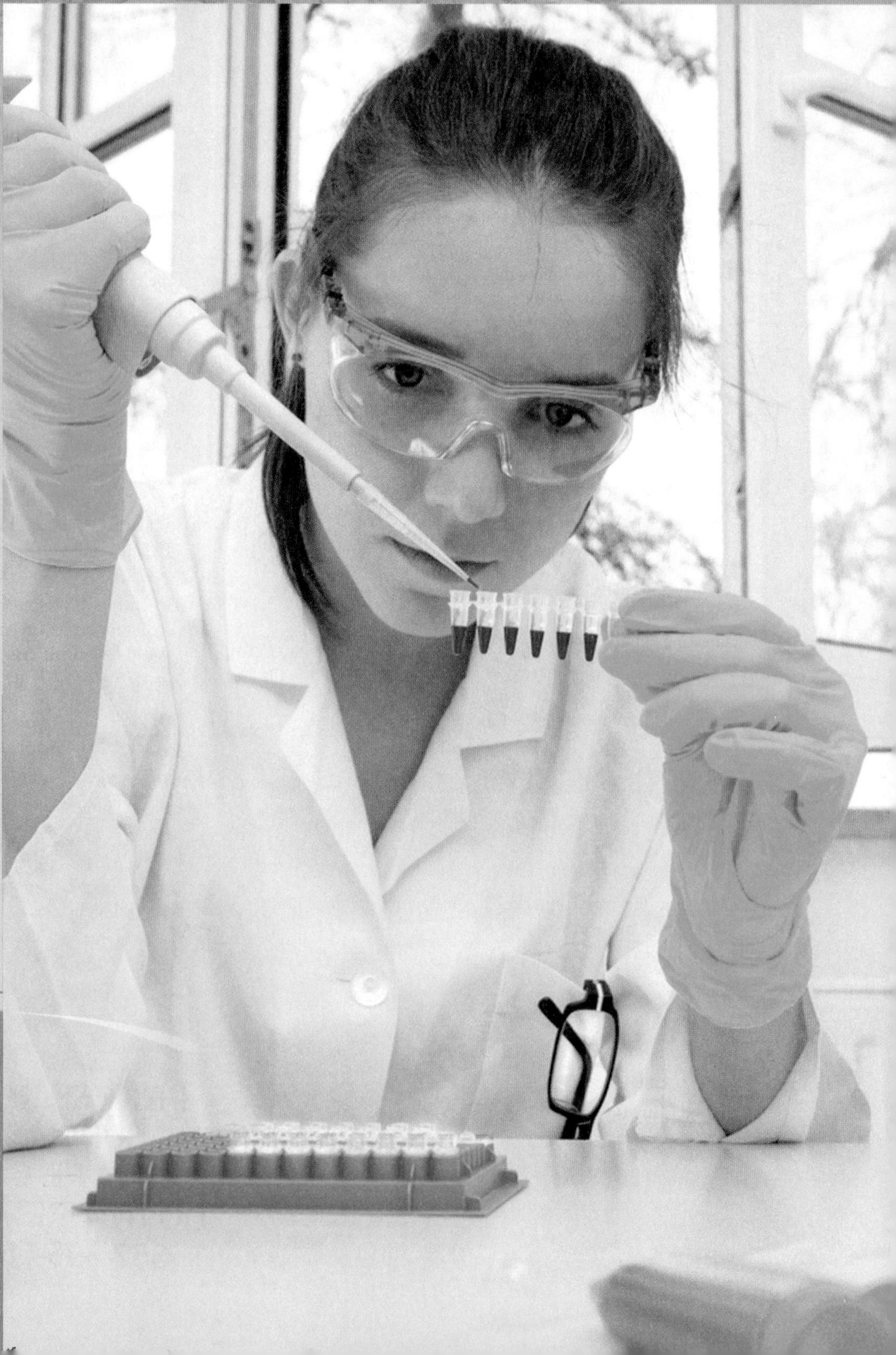

A scientist in a laboratory using a pipette to measure quantities of a liquid. anyaivanova/Shutterstock.com

Suppose you work for a consumer advocate organization and you want to test a company's advertising claims about the effectiveness of its antacid. The company claims that its product neutralizes 10 times as much stomach acid per tablet as its nearest competitor. How would you test the validity of this claim?

Or suppose that after graduation you go to work for a chemical company that makes methanol (methyl alcohol), a substance used as a starting material for the manufacture of products such as antifreeze and aviation fuels. You are working with an experienced chemist who is trying to improve the company's process for making methanol from the reaction of gaseous hydrogen with carbon monoxide gas. The first day on the job, you are instructed to order enough hydrogen and carbon monoxide to produce 6.0 kg of methanol in a test run. How would you determine how much carbon monoxide and hydrogen you should order?

After you study this chapter, you will be able to answer these questions.

Methanol is a starting material for some jet fuels. Royalty-Free Corbis

9-1 Information Given by Chemical Equations

OBJECTIVE **To understand the molecular and mass information given in a balanced equation.**

Reactions are what chemistry is really all about. Recall from Chapter 6 that chemical changes are actually rearrangements of atom groupings that can be described by chemical equations. These chemical equations tell us the identities (formulas) of the reactants and products and also show how much of each reactant and product participates in the reaction. The numbers (coefficients) in the balanced chemical equation enable us to determine just how much product we can get from a given quantity of reactants. It is important to recognize that the coefficients in a balanced equation give us the *relative* numbers of molecules. That is, we are interested in the *ratio* of the coefficients, not individual coefficients.

To illustrate this idea, consider a nonchemical analogy. Suppose you are in charge of making deli sandwiches at a fast-food restaurant. A particular type of sandwich requires 2 slices of bread, 3 slices of meat, and 1 slice of cheese. We can represent making the sandwich with the following equation:

$$2 \text{ slices bread} + 3 \text{ slices meat} + 1 \text{ slice cheese} \rightarrow 1 \text{ sandwich}$$

Your boss sends you to the store to get enough ingredients to make 50 sandwiches. How do you figure out how much of each ingredient to buy? Because you need enough to make 50 sandwiches, you multiply the preceding equation by 50.

$$50(2 \text{ slices bread}) + 50(3 \text{ slices meat}) + 50(1 \text{ slice cheese}) \rightarrow 50(1 \text{ sandwich})$$

That is

$$100 \text{ slices bread} + 150 \text{ slices meat} + 50 \text{ slices cheese} \rightarrow 50 \text{ sandwiches}$$

Notice that the numbers 100:150:50 correspond to the ratio 2:3:1, which represents the coefficients in the "balanced equation" of making a sandwich. If you were asked to make any number of sandwiches, it would be easy to use the original sandwich equation to determine how much of each ingredient you need.

The equation for a chemical reaction gives you the same type of information. It indicates the relative numbers of reactant and product molecules involved in the reaction. Using the equation permits us to determine the amounts of reactants needed to give a certain amount of product or to predict how much product we can make from a given quantity of reactants.

To illustrate how this idea works with a chemistry example, consider the reaction between gaseous carbon monoxide and hydrogen to produce liquid methanol, $CH_3OH(l)$. The reactants and products are

$$\text{Unbalanced: } \underbrace{CO(g) + H_2(g)}_{\text{Reactants}} \rightarrow \underbrace{CH_3OH(l)}_{\text{Product}}$$

Because atoms are just rearranged (not created or destroyed) in a chemical reaction, we must always balance a chemical equation. That is, we must choose coefficients that give the same number of each type of atom on both sides. Using the smallest set of integers that satisfies this condition gives the balanced equation

$$\text{Balanced: } CO(g) + 2H_2(g) \rightarrow CH_3OH(l)$$

CHECK Reactants: 1 C, 1 O, 4 H; Products: 1 C, 1 O, 4 H

Again, the coefficients in a balanced equation give the *relative* numbers of molecules. That is, we could multiply this balanced equation by any number and still have a balanced equation. For example, we could multiply by 12:

$$12[CO(g) + 2H_2(g) \rightarrow CH_3OH(l)]$$

to obtain

$$12CO(g) + 24H_2(g) \rightarrow 12CH_3OH(l)$$

This is still a balanced equation (check to be sure). Because 12 represents a dozen, we could even describe the reaction in terms of dozens:

$$1 \text{ dozen } CO(g) + 2 \text{ dozen } H_2(g) \rightarrow 1 \text{ dozen } CH_3OH(l)$$

We could also multiply the original equation by a very large number, such as 6.022×10^{23}:

$$6.022 \times 10^{23}[CO(g) + 2H_2(g) \rightarrow CH_3OH(l)]$$

which leads to the equation

$$6.022 \times 10^{23}\ CO(g) + 2(6.022 \times 10^{23})\ H_2(g) \rightarrow 6.022 \times 10^{23}\ CH_3OH(l)$$

One mole is 6.022×10^{23} units.

Just as 12 is called a dozen, chemists call 6.022×10^{23} a *mole* (abbreviated mol). Our equation, then, can be written in terms of moles:

$$1 \text{ mol } CO(g) + 2 \text{ mol } H_2(g) \rightarrow 1 \text{ mol } CH_3OH(l)$$

Various ways of interpreting this balanced chemical equation are given in Table 9.1.

Table 9.1 Information Conveyed by the Balanced Equation for the Production of Methanol

$CO(g)$	+	$2H_2(g)$	→	$CH_3OH(l)$
1 molecule CO	+	2 molecules H_2	→	1 molecule CH_3OH
1 dozen CO molecules	+	2 dozen H_2 molecules	→	1 dozen CH_3OH molecules
6.022×10^{23} CO molecules	+	$2(6.022 \times 10^{23})$ H_2 molecules	→	6.022×10^{23} CH_3OH molecules
1 mol CO molecules	+	2 mol H_2 molecules	→	1 mol CH_3OH molecules

Example 9.1 Relating Moles to Molecules in Chemical Equations

Propane, C_3H_8, is a fuel commonly used for cooking on gas grills and for heating in rural areas where natural gas is unavailable. Propane reacts with oxygen gas to produce heat and the products carbon dioxide and water. This combustion reaction is represented by the unbalanced equation

$$C_3H_8(g) + O_2(g) \rightarrow CO_2(g) + H_2O(g)$$

Give the balanced equation for this reaction, and state the meaning of the equation in terms of numbers of molecules and moles of molecules.

Roberto Westbook/Blend Images/Getty Images

Propane is often used as a fuel for outdoor grills.

SOLUTION Using the techniques explained in Chapter 6, we can balance the equation.

$$C_3H_8(g) + 5O_2(g) \rightarrow 3CO_2(g) + 4H_2O(g)$$

CHECK 3 C, 8 H, 10 O → 3 C, 8 H, 10 O

This equation can be interpreted in terms of molecules as follows:

1 molecule of C_3H_8 reacts with 5 molecules of O_2 to give 3 molecules of CO_2 plus 4 molecules of H_2O

or as follows in terms of moles (of molecules):

1 mole of C_3H_8 reacts with 5 moles of O_2 to give 3 moles of CO_2 plus 4 moles of H_2O ■

9-2 Mole–Mole Relationships

OBJECTIVE **To learn to use a balanced equation to determine relationships between moles of reactants and moles of products.**

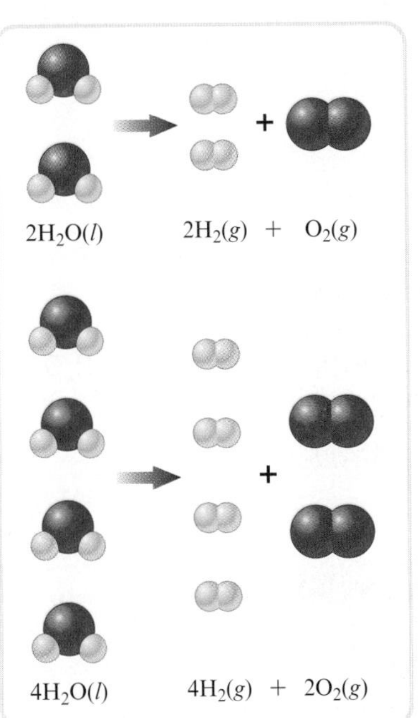

Now that we have discussed the meaning of a balanced chemical equation in terms of moles of reactants and products, we can use an equation to predict the moles of products that a given number of moles of reactants will yield. For example, consider the decomposition of water to give hydrogen and oxygen, which is represented by the following balanced equation:

$$2H_2O(l) \rightarrow 2H_2(g) + O_2(g)$$

This equation tells us that 2 moles of H_2O yields 2 moles of H_2 and 1 mole of O_2.

Now suppose that we have 4 moles of water. If we decompose 4 moles of water, how many moles of products do we get?

One way to answer this question is to multiply the entire equation by 2 (which will give us 4 moles of H_2O).

$$2[2H_2O(l) \rightarrow 2H_2(g) + O_2(g)]$$
$$4H_2O(l) \rightarrow 4H_2(g) + 2O_2(g)$$

Now we can state that

4 mol H_2O yields 4 mol H_2 plus 2 mol O_2

which answers the question of how many moles of products we get with 4 moles of H_2O.

Next, suppose we decompose 5.8 moles of water. What numbers of moles of products are formed in this process? We could answer this question by rebalancing the chemical equation as follows: First, we divide *all coefficients* of the balanced equation

$$2H_2O(l) \rightarrow 2H_2(g) + O_2(g)$$

by 2, to give

$$H_2O(l) \rightarrow H_2(g) + \tfrac{1}{2}O_2(g)$$

Now, because we have 5.8 moles of H_2O, we multiply this equation by 5.8.

$$5.8[H_2O(l) \rightarrow H_2(g) + \tfrac{1}{2}O_2(g)]$$

This gives

This equation with noninteger coefficients makes sense only if the equation means moles (of molecules) of the various reactants and products.

$$5.8H_2O(l) \rightarrow 5.8H_2(g) + 5.8(\tfrac{1}{2})O_2(g)$$
$$5.8H_2O(l) \rightarrow 5.8H_2(g) + 2.9O_2(g)$$

(Verify that this is a balanced equation.) Now we can state that

$$5.8 \text{ mol } H_2O \text{ yields } 5.8 \text{ mol } H_2 \text{ plus } 2.9 \text{ mol } O_2$$

This procedure of rebalancing the equation to obtain the number of moles involved in a particular situation always works, but it can be cumbersome. In Example 9.2 we will develop a more convenient procedure, which uses a conversion factor, or **mole ratio,** based on the balanced chemical equation.

Interactive Example 9.2 Determining Mole Ratios

What number of moles of O_2 will be produced by the decomposition of 5.8 moles of water?

SOLUTION

Where Are We Going?

We want to determine the number of moles of O_2 produced by the decomposition of 5.8 moles of H_2O.

What Do We Know?

- The balanced equation for the decomposition of water is

$$2H_2O \rightarrow 2H_2 + O_2$$

- We start with 5.8 moles of H_2O.

How Do We Get There?

Our problem can be diagrammed as follows:

5.8 mol H_2O → yields → ? mol O_2

To answer this question, we need to know the relationship between moles of H_2O and moles of O_2 in the balanced equation (conventional form):

$$2H_2O(l) \rightarrow 2H_2(g) + O_2(g)$$

From this equation we can state that

2 mol H_2O → yields → 1 mol O_2

The statement 2 mol H_2O = 1 mol O_2 is obviously not true in a literal sense, but it correctly expresses the chemical equivalence between H_2O and O_2.

which can be represented by the following equivalence statement:

$$2 \text{ mol } H_2O = 1 \text{ mol } O_2$$

We now want to use this equivalence statement to obtain the conversion factor (mole ratio) that we need. Because we want to go from moles of H_2O to moles of O_2, we need the mole ratio

$$\frac{1 \text{ mol } O_2}{2 \text{ mol } H_2O}$$

so that mol H_2O will cancel in the conversion from moles of H_2O to moles of O_2.

Math skill builder

For a review of equivalence statements and dimensional analysis, see Section 2-6.

$$5.8 \text{ \cancel{mol H_2O}} \times \frac{1 \text{ mol } O_2}{2 \text{ \cancel{mol H_2O}}} = 2.9 \text{ mol } O_2$$

So if we decompose 5.8 moles of H_2O, we will get 2.9 moles of O_2.

REALITY CHECK Note that this is the same answer we obtained earlier when we rebalanced the equation to give

$$5.8H_2O(l) \rightarrow 5.8H_2(g) + 2.9O_2(g)$$ ■

We saw in Example 9.2 that to determine the moles of a product that can be formed from a specified number of moles of a reactant, we can use the balanced equation to obtain the appropriate mole ratio. We will now extend these ideas in Example 9.3.

Interactive Example 9.3

Using Mole Ratios in Calculations

Calculate the number of moles of oxygen required to react exactly with 4.30 moles of propane, C_3H_8, in the reaction described by the following balanced equation:

$$C_3H_8(g) + 5O_2(g) \rightarrow 3CO_2(g) + 4H_2O(g)$$

SOLUTION

Where Are We Going?

We want to determine the number of moles of O_2 required to react with 4.30 moles of C_3H_8.

What Do We Know?

- The balanced equation for the reaction is

$$C_3H_8 + 5O_2 \rightarrow 3CO_2 + 4H_2O$$

- We start with 4.30 moles of C_3H_8.

How Do We Get There?

In this case the problem can be stated as follows:

To solve this problem, we need to consider the relationship between the reactants C_3H_8 and O_2. Using the balanced equation, we find that

$$1 \text{ mol } C_3H_8 \text{ requires } 5 \text{ mol } O_2$$

which can be represented by the equivalence statement

$$1 \text{ mol } C_3H_8 = 5 \text{ mol } O_2$$

This leads to the required mole ratio

$$\frac{5 \text{ mol } O_2}{1 \text{ mol } C_3H_8}$$

for converting from moles of C_3H_8 to moles of O_2. We construct the conversion ratio this way so that mol C_3H_8 cancels:

$$4.30 \cancel{\text{mol } C_3H_8} \times \frac{5 \text{ mol } O_2}{1 \cancel{\text{mol } C_3H_8}} = 21.5 \text{ mol } O_2$$

We can now answer the original question:

$$4.30 \text{ mol } C_3H_8 \text{ requires } 21.5 \text{ mol } O_2$$

REALITY CHECK According to the balanced equation, more O_2 is required (by moles) than C_3H_8 by a factor of 5. With about 4 moles of C_3H_8, we would expect about 20 moles of O_2, which is close to our answer.

SELF CHECK

Exercise 9.1 Calculate the moles of CO_2 formed when 4.30 moles of C_3H_8 reacts with the required 21.5 moles of O_2.

HINT Use the moles of C_3H_8, and obtain the mole ratio between C_3H_8 and CO_2 from the balanced equation.

See Problems 9.15 and 9.16. ■

9-3 Mass Calculations

OBJECTIVE **To learn to relate masses of reactants and products in a chemical reaction.**

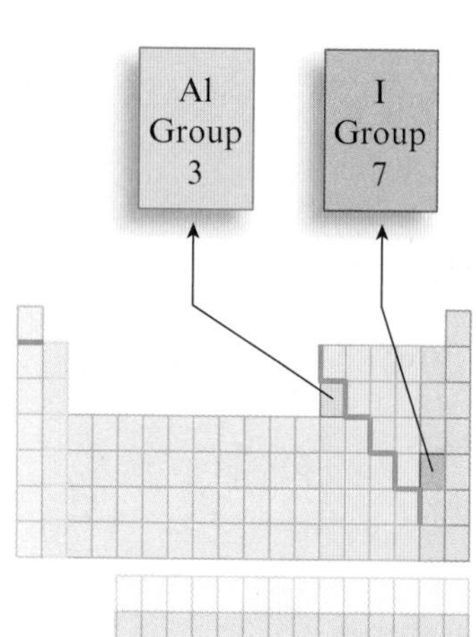

In the last section we saw how to use the balanced equation for a reaction to calculate the numbers of moles of reactants and products for a particular case. However, moles represent numbers of molecules, and we cannot count molecules directly. In chemistry we count by weighing. Therefore, in this section we will review the procedures for converting between moles and masses and will see how these procedures are applied to chemical calculations.

To develop these procedures we will consider the reaction between powdered aluminum metal and finely ground iodine to produce aluminum iodide. The balanced equation for this vigorous chemical reaction is

$$2\text{Al}(s) + 3\text{I}_2(s) \rightarrow 2\text{AlI}_3(s)$$

Suppose we have 35.0 g of aluminum. What mass of I_2 should we weigh out to react exactly with this amount of aluminum?

To answer this question, let's use the problem-solving strategy discussed in Chapter 8.

Where Are We Going?

We want to find the mass of iodine (I_2) that will react with 35.0 g of aluminum (Al). We know from the balanced equation that

$$2 \text{ mol Al requires } 3 \text{ mol } I_2$$

This can be written as the mole ratio

$$\frac{3 \text{ mol } I_2}{2 \text{ mol Al}}$$

© Cengage Learning

© Cengage Learning

Aluminum (*top left*) and iodine (*top right*) react vigorously to form aluminum iodide. The purple cloud results from excess iodine vaporized by the heat of the reaction (*bottom*).

We can use this ratio to calculate moles of I_2 needed from the moles of Al present. However, this leads us to two questions:

1. How many moles of Al are present?
2. How do we convert moles of I_2 to mass of I_2 as required by the problem?

We need to be able to convert from grams to moles and from moles to grams.

How Do We Get There?

The problem states that we have 35.0 g of aluminum, so we must convert from grams to moles of aluminum. This is something we already know how to do. Using the table of average atomic masses inside the front cover of this book, we find the atomic mass of aluminum to be 26.98. This means that 1 mole of aluminum has a mass of 26.98 g. We can use the equivalence statement

$$1 \text{ mol Al} = 26.98 \text{ g}$$

to find the moles of Al in 35.0 g.

$$35.0 \cancel{\text{g Al}} \times \frac{1 \text{ mol Al}}{26.98 \cancel{\text{g Al}}} = 1.30 \text{ mol Al}$$

Now that we have moles of Al, we can find the moles of I_2 required.

$$1.30 \cancel{\text{mol Al}} \times \frac{3 \text{ mol I}_2}{2 \cancel{\text{mol Al}}} = 1.95 \text{ mol I}_2$$

We now know the *moles* of I_2 required to react with the 1.30 moles of Al (35.0 g). The next step is to convert 1.95 moles of I_2 to grams so we will know how much to weigh out. We do this by using the molar mass of I_2. The atomic mass of iodine is 126.9 g (for 1 mole of I atoms), so the molar mass of I_2 is

$$2 \times 126.9 \text{ g/mol} = 253.8 \text{ g/mol} = \text{mass of 1 mol I}_2$$

Now we convert the 1.95 moles of I_2 to grams of I_2.

$$1.95 \cancel{\text{mol I}_2} \times \frac{253.8 \text{ g I}_2}{\cancel{\text{mol I}_2}} = 495 \text{ g I}_2$$

We have solved the problem. We need to weigh out 495 g of iodine (contains I_2 molecules) to react exactly with the 35.0 g of aluminum. We will further develop procedures for dealing with masses of reactants and products in Example 9.4.

REALITY CHECK We have determined that 495 g of I_2 is required to react with 35.0 g Al. Does this answer make sense? We know from the molar masses of Al and I_2 (26.98 g/mol and 253.8 g/mol) that the mass of 1 mole of I_2 is almost ten times as great as that of 1 mole of Al. We also know that we need a greater number of moles of I_2 compared with Al (by a 3:2 ratio). So, we should expect to get a mass of I_2 that is well over ten times as great as 35.0 g, and we did.

Interactive Example 9.4

Using Mass–Mole Conversions with Mole Ratios

Propane, C_3H_8, when used as a fuel, reacts with oxygen to produce carbon dioxide and water according to the following unbalanced equation:

$$C_3H_8(g) + O_2(g) \rightarrow CO_2(g) + H_2O(g)$$

What mass of oxygen will be required to react exactly with 96.1 g of propane?

SOLUTION

Where Are We Going?

We want to determine the mass of O_2 required to react exactly with 96.1 g C_3H_8.

What Do We Know?

- The unbalanced equation for the reaction is

$$C_3H_8 + O_2 \rightarrow CO_2 + H_2O.$$

- We start with 96.1 g C_3H_8.
- We know the atomic masses of carbon, hydrogen, and oxygen from the periodic table.

What Do We Need To Know?

- We need to know the balanced equation.
- We need the molar masses of O_2 and C_3H_8.

Always balance the equation for the reaction first.

How Do We Get There?

To deal with the amounts of reactants and products, we first need the balanced equation for this reaction:

$$C_3H_8(g) + 5O_2(g) \rightarrow 3CO_2(g) + 4H_2O(g)$$

Our problem, in schematic form, is

Math skill builder

Remember that to show the correct significant figures in each step, we are rounding off after each calculation. In doing problems, you should carry extra numbers, rounding off only at the end.

Using the ideas we developed when we discussed the aluminum–iodine reaction, we will proceed as follows:

1. We are given the number of grams of propane, so we must convert to moles of propane (C_3H_8).
2. Then we can use the coefficients in the balanced equation to determine the moles of oxygen (O_2) required.
3. Finally, we will use the molar mass of O_2 to calculate grams of oxygen.

We can sketch this strategy as follows:

$$C_3H_8(g) \quad + \quad 5O_2(g) \quad \rightarrow \quad 3CO_2(g) \quad + \quad 4H_2O(g)$$

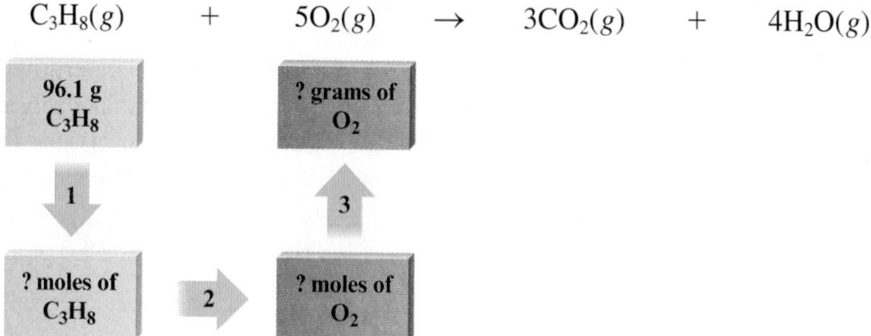

Thus the first question we must answer is, *How many moles of propane are present in 96.1 g of propane?* The molar mass of propane is 44.09 g (3 × 12.01 g + 8 × 1.008 g). The moles of propane present can be calculated as follows:

$$96.1 \text{ g } C_3H_8 \times \frac{1 \text{ mol } C_3H_8}{44.09 \text{ g } C_3H_8} = 2.18 \text{ mol } C_3H_8$$

Next we recognize that each mole of propane reacts with 5 moles of oxygen. This gives us the equivalence statement

$$1 \text{ mol } C_3H_8 = 5 \text{ mol } O_2$$

from which we construct the mole ratio

$$\frac{5 \text{ mol } O_2}{1 \text{ mol } C_3H_8}$$

that we need to convert from moles of propane molecules to moles of oxygen molecules.

$$2.18 \cancel{\text{mol } C_3H_8} \times \frac{5 \text{ mol } O_2}{1 \cancel{\text{mol } C_3H_8}} = 10.9 \text{ mol } O_2$$

Notice that the mole ratio is set up so that the moles of C_3H_8 cancel and the resulting units are moles of O_2.

Because the original question asked for the *mass* of oxygen needed to react with 96.1 g of propane, we must convert the 10.9 moles of O_2 to grams, using the molar mass of O_2 ($32.00 = 2 \times 16.00$ g).

$$10.9 \cancel{\text{mol } O_2} \times \frac{32.00 \text{ g } O_2}{1 \cancel{\text{mol } O_2}} = 349 \text{ g } O_2$$

Therefore, 349 g of oxygen is required to burn 96.1 g of propane. We can summarize this problem by writing out a "conversion string" that shows how the problem was done.

$$96.1 \cancel{\text{g } C_3H_8} \times \underbrace{\frac{1 \cancel{\text{mol } C_3H_8}}{44.09 \cancel{\text{g } C_3H_8}}}_{1} \times \underbrace{\frac{5 \cancel{\text{mol } O_2}}{1 \cancel{\text{mol } C_3H_8}}}_{2} \times \underbrace{\frac{32.00 \text{ g } O_2}{1 \cancel{\text{mol } O_2}}}_{3} = 349 \text{ g } O_2$$

Math skill builder
Use units as a check to see that you have used the correct conversion factors (mole ratios).

This is a convenient way to make sure the final units are correct. The procedure we have followed is summarized below.

$$C_3H_8(g) + 5O_2(g) \rightarrow 3CO_2(g) + 4H_2O(g)$$

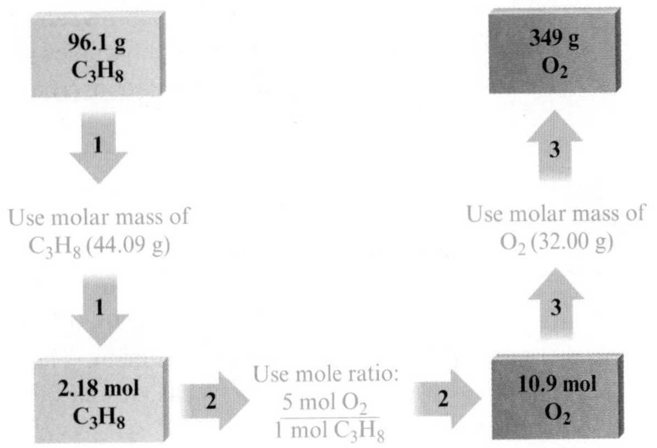

96.1 g C_3H_8 → (1) Use molar mass of C_3H_8 (44.09 g) → (1) 2.18 mol C_3H_8 → (2) Use mole ratio: $\frac{5 \text{ mol } O_2}{1 \text{ mol } C_3H_8}$ → (2) 10.9 mol O_2 → (3) Use molar mass of O_2 (32.00 g) → (3) 349 g O_2

REALITY CHECK According to the balanced equation, more O_2 is required (by moles) than C_3H_8 by a factor of 5. Because the molar mass of C_3H_8 is not much greater than that of O_2, we should expect that a greater mass of oxygen is required, and our answer confirms this.

SELF CHECK

Exercise 9.2 What mass of carbon dioxide is produced when 96.1 g of propane reacts with sufficient oxygen?

See Problems 9.23 through 9.26. ■

SELF CHECK

Exercise 9.3 Calculate the mass of water formed by the complete reaction of 96.1 g of propane with oxygen.

See Problems 9.23 through 9.26. ■

So far in this chapter, we have spent considerable time "thinking through" the procedures for calculating the masses of reactants and products in chemical reactions. We can summarize these procedures in the following steps:

Steps for Calculating the Masses of Reactants and Products in Chemical Reactions

Step 1 Balance the equation for the reaction.

Step 2 Convert the masses of reactants or products to moles.

Step 3 Use the balanced equation to set up the appropriate mole ratio(s).

Step 4 Use the mole ratio(s) to calculate the number of moles of the desired reactant or product.

Step 5 Convert from moles back to masses.

The process of using a chemical equation to calculate the relative masses of reactants and products involved in a reaction is called **stoichiometry** (pronounced stoi´ kē-ŏm´ i-trē). Chemists say that the balanced equation for a chemical reaction describes the stoichiometry of the reaction.

We will now consider a few more examples that involve chemical stoichiometry. Because real-world examples often involve very large or very small masses of chemicals that are most conveniently expressed by using scientific notation, we will deal with such a case in Example 9.5.

Critical Thinking

Your lab partner has made the observation that we always measure the mass of chemicals in lab, but then use mole ratios to balance equations. What if your lab partner decided to balance equations by using masses as coefficients? Is this possible? Why or why not?

Interactive Example 9.5

Stoichiometric Calculations: Using Scientific Notation

For a review of writing formulas of ionic compounds, see Chapter 5.

Solid lithium hydroxide has been used in space vehicles to remove exhaled carbon dioxide from the living environment. The products are solid lithium carbonate and liquid water. What mass of gaseous carbon dioxide can 1.00×10^3 g of lithium hydroxide absorb?

SOLUTION

Where Are We Going?

We want to determine the mass of carbon dioxide absorbed by 1.00×10^3 g of lithium hydroxide.

What Do We Know?

- The names of the reactants and products.
- We start with 1.00×10^3 g of lithium hydroxide.
- We can obtain the atomic masses from the periodic table.

What Do We Need To Know?

- We need to know the balanced equation for the reaction, but we first have to write the formulas for the reactants and products.
- We need the molar masses of lithium hydroxide and carbon dioxide.

Astronaut Sidney M. Gutierrez changes the lithium hydroxide canisters on the Space Shuttle *Columbia*.

How Do We Get There?

Step 1 Using the description of the reaction, we can write the unbalanced equation

$$\mathrm{LiOH}(s) + \mathrm{CO_2}(g) \rightarrow \mathrm{Li_2CO_3}(s) + \mathrm{H_2O}(l)$$

The balanced equation is

$$2\mathrm{LiOH}(s) + \mathrm{CO_2}(g) \rightarrow \mathrm{Li_2CO_3}(s) + \mathrm{H_2O}(l)$$

Check this for yourself.

Step 2 We convert the given mass of LiOH to moles, using the molar mass of LiOH, which is 6.941 g + 16.00 g + 1.008 g = 23.95 g.

$$1.00 \times 10^3 \text{ g LiOH} \times \frac{1 \text{ mol LiOH}}{23.95 \text{ g LiOH}} = 41.8 \text{ mol LiOH}$$

Step 3 The appropriate mole ratio is

$$\frac{1 \text{ mol } \mathrm{CO_2}}{2 \text{ mol LiOH}}$$

Step 4 Using this mole ratio, we calculate the moles of CO_2 needed to react with the given mass of LiOH.

$$41.8 \text{ mol LiOH} \times \frac{1 \text{ mol } \mathrm{CO_2}}{2 \text{ mol LiOH}} = 20.9 \text{ mol } \mathrm{CO_2}$$

Step 5 We calculate the mass of CO_2 by using its molar mass (44.01 g).

$$20.9 \text{ mol } \mathrm{CO_2} \times \frac{44.01 \text{ g } \mathrm{CO_2}}{1 \text{ mol } \mathrm{CO_2}} = 920. \text{ g } \mathrm{CO_2} = 9.20 \times 10^2 \text{ g } \mathrm{CO_2}$$

Thus 1.00×10^3 g of LiOH(s) can absorb 920. g of $CO_2(g)$.

We can summarize this problem as follows:

$$2\mathrm{LiOH}(s) + \mathrm{CO_2}(g) \rightarrow \mathrm{Li_2CO_3}(s) + \mathrm{H_2O}(l)$$

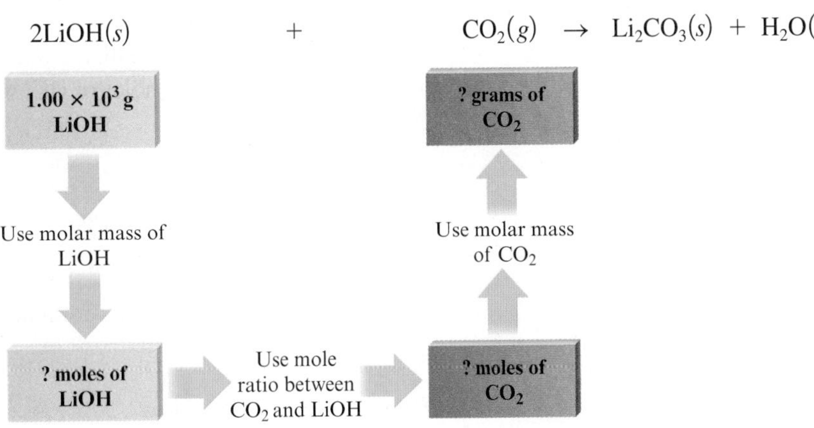

The conversion string is

$$1.00 \times 10^3 \text{ g LiOH} \times \frac{1 \text{ mol LiOH}}{23.95 \text{ g LiOH}} \times \frac{1 \text{ mol } \mathrm{CO_2}}{2 \text{ mol LiOH}} \times \frac{44.01 \text{ g } \mathrm{CO_2}}{1 \text{ mol } \mathrm{CO_2}} = 9.19 \times 10^2 \text{ g } \mathrm{CO_2}$$

Math skill builder

Carrying extra significant figures and rounding off only at the end gives an answer of 919 g CO_2.

REALITY CHECK According to the balanced equation, there is a 2:1 mole ratio of LiOH to CO_2. There is about a 1:2 molar mass ratio of LiOH:CO_2 (23.95:44.01). We should expect about the same mass of CO_2 as LiOH, and our answer confirms this (1000 g compared to 920 g).

SELF CHECK

Exercise 9.4 Hydrofluoric acid, an aqueous solution containing dissolved hydrogen fluoride, is used to etch glass by reacting with the silica, SiO_2, in the glass to produce gaseous silicon tetrafluoride and liquid water. The unbalanced equation is

$$HF(aq) + SiO_2(s) \rightarrow SiF_4(g) + H_2O(l)$$

a. Calculate the mass of hydrogen fluoride needed to react with 5.68 g of silica. *Hint:* Think carefully about this problem. What is the balanced equation for the reaction? What is given? What do you need to calculate? Sketch a map of the problem before you do the calculations.

b. Calculate the mass of water produced in the reaction described in part a.

See Problems 9.23 through 9.26. ■

Interactive Example 9.6

Stoichiometric Calculations: Comparing Two Reactions

Baking soda, $NaHCO_3$, is often used as an antacid. It neutralizes excess hydrochloric acid secreted by the stomach. The balanced equation for the reaction is

$$NaHCO_3(s) + HCl(aq) \rightarrow NaCl(aq) + H_2O(l) + CO_2(g)$$

Milk of magnesia, which is an aqueous suspension of magnesium hydroxide, $Mg(OH)_2$, is also used as an antacid. The balanced equation for the reaction is

$$Mg(OH)_2(s) + 2HCl(aq) \rightarrow 2H_2O(l) + MgCl_2(aq)$$

Which antacid can consume the most stomach acid, 1.00 g of $NaHCO_3$ or 1.00 g of $Mg(OH)_2$?

SOLUTION

Where Are We Going?

We want to compare the neutralizing power of two antacids, $NaHCO_3$ and $Mg(OH)_2$. In other words, how many moles of HCl will react with 1.00 g of each antacid?

What Do We Know?

- The balanced equations for the reactions.
- We start with 1.00 g each of $NaHCO_3$ and $Mg(OH)_2$.
- We can obtain atomic masses from the periodic table.

What Do We Need to Know?

- We need the molar masses of $NaHCO_3$ and $Mg(OH)_2$.

How Do We Get There?

The antacid that reacts with the larger number of moles of HCl is more effective because it will neutralize more moles of acid. A schematic for this procedure is

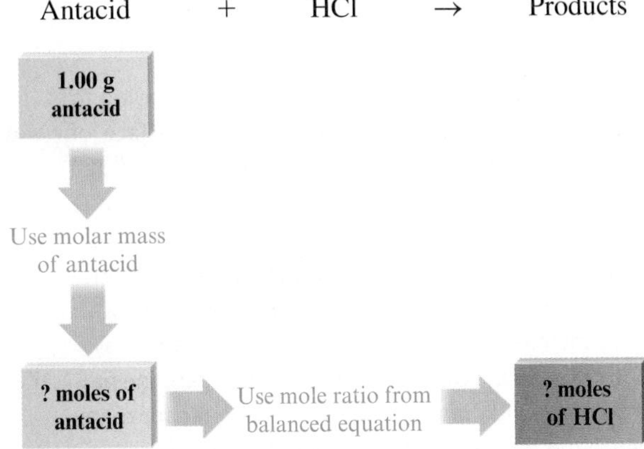

Notice that in this case we do not need to calculate how many grams of HCl react; we can answer the question with moles of HCl. We will now solve this problem for each antacid. Both of the equations are balanced, so we can proceed with the calculations.

Using the molar mass of $NaHCO_3$, which is 22.99 g + 1.008 g + 12.01 g + 3(16.00 g) = 84.01 g, we determine the moles of $NaHCO_3$ in 1.00 g of $NaHCO_3$.

$$1.00\ \cancel{\text{g NaHCO}_3} \times \frac{1\ \text{mol NaHCO}_3}{84.01\ \cancel{\text{g NaHCO}_3}} = 0.0119\ \text{mol NaHCO}_3$$

$$= 1.19 \times 10^{-2}\ \text{mol NaHCO}_3$$

Next we determine the moles of HCl, using the mole ratio $\frac{1\ \text{mol HCl}}{1\ \text{mol NaHCO}_3}$.

$$1.19 \times 10^{-2}\ \cancel{\text{mol NaHCO}_3} \times \frac{1\ \text{mol HCl}}{1\ \cancel{\text{mol NaHCO}_3}} = 1.19 \times 10^{-2}\ \text{mol HCl}$$

Thus 1.00 g of $NaHCO_3$ neutralizes 1.19×10^{-2} mole of HCl. We need to compare this to the number of moles of HCl that 1.00 g of $Mg(OH)_2$ neutralizes.

Using the molar mass of $Mg(OH)_2$, which is 24.31 g + 2(16.00 g) + 2(1.008 g) = 58.33 g, we determine the moles of $Mg(OH)_2$ in 1.00 g of $Mg(OH)_2$.

$$1.00\ \cancel{\text{g Mg(OH)}_2} \times \frac{1\ \text{mol Mg(OH)}_2}{58.33\ \cancel{\text{g Mg(OH)}_2}} = 0.0171\ \text{mol Mg(OH)}_2$$

$$= 1.71 \times 10^{-2}\ \text{mol Mg(OH)}_2$$

To determine the moles of HCl that react with this amount of $Mg(OH)_2$, we use the mole ratio $\frac{2\ \text{mol HCl}}{1\ \text{mol Mg(OH)}_2}$.

$$1.71 \times 10^{-2}\ \cancel{\text{mol Mg(OH)}_2} \times \frac{2\ \text{mol HCl}}{1\ \cancel{\text{mol Mg(OH)}_2}} = 3.42 \times 10^{-2}\ \text{mol HCl}$$

Therefore, 1.00 g of $Mg(OH)_2$ neutralizes 3.42×10^{-2} mole of HCl. We have already calculated that 1.00 g of $NaHCO_3$ neutralizes only 1.19×10^{-2} mole of HCl. Therefore, $Mg(OH)_2$ is a more effective antacid than $NaHCO_3$ on a mass basis.

CHEMISTRY IN FOCUS

Cars of the Future

There is a great deal of concern about how we are going to sustain our personal transportation system in the face of looming petroleum shortages (and the resultant high costs) and the challenges of global warming. The era of large gasoline-powered cars as the primary means of transportation in the United States seems to be drawing to a close. The fact that discoveries of petroleum are not keeping up with the rapidly increasing global demand for oil has caused skyrocketing prices. In addition, the combustion of gasoline produces carbon dioxide (about 1 lb of CO_2 per mile for many cars), which has been implicated in global warming.

So what will the car of the future in the United States be like? It seems that we are moving rapidly toward cars that have an electrical component as part of the power train. Hybrid cars, which use a small gasoline motor in conjunction with a powerful battery, have been quite successful. By supplementing the small gasoline engine, which would be inadequate by itself, with power from the battery, the typical hybrid gets 40 to 50 miles per gallon of gasoline. In this type of hybrid car, both the battery and the engine are used to power the wheels of the car as needed.

Another type of system that involves both a gasoline engine and a battery is the so-called "plug-in hybrid." In this car, the battery is the sole source of power to the car's wheels. The gasoline engine is only used to charge the battery as needed. One example of this type of car is the Chevrolet Volt, which is designed to run about 40 miles on each battery charge. The car can be plugged into a normal household electric outlet overnight to recharge the battery. For trips longer than 40 miles, the gasoline engine turns on to charge the battery.

Another type of "electrical car" being tested is one powered by a hydrogen–oxygen fuel cell. An example of such a car is the Honda FCX Clarity. The Clarity stores hydrogen in a tank that holds 4.1 kg of H_2 at a pressure of 5000 lb per square inch. The H_2 is sent to a fuel cell, where it reacts with oxygen from the air supplied by an air compressor. About 200 of these cars are going to be tested in Southern California in the next 3 years, leased to people who live near one of the three 24-hour public hydrogen stations. The Clarity gets about 72 miles per kilogram of hydrogen. One obvious advantage of a car powered by an H_2/O_2 fuel cell is that the combustion product is only H_2O. However, there is a catch (it seems there

SELF CHECK

Exercise 9.5 In Example 9.6 we answered one of the questions we posed in the introduction to this chapter. Now let's see if you can answer the other question posed there. Determine what mass of carbon monoxide and what mass of hydrogen are required to form 6.0 kg of methanol by the reaction

$$CO(g) + 2H_2(g) \rightarrow CH_3OH(l)$$

See Problem 9.39. ■

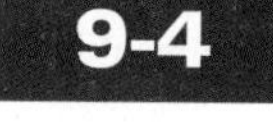

9-4 The Concept of Limiting Reactants

OBJECTIVE **To understand what is meant by the term "limiting reactant."**

Earlier in this chapter, we discussed making sandwiches. Recall that the sandwich-making process could be described as follows:

2 pieces bread + 3 slices meat + 1 slice cheese → 1 sandwich

In our earlier discussion, we always purchased the ingredients in the correct ratios so that we used all the components, with nothing left over.

Now assume that you came to work one day and found the following quantities of ingredients:

20 slices of bread
24 slices of meat
12 slices of cheese

is always a catch). Currently, 95% of hydrogen produced is obtained from natural gas (CH_4), and CO_2 is a by-product of this process. Intense research is now being conducted to find economically feasible ways to produce H_2 from water.

It appears that our cars of the future will have an electrical drive component. Whether it will involve a conventional battery or a fuel cell will depend on technological developments and costs.

Horizon Fuel Cell Technologies Pte. Ltd.

Even model cars are becoming "green." The H-racer from Horizon Fuel Cell Technologies uses a hydrogen–oxygen fuel cell.

Courtesy of American Honda Motor Co., Inc.

Honda FCX Clarity at a hydrogen refueling station.

How many sandwiches can you make? What will be left over?

To solve this problem, let's see how many sandwiches we can make with each component.

Bread: $$20 \text{ slices bread} \times \frac{1 \text{ sandwich}}{2 \text{ slices of bread}} = 10 \text{ sandwiches}$$

Meat: $$24 \text{ slices meat} \times \frac{1 \text{ sandwich}}{3 \text{ slices of meat}} = 8 \text{ sandwiches}$$

Cheese: $$12 \text{ slices cheese} \times \frac{1 \text{ sandwich}}{1 \text{ slice of cheese}} = 12 \text{ sandwiches}$$

How many sandwiches can you make? The answer is 8. When you run out of meat, you must stop making sandwiches. The meat is the limiting ingredient.

What do you have left over? Making 8 sandwiches requires 16 pieces of bread. You started with 20 pieces, so you have 4 pieces of bread left. You also used 8 pieces of cheese for the 8 sandwiches, so you have $12 - 8 = 4$ pieces of cheese left.

In this example, the ingredient present in the largest number (the meat) was actually the component that limited the number of sandwiches you could make. This situation arose because each sandwich required 3 slices of meat—more than the quantity required of any other ingredient.

You probably have been dealing with limiting-reactant problems for most of your life. For example, suppose a lemonade recipe calls for 1 cup of sugar for every 6 lemons. You have 12 lemons and 3 cups of sugar. Which ingredient is limiting, the lemons or the sugar?*

A Closer Look

When molecules react with each other to form products, considerations very similar to those involved in making sandwiches arise. We can illustrate these ideas with the reaction of $N_2(g)$ and $H_2(g)$ to form $NH_3(g)$:

$$N_2(g) + 3H_2(g) \rightarrow 2NH_3(g)$$

Consider the following container of $N_2(g)$ and $H_2(g)$:

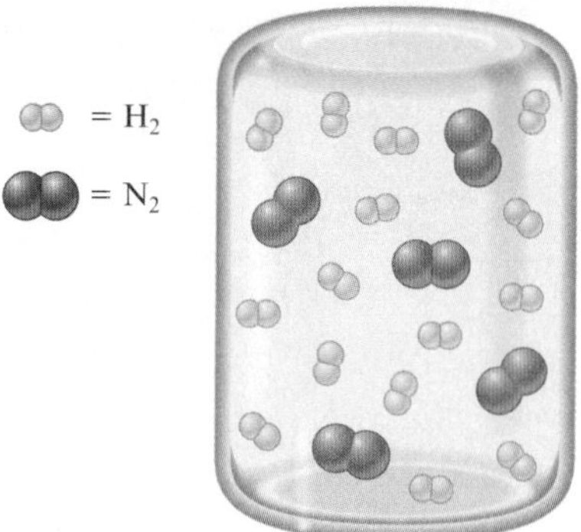

What will this container look like if the reaction between N_2 and H_2 proceeds to completion? To answer this question, you need to remember that each N_2 requires 3 H_2 molecules to form 2 NH_3. To make things more clear, we will circle groups of reactants:

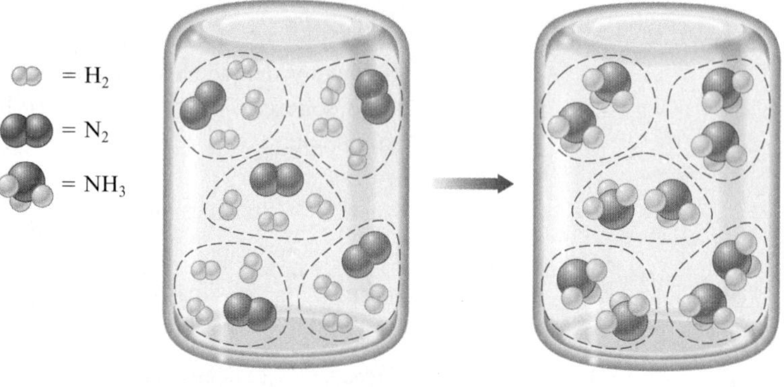

In this case, the mixture of N_2 and H_2 contained just the number of molecules needed to form NH_3 with nothing left over. That is, the ratio of the number of H_2 molecules to N_2 molecules was

$$\frac{15\ H_2}{5\ N_2} = \frac{3\ H_2}{1\ N_2}$$

*The ratio of lemons to sugar that the recipe calls for is 6 lemons to 1 cup of sugar. We can calculate the number of lemons required to "react with" the 3 cups of sugar as follows:

$$3\ \cancel{\text{cups sugar}} \times \frac{6\ \text{lemons}}{1\ \cancel{\text{cup sugar}}} = 18\ \text{lemons}$$

Thus 18 lemons would be required to use up 3 cups of sugar. However, we have only 12 lemons, so the lemons are limiting.

This ratio exactly matches the numbers in the balanced equation

$$3H_2(g) + N_2(g) \rightarrow 2NH_3(g).$$

This type of mixture is called a *stoichiometric mixture*—one that contains the relative amounts of reactants that matches the numbers in the balanced equation. In this case, all reactants will be consumed to form products.

Now consider another container of $N_2(g)$ and $H_2(g)$:

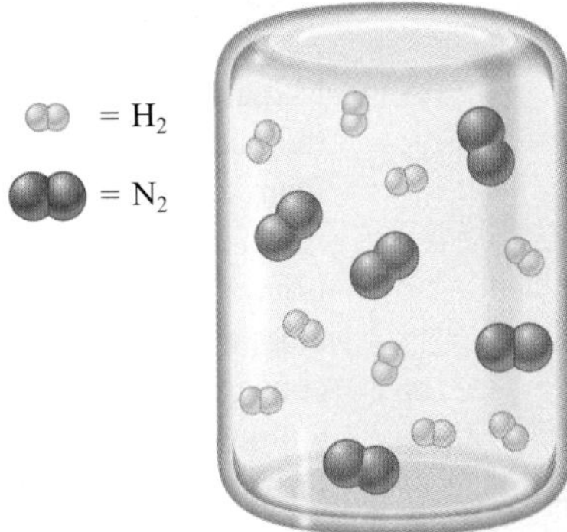

What will the container look like if the reaction between $N_2(g)$ and $H_2(g)$ proceeds to completion? Remember that each N_2 requires 3 H_2. Circling groups of reactants, we have

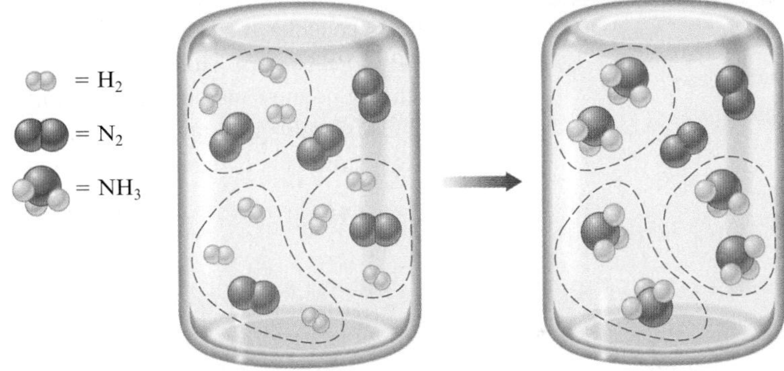

Before the reaction After the reaction

In this case, the hydrogen (H_2) is limiting. That is, the H_2 molecules are used up before all of the N_2 molecules are consumed. In this situation, the amount of hydrogen limits the amount of product (ammonia) that can form—hydrogen is the limiting reactant. Some N_2 molecules are left over in this case because the reaction runs out of H_2 molecules first.

To determine how much product can be formed from a given mixture of reactants, we have to look for the reactant that is limiting—the one that runs out first and thus limits the amount of product that can form.

In some cases, the mixture of reactants might be stoichiometric—that is, all reactants run out at the same time. In general, however, you cannot assume that a given mixture of reactants is a stoichiometric mixture, so you must determine whether one of the reactants is limiting.

The reactant that runs out first and thus limits the amounts of products that can form is called the **limiting reactant** or **limiting reagent.**

To this point, we have considered examples where the numbers of reactant molecules could be counted. In "real life" you can't count the molecules directly—you can't see them, and, even if you could, there would be far too many to count. Instead, you must count by weighing. We must therefore explore how to find the limiting reactant, given the masses of the reactants.

9-5

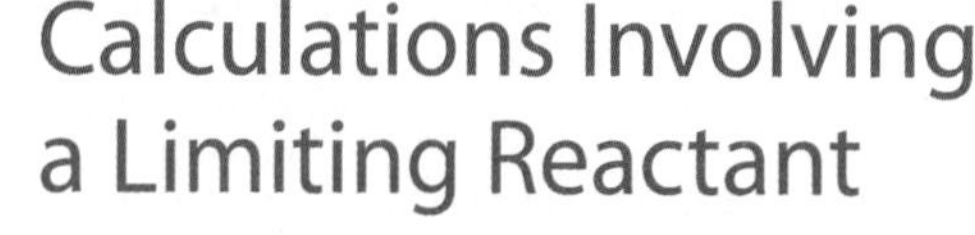

Calculations Involving a Limiting Reactant

OBJECTIVES

- **To learn to recognize the limiting reactant in a reaction.**
- **To learn to use the limiting reactant to do stoichiometric calculations.**

Andy Sotiriou/The Image Bank/Getty Images

Anhydrous ammonia tanks used as agricultural fertilizer.

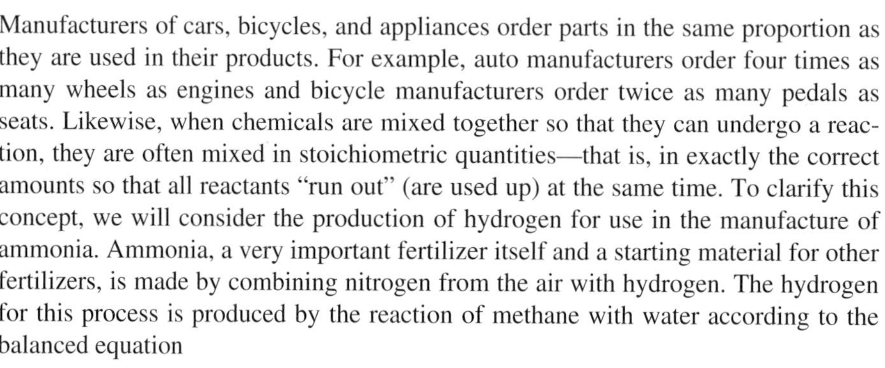

Manufacturers of cars, bicycles, and appliances order parts in the same proportion as they are used in their products. For example, auto manufacturers order four times as many wheels as engines and bicycle manufacturers order twice as many pedals as seats. Likewise, when chemicals are mixed together so that they can undergo a reaction, they are often mixed in stoichiometric quantities—that is, in exactly the correct amounts so that all reactants "run out" (are used up) at the same time. To clarify this concept, we will consider the production of hydrogen for use in the manufacture of ammonia. Ammonia, a very important fertilizer itself and a starting material for other fertilizers, is made by combining nitrogen from the air with hydrogen. The hydrogen for this process is produced by the reaction of methane with water according to the balanced equation

$$CH_4(g) + H_2O(g) \rightarrow 3H_2(g) + CO(g)$$

Let's consider the question, *What mass of water is required to react exactly with 249 g of methane?* That is, how much water will just use up all of the 249 g of methane, leaving no methane or water remaining?

This problem requires the same strategies we developed in the previous section. Again, drawing a map of the problem is helpful.

$$CH_4(g) \quad + \quad H_2O(g) \rightarrow 3H_2(g) + CO(g)$$

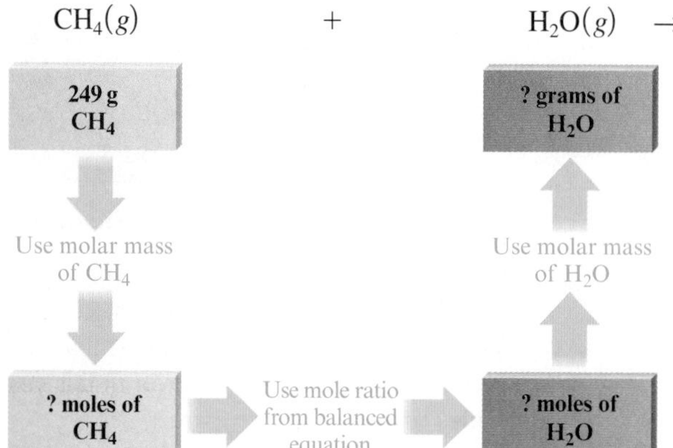

We first convert the mass of CH_4 to moles, using the molar mass of CH_4 (16.04 g/mol).

$$249 \text{ g } \cancel{CH_4} \times \frac{1 \text{ mol } CH_4}{16.04 \text{ g } \cancel{CH_4}} = 15.5 \text{ mol } CH_4$$

Because in the balanced equation 1 mole of CH_4 reacts with 1 mole of H_2O, we have

$$15.5 \cancel{\text{mol } CH_4} \times \frac{1 \text{ mol } H_2O}{1 \cancel{\text{mol } CH_4}} = 15.5 \text{ mol } H_2O$$

Therefore, 15.5 moles of H_2O will react exactly with the given mass of CH_4. Converting 15.5 moles of H_2O to grams of H_2O (molar mass = 18.02 g/mol) gives

$$15.5 \cancel{\text{mol } H_2O} \times \frac{18.02 \text{ g } H_2O}{1 \cancel{\text{mol } H_2O}} = 279 \text{ g } H_2O$$

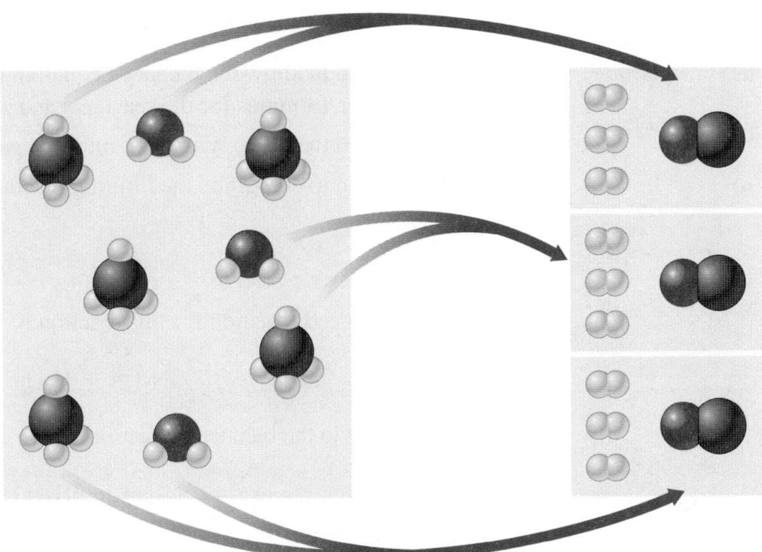

Figure 9.1 A mixture of $5CH_4$ and $3H_2O$ molecules undergoes the reaction $CH_4(g) + H_2O(g) \rightarrow 3H_2(g) + CO(g)$. Note that the H_2O molecules are used up first, leaving two CH_4 molecules unreacted.

This result means that if 249 g of methane is mixed with 279 g of water, both reactants will "run out" at the same time. The reactants have been mixed in stoichiometric quantities.

If, on the other hand, 249 g of methane is mixed with 300 g of water, the methane will be consumed before the water runs out. The water will be in *excess*. In this case, the quantity of products formed will be determined by the quantity of methane present. Once the methane is consumed, no more products can be formed, even though some water still remains. In this situation, the amount of methane *limits* the amount of products that can be formed. Recall from Section 9-4 that we call such a reactant the limiting reactant or the limiting reagent. There are two ways to determine the limiting reactant in a chemical reaction. One involves comparing the moles of reactants to see which runs out first. This concept is illustrated in Fig. 9.1. Note from this figure that because there are fewer water molecules than CH_4 molecules, the water is consumed first. After the water molecules are gone, no more products can form. So in this case water is the limiting reactant.

The reactant that is consumed first limits the amounts of products that can form.

A second method for determining which reactant in a chemical reaction is limiting is to consider the amounts of products that can be formed by completely consuming each reactant. The reactant that produces the smallest amount of product must run out first and thus be limiting. To see how this works, consider the discussion of both approaches in Example 9.7.

Interactive Example 9.7 Stoichiometric Calculations: Identifying the Limiting Reactant

Suppose 25.0 kg (2.50×10^4 g) of nitrogen gas and 5.00 kg (5.00×10^3 g) of hydrogen gas are mixed and reacted to form ammonia. Calculate the mass of ammonia produced when this reaction is run to completion.

SOLUTION

Where Are We Going?

We want to determine the mass of ammonia produced given the masses of both reactants.

What Do We Know?

- The names of the reactants and products.
- We start with 2.50×10^4 g of nitrogen gas and 5.00×10^3 g of hydrogen gas.
- We can obtain the atomic masses from the periodic table.

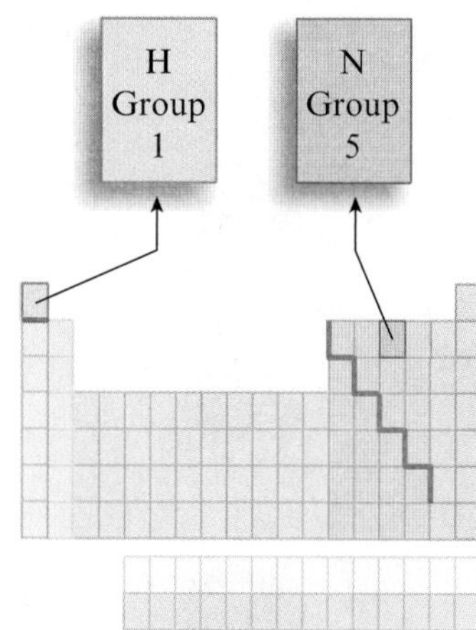

What Do We Need To Know?

- We need to know the balanced equation for the reaction, but we first have to write the formulas for the reactants and products.
- We need the molar masses of nitrogen gas, hydrogen gas, and ammonia.
- We need to determine the limiting reactant.

How Do We Get There?

The unbalanced equation for this reaction is

$$N_2(g) + H_2(g) \rightarrow NH_3(g)$$

which leads to the balanced equation

$$N_2(g) + 3H_2(g) \rightarrow 2NH_3(g)$$

This problem is different from the others we have done so far in that we are mixing *specified amounts of two reactants* together.

We first calculate the moles of the two reactants present:

$$2.50 \times 10^4 \cancel{\text{g N}_2} \times \frac{1 \text{ mol N}_2}{28.02 \cancel{\text{g N}_2}} = 8.92 \times 10^2 \text{ mol N}_2$$

$$5.00 \times 10^3 \cancel{\text{g H}_2} \times \frac{1 \text{ mol H}_2}{2.016 \cancel{\text{g H}_2}} = 2.48 \times 10^3 \text{ mol H}_2$$

A. First we will determine the limiting reactant by comparing the moles of reactants to see which is consumed first. That is, we must determine which is the limiting reactant in this experiment. To do so we must add a step to our normal procedure. We can map this process as follows:

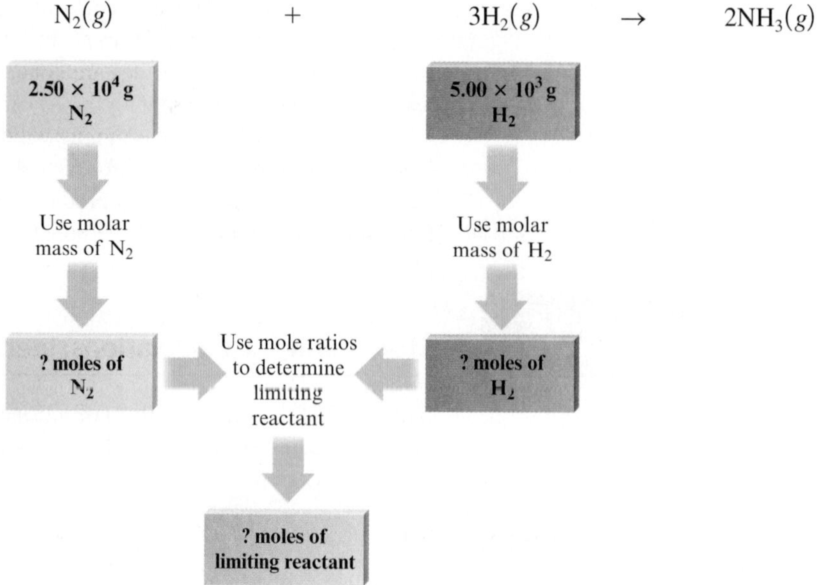

We will use the moles of the limiting reactant to calculate the moles and then the grams of the product.

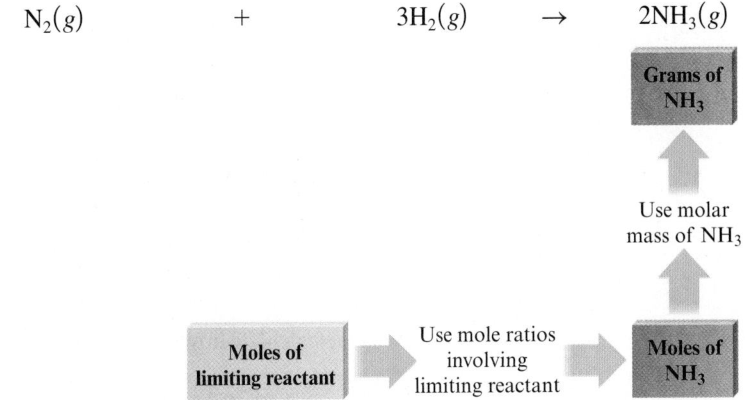

Now we must determine which reactant is limiting (will be consumed first). We have 8.92×10^2 moles of N_2. Let's determine *how many moles of H_2 are required to react with this much N_2*. Because 1 mole of N_2 reacts with 3 moles of H_2, the number of moles of H_2 we need to react completely with 8.92×10^2 moles of N_2 is determined as follows:

8.92 × 10² mol N₂ → $\frac{3 \text{ mol } H_2}{1 \text{ mol } N_2}$ → Moles of H_2 required

$$8.92 \times 10^2 \cancel{\text{mol } N_2} \times \frac{3 \text{ mol } H_2}{1 \cancel{\text{mol } N_2}} = 2.68 \times 10^3 \text{ mol } H_2$$

Is N_2 or H_2 the limiting reactant? The answer comes from the comparison

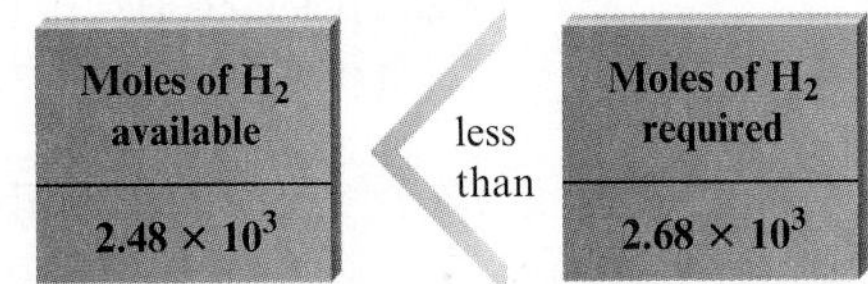

We see that 8.92×10^2 moles of N_2 requires 2.68×10^3 moles of H_2 to react completely. However, only 2.48×10^3 moles of H_2 is present. This means that the hydrogen will be consumed before the nitrogen runs out, so hydrogen is the *limiting reactant* in this particular situation.

Note that in our effort to determine the limiting reactant, we could have started instead with the given amount of hydrogen and calculated the moles of nitrogen required.

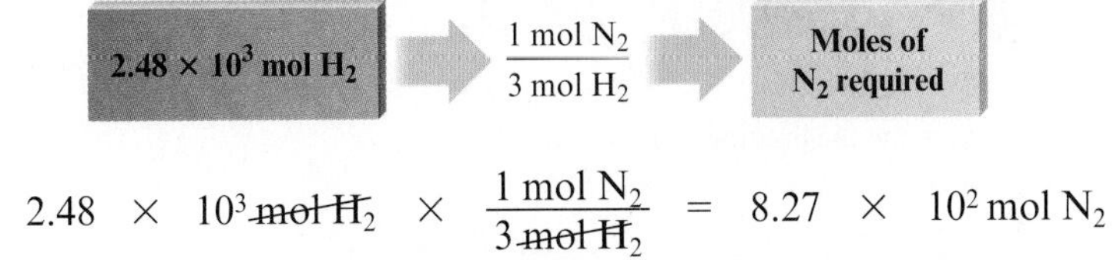

$$2.48 \times 10^3 \cancel{\text{mol } H_2} \times \frac{1 \text{ mol } N_2}{3 \cancel{\text{mol } H_2}} = 8.27 \times 10^2 \text{ mol } N_2$$

Thus 2.48×10^3 moles of H_2 requires 8.27×10^2 moles of N_2. Because 8.92×10^2 moles of N_2 is actually present, the nitrogen is in excess.

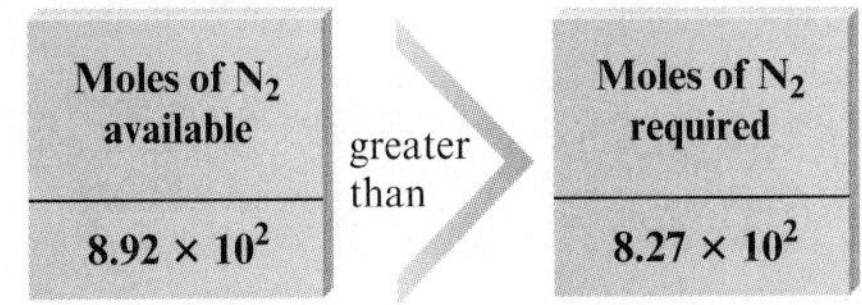

If nitrogen is in excess, hydrogen will "run out" first; again we find that hydrogen limits the amount of ammonia formed.

Because the moles of H_2 present are limiting, we must use this quantity to determine the moles of NH_3 that can form.

$$2.48 \times 10^3 \cancel{\text{mol H}_2} \times \frac{2 \text{ mol NH}_3}{3 \cancel{\text{mol H}_2}} = 1.65 \times 10^3 \text{ mol NH}_3$$

B. Alternately we can determine the limiting reactant by computing the moles of NH_3 that would be formed by the complete reaction of each N_2 and H_2.

Since 1 mole of N_2 produces 2 moles of NH_3, the amount of NH_3 that would be produced if all of the N_2 was used up is calculated as follows:

$$8.92 \times 10^2 \cancel{\text{mol N}_2} \times \frac{2 \text{ mol NH}_3}{1 \cancel{\text{mol N}_2}} = 1.78 \times 10^3 \text{ mol NH}_3$$

Next we calculate how much NH_3 would be produced if the H_2 was completely used up:

$$2.48 \times 10^3 \cancel{\text{mol H}_2} \times \frac{2 \text{ mol NH}_3}{3 \cancel{\text{mol H}_2}} = 1.65 \times 10^3 \text{ mol NH}_3$$

Because a smaller amount of NH_3 is produced from the H_2 than from the N_2, the amount of H_2 must be limiting. Thus, because the H_2 is the limiting reactant, the amount of NH_3 that can be produced is 1.65×10^3 moles, as we determined before.

Next we convert moles of NH_3 to mass of NH_3.

$$1.65 \times 10^3 \cancel{\text{mol NH}_3} \times \frac{17.03 \text{ g NH}_3}{1 \cancel{\text{mol NH}_3}} = 2.81 \times 10^4 \text{ g NH}_3 = 28.1 \text{ kg NH}_3$$

Therefore, 25.0 kg of N_2 and 5.00 kg of H_2 can form 28.1 kg of NH_3.

REALITY CHECK If neither reactant were limiting, we would expect an answer of 30.0 kg of NH_3 because mass is conserved (25.0 kg + 5.0 kg = 30.0 kg). Because one of the reactants (H_2 in this case) is limiting, the answer should be less than 30.0 kg, which it is. ■

The strategies used in Example 9.7 are summarized in Fig. 9.2.

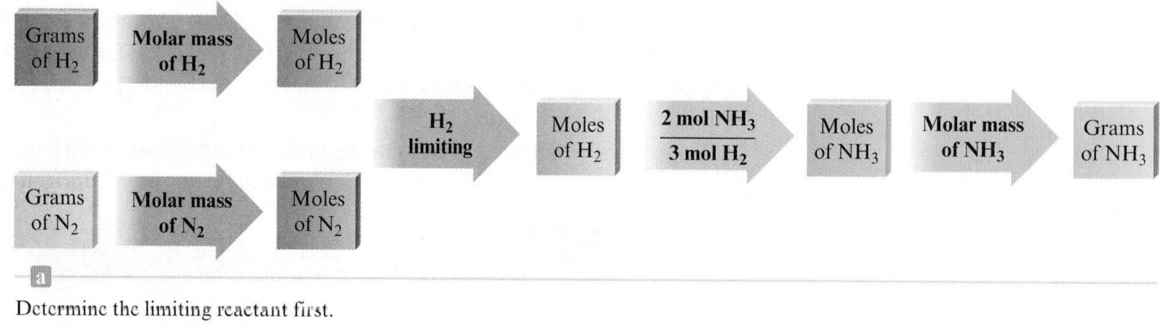

Determine the limiting reactant first.

Determine the possible number of moles of product first.

Figure 9.2 ▸ A map of the procedure used in Example 9.7.

The following list summarizes the steps to take in solving stoichiometry problems in which the amounts of two (or more) reactants are given.

Steps for Solving Stoichiometry Problems Involving Limiting Reactants

Step 1 Write and balance the equation for the reaction.

Step 2 Convert known masses of reactants to moles.

Step 3 Using the numbers of moles of reactants and the appropriate mole ratios, determine which reactant is limiting.

Step 4 Using the amount of the limiting reactant and the appropriate mole ratio, compute the number of moles of the desired product.

or

Step 3 Using the appropriate mole ratios, compute the numbers of moles of product formed if each reactant were consumed.

Step 4 Choose the least number of moles of product formed from Step 3.

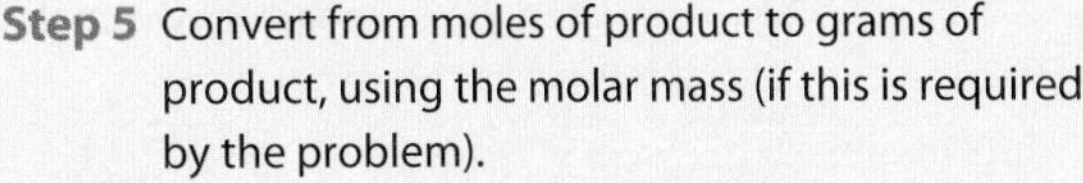

Step 5 Convert from moles of product to grams of product, using the molar mass (if this is required by the problem).

Interactive Example 9.8

Stoichiometric Calculations: Reactions Involving the Masses of Two Reactants

Nitrogen gas can be prepared by passing gaseous ammonia over solid copper(II) oxide at high temperatures. The other products of the reaction are solid copper and water vapor. How many grams of N_2 are formed when 18.1 g of NH_3 is reacted with 90.4 g of CuO?

SOLUTION

Ken O'Donoghue © Cengage Learning

Copper(II) oxide reacting with ammonia in a heated tube.

Where Are We Going?

We want to determine the mass of nitrogen produced given the masses of both reactants.

What Do We Know?

- The names or formulas of the reactants and products.
- We start with 18.1 g of NH_3 and 90.4 g of CuO.
- We can obtain the atomic masses from the periodic table.

What Do We Need To Know?

- We need to know the balanced equation for the reaction, but we first have to write the formulas for the reactants and products.
- We need the molar masses of NH_3, CuO, and N_2.
- We need to determine the limiting reactant.

How Do We Get There?

Step 1 From the description of the problem, we obtain the following balanced equation:

$$2NH_3(g) + 3CuO(s) \rightarrow N_2(g) + 3Cu(s) + 3H_2O(g)$$

Step 2 Next, from the masses of reactants available we must compute the moles of NH_3 (molar mass = 17.03 g) and of CuO (molar mass = 79.55 g).

$$18.1 \text{ g } NH_3 \times \frac{1 \text{ mol } NH_3}{17.03 \text{ g } NH_3} = 1.06 \text{ mol } NH_3$$

$$90.4 \text{ g CuO} \times \frac{1 \text{ mol CuO}}{79.55 \text{ g CuO}} = 1.14 \text{ mol CuO}$$

A. First we will determine the limiting reactant by comparing the moles of reactants to see which one is consumed first. We can then determine the number of moles of N_2 formed.

Step 3 To determine which reactant is limiting, we use the mole ratio between CuO and NH_3.

$$1.06 \text{ mol } NH_3 \times \frac{3 \text{ mol CuO}}{2 \text{ mol } NH_3} = 1.59 \text{ mol CuO}$$

Then we compare how much CuO we have with how much of it we need.

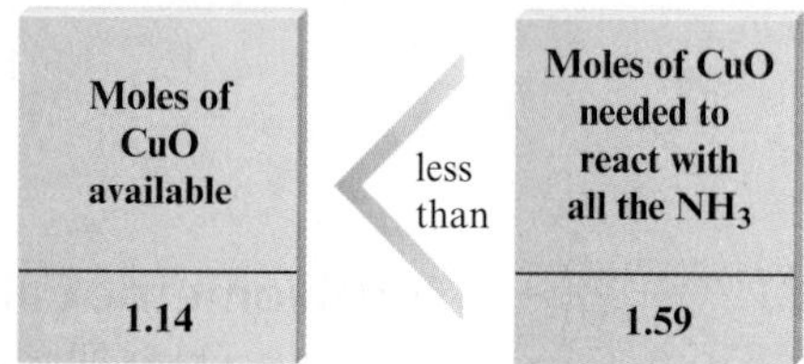

Therefore, 1.59 moles of CuO is required to react with 1.06 moles of NH_3, but only 1.14 moles of CuO is actually present. So the amount of CuO is limiting; CuO will run out before NH_3 does.

Step 4 CuO is the limiting reactant, so we must use the amount of CuO in calculating the amount of N_2 formed. Using the mole ratio between CuO and N_2 from the balanced equation, we have

$$1.14 \text{ mol CuO} \times \frac{1 \text{ mol } N_2}{3 \text{ mol CuO}} = 0.380 \text{ mol } N_2$$

B. Alternately we can determine the limiting reactant by computing the moles of N_2 that would be formed by complete combustion of NH_3 and CuO:

$$1.06 \text{ mol } NH_3 \times \frac{1 \text{ mol } N_2}{2 \text{ mol } NH_3} = 0.530 \text{ mol } N_2$$

$$1.14 \text{ mol CuO} \times \frac{1 \text{ mol } N_2}{3 \text{ mol CuO}} = 0.380 \text{ mol } N_2$$

As before, CuO is the limiting reactant, and we see that we produce 0.380 mole of N_2. Both methods lead us to the same final step.

Step 5 Using the molar mass of N_2 (28.02), we can now calculate the mass of N_2 produced.

$$0.380 \text{ mol } N_2 \times \frac{28.02 \text{ g } N_2}{1 \text{ mol } N_2} = 10.6 \text{ g } N_2$$

SELF CHECK

Exercise 9.6 Lithium nitride, an ionic compound containing the Li^+ and N^{3-} ions, is prepared by the reaction of lithium metal and nitrogen gas. Calculate the mass of lithium nitride formed from 56.0 g of nitrogen gas and 56.0 g of lithium in the unbalanced reaction

$$Li(s) + N_2(g) \rightarrow Li_3N(s)$$

See Problems 9.51 through 9.54. ■

9-6 Percent Yield

OBJECTIVE

To learn to calculate actual yield as a percentage of theoretical yield.

In the previous section we learned how to calculate the amount of products formed when specified amounts of reactants are mixed together. In doing these calculations, we used the fact that the amount of product is controlled by the limiting reactant. Products stop forming when one reactant runs out.

The amount of product calculated in this way is called the **theoretical yield** of that product. It is the amount of product predicted from the amounts of reactants used. For instance, in Example 9.8, 10.6 g of nitrogen represents the theoretical yield. This is the *maximum amount* of nitrogen that can be produced from the quantities of reactants used. Actually, however, the amount of product predicted (the theoretical yield) is seldom obtained. One reason for this is the presence of side reactions (other reactions that consume one or more of the reactants or products).

The *actual yield* of product, which is the amount of product *actually obtained,* is often compared to the theoretical yield. This comparison, usually expressed as a percentage, is called the **percent yield.** Percent yield is important as an indicator of the efficiency of a particular reaction.

$$\frac{\text{Actual yield}}{\text{Theoretical yield}} \times 100\% = \text{percent yield}$$

For example, *if* the reaction considered in Example 9.8 *actually* gave 6.63 g of nitrogen instead of the *predicted* 10.6 g, the percent yield of nitrogen would be

$$\frac{6.63 \text{ g } \cancel{N_2}}{10.6 \text{ g } \cancel{N_2}} \times 100\% = 62.5\%$$

Interactive Example 9.9

Stoichiometric Calculations: Determining Percent Yield

In Section 9-1, we saw that methanol can be produced by the reaction between carbon monoxide and hydrogen. Let's consider this process again. Suppose 68.5 kg (6.85×10^4 g) of $CO(g)$ is reacted with 8.60 kg (8.60×10^3 g) of $H_2(g)$.

a. Calculate the theoretical yield of methanol.

b. If 3.57×10^4 g of CH_3OH is actually produced, what is the percent yield of methanol?

SOLUTION (a)

Where Are We Going?

We want to determine the theoretical yield of methanol and the percent yield given an actual yield.

What Do We Know?

- From Section 9-1 we know the balanced equation is

$$2H_2 + CO \rightarrow CH_3OH$$

- We start with 6.85×10^4 g of CO and 8.60×10^3 g of H_2.
- We can obtain the atomic masses from the periodic table.

What Do We Need To Know?

- We need the molar masses of H_2, CO, and CH_3OH.
- We need to determine the limiting reactant.

How Do We Get There?

Step 1 The balanced equation is

$$2H_2(g) + CO(g) \rightarrow CH_3OH(l)$$

Step 2 Next we calculate the moles of reactants.

$$6.85 \times 10^4 \cancel{\text{g CO}} \times \frac{1 \text{ mol CO}}{28.01 \cancel{\text{g CO}}} = 2.45 \times 10^3 \text{ mol CO}$$

$$8.60 \times 10^3 \cancel{\text{g H}_2} \times \frac{1 \text{ mol H}_2}{2.016 \cancel{\text{g H}_2}} = 4.27 \times 10^3 \text{ mol H}_2$$

Step 3 Now we determine which reactant is limiting. Using the mole ratio between CO and H_2 from the balanced equation, we have

$$2.45 \times 10^3 \cancel{\text{mol CO}} \times \frac{2 \text{ mol H}_2}{1 \cancel{\text{mol CO}}} = 4.90 \times 10^3 \text{ mol H}_2$$

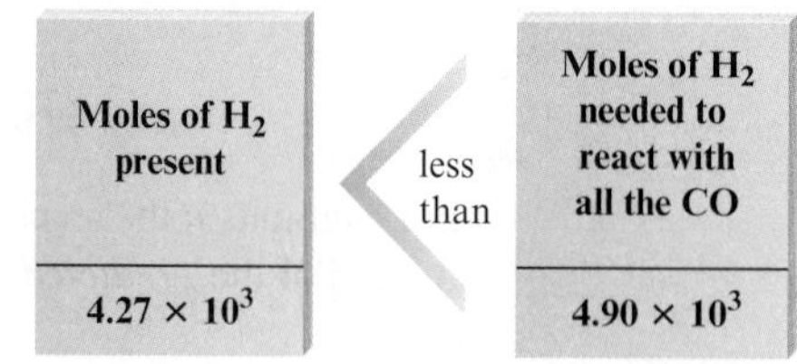

We see that 2.45×10^3 moles of CO requires 4.90×10^3 moles of H_2. Because only 4.27×10^3 moles of H_2 is actually present, *H_2 is limiting.*

Step 4 We must therefore use the amount of H_2 and the mole ratio between H_2 and CH_3OH to determine the maximum amount of methanol that can be produced in the reaction.

$$4.27 \times 10^3 \cancel{\text{mol H}_2} \times \frac{1 \text{ mol CH}_3\text{OH}}{2 \cancel{\text{mol H}_2}} = 2.14 \times 10^3 \text{ mol CH}_3\text{OH}$$

This represents the theoretical yield in moles.

Step 5 Using the molar mass of CH_3OH (32.04 g), we can calculate the theoretical yield in grams.

$$2.14 \times 10^3 \cancel{\text{mol CH}_3\text{OH}} \times \frac{32.04 \text{ g CH}_3\text{OH}}{1 \cancel{\text{mol CH}_3\text{OH}}} = 6.86 \times 10^4 \text{ g CH}_3\text{OH}$$

So, from the amounts of reactants given, the maximum amount of CH_3OH that can be formed is 6.86×10^4 g. This is the *theoretical yield.*

SOLUTION (b) The percent yield is

$$\frac{\text{Actual yield (grams)}}{\text{Theoretical yield (grams)}} \times 100\% = \frac{3.57 \times 10^4 \text{ g } \cancel{CH_3OH}}{6.86 \times 10^4 \text{ g } \cancel{CH_3OH}} \times 100\% = 52.0\%$$

SELF CHECK

Exercise 9.7 Titanium(IV) oxide is a white compound used as a coloring pigment. In fact, the pages in the books you read are white because of the presence of this compound in the paper. Solid titanium(IV) oxide can be prepared by reacting gaseous titanium(IV) chloride with oxygen gas. A second product of this reaction is chlorine gas.

$$TiCl_4(g) + O_2(g) \rightarrow TiO_2(s) + Cl_2(g)$$

a. Suppose 6.71×10^3 g of titanium(IV) chloride is reacted with 2.45×10^3 g of oxygen. Calculate the maximum mass of titanium(IV) oxide that can form.

b. If the percent yield of TiO_2 is 75%, what mass is actually formed?

See Problems 9.63 and 9.64. ■

CHAPTER 9 REVIEW

F directs you to the *Chemistry in Focus* feature in the chapter

Key Terms

- mole ratio (9-2)
- stoichiometry (9-3)
- limiting reactant (9-4)
- limiting reagent (9-4)
- theoretical yield (9-6)
- percent yield (9-6)

For Review

- A balanced chemical equation gives relative numbers (or moles) of reactant and product molecules that participate in a chemical reaction.
- Stoichiometric calculations involve using a balanced chemical equation to determine the amounts of reactants needed or products formed in a reaction.
- To convert between moles of reactants and moles of products, we use mole ratios derived from the balanced chemical equation.
- To calculate masses from the moles of reactants needed or products formed, we can use the molar masses of substances for finding the masses (g) needed or formed.
- Often, reactants in a chemical reaction are not present in stoichiometric quantities (i.e., they do not "run out" at the same time).
  - In this case, we must determine which reactant runs out first and thus limits the amount of products that can form—this is called the limiting reactant.
- The actual yield (amount produced) of a reaction is usually less than the maximum expected (theoretical yield).
- The actual yield is often expressed as a percentage of the theoretical yield:

$$\text{Percent yield} = \frac{\text{actual yield (g)}}{\text{theoretical yield (g)}} \times 100\%$$

Active Learning Questions

These questions are designed to be considered by groups of students in class. Often these questions work well for introducing a particular topic in class.

1. Relate Active Learning Question 2 from Chapter 2 to the concepts of chemical stoichiometry.
2. You are making cookies and are missing a key ingredient—eggs. You have plenty of the other ingredients, except that you have only 1.33 cups of butter and no eggs. You note that the recipe calls for 2 cups of butter and 3 eggs (plus the other ingredients) to make 6 dozen cookies. You telephone a friend and have him bring you some eggs.
   a. How many eggs do you need?
   b. If you use all the butter (and get enough eggs), how many cookies can you make?

   Unfortunately, your friend hangs up before you tell him how many eggs you need. When he arrives, he has a surprise for you—to save time he has broken the eggs in a bowl for you. You ask him how many he brought, and he replies, "All of them, but I spilled some on the way over." You weigh the eggs and find that they weigh 62.1 g. Assuming that an average egg weighs 34.21 g:

c. How much butter is needed to react with all the eggs?
d. How many cookies can you make?
e. Which will you have left over, eggs or butter?
f. How much is left over?
g. Relate this question to the concepts of chemical stoichiometry.

3. Nitrogen (N_2) and hydrogen (H_2) react to form ammonia (NH_3). Consider the mixture of N_2 ( ) and H_2 ( ) in a closed container as illustrated below:

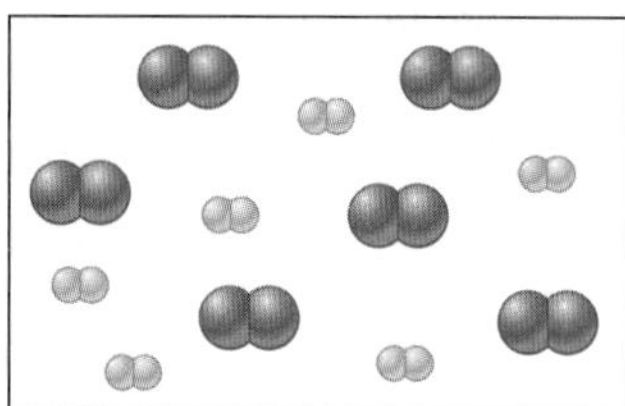

Assuming the reaction goes to completion, draw a representation of the product mixture. Explain how you arrived at this representation.

4. Which of the following equations best represents the reaction for Question 3?

a. $6N_2 + 6H_2 \rightarrow 4NH_3 + 4N_2$
b. $N_2 + H_2 \rightarrow NH_3$
c. $N + 3H \rightarrow NH_3$
d. $N_2 + 3H_2 \rightarrow 2NH_3$
e. $2N_2 + 6H_2 \rightarrow 4NH_3$

For choices you did not pick, explain what you feel is wrong with them, and justify the choice you did pick.

5. You know that chemical A reacts with chemical B. You react 10.0 g A with 10.0 g B. What information do you need to know to determine the amount of product that will be produced? Explain.

6. If 10.0 g of hydrogen gas is reacted with 10.0 g of oxygen gas according to the equation

$$2H_2 + O_2 \rightarrow 2H_2O$$

we should not expect to form 20.0 g of water. Why not? What mass of water can be produced with a complete reaction?

7. The limiting reactant in a reaction:

a. has the lowest coefficient in a balanced equation.
b. is the reactant for which you have the fewest number of moles.
c. has the lowest ratio: moles available/coefficient in the balanced equation.
d. has the lowest ratio: coefficient in the balanced equation/moles available.
e. None of the above.

For choices you did not pick, explain what you feel is wrong with them, and justify the choice you did pick.

8. Given the equation $3A + B \rightarrow C + D$, if 4 moles of A is reacted with 2 moles of B, which of the following is true?

a. The limiting reactant is the one with the higher molar mass.
b. A is the limiting reactant because you need 6 moles of A and have 4 moles.
c. B is the limiting reactant because you have fewer moles of B than moles of A.
d. B is the limiting reactant because three A molecules react with every one B molecule.
e. Neither reactant is limiting.

For choices you did not pick, explain what you feel is wrong with them, and justify the choice you did pick.

9. What happens to the weight of an iron bar when it rusts?

a. There is no change because mass is always conserved.
b. The weight increases.
c. The weight increases, but if the rust is scraped off, the bar has the original weight.
d. The weight decreases.

Justify your choice and, for choices you did not pick, explain what is wrong with them. Explain what it means for something to rust.

10. Consider the equation $2A + B \rightarrow A_2B$. If you mix 1.0 mole of A and 1.0 mole of B, how many moles of A_2B can be produced?

11. What is meant by the term *mole ratio*? Give an example of a mole ratio, and explain how it is used in solving a stoichiometry problem.

12. Which would produce a greater number of moles of product: a given amount of hydrogen gas reacting with an excess of oxygen gas to produce water, or the same amount of hydrogen gas reacting with an excess of nitrogen gas to make ammonia? Support your answer.

13. Consider a reaction represented by the following balanced equation

$$2A + 3B \rightarrow C + 4D$$

You find that it requires equal masses of A and B so that there are no reactants left over. Which of the following is true? Justify your choice.

a. The molar mass of A must be greater than the molar mass of B.
b. The molar mass of A must be less than the molar mass of B.
c. The molar mass of A must be the same as the molar mass of B.

14. Consider a chemical equation with two reactants forming one product. If you know the mass of each reactant, what else do you need to know to determine the mass of the product? Why isn't the mass necessarily the sum of the mass of the reactants? Provide a real example of such a reaction, and support your answer mathematically.

15. Consider the balanced chemical equation

$$A + 5B \rightarrow 3C + 4D$$

When equal masses of A and B are reacted, which is limiting, A or B? Justify your choice.

a. If the molar mass of A is greater than the molar mass of B, then A must be limiting.
b. If the molar mass of A is less than the molar mass of B, then A must be limiting.
c. If the molar mass of A is greater than the molar mass of B, then B must be limiting.
d. If the molar mass of A is less than the molar mass of B, then B must be limiting.

All even-numbered Questions and Problems have answers in the back of this book and solutions in the *Student Solutions Guide*.

16. Which of the following reaction mixtures would produce the greatest amount of product, assuming all went to completion? Justify your choice.

Each involves the reaction symbolized by the equation

$$2H_2 + O_2 \rightarrow 2H_2O$$

a. 2 moles of H_2 and 2 moles of O_2.
b. 2 moles of H_2 and 3 moles of O_2.
c. 2 moles of H_2 and 1 mole of O_2.
d. 3 moles of H_2 and 1 mole of O_2.
e. Each would produce the same amount of product.

17. Baking powder is a mixture of cream of tartar ($KHC_4H_4O_6$) and baking soda ($NaHCO_3$). When it is placed in an oven at typical baking temperatures (as part of a cake, for example), it undergoes the following reaction (CO_2 makes the cake rise):

$$KHC_4H_4O_6(s) + NaHCO_3(s) \rightarrow KNaC_4H_4O_6(s) + H_2O(g) + CO_2(g)$$

You decide to make a cake one day, and the recipe calls for baking powder. Unfortunately, you have no baking powder. You do have cream of tartar and baking soda, so you use stoichiometry to figure out how much of each to mix.

Of the following choices, which is the best way to make baking powder? The amounts given in the choices are in teaspoons (that is, you will use a teaspoon to measure the baking soda and cream of tartar). Justify your choice.

Assume a teaspoon of cream of tartar has the same mass as a teaspoon of baking soda.

a. Add equal amounts of baking soda and cream of tartar.
b. Add a bit more than twice as much cream of tartar as baking soda.
c. Add a bit more than twice as much baking soda as cream of tartar.
d. Add more cream of tartar than baking soda, but not quite twice as much.
e. Add more baking soda than cream of tartar, but not quite twice as much.

18. You have seven closed containers each with equal masses of chlorine gas (Cl_2). You add 10.0 g of sodium to the first sample, 20.0 g of sodium to the second sample, and so on (adding 70.0 g of sodium to the seventh sample). Sodium and chloride react to form sodium chloride according to the equation

$$2Na(s) + Cl_2(g) \rightarrow 2NaCl(s)$$

After each reaction is complete, you collect and measure the amount of sodium chloride formed. A graph of your results is shown below.

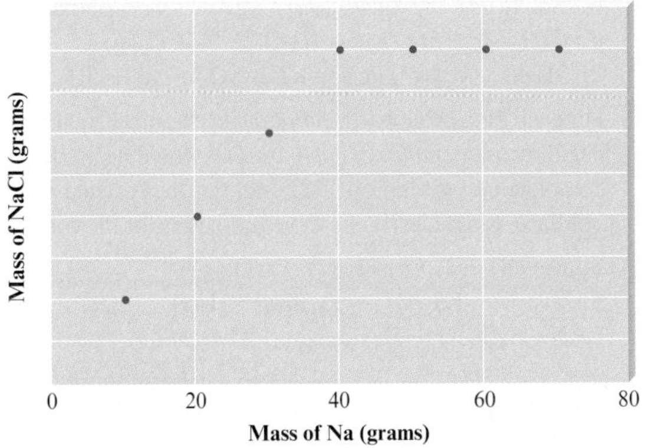

Answer the following questions:

a. Explain the shape of the graph.
b. Calculate the mass of NaCl formed when 20.0 g of sodium is used.
c. Calculate the mass of Cl_2 in each container.
d. Calculate the mass of NaCl formed when 50.0 g of sodium is used.
e. Identify the leftover reactant and determine its mass for parts b and d above.

19. You have a chemical in a sealed glass container filled with air. The setup is sitting on a balance as shown below. The chemical is ignited by means of a magnifying glass focusing sunlight on the reactant. After the chemical has completely burned, which of the following is true? Explain your answer.

a. The balance will read less than 250.0 g.
b. The balance will read 250.0 g.
c. The balance will read greater than 250.0 g.
d. Cannot be determined without knowing the identity of the chemical.

20. Consider an iron bar on a balance as shown.

As the iron bar rusts, which of the following is true? Explain your answer.

a. The balance will read less than 75.0 g.
b. The balance will read 75.0 g.
c. The balance will read greater than 75.0 g.
d. The balance will read greater than 75.0 g, but if the bar is removed, the rust scraped off, and the bar replaced, the balance will read 75.0 g.

21. Consider the reaction between NO(*g*) and $O_2(g)$ represented below.

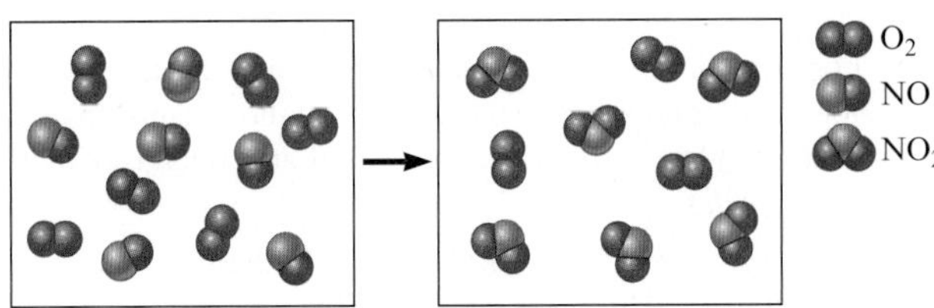

What is the balanced equation for this reaction, and what is the limiting reactant?

All even-numbered Questions and Problems have answers in the back of this book and solutions in the *Student Solutions Guide*.

Questions and Problems

9-1 Information Given by Chemical Equations

Questions

1. What do the coefficients of a balanced chemical equation tell us about the proportions in which atoms and molecules react on an individual (microscopic) basis?

2. The vigorous reaction between aluminum and iodine gives the balanced equation:

$$2Al(s) + 3I_2(s) \rightarrow 2AlI_3(s).$$

© Cengage Learning

What do the coefficients in this balanced chemical equation tell us about the proportions in which these substances react on a macroscopic (mole) basis?

3. Although *mass* is a property of matter we can conveniently measure in the laboratory, the coefficients of a balanced chemical equation are *not* directly interpreted on the basis of mass. Explain why.

4. Which of the following statements is *true* for the reaction of nitrogen gas with hydrogen gas to produce ammonia (NH_3)? Choose the *best* answer.

   a. Subscripts can be changed to balance this equation, just as they can be changed to balance the charges when writing the formula for an ionic compound.
   b. The nitrogen and hydrogen will not react until you have added the correct mole ratios.
   c. The mole ratio of nitrogen to hydrogen in the balanced equation is 1:2.
   d. Ammonia will not form unless 1 mole of nitrogen and 3 moles of hydrogen have been added.
   e. The balanced equation allows you to predict how much ammonia you will make based on the amount of nitrogen and hydrogen present.

Problems

5. For each of the following reactions, give the balanced equation for the reaction and state the meaning of the equation in terms of the numbers of *individual molecules* and in terms of *moles of molecules*.

   a. $PCl_3(l) + H_2O(l) \rightarrow H_3PO_3(aq) + HCl(g)$
   b. $XeF_2(g) + H_2O(l) \rightarrow Xe(g) + HF(g) + O_2(g)$
   c. $S(s) + HNO_3(aq) \rightarrow H_2SO_4(aq) + H_2O(l) + NO_2(g)$
   d. $NaHSO_3(s) \rightarrow Na_2SO_3(s) + SO_2(g) + H_2O(l)$

6. For each of the following reactions, give the balanced chemical equation for the reaction and state the meaning of the equation in terms of *individual molecules* and in terms of *moles* of molecules.

   a. $MnO_2(s) + Al(s) \rightarrow Mn(s) + Al_2O_3(s)$
   b. $B_2O_3(s) + CaF_2(s) \rightarrow BF_3(g) + CaO(s)$
   c. $NO_2(g) + H_2O(l) \rightarrow HNO_3(aq) + NO(g)$
   d. $C_6H_6(g) + H_2(g) \rightarrow C_6H_{12}(g)$

9-2 Mole–Mole Relationships

Questions

7. Consider the reaction represented by the chemical equation

$$C(s) + O_2(g) \rightarrow CO_2(g)$$

Since the coefficients of the balanced chemical equation are all equal to 1, we know that exactly 1 g of C will react with exactly 1 g of O_2. True or false? Explain.

8. For the balanced chemical equation for the combination reaction of sodium metal and chlorine gas

$$2Na(s) + Cl_2(g) \rightarrow 2NaCl(s)$$

explain why we know that 2 g of Na reacting with 1 g of Cl_2 will *not* result in the production of 2 g of NaCl.

9. Consider the balanced chemical equation

$$4Al(s) + 3O_2(g) \rightarrow 2Al_2O_3(s).$$

What mole ratio would you use to calculate how many moles of oxygen gas would be needed to react completely with a given number of moles of aluminum metal? What mole ratio would you use to calculate the number of moles of product that would be expected if a given number of moles of aluminum metal reacts completely?

10. Consider the balanced equation

$$CH_4(g) + 2O_2(g) \rightarrow CO_2(g) + 2H_2O(g)$$

What is the mole ratio that would enable you to calculate the number of moles of oxygen needed to react exactly with a given number of moles of $CH_4(g)$? What mole ratios would you use to calculate how many moles of each product form from a given number of moles of CH_4?

Problems

11. For each of the following balanced chemical equations, calculate how many *moles* of product(s) would be produced if 0.500 mole of the first reactant were to react completely.

   a. $CO_2(g) + 4H_2(g) \rightarrow CH_4(g) + 2H_2O(l)$
   b. $BaCl_2(aq) + 2AgNO_3(aq) \rightarrow 2AgCl(s) + Ba(NO_3)_2(aq)$
   c. $C_3H_8(g) + 5O_2(g) \rightarrow 4H_2O(l) + 3CO_2(g)$
   d. $3H_2SO_4(aq) + 2Fe(s) \rightarrow Fe_2(SO_4)_3(aq) + 3H_2(g)$

12. For each of the following *unbalanced* chemical equations, calculate how many *moles of each product* would be produced by the complete conversion of 0.125 mole of the reactant indicated in boldface. State clearly the mole ratio used for the conversion.

   a. $\mathbf{FeO}(s) + C(s) \rightarrow Fe(l) + CO_2(g)$
   b. $Cl_2(g) + \mathbf{KI}(aq) \rightarrow KCl(aq) + I_2(s)$
   c. $\mathbf{Na_2B_4O_7}(s) + H_2SO_4(aq) + H_2O(l) \rightarrow H_3BO_3(s) + Na_2SO_4(aq)$
   d. $\mathbf{CaC_2}(s) + H_2O(l) \rightarrow Ca(OH)_2(s) + C_2H_2(g)$

All even-numbered Questions and Problems have answers in the back of this book and solutions in the *Student Solutions Guide*.

13. For each of the following balanced chemical equations, calculate how many *grams* of the product(s) would be produced by complete reaction of 0.125 mole of the first reactant.

a. $AgNO_3(aq) + LiOH(aq) \rightarrow AgOH(s) + LiNO_3(aq)$
b. $Al_2(SO_4)_3(aq) + 3CaCl_2(aq) \rightarrow 2AlCl_3(aq) + 3CaSO_4(s)$
c. $CaCO_3(s) + 2HCl(aq) \rightarrow CaCl_2(aq) + CO_2(g) + H_2O(l)$
d. $2C_4H_{10}(g) + 13O_2(g) \rightarrow 8CO_2(g) + 10H_2O(g)$

14. For each of the following balanced chemical equations, calculate how many *moles* and how many *grams* of each product would be produced by the complete conversion of 0.50 mole of the reactant indicated in boldface. State clearly the mole ratio used for each conversion.

a. $\mathbf{NH_3}(g) + HCl(g) \rightarrow NH_4Cl(s)$
b. $CH_4(g) + \mathbf{4S}(s) \rightarrow CS_2(l) + 2H_2S(g)$
c. $\mathbf{PCl_3}(l) + 3H_2O(l) \rightarrow H_3PO_3(aq) + 3HCl(aq)$
d. $\mathbf{NaOH}(s) + CO_2(g) \rightarrow NaHCO_3(s)$

15. For each of the following *unbalanced* equations, indicate how many *moles* of the *second reactant* would be required to react exactly with *0.275 mole* of the *first reactant.* State clearly the mole ratio used for the conversion.

a. $Cl_2(g) + KI(aq) \rightarrow I_2(s) + KCl(aq)$
b. $Co(s) + P_4(s) \rightarrow Co_3P_2(s)$
c. $Zn(s) + HNO_3(aq) \rightarrow ZnNO_3(aq) + H_2(g)$
d. $C_5H_{12}(l) + O_2(g) \rightarrow CO_2(g) + H_2O(g)$

16. For each of the following *unbalanced* equations, indicate how many *moles* of the *first product* are produced if *0.625 mole* of the *second product* forms. State clearly the mole ratio used for each conversion.

a. $KO_2(s) + H_2O(l) \rightarrow O_2(g) + KOH(s)$
b. $SeO_2(g) + H_2Se(g) \rightarrow Se(s) + H_2O(g)$
c. $CH_3CH_2OH(l) + O_2(g) \rightarrow CH_3CHO(aq) + H_2O(l)$
d. $Fe_2O_3(s) + Al(s) \rightarrow Fe(l) + Al_2O_3(s)$

9-3 Mass Calculations

Questions

17. What quantity serves as the conversion factor between the mass of a sample and how many moles the sample contains?

18. What does it mean to say that the balanced chemical equation for a reaction describes the *stoichiometry* of the reaction?

Problems

19. Using the average atomic masses given inside the front cover of this book, calculate how many *moles* of each substance the following masses represent.

a. 4.15 g of silicon, Si
b. 2.72 mg of gold(III) chloride, $AuCl_3$
c. 1.05 kg of sulfur, S
d. 0.000901 g of iron(III) chloride, $FeCl_3$
e. 5.62×10^3 g of magnesium oxide, MgO

20. Using the average atomic masses given inside the front cover of this book, calculate the number of *moles* of each substance contained in the following *masses.*

a. 2.01×10^{-2} g of silver
b. 45.2 mg of ammonium sulfide
c. 61.7 μg of uranium
d. 5.23 kg of sulfur dioxide
e. 272 g of iron(III) nitrate

21. Using the average atomic masses given inside the front cover of this book, calculate the *mass in grams* of each of the following samples.

a. 2.17 moles of germanium, Ge
b. 4.24 mmol of lead(II) chloride (1 mmol = 1/1000 mol)
c. 0.0971 mole of ammonia, NH_3
d. 4.26×10^3 moles of hexane, C_6H_{14}
e. 1.71 moles of iodine monochloride, ICl

22. Using the average atomic masses given inside the front cover of this book, calculate the *mass in grams* of each of the following samples.

a. 0.341 mole of potassium nitride
b. 2.62 mmol of neon (1 mmol = 1/1000 mol)
c. 0.00449 mole of manganese(II) oxide
d. 7.18×10^5 moles of silicon dioxide
e. 0.000121 mole of iron(III) phosphate

23. For each of the following *unbalanced* equations, calculate how many *moles* of the second reactant would be required to react completely with 0.413 *moles* of the first reactant.

a. $Co(s) + F_2(g) \rightarrow CoF_3(s)$
b. $Al(s) + H_2SO_4(aq) \rightarrow Al_2(SO_4)_3(aq) + H_2(g)$
c. $K(s) + H_2O(l) \rightarrow KOH(aq) + H_2(g)$
d. $Cu(s) + O_2(g) \rightarrow Cu_2O(s)$

24. For each of the following *unbalanced* equations, calculate how many *moles* of the second reactant would be required to react completely with 0.557 *grams* of the first reactant.

a. $Al(s) + Br_2(l) \rightarrow AlBr_3(s)$
b. $Hg(s) + HClO_4(aq) \rightarrow Hg(ClO_4)_2(aq) + H_2(g)$
c. $K(s) + P(s) \rightarrow K_3P(s)$
d. $CH_4(g) + Cl_2(g) \rightarrow CCl_4(l) + HCl(g)$

25. For each of the following *unbalanced* equations, calculate how many *grams of each product* would be produced by complete reaction of 12.5 g of the reactant indicated in boldface. Indicate clearly the mole ratio used for the conversion.

a. $TiBr_4(g) + \mathbf{H_2}(g) \rightarrow Ti(s) + HBr(g)$
b. $\mathbf{SiH_4}(g) + NH_3(g) \rightarrow Si_3N_4(s) + H_2(g)$
c. $NO(g) + \mathbf{H_2}(g) \rightarrow N_2(g) + 2H_2O(l)$
d. $\mathbf{Cu_2S}(s) \rightarrow Cu(s) + S(g)$

26. Consider the following reaction:

$$PCl_3(s) + 3H_2O(l) \rightarrow H_3PO_3(aq) + 3HCl(aq)$$

What mass of H_2O is needed to completely react with 20.0 g of PCl_3?

27. "Smelling salts," which are used to revive someone who has fainted, typically contain ammonium carbonate, $(NH_4)_2CO_3$. Ammonium carbonate decomposes readily to form ammonia, carbon dioxide, and water. The strong odor of the ammonia usually restores consciousness in the person who has fainted. The unbalanced equation is

$$(NH_4)_2CO_3(s) \rightarrow NH_3(g) + CO_2(g) + H_2O(g)$$

Calculate the mass of ammonia gas that is produced if 1.25 g of ammonium carbonate decomposes completely.

All even-numbered Questions and Problems have answers in the back of this book and solutions in the *Student Solutions Guide.*

28. Calcium carbide, CaC_2, can be produced in an electric furnace by strongly heating calcium oxide (lime) with carbon. The unbalanced equation is

$$CaO(s) + C(s) \rightarrow CaC_2(s) + CO(g)$$

Calcium carbide is useful because it reacts readily with water to form the flammable gas acetylene, C_2H_2, which is used extensively in the welding industry. The unbalanced equation is

$$CaC_2(s) + H_2O(l) \rightarrow C_2H_2(g) + Ca(OH)_2(s)$$

What mass of acetylene gas, C_2H_2, would be produced by complete reaction of 3.75 g of calcium carbide?

29. When elemental carbon is burned in the open atmosphere, with plenty of oxygen gas present, the product is carbon dioxide.

$$C(s) + O_2(g) \rightarrow CO_2(g)$$

However, when the amount of oxygen present during the burning of the carbon is restricted, carbon monoxide is more likely to result.

$$2C(s) + O_2(g) \rightarrow 2CO(g)$$

What mass of each product is expected when a 5.00-g sample of pure carbon is burned under each of these conditions?

30. If baking soda (sodium hydrogen carbonate) is heated strongly, the following reaction occurs:

$$2NaHCO_3(s) \rightarrow Na_2CO_3(s) + H_2O(g) + CO_2(g)$$

Calculate the mass of sodium carbonate that will remain if a 1.52-g sample of sodium hydrogen carbonate is heated.

31. Although we usually think of substances as "burning" only in oxygen gas, the process of rapid oxidation to produce a flame may also take place in other strongly oxidizing gases. For example, when iron is heated and placed in pure chlorine gas, the iron "burns" according to the following (unbalanced) reaction:

$$Fe(s) + Cl_2(g) \rightarrow FeCl_3(s)$$

How many milligrams of iron(III) chloride result when 15.5 mg of iron is reacted with an excess of chlorine gas?

32. When yeast is added to a solution of glucose or fructose, the sugars are said to undergo *fermentation* and ethyl alcohol is produced.

$$C_6H_{12}O_6(aq) \rightarrow 2C_2H_5OH(aq) + 2CO_2(g)$$

This is the reaction by which wines are produced from grape juice. Calculate the mass of ethyl alcohol, C_2H_5OH, produced when 5.25 g of glucose, $C_6H_{12}O_6$, undergoes this reaction.

33. Sulfurous acid is unstable in aqueous solution and gradually decomposes to water and sulfur dioxide gas (which explains the choking odor associated with sulfurous acid solutions).

$$H_2SO_3(aq) \rightarrow H_2O(l) + SO_2(g)$$

If 4.25 g of sulfurous acid undergoes this reaction, what mass of sulfur dioxide is released?

34. Small quantities of oxygen gas can be generated in the laboratory by the decomposition of hydrogen peroxide. The unbalanced equation for the reaction is

$$H_2O_2(aq) \rightarrow H_2O(l) + O_2(g)$$

Calculate the mass of oxygen produced when 10.00 g of hydrogen peroxide decomposes.

35. Elemental phosphorus burns in oxygen with an intensely hot flame, producing a brilliant light and clouds of the oxide product. These properties of the combustion of phosphorus have led to its being used in bombs and incendiary devices for warfare.

$$P_4(s) + 5O_2(g) \rightarrow 2P_2O_5(s)$$

If 4.95 g of phosphorus is burned, what mass of oxygen does it combine with?

36. Although we tend to make less use of mercury these days because of the environmental problems created by its improper disposal, mercury is still an important metal because of its unusual property of existing as a liquid at room temperature. One process by which mercury is produced industrially is through the heating of its common ore cinnabar (mercuric sulfide, HgS) with lime (calcium oxide, CaO).

$$4HgS(s) + 4CaO(s) \rightarrow 4Hg(l) + 3CaS(s) + CaSO_4(s)$$

What mass of mercury would be produced by complete reaction of 10.0 kg of HgS?

37. Ammonium nitrate has been used as a high explosive because it is unstable and decomposes into several gaseous substances. The rapid expansion of the gaseous substances produces the explosive force.

$$NH_4NO_3(s) \rightarrow N_2(g) + O_2(g) + H_2O(g)$$

Calculate the mass of each product gas if 1.25 g of ammonium nitrate reacts.

38. If common sugars are heated too strongly, they char as they decompose into carbon and water vapor. For example, if sucrose (table sugar) is heated, the reaction is

$$C_{12}H_{22}O_{11}(s) \rightarrow 12C(s) + 11H_2O(g)$$

What mass of carbon is produced if 1.19 g of sucrose decomposes completely?

39. Thionyl chloride, $SOCl_2$, is used as a very powerful drying agent in many synthetic chemistry experiments in which the presence of even small amounts of water would be detrimental. The unbalanced chemical equation is

$$SOCl_2(l) + H_2O(l) \rightarrow SO_2(g) + HCl(g)$$

Calculate the mass of water consumed by complete reaction of 35.0 g of $SOCl_2$.

F 40. In the "Chemistry in Focus" segment *Cars of the Future,* the claim is made that the combustion of gasoline for some cars causes about 1 lb of CO_2 to be produced for each mile traveled.

Estimate the gas mileage of a car that produces about 1 lb of CO_2 per mile traveled. Assume gasoline has a density of 0.75 g/mL and is 100% octane (C_8H_{18}). While this last part is not true, it is close enough for an estimation. The reaction can be represented by the following *unbalanced* chemical equation:

$$C_8H_{18} + O_2 \rightarrow CO_2 + H_2O$$

All even-numbered Questions and Problems have answers in the back of this book and solutions in the *Student Solutions Guide*.

9-5 Calculations Involving a Limiting Reactant

Questions

41. Imagine you are chatting with a friend who has not yet taken a chemistry course. How would you explain the concept of *limiting reactant* to her? Your textbook uses the analogy of an automobile manufacturer ordering four wheels for each engine ordered as an example. Can you think of another analogy that might help your friend to understand the concept?

42. Explain how one determines which reactant in a process is the limiting reactant. Does this depend only on the masses of the reactant present? Is the mole ratio in which the reactants combine involved?

43. Consider the equation: $2A + B \rightarrow 5C$. If 10.0 g of A reacts with 5.00 g of B, how is the limiting reactant determined? Choose the *best* answer and explain.
    a. Choose the reactant with the smallest coefficient in the balanced chemical equation. So in this case, the limiting reactant is B.
    b. Choose the reactant with the smallest mass given. So in this case, the limiting reactant is B.
    c. The mass of each reactant must be converted to moles and then compared to the ratios in the balanced chemical equation. So in this case, the limiting reactant cannot be determined without the molar masses of A and B.
    d. The mass of each reactant must be converted to moles first. The reactant with the fewest moles present is the limiting reactant. So in this case, the limiting reactant cannot be determined without the molar masses of A and B.
    e. The mass of each reactant must be divided by their coefficients in the balanced chemical equation, and the smallest number present is the limiting reactant. So in this case, there is no limiting reactant because A and B are used up perfectly.

44. According to the law of conservation of mass, mass cannot be gained or destroyed in a chemical reaction. Why can't you simply add the masses of two reactants to determine the total mass of product? Choose the *best* answer and explain.
    a. One of the reactants could be present in excess, and not all of it will be used to make the product(s).
    b. The masses of the reactants must be converted to moles first and then added.
    c. Not all chemical reactions follow the law of conservation of mass, especially ones with mixed physical states present.
    d. The masses of the two reactants cannot be added until they are each multiplied by their coefficient in the balanced equation.
    e. It is only the molar masses that are conserved in chemical reactions, not the actual mass amounts given in the laboratory.

Problems

45. For each of the following *unbalanced* reactions, suppose exactly 5.00 g of *each reactant* is taken. Determine which reactant is limiting, and also determine what mass of the excess reagent will remain after the limiting reactant is consumed.
    a. $Na_2B_4O_7(s) + H_2SO_4(aq) + H_2O(l) \rightarrow H_3BO_3(s) + Na_2SO_4(aq)$
    b. $CaC_2(s) + H_2O(l) \rightarrow Ca(OH)_2(s) + C_2H_2(g)$
    c. $NaCl(s) + H_2SO_4(l) \rightarrow HCl(g) + Na_2SO_4(s)$
    d. $SiO_2(s) + C(s) \rightarrow Si(l) + CO(g)$

46. For each of the following *unbalanced* chemical equations, suppose that exactly 5.00 g of *each* reactant is taken. Determine which reactant is limiting, and calculate what mass of each product is expected (assuming that the limiting reactant is completely consumed).
    a. $S(s) + H_2SO_4(aq) \rightarrow SO_2(g) + H_2O(l)$
    b. $MnO_2(s) + H_2SO_4(l) \rightarrow Mn(SO_4)_2(s) + H_2O(l)$
    c. $H_2S(g) + O_2(g) \rightarrow SO_2(g) + H_2O(l)$
    d. $AgNO_3(aq) + Al(s) \rightarrow Ag(s) + Al(NO_3)_3(aq)$

47. For each of the following *unbalanced* chemical equations, suppose 10.0 g of *each* reactant is taken. Show by calculation which reactant is the limiting reagent. Calculate the mass of each product that is expected.
    a. $C_3H_8(g) + O_2(g) \rightarrow CO_2(g) + H_2O(g)$
    b. $Al(s) + Cl_2(g) \rightarrow AlCl_3(s)$
    c. $NaOH(s) + CO_2(g) \rightarrow Na_2CO_3(s) + H_2O(l)$
    d. $NaHCO_3(s) + HCl(aq) \rightarrow NaCl(aq) + H_2O(l) + CO_2(g)$

48. For each of the following *unbalanced* chemical equations, suppose that exactly 1.00 g of *each* reactant is taken. Determine which reactant is limiting, and calculate what mass of the product in boldface is expected (assuming that the limiting reactant is completely consumed).
    a. $CS_2(l) + O_2(g) \rightarrow \mathbf{CO_2}(g) + SO_2(g)$
    b. $NH_3(g) + CO_2(g) \rightarrow CN_2H_4O(s) + \mathbf{H_2O}(g)$
    c. $H_2(g) + MnO_2(s) \rightarrow MnO(s) + \mathbf{H_2O}(g)$
    d. $I_2(l) + Cl_2(g) \rightarrow \mathbf{ICl}(g)$

49. For each of the following *unbalanced* chemical equations, suppose 1.00 g of *each* reactant is taken. Show by calculation which reactant is limiting. Calculate the mass of each product that is expected.
    a. $UO_2(s) + HF(aq) \rightarrow UF_4(aq) + H_2O(l)$
    b. $NaNO_3(aq) + H_2SO_4(aq) \rightarrow Na_2SO_4(aq) + HNO_3(aq)$
    c. $Zn(s) + HCl(aq) \rightarrow ZnCl_2(aq) + H_2(g)$
    d. $B(OH)_3(s) + CH_3OH(l) \rightarrow B(OCH_3)_3(s) + H_2O(l)$

50. For each of the following *unbalanced* chemical equations, suppose that exactly 15.0 g of *each* reactant is taken. Determine which reactant is limiting, and calculate what mass of each product is expected. (Assume that the limiting reactant is completely consumed.)
    a. $Al(s) + HCl(aq) \rightarrow AlCl_3(aq) + H_2(g)$
    b. $NaOH(aq) + CO_2(g) \rightarrow Na_2CO_3(aq) + H_2O(l)$
    c. $Pb(NO_3)_2(aq) + HCl(aq) \rightarrow PbCl_2(s) + HNO_3(aq)$
    d. $K(s) + I_2(s) \rightarrow KI(s)$

All even-numbered Questions and Problems have answers in the back of this book and solutions in the *Student Solutions Guide*.

51. Lead(II) carbonate, also called "white lead," was formerly used as a pigment in white paints. However, because of its toxicity, lead can no longer be used in paints intended for residential homes. Lead(II) carbonate is prepared industrially by reaction of aqueous lead(II) acetate with carbon dioxide gas. The unbalanced equation is

$$Pb(C_2H_3O_2)_2(aq) + H_2O(l) + CO_2(g) \rightarrow PbCO_3(s) + HC_2H_3O_2(aq)$$

Suppose an aqueous solution containing 1.25 g of lead(II) acetate is treated with 5.95 g of carbon dioxide. Calculate the theoretical yield of lead carbonate.

52. Copper(II) sulfate has been used extensively as a fungicide (kills fungus) and herbicide (kills plants). Copper(II) sulfate can be prepared in the laboratory by reaction of copper(II) oxide with sulfuric acid. The unbalanced equation is

$$CuO(s) + H_2SO_4(aq) \rightarrow CuSO_4(aq) + H_2O(l)$$

If 2.49 g of copper(II) oxide is treated with 5.05 g of pure sulfuric acid, which reactant would limit the quantity of copper(II) sulfate that could be produced?

53. Lead(II) oxide from an ore can be reduced to elemental lead by heating in a furnace with carbon.

$$PbO(s) + C(s) \rightarrow Pb(l) + CO(g)$$

Calculate the expected yield of lead if 50.0 kg of lead oxide is heated with 50.0 kg of carbon.

54. If steel wool (iron) is heated until it glows and is placed in a bottle containing pure oxygen, the iron reacts spectacularly to produce iron(III) oxide.

$$Fe(s) + O_2(g) \rightarrow Fe_2O_3(s)$$

If 1.25 g of iron is heated and placed in a bottle containing 0.0204 mole of oxygen gas, what mass of iron(III) oxide is produced?

55. A common method for determining how much chloride ion is present in a sample is to precipitate the chloride from an aqueous solution of the sample with silver nitrate solution and then to weigh the silver chloride that results. The balanced net ionic reaction is

$$Ag^+(aq) + Cl^-(aq) \rightarrow AgCl(s)$$

Suppose a 5.45-g sample of pure sodium chloride is dissolved in water and is then treated with a solution containing 1.15 g of silver nitrate. Will this quantity of silver nitrate be capable of precipitating *all* the chloride ion from the sodium chloride sample?

56. Although many sulfate salts are soluble in water, calcium sulfate is not (Table 7.1). Therefore, a solution of calcium chloride will react with sodium sulfate solution to produce a precipitate of calcium sulfate. The balanced equation is

$$CaCl_2(aq) + Na_2SO_4(aq) \rightarrow CaSO_4(s) + 2NaCl(aq)$$

If a solution containing 5.21 g of calcium chloride is combined with a solution containing 4.95 g of sodium sulfate, which is the limiting reactant? Which reactant is present in excess?

57. Hydrogen peroxide is used as a cleaning agent in the treatment of cuts and abrasions for several reasons. It is an oxidizing agent that can directly kill many microorganisms; it decomposes upon contact with blood, releasing elemental oxygen gas (which inhibits the growth of anaerobic microorganisms); and it foams upon contact with blood, which provides a cleansing action. In the laboratory, small quantities of hydrogen peroxide can be prepared by the action of an acid on an alkaline earth metal peroxide, such as barium peroxide.

$$BaO_2(s) + 2HCl(aq) \rightarrow H_2O_2(aq) + BaCl_2(aq)$$

What amount of hydrogen peroxide should result when 1.50 g of barium peroxide is treated with 25.0 mL of hydrochloric acid solution containing 0.0272 g of HCl per mL?

58. Silicon carbide, SiC, is one of the hardest materials known. Surpassed in hardness only by diamond, it is sometimes known commercially as carborundum. Silicon carbide is used primarily as an abrasive for sandpaper and is manufactured by heating common sand (silicon dioxide, SiO_2) with carbon in a furnace.

$$SiO_2(s) + C(s) \rightarrow CO(g) + SiC(s)$$

What mass of silicon carbide should result when 1.0 kg of pure sand is heated with an excess of carbon?

9-6 Percent Yield

Questions

59. Your text talks about several sorts of "yield" when experiments are performed in the laboratory. Students often confuse these terms. Define, compare, and contrast what are meant by *theoretical* yield, *actual* yield, and *percent* yield.

60. The text explains that one reason why the actual yield for a reaction may be less than the theoretical yield is side reactions. Suggest some other reasons why the percent yield for a reaction might not be 100%.

61. According to his prelaboratory theoretical yield calculations, a student's experiment should have produced 1.44 g of magnesium oxide. When he weighed his product after reaction, only 1.23 g of magnesium oxide was present. What is the student's percent yield?

62. An air bag is deployed by utilizing the following reaction (the nitrogen gas produced inflates the air bag):

$$2NaN_3(s) \rightarrow 2Na(s) + 3N_2(g)$$

If 10.5 g of NaN_3 is decomposed, what theoretical mass of sodium should be produced? If only 2.84 g of sodium is actually collected, what is the percent yield?

Problems

63. The compound sodium thiosulfate pentahydrate, $Na_2S_2O_3 \cdot 5H_2O$, is important commercially to the photography business as "hypo," because it has the ability to dissolve unreacted silver salts from photographic film during development. Sodium thiosulfate pentahydrate can be produced by boiling elemental sulfur in an aqueous solution of sodium sulfite.

$$S_8(s) + Na_2SO_3(aq) + H_2O(l) \rightarrow Na_2S_2O_3 \cdot 5H_2O(s) \quad \text{(unbalanced)}$$

All even-numbered Questions and Problems have answers in the back of this book and solutions in the *Student Solutions Guide*.

What is the theoretical yield of sodium thiosulfate pentahydrate when 3.25 g of sulfur is boiled with 13.1 g of sodium sulfite? Sodium thiosulfate pentahydrate is very soluble in water. What is the percent yield of the synthesis if a student doing this experiment is able to isolate (collect) only 5.26 g of the product?

64. Alkali metal hydroxides are sometimes used to "scrub" excess carbon dioxide from the air in closed spaces (such as submarines and spacecraft). For example, lithium hydroxide reacts with carbon dioxide according to the unbalanced chemical equation

$$LiOH(s) + CO_2(g) \rightarrow Li_2CO_3(s) + H_2O(g)$$

Suppose a lithium hydroxide canister contains 155 g of $LiOH(s)$. What mass of $CO_2(g)$ will the canister be able to absorb? If it is found that after 24 hours of use the canister has absorbed 102 g of carbon dioxide, what percentage of its capacity has been reached?

65. Although they were formerly called the inert gases, at least the heavier elements of Group 8 do form relatively stable compounds. For example, xenon combines directly with elemental fluorine at elevated temperatures in the presence of a nickel catalyst.

$$Xe(g) + 2F_2(g) \rightarrow XeF_4(s)$$

What is the theoretical mass of xenon tetrafluoride that should form when 130. g of xenon is reacted with 100. g of F_2? What is the percent yield if only 145 g of XeF_4 is actually isolated?

66. Solid copper can be produced by passing gaseous ammonia over solid copper(II) oxide at high temperatures.

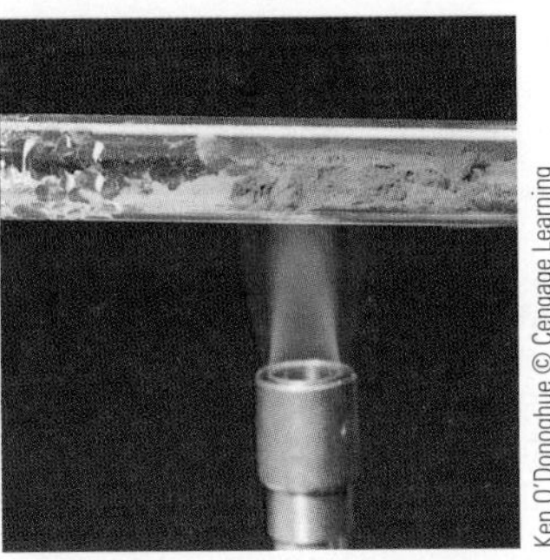
Ken O'Donoghue © Cengage Learning

The other products of the reaction are nitrogen gas and water vapor. The balanced equation for this reaction is:

$$2NH_3(g) + 3CuO(s) \rightarrow N_2(g) + 3Cu(s) + 3H_2O(g)$$

What is the theoretical yield of solid copper that should form when 18.1 g of NH_3 is reacted with 90.4 g of CuO? If only 45.3 g of copper is actually collected, what is the percent yield?

Additional Problems

67. Natural waters often contain relatively high levels of calcium ion, Ca^{2+}, and hydrogen carbonate ion (bicarbonate), HCO_3^-, from the leaching of minerals into the water. When such water is used commercially or in the home, heating of the water leads to the formation of solid calcium carbonate, $CaCO_3$, which forms a deposit ("scale") on the interior of boilers, pipes, and other plumbing fixtures.

$$Ca(HCO_3)_2(aq) \rightarrow CaCO_3(s) + CO_2(g) + H_2O(l)$$

If a sample of well water contains 2.0×10^{-3} mg of $Ca(HCO_3)_2$ per milliliter, what mass of $CaCO_3$ scale would 1.0 mL of this water be capable of depositing?

68. One process for the commercial production of baking soda (sodium hydrogen carbonate) involves the following reaction, in which the carbon dioxide is used in its solid form ("dry ice") both to serve as a source of reactant and to cool the reaction system to a temperature low enough for the sodium hydrogen carbonate to precipitate:

$$NaCl(aq) + NH_3(aq) + H_2O(l) + CO_2(s) \rightarrow NH_4Cl(aq) + NaHCO_3(s)$$

Because they are relatively cheap, sodium chloride and water are typically present in excess. What is the expected yield of $NaHCO_3$ when one performs such a synthesis using 10.0 g of ammonia and 15.0 g of dry ice, with an excess of NaCl and water?

69. A favorite demonstration among chemistry instructors, to show that the properties of a compound differ from those of its constituent elements, involves iron filings and powdered sulfur. If the instructor takes samples of iron and sulfur and just mixes them together, the two elements can be separated from one another with a magnet (iron is attracted to a magnet, sulfur is not). If the instructor then combines and *heats* the mixture of iron and sulfur, a reaction takes place and the elements combine to form iron(II) sulfide (which is not attracted by a magnet).

$$Fe(s) + S(s) \rightarrow FeS(s)$$

Suppose 5.25 g of iron filings is combined with 12.7 g of sulfur. What is the theoretical yield of iron(II) sulfide?

70. When the sugar glucose, $C_6H_{12}O_6$, is burned in air, carbon dioxide and water vapor are produced. Write the balanced chemical equation for this process, and calculate the theoretical yield of carbon dioxide when 1.00 g of glucose is burned completely.

71. When elemental copper is strongly heated with sulfur, a mixture of CuS and Cu_2S is produced, with CuS predominating.

$$Cu(s) + S(s) \rightarrow CuS(s)$$
$$2Cu(s) + S(s) \rightarrow Cu_2S(s)$$

What is the theoretical yield of CuS when 31.8 g of $Cu(s)$ is heated with 50.0 g of S? (Assume only CuS is produced in the reaction.) What is the percent yield of CuS if only 40.0 g of CuS can be isolated from the mixture?

72. Barium chloride solutions are used in chemical analysis for the quantitative precipitation of sulfate ion from solution.

$$Ba^{2+}(aq) + SO_4^{2-}(aq) \rightarrow BaSO_4(s)$$

Suppose a solution is known to contain on the order of 150 mg of sulfate ion. What mass of barium chloride should be added to guarantee precipitation of all the sulfate ion?

All even-numbered Questions and Problems have answers in the back of this book and solutions in the *Student Solutions Guide*.

73. The traditional method of analysis for the amount of chloride ion present in a sample is to dissolve the sample in water and then slowly to add a solution of silver nitrate. Silver chloride is very insoluble in water, and by adding a slight excess of silver nitrate, it is possible to effectively remove all chloride ion from the sample.

$$Ag^+(aq) + Cl^+(aq) \rightarrow AgCl(s)$$

Suppose a 1.054-g sample is known to contain 10.3% chloride ion by mass. What mass of silver nitrate must be used to completely precipitate the chloride ion from the sample? What mass of silver chloride will be obtained?

74. For each of the following reactions, give the balanced equation for the reaction and state the meaning of the equation in terms of numbers of *individual molecules* and in terms of *moles* of molecules.

a. $UO_2(s) + HF(aq) \rightarrow UF_4(aq) + H_2O(l)$
b. $NaC_2H_3O_2(aq) + H_2SO_4(aq) \rightarrow Na_2SO_4(aq) + HC_2H_3O_2(aq)$
c. $Mg(s) + HCl(aq) \rightarrow MgCl_2(aq) + H_2(g)$
d. $B_2O_3(s) + H_2O(l) \rightarrow B(OH)_3(aq)$

75. True or false? For the reaction represented by the balanced chemical equation

$$Mg(OH)_2(aq) + 2HCl(aq) \rightarrow 2H_2O(l) + MgCl_2(aq)$$

for 0.40 mole of $Mg(OH)_2$, 0.20 mol of HCl will be needed.

76. Consider the balanced equation

$$C_3H_8(g) + 5O_2(g) \rightarrow 3CO_2(g) + 4H_2O(g)$$

What mole ratio enables you to calculate the number of moles of oxygen needed to react exactly with a given number of moles of $C_3H_8(g)$? What mole ratios enable you to calculate how many moles of each product form from a given number of moles of C_3H_8?

77. For each of the following balanced reactions, calculate how many *moles of each product* would be produced by complete conversion of *0.50 mole* of the reactant indicated in boldface. Indicate clearly the mole ratio used for the conversion.

a. $\mathbf{2H_2O_2}(l) \rightarrow 2H_2O(l) + O_2(g)$
b. $\mathbf{2KClO_3}(s) \rightarrow 2KCl(s) + 3O_2(g)$
c. $\mathbf{2Al}(s) + 6HCl(aq) \rightarrow 2AlCl_3(aq) + 3H_2(g)$
d. $\mathbf{C_3H_8}(g) + 5O_2(g) \rightarrow 3CO_2(g) + 4H_2O(g)$

78. For each of the following balanced equations, indicate how many *moles of the product* could be produced by complete reaction of *1.00 g* of the reactant indicated in boldface. Indicate clearly the mole ratio used for the conversion.

a. $\mathbf{NH_3}(g) + HCl(g) \rightarrow NH_4Cl(s)$
b. $\mathbf{CaO}(s) + CO_2(g) \rightarrow CaCO_3(s)$
c. $\mathbf{4Na}(s) + O_2(g) \rightarrow 2Na_2O(s)$
d. $\mathbf{2P}(s) + 3Cl_2(g) \rightarrow 2PCl_3(l)$

79. Using the average atomic masses given inside the front cover of the text, calculate how many *moles* of each substance the following masses represent.

a. 4.21 g of copper(II) sulfate
b. 7.94 g of barium nitrate
c. 1.24 mg of water
d. 9.79 g of tungsten
e. 1.45 lb of sulfur
f. 4.65 g of ethyl alcohol, C_2H_5OH
g. 12.01 g of carbon

80. Using the average atomic masses given inside the front cover of the text, calculate the *mass in grams* of each of the following samples.

a. 5.0 moles of nitric acid
b. 0.000305 mole of mercury
c. 2.31×10^{-5} mole of potassium chromate
d. 10.5 moles of aluminum chloride
e. 4.9×10^4 moles of sulfur hexafluoride
f. 125 moles of ammonia
g. 0.01205 mole of sodium peroxide

81. For each of the following *incomplete* and *unbalanced* equations, indicate how many *moles* of the *second reactant* would be required to react completely with *0.145 mole* of the *first reactant.*

a. $BaCl_2(aq) + H_2SO_4(aq) \rightarrow$
b. $AgNO_3(aq) + NaCl(aq) \rightarrow$
c. $Pb(NO_3)_2(aq) + Na_2CO_3(aq) \rightarrow$
d. $C_3H_8(g) + O_2(g) \rightarrow$

82. One step in the commercial production of sulfuric acid, H_2SO_4, involves the conversion of sulfur dioxide, SO_2, into sulfur trioxide, SO_3.

$$2SO_2(g) + O_2(g) \rightarrow 2SO_3(g)$$

If 150 kg of SO_2 reacts completely, what mass of SO_3 should result?

83. Many metals occur naturally as sulfide compounds; examples include ZnS and CoS. Air pollution often accompanies the processing of these ores, because toxic sulfur dioxide is released as the ore is converted from the sulfide to the oxide by roasting (smelting). For example, consider the unbalanced equation for the roasting reaction for zinc:

$$ZnS(s) + O_2(g) \rightarrow ZnO(s) + SO_2(g)$$

How many kilograms of sulfur dioxide are produced when 1.0×10^2 kg of ZnS is roasted in excess oxygen by this process?

84. If sodium peroxide is added to water, elemental oxygen gas is generated:

$$Na_2O_2(s) + H_2O(l) \rightarrow NaOH(aq) + O_2(g)$$

Suppose 3.25 g of sodium peroxide is added to a large excess of water. What mass of oxygen gas will be produced?

85. When elemental copper is placed in a solution of silver nitrate, the following oxidation–reduction reaction takes place, forming elemental silver:

$$Cu(s) + 2AgNO_3(aq) \rightarrow Cu(NO_3)_2(aq) + 2Ag(s)$$

What mass of copper is required to remove all the silver from a silver nitrate solution containing 1.95 mg of silver nitrate?

86. When small quantities of elemental hydrogen gas are needed for laboratory work, the hydrogen is often generated by chemical reaction of a metal with acid. For example, zinc reacts with hydrochloric acid, releasing gaseous elemental hydrogen:

$$Zn(s) + 2HCl(aq) \rightarrow ZnCl_2(aq) + H_2(g)$$

What mass of hydrogen gas is produced when 2.50 g of zinc is reacted with excess aqueous hydrochloric acid?

All even-numbered Questions and Problems have answers in the back of this book and solutions in the *Student Solutions Guide.*

87. The gaseous hydrocarbon acetylene, C_2H_2, is used in welders' torches because of the large amount of heat released when acetylene burns with oxygen.

$$2C_2H_2(g) + 5O_2(g) \rightarrow 4CO_2(g) + 2H_2O(g)$$

How many grams of oxygen gas are needed for the complete combustion of 150 g of acetylene?

88. For each of the following *unbalanced* chemical equations, suppose exactly 5.0 g of each reactant is taken. Determine which reactant is limiting, and calculate what mass of each product is expected, assuming that the limiting reactant is completely consumed.

a. $Na(s) + Br_2(l) \rightarrow NaBr(s)$
b. $Zn(s) + CuSO_4(aq) \rightarrow ZnSO_4(aq) + Cu(s)$
c. $NH_4Cl(aq) + NaOH(aq) \rightarrow NH_3(g) + H_2O(l) + NaCl(aq)$
d. $Fe_2O_3(s) + CO(g) \rightarrow Fe(s) + CO_2(g)$

89. For each of the following *unbalanced* chemical equations, suppose 25.0 g of each reactant is taken. Show by calculation which reactant is limiting. Calculate the theoretical yield in grams of the product in boldface.

a. $C_2H_5OH(l) + O_2(g) \rightarrow \mathbf{CO_2}(g) + H_2O(l)$
b. $N_2(g) + O_2(g) \rightarrow \mathbf{NO}(g)$
c. $NaClO_2(aq) + Cl_2(g) \rightarrow ClO_2(g) + \mathbf{NaCl}(aq)$
d. $H_2(g) + N_2(g) \rightarrow \mathbf{NH_3}(g)$

90. Hydrazine, N_2H_4, emits a large quantity of energy when it reacts with oxygen, which has led to hydrazine's use as a fuel for rockets:

$$N_2H_4(l) + O_2(g) \rightarrow N_2(g) + 2H_2O(g)$$

How many moles of each of the gaseous products are produced when 20.0 g of pure hydrazine is ignited in the presence of 20.0 g of pure oxygen? How many grams of each product are produced?

91. Consider the following reaction for lighting a match:

$$3P_4(s) + 10KClO_3(s) \xrightarrow{\text{heat}} 10KCl(s) + 6P_2O_5(s)$$

a. If 3.50 moles of P_2O_5 were produced in this reaction, how many moles of KCl were produced?
b. If 3.50 moles of P_2O_5 were produced in this reaction, how many moles of P_4 were required?

92. Before going to lab, a student read in his lab manual that the percent yield for a difficult reaction to be studied was likely to be only 40.% of the theoretical yield. The student's prelab stoichiometric calculations predict that the theoretical yield should be 12.5 g. What is the student's actual yield likely to be?

ChemWork Problems

These multiconcept problems (and additional ones) are found interactively online with the same type of assistance a student would get from an instructor.

93. Consider the following unbalanced chemical equation for the combustion of pentane (C_5H_{12}):

$$C_5H_{12}(l) + O_2(g) \rightarrow CO_2(g) + H_2O(l)$$

If a 20.4-gram sample of pentane is burned in excess oxygen, what mass of water can be produced, assuming 100% yield?

94. A 0.4230-g sample of impure sodium nitrate (contains sodium nitrate plus inert ingredients) was heated, converting all the sodium nitrate to 0.2339 g of sodium nitrite and oxygen gas. Determine the percent of sodium nitrate in the original sample.

95. Consider the following *unbalanced* chemical equation.

$$LiOH(s) + CO_2(g) \rightarrow Li_2CO_3(s) + H_2O(l)$$

If 67.4 g of lithium hydroxide reacts with excess carbon dioxide, what mass of lithium carbonate will be produced?

96. Over the years, the thermite reaction has been used for welding railroad rails, in incendiary bombs, and to ignite solid fuel rocket motors. The reaction is

$$Fe_2O_3(s) + 2Al(s) \rightarrow 2Fe(l) + Al_2O_3(s)$$

a. What mass of iron(III) oxide must be used to produce 25.69 g of iron?
b. What mass of aluminum must be used to produce 25.69 g of iron?
c. What is the maximum mass of aluminum oxide that could be produced along with 25.69 g of iron?

97. Consider the following *unbalanced* chemical equation:

$$H_2S(g) + O_2(g) \rightarrow SO_2(g) + H_2O(g)$$

Determine the maximum number of moles of SO_2 produced from 8.0 moles of H_2S and 3.0 moles of O_2.

98. Ammonia gas reacts with sodium metal to form sodium amide ($NaNH_2$) and hydrogen gas. The *unbalanced* chemical equation for this reaction is as follows:

$$NH_3(g) + Na(s) \rightarrow NaNH_2(s) + H_2(g)$$

Assuming that you start with 32.8 g of ammonia gas and 16.6 g of sodium metal and assuming that the reaction goes to completion, determine the mass (in grams) of each product.

99. Sulfur dioxide gas reacts with sodium hydroxide to form sodium sulfite and water. The *unbalanced* chemical equation for this reaction is as follows:

$$SO_2(g) + NaOH(s) \rightarrow Na_2SO_3(s) + H_2O(l)$$

Assuming you react 38.3 g of sulfur dioxide with 32.8 g of sodium hydroxide and assuming that the reaction goes to completion, calculate the mass of each product formed.

100. The production capacity for acrylonitrile (C_3H_3N) in the United States is over 2 billion pounds per year. Acrylonitrile, the building block for polyacrylonitrile fibers and a variety of plastics, is produced from gaseous propylene, ammonia, and oxygen:

$$2C_3H_6(g) + 2NH_3(g) + 3O_2(g) \rightarrow 2C_3H_3N(g) + 6H_2O(g)$$

a. Assuming 100% yield, determine the mass of acrylonitrile which can be produced from the mixture below:

Mass	Reactant
5.23×10^2 g	propylene
5.00×10^2 g	ammonia
1.00×10^3 g	oxygen

b. What mass of water is formed from your mixture?
c. Calculate the mass (in grams) of each reactant after the reaction is complete.

All even-numbered Questions and Problems have answers in the back of this book and solutions in the *Student Solutions Guide*.

Questions

1. What does the average *atomic mass* of an element represent? What unit is used for average atomic mass? Express the atomic mass unit in grams. Why is the average atomic mass for an element typically *not* a whole number?
2. Perhaps the most important concept in introductory chemistry concerns what a *mole* of a substance represents. The mole concept will come up again and again in later chapters in this book. What does one mole of a substance represent on a microscopic, atomic basis? What does one mole of a substance represent on a macroscopic, mass basis? Why have chemists defined the mole in this manner?
3. How do we know that 16.00 g of oxygen contains the same number of atoms as does 12.01 g of carbon, and that 22.99 g of sodium contains the same number of atoms as each of these? How do we know that 106.0 g of Na_2CO_3 contains the same number of carbon atoms as does 12.01 g of carbon, but three times as many oxygen atoms as in 16.00 g of oxygen, and twice as many sodium atoms as in 22.99 g of sodium?
4. Define *molar mass*. Using H_3PO_4 as an example, calculate the molar mass from the atomic masses of the elements.
5. What is meant by the *percent composition* by mass for a compound? Describe in general terms how this information is obtained by experiment for new compounds. How can this information be calculated for known compounds?
6. Define, compare, and contrast what are meant by the *empirical* and *molecular* formulas for a substance. What does each of these formulas tell us about a compound? What information must be known for a compound before the molecular formula can be determined? Why is the molecular formula an *integer multiple* of the empirical formula?
7. When chemistry teachers prepare an exam question on determining the empirical formula of a compound, they usually take a known compound and calculate the percent composition of the compound from the formula. They then give students this percent composition data and have the students calculate the original formula. Using a compound of *your* choice, first use the molecular formula of the compound to calculate the percent composition of the compound. Then use this percent composition data to calculate the empirical formula of the compound.
8. Rather than giving students straight percent composition data for determining the empirical formula of a compound (see Question 7), sometimes chemistry teachers will try to emphasize the experimental nature of formula determination by converting the percent composition data into actual experimental masses. For example, the compound CH_4 contains 74.87% carbon by mass. Rather than giving students the data in this form, a teacher might say instead, "When 1.000 g of a compound was analyzed, it was found to contain 0.7487 g of carbon, with the remainder consisting of hydrogen." Using the compound you chose for Question 7, and the percent composition data you calculated, reword your data as suggested in this problem in terms of actual "experimental" masses. Then from these masses, calculate the empirical formula of your compound.
9. Balanced chemical equations give us information in terms of individual molecules reacting in the proportions indicated by the coefficients, and also in terms of macroscopic amounts (that is, moles). Write a balanced chemical equation of your choice, and interpret in words the meaning of the equation on the molecular and macroscopic levels.
10. Consider the *unbalanced* equation for the combustion of propane:

$$C_3H_8(g) + O_2(g) \rightarrow CO_2(g) + H_2O(g)$$

First, balance the equation. Then, for a given amount of propane, write the mole ratios that would enable you to calculate the number of moles of each product as well as the number of moles of O_2 that would be involved in a complete reaction. Finally, show how these mole ratios would be applied if 0.55 mole of propane is combusted.

11. In the practice of chemistry one of the most important calculations concerns the masses of products expected when particular masses of reactants are used in an experiment. For example, chemists judge the practicality and efficiency of a reaction by seeing how close the amount of product actually obtained is to the expected amount. Using a balanced chemical equation and an amount of starting material of your choice, summarize and illustrate the various steps needed in such a calculation for the expected amount of product.
12. What is meant by a *limiting reactant* in a particular reaction? In what way is the reaction "limited"? What does it mean to say that one or more of the reactants are present *in excess?* What happens to a reaction when the limiting reactant is used up?
13. For a balanced chemical equation of your choice, and using 25.0 g of each of the reactants in your equation, illustrate and explain how you would determine which reactant is the limiting reactant. Indicate *clearly* in your discussion how the choice of limiting reactant follows from your calculations.
14. What do we mean by the *theoretical yield* for a reaction? What is meant by the *actual yield?* Why might the actual yield for an experiment be *less* than the theoretical yield? Can the actual yield be *more* than the theoretical yield?

Problems

15. Consider 2.45-g samples of each of the following elements or compounds. Calculate the number of moles of the element or compound present in each sample.
    a. $Fe_2O_3(s)$
    b. $P_4(s)$
    c. $Cl_2(g)$
    d. $Hg_2O(s)$
    e. $HgO(s)$
    f. $Ca(NO_3)_2(s)$
    g. $C_3H_8(g)$
    h. $Al_2(SO_4)_3(s)$

All even-numbered Questions and Problems have answers in the back of this book and solutions in the *Student Solutions Guide*.

16. Calculate the percent by mass of the element whose symbol occurs *first* in the following compounds' formulas.

  a. $C_6H_6(l)$
  b. $Na_2SO_4(s)$
  c. $CS_2(l)$
  d. $AlCl_3(s)$
  e. $Cu_2O(s)$
  f. $CuO(s)$
  g. $Co_2O_3(s)$
  h. $C_6H_{12}O_6(s)$

17. A compound was analyzed and was found to have the following percent composition by mass: sodium, 43.38%; carbon, 11.33%; oxygen, 45.29%. Determine the empirical formula of the compound.

18. For each of the following balanced equations, calculate how many grams of each product would form if 12.5 g of the reactant listed *first* in the equation reacts completely (there is an excess of the second reactant).

  a. $SiC(s) + 2Cl_2(g) \rightarrow SiCl_4(l) + C(s)$
  b. $Li_2O(s) + H_2O(l) \rightarrow 2LiOH(aq)$
  c. $2Na_2O_2(s) + 2H_2O(l) \rightarrow 4NaOH(aq) + O_2(g)$
  d. $SnO_2(s) + 2H_2(g) \rightarrow Sn(s) + 2H_2O(l)$

19. For the reactions in Question 18, suppose that instead of an *excess* of the second reactant, only 5.00 g of the second reactant is available. Indicate which substance is the limiting reactant in each reaction.

20. Depending on the concentration of oxygen gas present when carbon is burned, either of two oxides may result.

$2C(s) + O_2(g) \rightarrow 2CO(g)$ (restricted amount of oxygen)

$C(s) + O_2(g) \rightarrow CO_2(g)$ (unrestricted amount of oxygen)

Suppose that experiments are performed in which duplicate 5.00-g samples of carbon are burned under both conditions. Calculate the theoretical yield of product for each experiment.

21. A traditional analysis for samples containing calcium ion was to precipitate the calcium ion with sodium oxalate ($Na_2C_2O_4$) solution and then to collect and weigh either the calcium oxalate itself or the calcium oxide produced by heating the oxalate precipitate:

$$Ca^{2+}(aq) + C_2O_4^{2-}(aq) \rightarrow CaC_2O_4(s)$$

Suppose a sample contained 0.1014 g of calcium ion. What theoretical yield of calcium oxalate would be expected? If only 0.2995 g of calcium oxalate is collected, what percentage of the theoretical yield does that represent?

All even-numbered Questions and Problems have answers in the back of this book and solutions in the *Student Solutions Guide.*

10 Energy

CHAPTER 10

A hummingbird exerts a great deal of energy in order to hover.

Image by © Raven Regan/Design Pics/Corbis

Energy is at the center of our very existence as individuals and as a society. The food that we eat furnishes the energy to live, work, and play, just as the coal and oil consumed by manufacturing and transportation systems power our modern industrialized civilization.

Huge quantities of carbon-based fossil fuels have been available for the taking. This abundance of fuels has led to a world society with a huge appetite for energy, consuming millions of barrels of petroleum every day. We are now dangerously dependent on the dwindling supplies of oil, and this dependence is an important source of tension among nations in today's world. In an incredibly short time we have moved from a period of ample and cheap supplies of petroleum to one of high prices and uncertain supplies. If our present standard of living is to be maintained, we must find alternatives to petroleum. To do this, we need to know the relationship between chemistry and energy, which we explore in this chapter.

Energy is a factor in all human activity. Tetra Images/Superstock

10-1 The Nature of Energy

OBJECTIVE **To understand the general properties of energy.**

Although energy is a familiar concept, it is difficult to define precisely. For our purposes we will define **energy** as *the ability to do work or produce heat.* We will define these terms below.

Energy can be classified as either potential or kinetic energy. **Potential energy** is energy due to position or composition. For example, water behind a dam has potential energy that can be converted to work when the water flows down through turbines, thereby creating electricity. Attractive and repulsive forces also lead to potential energy. The energy released when gasoline is burned results from differences in attractive forces between the nuclei and electrons in the reactants and products. The **kinetic energy** of an object is energy due to the motion of the object and depends on the mass of the object m and its velocity v: $\text{KE} = \frac{1}{2}mv^2$.

One of the most important characteristics of energy is that it is conserved. The **law of conservation of energy** states *that energy can be converted from one form to another but can be neither created nor destroyed.* That is, the energy of the universe is constant.

Although the energy of the universe is constant, it can be readily converted from one form to another. Consider the two balls in Fig. 10.1(a). Ball A, because of its initially higher position, has more potential energy than ball B. When ball A is released, it moves down the hill and strikes ball B. Eventually, the arrangement shown in Fig. 10.1(b) is achieved. What has happened in going from the initial to the final arrangement? The potential energy of A has decreased because its position was lowered. However, this energy cannot disappear. Where is the energy lost by A?

Initially, the potential energy of A is changed to kinetic energy as the ball rolls down the hill. Part of this energy is transferred to B, causing it to be raised to a higher final position. Thus the potential energy of B has been increased, which means that **work** (force acting over a distance) has been performed on B. Because the final position of B is lower than the original position of A, however, some of the energy is still unaccounted for. Both balls in their final positions are at rest, so the missing energy cannot be attributed to their motions.

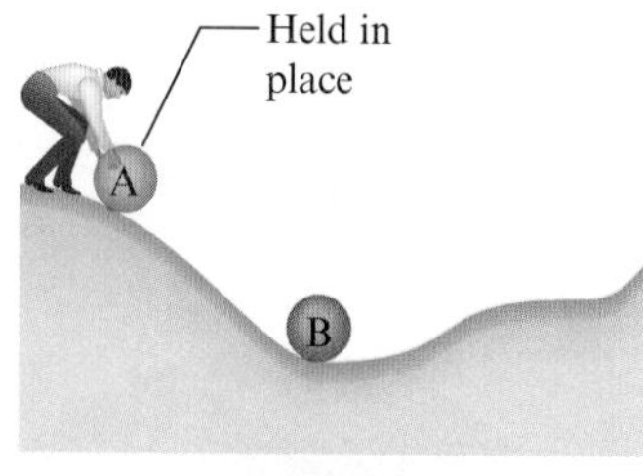

Initial

a

In the initial positions, ball A has a higher potential energy than ball B.

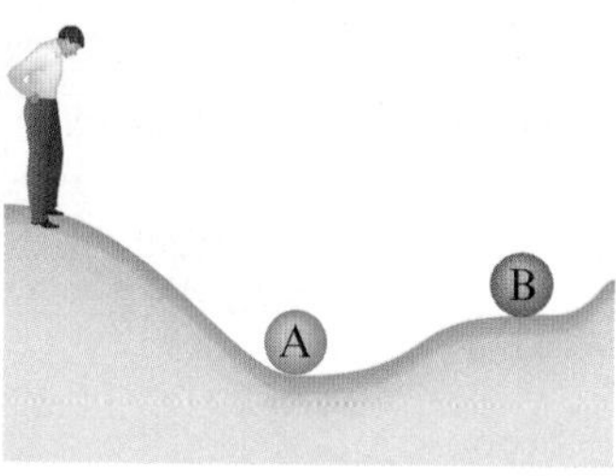

Final

b

After A has rolled down the hill, the potential energy lost by A has been converted to random motions of the components of the hill (frictional heating) and to an increase in the potential energy of B.

Figure 10.1

What has happened to the remaining energy? The answer lies in the interaction between the hill's surface and the ball. As ball A rolls down the hill, some of its kinetic energy is transferred to the surface of the hill as heat. This transfer of energy is called *frictional heating.* The temperature of the hill increases very slightly as the ball rolls down. Thus the energy stored in A in its original position (potential energy) is distributed to B through work and to the surface of the hill by heat.

Imagine that we perform this same experiment several times, varying the surface of the hill from very smooth to very rough. In rolling to the bottom of the hill (see Fig. 10.1), A always loses the same amount of energy because its position always changes by exactly the same amount. The way that this energy transfer is divided between work and heat, however, depends on the specific conditions—the *pathway.* For example, the surface of the hill might be so rough that the energy of A is expended completely through frictional heating: A is moving so slowly when it hits B that it cannot move B to the next level. In this case, no work is done. Regardless of the condition of the hill's surface, the *total energy* transferred will be constant, although the amounts of heat and work will differ. Energy change is independent of the pathway, whereas work and heat are both dependent on the pathway.

This brings us to a very important idea, the state function. A **state function** is a property of the system that changes independently of its pathway. Let's consider a nonchemical example. Suppose you are traveling from Chicago to Denver. Which of the following are state functions?

- Distance traveled
- Change in elevation

Because the distance traveled depends on the route taken (that is, the *pathway* between Chicago and Denver), it is *not* a state function. On the other hand, the change in elevation depends only on the difference between Denver's elevation (5280 ft) and Chicago's elevation (580 ft). The change in elevation is always 5280 ft − 580 ft = 4700 ft; it does not depend on the route taken between the two cities.

We can also learn about state functions from the example illustrated in Fig. 10.1. Because ball A always goes from its initial position on the hill to the bottom of the hill, its energy change is always the same, regardless of whether the hill is smooth or bumpy. This energy is a state function—a given change in energy is independent of the pathway of the process. In contrast, work and heat are *not* state functions. For a given change in the position of A, a smooth hill produces more work and less heat than a rough hill does. That is, for a given change in the position of A, the change in energy is always the same (state function) but the way the resulting energy is distributed as heat or work depends on the nature of the hill's surface (heat and work are not state functions).

Critical Thinking

What if energy were not conserved? How would this affect our lives?

10-2 Temperature and Heat

OBJECTIVE **To understand the concepts of temperature and heat.**

What does the temperature of a substance tell us about that substance? Put another way, how is warm water different from cold water? The answer lies in the motions of the water molecules. **Temperature** is a *measure of the random motions of the components of a substance.* That is, the H_2O molecules in warm water are moving around more rapidly than the H_2O molecules in cold water.

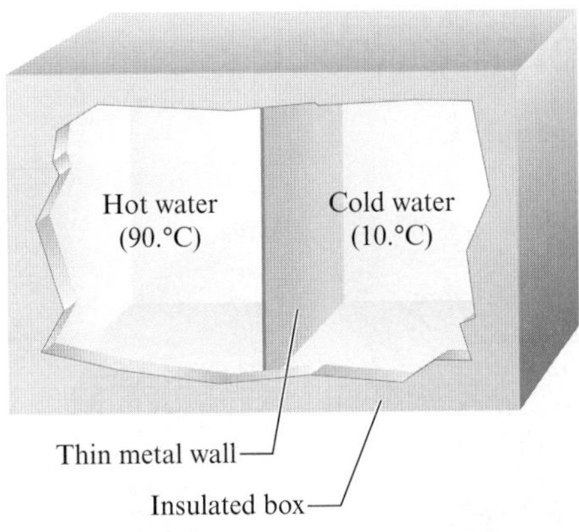

Figure 10.2 ▶ Equal masses of hot water and cold water separated by a thin metal wall in an insulated box.

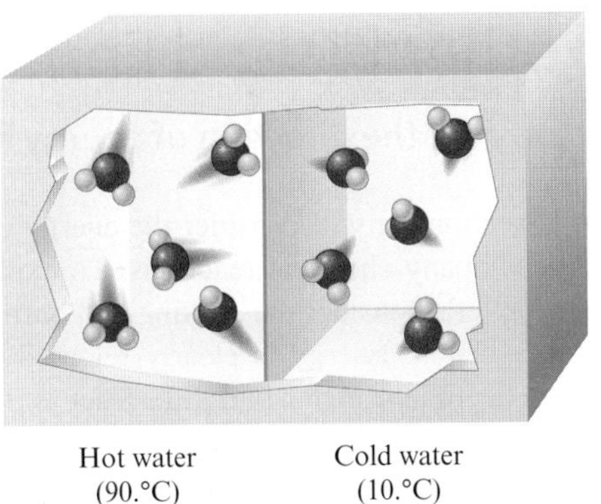

Figure 10.3 ▶ The H_2O molecules in hot water have much greater random motions than the H_2O molecules in cold water.

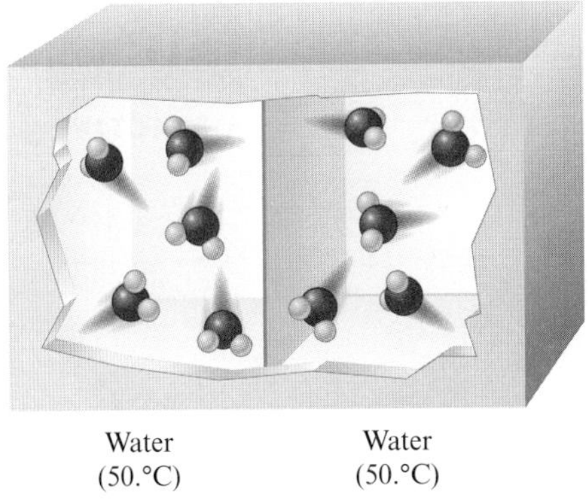

Figure 10.4 ▶ The water samples now have the same temperature (50. °C) and have the same random motions.

Consider an experiment in which we place 1.00 kg of hot water (90. °C) next to 1.00 kg of cold water (10. °C) in an insulated box. The water samples are separated from each other by a thin metal plate (Fig. 10.2). You already know what will happen: the hot water will cool down and the cold water will warm up.

Assuming that no energy is lost to the air, can we determine the final temperature of the two samples of water? Let's consider how to think about this problem.

First picture what is happening. Remember that the H_2O molecules in the hot water are moving faster than those in the cold water (Fig. 10.3). As a result, energy will be transferred through the metal wall from the hot water to the cold water. This energy transfer will cause the H_2O molecules in the hot water to slow down and the H_2O molecules in the cold water to speed up.

Thus we have a transfer of energy from the hot water to the cold water. This flow of energy is called heat. **Heat** can be defined as a *flow of energy due to a temperature difference.* What will eventually happen? The two water samples will reach the same temperature (Fig. 10.4). At this point, how does the energy lost by the hot water compare to the energy gained by the cold water? They must be the same (remember that energy is conserved).

We conclude that the final temperature is the average of the original temperatures:

$$T_{\text{final}} = \frac{T_{\substack{\text{hot}\\\text{initial}}} + T_{\substack{\text{cold}\\\text{initial}}}}{2} = \frac{90.\ °\text{C} + 10.\ °\text{C}}{2} = 50.\ °\text{C}$$

For the hot water, the temperature change is

$$\text{Change in temperature (hot)} = \Delta T_{\text{hot}} = 90.\ °\text{C} - 50.\ °\text{C} = 40.\ °\text{C}$$

The temperature change for the cold water is

$$\text{Change in temperature (cold)} = \Delta T_{\text{cold}} = 50.\ °\text{C} - 10.\ °\text{C} = 40.\ °\text{C}$$

In this example, the masses of hot water and cold water are equal. If they were unequal, this problem would be more complicated.

Let's summarize the ideas we have introduced in this section. Temperature is a measure of the random motions of the components of an object. Heat is a *flow* of energy due to a temperature difference. We say that the random motions of the components of an object constitute the *thermal energy* of that object. The flow of energy called heat is the way in which thermal energy is transferred from a hot object to a colder object.

10-3 Exothermic and Endothermic Processes

OBJECTIVE **To consider the direction of energy flow as heat.**

A burning match releases energy.
Elektra Vision AG/Jupiter Images

In this section we will consider the energy changes that accompany chemical reactions. To explore this idea, let's consider the striking and burning of a match. Energy is clearly released through heat as the match burns. To discuss this reaction, we divide the universe into two parts: the system and the surroundings. The **system** is the part of the universe on which we wish to focus attention; the **surroundings** include everything else in the universe. In this case we define the system as the reactants and products of the reaction. The surroundings consist of the air in the room and anything else other than the reactants and products.

When a process results in the evolution of heat, it is said to be **exothermic** (*exo-* is a prefix meaning "out of"); that is, energy flows *out of the system.* For example, in the combustion of a match, energy flows out of the system as heat. Processes that absorb energy from the surroundings are said to be **endothermic.** When the heat flow moves *into a system,* the process is endothermic. Boiling water to form steam is a common endothermic process.

Where does the energy, released as heat, come from in an exothermic reaction? The answer lies in the difference in potential energies between the products and the reactants. Which has lower potential energy, the reactants or the products? We know that total energy is conserved and that energy flows from the system into the surroundings in an exothermic reaction. Thus *the energy gained by the surroundings must be equal to the energy lost by the system.* In the combustion of a match, the burned match has lost potential energy (in this case potential energy stored in the bonds of the reactants), which was transferred through heat to the surroundings (Fig. 10.5). The heat flow into the surroundings results from a lowering of the potential energy of the reaction system. *In any exothermic reaction, some of the potential energy stored in the chemical bonds is converted to thermal energy (random kinetic energy) via heat.*

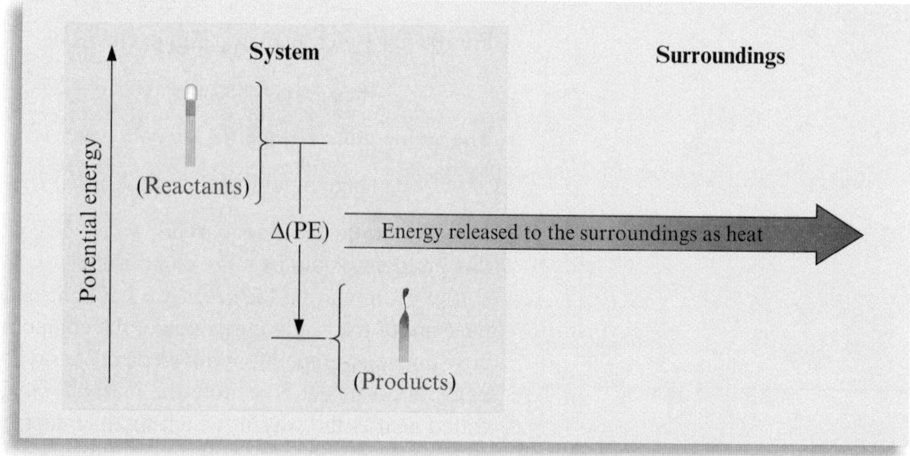

Figure 10.5 ▶ The energy changes accompanying the burning of a match.

10-4 Thermodynamics

OBJECTIVE To understand how energy flow affects internal energy.

The study of energy is called **thermodynamics.** The law of conservation of energy is often called the **first law of thermodynamics** and is stated as follows:

The energy of the universe is constant.

The **internal energy,** *E,* of a system can be defined most precisely as the sum of the kinetic and potential energies of all "particles" in the system. The internal energy of a system can be changed by a flow of work, heat, or both. That is,

$$\Delta E = q + w$$

where

Δ ("delta") means a change in the function that follows

q represents heat

w represents work

Thermodynamic quantities always consist of two parts: a *number,* giving the magnitude of the change, and a *sign,* indicating the direction of the flow. *The sign reflects the system's point of view.* For example, when a quantity of energy flows *into* the system via heat (an endothermic process), q is equal to $+x$, where the *positive* sign indicates that the *system's energy is increasing.* On the other hand, when energy flows *out of* the system via heat (an exothermic process), q is equal to $-x$, where the *negative* sign indicates that the *system's energy is decreasing.*

In this text the same conventions apply to the flow of work. If the system does work on the surroundings (energy flows out of the system), w is negative. If the surroundings do work on the system (energy flows into the system), w is positive. We define work from the system's point of view to be consistent for all thermodynamic quantities. That is, in this convention the signs of both q and w reflect what happens to the system; thus we use $\Delta E = q + w$.

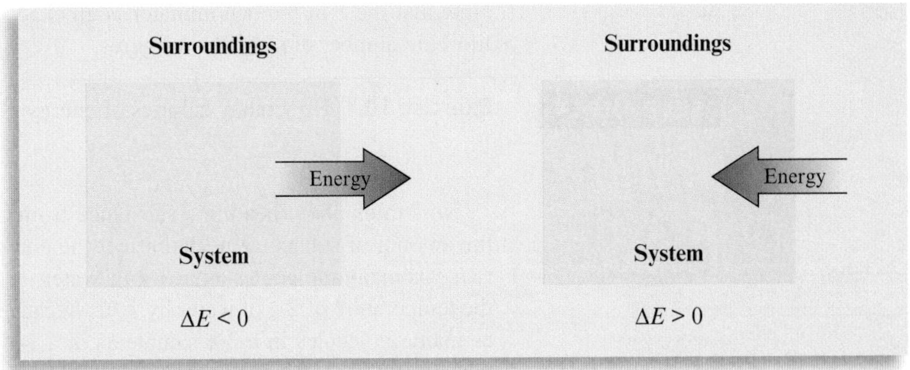

Critical Thinking

You are calculating ΔE in a chemistry problem. What if you confuse the system and the surroundings? How would this affect the magnitude of the answer you calculate? The sign?

10-5 Measuring Energy Changes

OBJECTIVE **To understand how heat is measured.**

© Cengage Learning

Diet drinks are now labeled as "low joule" instead of "low calorie" in European countries.

Earlier in this chapter we saw that when we heat a substance to a higher temperature, we increase the motions of the components of the substance—that is, we increase the thermal energy of the substance. Different materials respond differently to being heated. To explore this idea we need to introduce the common units of energy: the *calorie* and the *joule* (pronounced "jewel").

In the metric system the **calorie** is defined as the amount of energy (heat) required to raise the temperature of one gram of water by one Celsius degree. The "calorie" with which you are probably familiar is used to measure the energy content of food and is actually a kilocalorie (1000 calories), written with a capital *C* (Calorie) to distinguish it from the calorie used in chemistry. The **joule** (an SI unit) can be most conveniently defined in terms of the calorie:

$$1 \text{ calorie} = 4.184 \text{ joules}$$

or using the normal abbreviations

$$1 \text{ cal} = 4.184 \text{ J}$$

You need to be able to convert between calories and joules. We will consider that conversion process in Example 10.1.

Interactive Example 10.1 Converting Calories to Joules

Express 60.1 cal of energy in units of joules.

SOLUTION By definition 1 cal = 4.184 J, so the conversion factor needed is $\frac{4.184 \text{ J}}{1 \text{ cal}}$, and the result is

$$60.1 \cancel{\text{cal}} \times \frac{4.184 \text{ J}}{1 \cancel{\text{cal}}} = 251 \text{ J}$$

Note that the 1 in the denominator is an exact number by definition and so does not limit the number of significant figures.

SELF CHECK **Exercise 10.1** How many calories of energy correspond to 28.4 J?

See Problems 10.25 through 10.30. ■

Now think about heating a substance from one temperature to another. How does the amount of substance heated affect the energy required? In 2 g of water there are twice as many molecules as in 1 g of water. It takes twice as much energy to change the temperature of 2 g of water by 1 °C, because we must change the motions of twice as many molecules in a 2-g sample as in a 1-g sample. Also, as we would expect, it takes twice as much energy to raise the temperature of a given sample of water by 2 degrees as it does to raise the temperature by 1 degree.

Interactive Example 10.2 Calculating Energy Requirements

Determine the amount of energy (heat) in joules required to raise the temperature of 7.40 g water from 29.0 °C to 46.0 °C.

CHEMISTRY IN FOCUS

Coffee: Hot and Quick(lime)

Convenience and speed are the watchwords of our modern society. One new product that fits these requirements is a container of coffee that heats itself with no batteries needed. Consumers can now buy a 10-ounce container of Wolfgang Puck gourmet latte that heats itself to 145 °F in 6 minutes and stays hot for 30 minutes. What kind of chemical magic makes this happen? Pushing a button on the bottom of the container. This action allows water to mix with calcium oxide, or quicklime (see accompanying figure). The resulting reaction

$$CaO(s) + H_2O(l) \rightarrow Ca(OH)_2(s)$$

releases enough energy as heat to bring the coffee to a pleasant drinking temperature.

Other companies are experimenting with similar technology to heat liquids such as tea, hot chocolate, and soup.

A different reaction is now being used to heat MREs (meals ready-to-eat) for soldiers on the battlefield. In this case the energy to heat the meals is furnished by mixing magnesium iron oxide with water to produce an exothermic reaction.

Clearly, chemistry is "hot stuff."

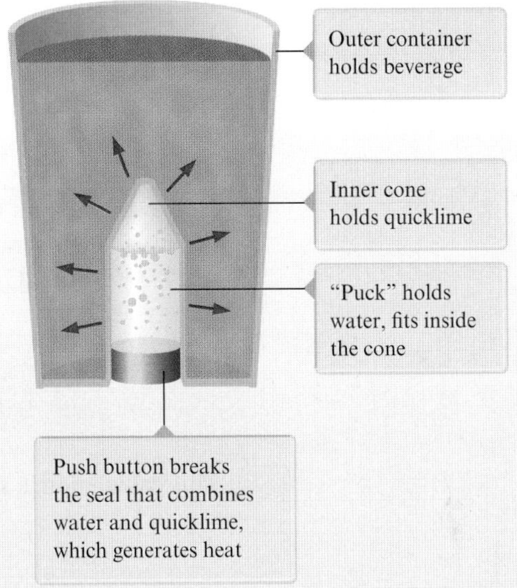

SOLUTION

Where Are We Going?

We want to determine the amount of energy (heat in joules) needed to increase the temperature of 7.40 g water from 29.0 °C to 46.0 °C.

What Do We Know?

- The mass of water is 7.40 g, and the temperature is increased from 29.0 °C to 46.0 °C.

What Information Do We Need?

- The amount of heat needed to raise 1.00 g water by 1.00 °C. From the text we see that 4.184 J of energy is required.

How Do We Get There?

In solving any kind of problem, it is often useful to draw a diagram that represents the situation. In this case, we have 7.40 g of water that is to be heated from 29.0 °C to 46.0 °C.

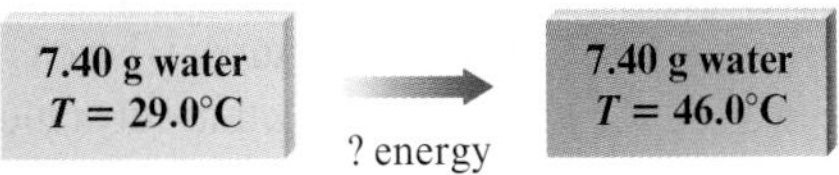

Our task is to determine how much energy is required to accomplish this change.

From the discussion in the text, we know that 4.184 J of energy is required to raise the temperature of *one* gram of water by *one* Celsius degree.

1.00 g water $T = 29.0°C$ → 4.184 J → **1.00 g water** $T = 30.0°C$

Because in our case we have 7.40 g of water instead of 1.00 g, it will take 7.40 × 4.184 J to raise the temperature by one degree.

7.40 g water $T = 29.0°C$ → (7.40 × 4.184) J → **7.40 g water** $T = 30.0°C$

However, we want to raise the temperature of our sample of water by more than 1 °C. In fact, the temperature change required is from 29.0 °C to 46.0 °C. This is a change of 17.0 °C (46.0 °C − 29.0 °C = 17.0 °C). Thus we will have to supply 17.0 times the energy necessary to raise the temperature of 7.40 g of water by 1 °C.

7.40 g water $T = 29.0°C$ → (17.0 × 7.40 × 4.184) J → **7.40 g water** $T = 46.0°C$

This calculation is summarized as follows:

$$4.184 \frac{\text{J}}{\text{g °C}} \times 7.40 \text{ g} \times 17.0 \text{ °C} = 526 \text{ J}$$

Energy per gram of water per degree of temperature × Actual grams of water × Actual temperature change = Energy required

Math skill builder

The result you will get on your calculator is 4.184 × 7.40 × 17.0 = 526.3472, which rounds off to 526.

We have shown that 526 J of energy (as heat) is required to raise the temperature of 7.40 g of water from 29.0 °C to 46.0 °C. Note that because 4.184 J of energy is required to heat 1 g of water by 1 °C, the units are J/g °C (joules per gram per Celsius degree).

REALITY CHECK The units (J) are correct, and the answer is reported to the correct number of significant figures (three).

SELF CHECK

Exercise 10.2 Calculate the joules of energy required to heat 454 g of water from 5.4 °C to 98.6 °C.

See Problems 10.31 through 10.36. ■

Table 10.1 The Specific Heat Capacities of Some Common Substances

Substance	Specific Heat Capacity (J/g °C)
water (*l*)* (liquid)	4.184
water (*s*) (ice)	2.03
water (*g*) (steam)	2.0
aluminum (*s*)	0.89
iron (*s*)	0.45
mercury (*l*)	0.14
carbon (*s*)	0.71
silver (*s*)	0.24
gold (*s*)	0.13

*The symbols (*s*), (*l*), and (*g*) indicate the solid, liquid, and gaseous states, respectively.

So far we have seen that the energy (heat) required to change the temperature of a substance depends on

1. The amount of substance being heated (number of grams)
2. The temperature change (number of degrees)

There is, however, another important factor: the identity of the substance.

Different substances respond differently to being heated. We have seen that 4.184 J of energy raises the temperature of 1 g of water by 1 °C. In contrast, this same amount of energy applied to 1 g of gold raises its temperature by approximately 32 °C! The point is that some substances require relatively large amounts of energy to change their temperatures, whereas others require relatively little. Chemists describe this difference by saying that substances have different heat capacities. *The amount of energy required to change the temperature of one gram of a substance by one Celsius degree* is called its **specific heat capacity** or, more commonly, its *specific heat*. The specific heat capacities for several substances are listed in Table 10.1. You can see from the table that the specific heat capacity for water is very high compared to those of the other substances listed. This is why lakes and seas are much slower to respond to cooling or heating than are the surrounding land masses.

CHEMISTRY IN FOCUS

Nature Has Hot Plants

The voodoo lily (Titan Arum) is a beautiful and seductive plant. The exotic-looking lily features an elaborate reproductive mechanism—a purple spike that can reach nearly 3 feet in length and is cloaked by a hoodlike leaf. But approach to the plant reveals bad news—it smells terrible!

Despite its antisocial odor, this putrid plant has fascinated biologists for many years because of its ability to generate heat. At the peak of its metabolic activity, the plant's blossom can be as much as 15 °C above its surrounding temperature. To generate this much heat, the metabolic rate of the plant must be close to that of a flying hummingbird!

What's the purpose of this intense heat production? For a plant faced with limited food supplies in the very competitive tropical climate where it grows, heat production seems like a great waste of energy. The answer to this mystery is that the voodoo lily is pollinated mainly by carrion-loving insects. Thus the lily prepares a malodorous mixture of chemicals characteristic of rotting meat, which it then "cooks" off into the surrounding air to attract flesh-feeding beetles and flies. Then, once the insects enter the pollination chamber, the high temperatures there (as high as 110 °F) cause the insects to remain very active to better carry out their pollination duties.

The voodoo lily is only one of many thermogenic (heat-producing) plants. These plants are of special interest to biologists because they provide opportunities to study metabolic reactions that are quite subtle in "normal" plants.

Titan Arum is reputedly the largest flower in the world.

Interactive Example 10.3 Calculations Involving Specific Heat Capacity

a. What quantity of energy (in joules) is required to heat a piece of iron weighing 1.3 g from 25 °C to 46 °C?

b. What is the answer in calories?

SOLUTION

Where Are We Going?

We want to determine the amount of energy (units of joules and calories) to increase the temperature of 1.3 g of iron from 25° C to 46° C.

What Do We Know?

- The mass of iron is 1.3 g, and the temperature is increased from 25° C to 46° C.

What Information Do We Need?

- We need the specific heat capacity of iron and the conversion factor between joules and calories.

How Do We Get There?

a. It is helpful to draw the following diagram to represent the problem.

1.3 g iron
T = 25°C

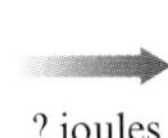

? joules

1.3 g iron
T = 46°C

From Table 10.1 we see that the specific heat capacity of iron is 0.45 J/g °C. That is, it takes 0.45 J to raise the temperature of a 1-g piece of iron by 1 °C.

1.0 g iron $T = 25°C$	→ 0.45 J	1.0 g iron $T = 26°C$

In this case our sample is 1.3 g, so 1.3 × 0.45 J is required for *each* degree of temperature increase.

1.3 g iron $T = 25°C$	→ (1.3 × 0.45) J	1.3 g iron $T = 26°C$

Because the temperature increase is 21 °C (46 °C − 25 °C = 21 °C), the total amount of energy required is

$$0.45 \frac{\text{J}}{\text{g °C}} \times 1.3 \text{ g} \times 21 \text{ °C} = 12 \text{ J}$$

Math skill builder

The result you will get on your calculator is 0.45 × 1.3 × 21 = 12.285, which rounds off to 12.

1.3 g iron $T = 25°C$	→ (21 × 1.3 × 0.45) J	1.3 g iron $T = 46°C$

Note that the final units are joules, as they should be.

b. To calculate this energy in calories, we can use the definition 1 cal = 4.184 J to construct the appropriate conversion factor. We want to change from joules to calories, so cal must be in the numerator and J in the denominator, where it cancels:

$$12 \cancel{\text{J}} \times \frac{1 \text{ cal}}{4.184 \cancel{\text{J}}} = 2.9 \text{ cal}$$

Remember that 1 in this case is an exact number by definition and therefore does not limit the number of significant figures (the number 12 is limiting here).

REALITY CHECK The units (joules and calories) are correct, and the answer is reported to the correct number of significant figures (two).

SELF CHECK

Exercise 10.3 A 5.63-g sample of solid gold is heated from 21 °C to 32 °C. How much energy (in joules and calories) is required?

See Problems 10.31 through 10.36. ■

Note that in Example 10.3, to calculate the energy (heat) required, we took the product of the specific heat capacity, the sample size in grams, and the change in temperature in Celsius degrees.

Energy (heat) required (Q)	=	Specific heat capacity (s)	×	Mass (m) in grams of sample	×	Change in temperature (ΔT) in °C

We can represent this by the following equation:

$$Q = s \times m \times \Delta T$$

Math skill builder

The symbol Δ (the Greek letter delta) is shorthand for "change in."

where

Q = energy (heat) required

s = specific heat capacity

m = mass of the sample in grams

ΔT = change in temperature in Celsius degrees

Firewalking: Magic or Science?

For millennia people have been amazed at the ability of Eastern mystics to walk across beds of glowing coals without any apparent discomfort. Even in the United States, thousands of people have performed feats of firewalking as part of motivational seminars. How can this be possible? Do firewalkers have supernatural powers?

Actually, there are good scientific explanations of why firewalking is possible. First, human tissue is mainly composed of water, which has a relatively large specific heat capacity. This means that a large amount of energy must be transferred from the coals to change significantly the temperature of the feet. During the brief contact between feet and coals involved in firewalking, there is relatively little time for energy flow, so the feet do not reach a high enough temperature to cause damage.

Also, although the surface of the coals has a very high temperature, the red-hot layer is very thin. Therefore, the quantity of energy available to heat the feet is smaller than might be expected.

Thus, although firewalking is impressive, there are several scientific reasons why anyone with the proper training should be able to do it on a properly prepared bed of coals. (Don't try this on your own!)

AP Photo/Itsuo Inouye

A group of firewalkers in Japan.

This equation always applies when a substance is being heated (or cooled) and no change of state occurs. Before you begin to use this equation, however, make sure you understand what it means.

Example 10.4 Specific Heat Capacity Calculations: Using the Equation

A 1.6-g sample of a metal that has the appearance of gold requires 5.8 J of energy to change its temperature from 23 °C to 41 °C. Is the metal pure gold?

SOLUTION

Where Are We Going?

We want to determine if a metal is gold.

What Do We Know?

- The mass of metal is 1.6 g, and 5.8 J of energy is required to increase the temperature from 23° C to 41° C.

What Information Do We Need?

- We need the specific heat capacity of gold.

How Do We Get There?

We can represent the data given in this problem by the following diagram:

1.6 g metal $T = 23°C$ → 5.8 J → 1.6 g metal $T = 41°C$

$$\Delta T = 41\ °C - 23\ °C = 18\ °C$$

Using the data given, we can calculate the value of the specific heat capacity for the metal and compare this value to the one for gold given in Table 10.1. We know that

$$Q = s \times m \times \Delta T$$

or, pictorially,

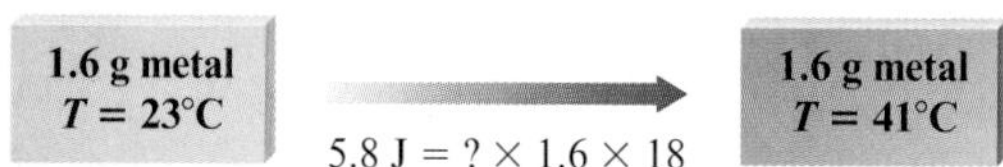

When we divide both sides of the equation

$$Q = s \times m \times \Delta T$$

by $m \times \Delta T$, we get

$$\frac{Q}{m \times \Delta T} = s$$

Thus, using the data given, we can calculate the value of s. In this case,

Q = energy (heat) required = 5.8 J

m = mass of the sample = 1.6 g

ΔT = change in temperature = 18 °C (41 °C − 23 °C = 18 °C)

Thus

$$s = \frac{Q}{m \times \Delta T} = \frac{5.8\ \text{J}}{(1.6\ \text{g})(18\ °\text{C})} = 0.20\ \text{J/g}\ °\text{C}$$

Math skill builder

The result you will get on your calculator is 5.8/(1.6 × 18) = 0.2013889, which rounds off to 0.20.

From Table 10.1, the specific heat capacity for gold is 0.13 J/g °C. Thus the metal must not be pure gold.

SELF CHECK

Exercise 10.4 A 2.8-g sample of pure metal requires 10.1 J of energy to change its temperature from 21 °C to 36 °C. What is this metal? (Use Table 10.1.)

See Problems 10.31 through 10.36. ■

10-6 Thermochemistry (Enthalpy)

OBJECTIVE **To consider the heat (enthalpy) of chemical reactions.**

We have seen that some reactions are exothermic (produce heat energy) and other reactions are endothermic (absorb heat energy). Chemists also like to know exactly how much energy is produced or absorbed by a given reaction. To make that process more convenient, we have invented a special energy function called **enthalpy,** which is designated by H. For a reaction occurring under conditions of constant pressure, the change in enthalpy (ΔH) is equal to the energy that flows as heat. That is,

$$\Delta H_p = \text{heat}$$

where the subscript "p" indicates that the process has occurred under conditions of constant pressure and Δ means "a change in." Thus the enthalpy change for a reaction (that occurs at constant pressure) is the same as the heat for that reaction.

Interactive Example 10.5

Enthalpy

When 1 mole of methane (CH_4) is burned at constant pressure, 890 kJ of energy is released as heat. Calculate ΔH for a process in which a 5.8-g sample of methane is burned at constant pressure.

SOLUTION

Where Are We Going?

We want to determine ΔH for the reaction of 5.8 g of methane (CH_4) with oxygen at constant pressure.

What Do We Know?

- When 1 mole of CH_4 is burned, 890 kJ of energy is released.
- We have 5.8 g of CH_4.

What Information Do We Need?

- Molar mass of methane, which we can get from the atomic masses of carbon (12.01 g/mol) and hydrogen (1.008 g/mol). The molar mass is 16.0 g/mol.

How Do We Get There?

At constant pressure, 890 kJ of energy per mole of CH_4 is produced as heat:

$$q_p = \Delta H = -890 \text{ kJ/mol CH}_4$$

Note that the minus sign indicates an exothermic process. In this case, a 5.8-g sample of CH_4 (molar mass = 16.0 g/mol) is burned. Since this amount is smaller than 1 mole, less than 890 kJ will be released as heat. The actual value can be calculated as follows:

$$5.8 \text{ g } \cancel{\text{CH}_4} \times \frac{1 \text{ mol CH}_4}{16.0 \text{ g } \cancel{\text{CH}_4}} = 0.36 \text{ mol CH}_4$$

and

$$0.36 \; \cancel{\text{mol CH}_4} \times \frac{-890 \text{ kJ}}{\cancel{\text{mol CH}_4}} = -320 \text{ kJ}$$

Thus, when a 5.8-g sample of CH_4 is burned at constant pressure,

$$\Delta H = \text{heat flow} = -320 \text{ kJ}$$

REALITY CHECK The mass of methane burned is less than 1 mole, so less than 890 kJ will be released as heat. The answer has two significant figures as required by the given quantities.

SELF CHECK

Exercise 10.5 The reaction that occurs in the heat packs used to treat sports injuries is

$$4\text{Fe}(s) + 3\text{O}_2(g) \rightarrow 2\text{Fe}_2\text{O}_3(s) \qquad \Delta H = -1652 \text{ kJ}$$

How much heat is released when 1.00 g of Fe(s) is reacted with excess $O_2(g)$?

See Problems 10.41 and 10.42. ■

Burning Calories

There is a growing concern in the United States about the increasing tendency for individuals to be overweight. Estimates indicate that two-thirds of the adults in the United States are overweight, and one-third are classified as obese. This is an alarming situation because obesity is associated with many serious diseases such as diabetes and heart disease. In an effort to reduce this problem, the U.S. government now requires national restaurant chains and grocery stores to post the calorie counts for the food they sell. The hope is that people will use the information to make better food choices for weight control.

All of this leads to the question of how the calorie content of food is determined. The process involves a special type of calorimeter called a *bomb calorimeter*. The food sample is enclosed in the bomb calorimeter and burned. The energy produced heats the water surrounding the calorimeter, and the amount of energy is determined by measuring the increase in temperature of the known amount of water. The "Calorie" used for food is equal to the kilocalorie used by the science community. Thus the number of kilocalories produced by burning the food is the "Calorie" content of the food.

How is food "burned" in a calorimeter? The exact composition of food can be determined from its ingredients. Each ingredient in its proper amount for the food is burned in a calorimeter, and the calories released are determined. The calorie count assigned to a particular food is totaled as the sum of the ingredients and is adjusted for the amount of energy the body will actually absorb (98% for fats and less for carbohydrates and proteins). Restaurants determine calorie counts from recipes. However, that means errors can occur, depending on how closely a chef follows a recipe and whether the portion size in the recipe is actually the size of the portion served in the restaurant. Although the "calorie" count may not be exact for restaurant foods, the listing on the menu will give people a chance to make better choices.

A menu board in a restaurant showing the calories for various choices.

Calorimetry

A **calorimeter** (Fig. 10.6) is a device used to determine the heat associated with a chemical reaction. The reaction is run in the calorimeter and the temperature change of the calorimeter is observed. Knowing the temperature change that occurs in the calorimeter and the heat capacity of the calorimeter enables us to calculate the heat energy released or absorbed by the reaction. Thus we can determine ΔH for the reaction.

Once we have measured the ΔH values for various reactions, we can use these data to *calculate* the ΔH values of other reactions. We will see how to carry out these calculations in the next section.

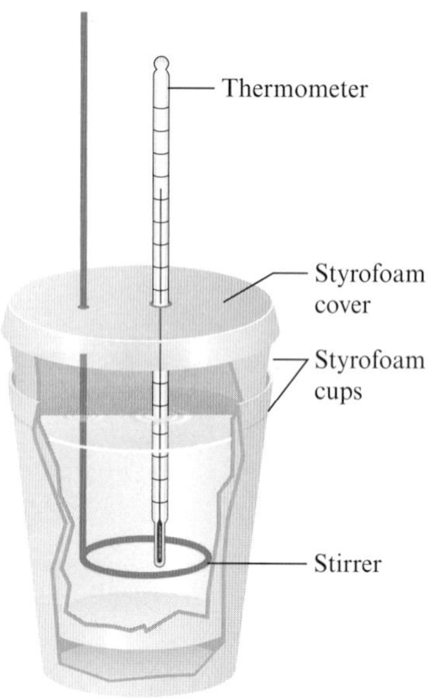

Figure 10.6 A coffee-cup calorimeter made of two Styrofoam cups.

10-7 Hess's Law

OBJECTIVE To understand Hess's law.

One of the most important characteristics of enthalpy is that it is a state function. That is, the change in enthalpy for a given process is independent of the pathway for the process. Consequently, *in going from a particular set of reactants to a particular set of products, the change in enthalpy is the same whether the reaction takes place in one step or in a series of steps.* This principle, which is known as **Hess's law,** can be illustrated by examining the oxidation of nitrogen to produce nitrogen dioxide. The overall reaction can be written in one step, where the enthalpy change is represented by ΔH_1.

$$N_2(g) + 2O_2(g) \rightarrow 2NO_2(g) \qquad \Delta H_1 = 68 \text{ kJ}$$

This reaction can also be carried out in two distinct steps, with the enthalpy changes being designated as ΔH_2 and ΔH_3:

$$N_2(g) + O_2(g) \rightarrow 2NO(g) \qquad \Delta H_2 = 180 \text{ kJ}$$

$$2NO(g) + O_2(g) \rightarrow 2NO_2(g) \qquad \Delta H_3 = -112 \text{ kJ}$$

$$\text{Net reaction:} \quad N_2(g) + 2O_2(g) \rightarrow 2NO_2(g) \qquad \Delta H_2 + \Delta H_3 = 68 \text{ kJ}$$

Note that the sum of the two steps gives the net, or overall, reaction and that

$$\Delta H_1 = \Delta H_2 + \Delta H_3 = 68 \text{ kJ}$$

The importance of Hess's law is that it allows us to *calculate* heats of reaction that might be difficult or inconvenient to measure directly in a calorimeter.

Characteristics of Enthalpy Changes

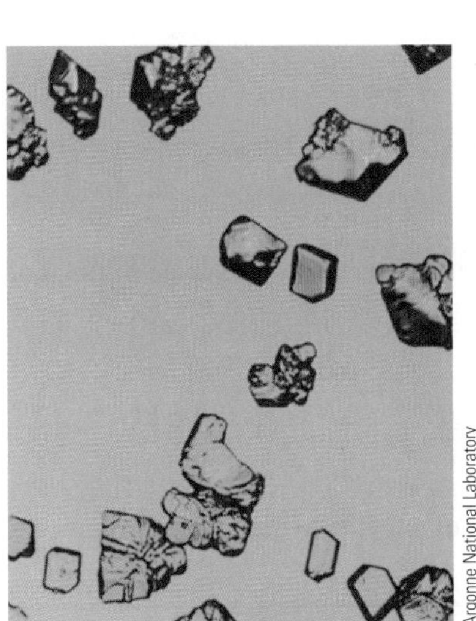
Argonne National Laboratory

Crystals of xenon tetrafluoride, the first reported binary compound containing a noble gas element.

To use Hess's law to compute enthalpy changes for reactions, it is important to understand two characteristics of ΔH for a reaction:

1. If a reaction is reversed, the sign of ΔH is also reversed.
2. The magnitude of ΔH is directly proportional to the quantities of reactants and products in a reaction. If the coefficients in a balanced reaction are multiplied by an integer, the value of ΔH is multiplied by the same integer.

Both these rules follow in a straightforward way from the properties of enthalpy changes. The first rule can be explained by recalling that the *sign* of ΔH indicates the *direction* of the heat flow at constant pressure. If the direction of the reaction is reversed, the direction of the heat flow also will be reversed. To see this, consider the preparation of xenon tetrafluoride, which was the first binary compound made from a noble gas:

$$Xe(g) + 2F_2(g) \rightarrow XeF_4(s) \qquad \Delta H = -251 \text{ kJ}$$

This reaction is exothermic, and 251 kJ of energy flows into the surroundings as heat. On the other hand, if the colorless XeF_4 crystals are decomposed into the elements, according to the equation

$$XeF_4(s) \rightarrow Xe(g) + 2F_2(g)$$

the opposite energy flow occurs because 251 kJ of energy must be added to the system to produce this endothermic reaction. Thus, for this reaction, $\Delta H = +251$ kJ.

The second rule comes from the fact that ΔH is an extensive property, depending on the amount of substances reacting. For example, since 251 kJ of energy is evolved for the reaction

$$Xe(g) + 2F_2(g) \rightarrow XeF_4(s)$$

then for a preparation involving twice the quantities of reactants and products, or

$$2Xe(g) + 4F_2(g) \rightarrow 2XeF_4(s)$$

twice as much heat would be evolved:

$$\Delta H = 2(-251 \text{ kJ}) = -502 \text{ kJ}$$

Critical Thinking

What if Hess's law were not true? What are some possible repercussions this would have?

Interactive Example 10.6

Hess's Law

Two forms of carbon are graphite, the soft, black, slippery material used in "lead" pencils and as a lubricant for locks, and diamond, the brilliant, hard gemstone. Using the enthalpies of combustion for graphite (−394 kJ/mol) and diamond (−396 kJ/mol), calculate ΔH for the conversion of graphite to diamond:

$$C_{graphite}(s) \rightarrow C_{diamond}(s)$$

SOLUTION

Where Are We Going?

We want to determine ΔH for the conversion of graphite to diamond.

What Do We Know?

The combustion reactions are

$$C_{graphite}(s) + O_2(g) \rightarrow CO_2(g) \qquad \Delta H = -394 \text{ kJ}$$
$$C_{diamond}(s) + O_2(g) \rightarrow CO_2(g) \qquad \Delta H = -396 \text{ kJ}$$

How Do We Get There?

Note that if we reverse the second reaction (which means we must change the sign of ΔH) and sum the two reactions, we obtain the desired reaction:

$$C_{graphite}(s) + O_2(g) \rightarrow CO_2(g) \qquad \Delta H = -394 \text{ kJ}$$
$$CO_2(g) \rightarrow C_{diamond}(s) + O_2(g) \qquad \Delta H = -(-396 \text{ kJ})$$
$$C_{graphite}(s) \rightarrow C_{diamond}(s) \qquad \Delta H = 2 \text{ kJ}$$

Thus 2 kJ of energy is required to change 1 mole of graphite to diamond. This process is endothermic.

SELF CHECK

Exercise 10.6 From the following information

$$S(s) + \tfrac{3}{2}O_2(g) \rightarrow SO_3(g) \qquad \Delta H = -395.2 \text{ kJ}$$
$$2SO_2(g) + O_2(g) \rightarrow 2SO_3(g) \qquad \Delta H = -198.2 \text{ kJ}$$

calculate ΔH for the reaction

$$S(s) + O_2(g) \rightarrow SO_2(g)$$

See Problems 10.45 through 10.48. ■

Quality Versus Quantity of Energy

OBJECTIVE **To see how the quality of energy changes as it is used.**

One of the most important characteristics of energy is that it is conserved. Thus the total energy content of the universe will always be what it is now. If that is the case, why are we concerned about energy? For example, why should we worry about conserving our petroleum supply? Surprisingly, the "energy crisis" is not about the *quantity* of energy, but rather about the *quality* of energy. To understand this idea, consider an automobile trip from Chicago to Denver. Along the way you would put gasoline into the car to get to Denver. What happens to that energy? The energy stored in the bonds of the gasoline and of the oxygen that reacts with it is changed to thermal energy, which is spread along the highway to Denver. The total quantity of energy remains the same as before the trip but the energy concentrated in the gasoline becomes widely distributed in the environment:

$$\text{gasoline}(l) + O_2(g) \rightarrow CO_2(g) + H_2O(l) + \text{energy}$$

gasoline: C_8H_{18} and other similar compounds

energy: Spread along the highway, heating the road and the air

Which energy is easier to use to do work: the concentrated energy in the gasoline or the thermal energy spread from Chicago to Denver? Of course, the energy concentrated in the gasoline is more convenient to use.

This example illustrates a very important general principle: when we utilize energy to do work, we degrade its usefulness. In other words, when we use energy the *quality* of that energy (its ease of use) is lowered.

In summary,

You may have heard someone mention the "heat death" of the universe. Eventually (many eons from now), all energy will be spread evenly throughout the universe and everything will be at the same temperature. At this point it will no longer be possible to do any work. The universe will be "dead."

We don't have to worry about the heat death of the universe anytime soon, of course, but we do need to think about conserving "quality" energy supplies. The energy stored in petroleum molecules got there over millions of years through plants and simple animals absorbing energy from the sun and using this energy to construct molecules. As these organisms died and became buried, natural processes changed them into the petroleum deposits we now access for our supplies of gasoline and natural gas.

Petroleum is highly valuable because it furnishes a convenient, concentrated source of energy. Unfortunately, we are using this fuel at a much faster rate than natural processes can replace it, so we are looking for new sources of energy. The most logical energy source is the sun. *Solar energy* refers to using the sun's energy directly to do productive work in our society. We will discuss energy supplies in the next section.

10-9 Energy and Our World

OBJECTIVE To consider the energy resources of our world.

Woody plants, coal, petroleum, and natural gas provide a vast resource of energy that originally came from the sun. By the process of photosynthesis, plants store energy that can be claimed by burning the plants themselves or the decay products that have been converted over millions of years to **fossil fuels.** Although the United States currently depends heavily on petroleum for energy, this dependency is a relatively recent phenomenon, as shown in Fig. 10.7. In this section we discuss some sources of energy and their effects on the environment.

Petroleum and Natural Gas

Although how they were produced is not completely understood, petroleum and natural gas were most likely formed from the remains of marine organisms that lived approximately 500 million years ago. Because of the way these substances were formed they are called *fossil fuels.* **Petroleum** is a thick, dark liquid composed mostly of compounds called *hydrocarbons* that contain carbon and hydrogen. (Carbon is unique among elements in the extent to which it can bond to itself to form chains of various lengths.) Table 10.2 gives the formulas and names for several common hydrocarbons. **Natural gas,** usually associated with petroleum deposits, consists mostly of methane, but it also contains significant amounts of ethane, propane, and butane.

Table 10.2 ▸ Names and Formulas for Some Common Hydrocarbons

Formula	Name
CH_4	Methane
C_2H_6	Ethane
C_3H_8	Propane
C_4H_{10}	Butane
C_5H_{12}	Pentane
C_6H_{14}	Hexane
C_7H_{16}	Heptane
C_8H_{18}	Octane

The composition of petroleum varies somewhat, but it includes mostly hydrocarbons having chains that contain from 5 to more than 25 carbons. To be used efficiently, the petroleum must be separated into fractions by boiling. The lighter molecules (having the lowest boiling points) can be boiled off, leaving the heavier ones behind. The commercial uses of various petroleum fractions are shown in Table 10.3.

At the end of the 20th century there was increasing concern that the global supplies of fossil fuels were being rapidly depleted. However, a new technology called *hydraulic fracturing* or *fracking* has changed the situation significantly. Fracking involves injecting a mixture of water, sand, and chemicals at high pressures through a well drilled deep into rock layers. The high pressures involved cause the rock layer to fracture, allowing deep pockets of petroleum and natural gas to escape and be recovered. This technique has the potential to produce huge amounts of previously unavailable petroleum. For example, it is estimated that recoverable supplies of natural gas in deep shale deposits amount to over 200 trillion cubic meters of natural gas. Thus the introduction of fracking has completely altered the thinking about future energy sources.

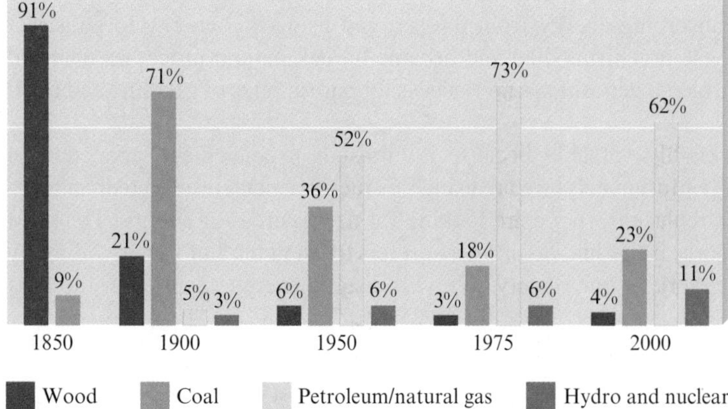

Figure 10.7 ▸ Energy sources used in the United States.

Table 10.3 ▸ Uses of the Various Petroleum Fractions

Petroleum Fraction in Terms of Numbers of Carbon Atoms	Major Uses
C_5–C_{10}	Gasoline
C_{10}–C_{18}	Kerosene Jet fuel
C_{15}–C_{25}	Diesel fuel Heating oil Lubricating oil
$>C_{25}$	Asphalt

The petroleum era began when the demand for lamp oil during the Industrial Revolution outstripped the traditional sources: animal fats and whale oil. In response to this increased demand, Edwin Drake drilled the first oil well in 1859 at Titusville, Pennsylvania. The petroleum from this well was refined to produce *kerosene* (fraction C_{10}–C_{18}), which served as an excellent lamp oil. *Gasoline* (fraction C_5–C_{10}) had limited use and was often discarded. This situation soon changed. The development of the electric light decreased the need for kerosene, and the advent of the "horseless carriage" with its gasoline-powered engine signaled the birth of the gasoline age.

As gasoline became more important, new ways were sought to increase the yield of gasoline obtained from each barrel of petroleum. William Burton invented a process at Standard Oil of Indiana called *pyrolytic (high-temperature) cracking.* In this process, the heavier molecules of the kerosene fraction are heated to about 700 °C, causing them to break (crack) into the smaller molecules of hydrocarbons in the gasoline fraction. As cars became larger, more efficient internal combustion engines were designed. Because of the uneven burning of the gasoline then available, these engines "knocked," producing unwanted noise and even engine damage. Intensive research to find additives that would promote smoother burning produced tetraethyl lead, $(C_2H_5)_4Pb$, a very effective "antiknock" agent.

The addition of tetraethyl lead to gasoline became a common practice, and by 1960, gasoline contained as much as 3 g of lead per gallon. As we have discovered so often in recent years, technological advances can produce environmental problems. To prevent air pollution from automobile exhaust, catalytic converters have been added to car exhaust systems. The effectiveness of these converters, however, is destroyed by lead. The use of leaded gasoline also greatly increased the amount of lead in the environment, where it can be ingested by animals and humans. For these reasons, the use of lead in gasoline has been phased out, requiring extensive (and expensive) modifications of engines and of the gasoline refining process.

Coal

Coal was formed from the remains of plants that were buried and subjected to high pressure and heat over long periods of time. Plant materials have a high content of cellulose, a complex molecule whose empirical formula is CH_2O but whose molar mass is approximately 500,000 g/mol. After the plants and trees that grew on the earth at various times and places died and were buried, chemical changes gradually lowered the oxygen and hydrogen content of the cellulose molecules. Coal "matures" through four stages: lignite, subbituminous, bituminous, and anthracite. Each stage has a higher carbon-to-oxygen and carbon-to-hydrogen ratio; that is, the relative carbon content gradually increases. Typical elemental compositions of the various coals are given in Table 10.4. The energy available from the combustion of a given mass of coal increases as the carbon content increases. Anthracite is the most valuable coal, and lignite is the least valuable.

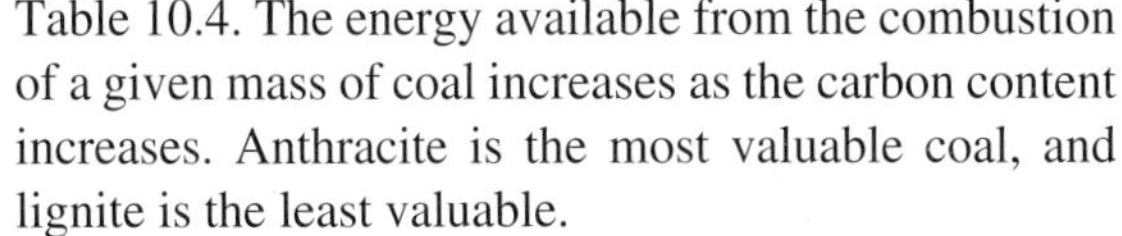

cbpix/Stockphoto.com

Table 10.4 ▸ Element Composition of Various Types of Coal

	Mass Percent of Each Element				
Type of Coal	C	H	O	N	S
Lignite	71	4	23	1	1
Subbituminous	77	5	16	1	1
Bituminous	80	6	8	1	5
Anthracite	92	3	3	1	1

Coal is an important and plentiful fuel in the United States, currently furnishing approximately 20% of our energy. As the supply of petroleum decreases, the share of the energy supply from coal could eventually increase to as high as 30%. However, coal is expensive and dangerous to mine underground, and the strip mining of fertile farmland in the Midwest or of scenic land in the West causes obvious problems. In addition, the burning of coal, especially high-sulfur coal, yields air pollutants such as sulfur dioxide, which, in turn, can lead to acid rain. However, even if coal were pure carbon, the carbon dioxide produced when it was burned would still have significant effects on the earth's climate.

Effects of Carbon Dioxide on Climate

The earth receives a tremendous quantity of radiant energy from the sun, about 30% of which is reflected back into space by the earth's atmosphere. The remaining energy passes through the atmosphere to the earth's surface. Some of this energy is absorbed by plants for photosynthesis and some by the oceans to evaporate water, but most of it is absorbed by soil, rocks, and water, increasing the temperature of the earth's surface. This energy is, in turn, radiated from the heated surface mainly as *infrared radiation,* often called *heat radiation.*

The atmosphere, like window glass, is transparent to visible light but does not allow all the infrared radiation to pass back into space. Molecules in the atmosphere, principally H_2O and CO_2, strongly absorb infrared radiation and radiate it back toward the earth, as shown in Fig. 10.8. A net amount of thermal energy is retained by the earth's atmosphere, causing the earth to be much warmer than it would be without its atmosphere. In a way, the atmosphere acts like the glass of a greenhouse, which is transparent to visible light but absorbs infrared radiation, thus raising the temperature inside the building. This **greenhouse effect** is seen even more spectacularly on Venus, where the dense atmosphere is thought to be responsible for the high surface temperature of that planet.

Thus the temperature of the earth's surface is controlled to a significant extent by the carbon dioxide and water content of the atmosphere. The effect of atmospheric moisture (humidity) is readily apparent in the Midwest, for example. In summer, when the humidity is high, the heat of the sun is retained well into the night, giving very high nighttime temperatures. In winter, the coldest temperatures always occur on clear nights, when the low humidity allows efficient radiation of energy back into space.

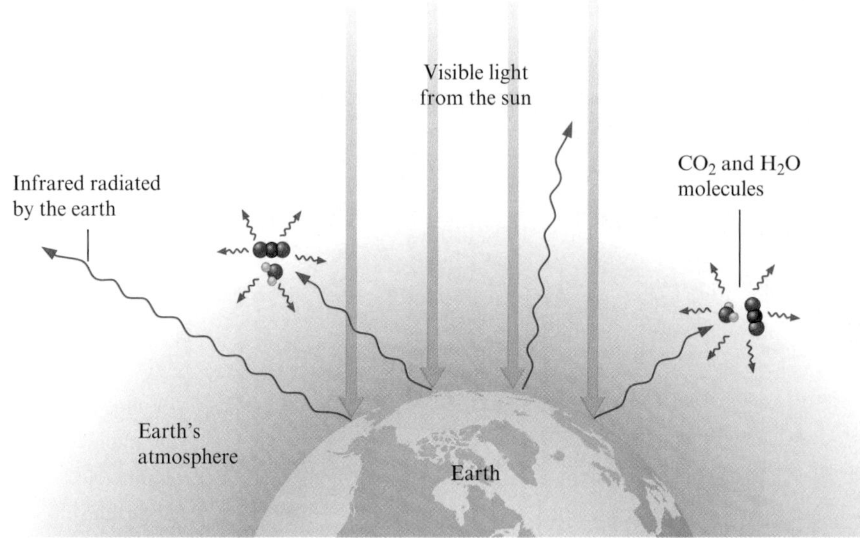

Figure 10.8 The earth's atmosphere is transparent to visible light from the sun. This visible light strikes the earth, and part of it is changed to infrared radiation. The infrared radiation from the earth's surface is strongly absorbed by CO_2, H_2O, and other molecules present in smaller amounts (for example, CH_4 and N_2O) in the atmosphere. In effect, the atmosphere traps some of the energy, acting like the glass in a greenhouse and keeping the earth warmer than it would otherwise be.

CHEMISTRY IN FOCUS

Seeing the Light

We are about to have a revolution in lighting. The incandescent light bulb developed by Thomas Edison in the late nineteenth century still dominates our lighting systems. However, this is about to change because Edison's light bulb is so inefficient: about 95% of the energy goes to heat instead of light. In the United States, 22% of total electricity production goes for lighting, for a cost of about $58 million. Globally, illumination consumes about 19% of electricity, and demand for lighting is expected to grow by 60% in the next 25 years. Given energy prices and the problems associated with global warming, we must find more efficient lighting devices.

In the short term, the answer appears to be compact fluorescent lights (CFLs). These bulbs, which have a screw-type base, draw only about 20% as much energy as incandescent bulbs for a comparable amount of light production. Although they cost four times as much, CFLs last ten times as long as incandescent bulbs. CFLs produce light from a type of compound called a phosphor that coats the inner walls of the bulb. The phosphor is mixed with a small amount of mercury (about 5 mg per bulb). When the bulb is turned on, a beam of electrons is produced. The electrons are absorbed by mercury atoms, which are caused to emit ultraviolet (UV) light. This UV light is absorbed by phosphor, which then emits visible light (a process called fluorescence). It is estimated that replacing all of the incandescent bulbs in our homes with CFLs would reduce our electrical demand in the United States by the equivalent of the power produced by 20 new 1000-MW nuclear power plants. This is a very significant savings.

Although the amount of mercury in each bulb is small (breaking a single CFL would not endanger a normal adult), recycling large numbers of CFLs does present potential pollution hazards. Research is now underway to find ways to alleviate this danger. For example, Professor Robert Hurt and his colleagues at Brown University have found that selenium prepared as tiny particles has a very high affinity for mercury and can be used in recycling operations to prevent dangerous occupational exposure to mercury.

Another type of lighting device that is now economical enough to be used widely is the light-emitting diode (LED). An LED is a solid-state semiconductor designed to emit visible light when its electrons fall to lower energy levels. The tiny glowing light that indicates an audio system or television is on is an LED. In recent years, LEDs have been used in traffic lights, turn signals on cars, flashlights, and street lights. The use of LEDs for holiday lighting is rapidly increasing. It is estimated that LEDs eventually will reduce energy consumption for holiday lighting by 90%. The light production of LEDs per amount of energy consumed has increased dramatically in recent months, and the costs are decreasing steadily. Although LED lights are more expensive than CFLs, they last more than 15 years. Thus dramatic changes are occurring in the methods for lighting, and we all need to do our part to make our lives more energy efficient.

A compact fluorescent light bulb (CFL).

The atmosphere's water content is controlled by the water cycle (evaporation and precipitation), and the average has remained constant over the years. However, as fossil fuels have been used more extensively, the carbon dioxide concentration has increased—up about 20% from 1880 to the present. Projections indicate that the carbon dioxide content of the atmosphere may be double in the twenty-first century what it was in 1880. This trend *could* increase the earth's average temperature by as much as 10 °C, causing dramatic changes in climate and greatly affecting the growth of food crops.

How well can we predict the long-term effects of carbon dioxide? Because weather has been studied for a period of time that is minuscule compared with the age of the earth, the factors that control the earth's climate in the long range are not clearly un-

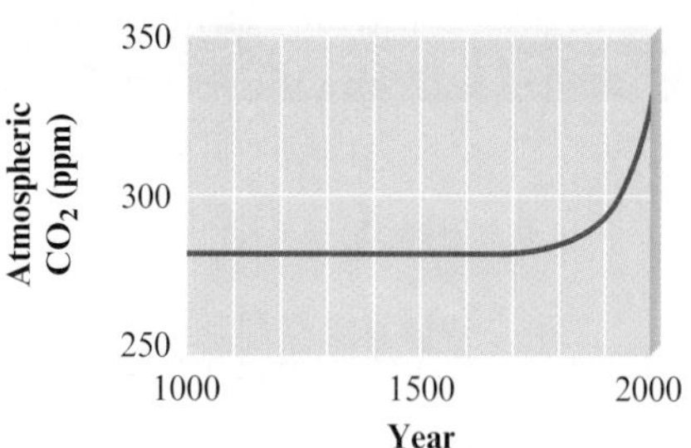

Figure 10.9 ▸ The atmospheric CO_2 concentration over the past 1000 years, based on ice core data and direct readings (since 1958). Note the dramatic increase in the past 100 years.

derstood. For example, we do not understand what causes the earth's periodic ice ages. So it is difficult to estimate the effects of the increasing carbon dioxide levels.

In fact, the variation in the earth's average temperature over the past century is somewhat confusing. In the northern latitudes during the past century, the average temperature rose by 0.8 °C over a period of 60 years, then cooled by 0.5 °C during the next 25 years, and finally warmed by 0.2 °C in the succeeding 15 years. Such fluctuations do not match the steady increase in carbon dioxide. However, in southern latitudes and near the equator during the past century, the average temperature showed a steady rise totaling 0.4 °C. This figure is in reasonable agreement with the predicted effect of the increasing carbon dioxide concentration over that period. Another significant fact is that the last 10 years of the twentieth century have been the warmest decade on record.

Although the exact relationship between the carbon dioxide concentration in the atmosphere and the earth's temperature is not known at present, one thing is clear: The increase in the atmospheric concentration of carbon dioxide is quite dramatic (Fig. 10.9). We must consider the implications of this increase as we consider our future energy needs.

New Energy Sources

As we search for the energy sources of the future, we need to consider economic, climatic, and supply factors. There are several potential energy sources: the sun (solar), nuclear processes (fission and fusion), biomass (plants), and synthetic fuels. Direct use of the sun's radiant energy to heat our homes and run our factories and transportation systems seems a sensible long-term goal. But what do we do now? Conservation of fossil fuels is one obvious step, but substitutes for fossil fuels also must be found. There is much research going on now to solve this problem.

Critical Thinking

A government study concludes that burning fossil fuels to power our automobiles causes too much pollution. What if Congress decided that all cars and trucks must be powered by batteries? Would this solve the air pollution problems caused by transportation?

10-10 Energy as a Driving Force

OBJECTIVE **To understand energy as a driving force for natural processes.**

A major goal of science is to understand why things happen as they do. In particular, we are interested in the driving forces of nature. Why do things occur in a particular direction? For example, consider a log that has burned in a fireplace, producing ashes and heat energy. If you are sitting in front of the fireplace, you would be very surprised to see the ashes begin to absorb heat from the air and reconstruct themselves into the log. It just doesn't happen. That is, the process that always occurs is

$$\text{log} + O_2(g) \rightarrow CO_2(g) + H_2O(g) + \text{ashes} + \text{energy}$$

The reverse of this process

$$CO_2(g) + H_2O(g) + \text{ashes} + \text{energy} \rightarrow \text{log} + O_2(g)$$

never happens.

Consider another example. A gas is trapped in one end of a vessel as shown below.

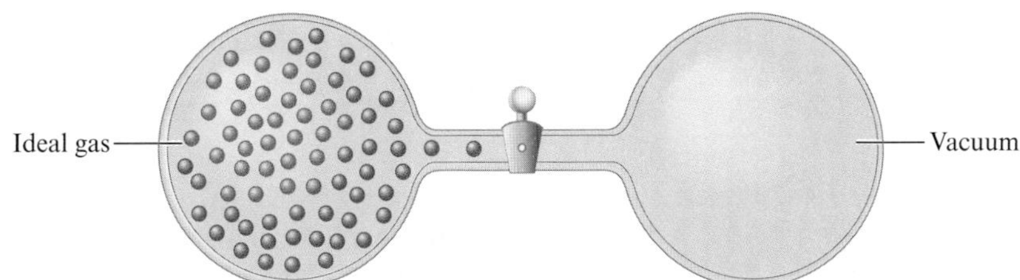

When the valve is opened, what always happens? The gas spreads evenly throughout the entire container.

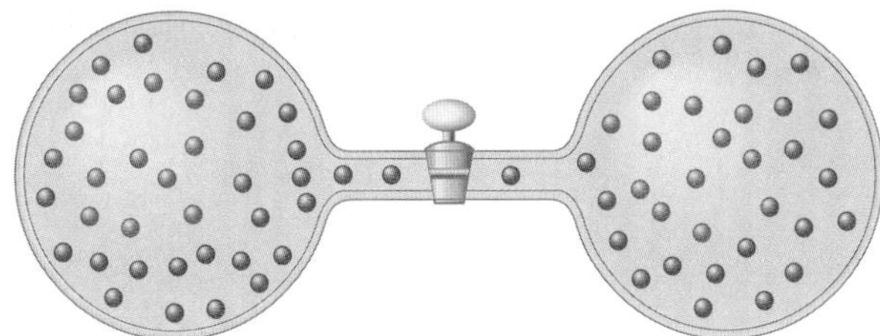

You would be very surprised to see the following process occur spontaneously:

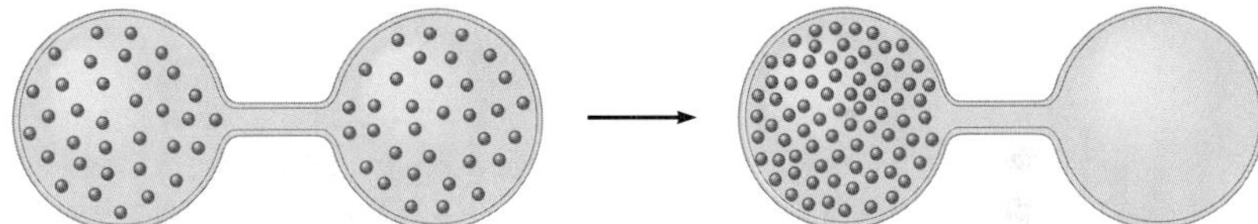

So, why does this process

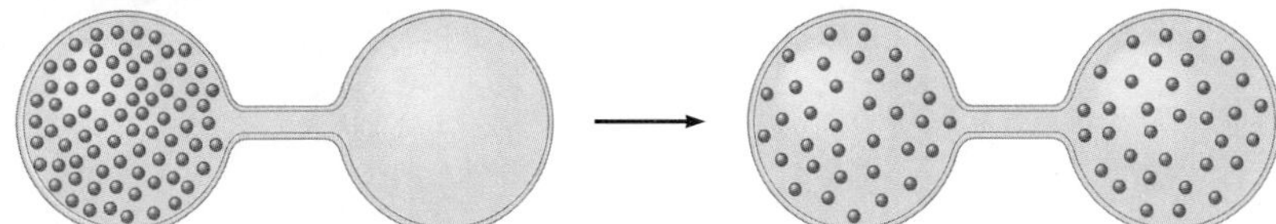

occur spontaneously but the reverse process

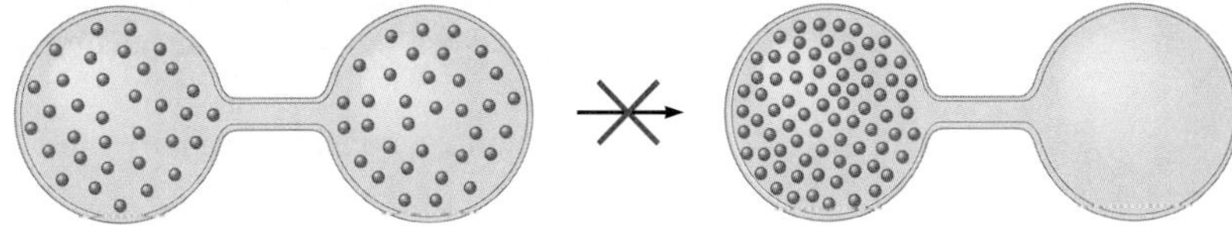

never occurs?

In many years of analyzing these and many other processes, scientists have discovered two very important driving forces:

- Energy spread
- Matter spread

Energy spread means that in a given process, concentrated energy is dispersed widely. This distribution happens every time an exothermic process occurs. For example, when a Bunsen burner burns, the energy stored in the fuel (natural gas—mostly methane) is dispersed into the surrounding air:

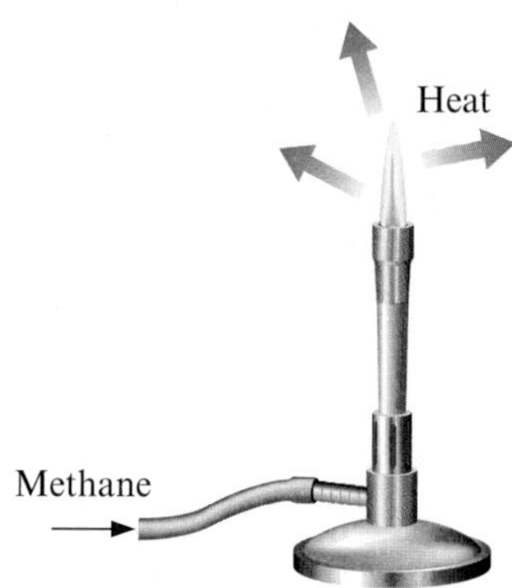

The energy that flows into the surroundings through heat increases the thermal motions of the molecules in the surroundings. In other words, this process increases the random motions of the molecules in the surroundings. *This always happens in every exothermic process.*

Matter spread means exactly what it says: the molecules of a substance are spread out and occupy a larger volume.

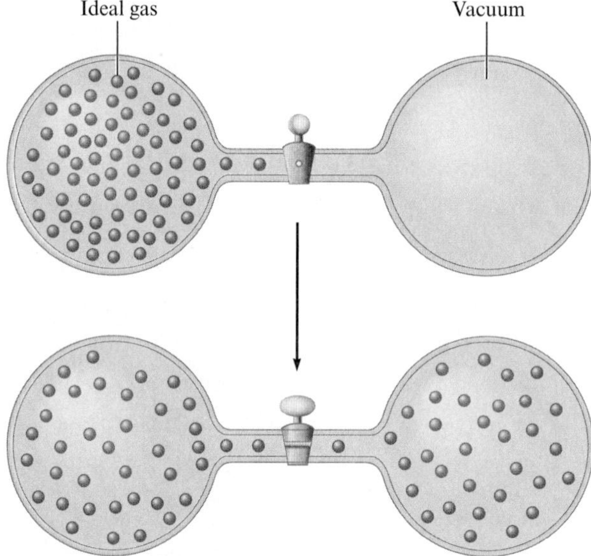

After looking at thousands of processes, scientists have concluded that these two factors are the important driving forces that cause events to occur. That is, processes are favored if they involve energy spread and matter spread.

Do these driving forces ever occur in opposition? Yes, they do—in many, many processes.

For example, consider ordinary table salt dissolving in water.

This process occurs spontaneously. You observe it every time you add salt to water to cook potatoes or pasta. Surprisingly, dissolving salt in water is *endothermic*. This process seems to go in the wrong direction—it involves energy concentration, not energy spread. Why does the salt dissolve? Because of matter spread. The Na^+ and Cl^- that are closely packed in the solid NaCl become spread around randomly in a much larger volume in the resulting solution. Salt dissolves in water because the favorable matter spread overcomes an unfavorable energy change.

Entropy

Entropy is a function we have invented to keep track of the natural tendency for the components of the universe to become disordered—entropy (designated by the letter S) is a measure of disorder or randomness. As randomness increases, S increases. Which has lower entropy, solid water (ice) or gaseous water (steam)? Remember that ice contains closely packed, ordered H_2O molecules, and steam has widely dispersed, randomly moving H_2O molecules (Fig. 10.10). Thus ice has more order and a lower value of S.

What do you suppose happens to the disorder of the universe as energy spread and matter spread occur during a process?

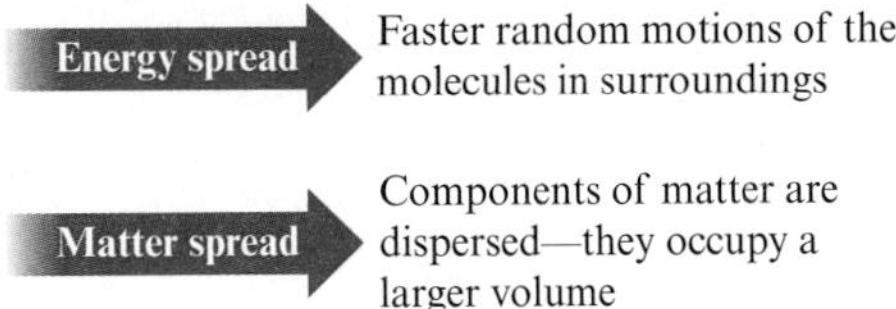

It seems clear that both energy spread and matter spread lead to greater entropy (greater disorder) in the universe. This idea leads to a very important conclusion that is summarized in the **second law of thermodynamics:**

The entropy of the universe is always increasing.

Figure 10.10 ▸ Comparing the entropies of ice and steam.

Critical Thinking

What if the first law of thermodynamics were true, but the second law were not? How would the world be different?

A **spontaneous process** is one that occurs in nature without outside intervention—it happens "on its own." The second law of thermodynamics helps us to understand why certain processes are spontaneous and others are not. It also helps us to understand the conditions necessary for a process to be spontaneous. For example, at 1 atm (1 atmosphere of pressure), ice will spontaneously melt above a temperature of 0 °C but not below this temperature. A process is spontaneous only if the entropy of the universe increases as a result of the process. That is, all processes that occur in the universe lead to a net increase in the disorder of the universe. As the universe "runs," it is always heading toward more disorder. We are plunging slowly but inevitably toward total randomness—the heat death of the universe. But don't despair; it will not happen soon.

CHAPTER 10 REVIEW

F directs you to the *Chemistry in Focus* feature in the chapter

Key Terms

energy (10-1)
potential energy (10-1)
kinetic energy (10-1)
law of conservation of energy (10-1)
work (10-1)
state function (10-1)
temperature (10-2)
heat (10-2)
system (10-3)
surroundings (10-3)
exothermic (10-3)
endothermic (10-3)
thermodynamics (10-4)
first law of thermodynamics (10-4)
internal energy (10-4)
calorie (10-5)
joule (10-5)
specific heat capacity (10-5)
enthalpy (10-6)
calorimeter (10-6)
Hess's law (10-7)
fossil fuels (10-9)
petroleum (10-9)
natural gas (10-9)
coal (10-9)
greenhouse effect (10-9)
energy spread (10-10)
matter spread (10-10)
entropy (10-10)
second law of thermodynamics (10-10)
spontaneous process (10-10)

For Review

- Energy is conserved.
- The law of conservation of energy states that energy is neither created nor destroyed in any process.
- In a process, energy can be changed from one form to another, but the amount of energy remains constant.
- Thermodynamics is the study of energy and its changes.
- Energy is classified as one of the following:
  - Kinetic energy—energy due to the motion of an object
  - Potential energy—energy due to the position or composition of an object
- Some functions, called state functions, depend only on the beginning and final states of the system, not on the specific pathway followed.
  - Energy is a state function.
  - Heat and work are not state functions.
- Temperature indicates the vigor of the random motions of the components of that substance.
- Thermal energy is the sum of the energy produced by the random motions of the components.
- Heat is a flow of energy between two objects due to a temperature difference between the objects.
  - An exothermic process is one in which energy as heat flows out of the system into the surroundings.
  - An endothermic process is one in which energy as heat flows into the system from the surroundings.
  - The common units for heat are calories and joules.

- Internal energy (E) is the sum of the kinetic and potential energy associated with an object.
  - Internal energy (E) can be changed by two types of energy flow:
    - Heat (q)
    - Work (w)
  - $\Delta E = q + w$
- Specific heat capacity is the energy required to change the temperature of a mass of 1 g of a substance by 1°C.

- For a process carried out at constant pressure, the change in enthalpy (ΔH) of that process is equal to the heat.

Characteristics of ΔH

- If a reaction is reversed, the sign of ΔH is also reversed.
- The magnitude of ΔH is directly proportional to the quantities of reactants and products in a reaction. If the coefficients in a balanced reaction are multiplied by an integer, the value of ΔH is multiplied by the same integer.

- Hess's law enables the calculation of the heat for a given reaction from known heats of related reactions.
- A calorimeter is a device used to measure the heat associated with a given chemical reaction.
- Although energy is conserved in every process, the quality (usefulness) of the energy decreases in every real process.
- Natural processes occur in the direction that leads to an increase in the disorder (increase in entropy) of the universe.
  - The principal driving forces for natural processes can be described in terms of "matter spread" and "energy spread."
- Our world has many sources of energy. The use of this energy affects the environment in various ways.

All even-numbered Questions and Problems have answers in the back of this book and solutions in the *Student Solutions Guide*.

Active Learning Questions

These questions are designed to be considered by groups of students in class. Often these questions work well for introducing a particular topic in class.

1. Look at Fig. 10.1 in your text. Ball A has stopped moving. However, energy must be conserved. So what happened to the energy of ball A?
2. A friend of yours reads that the process of water freezing is exothermic. This friend tells you that this can't be true because exothermic implies "hot," and ice is cold. Is the process of water freezing exothermic? If so, explain this process so your friend can understand it. If not, explain why not.
3. You place hot metal into a beaker of cold water.
   a. Eventually what is true about the temperature of the metal compared to that of the water? Explain why this is true.
   b. Label this process as endothermic or exothermic if we consider the system to be
      i. the metal. Explain.
      ii. the water. Explain.
4. What does it mean when the heat for a process is reported with a negative sign?
5. You place 100.0 g of a hot metal in 100.0 g of cold water. Which substance (metal or water) undergoes a larger temperature change? Why is this?
6. Explain why aluminum cans make good storage containers for soft drinks. Styrofoam cups can be used to keep coffee hot and cola cold. How can this be?
7. In Section 10-7, two characteristics of enthalpy changes for reactions are listed. What are these characteristics? Explain why these characteristics are true.
8. What is the difference between *quality* and *quantity* of energy? Are both conserved? Is either conserved?
9. What is meant by the term *driving forces?* Why are *matter spread* and *energy spread* considered to be driving forces?
10. Give an example of a process in which *matter spread* is a driving force and an example of a process in which *energy spread* is a driving force, and explain each. These examples should be different from the ones given in the text.
11. Explain in your own words what is meant by the term *entropy*. Explain how both *matter spread* and *energy spread* are related to the concept of entropy.
12. Consider the processes

$$H_2O(g) \rightarrow H_2O(l)$$
$$H_2O(l) \rightarrow H_2O(g)$$

   a. Which process is favored by energy spread? Explain.
   b. Which process is favored by matter spread? Explain.
   c. How does temperature affect which process is favored? Explain.
13. What if energy was not conserved? How would this affect our lives?
14. The internal energy of a system is said to be the sum of the kinetic and potential energies of all the particles in the system. Section 10-1 discusses *potential energy* and *kinetic energy* in terms of a ball on a hill. Explain *potential energy* and *kinetic energy* for a chemical reaction.
15. Hydrogen gas and oxygen gas react violently to form water.
   a. Which is lower in energy: a mixture of hydrogen gases, or water? Explain.
   b. Sketch an energy-level diagram (like Fig. 10.5) for this reaction and explain it.
16. Consider four 100.0-g samples of water, each in a separate beaker at 25.0 °C. Into each beaker you drop 10.0 g of a different metal that has been heated to 95.0 °C. Assuming no heat loss to the surroundings, which water sample will have the highest final temperature? Explain your answer.
   a. The water to which you have added aluminum (c = 0.89 J/g °C).
   b. The water to which you have added iron (c = 0.45 J/g °C).
   c. The water to which you have added copper (c = 0.20 J/g °C).
   d. The water to which you have added lead (c = 0.14 J/g °C).
   e. Because the masses of the metals are the same, the final temperatures would be the same.
17. For each of the following situations a–c, use the following choices i–iii to complete the statement "The final temperature of the water should be"
      i. Between 50 °C and 90 °C
      ii. 50 °C
      iii. Between 10 °C and 50 °C
   a. A 100.0-g sample of water at 90 °C is added to a 100.0-g sample of water at 10 °C.
   b. A 100.0-g sample of water at 90 °C is added to a 500.0-g sample of water at 10 °C.
   c. You have a Styrofoam cup with 50.0 g of water at 10 °C. You add a 50.0-g iron ball at 90 °C to the water.

18. How is Hess's law a restatement of the first law of thermodynamics?

19. Does the entropy of the system increase or decrease for each of the following? Explain.

a. the evaporation of alcohol
b. the freezing of water
c. dissolving NaCl in water

20. Predict the sign of $\Delta S°$ for each of the following changes.

a.

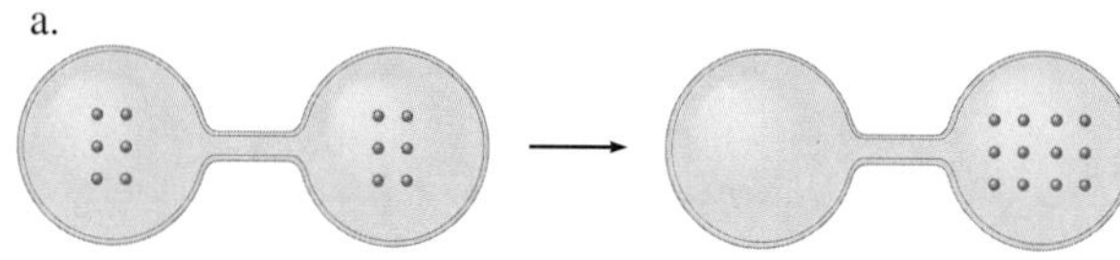

b. $AgCl(s) \rightarrow Ag^+(aq) + Cl^-(aq)$
c. $2H_2(g) + O_2 \rightarrow 2H_2O(l)$
d. $H_2O(l) \rightarrow H_2O(g)$

Questions and Problems

10-1 The Nature of Energy

Questions

1. ________ represents the ability to do work or to produce heat.
2. What is meant by *potential* energy? Give an example of an object or material that possesses potential energy.
3. What is the kinetic energy of a particle of mass m moving through space with velocity v?
4. The total energy of the universe is ________.
5. What is meant by a *state* function? Give an example.
6. In Fig. 10.1, what kind of energy does ball A possess initially when at rest at the top of the hill? What kind of energies are involved as ball A moves down the hill? What kind of energy does ball A possess when it reaches the bottom of the hill and *stops* moving after hitting ball B? Where did the energy gained by ball B, allowing it to move up the hill, come from?

10-2 Temperature and Heat

Questions

7. Students often confuse what is meant by *heat* and *temperature*. Define each. How are the two concepts related?
8. If you spilled a cup of freshly brewed *hot* tea on yourself, you would be burned. If you spilled the same quantity of *iced* tea on yourself, you would not be burned. Explain.
9. What does the *thermal energy* of an object represent?
10. How are the *temperature* of an object and the *thermal energy* of an object related?

10-3 Exothermic and Endothermic Processes

Questions

11. In studying heat flows for chemical processes, what do we mean by the terms *system* and *surroundings*?
12. When a chemical system evolves energy, where does the energy go?
13. The combustion of methane, CH_4, is an exothermic process. Therefore, the products of this reaction must possess (higher/lower) total potential energy than do the reactants.
14. Are the following processes exothermic or endothermic?

a. When solid KBr is dissolved in water, the solution gets colder.
b. Natural gas (CH_4) is burned in a furnace.
c. When concentrated sulfuric acid is added to water, the solution gets very hot.
d. Water is boiled in a teakettle.

10-4 Thermodynamics

Questions

15. What do we mean by *thermodynamics?* What is the *first law of thermodynamics?*
16. The ________ energy, E, of a system represents the sum of the kinetic and potential energies of all particles within the system.
17. Calculate ΔE for each of the following cases.

a. $q = +51$ kJ, $w = -15$ kJ
b. $q = +100.$ kJ, $w = -65$ kJ
c. $q = -65$ kJ, $w = -20.$ kJ

18. If q for a process is a negative number, then the system is (gaining/losing) energy.
19. For an endothermic process, q will have a (positive/negative) sign.
20. A system releases 125 kJ of heat, and 104 kJ of work is done on it. Calculate ΔE.

10-5 Measuring Energy Changes

Questions

21. How is the *calorie* defined? How does a *Calorie* differ from a *calorie?* How is the *joule* related to the calorie?
22. Write the conversion factors that would be necessary to perform each of the following conversions:

a. an energy given in calories to its equivalent in joules
b. an energy given in joules to its equivalent in calories
c. an energy given in calories to its equivalent in kilocalories
d. an energy given in kilojoules to its equivalent in joules

Problems

23. If 8.40 kJ of heat is needed to raise the temperature of a sample of metal from 15 °C to 20 °C, how many kilojoules of heat will be required to raise the temperature of the same sample of metal from 25 °C to 40 °C?
24. If it takes 654 J of energy to warm a 5.51-g sample of water, how much energy would be required to warm 55.1 g of water by the same amount?
25. Convert the following numbers of calories or kilocalories into joules and kilojoules (Remember: *kilo* means 1000.)

a. 75.2 kcal
b. 75.2 cal
c. 1.41×10^3 cal
d. 1.41 kcal

All even-numbered Questions and Problems have answers in the back of this book and solutions in the *Student Solutions Guide*.

26. Convert the following numbers of calories into kilocalories. (Remember: *kilo* means 1000.)
    a. 8254 cal
    b. 41.5 cal
    c. 8.231×10^3 cal
    d. 752,900 cal
27. Convert the following numbers of kilojoules into kilocalories. (Remember: *kilo* means 1000.)
    a. 652.1 kJ
    b. 1.00 kJ
    c. 4.184 kJ
    d. 4.351×10^3 kJ
28. Convert the following numbers of calories or kilocalories into joules or kilojoules.
    a. 7845 cal
    b. 4.55×10^4 cal
    c. 62.142 kcal
    d. 43,024 cal
29. Perform the indicated conversions.
    a. 625.2 cal into kilojoules
    b. 82.41 kJ into joules
    c. 52.61 kcal into joules
    d. 124.2 kJ into kilocalories
30. Perform the indicated conversions.
    a. 45.62 kcal into kilojoules
    b. 72.94 kJ into kilocalories
    c. 2.751 kJ into calories
    d. 5.721 kcal into joules
31. If 69.5 kJ of heat is applied to a 1012-g block of metal, the temperature of the metal increases by 11.4 °C. Calculate the specific heat capacity of the metal in J/g °C.
32. What quantity of heat energy must have been applied to a block of aluminum weighing 42.7 g if the temperature of the block of aluminum increased by 15.2 °C? (See Table 10.1.)
33. If 125 J of heat energy is applied to a block of silver weighing 29.3 g, by how many degrees will the temperature of the silver increase? (See Table 10.1.)
34. If 100. J of heat energy is applied to a 25-g sample of mercury, by how many degrees will the temperature of the sample of mercury increase? (See Table 10.1.)
35. What quantity of heat is required to raise the temperature of 55.5 g of gold from 20 °C to 45 °C? (See Table 10.1.)
36. The "Chemistry in Focus" segment *Coffee: Hot and Quick(lime)* discusses self-heating cups of coffee using the chemical reaction between quicklime, CaO(*s*), and water. Is this reaction endothermic or exothermic?
37. The "Chemistry in Focus" segment *Nature Has Hot Plants* discusses thermogenic, or heat-producing, plants. For some plants, enough heat is generated to increase the temperature of the blossom by 15 °C. About how much heat is required to increase the temperature of 1 L of water by 15 °C?
38. In the "Chemistry in Focus" segment *Firewalking: Magic or Science?*, it is claimed that one reason people can walk on hot coals is that human tissue is mainly composed of water. Because of this, a large amount of heat must be transferred from the coals to change significantly the temperature of the feet. How much heat must be transferred to 100.0 g of water to change its temperature by 35 °C?

All even-numbered Questions and Problems have answers in the back of this book and solutions in the *Student Solutions Guide.*

10-6 Thermochemistry (Enthalpy)

Questions

39. The enthalpy change for a reaction that occurs at constant pressure is (higher than/lower than/the same as) the heat for that reaction.
40. A ________ is a device used to determine the heat associated with a chemical reaction.

Problems

41. The enthalpy change for the reaction of hydrogen gas with fluorine gas to produce hydrogen fluoride is −542 kJ for the equation *as written*:

$$H_2(g) + F_2(g) \rightarrow 2HF(g) \qquad \Delta H = -542\ \text{kJ}$$

    a. What is the enthalpy change *per mole* of hydrogen fluoride produced?
    b. Is the reaction exothermic or endothermic as written?
    c. What would be the enthalpy change for the *reverse* of the given equation (that is, for the decomposition of HF into its constituent elements)?
42. For the reaction $S(s) + O_2(g) \rightarrow SO_2(g)$, $\Delta H = -296$ kJ per mole of SO_2 formed.
    a. Calculate the quantity of heat released when 1.00 g of sulfur is burned in oxygen.
    b. Calculate the quantity of heat released when 0.501 mole of sulfur is burned in air.
    c. What quantity of energy is required to break up 1 mole of $SO_2(g)$ into its constituent elements?
43. The "Chemistry in Focus" segment *Burning Calories* discusses calories in food. If a food is said to contain 350 calories per serving, determine this value in terms of joules.

10-7 Hess's Law

Questions

44. When ethanol (grain alcohol, C_2H_5OH) is burned in oxygen, approximately 1360 kJ of heat energy is released per mole of ethanol.

$$C_2H_5OH(l) + 3O_2(g) \rightarrow 2CO_2(g) + 3H_2O(g)$$

    a. What quantity of heat is released for each *gram* of ethanol burned?
    b. What is ΔH for the reaction *as written*?
    c. How much heat is released when sufficient ethanol is burned so as to produce 1 mole of water vapor?

Problems

45. Given the following hypothetical data:

$$X(g) + Y(g) \rightarrow XY(g) \text{ for which } \Delta H = a \text{ kJ}$$
$$X(g) + Z(g) \rightarrow XZ(g) \text{ for which } \Delta H = b \text{ kJ}$$

Calculate ΔH for the reaction

$$Y(g) + XZ(g) \rightarrow XY(g) + Z(g)$$

46. Given the following data:

$$C(s) + O_2(g) \rightarrow CO_2(g) \qquad \Delta H = -393 \text{ kJ}$$
$$2CO(g) + O_2(g) \rightarrow 2CO_2(g) \qquad \Delta H = -566 \text{ kJ}$$

Calculate ΔH for the reaction $2C(s) + O_2(g) \rightarrow CO(g)$.

47. Given the following data:

$$S(s) + \tfrac{3}{2}O_2(g) \rightarrow SO_3(g) \qquad \Delta H = -395.2 \text{ kJ}$$
$$2SO_2(g) + O_2(g) \rightarrow 2SO_3(g) \qquad \Delta H = -198.2 \text{ kJ}$$

Calculate ΔH for the reaction $S(s) + O_2(g) \rightarrow SO_2(g)$.

48. Given the following data:

$$C_2H_2(g) + \tfrac{5}{2}O_2(g) \rightarrow 2CO_2(g) + H_2O(l) \qquad \Delta H = -1300. \text{ kJ}$$
$$C(s) + O_2(g) \rightarrow CO_2(g) \qquad \Delta H = -394 \text{ kJ}$$
$$H_2(g) + \tfrac{1}{2}O_2(g) \rightarrow H_2O(l) \qquad \Delta H = -286 \text{ kJ}$$

Calculate ΔH for the reaction

$$2C(s) + H_2(g) \rightarrow C_2H_2(g)$$

10-8 Quality Versus Quantity of Energy

Questions

49. Consider the gasoline in your car's gas tank. What happens to the energy stored in the gasoline when you drive your car? Although the total energy in the universe remains constant, can the energy stored in the gasoline be reused once it is dispersed to the environment?

50. Although the total energy of the universe will remain constant, why will energy no longer be useful once everything in the universe is at the same temperature?

51. Why are petroleum products especially useful as sources of energy?

52. Why is the "quality" of energy decreasing in the universe?

10-9 Energy and Our World

Questions

53. Where did the energy stored in wood, coal, petroleum, and natural gas originally come from?

54. What does petroleum consist of? What are some "fractions" into which petroleum is refined? How are these fractions related to the sizes of the molecules involved?

55. What does natural gas consist of? Where is natural gas commonly found?

56. What was tetraethyl lead used for in the petroleum industry? Why is it no longer commonly used?

57. What are the four "stages" of coal formation? How do the four types of coal differ?

58. What is the "greenhouse effect"? Why is a certain level of greenhouse gases beneficial, but too high a level dangerous to life on earth? What is the most common greenhouse gas?

10-10 Energy as a Driving Force

Questions

59. A ________ is some factor that tends to make a process occur.

60. What is the second law of thermodynamics? How is this law related to energy spread and matter spread?

61. If a reaction occurs readily but has an endothermic heat of reaction, what must be the driving force for the reaction?

62. Does a double-displacement reaction such as

$$NaCl(aq) + AgNO_3(aq) \rightarrow AgCl(s) + NaNO_3(aq)$$

result in a matter spread or in a concentration of matter?

63. What do we mean by *entropy?* Why does the entropy of the universe increase during a spontaneous process?

64. A chunk of ice at room temperature melts, even though the process is endothermic. Why?

Additional Problems

65. In an endothermic reaction, do the reactants or the products have the lower potential energy?

66. Calculate the enthalpy change when 1.00 g of methane is burned in excess oxygen according to the reaction

$$CH_4(g) + 2O_2(g) \rightarrow CO_2(g) + H_2O(l) \qquad \Delta H = -891 \text{ kJ/mol}$$

67. Perform the indicated conversions.
    a. 85.21 cal into joules
    b. 672.1 J into calories
    c. 8.921 kJ into joules
    d. 556.3 cal into kilojoules

68. Calculate the amount of energy required (in calories) to heat 145 g of water from 22.3 °C to 75.0 °C.

69. It takes 1.25 kJ of energy to heat a certain sample of pure silver from 12.0 °C to 15.2 °C. Calculate the mass of the sample of silver.

70. What quantity of heat energy would have to be applied to a 25.1-g block of iron in order to raise the temperature of the iron sample by 17.5 °C? (See Table 10.1.)

71. The specific heat capacity of gold is 0.13 J/g °C. Calculate the specific heat capacity of gold in cal/g °C.

72. Calculate the amount of energy required (in joules) to heat 2.5 kg of water from 18.5 °C to 55.0 °C.

73. If 10. J of heat is applied to 5.0-g samples of each of the substances listed in Table 10.1, which substance's temperature will increase the most? Which substance's temperature will increase the least?

74. A 50.0-g sample of water at 100. °C is poured into a 50.0-g sample of water at 25 °C. What will be the final temperature of the water?

75. A 25.0-g sample of pure iron at 85 °C is dropped into 75 g of water at 20. °C. What is the final temperature of the water–iron mixture?

All even-numbered Questions and Problems have answers in the back of this book and solutions in the *Student Solutions Guide.*

76. If 7.24 kJ of heat is applied to a 952-g block of metal, the temperature increases by 10.7 °C. Calculate the specific heat capacity of the metal in J/g °C.

77. For each of the substances listed in Table 10.1, calculate the quantity of heat required to heat 150. g of the substance by 11.2 °C.

78. A system releases 213 kJ of heat and has a calculated ΔE of −45 kJ. How much work was done on the system?

79. Calculate ΔE for each of the following.

a. $q = -47$ kJ, $w = +88$ kJ
b. $q = +82$ kJ, $w = +47$ kJ
c. $q = +47$ kJ, $w = 0$
d. In which of these cases do the surroundings do work on the system?

80. Calculate the enthalpy change when 5.00 g of propane is burned with excess oxygen according to the reaction

$$C_3H_8(g) + 5O_2(g) \rightarrow 3CO_2(g) + 4H_2O(l) \qquad \Delta H = -2221 \text{ kJ/mol}$$

81. The overall reaction in commercial heat packs can be represented as

$$4Fe(s) + 3O_2(g) \rightarrow 2Fe_2O_3(s) \qquad \Delta H = -1652 \text{ kJ}$$

a. How much heat is released when 4.00 moles of iron is reacted with excess O_2?
b. How much heat is released when 1.00 mole of Fe_2O_3 is produced?
c. How much heat is released when 1.00 g iron is reacted with excess O_2?
d. How much heat is released when 10.0 g Fe and 2.00 g O_2 are reacted?

82. Consider the following equations:

$$3A + 6B \rightarrow 3D \qquad \Delta H = -403 \text{ kJ/mol}$$
$$E + 2F \rightarrow A \qquad \Delta H = -105.2 \text{ kJ/mol}$$
$$C \rightarrow E + 3D \qquad \Delta H = +64.8 \text{ kJ/mol}$$

Suppose the first equation is reversed and multiplied by $\frac{1}{6}$, the second and third equations are divided by 2, and the three adjusted equations are added. What is the net reaction and what is the overall heat of this reaction?

83. It has been determined that the body can generate 5500 kJ of energy during one hour of strenuous exercise. Perspiration is the body's mechanism for eliminating this heat. How many grams and how many liters of water would have to be evaporated through perspiration to rid the body of the heat generated during two hours of exercise? (The heat of vaporization of water is 40.6 kJ/mol.)

84. Liquid water turns to ice. Is this process endothermic or exothermic? Choose the *best* answer.

a. *Endothermic.* The water absorbed heat and got colder, thereby forming ice.
b. *Endothermic.* Energy in the form of heat was given off by the water, which became colder and formed ice.
c. *Exothermic.* The water released energy, slowing down the water molecules so that solid ice formed.
d. *Exothermic.* Heat was absorbed by the water, moving its molecules faster so that they condensed on an object and formed ice.
e. *Neither endothermic nor exothermic.* There was no energy transfer in or out of the water to form ice.

ChemWork Problems

These multiconcept problems (and additional ones) are found interactively online with the same type of assistance a student would get from an instructor.

85. Which of the following reactions is/are endothermic?

a. $CO_2(s) \rightarrow CO_2(g)$
b. $NH_3(g) \rightarrow NH_3(l)$
c. $2H_2(g) + O_2(g) \rightarrow 2H_2O(g)$
d. $H_2O(l) \rightarrow H_2O(s)$
e. $Cl_2(g) \rightarrow 2Cl(g)$

86. The specific heat capacity of graphite is 0.71 J/g °C. Calculate the energy required to raise the temperature of 2.4 moles of graphite by 25.0 °C.

87. A swimming pool, 10.0 m by 4.0 m, is filled with water to a depth of 3.0 m at a temperature of 20.2 °C. How much energy is required to raise the temperature of the water to 24.6 °C?

88. Consider the reaction

$$B_2H_6(g) + 3O_2(g) \rightarrow B_2O_3(s) + 3H_2O(g) \qquad \Delta H = -2035 \text{ kJ}$$

Calculate the amount of heat released when 54.0 g of diborane is combusted.

89. Calculate ΔH for the reaction

$$N_2H_4(l) + O_2(g) \rightarrow N_2(g) + 2H_2O(l)$$

given the following data:

Equation	ΔH (kJ)
$2NH_3(g) + 3N_2O(g) \rightarrow 4N_2(g) + 3H_2O(l)$	−1010
$N_2O(g) + 3H_2(g) \rightarrow N_2H_4(l) + H_2O(l)$	−317
$2NH_3(g) + \frac{1}{2}O_2(g) \rightarrow N_2H_4(l) + H_2O(l)$	−143
$H_2(g) + \frac{1}{2}O_2(g) \rightarrow H_2O(l)$	−286

All even-numbered Questions and Problems have answers in the back of this book and solutions in the *Student Solutions Guide.*

11 Modern Atomic Theory

CHAPTER 11

The Aurora Borealis. The colors are due to spectral emissions of nitrogen and oxygen.

John Hemmingsen/FlickrVision/Getty Images

The concept of atoms is a very useful one. It explains many important observations, such as why compounds always have the same composition (a specific compound always contains the same types and numbers of atoms) and how chemical reactions occur (they involve a rearrangement of atoms).

Once chemists came to "believe" in atoms, a logical question followed: What are atoms like? What is the structure of an atom? In Chapter 4 we learned to picture the atom with a positively charged nucleus composed of protons and neutrons at its center and electrons moving around the nucleus in a space very large compared to the size of the nucleus.

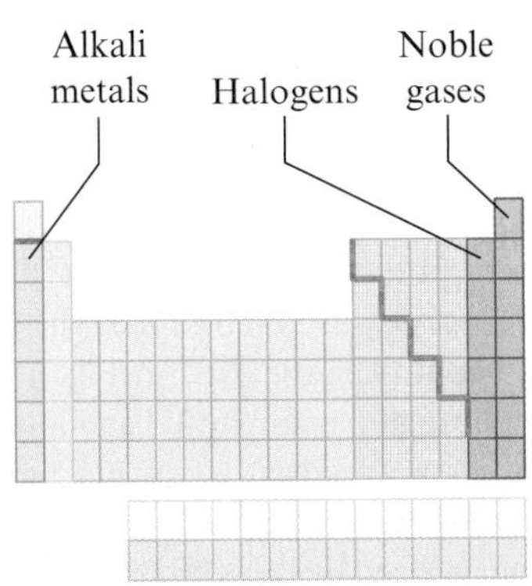

In this chapter we will look at atomic structure in more detail. In particular, we will develop a picture of the electron arrangements in atoms—a picture that allows us to account for the chemistry of the various elements. Recall from our discussion of the periodic table in Chapter 4 that, although atoms exhibit a great variety of characteristics, certain elements can be grouped together because they behave similarly. For example, fluorine, chlorine, bromine, and iodine (the halogens) show great chemical similarities. Likewise, lithium, sodium, potassium, rubidium, and cesium (the alkali metals) exhibit many similar properties, and the noble gases (helium, neon, argon, krypton, xenon, and radon) are all very nonreactive. Although the members of each of these groups of elements show great similarity *within* the group, the differences in behavior *between* groups are striking. In this chapter we will see that it is the way the electrons are arranged in various atoms that accounts for these facts.

Neon sign for a Chinese restaurant in New York City. Jeff Greenberg/Getty Images

Rutherford's Atom

OBJECTIVE **To describe Rutherford's model of the atom.**

Remember that in Chapter 4 we discussed the idea that an atom has a small positive core (called the nucleus) with negatively charged electrons moving around the nucleus in some way (Fig. 11.1). This concept of a *nuclear atom* resulted from Ernest Rutherford's experiments in which he bombarded metal foil with α particles (see Section 4.5). Rutherford and his coworkers were able to show that the nucleus of the atom is composed of positively charged particles called *protons* and neutral particles called *neutrons.* Rutherford also found that the nucleus is apparently very small compared to the size of the entire atom. The electrons account for the rest of the atom.

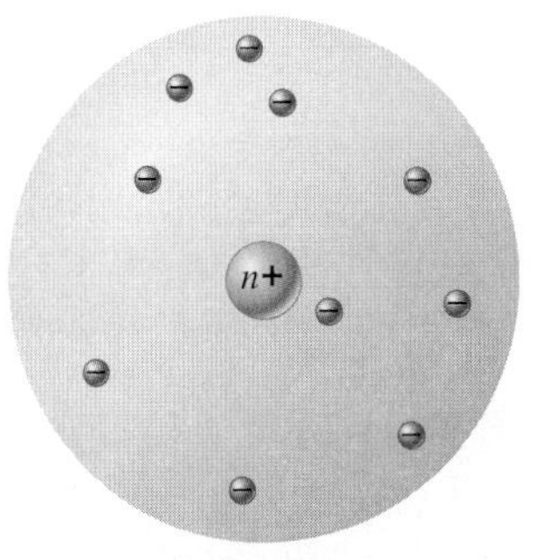

Figure 11.1 The Rutherford atom. The nuclear charge ($n+$) is balanced by the presence of n electrons moving in some way around the nucleus.

A major question left unanswered by Rutherford's work was, What are the electrons doing? That is, how are the electrons arranged and how do they move? Rutherford suggested that electrons might revolve around the nucleus like the planets revolve around the sun in our solar system. He couldn't explain, however, why the negative electrons aren't attracted into the positive nucleus, causing the atom to collapse.

At this point it became clear that more observations of the properties of atoms were needed to understand the structure of the atom more fully. To help us understand these observations, we need to discuss the nature of light and how it transmits energy.

11-2 Electromagnetic Radiation

OBJECTIVE To explore the nature of electromagnetic radiation.

If you hold your hand a few inches from a brightly glowing light bulb, what do you feel? Your hand gets warm. The "light" from the bulb somehow transmits energy to your hand. The same thing happens if you move close to the glowing embers of wood in a fireplace—you receive energy that makes you feel warm. The energy you feel from the sun is a similar example.

In all three of these instances, energy is being transmitted from one place to another by light—more properly called **electromagnetic radiation.** Many kinds of electromagnetic radiation exist, including the X rays used to make images of bones, the "white" light from a light bulb, the microwaves used to cook hot dogs and other food, and the radio waves that transmit voices and music. How do these various types of electromagnetic radiation differ from one another? To answer this question we need to talk about waves. To explore the characteristics of waves, let's think about ocean waves. In Fig. 11.2 a seagull is shown floating on the ocean and being raised and lowered by the motion of the water surface as waves pass by. Notice that the gull just moves up and down as the waves pass—it is not moved forward. A particular wave is characterized by three properties: *wavelength, frequency,* and *speed.*

Figure 11.2 ▸ A seagull floating on the ocean moves up and down as waves pass.

The **wavelength** (symbolized by the Greek letter lambda, λ) is the distance between two consecutive wave peaks (Fig. 11.3). The **frequency** of the wave (symbolized by the Greek letter nu, ν) indicates how many wave peaks pass a certain point per given time period. This idea can best be understood by thinking about how many times the seagull in Fig. 11.2 goes up and down per minute. The *speed* of a wave indicates how fast a given peak travels through the water.

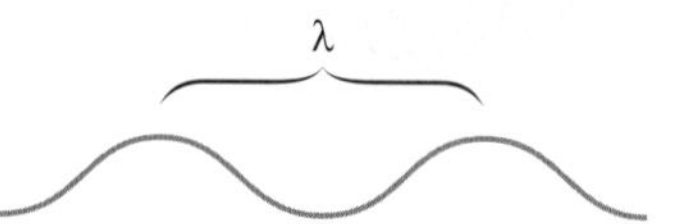

Figure 11.3 ▸ The wavelength of a wave is the distance between peaks.

Although it is more difficult to picture than water waves, light (electromagnetic radiation) also travels as waves. The various types of electromagnetic radiation (X rays, microwaves, and so on) differ in their wavelengths. The classes of electromagnetic radiation are shown in Fig. 11.4. Notice that X rays have very short wavelengths, whereas radio waves have very long wavelengths.

Radiation provides an important means of energy transfer. For example, the energy from the sun reaches the earth mainly in the forms of visible and ultraviolet radiation. The glowing coals of a fireplace transmit heat energy by infrared radiation. In a microwave oven, the water molecules in food absorb microwave radiation, which increases their mo-

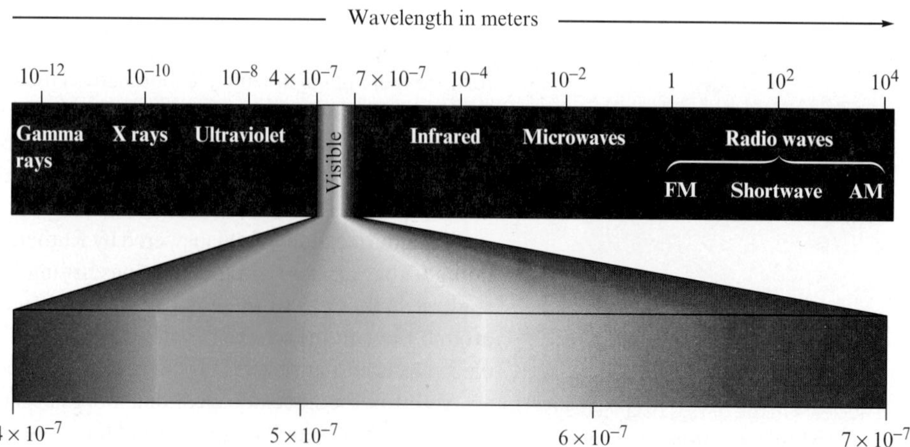

Figure 11.4 ▸ The different wavelengths of electromagnetic radiation.

Light as a Sex Attractant

Parrots, which are renowned for their vibrant colors, apparently have a secret weapon that enhances their colorful appearance—a phenomenon called *fluorescence.* Fluorescence occurs when a substance absorbs ultraviolet (UV) light, which is invisible to the human eye, and converts it to visible light. This phenomenon is widely used in interior lighting in which long tubes are coated with a fluorescent substance. The fluorescent coating absorbs UV light (produced in the interior of the tube) and emits intense white light, which consists of all wavelengths of visible light.

Interestingly, scientists have shown that parrots have fluorescent feathers that are used to attract the opposite sex. Note in the accompanying photos that a budgerigar parrot has certain feathers that produce fluorescence. Kathryn E. Arnold of the University of Queensland in Australia examined the skins of 700 Australian parrots from museum collections and found that the feathers that showed fluorescence were always display feathers—ones that were fluffed or waggled during courtship. To test her theory that fluorescence is a significant aspect of parrot romance, Arnold studied the behavior of a parrot toward birds of the opposite sex. In some cases, the potential mate had a UV-blocking substance applied to its feathers, blocking its fluorescence. Arnold's study revealed that parrots always preferred partners that showed fluorescence over those in which the fluorescence was blocked. Perhaps on your next date you might consider wearing a shirt with some fluorescent decoration!

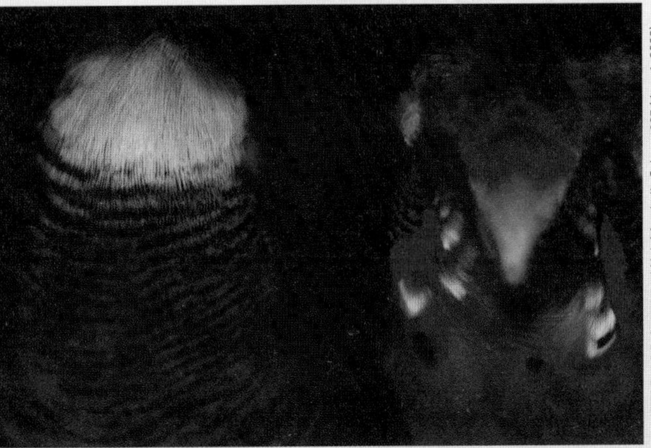

Arnold, K.E., I.P.F. Owens and N.J. Marshall. *Science* 295 (Jan. 4, 2002): 92/Justin Marshall

The back and front of a budgerigar parrot. In the photo at the right, the same parrot is seen under ultraviolet light.

tions; this energy is then transferred to other types of molecules by collisions, increasing the food's temperature.

Thus we visualize electromagnetic radiation ("light") as a wave that carries energy through space. Sometimes, however, light doesn't behave as though it were a wave. That is, electromagnetic radiation can sometimes have properties that are characteristic of particles. (You will learn more about this idea in later courses.) Another way to think of a beam of light traveling through space, then, is as a stream of tiny packets of energy called **photons.**

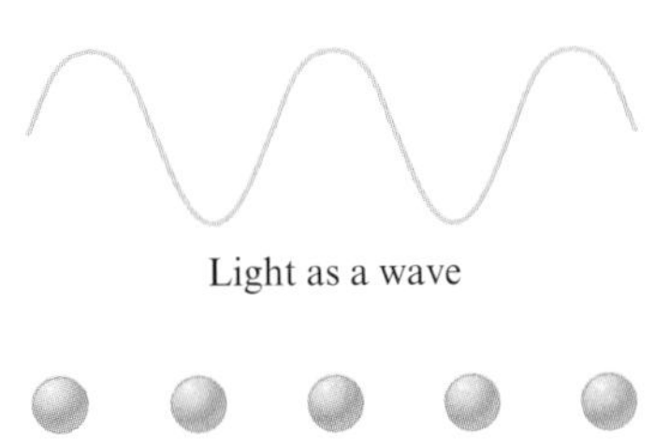

Figure 11.5 Electromagnetic radiation (a beam of light) can be pictured in two ways: as a wave and as a stream of individual packets of energy called photons.

What is the exact nature of light? Does it consist of waves or is it a stream of particles of energy? It seems to be both (Fig. 11.5). This situation is often referred to as the wave–particle nature of light.

Different wavelengths of electromagnetic radiation carry different amounts of energy. For example, the photons that correspond to red light carry less energy than the photons that correspond to blue light. In general, the longer the wavelength of light, the lower the energy of its photons (Fig. 11.6).

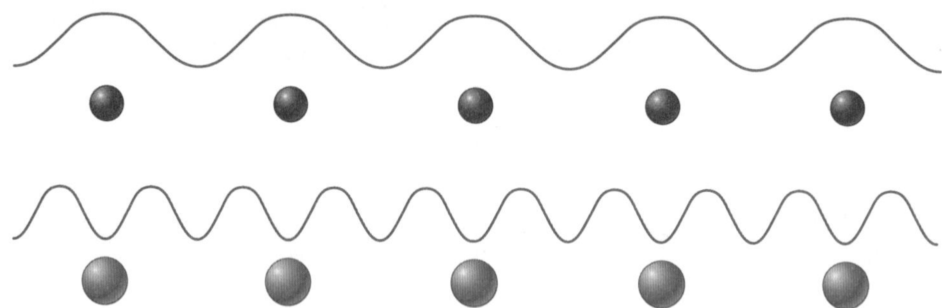

Figure 11.6 ▸ A photon of red light (relatively long wavelength) carries less energy than does a photon of blue light (relatively short wavelength).

11-3 Emission of Energy by Atoms

OBJECTIVE To understand how atoms emit light.

Consider the results of the experiment sometimes referred to as a "flame test." This experiment is run by dissolving compounds containing the Li^+ ion, the Cu^{2+} ion, and the Na^+ ion in separate dishes containing methyl alcohol (with a little water added to help dissolve the compounds). The solutions are then set on fire. Notice the brilliant colors that result. The solution containing Li^+ gives a beautiful, deep-red color, while the Cu^{2+} solution burns green. Notice that the Na^+ solution burns with a yellow–orange color, a color that should look familiar to you from the lights used in many parking lots. The color of these "sodium vapor lights" arises from the same source (the sodium atom) as the color of the burning solution containing Na^+ ions.

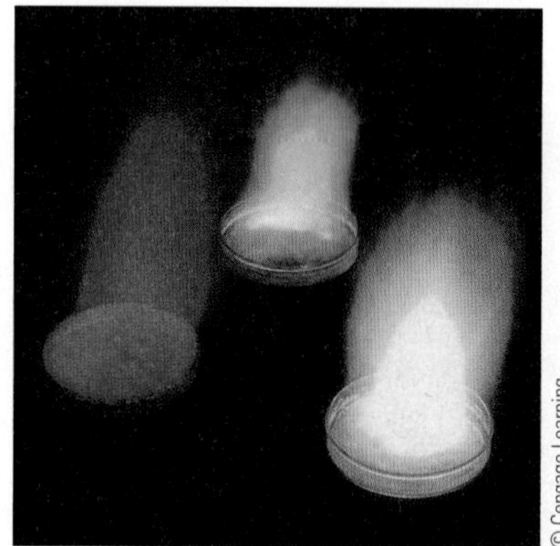

When salts containing Li^+, Cu^{2+}, and Na^+ dissolved in methyl alcohol are set on fire, brilliant colors result: Li^+, red; Cu^{2+}, green; and Na^+, yellow.

As we will see in more detail in the next section, the colors of these flames result from atoms in these solutions releasing energy by emitting visible light of specific wavelengths (that is, specific colors). The heat from the flame causes the atoms to absorb energy—we say that the atoms become *excited.* Some of this excess energy is then released in the form of light. The atom moves to a lower energy state as it emits a photon of light.

Lithium emits red light because its energy change corresponds to photons of red light (Fig. 11.7). Copper emits green light because it undergoes a different energy change than lithium; the energy change for copper corresponds to the energy of a photon of green light. Likewise, the energy change for sodium corresponds to a photon with a yellow–orange color.

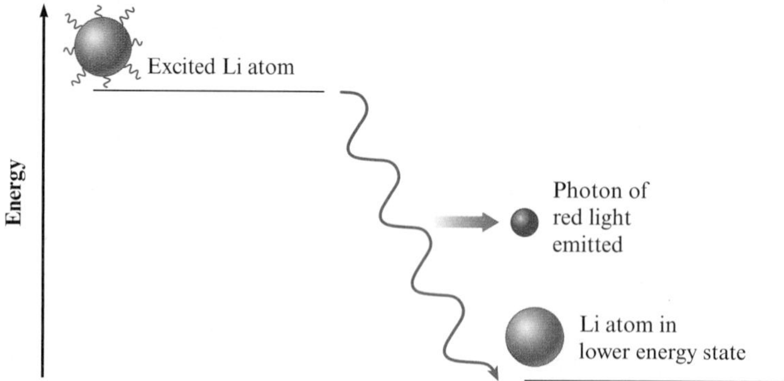

Figure 11.7 ▸ An excited lithium atom emitting a photon of red light to drop to a lower energy state.

To summarize, we have the following situation. When atoms receive energy from some source—they become excited—they can release this energy by emitting light. The emitted energy is carried away by a photon. Thus the energy of the photon corresponds exactly to the energy change experienced by the emitting atom. High-energy photons correspond to short-wavelength light and low-energy photons correspond to long-wavelength light. The photons of red light therefore carry less energy than the photons of blue light because red light has a longer wavelength than blue light does.

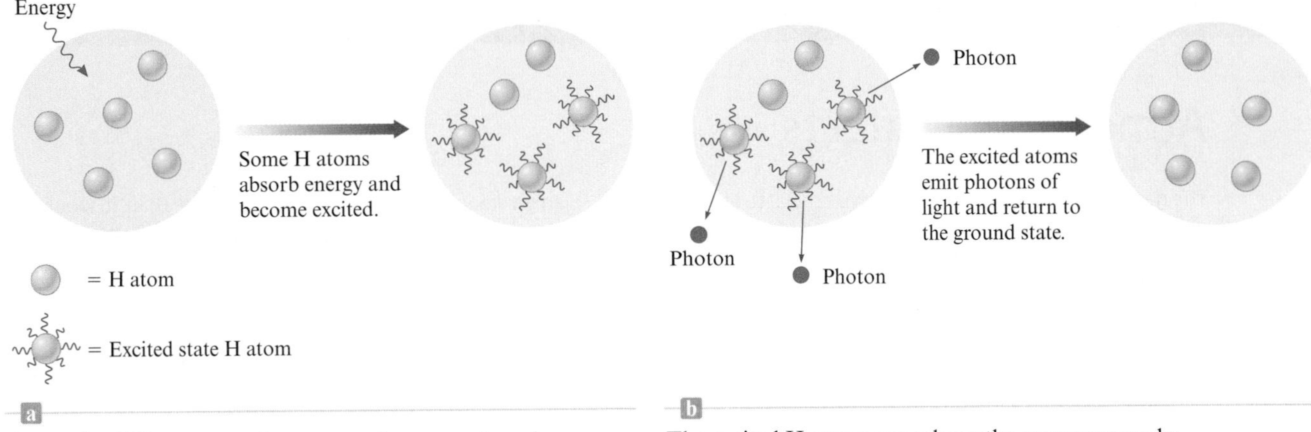

a A sample of H atoms receives energy from an external source, which causes some of the atoms to become excited (to possess excess energy).

b The excited H atoms can release the excess energy by emitting photons. The energy of each emitted photon corresponds exactly to the energy lost by each excited atom.

Figure 11.8

11-4 The Energy Levels of Hydrogen

OBJECTIVE **To understand how the emission spectrum of hydrogen demonstrates the quantized nature of energy.**

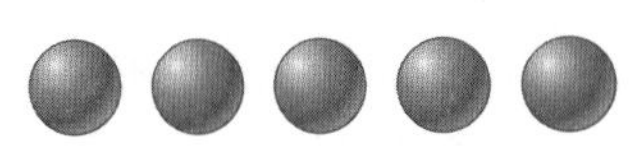

Each photon of blue light carries a larger quantity of energy than a photon of red light.

A particular color (wavelength) of light carries a particular amount of energy per photon.

As we learned in the last section, an atom with excess energy is said to be in an *excited state*. An excited atom can release some or all of its excess energy by emitting a photon (a "particle" of electromagnetic radiation) and thus move to a lower energy state. The lowest possible energy state of an atom is called its *ground state*.

We can learn a great deal about the energy states of hydrogen atoms by observing the photons they emit. To understand the significance of this, you need to remember that the *different wavelengths of light carry different amounts of energy per photon.* Recall that a beam of red light has lower-energy photons than a beam of blue light.

When a hydrogen atom absorbs energy from some outside source, it uses this energy to enter an excited state. It can release this excess energy (go back to a lower state) by emitting a photon of light (Fig. 11.8). We can picture this process in terms of the energy-level diagram shown in Fig. 11.9. The important point here is that *the energy contained in the photon corresponds to the change in energy that the atom experiences* in going from the excited state to the lower state.

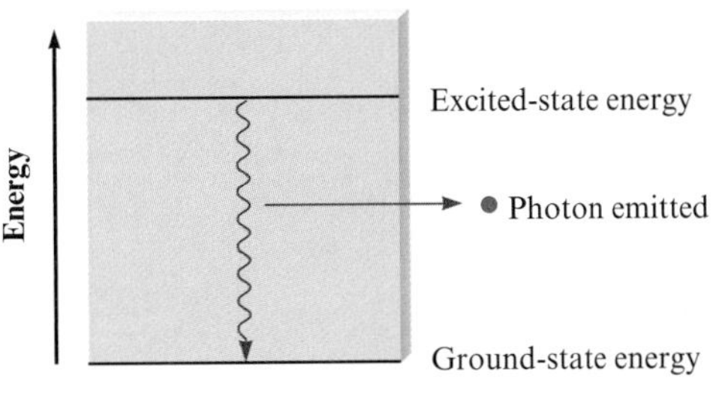

Figure 11.9 When an excited H atom returns to a lower energy level, it emits a photon that contains the energy released by the atom. Thus the energy of the photon corresponds to the difference in energy between the two states.

Consider the following experiment. Suppose we take a sample of H atoms and put a lot of energy into the system (as represented in Fig. 11.8). When we study the photons of visible light emitted, we see only certain colors (Fig. 11.10). That is,

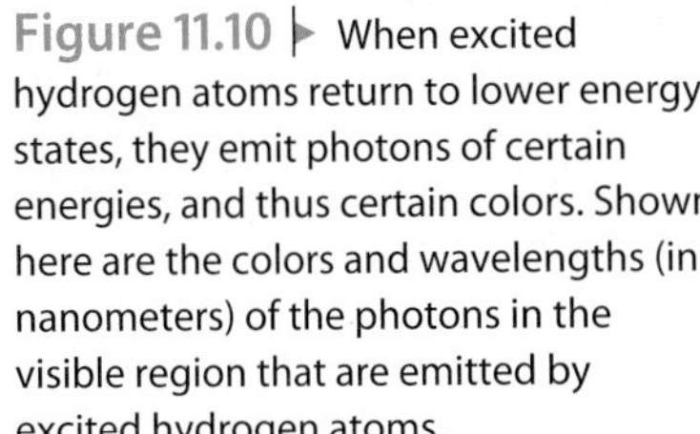

Figure 11.10 When excited hydrogen atoms return to lower energy states, they emit photons of certain energies, and thus certain colors. Shown here are the colors and wavelengths (in nanometers) of the photons in the visible region that are emitted by excited hydrogen atoms.

Atmospheric Effects

The gaseous atmosphere of the earth is crucial to life in many different ways. One of the most important characteristics of the atmosphere is the way its molecules absorb radiation from the sun.

If it weren't for the protective nature of the atmosphere, the sun would "fry" us with its high-energy radiation. We are protected by the atmospheric ozone, a form of oxygen consisting of O_3 molecules, which absorbs high-energy radiation and thus prevents it from reaching the earth. This explains why we are so concerned that chemicals released into the atmosphere are destroying this high-altitude ozone.

The atmosphere also plays a central role in controlling the earth's temperature, a phenomenon called the *greenhouse effect.* The atmospheric gases CO_2, H_2O, CH_4, N_2O, and others do not absorb light in the visible region. Therefore, the visible light from the sun passes through the atmosphere to warm the earth. In turn, the earth radiates this energy back toward space as infrared radiation. (For example, think of the heat radiated from black asphalt on a hot summer day.) But the gases listed earlier are strong *absorbers* of *infrared* waves, and they reradiate some of this energy back toward the earth as shown in Fig. 11.13. Thus these gases act as an insulating blanket keeping the earth much warmer than it would be without them. (If these gases were not present, all of the heat the earth radiates would be lost into space.)

However, there is a problem. When we burn fossil fuels (coal, petroleum, and natural gas), one of the products is CO_2. Because we use such huge quantities of fossil fuels, the CO_2 content in the atmosphere is increasing gradually but significantly. This should cause the earth to get warmer, eventually changing the weather patterns on the earth's surface and melting the polar ice caps, which would flood many low-lying areas.

Because the natural forces that control the earth's temperature are not very well understood at this point, it is difficult to decide whether the greenhouse warming has already started. But most scientists think it has. For example, the 1980s and 1990s were among the warmest years the earth has experienced since people started keeping records. Also, studies at the Scripps Institution of Oceanography indicate that the average temperatures of surface waters in the world's major oceans have risen since the 1960s in close agreement with the predictions of models based on the increase in CO_2 concentrations. Studies also show that Arctic sea ice, the Greenland Ice Sheet,

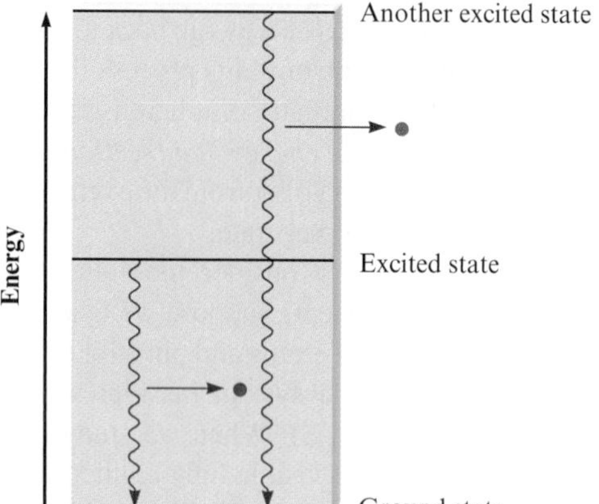

Figure 11.11 ▸ Hydrogen atoms have several excited-state energy levels. The color of the photon emitted depends on the energy change that produces it. A larger energy change may correspond to a blue photon, whereas a smaller change may produce a red photon.

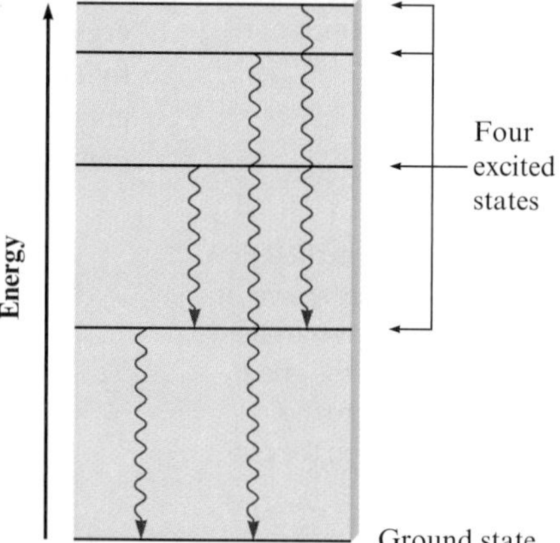

Figure 11.12 ▸ Each photon emitted by an excited hydrogen atom corresponds to a particular energy change in the hydrogen atom. In this diagram the horizontal lines represent discrete energy levels present in the hydrogen atom. A given H atom can exist in any of these energy states and can undergo energy changes to the ground state as well as to other excited states.

and various glaciers are melting much faster in recent years. These changes indicate that global warming is occurring.

The greenhouse effect is something we must watch closely. Controlling it may mean lowering our dependence on fossil fuels and increasing our reliance on nuclear, solar, or other power sources. In recent years, the trend has been in the opposite direction.

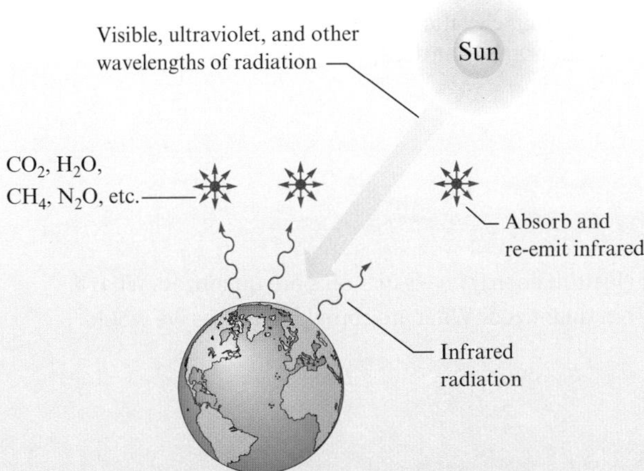

Figure 11.13 Certain gases in the earth's atmosphere absorb and re-emit some of the infrared (heat) radiation produced by the earth. This keeps the earth warmer than it would be otherwise.

A composite satellite image of the earth's biomass constructed from the radiation given off by living matter over a multiyear period.

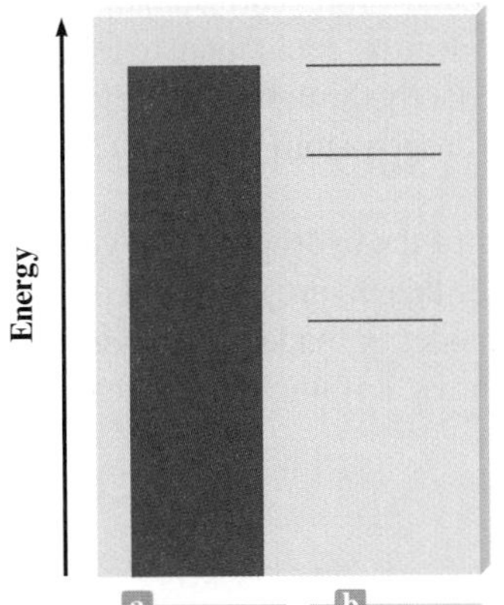

Figure 11.14 (a) Continuous energy levels. Any energy value is allowed. (b) Discrete (quantized) energy levels. Only certain energy states are allowed.

only certain types of photons are produced. We don't see all colors, which would add up to give "white light"; we see only selected colors. This is a very significant result. Let's discuss carefully what it means.

Because only certain photons are emitted, we know that only certain energy changes are occurring (Fig. 11.11). This means that the hydrogen atom must have *certain discrete energy levels* (Fig. 11.12). Excited hydrogen atoms *always* emit photons with the same discrete colors (wavelengths)—those shown in Fig. 11.10. They *never* emit photons with energies (colors) in between those shown. So we can conclude that all hydrogen atoms have the same set of discrete energy levels. We say the energy levels of hydrogen are **quantized energy levels.** That is, only *certain values are allowed.* Scientists have found that the energy levels of *all* atoms are quantized.

The quantized nature of the energy levels in atoms was a surprise when scientists discovered it. It had been assumed previously that an atom could exist at any energy level. That is, everyone had assumed that atoms could have a continuous set of energy levels rather than only certain discrete values (Fig. 11.14). A useful analogy here is the contrast between the elevations allowed by a ramp, which vary continuously, and those allowed by a set of steps, which are discrete (Fig. 11.15). The discovery of the quantized nature of energy has radically changed our view of the atom, as we will see in the next few sections.

a
A ramp varies continously in elevation.

b
A flight of stairs allows only certain elevations; the elevations are quantized.

Figure 11.15 ▶ The difference between continuous and quantized energy levels can be illustrated by comparing a flight of stairs with a ramp.

Critical Thinking

We now have evidence that electron energy levels in atoms are quantized. What if energy levels in atoms were not quantized? What are some differences we would notice?

11-5 The Bohr Model of the Atom

OBJECTIVE **To learn about Bohr's model of the hydrogen atom.**

In 1911 at the age of 25, Niels Bohr (Fig. 11.16) received his Ph.D. in physics. He was convinced that the atom could be pictured as a small positive nucleus with electrons orbiting around it.

Over the next two years, Bohr constructed a model of the hydrogen atom with quantized energy levels that agreed with the hydrogen emission results we have just discussed. Bohr pictured the electron moving in circular orbits corresponding to the various allowed energy levels. He suggested that the electron could jump to a different orbit by absorbing or emitting a photon of light with exactly the correct energy content. Thus, in the Bohr atom, the energy levels in the hydrogen atom represented certain allowed circular orbits (Fig. 11.17).

At first Bohr's model appeared very promising. It fit the hydrogen atom very well. However, when this model was applied to atoms other than hydrogen, it did not work. In fact, further experiments showed that the Bohr model is fundamentally incorrect. Although the Bohr model paved the way for later theories, it is important to realize that

The Niels Bohr Archive, Copenhagen

Figure 11.16 ▶ Niels Hendrik David Bohr (1885–1962) as a boy lived in the shadow of his younger brother Harald, who played on the 1908 Danish Olympic Soccer Team and later became a distinguished mathematician. In school, Bohr received his poorest marks in composition and struggled with writing during his entire life. In fact, he wrote so poorly that he was forced to dictate his Ph.D. thesis to his mother. He is one of the very few people who felt the need to write rough drafts of postcards. Nevertheless, Bohr was a brilliant physicist. After receiving his Ph.D. in Denmark, he constructed a quantum model for the hydrogen atom by the time he was 27. Even though his model later proved to be incorrect, Bohr remained a central figure in the drive to understand the atom. He was awarded the Nobel Prize in physics in 1922.

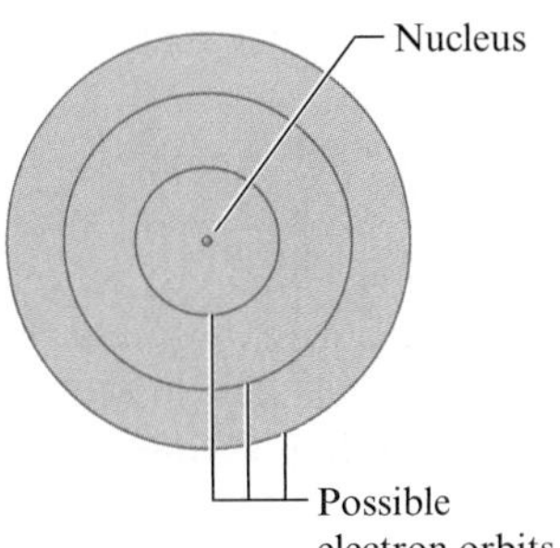

Figure 11.17 ▸ The Bohr model of the hydrogen atom represented the electron as restricted to certain circular orbits around the nucleus.

the current theory of atomic structure is not the same as the Bohr model. Electrons do *not* move around the nucleus in circular orbits like planets orbiting the sun. Surprisingly, as we shall see later in this chapter, we do not know exactly how the electrons move in an atom.

11-6 The Wave Mechanical Model of the Atom

OBJECTIVE To understand how the electron's position is represented in the wave mechanical model.

Louis Victor de Broglie

By the mid-1920s it had become apparent that the Bohr model was incorrect. Scientists needed to pursue a totally new approach. Two young physicists, Louis Victor de Broglie from France and Erwin Schrödinger from Austria, suggested that because light seems to have both wave and particle characteristics (it behaves simultaneously as a wave and as a stream of particles), the electron might also exhibit both of these characteristics. Although everyone had assumed that the electron was a tiny particle, these scientists said it might be useful to find out whether it could be described as a wave.

When Schrödinger carried out a mathematical analysis based on this idea, he found that it led to a new model for the hydrogen atom that seemed to apply equally well to other atoms—something Bohr's model failed to do. We will now explore a general picture of this model, which is called the **wave mechanical model** of the atom.

In the Bohr model, the electron was assumed to move in circular orbits. In the wave mechanical model, on the other hand, the electron states are described by orbitals. *Orbitals are nothing like orbits.* To approximate the idea of an orbital, picture a single male firefly in a room in the center of which an open vial of female sex-attractant hormones is suspended. The room is extremely dark and there is a camera in one corner with its shutter open. Every time the firefly "flashes," the camera records a pinpoint of light and thus the firefly's position in the room at that moment. The firefly senses the sex attractant, and as you can imagine, it spends a lot of time at or close to it. However, now and then the insect flies randomly around the room.

When the film is taken out of the camera and developed, the picture will probably look like Fig. 11.18. Because a picture is brightest where the film has been exposed to the most light, the color intensity at any given point tells us how often the firefly visited

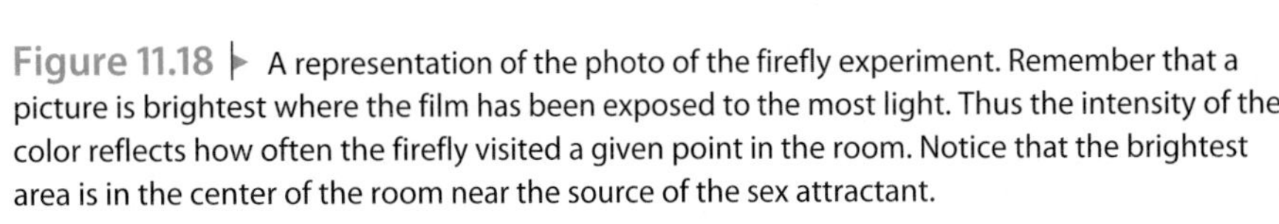

Figure 11.18 ▸ A representation of the photo of the firefly experiment. Remember that a picture is brightest where the film has been exposed to the most light. Thus the intensity of the color reflects how often the firefly visited a given point in the room. Notice that the brightest area is in the center of the room near the source of the sex attractant.

a given point in the room. Notice that, as we might expect, the firefly spent the most time near the room's center.

Now suppose you are watching the firefly in the dark room. You see it flash at a given point far from the center of the room. Where do you expect to see it next? There is really no way to be sure. The firefly's flight path is not precisely predictable. However, if you had seen the time-exposure picture of the firefly's activities (Fig. 11.18), you would have some idea where to look next. Your best chance would be to look more toward the center of the room. Figure 11.18 suggests there is the highest probability (the highest odds, the greatest likelihood) of finding the firefly at any particular moment near the center of the room. You *can't be sure* the firefly will fly toward the center of the room, but it *probably* will. So the time-exposure picture is a kind of "probability map" of the firefly's flight pattern.

Figure 11.19 The probability map, or orbital, that describes the hydrogen electron in its lowest possible energy state. The more intense the color of a given dot, the more likely it is that the electron will be found at that point. We have no information about when the electron will be at a particular point or about how it moves. Note that the probability of the electron's presence is highest closest to the positive nucleus (located at the center of this diagram), as might be expected.

According to the wave mechanical model, the electron in the hydrogen atom can be pictured as being something like this firefly. Schrödinger found that he could not precisely describe the electron's path. His mathematics enabled him only to predict the probabilities of finding the electron at given points in space around the nucleus. In its ground state the hydrogen electron has a probability map like that shown in Fig. 11.19. The more intense the color at a particular point, the more probable that the electron will be found at that point at a given instant. The model gives *no information about when* the electron occupies a certain point in space or *how it moves.* In fact, we have good reasons to believe that we can *never know* the details of electron motion, no matter how sophisticated our models may become. But one thing we feel confident about is that the electron *does not* orbit the nucleus in circles as Bohr suggested.

11-7 The Hydrogen Orbitals

OBJECTIVE To learn about the shapes of orbitals designated by *s*, *p*, and *d*.

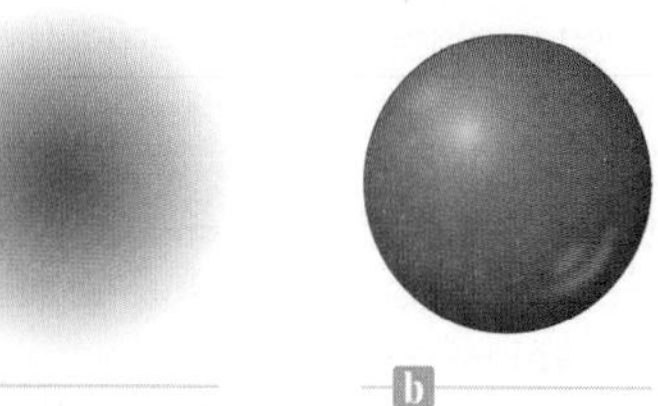

Figure 11.20 (a) The hydrogen 1*s* orbital. (b) The size of the orbital is defined by a sphere that contains 90% of the total electron probability. That is, the electron can be found *inside* this sphere 90% of the time. The 1*s* orbital is often represented simply as a sphere. However, the most accurate picture of the orbital is the probability map represented in (a).

The probability map for the hydrogen electron shown in Fig. 11.19 is called an **orbital.** Although the probability of finding the electron decreases at greater distances from the nucleus, the probability of finding it even at great distances from the nucleus never becomes exactly zero. A useful analogy might be the lack of a sharp boundary between the earth's atmosphere and "outer space." The atmosphere fades away gradually, but there are always a few molecules present. Because the edge of an orbital is "fuzzy," an orbital does not have an exactly defined size. So chemists arbitrarily define its size as the sphere that contains 90% of the total electron probability [Fig. 11.20(b)]. This means that the electron spends 90% of the time inside this surface and 10% somewhere outside this surface. (Note that we are *not* saying the electron travels only on the *surface* of the sphere.) The orbital represented in Fig. 11.20 is named the 1*s* orbital, and it describes the hydrogen electron's lowest energy state (the ground state).

In Section 11-4 we saw that the hydrogen atom can absorb energy to transfer the electron to a higher energy state (an excited state). In terms of the obsolete Bohr model, this meant the electron was transferred to an orbit with a larger radius. In the wave mechanical model, these higher energy states correspond to different kinds of orbitals with different shapes.

At this point we need to stop and consider how the hydrogen atom is organized. Remember, we showed earlier that the hydrogen atom has discrete energy levels. We call these levels **principal energy levels** and label them with integers (Fig. 11.21). Next we find that each of these levels is subdivided into **sublevels.** The following analogy should help you understand this. Picture an inverted triangle (Fig. 11.22). We divide the principal levels into various numbers of sublevels. Principal level 1 consists of one sublevel, principal level 2 has two sublevels, principal level 3 has three sublevels, and principal level 4 has four sublevels.

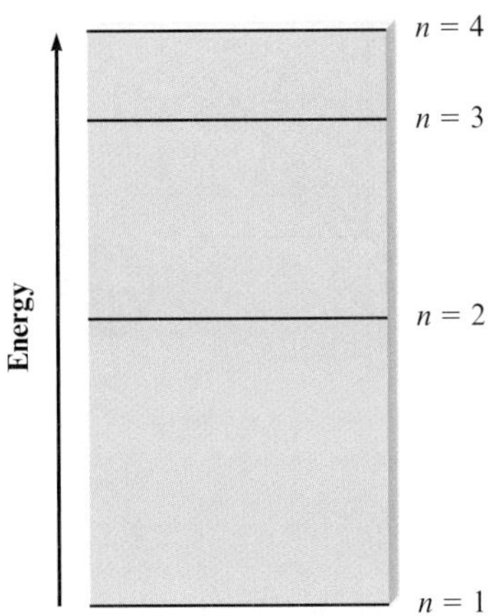

Figure 11.21 The first four principal energy levels in the hydrogen atom. Each level is assigned an integer, n.

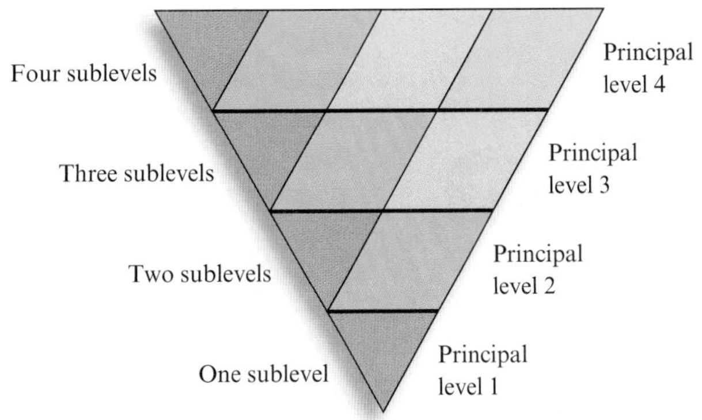

Figure 11.22 An illustration of how principal levels can be divided into sublevels.

Like our triangle, the principal energy levels in the hydrogen atom contain sublevels. As we will see presently, these sublevels contain spaces for the electron that we call orbitals. Principal energy level 1 consists of just one sublevel, or one type of orbital. The spherical shape of this orbital is shown in Fig. 11.20. We label this orbital 1*s*. The number 1 is for the principal energy level, and *s* is a shorthand way to label a particular sublevel (type of orbital).

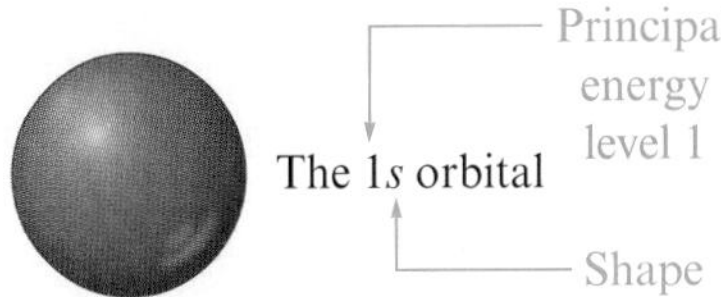

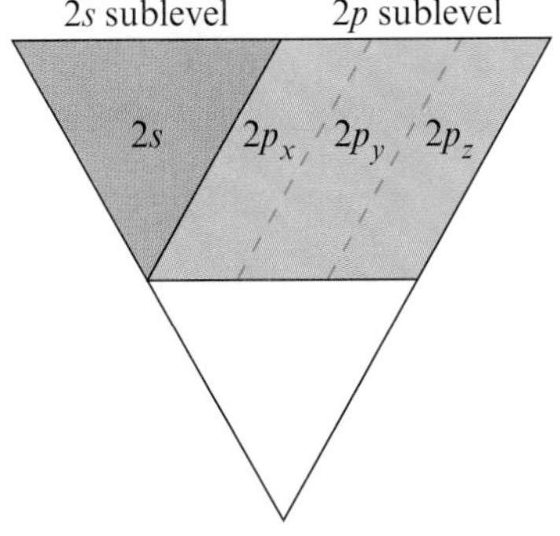

Figure 11.23 Principal level 2 shown divided into the 2*s* and 2*p* sublevels.

Principal energy level 2 has two sublevels. (Note the correspondence between the principal energy level number and the number of sublevels.) These sublevels are labeled 2*s* and 2*p*. The 2*s* sublevel consists of one orbital (called the 2*s*), and the 2*p* sublevel consists of three orbitals (called $2p_x$, $2p_y$, and $2p_z$). Let's return to the inverted triangle to illustrate this. Figure 11.23 shows principal level 2 divided into the sublevels 2*s* and 2*p* (which is subdivided into $2p_x$, $2p_y$, and $2p_z$). The orbitals have the shapes shown in Figs. 11.24 and 11.25. The 2*s* orbital is spherical like the 1*s* orbital but larger in size (Fig. 11.24). The three 2*p* orbitals are not spherical but have two "lobes." These orbitals are shown in Fig. 11.25 both as electron probability maps and as surfaces that contain 90% of the total electron probability. Notice that the label *x*, *y*, or *z* on a given 2*p* orbital tells along which axis the lobes of that orbital are directed.

What we have learned so far about the hydrogen atom is summarized in Fig. 11.26. Principal energy level 1 has one sublevel, which contains the 1*s* orbital. Principal energy level 2 contains two sublevels, one of which contains the 2*s* orbital and one of which contains the 2*p* orbitals (three of them). Note that each orbital is designated by a symbol or label. We summarize the information given by this label in the following box.

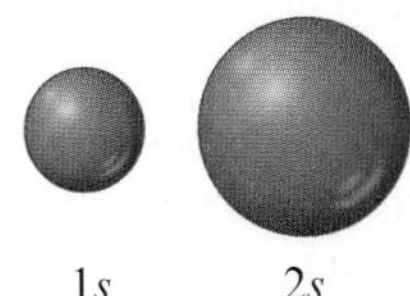

Figure 11.24 The relative sizes of the 1*s* and 2*s* orbitals of hydrogen.

Orbital Labels

1. The number tells the principal energy level.
2. The letter tells the shape. The letter *s* means a spherical orbital; the letter *p* means a two-lobed orbital. The *x*, *y*, or *z* subscript on a *p* orbital label tells along which of the coordinate axes the two lobes lie.

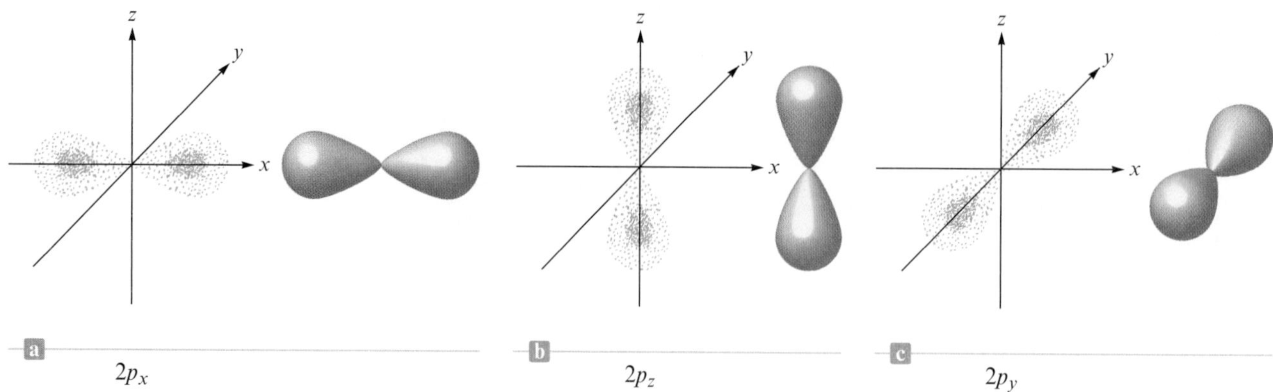

Figure 11.25 ▸ The three 2*p* orbitals: (a) $2p_x$, (b) $2p_z$, (c) $2p_y$. The *x, y,* or *z* label indicates along which axis the two lobes are directed. Each orbital is shown both as a probability map and as a surface that encloses 90% of the electron probability.

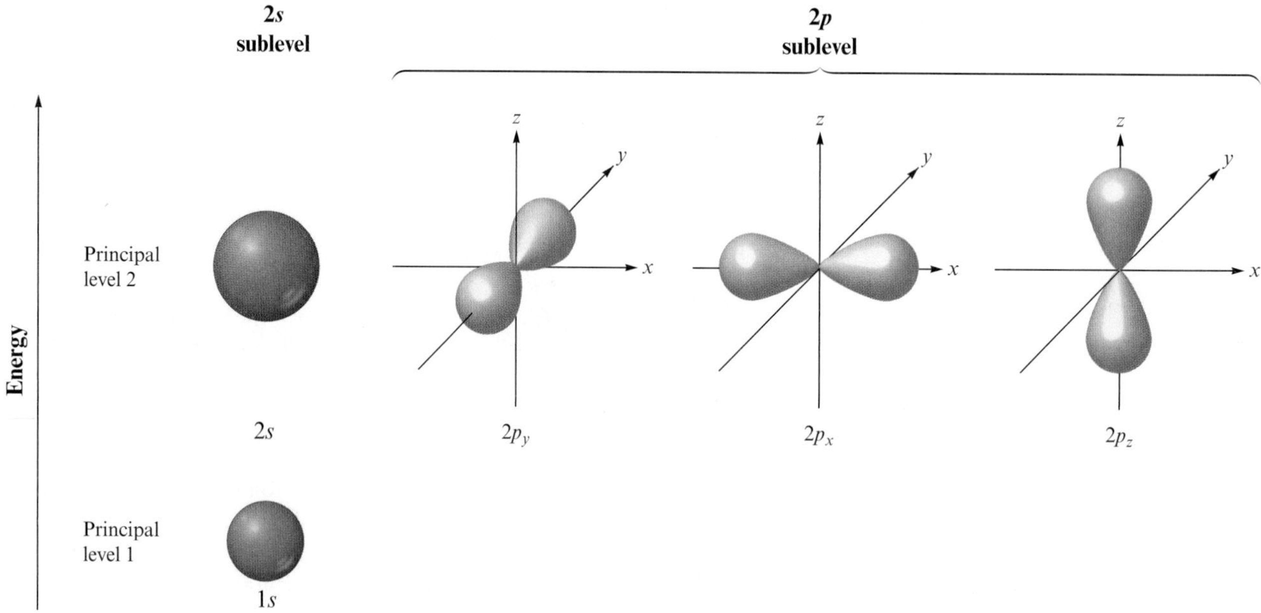

Figure 11.26 ▸ A diagram of principal energy levels 1 and 2 showing the shapes of orbitals that compose the sublevels.

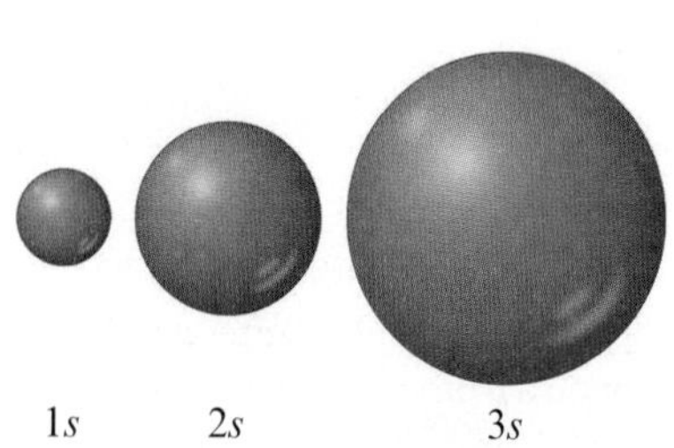

Figure 11.27 ▸ The relative sizes of the spherical 1*s*, 2*s*, and 3*s* orbitals of hydrogen.

One important characteristic of orbitals is that as the level number increases, the average distance of the electron in that orbital from the nucleus also increases. That is, when the hydrogen electron is in the 1*s* orbital (the ground state), it spends most of its time much closer to the nucleus than when it occupies the 2*s* orbital (an excited state).

You may be wondering at this point why hydrogen, which has only one electron, has more than one orbital. It is best to think of an orbital as a *potential space* for an electron. The hydrogen electron can occupy only a single orbital at a time, but the other orbitals are still available should the electron be transferred into one of them. For example, when a hydrogen atom is in its ground state (lowest possible energy state), the electron is in the 1*s* orbital. By adding the correct amount of energy (for example, a specific photon of light), we can excite the electron to the 2*s* orbital or to one of the 2*p* orbitals.

So far we have discussed only two of hydrogen's energy levels. There are many others. For example, level 3 has three sublevels (Fig. 11.22), which we label 3*s*, 3*p*, and 3*d*. The 3*s* sublevel contains a single 3*s* orbital, a spherical orbital larger than 1*s* and 2*s* (Fig. 11.27). Sublevel 3*p* contains three orbitals: $3p_x$, $3p_y$, and $3p_z$, which are shaped like the 2*p* orbitals except that they are larger. The 3*d* sublevel contains five 3*d*

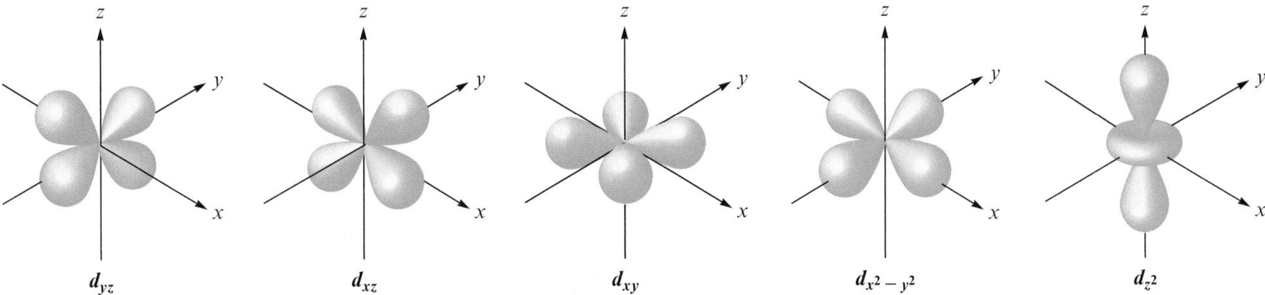

Figure 11.28 The shapes and labels of the five $3d$ orbitals.

orbitals with the shapes and labels shown in Fig. 11.28. (You do not need to memorize the $3d$ orbital shapes and labels. They are shown for completeness.)

Notice as you compare levels 1, 2, and 3 that a new type of orbital (sublevel) is added in each principal energy level. (Recall that the p orbitals are added in level 2 and the d orbitals in level 3.) This makes sense because in going farther out from the nucleus, there is more space available and thus room for more orbitals.

It might help you to understand that the number of orbitals increases with the principal energy level if you think of a theater in the round. Picture a round stage with circular rows of seats surrounding it. The farther from the stage a row of seats is, the more seats it contains because the circle is larger. Orbitals divide up the space around a nucleus somewhat like the seats in this circular theater. The greater the distance from the nucleus, the more space there is and the more orbitals we find.

The pattern of increasing numbers of orbitals continues with level 4. Level 4 has four sublevels labeled $4s$, $4p$, $4d$, and $4f$. The $4s$ sublevel has a single $4s$ orbital. The $4p$ sublevel contains three orbitals ($4p_x$, $4p_y$, and $4p_z$). The $4d$ sublevel has five $4d$ orbitals. The $4f$ sublevel has seven $4f$ orbitals.

The $4s$, $4p$, and $4d$ orbitals have the same shapes as the earlier s, p, and d orbitals, respectively, but are larger. We will not be concerned here with the shapes of the f orbitals.

The Wave Mechanical Model: Further Development

OBJECTIVES

- **To review the energy levels and orbitals of the wave mechanical model of the atom.**
- **To learn about electron spin.**

A model for the atom is of little use if it does not apply to all atoms. The Bohr model was discarded because it could be applied only to hydrogen. The wave mechanical model can be applied to all atoms in basically the same form as the one we have just used for hydrogen. In fact, the major triumph of this model is its ability to explain the periodic table of the elements. Recall that the elements on the periodic table are arranged in vertical groups, which contain elements that typically show similar chemical properties. The wave mechanical model of the atom allows us to explain, based on electron arrangements, why these similarities occur. We will see in due time how this is done.

Remember that an atom has as many electrons as it has protons to give it a zero overall charge. Therefore, all atoms beyond hydrogen have more than one electron. Before we can consider the atoms beyond hydrogen, we must describe one more property of electrons that determines how they can be arranged in an atom's orbitals. This property is spin. Each electron appears to be spinning as a top spins on its axis. Like the top, an electron can spin only in one of two directions. We often represent spin with an arrow: either ↑ or ↓. One arrow represents the electron spinning in the one direction, and the other represents the electron spinning in the opposite direction. For our purposes, what is most important about electron spin is that two electrons must have *opposite* spins to occupy the same orbital. That is, two electrons that have the same spin cannot occupy the same orbital. This leads to the **Pauli exclusion principle:** an atomic orbital can hold a maximum of two electrons, and those two electrons must have opposite spins.

Before we apply the wave mechanical model to atoms beyond hydrogen, we will summarize the model for convenient reference.

Principal Components of the Wave Mechanical Model of the Atom

1. Atoms have a series of energy levels called **principal energy levels,** which are designated by whole numbers symbolized by *n*; *n* can equal 1, 2, 3, 4, . . . Level 1 corresponds to $n = 1$, level 2 corresponds to $n = 2$, and so on.
2. The energy of the level increases as the value of *n* increases.
3. Each principal energy level contains one or more *types* of orbitals, called **sublevels.**
4. The number of sublevels present in a given principal energy level equals *n*. For example, level 1 contains one sublevel (1*s*); level 2 contains two sublevels (two types of orbitals), the 2*s* orbital and the three 2*p* orbitals; and so on. These are summarized in the following table. The number of each type of orbital is shown in parentheses.

n	Sublevels (Types of Orbitals) Present
1	1*s*(1)
2	2*s*(1) 2*p*(3)
3	3*s*(1) 3*p*(3) 3*d*(5)
4	4*s*(1) 4*p*(3) 4*d*(5) 4*f*(7)

5. The *n* value is always used to label the orbitals of a given principal level and is followed by a letter that indicates the type (shape) of the orbital. For example, the designation 3*p* means an orbital in level 3 that has two lobes (a *p* orbital always has two lobes).
6. An orbital can be empty or it can contain one or two electrons, but never more than two. If two electrons occupy the same orbital, they must have opposite spins.
7. The shape of an orbital does not indicate the details of electron movement. It indicates the probability distribution for an electron residing in that orbital.

Interactive Example 11.1

Understanding the Wave Mechanical Model of the Atom

Indicate whether each of the following statements about atomic structure is true or false.

a. An *s* orbital is always spherical in shape.

b. The 2*s* orbital is the same size as the 3*s* orbital.

c. The number of lobes on a *p* orbital increases as *n* increases. That is, a 3*p* orbital has more lobes than a 2*p* orbital.

d. Level 1 has one *s* orbital, level 2 has two *s* orbitals, level 3 has three *s* orbitals, and so on.

e. The electron path is indicated by the surface of the orbital.

SOLUTION

a. True. The size of the sphere increases as n increases, but the shape is always spherical.

b. False. The 3*s* orbital is larger (the electron is farther from the nucleus on average) than the 2*s* orbital.

c. False. A *p* orbital always has two lobes.

d. False. Each principal energy level has only one *s* orbital.

e. False. The electron is *somewhere inside* the orbital surface 90% of the time. The electron does not move around *on* this surface.

SELF CHECK

Exercise 11.1 Define the following terms.

a. Bohr orbits

b. orbitals

c. orbital size

d. sublevel

See Problems 11.37 through 11.44. ■

Electron Arrangements in the First Eighteen Atoms on the Periodic Table

OBJECTIVES

- **To understand how the principal energy levels fill with electrons in atoms beyond hydrogen.**
- **To learn about valence electrons and core electrons.**

We will now describe the electron arrangements in atoms with $Z = 1$ to $Z = 18$ by placing electrons in the various orbitals in the principal energy levels, starting with $n = 1$, and then continuing with $n = 2$, $n = 3$, and so on. For the first eighteen elements, the individual sublevels fill in the following order: 1*s*, then 2*s*, then 2*p*, then 3*s*, then 3*p*.

The most attractive orbital to an electron in an atom is always the 1*s*, because in this orbital the negatively charged electron is closer to the positively charged nucleus than in any other orbital. That is, the 1*s* orbital involves the space around the nucleus that is closest to the nucleus. As n increases, the orbital becomes larger—the electron, on average, occupies space farther from the nucleus.

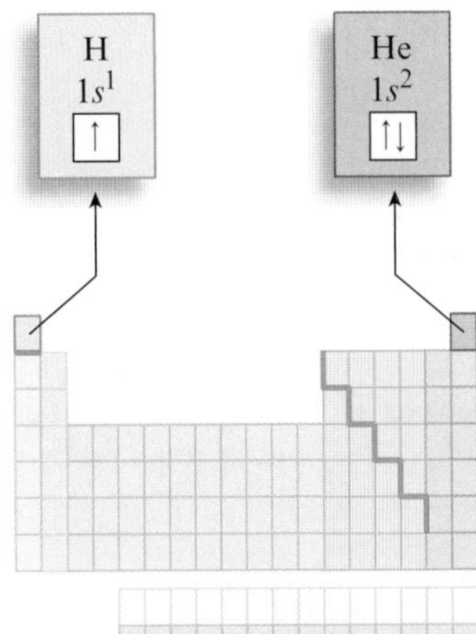

So in its ground state hydrogen has its lone electron in the 1*s* orbital. This is commonly represented in two ways. First, we say that hydrogen has the electron arrangement, or **electron configuration,** $1s^1$. This just means there is one electron in the 1*s* orbital. We can also represent this configuration by using an **orbital diagram,** also called a **box diagram,** in which orbitals are represented by boxes grouped by sublevel with small arrows indicating the electrons. For *hydrogen,* the electron configuration and box diagram are

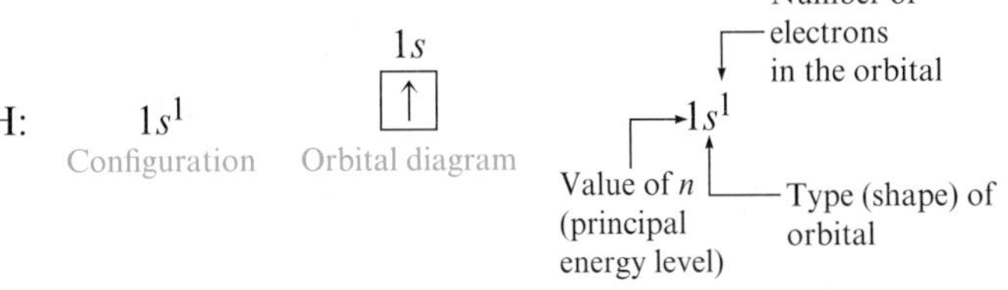

The arrow represents an electron spinning in a particular direction. The next element is *helium,* $Z = 2$. It has two protons in its nucleus and so has two electrons. Because the $1s$ orbital is the most desirable, both electrons go there but with opposite spins. For helium, the electron configuration and box diagram are

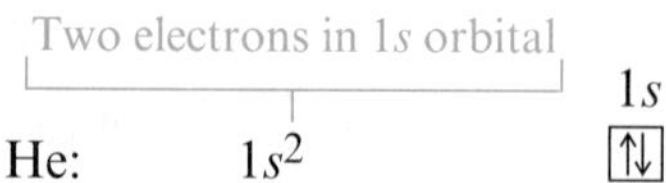

The opposite electron spins are shown by the opposing arrows in the box.

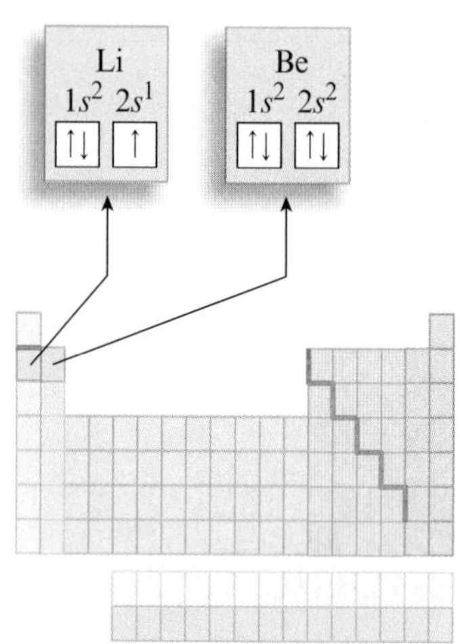

Lithium ($Z = 3$) has three electrons, two of which go into the $1s$ orbital. That is, two electrons fill that orbital. The $1s$ orbital is the only orbital for $n = 1$, so the third electron must occupy an orbital with $n = 2$—in this case the $2s$ orbital. This gives a $1s^22s^1$ configuration. The electron configuration and box diagram are

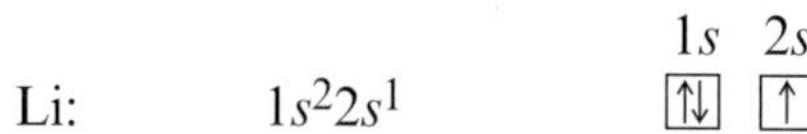

The next element, *beryllium,* has four electrons, which occupy the $1s$ and $2s$ orbitals with opposite spins.

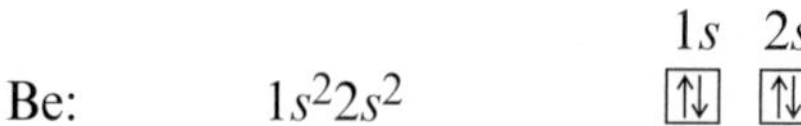

Boron has five electrons, four of which occupy the $1s$ and $2s$ orbitals. The fifth electron goes into the second type of orbital with $n = 2$, one of the $2p$ orbitals.

B: $1s^22s^22p^1$ — $1s$ [↑↓] $2s$ [↑↓] $2p$ [↑| |]

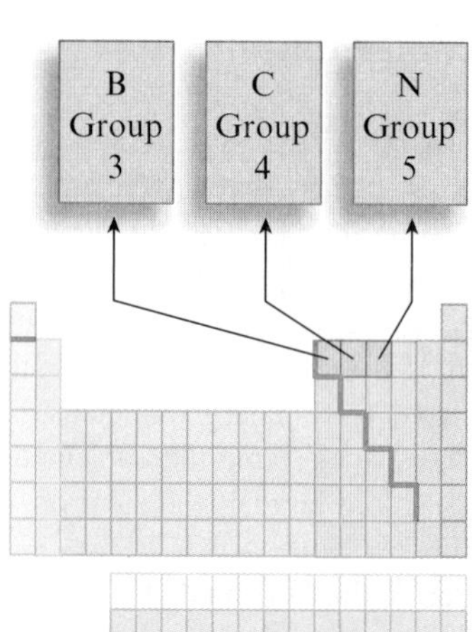

Because all the $2p$ orbitals have the same energy, it does not matter which $2p$ orbital the electron occupies.

Carbon, the next element, has six electrons: two electrons occupy the $1s$ orbital, two occupy the $2s$ orbital, and two occupy $2p$ orbitals. There are three $2p$ orbitals, so each of the mutually repulsive electrons occupies a different $2p$ orbital. For reasons we will not consider, in the separate $2p$ orbitals the electrons have the same spin.

The configuration for carbon could be written $1s^22s^22p^12p^1$ to indicate that the electrons occupy separate $2p$ orbitals. However, the configuration is usually given as $1s^22s^22p^2$, and it is understood that the electrons are in different $2p$ orbitals.

C: $1s^22s^22p^2$ — $1s$ [↑↓] $2s$ [↑↓] $2p$ [↑|↑|]

Note the like spins for the unpaired electrons in the $2p$ orbitals.

The configuration for *nitrogen,* which has seven electrons, is $1s^22s^22p^3$. The three electrons in $2p$ orbitals occupy separate orbitals and have like spins.

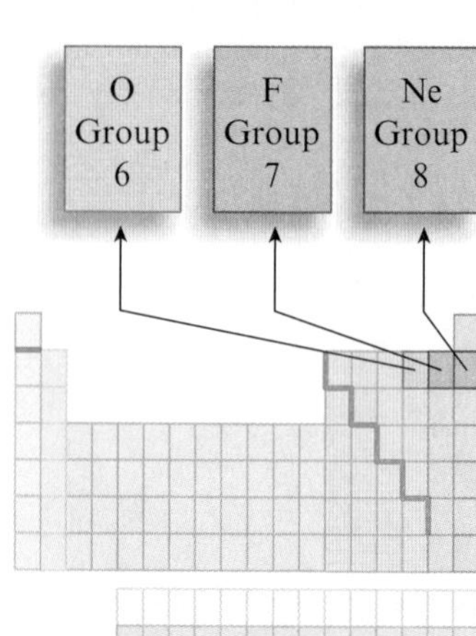

N: $1s^22s^22p^3$ — $1s$ [↑↓] $2s$ [↑↓] $2p$ [↑|↑|↑]

The configuration for *oxygen,* which has eight electrons, is $1s^22s^22p^4$. One of the $2p$ orbitals is now occupied by a pair of electrons with opposite spins, as required by the Pauli exclusion principle.

O: $1s^22s^22p^4$ — $1s$ [↑↓] $2s$ [↑↓] $2p$ [↑↓|↑|↑]

H $1s^1$							He $1s^2$
Li $2s^1$	Be $2s^2$	B $2p^1$	C $2p^2$	N $2p^3$	O $2p^4$	F $2p^5$	Ne $2p^6$
Na $3s^1$	Mg $3s^2$	Al $3p^1$	Si $3p^2$	P $3p^3$	S $3p^4$	Cl $3p^5$	Ar $3p^6$

Figure 11.29 ▸ The electron configurations in the sublevel last occupied for the first eighteen elements.

The electron configurations and orbital diagrams for *fluorine* (nine electrons) and *neon* (ten electrons) are

		1s	2s	2p
F:	$1s^22s^22p^5$	⇅	⇅	⇅ ⇅ ↑
Ne:	$1s^22s^22p^6$	⇅	⇅	⇅ ⇅ ⇅

With neon, the orbitals with $n = 1$ and $n = 2$ are completely filled.

For *sodium,* which has eleven electrons, the first ten electrons occupy the 1*s*, 2*s*, and 2*p* orbitals, and the eleventh electron must occupy the first orbital with $n = 3$, the 3*s* orbital. The electron configuration for sodium is $1s^22s^22p^63s^1$. To avoid writing the inner-level electrons, we often abbreviate the configuration $1s^22s^22p^63s^1$ as $[Ne]3s^1$, where [Ne] represents the electron configuration of neon, $1s^22s^22p^6$.

The orbital diagram for sodium is

1s	2s	2p	3s
⇅	⇅	⇅ ⇅ ⇅	↑

The next element, *magnesium,* $Z = 12$, has the electron configuration $1s^22s^22p^63s^2$, or $[Ne]3s^2$.

The next six elements, *aluminum* through *argon,* have electron configurations obtained by filling the 3*p* orbitals one electron at a time. Figure 11.29 summarizes the electron configurations of the first eighteen elements by giving the number of electrons in the type of orbital (sublevel) occupied last.

Interactive Example 11.2

Writing Orbital Diagrams

Write the orbital diagram for magnesium.

SOLUTION Magnesium ($Z = 12$) has twelve electrons that are placed successively in the 1*s*, 2*s*, 2*p*, and 3*s* orbitals to give the electron configuration $1s^22s^22p^63s^2$. The orbital diagram is

1s	2s	2p	3s
⇅	⇅	⇅ ⇅ ⇅	⇅

Only occupied orbitals are shown here.

SELF CHECK

Exercise 11.2 Write the complete electron configuration and the orbital diagram for each of the elements aluminum through argon.

See Problems 11.49 through 11.54. ■

CHEMISTRY IN FOCUS

A Magnetic Moment

An anesthetized frog lies in the hollow core of an electromagnet. As the current in the coils of the magnet is increased, the frog magically rises and floats in midair (see photo). How can this happen? Is the electromagnet an antigravity machine? In fact, there is no magic going on here. This phenomenon demonstrates the magnetic properties of all matter. We know that iron magnets attract and repel each other depending on their relative orientations. Is a frog magnetic like a piece of iron? If a frog lands on a steel manhole cover, will it be trapped there by magnetic attractions? Of course not. The magnetism of the frog, as with most objects, shows up only in the presence of a strong inducing magnetic field. In other words, the powerful electromagnet surrounding the frog in the experiment described here *induces* a magnetic field in the frog that opposes the inducing field. The opposing magnetic field in the frog repels the inducing field, and the frog lifts up until the magnetic force is balanced by the gravitational pull on its body. The frog then "floats" in air.

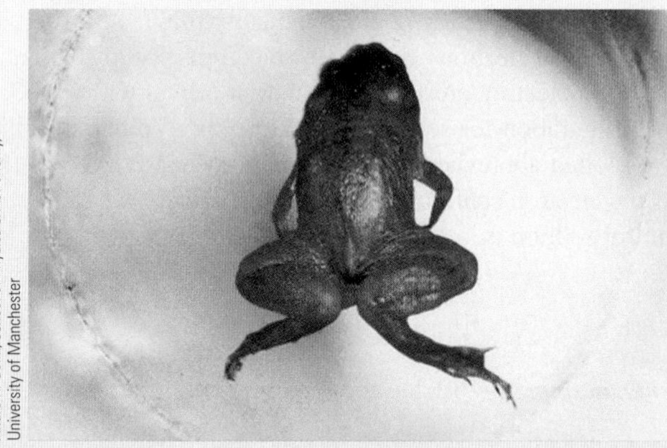

A live frog levitated in a magnetic field.

How can a frog be magnetic if it is not made of iron? It's the electrons. Frogs are composed of cells containing many kinds of molecules. Of course, these molecules are made of atoms—carbon atoms, nitrogen atoms, oxygen atoms, and other types. Each of these atoms contains electrons that are moving around the atomic nuclei. When these electrons sense a strong magnetic field, they respond by moving in a fashion that produces magnetic fields aligned to oppose the inducing field. This phenomenon is called *diamagnetism*.

All substances, animate and inanimate, because they are made of atoms, exhibit diamagnetism. Andre Geim and his colleagues at the University of Nijmegan, the Netherlands, have levitated frogs, grasshoppers, plants, and water droplets, among other objects. Geim says that, given a large enough electromagnet, even humans can be levitated. He notes, however, that constructing a magnet strong enough to float a human would be very expensive, and he sees no point in it. Geim does point out that inducing weightlessness with magnetic fields may be a good way to pretest experiments on weightlessness intended as research for future space flights—to see if the ideas fly as well as the objects.

At this point it is useful to introduce the concept of **valence electrons**—that is, *the electrons in the outermost (highest) principal energy level of an atom.* For example, nitrogen, which has the electron configuration $1s^22s^22p^3$, has electrons in principal levels 1 and 2. Therefore, level 2 (which has $2s$ and $2p$ sublevels) is the valence level of nitrogen, and the $2s$ and $2p$ electrons are the valence electrons. For the sodium atom (electron configuration $1s^22s^22p^63s^1$, or $[Ne]3s^1$), the valence electron is the electron in the $3s$ orbital, because in this case principal energy level 3 is the outermost level that contains an electron. The valence electrons are the most important electrons to chemists because, being the outermost electrons, they are the ones involved when atoms attach to each other (form bonds), as we will see in the next chapter. The inner electrons, which are known as **core electrons,** are not involved in bonding atoms to each other.

Note in Fig. 11.29 that a very important pattern is developing: except for helium, *the atoms of elements in the same group (vertical column of the periodic table) have the same number of electrons in a given type of orbital* (sublevel), except that the orbitals are in different principal energy levels. Remember that the elements were originally organized into groups on the periodic table on the basis of similarities in chemical properties. Now we understand the reason behind these groupings. Elements with the same valence electron arrangement show very similar chemical behavior.

Electron Configurations and the Periodic Table

OBJECTIVE **To learn about the electron configurations of atoms with *Z* greater than 18.**

In the previous section we saw that we can describe the atoms beyond hydrogen by simply filling the atomic orbitals starting with level $n = 1$ and working outward in order. This works fine until we reach the element *potassium* ($Z = 19$), which is the next element after argon. Because the $3p$ orbitals are fully occupied in argon, we might expect the next electron to go into a $3d$ orbital (recall that for $n = 3$ the sublevels are $3s$, $3p$, and $3d$). However, experiments show that the chemical properties of potassium are very similar to those of lithium and sodium. Because we have learned to associate similar chemical properties with similar valence-electron arrangements, we predict that the valence-electron configuration for potassium is $4s^1$, resembling sodium ($3s^1$) and lithium ($2s^1$). That is, we expect the last electron in potassium to occupy the $4s$ orbital instead of one of the $3d$ orbitals. This means that principal energy level 4 begins to fill before level 3 has been completed. This conclusion is confirmed by many types of experiments. So the electron configuration of potassium is

$$\text{K: } 1s^2 2s^2 2p^6 3s^2 3p^6 4s^1, \text{ or } [\text{Ar}]4s^1$$

The next element is *calcium,* with an additional electron that also occupies the $4s$ orbital.

$$\text{Ca: } 1s^2 2s^2 2p^6 3s^2 3p^6 4s^2, \text{ or } [\text{Ar}]4s^2$$

The $4s$ orbital is now full.

After calcium the next electrons go into the $3d$ orbitals to complete principal energy level 3. The elements that correspond to filling the $3d$ orbitals are called transition metals. Then the $4p$ orbitals fill. Figure 11.30 gives partial electron configurations for the elements potassium through krypton.

Note from Fig. 11.30 that all of the transition metals have the general configuration $[\text{Ar}]4s^2 3d^n$ except chromium ($4s^1 3d^5$) and copper ($4s^1 3d^{10}$). The reasons for these exceptions are complex and will not be discussed here.

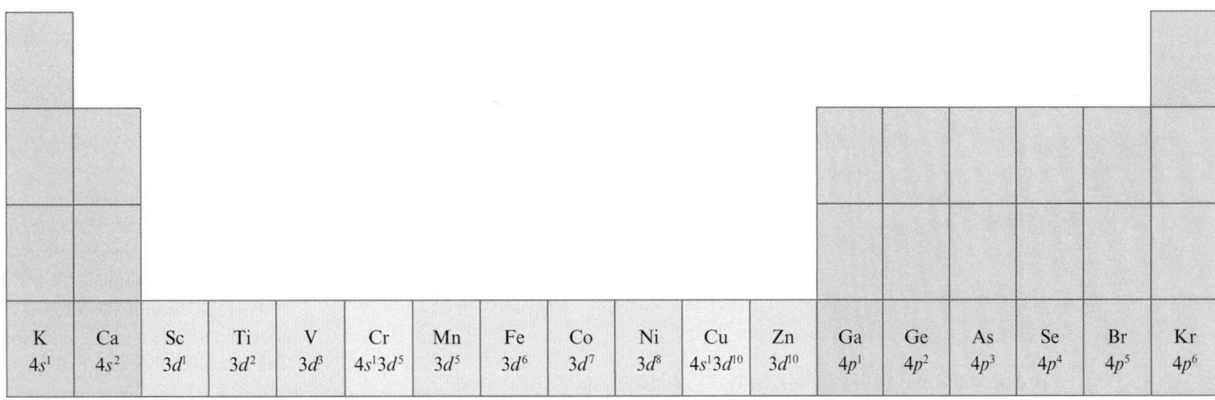

Figure 11.30 ▶ Partial electron configurations for the elements potassium through krypton. The transition metals shown in green (scandium through zinc) have the general configuration $[\text{Ar}]4s^2 3d^n$, except for chromium and copper.

Instead of continuing to consider the elements individually, we will now look at the overall relationship between the periodic table and orbital filling. Figure 11.31 shows which type of orbital is filling in each area of the periodic table. Note the points in the "Orbital Filling" box.

Orbital Filling

1. In a principal energy level that has *d* orbitals, the *s* orbital from the *next* level fills before the *d* orbitals in the current level. That is, the $(n + 1)s$ orbitals always fill before the *nd* orbitals. For example, the 5*s* orbitals fill for rubidium and strontium before the 4*d* orbitals fill for the second row of transition metals (yttrium through cadmium).
2. After lanthanum, which has the electron configuration $[Xe]6s^25d^1$, a group of fourteen elements called the **lanthanide series,** or the lanthanides, occurs. This series of elements corresponds to the filling of the seven 4*f* orbitals.
3. After actinium, which has the configuration $[Rn]7s^26d^1$, a group of fourteen elements called the **actinide series,** or the actinides, occurs. This series corresponds to the filling of the seven 5*f* orbitals.
4. Except for helium, the group numbers indicate the sum of electrons in the *ns* and *np* orbitals in the highest principal energy level that contains electrons (where *n* is the number that indicates a particular principal energy level). These electrons are the valence electrons, the electrons in the outermost principal energy level of a given atom.

To help you further understand the connection between orbital filling and the periodic table, Fig. 11.32 shows the orbitals in the order in which they fill.

A periodic table is almost always available to you. If you understand the relationship between the electron configuration of an element and its position on the periodic table, you can figure out the expected electron configuration of any atom.

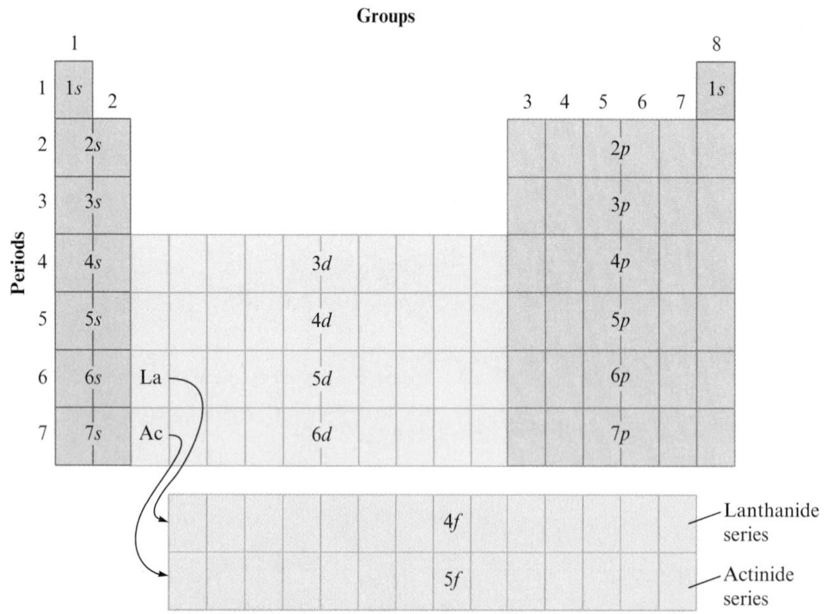

Figure 11.31 ▸ The orbitals being filled for elements in various parts of the periodic table. Note that in going along a horizontal row (a period), the $(n + 1)s$ orbital fills before the *nd* orbital. The group label indicates the number of valence electrons (the number of *s* plus the number of *p* electrons in the highest occupied principal energy level) for the elements in each group.

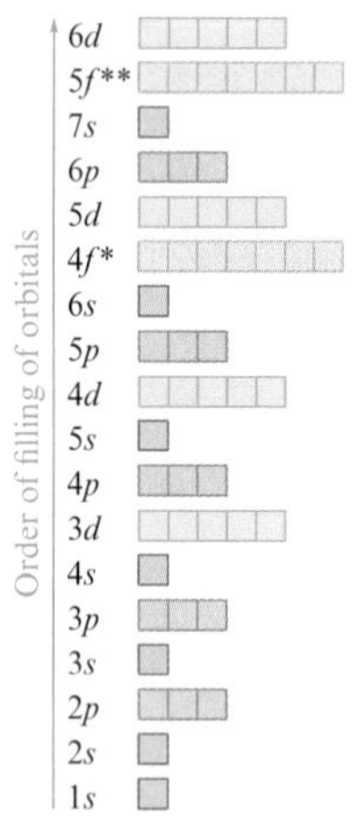

*After the 6*s* orbital is full, one electron goes into a 5*d* orbital. This corresponds to the element lanthanum ($[Xe]6s^25d^1$). After lanthanum, the 4*f* orbitals fill with electrons.

**After the 7*s* orbital is full, one electron goes into 6*d*. This is actinium ($[Rn]7s^26d^1$). The 5*f* orbitals then fill.

Figure 11.32 ▸ A box diagram showing the order in which orbitals fill to produce the atoms in the periodic table. Each box can hold two electrons.

CHEMISTRY IN FOCUS

The Chemistry of Bohrium

One of the best uses of the periodic table is to predict the properties of newly discovered elements. For example, the artificially synthesized element bohrium ($Z = 107$) is found in the same family as manganese, technetium, and rhenium and is expected to show chemistry similar to these elements. The problem, of course, is that only a few atoms of bohrium (Bh) can be made at a time, and the atoms exist for only a very short time (about 17 seconds). It's a real challenge to study the chemistry of an element under these conditions. However, a team of nuclear chemists led by Heinz W. Gaggeler of the University of Bern in Switzerland isolated six atoms of ^{267}Bh and prepared the compound BhO_3Cl. Analysis of the decay products of this compound helped define the thermochemical properties of BhO_3Cl and showed that bohrium seems to behave as might be predicted from its position in the periodic table.

Interactive Example 11.3

Determining Electron Configurations

Using Fig. 11.34, give the electron configurations for sulfur (S), gallium (Ga), hafnium (Hf), and radium (Ra).

SOLUTION

Sulfur is element 16 and resides in Period 3, where the 3*p* orbitals are being filled (Fig. 11.33). Because sulfur is the fourth among the "3*p* elements," it must have four 3*p* electrons. Sulfur's electron configuration is

$$\text{S: } 1s^2 2s^2 2p^6 3s^2 3p^4\text{, or [Ne]}3s^2 3p^4$$

Gallium is element 31 in Period 4 just after the transition metals (Fig. 11.33). It is the first element in the "4*p* series" and has a $4p^1$ arrangement. Gallium's electron configuration is

$$\text{Ga: } 1s^2 2s^2 2p^6 3s^2 3p^6 4s^2 3d^{10} 4p^1\text{, or [Ar]}4s^2 3d^{10} 4p^1$$

Hafnium is element 72 and is found in Period 6, as shown in Fig. 11.33. Note that it occurs just after the lanthanide series (Fig. 11.31). Thus the 4*f* orbitals are already filled. Hafnium is the second member of the 5*d* transition series and has two 5*d* electrons. Its electron configuration is

$$\text{Hf: } 1s^2 2s^2 2p^6 3s^2 3p^6 4s^2 3d^{10} 4p^6 5s^2 4d^{10} 5p^6 6s^2 4f^{14} 5d^2\text{, or [Xe]}6s^2 4f^{14} 5d^2$$

Radium is element 88 and is in Period 7 (and Group 2), as shown in Fig. 11.33. Thus radium has two electrons in the 7*s* orbital, and its electron configuration is

$$\text{Ra: } 1s^2 2s^2 2p^6 3s^2 3p^6 4s^2 3d^{10} 4p^6 5s^2 4d^{10} 5p^6 6s^2 4f^{14} 5d^{10} 6p^6 7s^2\text{, or [Rn]}7s^2$$

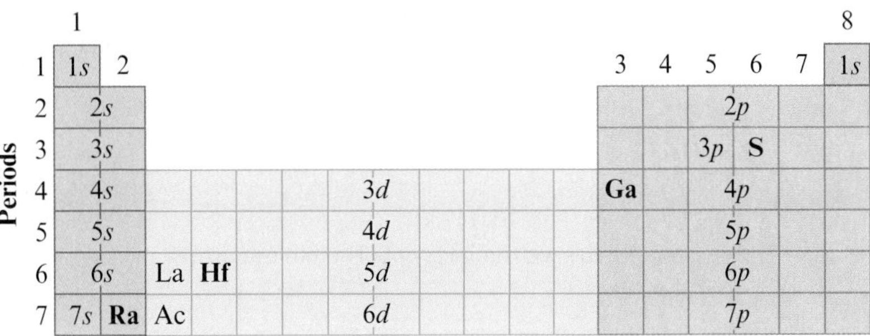

Figure 11.33 The positions of the elements considered in Example 11.3.

SELF CHECK

Exercise 11.3 Using Fig. 11.34, predict the electron configurations for fluorine, silicon, cesium, lead, and iodine. If you have trouble, use Fig. 11.31.

See Problems 11.59 through 11.68. ■

Summary of the Wave Mechanical Model and Valence-Electron Configurations

The concepts we have discussed in this chapter are very important. They allow us to make sense of a good deal of chemistry. When it was first observed that elements with similar properties occur periodically as the atomic number increases, chemists wondered why. Now we have an explanation. The wave mechanical model pictures the electrons in an atom as arranged in orbitals, with each orbital capable of holding two electrons. As we build up the atoms, the same types of orbitals recur in going from one principal energy level to another. This means that particular valence-electron configurations recur periodically. For reasons we will explore in the next chapter, elements

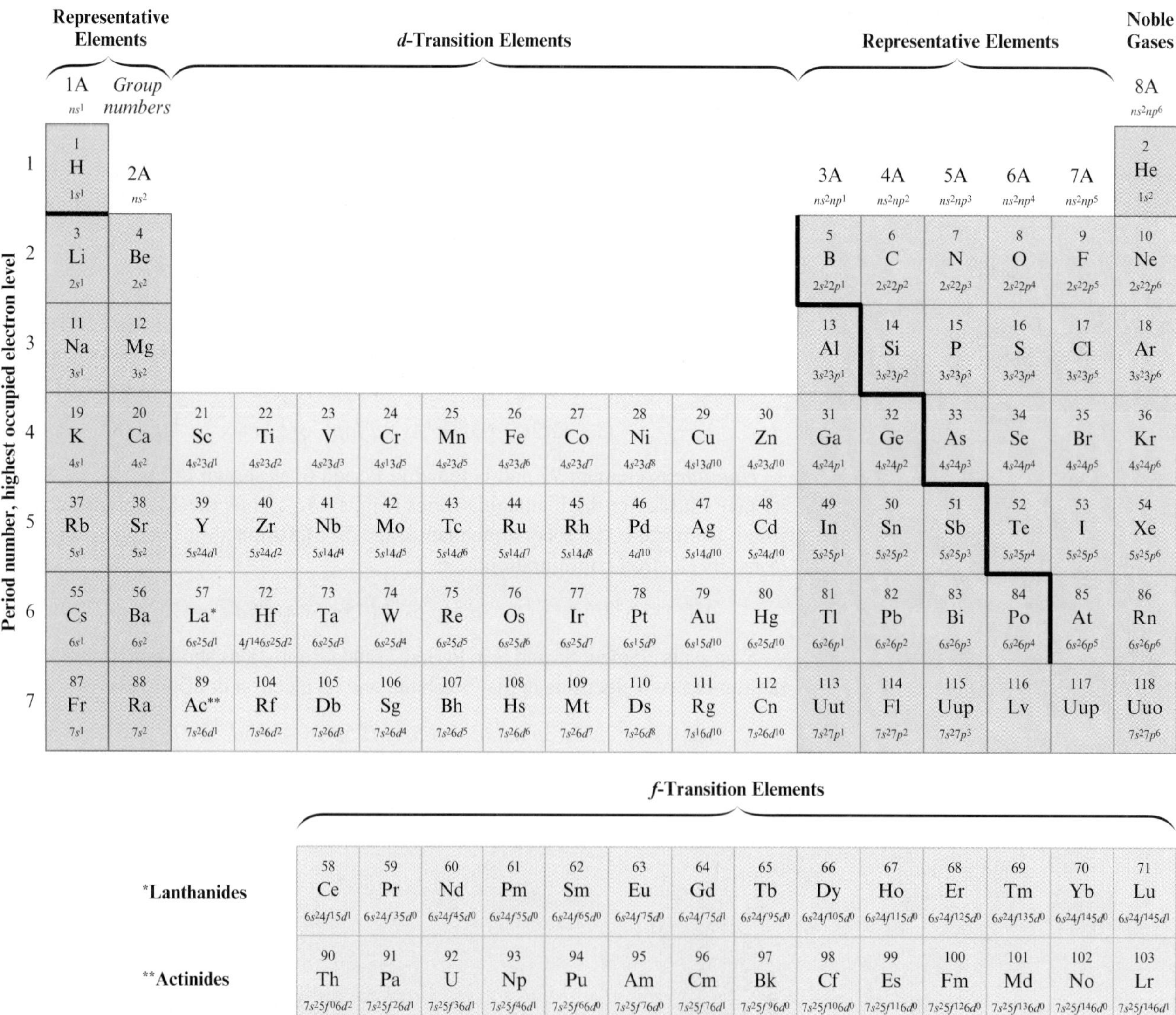

Figure 11.34 ▶ The periodic table with atomic symbols, atomic numbers, and partial electron configurations.

with a particular type of valence configuration all show very similar chemical behavior. Thus groups of elements, such as the alkali metals, show similar chemistry because all the elements in that group have the same type of valence-electron arrangement. This concept, which explains so much chemistry, is the greatest contribution of the wave mechanical model to modern chemistry.

For reference, the valence-electron configurations for all the elements are shown on the periodic table in Fig. 11.34. Note the following points:

1. The group labels for Groups 1, 2, 3, 4, 5, 6, 7, and 8 indicate the *total number* of valence electrons for the atoms in these groups. For example, all the elements in Group 5 have the configuration ns^2np^3. (Any d electrons present are always in the next lower principal energy level than the valence electrons and so are not counted as valence electrons.)
2. The elements in Groups 1, 2, 3, 4, 5, 6, 7, and 8 are often called the **main-group elements,** or **representative elements.** Remember that every member of a given group (except for helium) has the same valence-electron configuration, except that the electrons are in different principal energy levels.
3. We will not be concerned in this text with the configurations for the f-transition elements (lanthanides and actinides), although they are included in Fig. 11.34.

Critical Thinking

You have learned that each orbital can hold two electrons and this pattern is evident on the periodic table. What if each orbital could hold three electrons? How would this change the appearance of the periodic table? For example, what would be the atomic numbers of the noble gases?

11-11 Atomic Properties and the Periodic Table

OBJECTIVE **To understand the general trends in atomic properties in the periodic table.**

With all of this talk about electron probability and orbitals, we must not lose sight of the fact that chemistry is still fundamentally a science based on the observed properties of substances. We know that wood burns, steel rusts, plants grow, sugar tastes sweet, and so on because we *observe* these phenomena. The atomic theory is an attempt to help us understand why these things occur. If we understand why, we can hope to better control the chemical events that are so crucial in our daily lives.

In the next chapter we will see how our ideas about atomic structure help us understand how and why atoms combine to form compounds. As we explore this, and as we use theories to explain other types of chemical behavior later in the text, it is important that we distinguish the observation (steel rusts) from the attempts to explain why the observed event occurs (theories). The observations remain the same over the decades, but the theories (our explanations) change as we gain a clearer understanding of how nature operates. A good example of this is the replacement of the Bohr model for atoms by the wave mechanical model.

Because the observed behavior of matter lies at the heart of chemistry, you need to understand thoroughly the characteristic properties of the various elements and the trends (systematic variations) that occur in those properties. To that end, we will now consider some especially important properties of atoms and see how they vary, horizontally and vertically, on the periodic table.

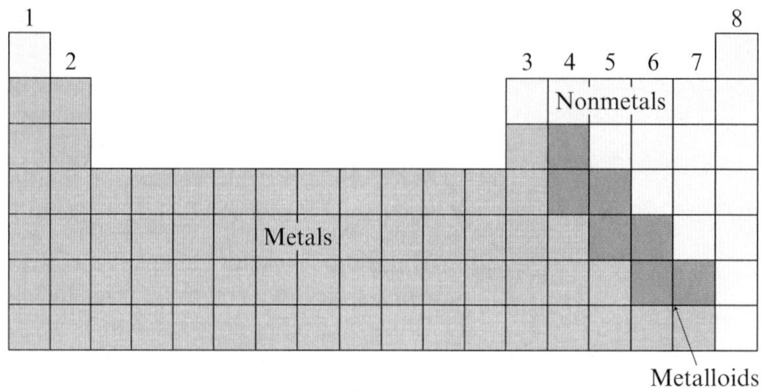

Figure 11.35 The classification of elements as metals, nonmetals, and metalloids.

Gold leaf being applied to Chichester Cathedral's weather vane in Sussex, England. Chris Ison/PA Wire URN: 11984321 (Press Association via AP Images)

Metals and Nonmetals

The most fundamental classification of the chemical elements is into metals and nonmetals. **Metals** typically have the following physical properties: a lustrous appearance, the ability to change shape without breaking (they can be pulled into a wire or pounded into a thin sheet), and excellent conductivity of heat and electricity. **Nonmetals** typically do not have these physical properties, although there are some exceptions. (For example, solid iodine is lustrous; the graphite form of carbon is an excellent conductor of electricity; and the diamond form of carbon is an excellent conductor of heat.) However, it is the *chemical* differences between metals and nonmetals that interest us the most: *metals tend to lose electrons to form positive ions, and nonmetals tend to gain electrons to form negative ions.* When a metal and a nonmetal react, a transfer of one or more electrons from the metal to the nonmetal often occurs.

Most of the elements are classified as metals, as is shown in Fig. 11.35. Note that the metals are found on the left side and at the center of the periodic table. The relatively few nonmetals are in the upper-right corner of the table. A few elements exhibit both metallic and nonmetallic behavior; they are classified as **metalloids** or semimetals.

It is important to understand that simply being classified as a metal does not mean that an element behaves exactly like all other metals. For example, some metals can lose one or more electrons much more easily than others. In particular, cesium can give up its outermost electron (a 6*s* electron) more easily than can lithium (a 2*s* electron). In fact, for the alkali metals (Group 1) the ease of giving up an electron varies as follows:

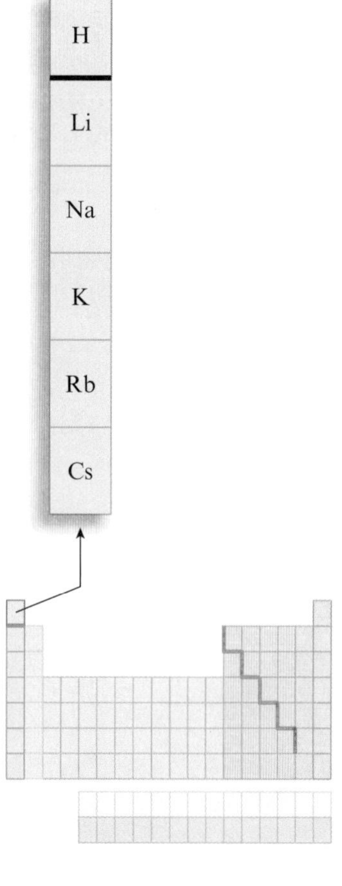

$$\text{Cs} > \text{Rb} > \text{K} > \text{Na} > \text{Li}$$

Loses an electron most easily

Note that as we go down the group, the metals become more likely to lose an electron. This makes sense because as we go down the group, the electron being removed resides, on average, farther and farther from the nucleus. That is, the 6*s* electron lost from Cs is much farther from the attractive positive

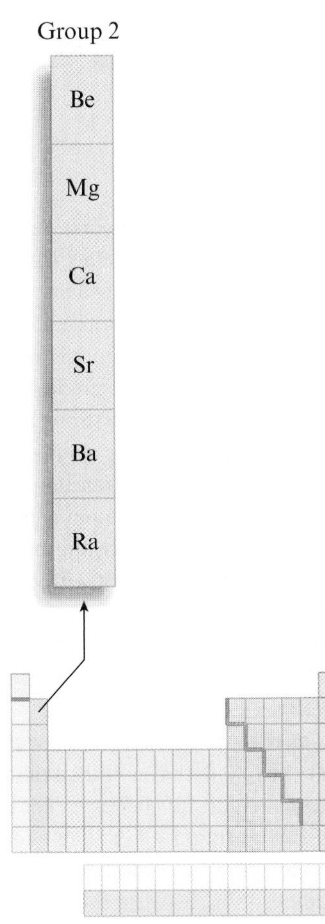

nucleus—and so is easier to remove—than the 2*s* electron that must be removed from a lithium atom.

The same trend is also seen in the Group 2 metals (alkaline earth metals): the farther down in the group the metal resides, the more likely it is to lose an electron.

Just as metals vary somewhat in their properties, so do nonmetals. In general, the elements that can most effectively pull electrons from metals occur in the upper-right corner of the periodic table.

As a general rule, we can say that the most chemically active metals appear in the lower-left region of the periodic table, whereas the most chemically active nonmetals appear in the upper-right region. The properties of the semimetals, or metalloids, lie between the metals and the nonmetals, as might be expected.

Ionization Energies

The **ionization energy** of an atom is the energy required to remove an electron from an individual atom in the gas phase:

$$M(g) \xrightarrow[\text{energy}]{\text{Ionization}} M^{+}(g) + e^{-}$$

As we have noted, the most characteristic chemical property of a metal atom is losing electrons to nonmetals. Another way of saying this is to say that *metals have relatively low ionization energies*—a relatively small amount of energy is needed to remove an electron from a typical metal.

Recall that metals at the bottom of a group lose electrons more easily than those at the top. In other words, ionization energies tend to decrease in going from the top to the bottom of a group.

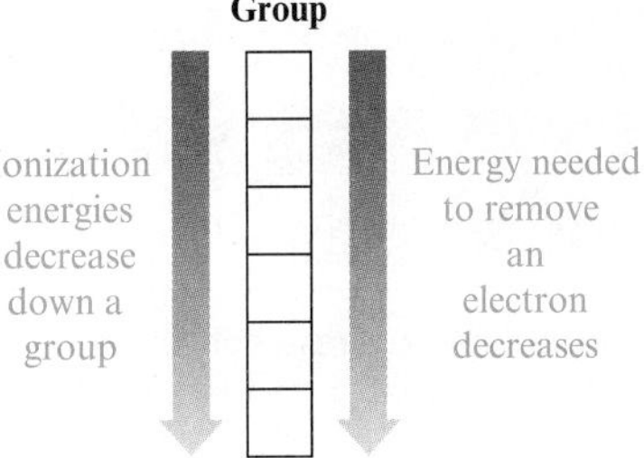

In contrast to metals, nonmetals have relatively large ionization energies. Nonmetals tend to gain, not lose, electrons. Recall that metals appear on the left side of the periodic table and nonmetals appear on the right. Thus it is not surprising that ionization energies tend to increase from left to right across a given period on the periodic table.

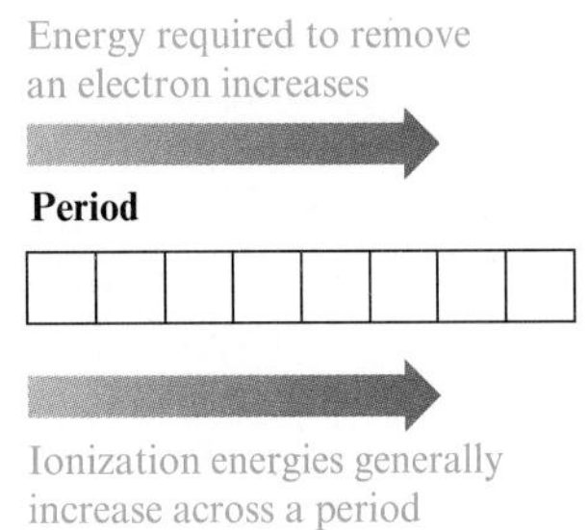

In general, the elements that appear in the lower-left region of the periodic table have the lowest ionization energies (and are therefore the most chemically active metals). On the other hand, the elements with the highest ionization energies (the most chemically active nonmetals) occur in the upper-right region of the periodic table.

CHEMISTRY IN FOCUS

Fireworks

The art of using mixtures of chemicals to produce explosives is an ancient one. Black powder—a mixture of potassium nitrate, charcoal, and sulfur—was being used in China well before A.D. 1000, and it has been used through the centuries in military explosives, in construction blasting, and for fireworks.

Before the nineteenth century, fireworks were confined mainly to rockets and loud bangs. Orange and yellow colors came from the presence of charcoal and iron filings. However, with the great advances in chemistry in the nineteenth century, new compounds found their way into fireworks. Salts of copper, strontium, and barium added brilliant colors. Magnesium and aluminum metals gave a dazzling white light.

How do fireworks produce their brilliant colors and loud bangs? Actually, only a handful of different chemicals are responsible for most of the spectacular effects. To produce the noise and flashes, an oxidizer (something with a strong affinity for electrons) is reacted with a metal such as magnesium or aluminum mixed with sulfur. The resulting reaction produces a brilliant flash, which is due to the aluminum or magnesium burning, and a loud report is produced by the rapidly expanding gases. For a color effect, an element with a colored flame is included.

Yellow colors in fireworks are due to sodium. Strontium salts give the red color familiar from highway safety flares. Barium salts give a green color.

Although you might think that the chemistry of fireworks is simple, achieving the vivid white flashes and the brilliant colors requires complex combinations of chemicals. For example, because the white flashes produce high flame temperatures, the colors tend to wash out. Another problem arises from the use of sodium salts. Because sodium produces an extremely bright yellow color, sodium salts cannot be used when other colors are desired. In short, the manufacture of fireworks that produce the desired effects and are also safe to handle requires very careful selection of chemicals.*

*The chemical mixtures in fireworks are very dangerous. *Do not* experiment with chemicals on your own.

Eyewire/Alamy Images

These brightly colored fireworks are the result of complex mixtures of chemicals.

Atomic Size

The sizes of atoms vary as shown in Fig. 11.36. Notice that atoms get larger as we go down a group on the periodic table and that they get smaller as we go from left to right across a period.

We can understand the increase in size that we observe as we go down a group by remembering that as the principal energy level increases, the average distance of the electrons from the nucleus also increases. So atoms get bigger as electrons are added to larger principal energy levels.

Explaining the decrease in **atomic size** across a period requires a little thought about the atoms in a given row (period) of the periodic table. Recall that the atoms in a particular period all have their outermost electrons in a given principal energy level. That is, the atoms in Period 1 have their outer electrons in the 1*s* orbital (principal energy level 1), the atoms in Period 2 have their outermost electrons in principal energy level 2 (2*s* and 2*p* orbitals), and so on (see Fig. 11.31). Because all the orbitals in a given principal energy level are expected to be the same size, we might expect the atoms in a given period to be the same size. However, remember that the number of protons in

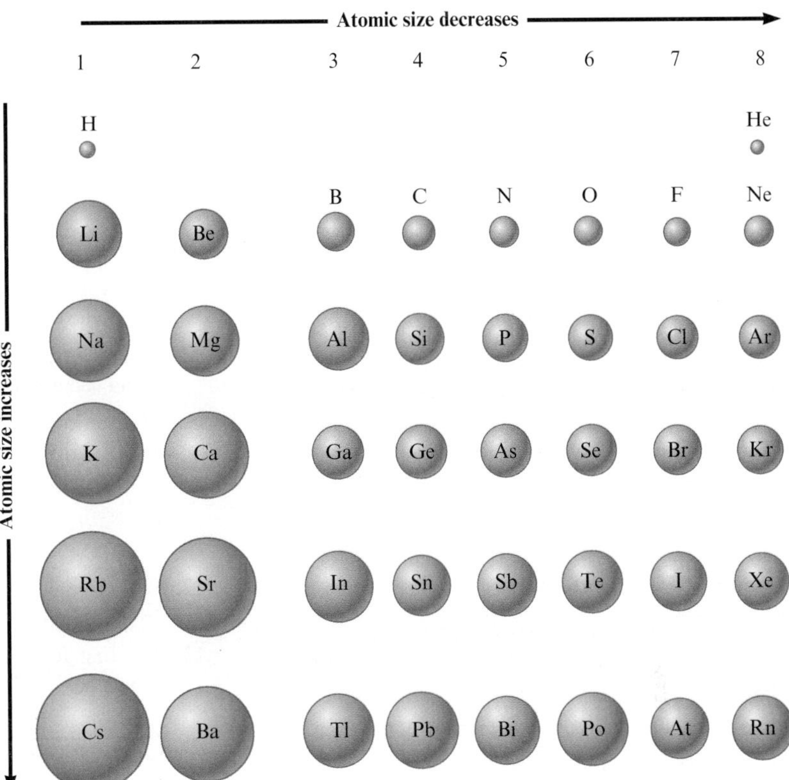

Figure 11.36 ▸ Relative atomic sizes for selected atoms. Note that atomic size increases down a group and decreases across a period.

the nucleus increases as we move from atom to atom in the period. The resulting increase in positive charge on the nucleus tends to pull the electrons closer to the nucleus. So instead of remaining the same size across a period as electrons are added in a given principal energy level, the atoms get smaller as the electron "cloud" is drawn in by the increasing nuclear charge.

CHAPTER 11 REVIEW

F directs you to the *Chemistry in Focus* feature in the chapter

Key Terms

electromagnetic radiation (11-2)
wavelength (11-2)
frequency (11-2)
photons (11-2)
quantized energy levels (11-4)
wave mechanical model (11-6)
orbital (11-7)
principal energy levels (11-7)
sublevels (11-7)
Pauli exclusion principle (11-8)
electron configuration (11-9)
orbital diagram (11-9)
box diagram (11-9)
valence electrons (11-9)
core electrons (11-9)
lanthanide series (11-10)
actinide series (11-10)
main-group elements (11-10)
representative elements (11-10)
metals (11-11)
nonmetals (11-11)
metalloids (11-11)
ionization energy (11-11)
atomic size (11-11)

For Review

- Rutherford's atom consists of a tiny, dense nucleus at the center and electrons that occupy most of the volume of the atom.
- Electromagnetic radiation
  - Characterized by its wavelength and frequency
  - Can be thought of as a stream of packets of energy called photons
  - Atoms can gain energy by absorbing a photon or lose energy by emitting a photon.
  - The energy of a photon is equal to $h\nu$, where $h = 6.626 \times 10^{-34}$ J·s.
- The hydrogen atom can emit only certain energies as it changes from a higher to a lower energy.
- Hydrogen has quantized energy levels.

- The Bohr model assumed electrons travel around the nucleus in circular orbits, which is incorrect.
- The wave mechanical model assumes the electron has both particle and wave properties and describes electrons as occupying orbitals.
  - The orbitals are different from the Bohr orbits.
  - Probability maps indicate the likelihood of finding the electron at a given point in space.
  - The size of an atom can be described by a surface that contains 90% of the total electron probability.
- Atomic energy levels are broken down into principal levels (n), which contain various numbers of sublevels.
  - The sublevels represent various types of orbitals (s, p, d, f), which have different shapes.
  - The number of sublevels increases as n increases.
- A given atom has Z protons in its nucleus and Z electrons surrounding the nucleus.
- The electrons occupy atomic orbitals starting with the lowest energy (the orbital closest to the nucleus).
- The Pauli exclusion principle states that an orbital can hold only two electrons with opposite spins.
- The electrons in the highest energy level are called valence electrons.
- The electron arrangement for a given atom explains its position on the periodic table.
- Atomic size generally increases down a group of the periodic table and decreases across a period.
- Ionization energy generally decreases down a group and increases across a period.

Active Learning Questions

These questions are designed to be considered by groups of students in class. Often these questions work well for introducing a particular topic in class.

1. How does probability fit into the description of the atom?
2. What is meant by an *orbital?*
3. Account for the fact that the line that separates the metals from the nonmetals on the periodic table is diagonal downward to the right instead of horizontal or vertical.
4. Consider the following statements: "The ionization energy for the potassium atom is negative because when K loses an electron to become K^+, it achieves a noble gas electron configuration." Indicate everything that is correct in this statement. Indicate everything that is incorrect. Correct the mistaken information and explain the error.
5. In going across a row of the periodic table, protons and electrons are added and ionization energy generally increases. In going down a column of the periodic table, protons and electrons are also being added but ionization energy generally decreases. Explain.
6. Which is larger, the H 1s orbital or the Li 1s orbital? Why? Which has the larger radius, the H atom or the Li atom? Why?
7. True or false? The hydrogen atom has a 3s orbital. Explain.
8. Differentiate among the terms *energy level, sublevel,* and *orbital.*
9. Make sense of the fact that metals tend to lose electrons and nonmetals tend to gain electrons. Use the periodic table to support your answer.
10. Show how using the periodic table helps you find the expected electron configuration of any element.

For Questions 11–13, you will need to consider ionizations beyond the first ionization energy. For example, the second ionization energy is the energy to remove a second electron from an element.

11. Compare the first ionization energy of helium to its second ionization energy, remembering that both electrons come from the 1s orbital.
12. Which would you expect to have a larger second ionization energy, lithium or beryllium? Why?
13. The first four ionization energies for elements X and Y are shown below. The units are not kJ/mol.

	X	Y
first	170	200
second	350	400
third	1800	3500
fourth	2500	5000

Identify the elements X and Y. There may be more than one answer, so explain completely.

14. Explain what is meant by the term "excited state" as it applies to an electron. Is an electron in an excited state higher or lower in energy than an electron in the ground state? Is an electron in an excited state more or less stable than an electron in the ground state?
15. What does it mean when we say energy levels are *quantized?*
16. What evidence do we have that energy levels in an atom are quantized? State and explain the evidence.
17. Explain the hydrogen emission spectrum. Why is it significant that the color emitted is not white? How does the emission spectrum support the idea of quantized energy levels?
18. There are an infinite number of allowed transitions in the hydrogen atom. Why don't we see more lines in the emission spectrum for hydrogen?
19. You have learned that each orbital is allowed two electrons, and this pattern is evident on the periodic table. What if each orbital was allowed three electrons? How would this change the appearance of the periodic table? For example, what would be the atomic numbers of the noble gases?
20. Atom A has valence electrons that are lower in energy than the valence electrons of Atom B. Which atom has the higher ionization energy? Explain.

All even-numbered Questions and Problems have answers in the back of this book and solutions in the *Student Solutions Guide.*

21. Consider the following waves representing electromagnetic radiation:

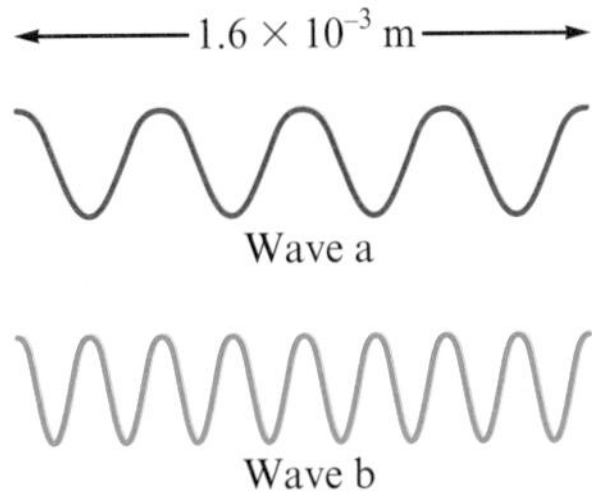

Which wave has the longer wavelength? Calculate the wavelength. Which wave has the higher frequency and photon energy? Calculate these values. Which wave has the greater velocity? What type of electromagnetic radiation does each wave represent?

All even-numbered Questions and Problems have answers in the back of this book and solutions in the *Student Solutions Guide*.

Questions and Problems

11-1 Rutherford's Atom

Questions

1. An atom has a small ________ charged core called the nucleus, with ________ charged electrons moving in the space around the nucleus.
2. What questions were left unanswered by Rutherford's experiments?

11-2 Electromagnetic Radiation

Questions

3. What is *electromagnetic radiation?* At what speed does electromagnetic radiation travel?
4. How are the different types of electromagnetic radiation similar? How do they differ?
5. What does the *wavelength* of electromagnetic radiation represent? How is the wavelength of radiation related to the *energy* of the photons of the radiation?
6. What do we mean by the *frequency* of electromagnetic radiation? Is the frequency the same as the *speed* of the electromagnetic radiation?
7. (F) The "Chemistry in Focus" segment *Light as a Sex Attractant* discusses fluorescence. In fluorescence, ultraviolet radiation is absorbed and intense white visible light is emitted. Is ultraviolet radiation a higher or a lower energy radiation than visible light?
8. (F) The "Chemistry in Focus" segment *Atmospheric Effects* discusses the greenhouse effect. How do the greenhouse gases CO_2, H_2O, and CH_4 have an effect on the temperature of the atmosphere?

11-3 Emission of Energy by Atoms

Questions

9. When lithium salts are heated in a flame, they emit red light. When copper salts are heated in a flame in the same manner, they emit green light. Why do we know that lithium salts will never emit green light, and copper salts will never emit red light?
10. The energy of a photon of visible light emitted by an excited atom is ________ the energy change that takes place within the atom itself.

11-4 The Energy Levels of Hydrogen

Questions

11. What does the *ground state* of an atom represent?
12. When an atom in an excited state returns to its ground state, what happens to the excess energy of the atom?
13. How is the energy carried per photon of light related to the wavelength of the light? Does short-wavelength light carry more energy or less energy than long-wavelength light?
14. When an atom ________ energy from outside, the atom goes from a lower energy state to a higher energy state.
15. Describe briefly why the study of electromagnetic radiation has been important to our understanding of the arrangement of electrons in atoms.
16. What does it mean to say that the hydrogen atom has *discrete energy levels?* How is this fact reflected in the radiation that excited hydrogen atoms emit?
17. Because a given element's atoms emit only certain photons of light, only certain ________ are occurring in those particular atoms.
18. How does the energy possessed by an emitted photon compare to the difference in energy levels that gave rise to the emission of the photon?
19. The energy levels of hydrogen (and other atoms) are said to be ________, which means that only certain energy values are allowed.
20. When a tube containing hydrogen atoms is energized by passing several thousand volts of electricity into the tube, the hydrogen emits light that, when passed through a prism, resolves into the "bright line" spectrum shown in Fig. 11.10. Why do hydrogen atoms emit bright lines of specific wavelengths rather than a continuous spectrum?

11-5 The Bohr Model of the Atom

Questions

21. What are the essential points of Bohr's theory of the structure of the hydrogen atom?
22. According to Bohr, what happens to the electron when a hydrogen atom absorbs a photon of light of sufficient energy?
23. How does the Bohr theory account for the observed phenomenon of the emission of discrete wavelengths of light by excited atoms?
24. Why was Bohr's theory for the hydrogen atom initially accepted, and why was it ultimately discarded?

11-6 The Wave Mechanical Model of the Atom

Questions

25. What major assumption (that was analogous to what had already been demonstrated for electromagnetic radiation) did de Broglie and Schrödinger make about the motion of tiny particles?

26. Discuss briefly the difference between an orbit (as described by Bohr for hydrogen) and an orbital (as described by the more modern, wave mechanical picture of the atom).

27. Why was Schrödinger not able to describe exactly the pathway an electron takes as it moves through the space of an atom?

28. Section 11-6 uses a "firefly" analogy to illustrate how the wave mechanical model for the atom differs from Bohr's model. Explain this analogy.

11-7 The Hydrogen Orbitals

Questions

29. Your text describes the probability map for an *s* orbital using an analogy to the earth's atmosphere. Explain this analogy.

30. Consider the following representation of a set of *p* orbitals for an atom:

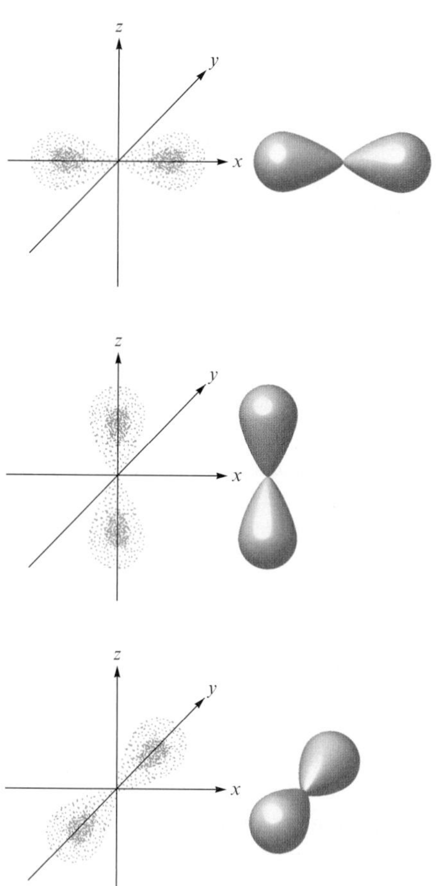

Which of the following statements is(are) true?

a. The areas represented by the *p* orbitals are positively charged clouds with negatively charged electrons embedded within these clouds.
b. The nucleus is located at the center point of each axis.
c. The electrons move along the elliptical paths as indicated by the *p* orbitals above.
d. The atom could not be hydrogen since its electron is found in the 1*s* orbital.

31. What are the differences between the 2*s* orbital and the 1*s* orbital of hydrogen? How are they similar?

32. What overall shape do the 2*p* and 3*p* orbitals have? How do the 2*p* orbitals differ from the 3*p* orbitals? How are they similar?

33. The higher the principal energy level, *n*, the (closer to/farther from) the nucleus is the electron.

34. When the electron in hydrogen is in the $n = 3$ principal energy level, the atom is in a/an ________ state.

35. Although a hydrogen atom has only one electron, the hydrogen atom possesses a complete set of available orbitals. What purpose do these additional orbitals serve?

36. Complete the following table.

Value of *n*	Possible Sublevels
1	________
2	________
3	________
4	________

11-8 The Wave Mechanical Model: Further Development

Questions

37. When describing the electrons in an orbital, we use arrows pointing upward and downward (↑ and ↓) to indicate what property?

38. Why can only two electrons occupy a particular orbital? What is this idea called?

39. How does the *energy* of a principal energy level depend on the value of *n*? Does a higher value of *n* mean a higher or lower energy?

40. The number of sublevels in a principal energy level (increases/decreases) as *n* increases.

41. According to the Pauli exclusion principle, a given orbital can contain only ________ electrons.

42. According to the Pauli exclusion principle, the electrons within a given orbital must have ________ spins.

43. Which of the following orbital designations is(are) possible?

a. 1*s* c. 2*d*
b. 2*p* d. 4*f*

44. Which of the following orbital designations is(are) *not* possible?

a. 2*f* c. 1*d*
b. 4*s* d. 5*p*

11-9 Electron Arrangements in the First Eighteen Atoms on the Periodic Table

Questions

45. Which orbital is the *first* to be filled in any atom? Why?

46. When a hydrogen atom is in its ground state, in which orbital is its electron found? Why?

47. Where are the *valence electrons* found in an atom, and why are these particular electrons most important to the chemical properties of the atom?

All even-numbered Questions and Problems have answers in the back of this book and solutions in the *Student Solutions Guide*.

48. How are the electron arrangements in a given group (vertical column) of the periodic table related? How is this relationship manifested in the properties of the elements in the given group?

Problems

49. Write the full electron configuration ($1s^22s^2$, etc.) for each of the following elements.

a. magnesium, $Z = 12$
b. lithium, $Z = 3$
c. oxygen, $Z = 8$
d. sulfur, $Z = 16$

50. To which element does each of the following electron configurations correspond?

a. $1s^22s^22p^63s^23p^2$
b. $1s^22s^2$
c. $1s^22s^22p^6$
d. $1s^22s^22p^63s^23p^6$

51. Write the full electron configuration ($1s^22s^2$, etc.) for each of the following elements.

a. phosphorus, $Z = 15$
b. calcium, $Z = 20$
c. potassium, $Z = 19$
d. boron, $Z = 5$

52. To which element does each of the following electron configurations correspond?

a. $1s^22s^22p^63s^23p^64s^23d^{10}4p^4$
b. $1s^22s^22p^63s^23p^64s^23d^1$
c. $1s^22s^22p^63s^23p^4$
d. $1s^22s^22p^63s^23p^64s^23d^{10}4p^65s^24d^{10}5p^5$

53. Write the complete orbital diagram for each of the following elements, using boxes to represent orbitals and arrows to represent electrons.

a. helium, $Z = 2$
b. neon, $Z = 10$
c. krypton, $Z = 36$
d. xenon, $Z = 54$

54. Write the complete orbital diagram for each of the following elements, using boxes to represent orbitals and arrows to represent electrons.

a. aluminum, $Z = 13$
b. phosphorus, $Z = 15$
c. bromine, $Z = 35$
d. argon, $Z = 18$

F 55. The "Chemistry in Focus" segment *A Magnetic Moment* discusses the ability to levitate a frog in a magnetic field because electrons, when sensing a strong magnetic field, respond by opposing it. This is called *diamagnetism.* Atoms that are diamagnetic have all paired electrons. Which columns among the representative elements in the periodic table consist of diamagnetic atoms? Consider orbital diagrams when answering this question.

56. For each of the following, give an atom and its complete electron configuration that would be expected to have the indicated number of valence electrons.

a. one
b. three
c. five
d. seven

11-10 Electron Configurations and the Periodic Table

Questions

57. Why do we believe that the valence electrons of calcium and potassium reside in the 4*s* orbital rather than in the 3*d* orbital?

58. Would you expect the valence electrons of rubidium and strontium to reside in the 5*s* or the 4*d* orbitals? Why?

Problems

59. Using the symbol of the previous noble gas to indicate the core electrons, write the electron configuration for each of the following elements.

a. arsenic, $Z = 33$
b. titanium, $Z = 22$
c. strontium, $Z = 38$
d. chlorine, $Z = 17$

60. To which element does each of the following abbreviated electron configurations refer?

a. $[Ne]3s^23p^1$
b. $[Ar]4s^1$
c. $[Ar]4s^23d^{10}4p^5$
d. $[Kr]5s^24d^{10}5p^2$

61. Using the symbol of the previous noble gas to indicate the core electrons, write the electron configuration for each of the following elements.

a. scandium, $Z = 21$
b. yttrium, $Z = 39$
c. lanthanum, $Z = 57$
d. actinium, $Z = 89$

62. How many valence electrons does each of the following atoms have?

a. rubidium, $Z = 37$
b. arsenic, $Z = 33$
c. aluminum, $Z = 13$
d. nickel, $Z = 28$

63. How many 3*d* electrons are found in each of the following elements?

a. nickel, $Z = 28$
b. vanadium, $Z = 23$
c. manganese, $Z = 25$
d. iron, $Z = 26$

64. Based on the elements' locations on the periodic table, how many 4*d* electrons would be predicted for each of the following elements?

a. ruthenium, $Z = 44$
b. palladium, $Z = 46$
c. tin, $Z = 50$
d. iron, $Z = 26$

65. For each of the following elements, indicate which set of orbitals is filled last.

a. radium, $Z = 88$
b. iodine, $Z = 53$
c. gold, $Z = 79$
d. lead, $Z = 82$

66. Write the valence-electron configuration of each of the following elements, basing your answer on the element's location on the periodic table.

a. uranium, $Z = 92$
b. manganese, $Z = 25$
c. mercury, $Z = 80$
d. francium, $Z = 87$

67. Write the valence shell electron configuration of each of the following elements, basing your answer on the element's location on the periodic table.

a. rubidium, $Z = 37$
b. barium, $Z = 56$
c. titanium, $Z = 22$
d. germanium, $Z = 32$

F 68. The "Chemistry in Focus" segment *The Chemistry of Bohrium* discusses element 107, bohrium (Bh). What is the expected electron configuration of Bh?

All even-numbered Questions and Problems have answers in the back of this book and solutions in the *Student Solutions Guide.*

11-11 Atomic Properties and the Periodic Table

Questions

69. What are some of the physical properties that distinguish the metallic elements from the nonmetals? Are these properties absolute, or do some nonmetallic elements exhibit some metallic properties (and vice versa)?

70. What types of ions do the metals and the nonmetallic elements form? Do the metals lose or gain electrons in doing this? Do the nonmetallic elements gain or lose electrons in doing this?

71. Give some similarities that exist among the elements of Group 1.

72. Give some similarities that exist among the elements of Group 7.

73. Which of the following elements most easily gives up electrons during reactions: Li, K, or Cs? Explain your choice.

74. Which elements in a given period (horizontal row) of the periodic table lose electrons most easily? Why?

75. Where are the most nonmetallic elements located on the periodic table? Why do these elements pull electrons from metallic elements so effectively during a reaction?

76. Why do the metallic elements of a given period (horizontal row) typically have much lower ionization energies than do the nonmetallic elements of the same period?

77. What are the *metalloids?* Where are the metalloids found on the periodic table?

78. The "Chemistry in Focus" segment *Fireworks* discusses some of the chemicals that give rise to the colors of fireworks. How do these colors support the existence of quantized energy levels in atoms?

Problems

79. In each of the following groups, which element is least reactive?

a. Group 1
b. Group 7
c. Group 2
d. Group 6

80. In each of the following sets of elements, which element would be expected to have the highest ionization energy?

a. Cs, K, Li
b. Ba, Sr, Ca
c. I, Br, Cl
d. Mg, Si, S

81. Arrange the following sets of elements in order of increasing atomic size.

a. Sn, Xe, Rb, Sr
b. Rn, He, Xe, Kr
c. Pb, Ba, Cs, At

82. In each of the following sets of elements, indicate which element has the smallest atomic size.

a. Na, K, Rb
b. Na, Si, S
c. N, P, As
d. N, O, F

Additional Problems

83. Consider the bright line spectrum of hydrogen shown in Fig. 11.10. Which line in the spectrum represents photons with the highest energy? With the lowest energy?

84. The speed at which electromagnetic radiation moves through a vacuum is called the ________.

85. The portion of the electromagnetic spectrum between wavelengths of approximately 400 and 700 nanometers is called the ________ region.

86. A beam of light can be thought of as consisting of a stream of light particles called ________.

87. The lowest possible energy state of an atom is called the ________ state.

88. The energy levels of hydrogen (and other atoms) are ________, which means that only certain values of energy are allowed.

89. According to Bohr, the electron in the hydrogen atom moved around the nucleus in circular paths called ________.

90. In the modern theory of the atom, a(n) ________ represents a region of space in which there is a high probability of finding an electron.

91. Electrons found in the outermost principal energy level of an atom are referred to as ________ electrons.

92. An element with partially filled *d* orbitals is called a(n) ________.

93. The ________ of electromagnetic radiation represents the number of waves passing a given point in space each second.

94. Only two electrons can occupy a given orbital in an atom, and to be in the same orbital, they must have opposite ________.

95. One bit of evidence that the present theory of atomic structure is "correct" lies in the magnetic properties of matter. Atoms with *unpaired* electrons are attracted by magnetic fields and thus are said to exhibit *paramagnetism.* The degree to which this effect is observed is directly related to the *number* of unpaired electrons present in the atom. On the basis of the electron orbital diagrams for the following elements, indicate which atoms would be expected to be paramagnetic, and tell how many unpaired electrons each atom contains.

a. phosphorus, $Z = 15$
b. iodine, $Z = 53$
c. germanium, $Z = 32$

96. Without referring to your textbook or a periodic table, write the full electron configuration, the orbital box diagram, and the noble gas shorthand configuration for the elements with the following atomic numbers.

a. $Z = 19$
b. $Z = 22$
c. $Z = 14$
d. $Z = 26$
e. $Z = 30$

97. Without referring to your textbook or a periodic table, write the full electron configuration, the orbital box diagram, and the noble gas shorthand configuration for the elements with the following atomic numbers.

a. $Z = 21$
b. $Z = 15$
c. $Z = 36$
d. $Z = 38$
e. $Z = 30$

All even-numbered Questions and Problems have answers in the back of this book and solutions in the *Student Solutions Guide.*

All even-numbered Questions and Problems have answers in the back of this book and solutions in the *Student Solutions Guide.*

98. Write the general valence configuration (for example, ns^1 for Group 1) for the group in which each of the following elements is found.

a. barium, Z = 56
b. bromine, Z = 35
c. tellurium, Z = 52
d. potassium, Z = 19
e. sulfur, Z = 16

99. How many valence electrons does each of the following atoms have?

a. titanium, Z = 22
b. iodine, Z = 53
c. radium, Z = 88
d. manganese, Z = 25

100. In the text (Section 11-6) it was mentioned that current theories of atomic structure suggest that all matter and all energy demonstrate both particle-like and wave-like properties under the appropriate conditions, although the wave-like nature of matter becomes apparent only in very small and very fast-moving particles. The relationship between wavelength (λ) observed for a particle and the mass and velocity of that particle is called the de Broglie relationship. It is

$$\lambda = h/mv$$

in which h is Planck's constant (6.63×10^{-34} J · s),* m represents the mass of the particle in kilograms, and v represents the velocity of the particle in meters per second. Calculate the "de Broglie wavelength" for each of the following, and use your numerical answers to explain why macroscopic (large) objects are not ordinarily discussed in terms of their "wave-like" properties.

a. an electron moving at 0.90 times the speed of light
b. a 150-g ball moving at a speed of 10. m/s
c. a 75-kg person walking at a speed of 2.0 km/h

101. Light waves move through space at a speed of ________ meters per second.

102. How do we know that the energy levels of the hydrogen atom are not *continuous,* as physicists originally assumed?

103. How does the attractive force that the nucleus exerts on an electron change with the principal energy level of the electron?

104. How many *unpaired* electrons does cobalt contain in its ground state?

105. A student writes the electron configuration of carbon (Z = 6) as $1s^3 2s^3$. Explain to him what is *wrong* with this configuration.

106. Given the following valence-electron orbital level diagram and the description, identify the element or ion.

a. a ground-state atom

3s [↑↓]   3p [↑↓ | ↑ | ↑]

b. an atom in an excited state (assume two electrons occupy the 1s orbital)

2s [↑]   2p [↑↓ | ↑ | ↑]

c. a ground-state ion with a charge of −1

4s [↑↓]   4p [↑↓ | ↑↓ | ↑]

107. Why do we believe that the three electrons in the 2p sublevel of nitrogen occupy different orbitals?

108. Write the full electron configuration ($1s^2 2s^2$, etc.) for each of the following elements.

a. bromine, Z = 35
b. xenon, Z = 54
c. barium, Z = 56
d. selenium, Z = 34

109. Write the complete orbital diagram for each of the following elements, using boxes to represent orbitals and arrows to represent electrons.

a. scandium, Z = 21
b. sulfur, Z = 16
c. potassium, Z = 19
d. nitrogen, Z = 7

110. How many valence electrons does each of the following atoms have?

a. nitrogen, Z = 7
b. chlorine, Z = 17
c. sodium, Z = 11
d. aluminum, Z = 13

111. What name is given to the series of ten elements in which the electrons are filling the 3*d* sublevel?

112. Write the general valence configuration (for example, ns^1 for Group 1) for the group in which each of the following elements is found.

a. nitrogen, Z = 7
b. francium, Z = 87
c. chlorine, Z = 17
d. selenium, Z = 34
e. magnesium, Z = 12

113. Using the symbol of the previous noble gas to indicate core electrons, write the valence shell electron configuration for each of the following elements.

a. titanium, Z = 22
b. selenium, Z = 34
c. antimony, Z = 51
d. strontium, Z = 38

114. Arrange the following atoms in order of *increasing* size (assuming all atoms are in their ground states).

a. $[Kr]5s^2 4d^{10} 5p^6$
b. $[Kr]5s^2 4d^{10} 5p^1$
c. $[Kr]5s^2 4d^{10} 5p^3$

115. Write the shorthand valence shell electron configuration of each of the following elements, basing your answer on the element's location on the periodic table.

a. nickel, Z = 28
b. niobium, Z = 41
c. hafnium, Z = 72
d. astatine, Z = 85

116. Metals have relatively (low/high) ionization energies, whereas nonmetals have relatively (high/low) ionization energies.

117. In each of the following sets of elements, indicate which element shows the most active chemical behavior.

a. B, Al, In   b. Na, Al, S   c. B, C, F

118. In each of the following sets of elements, indicate which element has the smallest atomic size.

a. Ba, Ca, Ra   b. P, Si, Al   c. Rb, Cs, K

*Note that s is the abbreviation for "seconds."

ChemWork Problems

These multiconcept problems (and additional ones) are found interactively online with the same type of assistance a student would get from an instructor.

119. Determine the maximum number of electrons that can have each of the following designations: $2f$, $2d_{xy}$, $3p$, $5d_{yz}$, and $4p$.

120. Which of the following statements is(are) *true*?

a. The $2s$ orbital in the hydrogen atom is larger than the $3s$ orbital also in the hydrogen atom.
b. The Bohr model of the hydrogen atom has been found to be incorrect.
c. The hydrogen atom has quantized energy levels.
d. An orbital is the same as a Bohr orbit.
e. The third energy level has three sublevels, the s, p, and d sublevels.

121. Give the electron configurations for the following atoms. Do not use the noble gas notation. Write out the complete electron configuration.

Element	Electron Configuration
Ca	________
B	________
H	________
S	________
Be	________

122. Identify the following three elements.

a. The ground-state electron configuration is $[Kr]5s^24d^{10}5p^4$.
b. The ground-state electron configuration is $[Ar]4s^23d^{10}4p^2$.
c. An excited state of this element has the electron configuration $1s^22s^22p^43s^1$.

123. Give the electron configurations for the following atoms. Use the noble gas notation.

Element	Electron Configuration
K	________
Be	________
Zr	________
Se	________
C	________

124. Compare the atomic sizes of each pair of atoms. State the larger atom for each pair.

Pair	Symbol for Larger Atom
F and B	________
C and N	________
B and Al	________

125. Compare the ionization energies of each pair of atoms. State the atom with the larger ionization energy for each pair.

Pair	Symbol of Atom with the Larger Ionization Energy
He and Kr	________
Na and Al	________
Cl and I	________

126. Three elements have the electron configurations $1s^22s^22p^63s^2$, $1s^22s^22p^63s^23p^4$, and $1s^22s^22p^63s^23p^64s^2$. The first ionization energies of these elements (not in the same order) are 0.590, 0.999, and 0.738 MJ/mol. The atomic radii are 104, 160, and 197 pm. Identify the three elements, and match the appropriate values of ionization energy and atomic radius to each configuration. Complete the following table with the correct information.

Electron Configuration	Element Symbol	First Ionization Energy (MJ/mol)	Atomic Radius (pm)
$1s^22s^22p^63s^2$	________	________	________
$1s^22s^22p^63s^23p^4$	________	________	________
$1s^22s^22p^63s^23p^64s^2$	________	________	________

All even-numbered Questions and Problems have answers in the back of this book and solutions in the *Student Solutions Guide*.

Chemical Bonding

CHAPTER 12

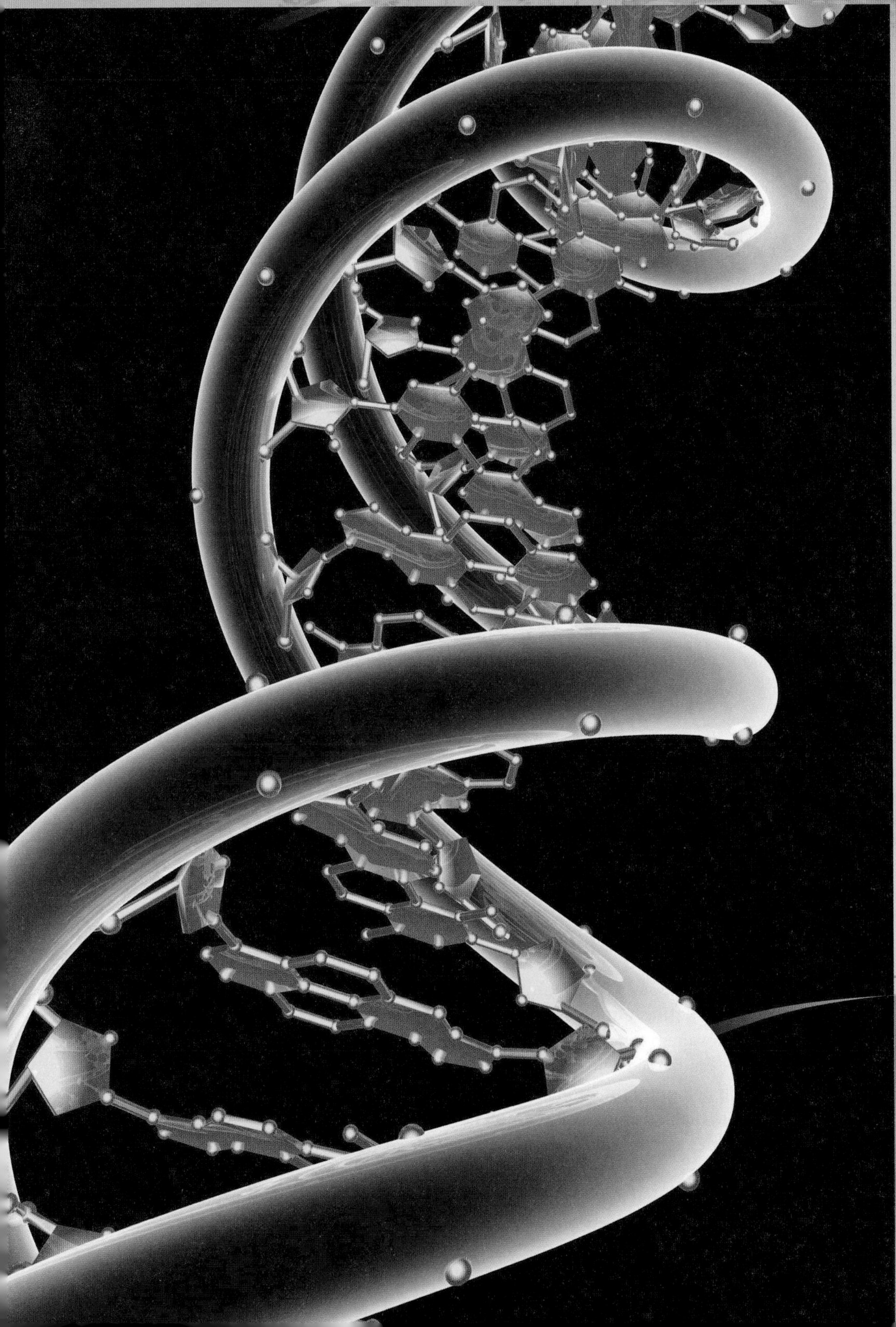

A representation of a segment of the DNA molecule. Science Photo Library/Superstock

The world around us is composed almost entirely of compounds and mixtures of compounds. Rocks, coal, soil, petroleum, trees, and human beings are all complex mixtures of chemical compounds in which different kinds of atoms are bound together. Most of the pure elements found in the earth's crust also contain many atoms bound together. In a gold nugget each gold atom is bound to many other gold atoms, and in a diamond many carbon atoms are bonded very strongly to each other. Substances composed of unbound atoms do exist in nature, but they are very rare. (Examples include the argon atoms in the atmosphere and the helium atoms found in natural gas reserves.)

The manner in which atoms are bound together has a profound effect on the chemical and physical properties of substances. For example, both graphite and diamond are composed solely of carbon atoms. However, graphite is a soft, slippery material used as a lubricant in locks, and diamond is one of the hardest materials known, valuable both as a gemstone and in industrial cutting tools. Why do these materials, both composed solely of carbon atoms, have such different properties? The answer lies in the different ways in which the carbon atoms are bound to each other in these substances.

Molecular bonding and structure play the central role in determining the course of chemical reactions, many of which are vital to our survival. Most reactions in biological systems are very sensitive to the structures of the participating molecules; in fact, very subtle differences in shape sometimes serve to channel the chemical reaction one way rather than another. Molecules that act as drugs must have exactly the right structure to perform their functions correctly. Structure also plays a central role in our senses of smell and taste. Substances have a particular odor because they fit into the specially shaped receptors in our nasal passages. Taste is also dependent on molecular shape, as we discuss in the "Chemistry in Focus" segment *Taste—It's the Structure That Counts.*

To understand the behavior of natural materials, we must understand the nature of chemical bonding and the factors that control the structures of compounds. In this chapter, we will present various classes of compounds that illustrate the different types of bonds. We will then develop models to describe the structure and bonding that characterize the materials found in nature.

Diamond, composed of carbon atoms bonded together to produce one of the hardest materials known, makes a beautiful gemstone. hacohob/Shutterstock.com

12-1 Types of Chemical Bonds

OBJECTIVES

- **To learn about ionic and covalent bonds and explain how they are formed.**
- **To learn about the polar covalent bond.**

A water molecule.

What is a chemical bond? Although there are several possible ways to answer this question, we will define a **bond** as a force that holds groups of two or more atoms together and makes them function as a unit. For example, in water the fundamental unit is the H—O—H molecule, which we describe as being held together by the two O—H bonds. We can obtain information about the strength of a bond by measuring the energy required to break the bond, the **bond energy.**

Atoms can interact with one another in several ways to form aggregates. We will consider specific examples to illustrate the various types of chemical bonds.

In Chapter 7 we saw that when solid sodium chloride is dissolved in water, the resulting solution conducts electricity, a fact that convinces chemists that sodium chloride is composed of Na^+ and Cl^- ions. Thus, when sodium and chlorine react to form sodium chloride, electrons are transferred from the sodium atoms to the chlorine atoms to form Na^+ and Cl^- ions, which then aggregate to form solid sodium chloride. The resulting solid sodium chloride is a very sturdy material; it has a melting point of approximately 800 °C. The strong bonding forces present in sodium chloride result from the attractions among the closely packed, oppositely charged ions. This is an example of **ionic bonding.** Ionic substances are formed when an atom that loses electrons relatively easily reacts with an atom that has a high affinity for electrons. In other words, an **ionic compound** results when a metal reacts with a nonmetal.

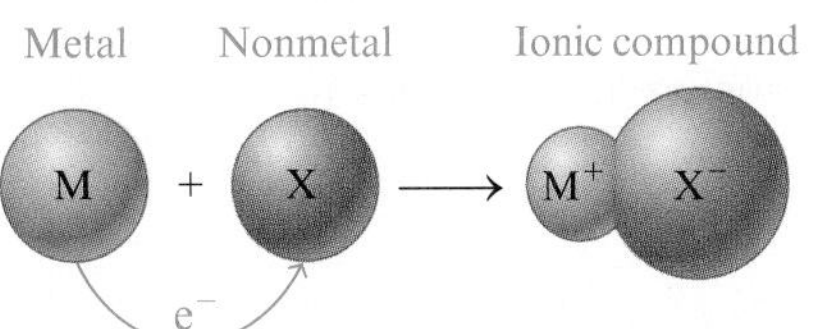

We have seen that a bonding force develops when two very different types of atoms react to form oppositely charged ions. But how does a bonding force develop between two identical atoms? Let's explore this situation by considering what happens when two hydrogen atoms are brought close together, as shown in Fig. 12.1. When hydrogen atoms are close together, the two electrons are simultaneously attracted to both nuclei. Note in Fig. 12.1(b) how the electron probability increases between the two nuclei indicating that the electrons are shared by the two nuclei.

The type of bonding we encounter in the hydrogen molecule and in many other molecules where *electrons are shared by nuclei* is called **covalent bonding.** Note that in the H_2 molecule the electrons reside primarily in the space between the two nuclei, where they are attracted simultaneously by both protons. Although we will not go into detail about it here, the increased attractive forces in this area lead to the formation of the H_2 molecule from the two separated hydrogen atoms. When we say that a bond is formed between the hydrogen atoms, we mean that the H_2 molecule is more stable than two separated hydrogen atoms by a certain quantity of energy (the bond energy).

So far we have considered two extreme types of bonding. In ionic bonding, the participating atoms are so different that one or more electrons are transferred to form oppositely charged ions. The bonding results from the attractions among these ions. In covalent bonding, two identical atoms share electrons equally. The bonding results from the mutual attraction of the two nuclei for the shared electrons. Between these extremes are intermediate cases in which the atoms are not so different that electrons are completely transferred but are different enough so that unequal sharing of electrons results,

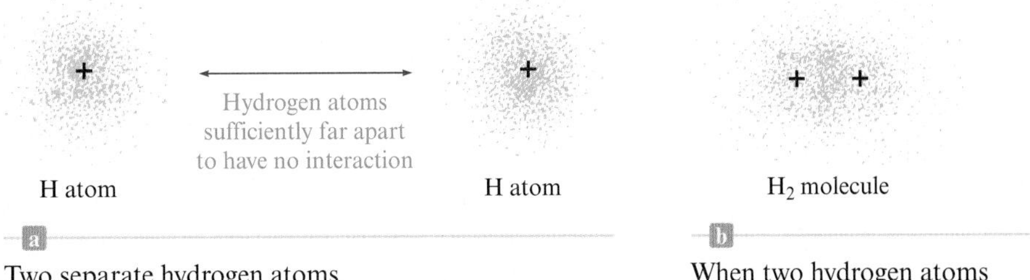

a Two separate hydrogen atoms.

b When two hydrogen atoms come close together, the two electrons are attracted simultaneously by both nuclei. This produces the bond. Note the relatively large electron probability between the nuclei, indicating sharing of the electrons.

Figure 12.1 The formation of a bond between two hydrogen atoms.

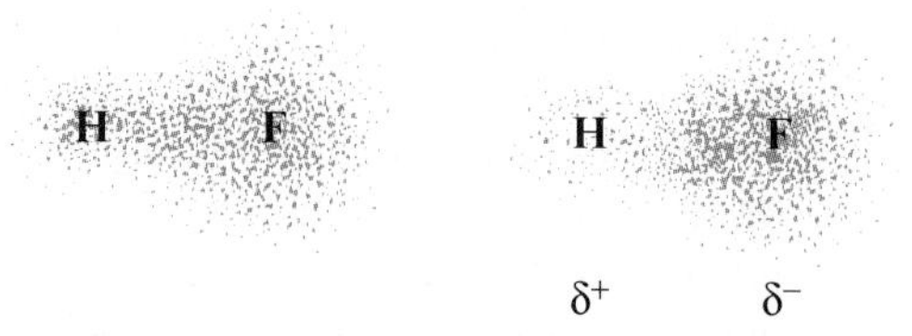

H F

H F

δ^+ δ^-

a

What the probability map would be like if the two electrons in the H—F bond were shared equally.

b

The actual situation, where the shared pair spends more time close to the fluorine atom than to the hydrogen atom. This gives fluorine a slight excess of negative charge and the hydrogen a slight deficit of negative charge (a slight positive charge).

Figure 12.2 ▸ Probability representations of the electron sharing in HF.

forming what is called a **polar covalent bond.** The hydrogen fluoride (HF) molecule contains this type of bond, which produces the following charge distribution,

$$\underset{\delta^+}{H}—\underset{\delta^-}{F}$$

where δ (delta) is used to indicate a partial or fractional charge.

The most logical explanation for the development of *bond polarity* (the partial positive and negative charges on the atoms in such molecules as HF) is that the electrons in the bonds are not shared equally. For example, we can account for the polarity of the HF molecule by assuming that the fluorine atom has a stronger attraction than the hydrogen atom for the shared electrons (Fig. 12.2). Because bond polarity has important chemical implications, we find it useful to assign a number that indicates an atom's ability to attract shared electrons. In the next section we show how this is done.

12-2 Electronegativity

OBJECTIVE To understand the nature of bonds and their relationship to electronegativity.

We saw in the previous section that when a metal and a nonmetal react, one or more electrons are transferred from the metal to the nonmetal to give ionic bonding. On the other hand, two identical atoms react to form a covalent bond in which electrons are shared equally. When *different* nonmetals react, a bond forms in which electrons are shared *unequally,* giving a polar covalent bond. The unequal sharing of electrons between two atoms is described by a property called **electronegativity:** *the relative ability of an atom in a molecule to attract shared electrons to itself.*

Chemists determine electronegativity values for the elements (Fig. 12.3) by measuring the polarities of the bonds between various atoms. Note that electronegativity generally increases going from left to right across a period and decreases going down a group for the representative elements. The range of electronegativity values is from 4.0 for fluorine to 0.7 for cesium and francium. Remember, the higher the atom's electronegativity value, the closer the shared electrons tend to be to that atom when it forms a bond.

The polarity of a bond depends on the *difference* between the electronegativity values of the atoms forming the bond. If the atoms have very similar electronegativities, the electrons are shared almost equally and the bond shows little polarity. If the atoms have very different electronegativity values, a very polar bond is formed. In extreme cases one or more electrons are actually transferred, forming ions and an ionic bond. For example, when an element from Group 1 (electronegativity values of about

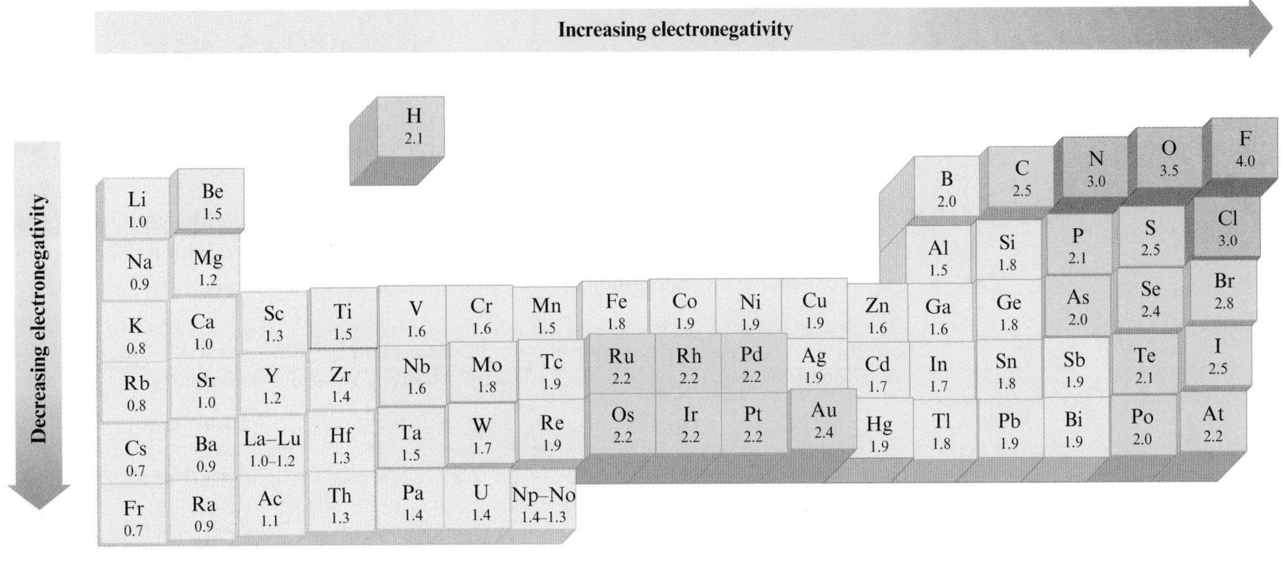

< 1.5
1.5–1.9
2.0–2.9
3.0–4.0

Figure 12.3 ▸ Electronegativity values for selected elements. Note that electronegativity generally increases across a period and decreases down a group. Note also that metals have relatively low electronegativity values and that nonmetals have relatively high values.

Table 12.1 ▸ The Relationship Between Electronegativity and Bond Type

Electronegativity Difference Between the Bonding Atoms	Bond Type	Covalent Character	Ionic Character
Zero	Covalent	↑ Increases	↓ Increases
↓	↓		
Intermediate	Polar covalent		
↓	↓		
Large	Ionic		

0.8) reacts with an element from Group 7 (electronegativity values of about 3), ions are formed and an ionic substance results.

The relationship between electronegativity and bond type is shown in Table 12.1. The various types of bonds are summarized in Fig. 12.4.

Critical Thinking

We use differences in electronegativity to account for certain properties of bonds. What if all atoms had the same electronegativity values? How would bonding between atoms be affected? What are some differences we would notice?

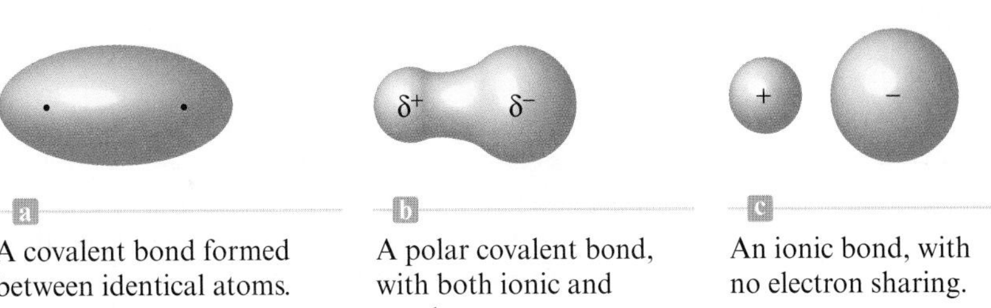

a A covalent bond formed between identical atoms.

b A polar covalent bond, with both ionic and covalent components.

c An ionic bond, with no electron sharing.

Figure 12.4 ▸ The three possible types of bonds.

Interactive Example 12.1

Using Electronegativity to Determine Bond Polarity

Using the electronegativity values given in Fig. 12.3, arrange the following bonds in order of increasing polarity: H—H, O—H, Cl—H, S—H, and F—H.

SOLUTION The polarity of the bond increases as the difference in electronegativity increases. From the electronegativity values in Fig. 12.3, the following variation in bond polarity is expected (the electronegativity value appears below each element).

Bond	Electronegativity Values	Difference in Electronegativity Values	Bond Type	Polarity
H—H	(2.1)(2.1)	2.1 − 2.1 = 0	Covalent	Increasing ↓
S—H	(2.5)(2.1)	2.5 − 2.1 = 0.4	Polar covalent	
Cl—H	(3.0)(2.1)	3.0 − 2.1 = 0.9	Polar covalent	
O—H	(3.5)(2.1)	3.5 − 2.1 = 1.4	Polar covalent	
F—H	(4.0)(2.1)	4.0 − 2.1 = 1.9	Polar covalent	

Therefore, in order of increasing polarity, we have

H—H S—H Cl—H O—H F—H

Least polar → Most polar

SELF CHECK

Exercise 12.1 For each of the following pairs of bonds, choose the bond that will be more polar.

a. H—P, H—C

b. O—F, O—I

c. N—O, S—O

d. N—H, Si—H

See Problems 12.17 through 12.20. ■

12-3 Bond Polarity and Dipole Moments

OBJECTIVE **To understand bond polarity and how it is related to molecular polarity.**

We saw in Section 12-1 that hydrogen fluoride has a positive end and a negative end. A molecule such as HF that has a center of positive charge and a center of negative charge is said to have a **dipole moment.** The dipolar character of a molecule is often represented by an arrow. This arrow points toward the negative charge center, and its tail indicates the positive center of charge:

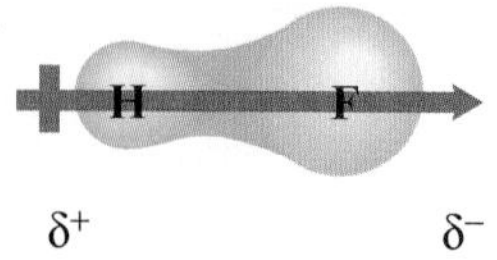

Any diatomic (two-atom) molecule that has a polar bond has a dipole moment. Some polyatomic (more than two atoms) molecules also have dipole moments. For example, because the oxygen atom in the water molecule has a greater electronegativ-

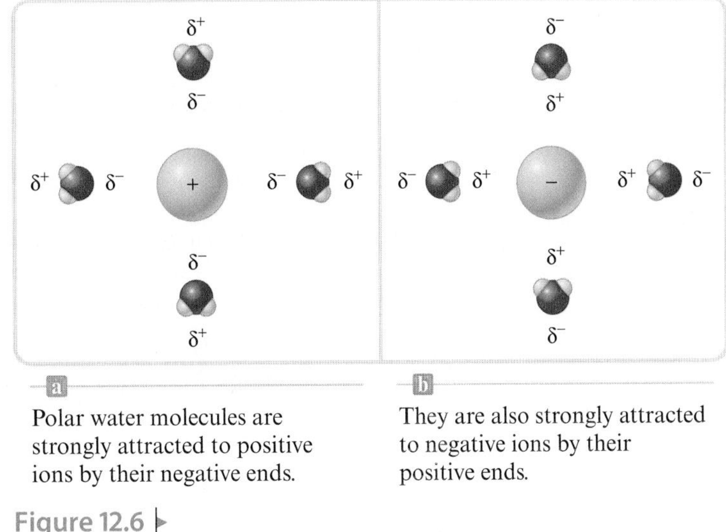

a Polar water molecules are strongly attracted to positive ions by their negative ends.

b They are also strongly attracted to negative ions by their positive ends.

Figure 12.6

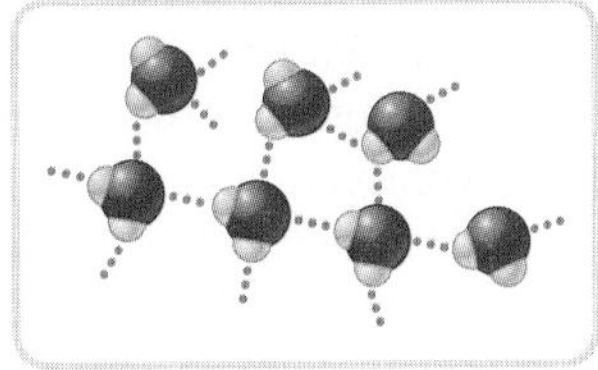

Figure 12.7 Polar water molecules are strongly attracted to each other.

δ^+ H

O $2\delta^-$

H δ^+

a The charge distribution in the water molecule. The oxygen has a charge of $2\delta^-$ because it pulls δ^- of charge from each hydrogen atom ($\delta^- + \delta^- = 2\delta^-$).

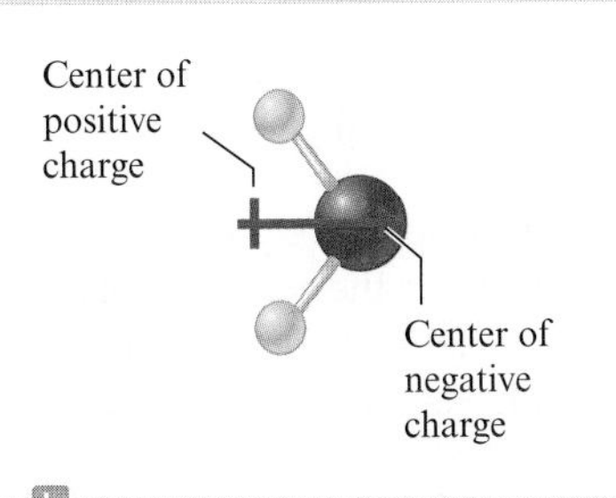

b The water molecule behaves as if it had a positive end and a negative end, as indicated by the arrow.

Figure 12.5

ity than the hydrogen atoms, the electrons are not shared equally. This results in a charge distribution (Fig. 12.5) that causes the molecule to behave as though it had two centers of charge—one positive and one negative. So the water molecule has a dipole moment.

The fact that the water molecule is polar (has a dipole moment) has a profound impact on its properties. In fact, it is not overly dramatic to state that the polarity of the water molecule is crucial to life as we know it on earth. Because water molecules are polar, they can surround and attract both positive and negative ions (Fig. 12.6). These attractions allow ionic materials to dissolve in water. Also, the polarity of water molecules causes them to attract each other strongly (Fig. 12.7). This means that much energy is required to change water from a liquid to a gas (the molecules must be separated from each other to undergo this change of state). Therefore, it is the polarity of the water molecule that causes water to remain a liquid at the temperatures on the earth's surface. If it were nonpolar, water would be a gas and the oceans would be empty.

12-4 Stable Electron Configurations and Charges on Ions

OBJECTIVES

- To learn about stable electron configurations.
- To learn to predict the formulas of ionic compounds.

We have seen many times that when a metal and a nonmetal react to form an ionic compound, the metal atom loses one or more electrons to the nonmetal. In Chapter 5, where binary ionic compounds were introduced, we saw that in these reactions, Group 1 metals always form 1+ cations, Group 2 metals always form 2+ cations, and aluminum in Group 3 always forms a 3+ cation. For the nonmetals, the Group 7 elements

Table 12.2 ▶ The Formation of Ions by Metals and Nonmetals

Group	Ion Formation	Electron Configuration: Atom		Ion
1	$Na \rightarrow Na^+ + e^-$	$[Ne]3s^1$	e^- lost $\longrightarrow$	[Ne]
2	$Mg \rightarrow Mg^{2+} + 2e^-$	$[Ne]3s^2$	$2e^-$ lost $\longrightarrow$	[Ne]
3	$Al \rightarrow Al^{3+} + 3e^-$	$[Ne]3s^23p^1$	$3e^-$ lost $\longrightarrow$	[Ne]
6	$O + 2e^- \rightarrow O^{2-}$	$[He]2s^22p^4 + 2e^-$	$\rightarrow [He]2s^22p^6 =$	[Ne]
7	$F + e^- \rightarrow F^-$	$[He]2s^22p^5 + e^-$	$\rightarrow [He]2s^22p^6 =$	[Ne]

always form 1− anions, and the Group 6 elements always form 2− anions. This is further illustrated in Table 12.2.

Notice something very interesting about the ions in Table 12.2: they all have the electron configuration of neon, a noble gas. That is, sodium loses its one valence electron (the 3*s*) to form Na^+, which has an [Ne] electron configuration. Likewise, Mg loses its two valence electrons to form Mg^{2+}, which also has an [Ne] electron configuration. On the other hand, the nonmetal atoms gain just the number of electrons needed for them to achieve the noble gas electron configuration. The O atom gains two electrons and the F atom gains one electron to give O^{2-} and F^-, respectively, both of which have the [Ne] electron configuration. We can summarize these observations as follows:

Electron Configurations of Ions

1. Representative (main-group) metals form ions by losing enough electrons to achieve the configuration of the previous noble gas (that is, the noble gas that occurs before the metal in question on the periodic table). For example, note that neon is the noble gas previous to sodium and magnesium. Similarly, helium is the noble gas previous to lithium and beryllium.
2. Nonmetals form ions by gaining enough electrons to achieve the configuration of the next noble gas (that is, the noble gas that follows the element in question on the periodic table). For example, note that neon is the noble gas that follows oxygen and fluorine, and argon is the noble gas that follows sulfur and chlorine.

This brings us to an important general principle. In observing millions of stable compounds, chemists have learned that **in almost all stable chemical compounds of the representative elements, all of the atoms have achieved a noble gas electron configuration.** The importance of this observation cannot be overstated. It forms the basis for all of our fundamental ideas about why and how atoms bond to each other.

We have already seen this principle operating in the formation of ions (see Table 12.2). We can summarize this behavior as follows: when representative metals and nonmetals react, they transfer electrons in such a way that both the cation and the anion have noble gas electron configurations.

On the other hand, when nonmetals react with each other, they share electrons in ways that lead to a noble gas electron configuration for each atom in the resulting molecule. For example, oxygen ($[He]2s^22p^4$), which needs two more electrons to achieve an [Ne] configuration, can get these electrons by combining with two H atoms (each of which has one electron),

O: [He] 2s ↑↓ 2p ↑↓ ↑↓ ↑↓
H H

to form water, H_2O. This fills the valence orbitals of oxygen.

In addition, each H shares two electrons with the oxygen atom,

O
H H

which fills the H 1*s* orbital, giving it a $1s^2$ or [He] electron configuration. We will have much more to say about covalent bonding in Section 12-6.

At this point let's summarize the ideas we have introduced so far.

Electron Configurations and Bonding

1. When a *nonmetal and a Group 1, 2, or 3 metal* react to form a binary ionic compound, the ions form in such a way that the valence-electron configuration of the *nonmetal* is *completed* to achieve the configuration of the *next* noble gas, and the valence orbitals of the *metal* are *emptied* to achieve the configuration of the *previous* noble gas. In this way both ions achieve noble gas electron configurations.
2. When *two nonmetals* react to form a covalent bond, they share electrons in a way that completes the valence-electron configurations of both atoms. That is, both nonmetals attain noble gas electron configurations by sharing electrons.

Predicting Formulas of Ionic Compounds

Now that we know something about the electron configuration of atoms, we can explain *why* these various ions are formed. To show how to predict what ions form when a metal reacts with a nonmetal, we will consider the formation of an ionic compound from calcium and oxygen. We can predict what compound will form by considering the valence-electron configurations of the following two atoms:

$$\text{Ca: } [\text{Ar}]4s^2$$
$$\text{O: } [\text{He}]2s^2 2p^4$$

From Fig. 12.3 we see that the electronegativity of oxygen (3.5) is much greater than that of calcium (1.0), giving a difference of 2.5. Because of this large difference, electrons are transferred from calcium to oxygen to form an oxygen anion and a calcium cation. How many electrons are transferred? We can base our prediction on the observation that noble gas configurations are the most stable. Note that oxygen needs two electrons to fill its valence orbitals (2*s* and 2*p*) and achieve the configuration of neon ($1s^2 2s^2 2p^6$), which is the next noble gas.

$$\text{O} + 2e^- \rightarrow \text{O}^{2-}$$
$$[\text{He}]2s^2 2p^4 + 2e^- \rightarrow [\text{He}]2s^2 2p^6, \text{ or } [\text{Ne}]$$

And by losing two electrons, calcium can achieve the configuration of argon (the previous noble gas).

$$\text{Ca} \rightarrow \text{Ca}^{2+} + 2e^-$$
$$[\text{Ar}]4s^2 \rightarrow [\text{Ar}] + 2e^-$$

Table 12.3 ▸ Common Ions with Noble Gas Configurations in Ionic Compounds

Group 1	Group 2	Group 3	Group 6	Group 7	Electron Configuration
Li^+	Be^{2+}				[He]
Na^+	Mg^{2+}	Al^{3+}	O^{2-}	F^-	[Ne]
K^+	Ca^{2+}		S^{2-}	Cl^-	[Ar]
Rb^+	Sr^{2+}		Se^{2-}	Br^-	[Kr]
Cs^+	Ba^{2+}		Te^{2-}	I^-	[Xe]

Two electrons are therefore transferred as follows:

$$Ca + O \rightarrow Ca^{2+} + O^{2-}$$

$2e^-$

To predict the formula of the ionic compound, we use the fact that chemical compounds are always electrically neutral—they have the same total quantities of positive and negative charges. In this case we must have equal numbers of Ca^{2+} and O^{2-} ions, and the empirical formula of the compound is CaO.

The same principles can be applied to many other cases. For example, consider the compound formed from aluminum and oxygen. Aluminum has the electron configuration $[Ne]3s^23p^1$. To achieve the neon configuration, aluminum must lose three electrons, forming the Al^{3+} ion.

$$Al \rightarrow Al^{3+} + 3e^-$$
$$[Ne]3s^23p^1 \rightarrow [Ne] + 3e^-$$

$3 \times (2-)$ balances $2 \times (3+)$.

Therefore, the ions will be Al^{3+} and O^{2-}. Because the compound must be electrically neutral, there will be three O^{2-} ions for every two Al^{3+} ions, and the compound has the empirical formula Al_2O_3.

Table 12.3 shows common elements that form ions with noble gas electron configurations in ionic compounds.

Notice that our discussion in this section refers to metals in Groups 1, 2, and 3 (the representative metals). The transition metals exhibit more complicated behavior (they form a variety of ions), which we will not be concerned with in this text.

Ionic Bonding and Structures of Ionic Compounds

OBJECTIVES

- **To learn about ionic structures.**
- **To understand factors governing ionic size.**

When metals and nonmetals react, the resulting ionic compounds are very stable; large amounts of energy are required to "take them apart." For example, the melting point of sodium chloride is approximately 800 °C. The strong bonding in these ionic compounds results from the attractions among the oppositely charged cations and anions.

We write the formula of an ionic compound such as lithium fluoride simply as LiF, but this is really the empirical, or simplest, formula. The actual solid contains huge and equal numbers of Li^+ and F^- ions packed together in a way that maximizes the attractions of the oppositely charged ions. A representative part of the lithium fluoride structure is shown in Fig. 12.8(a). In this structure the larger F^- ions are packed together like hard spheres, and the much smaller Li^+ ions are interspersed regularly among the

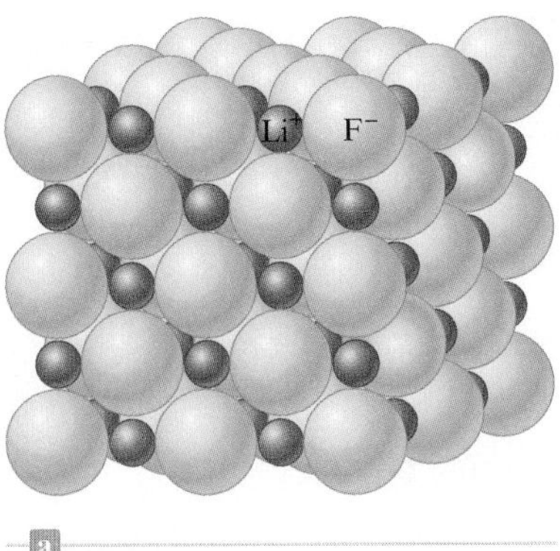

a This structure represents the ions as packed spheres.

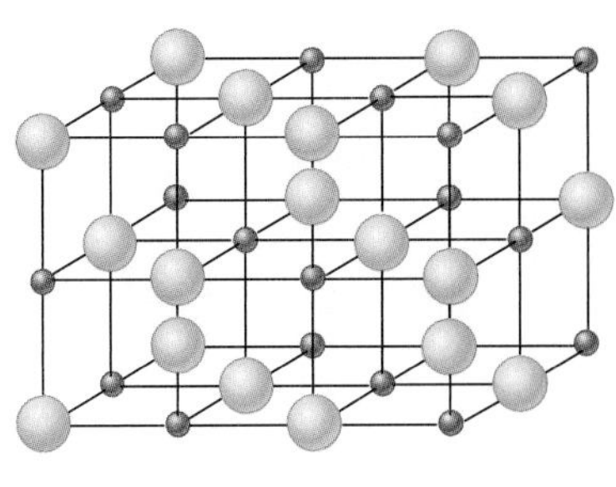

b This structure shows the positions (centers) of the ions. The spherical ions are packed in the way that maximizes the ionic attractions.

Figure 12.8 ▸ The structure of lithium fluoride.

When spheres are packed together, they do not fill up all of the space. The spaces (holes) that are left can be occupied by smaller spheres.

F^- ions. The structure shown in Fig. 12.8(b) represents only a tiny part of the actual structure, which continues in all three dimensions with the same pattern as that shown.

The structures of virtually all binary ionic compounds can be explained by a model that involves packing the ions as though they were hard spheres. The larger spheres (usually the anions) are packed together, and the small ions occupy the interstices (spaces or holes) among them.

To understand the packing of ions it helps to realize that *a cation is always smaller than the parent atom, and an anion is always larger than the parent atom.* This makes sense because when a metal loses all of its valence electrons to form a cation, it gets much smaller. On the other hand, in forming an anion, a nonmetal gains enough electrons to achieve the next noble gas electron configuration and so becomes much larger. The relative sizes of the Group 1 and Group 7 atoms and their ions are shown in Fig. 12.9.

Critical Thinking

Ions have different radii than their parent atoms. What if ions stayed the same size as their parent atoms? How would this affect the structure of ionic compounds?

Ionic Compounds Containing Polyatomic Ions

So far in this chapter we have discussed only binary ionic compounds, which contain ions derived from single atoms. However, many compounds contain polyatomic ions: charged species composed of several atoms. For example, ammonium nitrate contains the NH_4^+ and NO_3^- ions. These ions with their opposite charges attract each other in the same way as do the simple ions in binary ionic compounds. However, the *individual* polyatomic ions are held together by covalent bonds, with all of the atoms behaving as a unit. For example, in the ammonium ion, NH_4^+, there are four N—H covalent bonds. Likewise the nitrate ion, NO_3^-, contains three covalent N—O bonds. Thus, although ammonium nitrate is an ionic compound because it contains the NH_4^+ and NO_3^- ions, it also contains covalent bonds in the individual polyatomic ions. When ammonium nitrate is dissolved in water, it behaves as a strong electrolyte like the binary ionic compounds sodium chloride and potassium bromide. As we saw in Chapter 7, this occurs because when an ionic solid dissolves, the ions are freed to move independently and can conduct an electric current.

The common polyatomic ions, which are listed in Table 5.4, are all held together by covalent bonds.

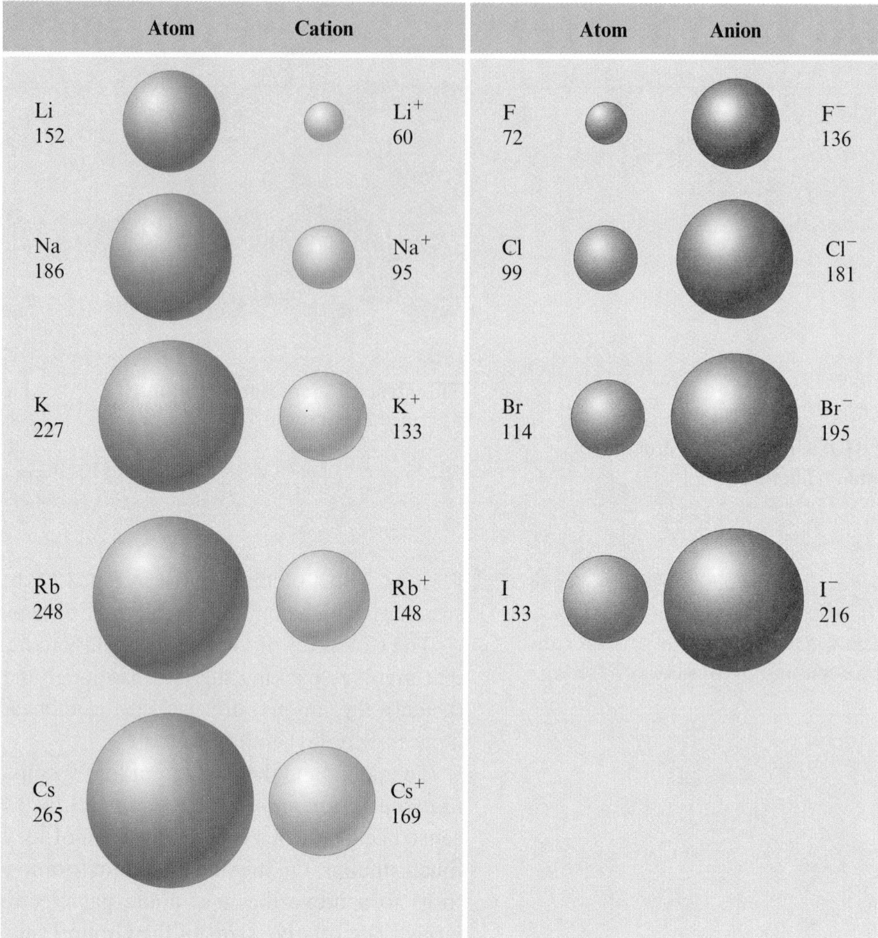

Figure 12.9 ▸ Relative sizes of some ions and their parent atoms. Note that cations are smaller and anions are larger than their parent atoms. The sizes (radii) are given in units of picometers (1 pm = 10^{-12} m).

12-6 Lewis Structures

OBJECTIVE To learn to write Lewis structures.

Remember that the electrons in the highest principal energy level of an atom are called the valence electrons.

Bonding involves just the valence electrons of atoms. Valence electrons are transferred when a metal and a nonmetal react to form an ionic compound. Valence electrons are shared between nonmetals in covalent bonds.

The **Lewis structure** is a representation of a molecule that shows how the valence electrons are arranged among the atoms in the molecule. These representations are named after G. N. Lewis, who conceived the idea while lecturing to a class of general chemistry students in 1902. The rules for writing Lewis structures are based on observations of many molecules from which chemists have learned that the *most important requirement for the formation of a stable compound is that the atoms achieve noble gas electron configurations.*

We have already seen this rule operate in the reaction of metals and nonmetals to form binary ionic compounds. An example is the formation of KBr, where the K^+ ion has the [Ar] electron configuration and the Br^- ion has the [Kr] electron configuration. In writing Lewis structures, *we include only the valence electrons.* Using dots to represent valence electrons, we write the Lewis structure for KBr as follows:

$$K^+ \qquad [:\ddot{\underset{..}{Br}}:]^-$$

Noble gas configuration [Ar] — Noble gas configuration [Kr]

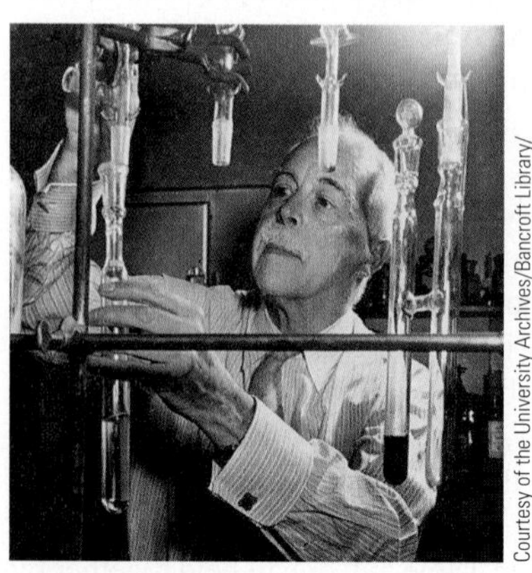
Courtesy of the University Archives/Bancroft Library/ University of California, Berkeley #UARC PIC 13:596

G. N. Lewis in his lab.

No dots are shown on the K^+ ion because it has lost its only valence electron (the 4*s* electron). The Br^- ion is shown with eight electrons because it has a filled valence shell.

Next we will consider Lewis structures for molecules with covalent bonds, involving nonmetals in the first and second periods. The principle of achieving a noble gas electron configuration applies to these elements as follows:

1. Hydrogen forms stable molecules where it shares two electrons. That is, it follows a **duet rule.** For example, when two hydrogen atoms, each with one electron, combine to form the H_2 molecule, we have

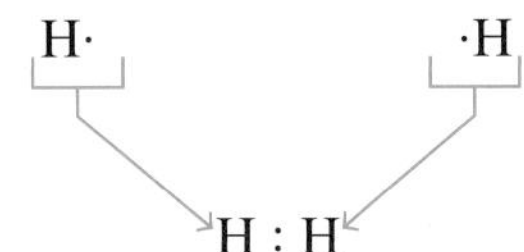

By sharing electrons, each hydrogen in H_2 has, in effect, two electrons; that is, each hydrogen has a filled valence shell.

H [↑] 1*s*
H [↓] 1*s*
→ H_2 [↑↓]
[He] configuration

2. Helium does not form bonds because its valence orbital is already filled; it is a noble gas. Helium has the electron configuration $1s^2$ and can be represented by the Lewis structure

He:
[He] configuration

3. The second-row nonmetals carbon through fluorine form stable molecules when they are surrounded by enough electrons to fill the valence orbitals—that is, the one 2*s* and the three 2*p* orbitals. Eight electrons are required to fill these orbitals, so these elements typically obey the **octet rule;** they are surrounded by eight electrons. An example is the F_2 molecule, which has the following Lewis structure:

Carbon, nitrogen, oxygen, and fluorine almost always obey the octet rule in stable molecules.

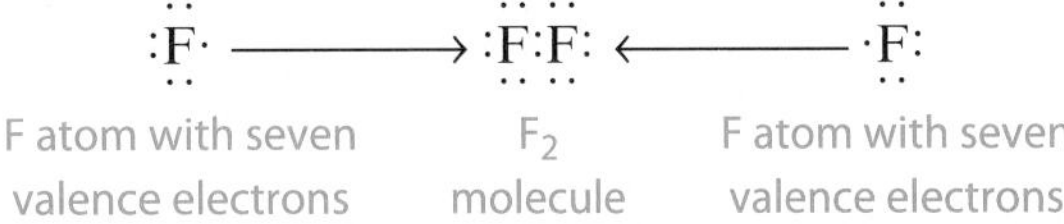

Note that each fluorine atom in F_2 is, in effect, surrounded by eight valence electrons, two of which are shared with the other atom. This is a **bonding pair** of electrons, as we discussed earlier. Each fluorine atom also has three pairs of electrons that are not involved in bonding. These are called **lone pairs** or **unshared pairs.**

4. Neon does not form bonds because it already has an octet of valence electrons (it is a noble gas). The Lewis structure is

:Ne:

Note that only the valence electrons ($2s^22p^6$) of the neon atom are represented by the Lewis structure. The $1s^2$ electrons are core electrons and are not shown.

Next we want to develop some general procedures for writing Lewis structures for molecules. Remember that Lewis structures involve only the valence electrons of atoms, so before we proceed, we will review the relationship of an element's position on the periodic table to the number of valence electrons it has. Recall that the group number gives the total number of valence electrons. For example, all Group 6 elements have six valence electrons (valence configuration ns^2np^4).

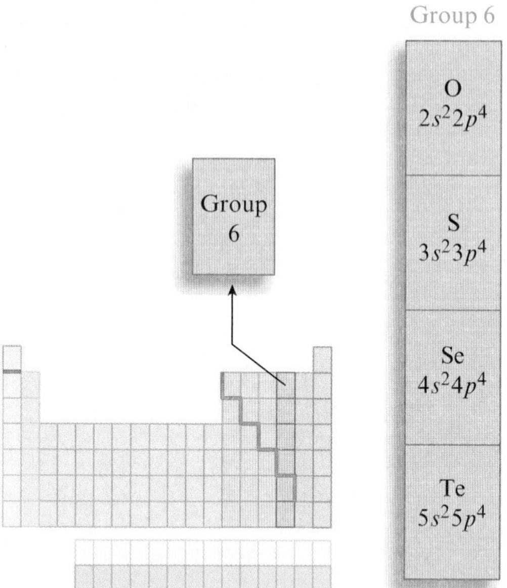

Similarly, all Group 7 elements have seven valence electrons (valence configuration ns^2np^5).

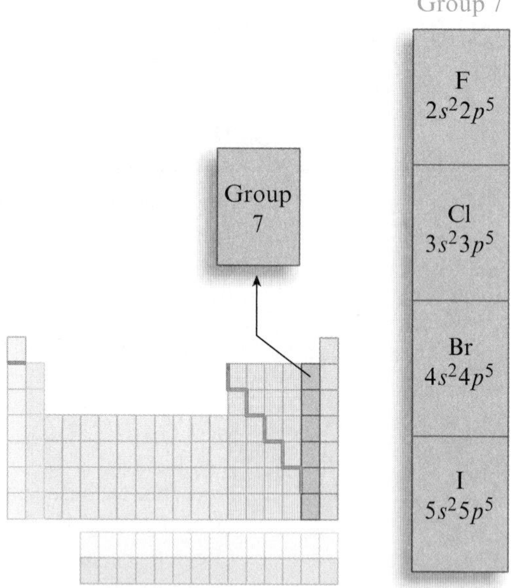

In writing the Lewis structure for a molecule, we need to keep the following things in mind:

1. We must include all the valence electrons from all atoms. The total number of electrons available is the sum of all the valence electrons from all the atoms in the molecule.
2. Atoms that are bonded to each other share one or more pairs of electrons.
3. The electrons are arranged so that each atom is surrounded by enough electrons to fill the valence orbitals of that atom. This means two electrons for hydrogen and eight electrons for second-row nonmetals.

CHEMISTRY IN FOCUS

To Bee or Not to Bee

One of the problems we face in modern society is how to detect illicit substances, such as drugs and explosives, in a convenient, accurate manner. Trained dogs are often used for this purpose because of their acute sense of smell. Now several researchers are trying to determine whether insects, such as honeybees and wasps, can be even more effective chemical detectors. In fact, studies have shown that bees can be trained in just a few minutes to detect the smell of almost any chemical.

Scientists at Los Alamos National Laboratory in New Mexico are designing a portable device using bees that possibly could be used to sniff out drugs and bombs at airports, border crossings, and schools. They call their study the Stealthy Insect Sensor Project. The Los Alamos project is based on the idea that bees can be trained to associate the smell of a particular chemical with a sugary treat. Bees stick out their "tongues" when they detect a food source. By pairing a drop of sugar water with the scent of TNT (trinitrotoluene) or C-4 (composition 4) plastic explosive about six times, the bees can be trained to extend their proboscis at a whiff of the chemical alone. The bee bomb detector is about half the size of a shoe box and weighs 4 lb. Inside the box, bees are lined up in a row and strapped into straw-like tubes, then exposed to puffs of air as a camera monitors their reactions. The signals from the video camera are sent to a computer, which analyzes the bees' behavior and signals when the bees respond to the particular scent they have been trained to detect.

A project at the University of Georgia uses tiny parasitic wasps as a chemical detector. Wasps do not extend their tongues when they detect a scent. Instead, they communicate the discovery of a scent by body movements that the scientists call "dances." The device, called the Wasp Hound, contains a team of wasps in a hand-held ventilated cartridge that has a fan at one end to draw in air from outside. If the scent is one the wasps do not recognize, they continue flying randomly. However, if the scent is one the wasps have been conditioned to recognize, they cluster around the opening. A video camera paired with a computer analyzes their behavior and signals when a scent is detected.

The insect sensors are now undergoing field trials, which typically compare the effectiveness of insects to that of trained dogs. Initial results appear promising, but the effectiveness of these devices remains to be proved.

A honeybee receives a fragrant reminder of its target scent each morning and responds by sticking out its proboscis.

The best way to make sure we arrive at the correct Lewis structure for a molecule is to use a systematic approach. We will use the approach summarized by the following rules.

Steps for Writing Lewis Structures

Step 1 Obtain the sum of the valence electrons from all of the atoms. Do not worry about keeping track of which electrons come from which atoms. It is the *total* number of valence electrons that is important.

Step 2 Use one pair of electrons to form a bond between each pair of bound atoms. For convenience, a line (instead of a pair of dots) is often used to indicate each pair of bonding electrons.

Step 3 Arrange the remaining electrons to satisfy the duet rule for hydrogen and the octet rule for each second-row element.

To see how these rules are applied, we will write the Lewis structures of several molecules.

Interactive Example 12.2 Writing Lewis Structures: Simple Molecules

Write the Lewis structure of the water molecule.

SOLUTION We will follow the *Steps for Writing Lewis Structures.*

Step 1 Find the sum of the *valence* electrons for H_2O.

$$\underset{\substack{\text{H}\\ \text{(Group 1)}}}{1} + \underset{\substack{\text{H}\\ \text{(Group 1)}}}{1} + \underset{\substack{\text{O}\\ \text{(Group 6)}}}{6} = 8 \text{ valence electrons}$$

Step 2 Using a pair of electrons per bond, we draw in the two O—H bonds, using a line to indicate each pair of bonding electrons.

H—O—H

Note that

H—O—H represents H : O : H

Step 3 We arrange the remaining electrons around the atoms to achieve a noble gas electron configuration for each atom. Four electrons have been used in forming the two bonds, so four electrons (8 − 4) remain to be distributed. Each hydrogen is satisfied with two electrons (duet rule), but oxygen needs eight electrons to have a noble gas electron configuration. So the remaining four electrons are added to oxygen as two lone pairs. Dots are used to represent the lone pairs.

H—Ö—H (with two lone pairs on O) Lone pairs

H—Ö—H might also be drawn as H:Ö:H

This is the correct Lewis structure for the water molecule. Each hydrogen shares two electrons, and the oxygen has four electrons and shares four to give a total of eight.

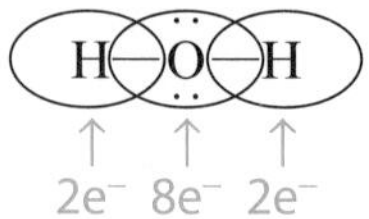

Note that a line is used to represent a shared pair of electrons (bonding electrons) and dots are used to represent unshared pairs.

SELF CHECK **Exercise 12.2** Write the Lewis structure for HCl.

See Problems 12.59 through 12.62. ■

CHEMISTRY IN FOCUS

Hiding Carbon Dioxide

As we discussed in Chapter 11 (see "Chemistry in Focus" segment *Atmospheric Effects*), global warming seems to be a reality. At the heart of this issue is the carbon dioxide produced by society's widespread use of fossil fuels. For example, in the United States, CO_2 makes up 81% of greenhouse gas emissions. Thirty percent of this CO_2 comes from coal-fired power plants used to produce electricity. One way to solve this problem would be to phase out coal-fired power plants. However, this outcome is not likely because the United States possesses so much coal (at least a 250-year supply) and coal is so cheap (about $0.03 per pound). Recognizing this fact, the U.S. government has instituted a research program to see if the CO_2 produced at power plants can be captured and sequestered (stored) underground in deep geologic formations. The factors that need to be explored to determine whether sequestration is feasible are the capacities of underground storage sites and the chances that the sites will leak.

The injection of CO_2 into the earth's crust is already being undertaken by various oil companies. Since 1996, the Norwegian oil company Statoil has separated more than 1 million tons of CO_2 annually from natural gas and pumped it into a saltwater aquifer beneath the floor of the North Sea. In western Canada a group of oil companies has injected CO_2 from a North Dakota synthetic fuels plant into oil fields in an effort to increase oil recovery. The oil companies expect to store 22 million tons of CO_2 there and to produce 130 million barrels of oil over the next 20 years.

Sequestration of CO_2 has great potential as one method for decreasing the rate of global warming. Only time will tell whether it will work.

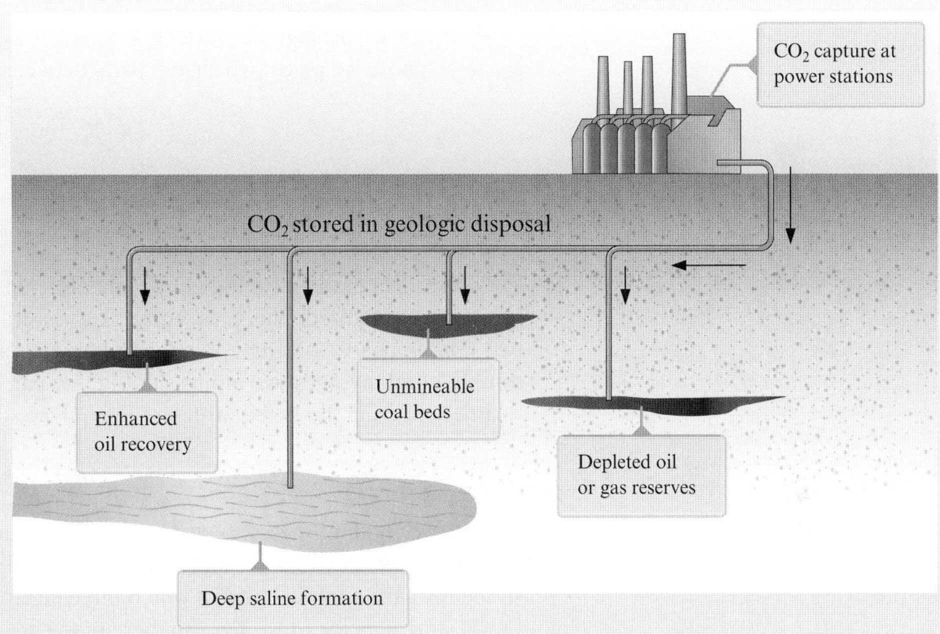

12-7 Lewis Structures of Molecules with Multiple Bonds

OBJECTIVE **To learn how to write Lewis structures for molecules with multiple bonds.**

Now let's write the Lewis structure for carbon dioxide.

Step 1 Summing the valence electrons gives

$$\underset{\substack{\uparrow \\ \text{C} \\ \text{(Group 4)}}}{4} + \underset{\substack{\uparrow \\ \text{O} \\ \text{(Group 6)}}}{6} + \underset{\substack{\uparrow \\ \text{O} \\ \text{(Group 6)}}}{6} = 16$$

O—C—O
represents
O:C:O

Step 2 Form a bond between the carbon and each oxygen:

O—C—O

Step 3 Next, distribute the remaining electrons to achieve noble gas electron configurations on each atom. In this case twelve electrons (16 − 4) remain after the bonds are drawn. The distribution of these electrons is determined by a trial-and-error process. We have six pairs of electrons to distribute. Suppose we try three pairs on each oxygen to give

:Ö—C—Ö:
represents
:Ö:C:Ö:

:Ö—C—Ö:

Is this correct? To answer this question we need to check two things:

1. The total number of electrons. There are sixteen valence electrons in this structure, which is the correct number.
2. The octet rule for each atom. Each oxygen has eight electrons around it, but the carbon has only four. This cannot be the correct Lewis structure.

How can we arrange the sixteen available electrons to achieve an octet for each atom? Suppose we place two shared pairs between the carbon and each oxygen:

Ö=C=Ö
represents
Ö::C::Ö

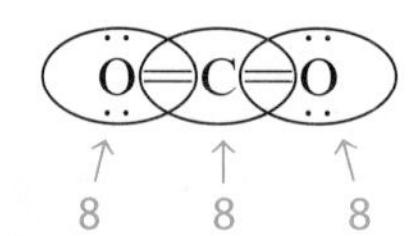

Now each atom is surrounded by eight electrons, and the total number of electrons is sixteen, as required. This is the correct Lewis structure for carbon dioxide, which has two *double* bonds. A **single bond** involves two atoms sharing one electron pair. A **double bond** involves two atoms sharing two pairs of electrons.

:O≡C—Ö:
represents
:O:::C:Ö:

In considering the Lewis structure for CO_2, you may have come up with

:O≡C—Ö: or :Ö—C≡O:

Note that both of these structures have the required sixteen electrons and that both have octets of electrons around each atom (verify this for yourself). Both of these structures have a **triple bond** in which three electron pairs are shared. Are these valid Lewis structures for CO_2? Yes. So there really are three Lewis structures for CO_2:

:Ö—C≡O: Ö=C=Ö :O≡C—Ö:

This brings us to a new term, **resonance.** A molecule shows resonance when *more than one Lewis structure can be drawn for the molecule.* In such a case we call the various Lewis structures **resonance structures.**

Of the three resonance structures for CO_2 shown above, the one in the center with two double bonds most closely fits our experimental information about the CO_2 molecule. In this text we will not be concerned about how to choose which resonance structure for a molecule gives the "best" description of that molecule's properties.

Next let's consider the Lewis structure of the CN^- (cyanide) ion.

Step 1 Summing the valence electrons, we have

CN^-

4 + 5 + 1 = 10

Note that the negative charge means an extra electron must be added.

Step 2 Draw a single bond (C—N).

CHEMISTRY IN FOCUS

Broccoli—Miracle Food?

Eating the right foods is critical to our health. In particular, certain vegetables, although they do not enjoy a very jazzy image, seem especially important. A case in point is broccoli, a vegetable with a humble reputation that packs a powerful chemistry wallop.

Broccoli contains a chemical called sulforaphane, which has the following Lewis structure:

$$CH_3—\underset{\underset{:\ddot{O}:}{\|}}{S}—(CH_2)_4—\ddot{N}=C=\dot{\underset{\cdot}{S}}:$$

Experiments indicate that sulforaphane furnishes protection against certain cancers by increasing the production of enzymes (called phase 2 enzymes) that "mop up" reactive molecules that can harm DNA. Sulforaphane also seems to combat bacteria. For example, among the most common harmful bacteria in humans is *Helicobacter pylori (H. pylori)*, which has been implicated in the development of several diseases of the stomach, including inflammation, cancer, and ulcers. Antibiotics are clearly the best treatment for *H. pylori* infections. However, especially in developing countries, where *H. pylori* is rampant, antibiotics are often too expensive to be available to the general population. In addition, the bacteria sometimes evade antibiotics by "hiding" in cells on the stomach walls and then reemerging after treatment ends.

Studies at Johns Hopkins in Baltimore and Vandoeuvre-les Nancy in France have shown that sulforaphane kills *H. pylori* (even when it has taken refuge in stomach-wall cells) at concentrations that are achievable by eating broccoli. The scientists at Johns Hopkins also found that sulforaphane seems to inhibit stomach cancer in mice. Although there are no guarantees that broccoli will keep you healthy, it might not hurt to add it to your diet.

Squared Studio/PhotoDisc/Getty Images

Step 3 Next, we distribute the remaining electrons to achieve a noble gas configuration for each atom. Eight electrons remain to be distributed. We can try various possibilities, such as

$$\underset{\cdot\cdot}{\overset{\cdot\cdot}{C}}—\underset{\cdot\cdot}{\overset{\cdot\cdot}{N}} \quad \text{or} \quad :\underset{\cdot\cdot}{\overset{\cdot\cdot}{C}}—N: \quad \text{or} \quad :C—\underset{\cdot\cdot}{\overset{\cdot\cdot}{N}}:$$

These structures are incorrect. To show why none is a valid Lewis structure, count the electrons around the C and N atoms. In the left structure, neither atom satisfies the octet rule. In the center structure, C has eight electrons but N has only four. In the right structure, the opposite is true. Remember that both atoms must simultaneously satisfy the octet rule. Therefore, the correct arrangement is

$$:C\equiv N:$$

$:C\equiv N:$ represents $:C:::N:$

(Satisfy yourself that both carbon and nitrogen have eight electrons.) In this case we have a triple bond between C and N, in which three electron pairs are shared. Because this is an anion, we indicate the charge outside of square brackets around the Lewis structure.

$$[:C\equiv N:]^-$$

In summary, sometimes we need double or triple bonds to satisfy the octet rule. Writing Lewis structures is a trial-and-error process. Start with single bonds between the bonded atoms and add multiple bonds as needed.

We will write the Lewis structure for NO_2^- in Example 12.3 to make sure the procedures for writing Lewis structures are clear.

Interactive Example 12.3

Writing Lewis Structures: Resonance Structures

Write the Lewis structure for the NO_2^- anion.

SOLUTION

Step 1 Sum the valence electrons for NO_2^-.

Valence electrons: 6 + 5 + 6 + 1 = 18 electrons
(O N O −1 charge)

Step 2 Put in single bonds.

O—N—O

Step 3 Satisfy the octet rule. In placing the electrons, we find there are two Lewis structures that satisfy the octet rule:

$[\ddot{\underset{..}{O}}{=}\ddot{N}{-}\ddot{\underset{..}{O}}{:}]^-$ and $[{:}\ddot{\underset{..}{O}}{-}\ddot{N}{=}\ddot{\underset{..}{O}}]^-$

Verify that each atom in these structures is surrounded by an octet of electrons. Try some other arrangements to see whether other structures exist in which the eighteen electrons can be used to satisfy the octet rule. It turns out that these are the only two that work. Note that this is another case where resonance occurs; there are two valid Lewis structures.

SELF CHECK

Exercise 12.3 Ozone is a very important constituent of the atmosphere. At upper levels it protects us by absorbing high-energy radiation from the sun. Near the earth's surface it produces harmful air pollution. Write the Lewis structure for ozone, O_3.

See Problems 12.63 through 12.68. ■

Now let's consider a few more cases in Example 12.4.

Interactive Example 12.4

Writing Lewis Structures: Summary

You may wonder how to decide which atom is the central atom in molecules of binary compounds. In cases where there is one atom of a given element and several atoms of a second element, the single atom is almost always the central atom of the molecule.

Give the Lewis structure for each of the following:

a. HF
b. N_2
c. NH_3
d. CH_4
e. CF_4
f. NO^+
g. NO_3^-

SOLUTION

In each case we apply the three steps for writing Lewis structures. Recall that lines are used to indicate shared electron pairs and that dots are used to indicate nonbonding pairs (lone pairs). The table following the *Self Check* exercises summarizes our results.

SELF CHECK

Exercise 12.4 Write the Lewis structures for the following molecules:

a. NF_3
b. O_2
c. CO
d. PH_3
e. H_2S
f. SO_4^{2-}
g. NH_4^+
h. ClO_3^-
i. SO_2

See Problems 12.55 through 12.68. ■

Molecule or Ion	Total Valence Electrons	Draw Single Bonds	Calculate Number of Electrons Remaining	Use Remaining Electrons to Achieve Noble Gas Configurations	Check: Atom	Check: Electrons
a. HF	1 + 7 = 8	H—F	8 − 2 = 6	H—F: (with lone pairs on F)	H F	2 8
b. N_2	5 + 5 = 10	N—N	10 − 2 = 8	:N≡N:	N	8
c. NH_3	5 + 3(1) = 8	H—N—H H (below N)	8 − 6 = 2	H—N—H with H below N and lone pair on N	H N	2 8
d. CH_4	4 + 4(1) = 8	H—C—H with H above and below C	8 − 8 = 0	H—C—H with H above and below C	H C	2 8
e. CF_4	4 + 4(7) = 32	F—C—F with F above and below C	32 − 8 = 24	:F—C—F: with :F: above and below C (each F with three lone pairs)	F C	8 8
f. NO^+	5 + 6−1 = 10	N—O	10 − 2 = 8	$[:N{\equiv}O:]^+$	N O	8 8
g. NO_3^-	5 + 3(6)+1 = 24	[O—N(—O)—O] (N bonded to three O)	24 − 6 = 18	$[$:O: single-bonded to N; N=O (lower left); N—O (lower right)$]^-$	N O	8 8
				NO_3^- shows resonance: $[$:O: single-bonded to N; N—O (lower left); N=O (lower right)$]^-$	N O	8 8
				$[$:O: double-bonded to N; N—O (lower left); N—O (lower right)$]^-$	N O	8 8

Remember, when writing Lewis structures, you don't have to worry about which electrons come from which atoms in a molecule. It is best to think of a molecule as a new entity that uses all the available valence electrons from the various atoms to achieve the strongest possible bonds. Think of the valence electrons as belonging to the molecule, rather than to the individual atoms. Simply distribute all the valence electrons so that noble gas electron configurations are obtained for each atom, without regard to the origin of each particular electron.

Some Exceptions to the Octet Rule

The idea that covalent bonding can be predicted by achieving noble gas electron configurations for all atoms is a simple and very successful idea. The rules we have used for Lewis structures describe correctly the bonding in most molecules. However, with such a simple model, some exceptions are inevitable. Boron, for example, tends to form compounds in which the boron atom has fewer than eight electrons around it—that is, it does not have a complete octet. Boron trifluoride, BF_3, a gas at normal temperatures and pressures, reacts very energetically with molecules such as water and ammonia that have unshared electron pairs (lone pairs).

H₂O: (lone pairs on O) ← Lone pairs → :NH₃

The violent reactivity of BF_3 with electron-rich molecules arises because the boron atom is electron-deficient. The Lewis structure that seems most consistent with the properties of BF_3 (twenty-four valence electrons) is

$$\text{:}\ddot{\text{F}}\text{:} - \text{B}(-\ddot{\text{F}}\text{:})_2$$

Note that in this structure the boron atom has only six electrons around it. The octet rule for boron could be satisfied by drawing a structure with a double bond between the boron and one of the fluorines. However, experiments indicate that each B—F bond is a single bond in accordance with the above Lewis structure. This structure is also consistent with the reactivity of BF_3 with electron-rich molecules. For example, BF_3 reacts vigorously with NH_3 to form H_3NBF_3.

$$\text{H}_3\text{N:} + \text{BF}_3 \rightarrow \text{H}_3\text{N—BF}_3$$

Note that in the product H_3NBF_3, which is very stable, boron has an octet of electrons.

It is also characteristic of beryllium to form molecules where the beryllium atom is electron-deficient.

The compounds containing the elements carbon, nitrogen, oxygen, and fluorine are accurately described by Lewis structures in the vast majority of cases. However, there are a few exceptions. One important example is the oxygen molecule, O_2. The following Lewis structure that satisfies the octet rule can be drawn for O_2 (see Self Check Exercise 12.4).

$$\ddot{\text{O}}=\ddot{\text{O}}$$

However, this structure does not agree with the *observed behavior* of oxygen. For example, the photos in Fig. 12.10 show that when liquid oxygen is poured between the poles of a strong magnet, it "sticks" there until it boils away. This provides clear evidence that oxygen is paramagnetic—that is, it contains unpaired electrons. However, the above Lewis structure shows only pairs of electrons. That is, no unpaired electrons are shown. There is no simple Lewis structure that satisfactorily explains the paramagnetism of the O_2 molecule.

Any molecule that contains an odd number of electrons does not conform to our rules for Lewis structures. For example, NO and NO_2 have eleven and seventeen valence electrons, respectively, and conventional Lewis structures cannot be drawn for these cases.

Even though there are exceptions, most molecules can be described by Lewis structures in which all the atoms have noble gas electron configurations, and this is a very useful model for chemists.

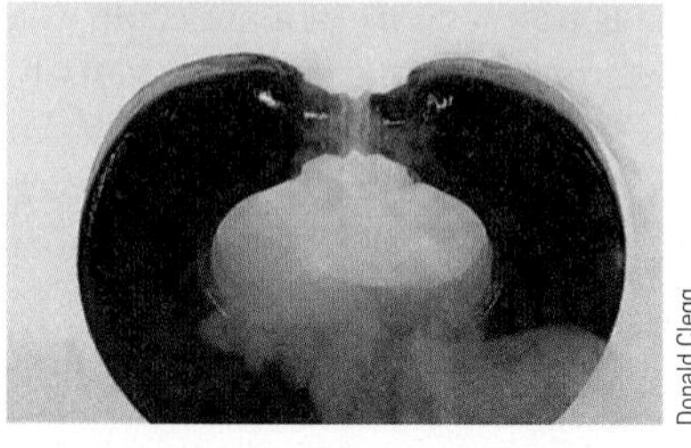

Figure 12.10 When liquid oxygen is poured between the poles of a magnet, it "sticks" until it boils away. This shows that the O_2 molecule has unpaired electrons (is paramagnetic).

12-8 Molecular Structure

OBJECTIVE To understand molecular structure and bond angles.

So far in this chapter we have considered the Lewis structures of molecules. These structures represent the arrangement of the *valence electrons* in a molecule. We use the word *structure* in another way when we talk about the **molecular structure** or **geometric structure** of a molecule. These terms refer to the three-dimensional arrangement of the *atoms* in a molecule. For example, the water molecule is known to have the molecular structure

O
H H

which is often called "bent" or "V-shaped." To describe the structure more precisely, we often specify the **bond angle.** For the H_2O molecule the bond angle is about 105°.

O
H H
~105°

a Computer graphic of a linear molecule containing three atoms

b Computer graphic of a trigonal planar molecule

c Computer graphic of a tetrahedral molecule

Frank Cox © Cengage Learning

On the other hand, some molecules exhibit a **linear structure** (all atoms in a line). An example is the CO_2 molecule.

O—C—O
180°

Note that a linear molecule has a 180° bond angle.

A third type of molecular structure is illustrated by BF_3, which is planar or flat (all four atoms in the same plane) with 120° bond angles.

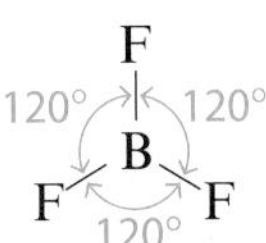

The name usually given to this structure is **trigonal planar structure,** although triangular might seem to make more sense.

Another type of molecular structure is illustrated by methane, CH_4. This molecule has the molecular structure shown in Fig. 12.11, which is called a **tetrahedral structure** or a **tetrahedron.** The dashed lines shown connecting the H atoms define the four identical triangular faces of the tetrahedron.

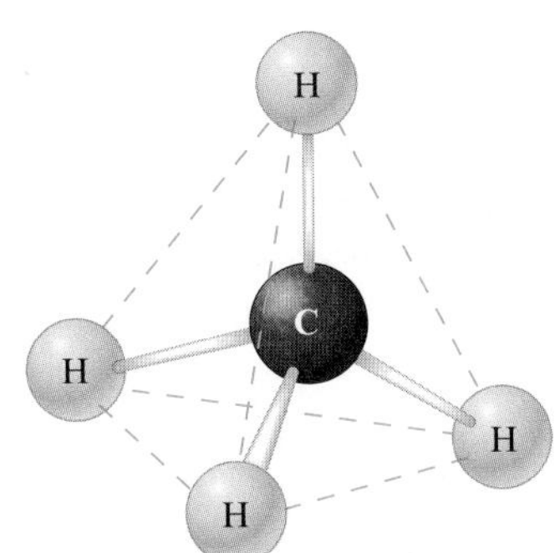

Figure 12.11 The tetrahedral molecular structure of methane. This representation is called a ball-and-stick model; the atoms are represented by balls and the bonds by sticks. The dashed lines show the outline of the tetrahedron.

In the next section we will discuss these various molecular structures in more detail. In that section we will learn how to predict the molecular structure of a molecule by looking at the molecule's Lewis structure.

12-9 Molecular Structure: The VSEPR Model

OBJECTIVE **To learn to predict molecular geometry from the number of electron pairs.**

The structures of molecules play a very important role in determining their properties. For example, as we see in the "Chemistry in Focus" segment *Taste—It's the Structure That Counts,* taste is directly related to molecular structure. Structure is particularly important for biological molecules; a slight change in the structure of a large biomolecule can completely destroy its usefulness to a cell and may even change the cell from a normal one to a cancerous one.

Many experimental methods now exist for determining the molecular structure of a molecule—that is, the three-dimensional arrangement of the atoms. These methods must be used when accurate information about the structure is required. However, it is often useful to be able to predict the *approximate* molecular structure of a molecule. In this section we consider a simple model that allows us to do this. This model, called the **valence shell electron pair repulsion (VSEPR) model,** is useful for predicting the molecular structures of molecules formed from nonmetals. The main idea of this model is that *the structure around a given atom is determined by minimizing repulsions between electron pairs.* This means that the bonding and nonbonding electron pairs (lone pairs) around a given atom are positioned *as far apart as possible.* To see how this model works, we will first consider the molecule $BeCl_2$, which has the following Lewis structure (it is an exception to the octet rule):

$$:\ddot{\underset{..}{Cl}}—Be—\ddot{\underset{..}{Cl}}:$$

Note that there are two pairs of electrons around the beryllium atom. What arrangement of these electron pairs allows them to be as far apart as possible to minimize the repulsions? The best arrangement places the pairs on opposite sides of the beryllium atom at 180° from each other.

—Be—
180°

This is the maximum possible separation for two electron pairs. Now that we have determined the optimal arrangement of the electron pairs around the central atom, we can specify the molecular structure of $BeCl_2$—that is, the positions of the atoms. Because each electron pair on beryllium is shared with a chlorine atom, the molecule has a **linear structure** with a 180° bond angle.

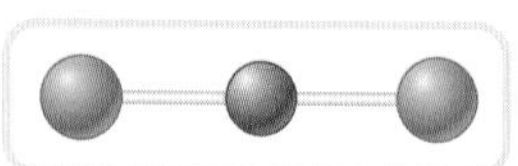

$$:\ddot{\underset{..}{Cl}}—Be—\ddot{\underset{..}{Cl}}:$$
180°

Whenever two pairs of electrons are present around an atom, they should always be placed at an angle of 180° to each other to give a linear arrangement.

Next let's consider BF_3, which has the following Lewis structure (it is another exception to the octet rule):

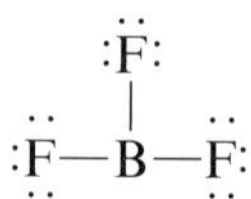

CHEMISTRY IN FOCUS

Taste—It's the Structure That Counts

Why do certain substances taste sweet, sour, bitter, or salty? Of course, it has to do with the taste buds on our tongues. But how do these taste buds work? For example, why does sugar taste sweet to us? The answer to this question remains elusive, but it does seem clear that sweet taste depends on how certain molecules fit the "sweet receptors" in our taste buds.

One of the mysteries of the sweet taste sensation is the wide variety of molecules that taste sweet. For example, the many types of sugars include glucose and sucrose (table sugar). The first artificial sweetener was probably the Romans' sapa (see "Chemistry in Focus" *Sugar of Lead* in Chapter 5), made by boiling wine in lead vessels to produce a syrup that contained lead acetate, $Pb(C_2H_3O_2)_2$, called sugar of lead because of its sweet taste. Other widely used modern artificial sweeteners include saccharin, aspartame, sucralose, and steviol, whose structures are shown in the accompanying figure. The structure of steviol is shown in simplified form. Each vertex represents a carbon atom, and not all of the hydrogen atoms are shown. Note the great disparity of structures for these sweet-tasting molecules. It's certainly not obvious which structural features trigger a sweet sensation when these molecules interact with the taste buds.

The pioneers in relating structure to sweet taste were two chemists, Robert S. Shallenberger and Terry E. Acree of Cornell University, who almost thirty years ago suggested that all sweet-tasting substances must contain a common feature they called a glycophore. They postulated that a glycophore always contains an atom or group of atoms that have available electrons located near a hydrogen atom attached to a relatively electronegative atom. Murray Goodman, a chemist at the University of California at San Diego, expanded the definition of a glycophore to include a hydrophobic ("water-hating") region. Goodman finds that a "sweet molecule" tends to be L-shaped with positively and negatively charged regions on the upright of the L and a hydrophobic region on the base of the L. For a molecule to be sweet, the L must be planar. If the L is twisted in one direction, the molecule has a bitter taste. If the molecule is twisted in the other direction, the molecule is tasteless.

The latest model for the sweet-taste receptor, proposed by Piero Temussi of the University of Naples, postulates that there are four binding sites on the receptor that can be occupied independently. Small sweet-tasting molecules might bind to one of the sites, while a large molecule would bind to more than one site simultaneously.

So the search goes on for a better artificial sweetener. One thing's for sure; it all has to do with molecular structure.

Saccharin

Sucralose

Aspartame (NutraSweet™)

Steviol

Note: the Lewis structures above are all drawn without lone pairs of electrons.

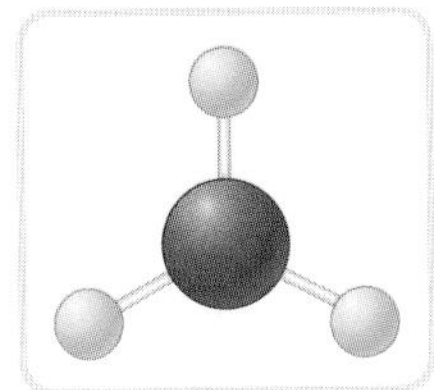

Here the boron atom is surrounded by three pairs of electrons. What arrangement minimizes the repulsions among three pairs of electrons? Here the greatest distance between electron pairs is achieved by angles of 120°.

120° 120°
B
120°

Because each of the electron pairs is shared with a fluorine atom, the molecular structure is

F
120° 120°
B
F F
120°

or

F
B
F - - - - F

This is a planar (flat) molecule with a triangular arrangement of F atoms, commonly described as a trigonal planar structure. *Whenever three pairs of electrons are present around an atom, they should always be placed at the corners of a triangle (in a plane at angles of 120° to each other).*

Next let's consider the methane molecule, which has the Lewis structure

```
    H
    |               H
H — C — H    or   H:C:H
    |               H
    H
```

There are four pairs of electrons around the central carbon atom. What arrangement of these electron pairs best minimizes the repulsions? First we try a square planar arrangement:

C 90°

The carbon atom and the electron pairs are all in a plane represented by the surface of a page, and the angles between the pairs are all 90°.

Is there another arrangement with angles greater than 90° that would put the electron pairs even farther away from each other? The answer is yes. We can get larger angles than 90° by using the following three-dimensional structure, which has angles of approximately 109.5°.

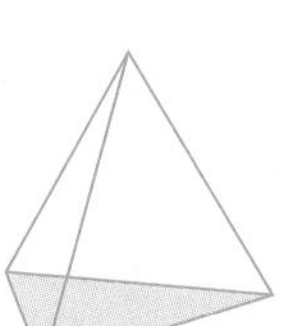

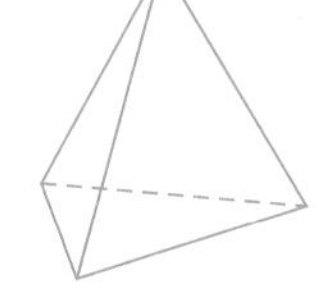

A tetrahedron has four equal triangular faces.

In this drawing the wedge indicates a position above the surface of a page and the dashed lines indicate positions behind that surface. The solid line indicates a position on the surface of a page. The figure formed by connecting the lines is called a tetrahedron, so we call this arrangement of electron pairs the **tetrahedral arrangement.**

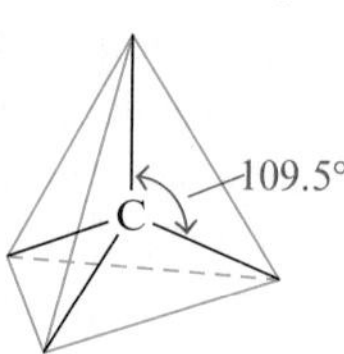

This is the maximum possible separation of four pairs around a given atom. *Whenever four pairs of electrons are present around an atom, they should always be placed at the corners of a tetrahedron (the tetrahedral arrangement).*

Now that we have the arrangement of electron pairs that gives the least repulsion, we can determine the positions of the atoms and thus the molecular structure of CH_4. In methane each of the four electron pairs is shared between the carbon atom and a hydrogen atom. Thus the hydrogen atoms are placed as shown in Fig. 12.12, and the molecule has a tetrahedral structure with the carbon atom at the center.

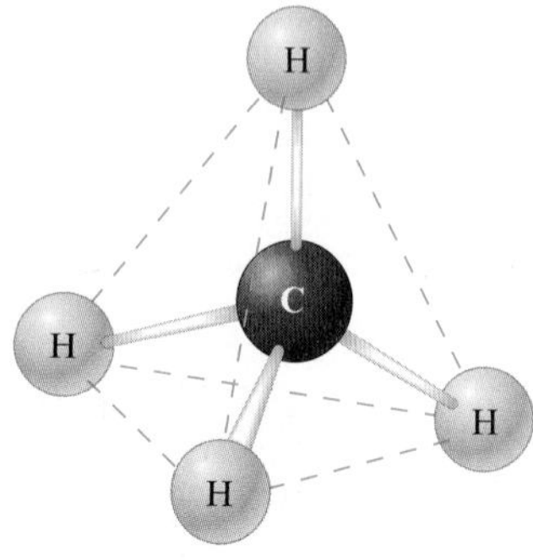

Figure 12.12 ▶ The molecular structure of methane. The tetrahedral arrangement of electron pairs produces a tetrahedral arrangement of hydrogen atoms.

Critical Thinking

You have seen that molecules with four electron pairs, such as CH_4, are tetrahedral. What if a molecule had six electron pairs, like SF_6? Predict the geometry and bond angles of SF_6.

Recall that the main idea of the VSEPR model is to find the arrangement of electron pairs around the central atom that minimizes the repulsions. Then we can determine the *molecular structure* by knowing how the electron pairs are shared with the peripheral atoms. A systematic procedure for using the VSEPR model to predict the structure of a molecule is outlined below.

Steps for Predicting Molecular Structure Using the VSEPR Model

Step 1 Draw the Lewis structure for the molecule.

Step 2 Count the electron pairs and arrange them in the way that minimizes repulsion (that is, put the pairs as far apart as possible).

Step 3 Determine the positions of the atoms from the way the electron pairs are shared.

Step 4 Determine the name of the molecular structure from the positions of the atoms.

Interactive Example 12.5 Predicting Molecular Structure Using the VSEPR Model, I

Ammonia, NH_3, is used as a fertilizer (injected into the soil) and as a household cleaner (in aqueous solution). Predict the structure of ammonia using the VSEPR model.

SOLUTION

Step 1 Draw the Lewis structure.

```
H—N̈—H
   |
   H
```

Step 2 Count the pairs of electrons and arrange them to minimize repulsions. The NH_3 molecule has four pairs of electrons around the N atom: three bonding pairs and one nonbonding pair. From the discussion of the methane molecule, we know that the best arrangement of four electron pairs is the tetrahedral structure shown in Fig. 12.13(a).

Step 3 Determine the positions of the atoms. The three H atoms share electron pairs as shown in Fig. 12.13(b).

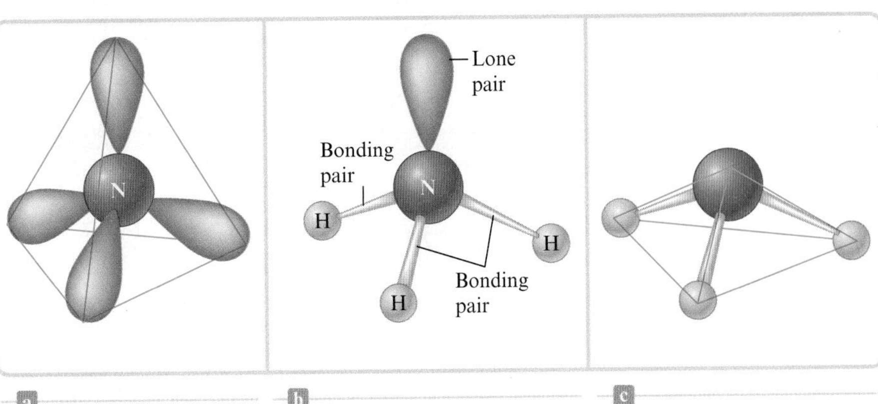

Figure 12.13 The structure of ammonia.

(a) The tetrahedral arrangement of electron pairs around the nitrogen atom in the ammonia molecule.

(b) Three of the electron pairs around nitrogen are shared with hydrogen atoms as shown, and one is a lone pair.

(c) The NH_3 molecule has the trigonal pyramid structure (a pyramid with a triangle as a base).

Step 4 Name the molecular structure. It is very important to recognize that the name of the molecular structure is always based on the *positions of the atoms. The placement of the electron pairs determines the structure, but the name is based on the positions of the atoms.* Thus it is incorrect to say that the NH_3 molecule is tetrahedral. It has a tetrahedral arrangement of electron pairs but *not* a tetrahedral arrangement of atoms. The molecular structure of ammonia is a **trigonal pyramid** (one side is different from the other three) rather than a tetrahedron. ■

Interactive Example 12.6

Predicting Molecular Structure Using the VSEPR Model, II

Describe the molecular structure of the water molecule.

SOLUTION

Step 1 The Lewis structure for water is

$$H{-}\ddot{\underset{\cdot\cdot}{O}}{-}H$$

Step 2 There are four pairs of electrons: two bonding pairs and two nonbonding pairs. To minimize repulsions, these are best arranged in a tetrahedral structure as shown in Fig. 12.14(a).

Step 3 Although H_2O has a tetrahedral arrangement of *electron pairs,* it is *not a tetrahedral molecule.* The *atoms* in the H_2O molecule form a V shape, as shown in Fig. 12.14(b) and (c).

Step 4 The molecular structure is called V-shaped or bent.

SELF CHECK

Exercise 12.5 Predict the arrangement of electron pairs around the central atom. Then sketch and name the molecular structure for each of the following molecules or ions.

a. NH_4^+
b. SO_4^{2-}
c. NF_3
d. H_2S
e. ClO_3^-
f. BeF_2

See Problems 12.81 through 12.84. ■

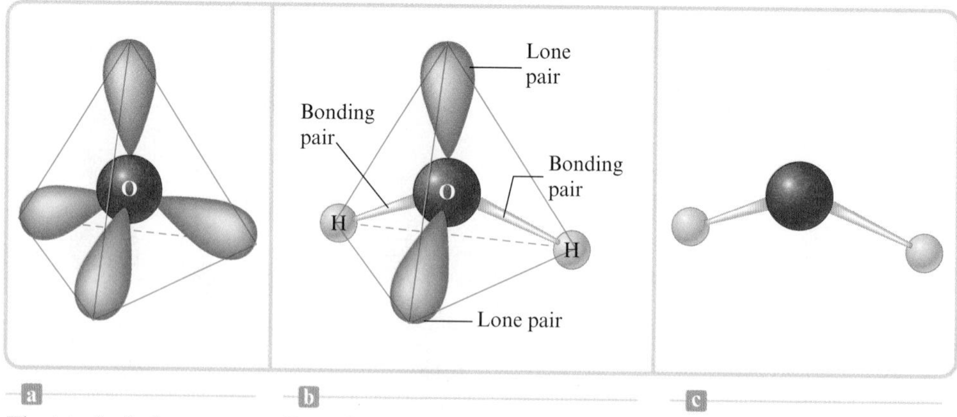

Figure 12.14 ▸ The structure of water.

a The tetrahedral arrangement of the four electron pairs around oxygen in the water molecule.

b Two of the electron pairs are shared between oxygen and the hydrogen atoms, and two are lone pairs.

c The V-shaped molecular structure of the water molecule.

The various molecules we have considered are summarized in Table 12.4 below. Note the following general rules.

Rules for Predicting Molecular Structure Using the VSEPR Model

1. Two pairs of electrons on a central atom in a molecule are always placed 180° apart. This is a linear arrangement of pairs.
2. Three pairs of electrons on a central atom in a molecule are always placed 120° apart in the same plane as the central atom. This is a trigonal planar (triangular) arrangement of pairs.
3. Four pairs of electrons on a central atom in a molecule are always placed 109.5° apart. This is a tetrahedral arrangement of electron pairs.
4. When *every pair* of electrons on the central atom is *shared* with another atom, the molecular structure has the same name as the arrangement of electron pairs.

Number of Pairs	Name of Arrangement
2	linear
3	trigonal planar
4	tetrahedral

5. When one or more of the electron pairs around a central atom are unshared (lone pairs), the name for the molecular structure is *different* from that for the arrangement of electron pairs (see rows 4 and 5 in Table 12.4).

Table 12.4 Arrangements of Electron Pairs and the Resulting Molecular Structures for Two, Three, and Four Electron Pairs

Number of Electron Pairs	Bonds	Electron Pair Arrangement	Ball-and-Stick Model	Molecular Structure	Partial Lewis Structure	Example Ball-and-Stick Model
2	2	Linear	 180°	Linear	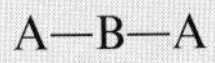 A—B—A	 Cl Be Cl
3	3	Trigonal planar (triangular)	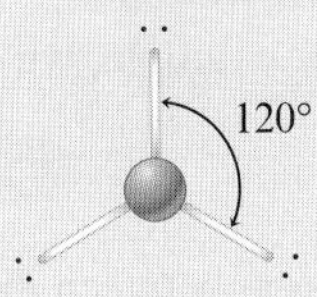 120°	Trigonal planar (triangular)	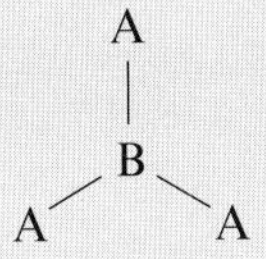 A, B, A, A	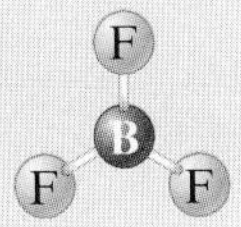 F, B, F, F
4	4	Tetrahedral	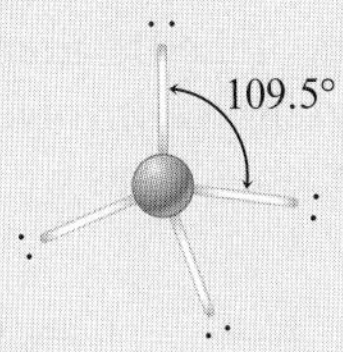 109.5°	Tetrahedral	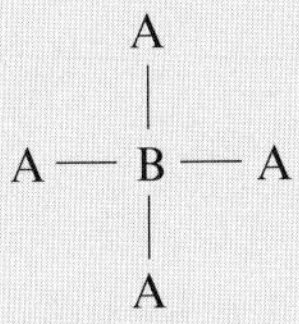 A, A—B—A, A	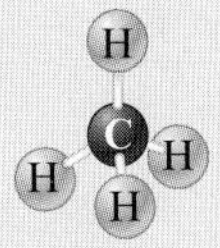 H, C, H, H, H
4	3	Tetrahedral	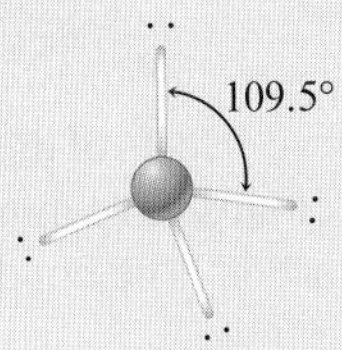 109.5°	Trigonal pyramid	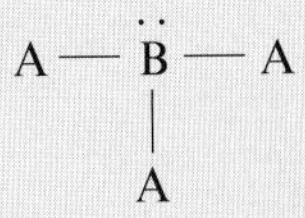 A—B̈—A, A	 N, H, H, H
4	2	Tetrahedral	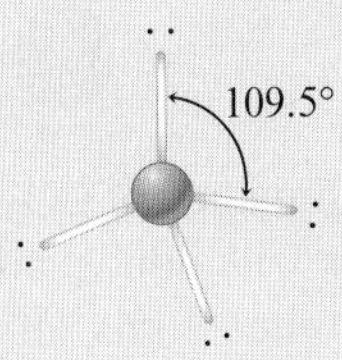 109.5°	Bent or V-shaped	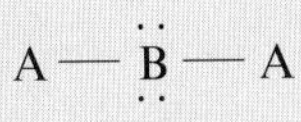 A—B—A (with two lone pairs on B)	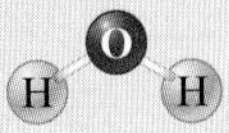 O, H, H

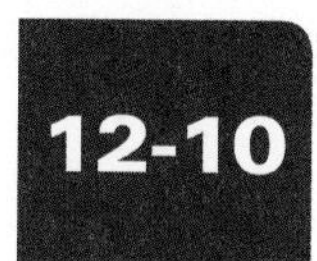

Molecular Structure: Molecules with Double Bonds

OBJECTIVE **To learn to apply the VSEPR model to molecules with double bonds.**

Up to this point we have applied the VSEPR model only to molecules (and ions) that contain single bonds. In this section we will show that this model applies equally well to species with one or more double bonds. We will develop the procedures for dealing with molecules with double bonds by considering examples whose structures are known.

First we will examine the structure of carbon dioxide, a substance that may be contributing to the warming of the earth. The carbon dioxide molecule has the Lewis structure

$$\underset{\cdot\cdot}{\ddot{O}}=C=\underset{\cdot\cdot}{\ddot{O}}$$

as discussed in Section 12-7. Carbon dioxide is known by experiment to be a linear molecule. That is, it has a 180° bond angle.

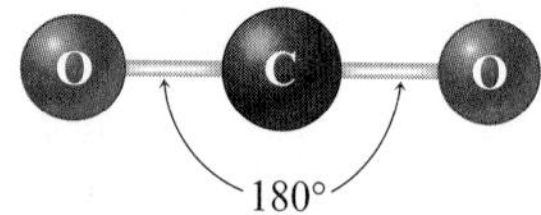

Recall from Section 12-9 that two electron pairs around a central atom can minimize their mutual repulsions by taking positions on opposite sides of the atom (at 180° from each other). This causes a molecule like $BeCl_2$, which has the Lewis structure

$$:\underset{\cdot\cdot}{\ddot{Cl}}-Be-\underset{\cdot\cdot}{\ddot{Cl}}:$$

to have a linear structure. Now recall that CO_2 has two double bonds and is known to be linear, so the double bonds must be at 180° from each other. Therefore, we conclude that each double bond in this molecule acts *effectively* as one repulsive unit. This conclusion makes sense if we think of a bond in terms of an electron density "cloud" between two atoms. For example, we can picture the single bonds in $BeCl_2$ as follows:

The minimum repulsion between these two electron density clouds occurs when they are on opposite sides of the Be atom (180° angle between them).

Each double bond in CO_2 involves the sharing of four electrons between the carbon atom and an oxygen atom. Thus we might expect the bonding cloud to be "fatter" than for a single bond:

However, the repulsive effects of these two clouds produce the same result as for single bonds; the bonding clouds have minimum repulsions when they are positioned on opposite sides of the carbon. The bond angle is 180°, and so the molecule is linear:

In summary, examination of CO_2 leads us to the conclusion that in using the VSEPR model for molecules with double bonds, each double bond should be treated the same as a single bond. In other words, although a double bond involves four electrons, these electrons are restricted to the space between a given pair of atoms. Therefore, these four electrons do not function as two independent pairs but are "tied together" to form one effective repulsive unit.

We reach this same conclusion by considering the known structures of other molecules that contain double bonds. For example, consider the ozone molecule, which has eighteen valence electrons and exhibits two resonance structures:

$$:\ddot{\underset{..}{O}}-\ddot{O}=\ddot{O}: \longleftrightarrow :\ddot{O}=\ddot{O}-\ddot{\underset{..}{O}}:$$

The ozone molecule is known to have a bond angle close to 120°. Recall that 120° angles represent the minimum repulsion for three pairs of electrons.

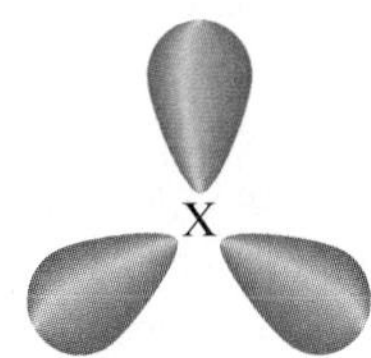

This indicates that the double bond in the ozone molecule is behaving as one effective repulsive unit:

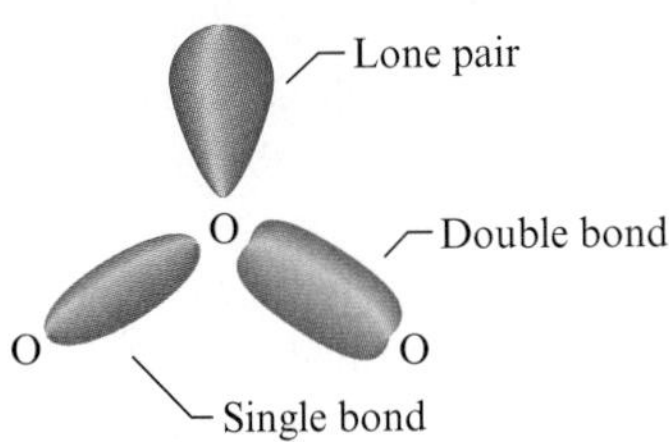

These and other examples lead us to the following rule: *When using the VSEPR model to predict the molecular geometry of a molecule, a double bond is counted the same as a single electron pair.*

Thus CO_2 has two "effective pairs" that lead to its linear structure, whereas O_3 has three "effective pairs" that lead to its bent structure with a 120° bond angle. Therefore, to use the VSEPR model for molecules (or ions) that have double bonds, we use the same steps as those given in Section 12-9, but we count any double bond the same as a single electron pair. Although we have not shown it here, triple bonds also count as one repulsive unit in applying the VSEPR model.

Interactive Example 12.7

Predicting Molecular Structure Using the VSEPR Model, III

Predict the structure of the nitrate ion.

SOLUTION

Step 1 The Lewis structures for NO_3^- are

$$\left[\begin{array}{c} :O: \\ \| \\ N \\ / \quad \backslash \\ :\ddot{\underset{..}{O}} \quad \ddot{\underset{..}{O}}: \end{array}\right]^- \leftrightarrow \left[\begin{array}{c} :\ddot{O}: \\ | \\ N \\ /\!/ \quad \backslash \\ \ddot{O} \quad \ddot{\underset{..}{O}}: \end{array}\right]^- \leftrightarrow \left[\begin{array}{c} :\ddot{O}: \\ | \\ N \\ / \quad \backslash\!\backslash \\ :\ddot{\underset{..}{O}} \quad \ddot{O} \end{array}\right]^-$$

Step 2 In each resonance structure there are effectively three pairs of electrons: the two single bonds and the double bond (which counts as one pair). These three "effective pairs" will require a trigonal planar arrangement (120° angles).

Step 3 The atoms are all in a plane, with the nitrogen at the center and the three oxygens at the corners of a triangle (trigonal planar arrangement).

Step 4 The NO_3^- ion has a trigonal planar structure. ■

CHAPTER 12 REVIEW

F directs you to the *Chemistry in Focus* feature in the chapter

Key Terms

bond (12-1)
bond energy (12-1)
ionic bonding (12-1)
ionic compound (12-1)
covalent bonding (12-1)
polar covalent bond (12-1)
electronegativity (12-2)
dipole moment (12-3)
Lewis structure (12-6)
duet rule (12-6)
octet rule (12-6)
bonding pair (12-6)
lone pairs (12-6)
unshared pairs (12-6)
single bond (12-7)
double bond (12-7)
triple bond (12-7)
resonance (12-7)
resonance structures (12-7)
molecular structure (12-8)
geometric structure (12-8)
bond angle (12-8)
linear structure (12-8)
trigonal planar structure (12-8)
tetrahedral structure (12-8)
valence shell electron pair repulsion (VSEPR) model (12-9)
tetrahedral arrangement (12-9)
trigonal pyramid (12-9)

For Review

- Chemical bonds hold groups of atoms together to form molecules and ionic solids.
- Bonds are classified as
  - Ionic: Formed when one or more electrons are transferred to form positive and negative ions
  - Covalent: Electrons are shared equally between identical atoms
    - Polar covalent bond: Unequal electron sharing between different atoms
- Electronegativity is the relative ability of an atom to attract the electrons shared with another atom in a bond.
- The difference in electronegativity of the atoms forming a bond determines the polarity of that bond.
- In stable compounds, atoms tend to achieve the electron configuration of the nearest noble gas atom.
- In ionic compounds,
  - Nonmetals tend to gain electrons to reach the electron configuration of the next noble gas atom.
  - Metals tend to lose electrons to reach the electron configuration of the previous noble gas atom.
- Ions group together to form compounds that are electrically neutral.
- In covalent compounds, nonmetals share electrons so that both atoms achieve noble gas configurations.
- Lewis structures represent the valence electron arrangements of the atoms in a compound.
- The rules for drawing Lewis structures recognize the importance of noble gas electron configurations.
  - Duet rule for hydrogen
  - Octet rule for most other atoms
- Some molecules have more than one valid Lewis structure, called resonance.
- Some molecules violate the octet rule for the component atoms.
  - Examples are BF_3, NO_2, and NO.
- Molecular structure describes how the atoms in a molecule are arranged in space.
- Molecular structure can be predicted by using the valence shell electron pair repulsion (VSEPR) model.

All even-numbered Questions and Problems have answers in the back of this book and solutions in the *Student Solutions Guide.*

Active Learning Questions

These questions are designed to be considered by groups of students in class. Often these questions work well for introducing a particular topic in class.

1. Using only the periodic table, predict the most stable ion for Na, Mg, Al, S, Cl, K, Ca, and Ga. Arrange these elements from largest to smallest radius and explain why the radius varies as it does.
2. Write the proper charges so that an alkali metal, a noble gas, and a halogen have the same electron configurations. What is the number of protons in each? The number of electrons in each? Arrange them from smallest to largest radii and explain your ordering rationale.
3. What is meant by a *chemical bond?*
4. Why do atoms form bonds with one another? What can make a molecule favored compared with the lone atoms?
5. How does a bond between Na and Cl differ from a bond between C and O? What about a bond between N and N?
6. In your own words, what is meant by the term *electronegativity?* What are the trends across and down the periodic table for electronegativity? Explain them, and describe how they are consistent with trends of ionization energy and atomic radii.
7. Explain the difference between ionic bonding and covalent bonding. How can we use the periodic table to help us determine the type of bonding between atoms?
8. True or false? In general, a larger atom has a smaller electronegativity. Explain.
9. Why is there an octet rule (and what does *octet* mean) in writing Lewis structures?
10. Does a Lewis structure tell which electrons came from which atoms? Explain.
11. If lithium and fluorine react, which has more attraction for an electron? Why?
12. In a bond between fluorine and iodine, which has more attraction for an electron? Why?

13. We use differences in electronegativity to account for certain properties of bonds. What if all atoms had the same electronegativity values? How would bonding between atoms be affected? What are some differences we would notice?
14. Explain how you can use the periodic table to predict the formula of compounds.
15. Why do we only consider the valence electrons in drawing Lewis structures?
16. How do we determine the total number of valence electrons for an ion? Provide an example of an anion and a cation, and explain your answer.
17. What is the main idea in the valence shell electron pair repulsion (VSEPR) theory?
18. The molecules NH_3 and BF_3 have the same general formula (AB_3) but different shapes.
    a. Find the shape of each of the above molecules.
    b. Provide more examples of real molecules that have the same general formulas but different shapes.
19. How do we deal with multiple bonds in VSEPR theory?
20. In Section 12-10 of your text, the term "effective pairs" is used. What does this mean?
21. Consider the ions Sc^{3+}, Cl^-, K^+, Ca^{2+}, and S^{2-}. Match these ions to the following pictures that represent the relative sizes of the ions.

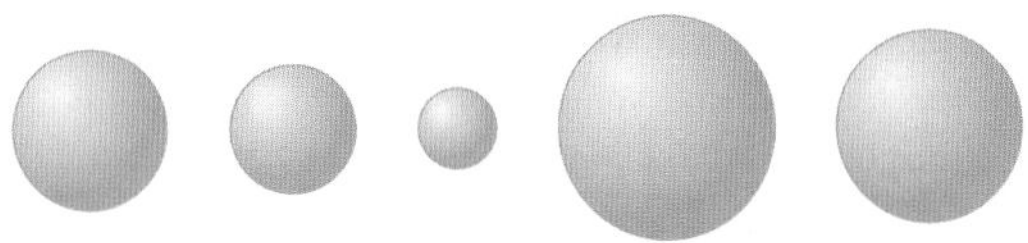

22. Write the name of each of the following shapes of molecules.

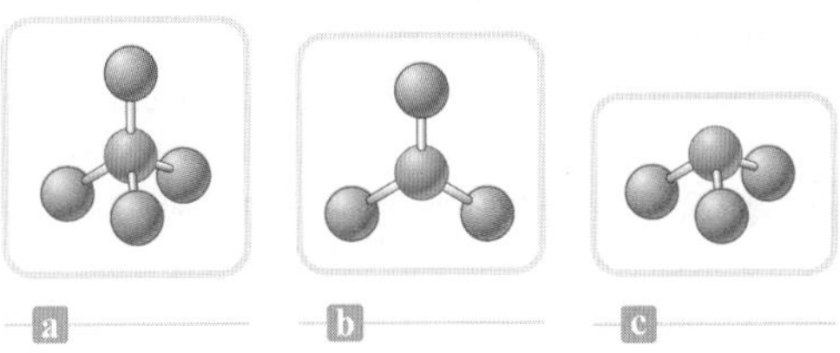

Questions and Problems

12-1 Types of Chemical Bonds

Questions

1. In general terms, what is a chemical *bond?*
2. What does the *bond energy* of a chemical bond represent?
3. What sorts of elements react to form *ionic* compounds?
4. In general terms, what is a *covalent* bond?
5. Describe the type of bonding that exists in the $Cl_2(g)$ molecule. How does this type of bonding differ from that found in the $HCl(g)$ molecule? How is it similar?
6. Compare and contrast the bonding found in the $H_2(g)$ and $HF(g)$ molecules with that found in $NaF(s)$.

12-2 Electronegativity

Questions

7. The relative ability of an atom in a molecule to attract electrons to itself is called the atom's ________.
8. What does it mean to say that a bond is *polar?* Give two examples of molecules with *polar* bonds. Indicate in your examples the direction of the polarity.
9. A bond between atoms having a (small/large) difference in electronegativity will be ionic.
10. What factor determines the relative level of polarity of a polar covalent bond?

Problems

11. In each of the following groups, which element is the most electronegative? Which is the least electronegative?
    a. K, Na, H
    b. F, Br, Na
    c. B, N, F
12. In each of the following groups, which element is the most electronegative? Which is the least electronegative?
    a. Cs, Ba, At
    b. Ba, Sr, Ra
    c. O, Rb, Mg
13. On the basis of the electronegativity values given in Fig. 12.3, indicate whether each of the following bonds would be expected to be ionic, covalent, or polar covalent.
    a. O—O
    b. Al—O
    c. B—O
14. On the basis of the electronegativity values given in Fig. 12.3, indicate whether each of the following bonds would be expected to be covalent, polar covalent, or ionic.
    a. S—S
    b. S—H
    c. S—K
15. Which of the following molecules contain polar covalent bonds?
    a. water, H_2O
    b. carbon monoxide, CO
    c. fluorine, F_2
    d. nitrogen, N_2
16. Which of the following molecules contain polar covalent bonds?
    a. phosphorus, P_4
    b. oxygen, O_2
    c. ozone, O_3
    d. hydrogen fluoride, HF
17. On the basis of the electronegativity values given in Fig. 12.3, indicate which is the more polar bond in each of the following pairs.
    a. H—F or H—Cl
    b. H—Cl or H—I
    c. H—Br or H—Cl
    d. H—I or H—Br

18. On the basis of the electronegativity values given in Fig. 12.3, indicate which is the more polar bond in each of the following pairs.
   a. O—Cl or O—Br
   b. N—O or N—F
   c. P—S or P—O
   d. H—O or H—N
19. Which bond in each of the following pairs has the greater ionic character?
   a. Na—F or Na—I
   b. Ca—S or Ca—O
   c. Li—Cl or Cs—Cl
   d. Mg—N or Mg—P
20. Which bond in each of the following pairs has less ionic character?
   a. Na—O or Na—N
   b. K—S or K—P
   c. Na—Cl or K—Cl
   d. Na—Cl or Mg—Cl

12-3 Bond Polarity and Dipole Moments

Questions

21. What is a *dipole moment?* Give four examples of molecules that possess dipole moments, and draw the direction of the dipole as shown in Section 12-3.
22. Why is the presence of a dipole moment in the water molecule so important? What are some properties of water that are determined by its polarity?

Problems

23. In each of the following diatomic molecules, which end of the molecule is negative relative to the other end?
   a. hydrogen chloride, HCl
   b. carbon monoxide, CO
   c. bromine monofluoride, BrF
24. In each of the following diatomic molecules, which end of the molecule is positive relative to the other end?
   a. hydrogen fluoride, HF
   b. chlorine monofluoride, ClF
   c. iodine monochloride, ICl
25. For each of the following bonds, draw a figure indicating the direction of the bond dipole, including which end of the bond is positive and which is negative.
   a. C—F
   b. Si—C
   c. C—O
   d. B—C
26. For each of the following bonds, draw a figure indicating the direction of the bond dipole, including which end of the bond is positive and which is negative.
   a. S—P
   b. S—F
   c. S—Cl
   d. S—Br
27. For each of the following bonds, draw a figure indicating the direction of the bond dipole, including which end of the bond is positive and which is negative.
   a. Si—H
   b. P—H
   c. S—H
   d. Cl—H
28. For each of the following bonds, draw a figure indicating the direction of the bond dipole, including which end of the bond is positive and which is negative.
   a. H—C
   b. N—O
   c. N—S
   d. N—C

12-4 Stable Electron Configurations and Charges on Ions

Questions

29. What does it mean when we say that in forming bonds, atoms try to achieve an electron configuration analogous to a noble gas?
30. The metallic elements lose electrons when reacting, and the resulting positive ions have an electron configuration analogous to the ________ noble gas element.
31. Nonmetals form negative ions by (losing/gaining) enough electrons to achieve the electron configuration of the next noble gas.
32. Explain how the atoms in *covalent* molecules achieve electron configurations similar to those of the noble gases. How does this differ from the situation in ionic compounds?

Problems

33. Which simple ion would each of the following elements be expected to form? What noble gas has an analogous electron configuration to each of the ions?
   a. chlorine, $Z = 17$
   b. strontium, $Z = 38$
   c. oxygen, $Z = 8$
   d. rubidium, $Z = 37$
34. Which simple ion would each of the following elements be expected to form? Which noble gas has an analogous electron configuration to each of the ions?
   a. bromine, $Z = 35$
   b. cesium, $Z = 55$
   c. phosphorus, $Z = 15$
   d. sulfur, $Z = 16$
35. For each of the following numbers of electrons, give the formula of a *positive ion* that would have that number of electrons, and write the complete electron configuration for each ion.
   a. 10 electrons
   b. 2 electrons
   c. 18 electrons
   d. 36 electrons
36. Give the formula of a *negative* ion that would have the same number of electrons as each of the following *positive* ions.
   a. K^+
   b. Mg^{2+}
   c. Sr^{2+}
   d. Cs^+
37. On the basis of their electron configurations, predict the formula of the simple binary ionic compounds likely to form when the following pairs of elements react with each other.
   a. aluminum, Al, and sulfur, S
   b. radium, Ra, and oxygen, O
   c. calcium, Ca, and fluorine, F
   d. cesium, Cs, and nitrogen, N
   e. rubidium, Rb, and phosphorus, P
38. On the basis of their electron configurations, predict the formula of the simple binary ionic compound likely to form when the following pairs of elements react with each other.
   a. aluminum and bromine
   b. aluminum and oxygen
   c. aluminum and phosphorus
   d. aluminum and hydrogen

All even-numbered Questions and Problems have answers in the back of this book and solutions in the *Student Solutions Guide*.

39. Name the noble gas atom that has the same electron configuration as each of the ions in the following compounds.
    a. barium sulfide, BaS
    b. strontium fluoride, SrF_2
    c. magnesium oxide, MgO
    d. aluminum sulfide, Al_2S_3
40. Atoms form ions so as to achieve electron configurations similar to those of the noble gases. For the following pairs of noble gas configurations, give the formulas of two simple ionic compounds that would have comparable electron configurations.
    a. [He] and [Ne]
    b. [Ne] and [Ne]
    c. [He] and [Ar]
    d. [Ne] and [Ar]

12-5 Ionic Bonding and Structures of Ionic Compounds

Questions

41. Is the formula we write for an ionic compound the *molecular* formula or the *empirical* formula? Why?
42. Describe in general terms the structure of ionic solids such as NaCl. How are the ions packed in the crystal?
43. Why are cations always smaller than the atoms from which they are formed?
44. Why are anions always larger than the atoms from which they are formed?

Problems

45. For each of the following pairs, indicate which species is smaller. Explain your reasoning in terms of the electron structure of each species.
    a. H or H^-
    b. N or N^{3-}
    c. Al or Al^{3+}
    d. F or Cl
46. For each of the following pairs, indicate which species is larger. Explain your reasoning in terms of the electron structure of each species.
    a. Mg^{2+} or Mg
    b. Ca^{2+} or K^+
    c. Rb^+ or Br^-
    d. Se^{2-} or Se
47. For each of the following pairs, indicate which is smaller.
    a. Fe or Fe^{3+}
    b. Cl or Cl^-
    c. Al^{3+} or Na^+
48. For each of the following pairs, indicate which is larger.
    a. I or F
    b. F or F^-
    c. Na^+ or F^-

12-6 and 12-7 Lewis Structures

Questions

49. Why are the *valence* electrons of an atom the only electrons likely to be involved in bonding to other atoms?
50. Explain what the "duet" and "octet" rules are and how they are used to describe the arrangement of electrons in a molecule.
51. What type of structure must each atom in a compound usually exhibit for the compound to be stable?
52. When elements in the second and third periods occur in compounds, what number of electrons in the valence shell represents the most stable electron arrangement? Why?

Problems

53. How many electrons are involved when two atoms in a molecule are connected by a "double bond"? Write the Lewis structure of a molecule containing a double bond.
54. What does it mean when two atoms in a molecule are connected by a "triple bond"? Write the Lewis structure of a molecule containing a triple bond.
55. Write the simple Lewis structure for each of the following atoms.
    a. I ($Z = 53$)
    b. Al ($Z = 13$)
    c. Xe ($Z = 54$)
    d. Sr ($Z = 38$)
56. Write the simple Lewis structure for each of the following atoms.
    a. Mg ($Z = 12$)
    b. Br ($Z = 35$)
    c. S ($Z = 16$)
    d. Si ($Z = 14$)
57. Give the *total* number of valence electrons in each of the following molecules.
    a. N_2O
    b. B_2H_6
    c. C_3H_8
    d. NCl_3
58. Give the *total* number of valence electrons in each of the following molecules.
    a. B_2O_3
    b. CO_2
    c. C_2H_6O
    d. NO_2
59. Write a Lewis structure for each of the following simple molecules. Show all bonding valence electron pairs as lines and all nonbonding valence electron pairs as dots.
    a. NBr_3
    b. HF
    c. CBr_4
    d. C_2H_2
60. Write a Lewis structure for each of the following simple molecules. Show all bonding valence electrons pairs as lines and all nonbonding valence electron pairs as dots.
    a. H_2S
    b. SiF_4
    c. C_2H_4
    d. C_3H_8
61. Write a Lewis structure for each of the following simple molecules. Show all bonding valence electron pairs as lines and all nonbonding valence electron pairs as dots.
    a. C_2H_6
    b. NF_3
    c. C_4H_{10}
    d. $SiCl_4$
62. Write a Lewis structure for each of the following molecules. Show all bonding valence electron pairs as lines and all nonbonding valence electron pairs as dots.
    a. PCl_3
    b. $CHCl_3$
    c. $C_2H_4Cl_2$
    d. N_2H_4
63. (F) The "Chemistry in Focus" segment *Broccoli—Miracle Food?* discusses the health benefits of eating broccoli and gives a Lewis structure for sulforaphane, a chemical in broccoli. Draw possible resonance structures for sulforaphane.
64. (F) The "Chemistry in Focus" segment *Hiding Carbon Dioxide* discusses attempts at sequestering (storing) underground CO_2 produced at power plants so as to diminish the greenhouse effect. Draw all resonance structures of the CO_2 molecule.

65. Write a Lewis structure for each of the following polyatomic ions. Show all bonding valence electron pairs as lines and all nonbonding valence electron pairs as dots. For those ions that exhibit resonance, draw the various possible resonance forms.
   a. sulfate ion, SO_4^{2-}
   b. phosphate ion, PO_4^{3-}
   c. sulfite ion, SO_3^{2-}

66. Write a Lewis structure for each of the following polyatomic ions. Show all bonding valence electron pairs as lines and all nonbonding valence electron pairs as dots. For those ions that exhibit resonance, draw the various possible resonance forms.
   a. chlorate ion, ClO_3^-
   b. peroxide ion, O_2^{2-}
   c. acetate ion, $C_2H_3O_2^-$

67. Write a Lewis structure for each of the following polyatomic ions. Show all bonding valence electron pairs as lines and all nonbonding valence electron pairs as dots. For those ions that exhibit resonance, draw the various possible resonance forms.
   a. chlorite ion, ClO_2^-
   b. perbromate ion, BrO_4^-
   c. cyanide ion, CN^-

68. Write a Lewis structure for each of the following polyatomic ions. Show all bonding valence electron pairs as lines and all nonbonding valence electron pairs as dots. For those ions that exhibit resonance, draw the various possible resonance forms.
   a. carbonate ion, CO_3^{2-}
   b. ammonium ion, NH_4^+
   c. hypochlorite ion, ClO^-

12-8 Molecular Structure

Questions

69. What is the geometric structure of the water molecule? How many pairs of valence electrons are there on the oxygen atom in the water molecule? What is the approximate H—O—H bond angle in water?

70. What is the geometric structure of the ammonia molecule? How many pairs of electrons surround the nitrogen atom in NH_3? What is the approximate H—N—H bond angle in ammonia?

71. What is the geometric structure of the boron trifluoride molecule, BF_3? How many pairs of valence electrons are present on the boron atom in BF_3? What are the approximate F—B—F bond angles in BF_3?

72. What is the geometric structure of the SiF_4 molecule? How many pairs of valence electrons are present on the silicon atom of SiF_4? What are the approximate F—Si—F bond angles in SiF_4?

12-9 and 12-10 Molecular Structure: The VSEPR Model

Questions

73. Why is the geometric structure of a molecule important, especially for biological molecules?

74. What general principles determine the molecular structure (shape) of a molecule?

75. How is the structure around a given atom related to repulsion between valence electron pairs on the atom?

76. Why are all diatomic molecules *linear,* regardless of the number of valence electron pairs on the atoms involved?

77. Although the valence electron pairs in ammonia have a tetrahedral arrangement, the overall geometric structure of the ammonia molecule is *not* described as being tetrahedral. Explain.

78. Although both the BF_3 and NF_3 molecules contain the same number of atoms, the BF_3 molecule is flat, whereas the NF_3 molecule is trigonal pyramidal. Explain.

Problems

79. For the indicated atom in each of the following molecules or ions, give the number and arrangement of the electron pairs around that atom.
   a. As in AsO_4^{3-}
   b. Se in SeO_4^{2-}
   c. S in H_2S

80. For the indicated atom in each of the following molecules or ions, give the number and arrangement of the electron pairs around that atom.
   a. S in SO_3^{2-}
   b. S in HSO_3^-
   c. S in HS^-

81. Using the VSEPR theory, predict the molecular structure of each of the following molecules.
   a. NCl_3 b. H_2Se c. $SiCl_4$

82. Using the VSEPR theory, predict the molecular structure of each of the following molecules.
   a. CBr_4 b. PH_3 c. OCl_2

83. Using the VSEPR theory, predict the molecular structure of each of the following polyatomic ions.
   a. sulfate ion, SO_4^{2-}
   b. phosphate ion, PO_4^{3-}
   c. ammonium ion, NH_4^+

84. Using the VSEPR theory, predict the molecular structure of each of the following polyatomic ions.
   a. dihydrogen phosphate ion, $H_2PO_4^-$
   b. perchlorate ion, ClO_4^-
   c. sulfite ion, SO_3^{2-}

85. For each of the following molecules or ions, indicate the bond angle expected between the central atom and any two adjacent hydrogen atoms.
   a. H_2O b. NH_3 c. NH_4^+ d. CH_4

86. For each of the following molecules or ions, indicate the bond angle expected between the central atom and any two adjacent chlorine atoms.
   a. Cl_2O b. NCl_3 c. CCl_4 d. C_2Cl_4

F 87. The "Chemistry in Focus" segment *Taste—It's the Structure That Counts* discusses artificial sweeteners. What are the expected bond angles around the nitrogen atom in aspartame?

All even-numbered Questions and Problems have answers in the back of this book and solutions in the *Student Solutions Guide.*

88. For each of the following molecules, predict both the molecular structure and bond angles around the central atom.

a. SeS_2
b. SeS_3
c. SO_2
d. CS_2

Additional Problems

89. What is *resonance?* Give three examples of molecules or ions that exhibit resonance, and draw Lewis structures for each of the possible resonance forms.

90. When two atoms share two pairs of electrons, a(n) ________ bond is said to exist between them.

91. The geometric arrangement of electron pairs around a given atom is determined principally by the tendency to minimize ________ between the electron pairs.

92. Choose the bond that is the least polar. Explain your reasoning.

a. P—S
b. C—O
c. N—N
d. Sr—O
e. Fe—P

93. In each case, which of the following pairs of bonded elements forms the more polar bond?

a. Br—Cl or Br—F
b. As—S or As—O
c. Pb—C or Pb—Si

94. What do we mean by the *bond energy* of a chemical bond?

95. A(n) ________ chemical bond represents the equal sharing of a pair of electrons between two nuclei.

96. For each of the following pairs of elements, identify which element would be expected to be more electronegative. It should not be necessary to look at a table of actual electronegativity values.

a. Be or Ba
b. N or P
c. F or Cl

97. On the basis of the electronegativity values given in Fig. 12.3, indicate whether each of the following bonds would be expected to be ionic, covalent, or polar covalent.

a. H—O
b. O—O
c. H—H
d. H—Cl

98. Which of the following molecules contain polar covalent bonds?

a. carbon monoxide, CO
b. chlorine, Cl_2
c. iodine monochloride, ICl
d. phosphorus, P_4

99. On the basis of the electronegativity values given in Fig. 12.3, indicate which is the more polar bond in each of the following pairs.

a. N—P or N—O
b. N—C or N—O
c. N—S or N—C
d. N—F or N—S

100. In each of the following molecules, which end of the molecule is negative relative to the other end?

a. carbon monoxide, CO
b. iodine monobromide, IBr
c. hydrogen iodide, HI

101. For each of the following bonds, draw a figure indicating the direction of the bond dipole, including which end of the bond is positive and which is negative.

a. N—Cl
b. N—P
c. N—S
d. N—C

102. Which noble gases correspond to the same electron configuration for each of the ions in the compound calcium nitride (in the order as written)?

a. Kr; Ne
b. Kr; He
c. Ar; He
d. Ar; Ne
e. Ar; Ar

103. What simple ion does each of the following elements most commonly form?

a. sodium
b. iodine
c. potassium
d. calcium
e. sulfur
f. magnesium
g. aluminum
h. nitrogen

104. On the basis of their electron configurations, predict the formula of the simple binary ionic compound likely to form when the following pairs of elements react with each other.

a. sodium, Na, and selenium, Se
b. rubidium, Rb, and fluorine, F
c. potassium, K, and tellurium, Te
d. barium, Ba, and selenium, Se
e. potassium, K, and astatine, At
f. francium, Fr, and chlorine, Cl

105. Which noble gas has the same electron configuration as each of the ions in the following compounds?

a. calcium bromide, $CaBr_2$
b. aluminum selenide, Al_2Se_3
c. strontium oxide, SrO
d. potassium sulfide, K_2S

106. For each of the following pairs, indicate which is smaller.

a. Rb^+ or Na^+
b. Mg^{2+} or Al^{3+}
c. F^- or I^-
d. Na^+ or K^+

107. Write the Lewis structure for each of the following atoms.

a. He ($Z = 2$)
b. Br ($Z = 35$)
c. Sr ($Z = 38$)
d. Ne ($Z = 10$)
e. I ($Z = 53$)
f. Ra ($Z = 88$)

108. What is the *total* number of *valence* electrons in each of the following molecules?

a. HNO_3
b. H_2SO_4
c. H_3PO_4
d. $HClO_4$

109. Write a Lewis structure for each of the following simple molecules. Show all bonding valence electron pairs as lines and all nonbonding valence electron pairs as dots.

a. GeH_4
b. ICl
c. NI_3
d. PF_3

110. Write a Lewis structure for each of the following simple molecules. Show all bonding valence electron pairs as lines and all nonbonding valence electron pairs as dots.

a. N_2H_4
b. C_2H_6
c. NCl_3
d. $SiCl_4$

All even-numbered Questions and Problems have answers in the back of this book and solutions in the *Student Solutions Guide.*

111. Write a Lewis structure for each of the following simple molecules. Show all bonding valence electron pairs as lines and all nonbonding valence electron pairs as dots. For those molecules that exhibit resonance, draw the various possible resonance forms.

a. SO_2
b. N_2O (N in center)
c. O_3

112. Write a Lewis structure for each of the following polyatomic ions. Show all bonding valence electron pairs as lines and all nonbonding valence electron pairs as dots. For those ions that exhibit resonance, draw the various possible resonance forms.

a. nitrate ion
b. carbonate ion
c. ammonium ion

113. Why is the molecular structure of H_2O nonlinear, whereas that of BeF_2 is linear, even though both molecules consist of three atoms?

114. For the indicated atom in each of the following molecules, give the number and the arrangement of the electron pairs around that atom.

a. C in CCl_4
b. Ge in GeH_4
c. B in BF_3

115. Using the VSEPR theory, predict the molecular structure of each of the following molecules.

a. Cl_2O b. OF_2 c. $SiCl_4$

116. Using the VSEPR theory, predict the molecular structure of each of the following polyatomic ions.

a. chlorate ion
b. chlorite ion
c. perchlorate ion

117. For each of the following molecules, indicate the bond angle expected between the central atom and any two adjacent chlorine atoms.

a. Cl_2O
b. CCl_4
c. $BeCl_2$
d. BCl_3

118. Using the VSEPR theory, predict the molecular structure of each of the following molecules or ions containing multiple bonds.

a. SO_2
b. SO_3
c. HCO_3^- (hydrogen is bonded to oxygen)
d. HCN

119. Using the VSEPR theory, predict the molecular structure of each of the following molecules or ions containing multiple bonds.

a. CO_3^{2-}
b. HNO_3 (hydrogen is bonded to oxygen)
c. NO_2^-
d. C_2H_2

120. Explain briefly how substances with ionic bonding differ in properties from substances with covalent bonding.

121. Explain the difference between a covalent bond formed between two atoms of the same element and a covalent bond formed between atoms of two different elements.

ChemWork Problems

These multiconcept problems (and additional ones) are found interactively online with the same type of assistance a student would get from an instructor.

122. Classify the bonding in each of the following molecules as ionic, polar covalent, or nonpolar covalent.

a. H_2
b. K_3P
c. NaI
d. SO_2
e. HF
f. CCl_4
g. CF_4
h. K_2S

123. Compare the electronegativities of each pair of atoms. State the element of each pair that has the greater electronegativity.

Pair	Symbol for Element with Greater Electronegativity
P and Cl	______
Ca and N	______
N and As	______

124. List the bonds P—Cl, P—F, O—F, and Si—F from least polar to most polar.

125. Arrange the atoms and/or ions in the following groups in order of decreasing size.

a. O, O^-, O^{2-}
b. Fe^{2+}, Ni^{2+}, Zn^{2+}
c. Ca^{2+}, K^+, Cl^-

126. Write electron configurations for the most stable ion formed by each of the following elements. Do not use the noble gas notation. Write out the complete electron configuration.

Element	Electron Configuration of the Most Stable Ion
Na	______
K	______
Li	______
Cs	______

127. Which of the following compounds or ions exhibit resonance?

a. O_3
b. CNO^-
c. AsI_3
d. CO_3^{2-}
e. AsF_3

128. The formulas of several chemical substances are given in the table below. For each substance in the table, give its chemical name and predict its molecular structure.

Formula	Compound Name	Molecular Structure
CO_2	______	______
NH_3	______	______
SO_3	______	______
H_2O	______	______
ClO_4^-	______	______

All even-numbered Questions and Problems have answers in the back of this book and solutions in the *Student Solutions Guide.*

Questions

1. What is *potential* energy? What is *kinetic* energy? What do we mean by the *law of conservation of energy?* What do scientists mean by *work?* Explain what scientists mean by a *state function* and give an example of one.

2. What does *temperature* measure? Are the molecules in a beaker of warm water moving at the same speed as the molecules in a beaker of cold water? Explain. What is *heat?* Is *heat* the same as *temperature?*

3. When describing a reaction, a chemist might refer to the *system* and the *surroundings*. Explain each of these terms. If a reaction is *endothermic,* does heat travel from the surroundings into the system, or from the system into the surroundings? Suppose a reaction between ionic solutes is performed in aqueous solution, and the temperature of the solution increases. Is the reaction exothermic or endothermic? Explain.

4. What is the study of energy and energy changes called? What is the "first law" of thermodynamics and what does it mean? What do scientists mean by the *internal energy* of a system? Is the *internal energy* the same as *heat?*

5. How is the *calorie* defined? Is the *thermodynamic calorie* the same as the *Calorie* we are careful of when planning our diets? Although the calorie is our "working unit" of energy (based on its experimental definition), the SI unit of energy is the *joule*. How are joules and calories related? What does the *specific heat capacity* of a substance represent? What common substance has a relatively high specific heat capacity, which makes it useful for cooling purposes?

6. What is the *enthalpy* change for a process? Is enthalpy a state function? In what experimental apparatus are enthalpy changes measured?

7. Hess's law is often confusing to students. Imagine you are talking to a friend who has not taken any science courses. Using the reactions

$$P_4(s) + 6Cl_2(g) \rightarrow 4PCl_3(g) \qquad \Delta H = -2.44 \times 10^3 \text{ kJ}$$

$$4PCl_5(g) \rightarrow P_4(s) + 10Cl_2(g) \qquad \Delta H = 3.43 \times 10^3 \text{ kJ}$$

Explain to your friend how Hess's law can be used to calculate the enthalpy change for the reaction

$$PCl_5(g) \rightarrow PCl_3(g) + Cl_2(g)$$

8. The first law of thermodynamics indicates that the total energy content of the universe is constant. If this is true, why do we worry about "energy conservation"? What do we mean by the *quality* of energy, rather than the *quantity?* Give an example. Although the quantity of energy in the universe may be constant, is the *quality* of that energy changing?

9. What do *petroleum* and *natural gas* consist of? Indicate some petroleum "fractions" and explain what they are used for. What does it mean to "crack" petroleum and why is this done? What was tetraethyl lead used for, and why has its use been drastically reduced? What is the *greenhouse effect,* and why are scientists concerned about it?

10. What is a *driving force?* Name two common and important driving forces, and give an example of each. What is *entropy?* Although the total *energy* of the universe is constant, is the *entropy* of the universe constant? What is a spontaneous process?

11. Suppose we have separate 25-g samples of iron, silver, and gold. If 125 J of heat energy is applied separately to each of the three samples, show by calculation which sample will end up at the highest temperature.

12. Methane, CH_4, is the major component of natural gas. Methane burns in air, releasing approximately 890 kJ of heat energy per mole.

$$CH_4(g) + 2O_2(g) \rightarrow CO_2(g) + 2H_2O(g)$$

   a. What quantity of heat is released if 0.521 mole of methane is burned?
   b. What quantity of heat is released if 1.25 g of methane is burned?
   c. What quantity of methane must have reacted if 1250 kJ of heat energy was released?

13. What is *electromagnetic radiation?* Give some examples of such radiation. Explain what the *wavelength* (λ) and *frequency* (ν) of electromagnetic radiation represent. Sketch a representation of a wave and indicate on your drawing one wavelength of the wave. At what speed does electromagnetic radiation move through space? How is this speed related to λ and ν?

14. Explain what it means for an atom to be in an *excited state* and what it means for an atom to be in its *ground state.* How does an excited atom *return* to its ground state? What is a *photon?* How is the wavelength (color) of light related to the energy of the photons being emitted by an atom? How is the energy of the photons being *emitted* by an atom related to the energy changes taking place *within* the atom?

15. Do atoms in excited states emit radiation randomly, at any wavelength? Why? What does it mean to say that the hydrogen atom has only certain *discrete energy levels* available? How do we know this? Why was the quantization of energy levels surprising to scientists when it was first discovered?

16. Describe Bohr's model of the hydrogen atom. How did Bohr envision the relationship between the electron and the nucleus of the hydrogen atom? How did Bohr's model explain the emission of only discrete wavelengths of light by excited hydrogen atoms? Why did Bohr's model not stand up as more experiments were performed using elements other than hydrogen?

17. Schrödinger and de Broglie suggested a "wave–particle duality" for small particles—that is, if electromagnetic radiation showed some particle-like properties, then perhaps small particles might exhibit some wave-like properties. Explain. How does the wave mechanical picture of the atom fundamentally differ from the Bohr model? How do wave mechanical *orbitals* differ from Bohr's *orbits?* What does it mean to say that an orbital represents a probability map for an electron?

All even-numbered Questions and Problems have answers in the back of this book and solutions in the *Student Solutions Guide.*

18. Describe the general characteristics of the first (lowest-energy) hydrogen atomic orbital. How is this orbital designated symbolically? Does this orbital have a sharp "edge"? Does the orbital represent a surface upon which the electron travels at all times?

19. Use the wave mechanical picture of the hydrogen atom to describe what happens when the atom absorbs energy and moves to an "excited" state. What do the *principal energy levels* and their sublevels represent for a hydrogen atom? How do we designate specific principal energy levels and sublevels in hydrogen?

20. Describe the sublevels and orbitals that constitute the third and fourth principal energy levels of hydrogen. How is each of the orbitals designated and what are the general shapes of their probability maps?

21. Describe *electron spin.* How does electron spin affect the total number of electrons that can be accommodated in a given orbital? What does the *Pauli exclusion principle* tell us about electrons and their spins?

22. Summarize the postulates of the wave mechanical model of the atom.

23. List the *order* in which the orbitals are filled as the atoms beyond hydrogen are built up. How many electrons overall can be accommodated in the first and second principal energy levels? How many electrons can be placed in a given *s* subshell? In a given *p* subshell? In a specific *p* orbital? Why do we assign unpaired electrons in the 2*p* orbitals of carbon, nitrogen, and oxygen?

24. Which are the *valence* electrons in an atom? Choose three elements and write their electron configurations, circling the valence electrons in the configurations. Why are the valence electrons more important to an atom's chemical properties than are the core electrons or the nucleus?

25. Sketch the overall shape of the periodic table and indicate the general regions of the table that represent the various *s, p, d,* and *f* orbitals being filled. How is an element's position in the periodic table related to its chemical properties?

26. Using the general periodic table you developed in Question 25, show how the valence-electron configuration of most of the elements can be written just by knowing the relative *location* of the element on the table. Give specific examples.

27. What are the *representative elements?* In what region(s) of the periodic table are these elements found? In what general area of the periodic table are the *metallic* elements found? In what general area of the table are the *nonmetals* found? Where in the table are the *metalloids* located?

28. You have learned how the properties of the elements vary *systematically,* corresponding to the electron structures of the elements being considered. Discuss how the *ionization energies* and *atomic sizes* of elements vary, both within a vertical group (family) of the periodic table and within a horizontal row (period).

29. In general, what do we mean by a *chemical bond?* What does the *bond energy* tell us about the strength of a chemical bond? Name the principal types of chemical bonds.

30. What do we mean by *ionic* bonding? Give an example of a substance whose particles are held together by ionic bonding. What experimental evidence do we have for the existence of ionic bonding? In general, what types of substances react to produce compounds having ionic bonding?

31. What do we mean by *covalent* bonding and *polar covalent* bonding? How are these two bonding types similar and how do they differ? What circumstance must exist for a bond to be purely covalent? How does a polar covalent bond differ from an ionic bond?

32. What is meant by *electronegativity*? How is the difference in electronegativity between two bonded atoms related to the polarity of the bond? Using Fig. 12.3, give an example of a bond that would be nonpolar and of a bond that would be highly polar.

33. What does it mean to say that a molecule has a dipole moment? What is the *difference* between a polar bond and a polar molecule (one that has a dipole moment)? Give an example of a molecule that has polar bonds and that has a dipole moment. Give an example of a molecule that has polar bonds, but that does *not* have a dipole moment. What are some implications of the fact that water has a dipole moment?

34. How is the attainment of a noble gas electron configuration important to our ideas of how atoms bond to each other? When atoms of a metal react with atoms of a nonmetal, what type of electron configurations do the resulting ions attain? Explain how the atoms in a covalently bonded compound can attain noble gas electron configurations.

35. Give evidence that ionic bonds are very strong. Does an ionic substance contain discrete molecules? With what general type of structure do ionic compounds occur? Sketch a representation of a general structure for an ionic compound. Why is a cation always smaller and an anion always larger than the respective parent atom? Describe the bonding in an ionic compound containing polyatomic ions.

36. Why does a Lewis structure for a molecule show only the valence electrons? What is the most important factor for the formation of a stable compound? How do we use this requirement when writing Lewis structures?

37. In writing Lewis structures for molecules, what is meant by the *duet rule?* To which element does the duet rule apply? What do we mean by the *octet rule?* Why is attaining an octet of electrons important for an atom when it forms bonds to other atoms? What is a bonding *pair* of electrons? What is a nonbonding (or *lone*) pair of electrons?

38. For three simple molecules of your own choice, *apply* the rules for writing Lewis structures. Write your discussion as if you are explaining the method to someone who is *not* familiar with Lewis structures.

39. What does a *double* bond between two atoms represent in terms of the number of electrons shared? What does a *triple* bond represent? When writing a Lewis structure, explain how we recognize when a molecule must contain double or triple bonds. What are *resonance structures?*

All even-numbered Questions and Problems have answers in the back of this book and solutions in the *Student Solutions Guide.*

40. Although many simple molecules fulfill the octet rule, some common molecules are exceptions to this rule. Give three examples of molecules whose Lewis structures are exceptions to the octet rule.

41. What do we mean by the *geometric structure* of a molecule? Draw the geometric structures of at least four simple molecules of your choosing and indicate the bond angles in the structures. Explain the main ideas of the *valence shell electron pair repulsion (VSEPR) theory*. Using several examples, explain how you would *apply* the VSEPR theory to predict their geometric structures.

42. What bond angle results when there are only two valence electron pairs around an atom? What bond angle results when there are three valence pairs? What bond angle results when there are four pairs of valence electrons around the central atom in a molecule? Give examples of molecules containing these bond angles.

43. How do we predict the geometric structure of a molecule whose Lewis structure indicates that the molecule contains a double or triple bond? Give an example of such a molecule, write its Lewis structure, and show how the geometric shape is derived.

44. Write the electron configuration for the following atoms, using the appropriate noble gas to abbreviate the configuration of the core electrons.

a. Sr, $Z = 38$
b. Al, $Z = 13$
c. Cl, $Z = 17$
d. K, $Z = 19$
e. S, $Z = 16$
f. As, $Z = 33$

45. Based on the electron configuration of the simple ions that the pairs of elements given below would be expected to form, predict the formula of the simple binary compound that would be formed by each pair.

a. Al and F
b. Li and N
c. Ca and S
d. Mg and P
e. Al and O
f. K and S

46. Draw the Lewis structure for each of the following molecules or ions. Indicate the number and spatial orientation of the electron pairs around the boldface atom in each formula. Predict the simple geometric structure of each molecule or ion, and indicate the approximate bond angles around the boldface atom.

a. $H_2\mathbf{O}$
b. $\mathbf{P}H_3$
c. $\mathbf{C}Br_4$
d. $\mathbf{Cl}O_4^-$
e. $\mathbf{B}F_3$
f. $\mathbf{Be}F_2$

Gases

13

CHAPTER 13

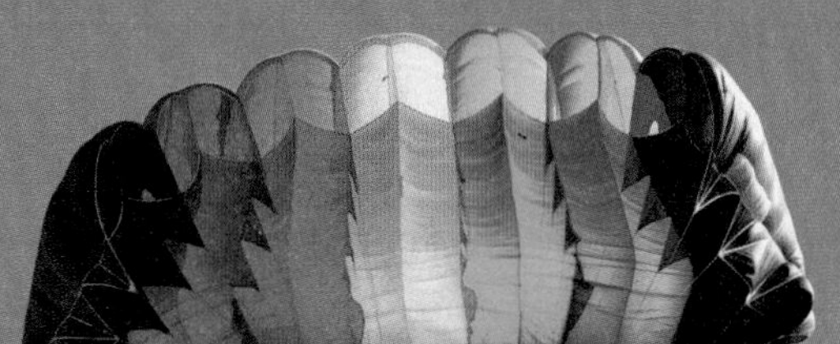

The atmosphere, consisting mostly of nitrogen and oxygen molecules, provides the air resistance necessary to keep the skydivers from descending too quickly.

Danshutter/Shutterstock.com

We live immersed in a gaseous solution. The earth's atmosphere is a mixture of gases that consists mainly of elemental nitrogen, N_2, and oxygen, O_2. The atmosphere both supports life and acts as a waste receptacle for the exhaust gases that accompany many industrial processes. The chemical reactions of these waste gases in the atmosphere lead to various types of pollution, including smog and acid rain. The two main sources of pollution are transportation and the production of electricity. The combustion of fuel in vehicles produces CO, CO_2, NO, and NO_2, along with unburned fragments of the petroleum used as fuel. The combustion of coal and petroleum in power plants produces NO_2 and SO_2 in the exhaust gases. These mixtures of chemicals can be activated by absorbing light to produce the photochemical smog that afflicts most large cities. The SO_2 in the air reacts with oxygen to produce SO_3 gas, which combines with water in the air to produce droplets of sulfuric acid (H_2SO_4), a major component of acid rain.

The gases in the atmosphere also shield us from harmful radiation from the sun and keep the earth warm by reflecting heat radiation back toward the earth. In fact, there is now great concern that an increase in atmospheric carbon dioxide, a product of the combustion of fossil fuels, is causing a dangerous warming of the earth. (See "Chemistry in Focus" *Atmospheric Effects* in Chapter 11.)

In this chapter we will look carefully at the properties of gases. First, we will see how measurements of gas properties lead to various types of laws—statements that show how the properties are related to each other. Then we will construct a model to explain why gases behave as they do. This model will show how the behavior of the individual particles of a gas leads to the observed properties of the gas itself (a collection of many, many particles).

The study of gases provides an excellent example of the scientific method in action. It illustrates how observations lead to natural laws, which in turn can be accounted for by models.

The *Breitling Orbiter 3,* shown over the Swiss Alps, recently completed a nonstop trip around the world. Gamma-Rapho via Getty Images

13-1 Pressure

OBJECTIVES

- **To learn about atmospheric pressure and how barometers work.**
- **To learn the various units of pressure.**

A gas uniformly fills any container, is easily compressed, and mixes completely with any other gas (see Section 3-1). One of the most obvious properties of a gas is that it exerts pressure on its surroundings. For example, when you blow up a balloon, the air inside pushes against the elastic sides of the balloon and keeps it firm.

The gases most familiar to us form the earth's atmosphere. The pressure exerted by this gaseous mixture that we call air can be dramatically demonstrated by the experiment shown in Fig. 13.1. A small volume of water is placed in a metal can and the water is boiled, which fills the can with steam. The can is then sealed and allowed to cool. Why does the can collapse as it cools? It is the atmospheric pressure that crumples the can. When the can is cooled after being sealed so that no air can flow in, the water vapor (steam) inside the can condenses to a very small volume of liquid water. As a gas, the water vapor filled the can, but when it is condensed to a liquid, the liquid does not come close to filling the can. The H_2O molecules formerly present as a gas are now collected in a much smaller volume of liquid, and there are very few molecules

Dry air (air from which the water vapor has been removed) is 78.1% N_2 molecules, 20.9% O_2 molecules, 0.9% Ar atoms, and 0.03% CO_2 molecules, along with smaller amounts of Ne, He, CH_4, Kr, and other trace components.

As a gas, water occupies 1200 times as much space as it does as a liquid at 25 °C and atmospheric pressure.

Soon after Torricelli died, a German physicist named Otto von Guericke invented an air pump. In a famous demonstration for the King of Prussia in 1683, Guericke placed two hemispheres together, pumped the air out of the resulting sphere through a valve, and showed that teams of horses could not pull the hemispheres apart. Then, after secretly opening the air valve, Guericke easily separated the hemispheres by hand. The King of Prussia was so impressed that he awarded Guericke a lifetime pension!

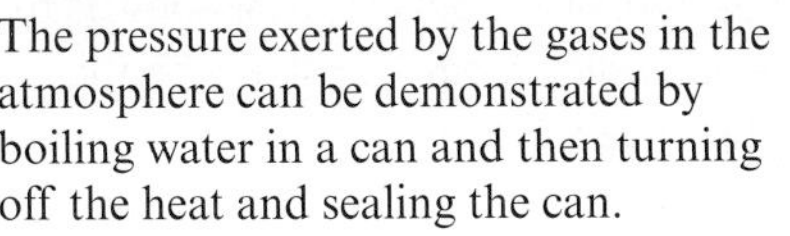

a The pressure exerted by the gases in the atmosphere can be demonstrated by boiling water in a can and then turning off the heat and sealing the can.

b As the can cools, the water vapor condenses, lowering the gas pressure inside the can. This causes the can to crumple.

Figure 13.1

of gas left to exert pressure outward and counteract the air pressure. As a result, the pressure exerted by the gas molecules in the atmosphere smashes the can.

A device that measures atmospheric pressure, the **barometer,** was invented in 1643 by an Italian scientist named Evangelista Torricelli (1608–1647), who had been a student of the famous astronomer Galileo. Torricelli's barometer is constructed by filling a glass tube with liquid mercury and inverting it in a dish of mercury, as shown in Fig. 13.2. Notice that a large quantity of mercury stays in the tube. In fact, at sea level the height of this column of mercury averages 760 mm. Why does this mercury stay in the tube, seemingly in defiance of gravity? Figure 13.2 illustrates how the pressure exerted by the atmospheric gases on the surface of mercury in the dish keeps the mercury in the tube.

Atmospheric pressure results from the mass of the air being pulled toward the center of the earth by gravity—in other words, it results from the weight of the air. Changing weather conditions cause the atmospheric pressure to vary, so the height of the column of Hg supported by the atmosphere at sea level varies; it is not always 760 mm. The meteorologist who says a "low" is approaching means that the atmospheric pressure is going to decrease. This condition often occurs in conjunction with a storm.

Atmospheric pressure also varies with altitude. For example, when Torricelli's experiment is done in Breckenridge, Colorado (elevation 9600 feet), the atmosphere supports a column of mercury only about 520 mm high because the air is "thinner." That is, there is less air pushing down on the earth's surface at Breckenridge than at sea level.

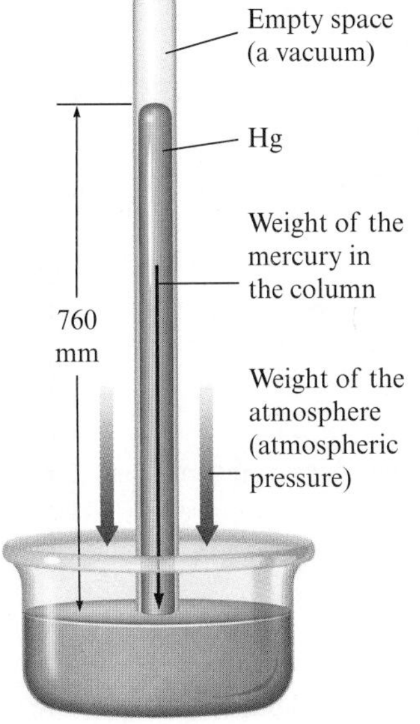

Figure 13.2 When a glass tube is filled with mercury and inverted in a dish of mercury at sea level, the mercury flows out of the tube until a column approximately 760 mm high remains (the height varies with atmospheric conditions). Note that the pressure of the atmosphere balances the weight of the column of mercury in the tube.

Breath Fingerprinting

Medical science is always looking for noninvasive ways to diagnose diseases. One very promising method involves analysis of a person's breath. This is really not a new idea. The concept traces way back to Hippocrates, who suggested a link between breath odor and disease as early as 400 B.C. More recently doctors have been aware that patients suffering from kidney and liver failure have a distinctive smell in their breaths. Modern technology is now enabling scientists to identify specific compounds in patient's breath that are directly associated with particular diseases.

However, breath analysis is not easy. A person's breath is chemically complex, containing a myriad of substances. Many of these substances result from body chemistry, but many others are present because of the person's diet, medications taken, and substances in inhaled ambient air. In spite of the difficulties, progress is being made. One pioneer in this area is Cristina Davis, a Professor of Mechanical and Aerospace Engineering at the University of California. Professor Davis is developing a portable pediatric asthma monitor that determines the level of inflammation by measuring the level of nitric oxide in a patient's breath. In another effort, Peter Mazzone of the Cleveland Clinic is testing a breath sensor that is able to diagnose lung cancer with 80% accuracy. Tests are now underway using a more sensitive device that should produce even greater accuracy. Researchers at various facilities are studying breath tests for other types of cancer, such as colon cancer and breast cancer.

It appears that important developments in breath analysis are on the horizon.

Units of Pressure

Mercury is used to measure pressure because of its high density. By way of comparison, the column of water required to measure a given pressure would be 13.6 times as high as a mercury column used for the same purpose.

Because instruments used for measuring pressure (Fig. 13.3) often contain mercury, the most commonly used units for pressure are based on the height of the mercury column (in millimeters) that the gas pressure can support. The unit **millimeters of mercury (mm Hg)** is often called the **torr** in honor of Torricelli. The terms *torr* and *mm Hg* are used interchangeably by chemists. A related unit for pressure is the **standard atmosphere** (abbreviated atm).

$$1 \text{ standard atmosphere} = 1.000 \text{ atm} = 760.0 \text{ mm Hg} = 760.0 \text{ torr}$$

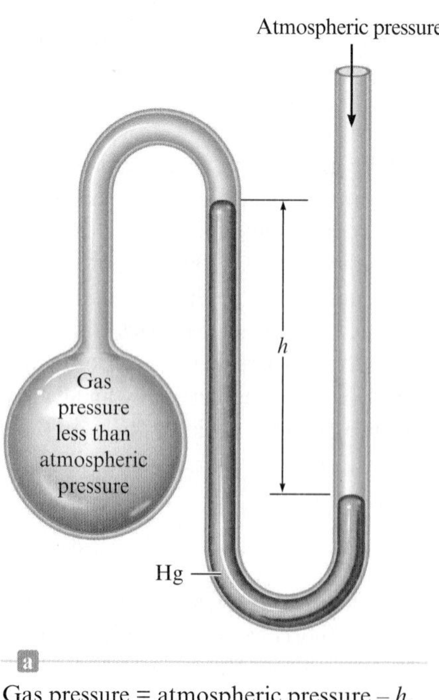

a Gas pressure = atmospheric pressure − *h*.

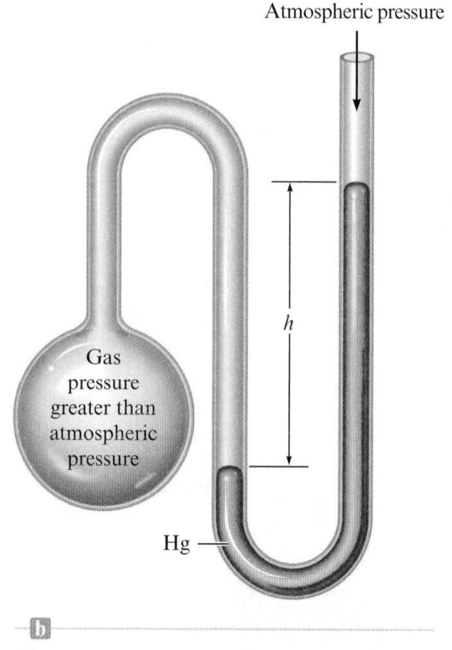

b Gas pressure = atmospheric pressure + *h*.

Figure 13.3 A device (called a manometer) for measuring the pressure of a gas in a container. The pressure of the gas is equal to *h* (the difference in mercury levels) in units of torr (equivalent to mm Hg).

The SI unit for pressure is the **pascal** (abbreviated Pa).

$$1 \text{ standard atmosphere} = 101{,}325 \text{ Pa}$$

Thus 1 atmosphere is about 100,000 or 10^5 pascals. Because the pascal is so small we will use it sparingly in this book. A unit of pressure that is employed in the engineering sciences and that we use for measuring tire pressure is pounds per square inch, abbreviated psi.

1.000 atm
760.0 mm Hg
760.0 torr
14.69 psi
101,325 Pa

$$1.000 \text{ atm} = 14.69 \text{ psi}$$

Sometimes we need to convert from one unit of pressure to another. We do this by using conversion factors. The process is illustrated in Example 13.1.

Interactive Example 13.1

Pressure Unit Conversions

The pressure of the air in a tire is measured to be 28 psi. Represent this pressure in atmospheres, torr, and pascals.

SOLUTION

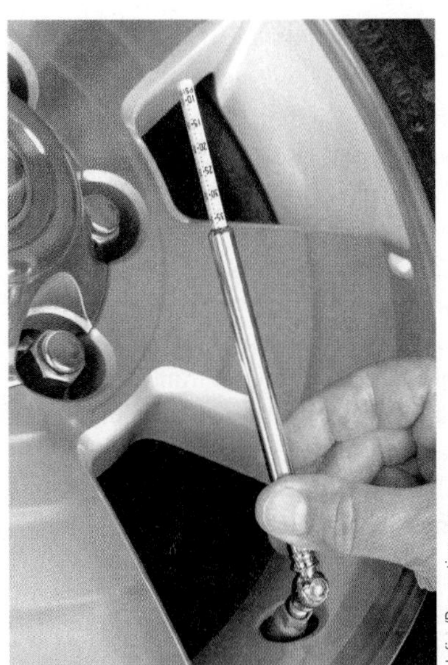
Kelpfish/Dreamstime.com

Checking the air pressure in a tire.

Where Are We Going?

We want to convert from units of pounds per square inch to units of atmospheres, torr, and pascals.

What Do We Know?

- 28 psi

What Information Do We Need?

- We need the equivalence statements for the units.

How Do We Get There?

To convert from pounds per square inch to atmospheres, we need the equivalence statement

$$1.000 \text{ atm} = 14.69 \text{ psi}$$

which leads to the conversion factor

$$\frac{1.000 \text{ atm}}{14.69 \text{ psi}}$$

$$28 \cancel{\text{psi}} \times \frac{1.000 \text{ atm}}{14.69 \cancel{\text{psi}}} = 1.9 \text{ atm}$$

To convert from atmospheres to torr, we use the equivalence statement

$$1.000 \text{ atm} = 760.0 \text{ torr}$$

which leads to the conversion factor

$$\frac{760.0 \text{ torr}}{1.000 \text{ atm}}$$

$$1.9 \cancel{\text{atm}} \times \frac{760.0 \text{ torr}}{1.000 \cancel{\text{atm}}} = 1.4 \times 10^3 \text{ torr}$$

Math skill builder

$1.9 \times 760.0 = 1444$

$1444 \Rightarrow 1400 = 1.4 \times 10^3$

Round off

To change from torr to pascals, we need the equivalence statement

$$1.000 \text{ atm} = 101{,}325 \text{ Pa}$$

which leads to the conversion factor

$$\frac{101{,}325 \text{ Pa}}{1.000 \text{ atm}}$$

Math skill builder

1.9 × 101,325 = 192,517.5

192,517.5 ⇨ 190,000 = 1.9 × 10^5

Round off

$$1.9 \cancel{\text{atm}} \times \frac{101{,}325 \text{ Pa}}{1.000 \cancel{\text{atm}}} = 1.9 \times 10^5 \text{ Pa}$$

REALITY CHECK The units on the answers are the units required.

SELF CHECK

Exercise 13.1 On a summer day in Breckenridge, Colorado, the atmospheric pressure is 525 mm Hg. What is this air pressure in atmospheres?

See Problems 13.7 through 13.12. ■

13-2 Pressure and Volume: Boyle's Law

OBJECTIVES

- **To understand the law that relates the pressure and volume of a gas.**
- **To do calculations involving this law.**

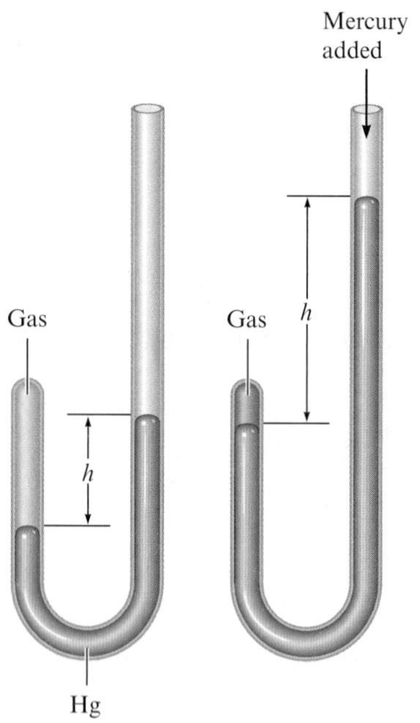

Mercury added

Gas

Gas

h

h

Hg

Figure 13.4 A J-tube similar to the one used by Boyle. The pressure on the trapped gas can be changed by adding or withdrawing mercury.

The first careful experiments on gases were performed by the Irish scientist Robert Boyle (1627–1691). Using a J-shaped tube closed at one end (Fig. 13.4), which he reportedly set up in the multi-story entryway of his house, Boyle studied the relationship between the pressure of the trapped gas and its volume. Representative values from Boyle's experiments are given in Table 13.1. The units given for the volume (cubic inches) and pressure (inches of mercury) are the ones Boyle used. Keep in mind that the metric system was not in use at this time.

First let's examine Boyle's observations (Table 13.1) for general trends. Note that as the pressure increases, the volume of the trapped gas decreases. In fact, if you compare the data from experiments 1 and 4, you can see that as the pressure is doubled (from 29.1 to 58.2), the volume of the gas is halved (from 48.0 to 24.0). The same relationship can be seen in experiments 2 and 5 and in experiments 3 and 6 (approximately).

The fact that the constant is sometimes 1.40 × 10^3 instead of 1.41 × 10^3 is due to experimental error (uncertainties in measuring the values of P and V).

Table 13.1 A Sample of Boyle's Observations (moles of gas and temperature both constant)

			Pressure × Volume (in. Hg) × (in.³)	
Experiment	Pressure (in. Hg)	Volume (in.³)	Actual	Rounded*
1	29.1	48.0	1396.8	1.40 × 10^3
2	35.3	40.0	1412.0	1.41 × 10^3
3	44.2	32.0	1414.4	1.41 × 10^3
4	58.2	24.0	1396.8	1.40 × 10^3
5	70.7	20.0	1414.0	1.41 × 10^3
6	87.2	16.0	1395.2	1.40 × 10^3
7	117.5	12.0	1410.0	1.41 × 10^3

*Three significant figures are allowed in the product because both of the numbers that are multiplied together have three significant figures.

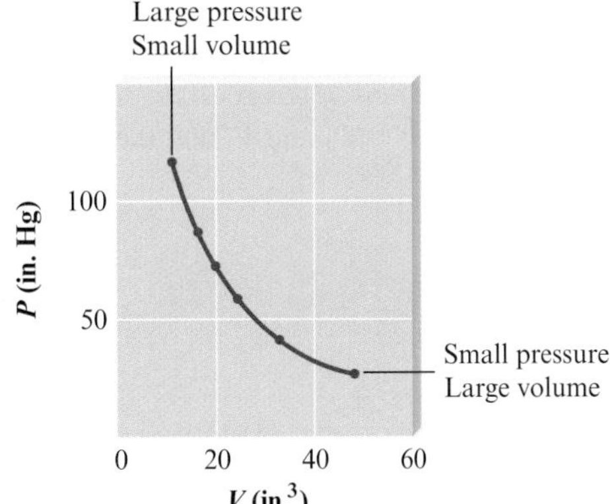

Figure 13.5 A plot of P vs. V from Boyle's data in Table 13.1.

We can see the relationship between the volume of a gas and its pressure more clearly by looking at the product of the values of these two properties ($P \times V$) using Boyle's observations. This product is shown in the last column of Table 13.1. Note that for all the experiments,

$$P \times V = 1.4 \times 10^3 \text{ (in Hg)} \times \text{in.}^3$$

with only a slight variation due to experimental error. Other similar measurements on gases show the same behavior. This means that the relationship of the pressure and volume of a gas can be expressed in words as

pressure times volume equals a constant

or in terms of an equation as

$$PV = k$$

For Boyle's law to hold, the amount of gas (moles) must not be changed. The temperature must also be constant.

which is called **Boyle's law,** where k is a constant at a specific temperature for a given amount of gas. For the data we used from Boyle's experiment, $k = 1.41 \times 10^3$ (in. Hg) $\times$ in.^{3}.

It is often easier to visualize the relationships between two properties if we make a graph. Figure 13.5 uses the data given in Table 13.1 to show how pressure is related to volume. This relationship, called a plot or a graph, shows that V decreases as P increases. When this type of relationship exists, we say that volume and pressure are inversely related or *inversely proportional;* when one increases, the other decreases. Boyle's law is illustrated by the gas samples in Fig. 13.6.

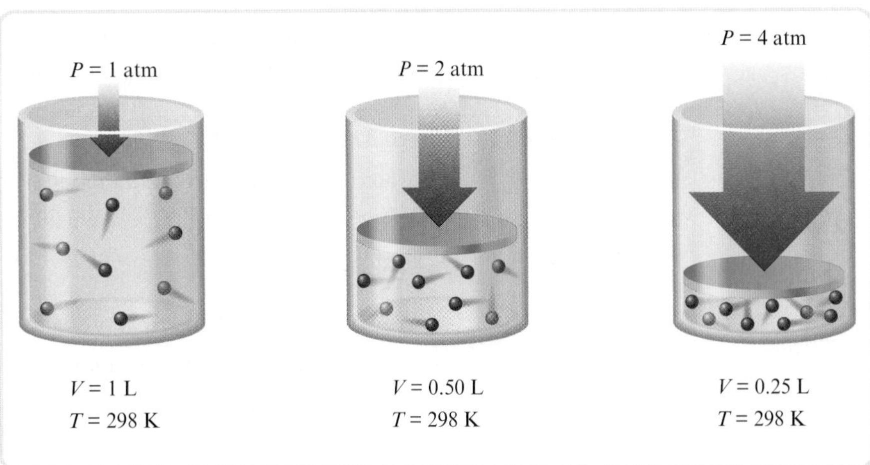

Figure 13.6 Illustration of Boyle's law. These three containers contain the same number of molecules. At 298 K, $P \times V = 1$ L atm in all three containers.

Boyle's law means that if we know the volume of a gas at a given pressure, we can predict the new volume if the pressure is changed, *provided that neither the temperature nor the amount of gas is changed.* For example, if we represent the original pressure and volume as P_1 and V_1 and the final values as P_2 and V_2, using Boyle's law we can write

$$P_1V_1 = k$$

and

$$P_2V_2 = k$$

We can also say

$$P_1V_1 = k = P_2V_2$$

or simply

$$P_1V_1 = P_2V_2$$

This is really another way to write Boyle's law. We can solve for the final volume (V_2) by dividing both sides of the equation by P_2.

$$\frac{P_1V_1}{P_2} = \frac{P_2V_2}{P_2}$$

Canceling the P_2 terms on the right gives

$$\frac{P_1}{P_2} \times V_1 = V_2$$

or

$$V_2 = V_1 \times \frac{P_1}{P_2}$$

This equation tells us that we can calculate the new gas volume (V_2) by multiplying the original volume (V_1) by the ratio of the original pressure to the final pressure (P_1/P_2), as illustrated in Example 13.2.

Interactive Example 13.2

Calculating Volume Using Boyle's Law

Freon-12 (the common name for the compound CCl_2F_2) was widely used in refrigeration systems but has now been replaced by other compounds that do not lead to the breakdown of the protective ozone in the upper atmosphere. Consider a 1.5-L sample of gaseous CCl_2F_2 at a pressure of 56 torr. If pressure is changed to 150 torr at a constant temperature,

a. Will the volume of the gas increase or decrease?

b. What will be the new volume of the gas?

SOLUTION

Where Are We Going?

We want to determine if the volume will increase or decrease when the pressure is changed, and we want to calculate the new volume.

What Do We Know?

- We know the initial and final pressures and the initial volume.
- The amount of gas and temperature are held constant.
- Boyle's law: $P_1V_1 = P_2V_2$.

How Do We Get There?

a. As the first step in a gas law problem, always write down the information given, in the form of a table showing the initial and final conditions.

Initial Conditions	**Final Conditions**
$P_1 = 56$ torr	$P_2 = 150$ torr
$V_1 = 1.5$ L	$V_2 = ?$

Drawing a picture also is often helpful. Notice that the pressure is increased from 56 torr to 150 torr, so the volume must decrease:

$P_1V_1 \Rightarrow P_2V_2$

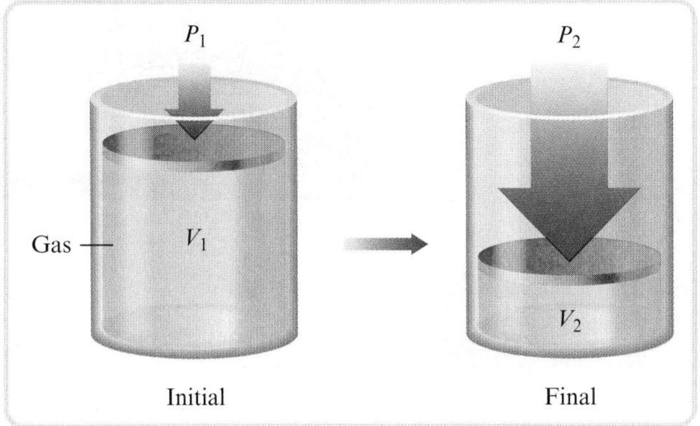

We can verify this by using Boyle's law in the form

$$V_2 = V_1 \times \frac{P_1}{P_2}$$

Note that V_2 is obtained by "correcting" V_1 using the ratio P_1/P_2. Because P_1 is less than P_2, the ratio P_1/P_2 is a fraction that is less than 1. Thus V_2 must be a fraction of (smaller than) V_1; the volume decreases.

b. We calculate V_2 as follows:

$$V_2 = V_1 \times \frac{P_1}{P_2} = \underset{\uparrow\,V_1}{1.5\text{ L}} \times \frac{\overset{P_1\,\downarrow}{56\text{ torr}}}{\underset{\uparrow\,P_2}{150\text{ torr}}} = 0.56\text{ L}$$

REALITY CHECK Because the pressure increases, we expect the volume to decrease. The pressure increased by almost a factor of three, and the volume decreased by about a factor of three.

SELF CHECK

Exercise 13.2 A sample of neon to be used in a neon sign has a volume of 1.51 L at a pressure of 635 torr. Calculate the volume of the gas after it is pumped into the glass tubes of the sign, where it shows a pressure of 785 torr.

See Problems 13.21 and 13.22. ■

Dave Jacobs/Stone/Getty Images

Neon signs in Hong Kong.

Interactive Example 13.3 Calculating Pressure Using Boyle's Law

In an automobile engine the gaseous fuel–air mixture enters the cylinder and is compressed by a moving piston before it is ignited. In a certain engine the initial cylinder volume is 0.725 L. After the piston moves up, the volume is 0.075 L. The fuel–air mixture initially has a pressure of 1.00 atm. Calculate the pressure of the compressed fuel–air mixture, assuming that both the temperature and the amount of gas remain constant.

SOLUTION

Where Are We Going?

We want to determine the new pressure of a fuel–air mixture that has undergone a volume change.

What Do We Know?

- We know the initial and final volumes and the initial pressure.
- The amount of gas and temperature are held constant.
- Boyle's law: $P_1V_1 = P_2V_2$.

How Do We Get There?

We summarize the given information in the following table:

Initial Conditions	Final Conditions
$P_1 = 1.00$ atm	$P_2 = ?$
$V_1 = 0.725$ L	$V_2 = 0.075$ L

Then we solve Boyle's law in the form $P_1V_1 = P_2V_2$ for P_2 by dividing both sides by V_2 to give the equation

$$P_2 = P_1 \times \frac{V_1}{V_2} = 1.00 \text{ atm} \times \frac{0.725 \cancel{\text{L}}}{0.075 \cancel{\text{L}}} = 9.7 \text{ atm}$$

Math skill builder

$$P_1V_1 = P_2V_2$$
$$\frac{P_1V_1}{V_2} = \frac{P_2\cancel{V_2}}{\cancel{V_2}}$$
$$P_1 \times \frac{V_1}{V_2} = P_2$$
$$\frac{0.725}{0.075} = 9.666\ldots$$
$$9.666 \Rightarrow 9.7$$
Round off

REALITY CHECK Because the volume decreases, we expect the pressure to increase. The volume decreased by about a factor of 10, and the pressure increased by about a factor of 10. ■

13-3 Volume and Temperature: Charles's Law

OBJECTIVES

- **To learn about absolute zero.**
- **To learn about the law relating the volume and temperature of a sample of gas at constant moles and pressure, and to do calculations involving that law.**

John A. Rizzo/PhotoDisc/Getty Images

The air in a balloon expands when it is heated. This means that some of the air escapes from the balloon, lowering the air density inside and thus making the balloon buoyant.

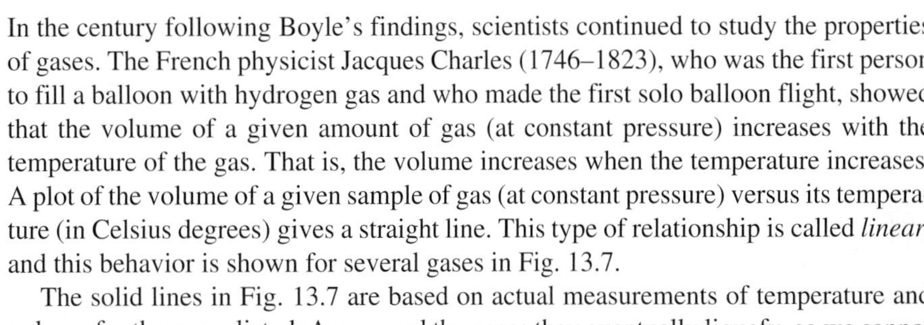

In the century following Boyle's findings, scientists continued to study the properties of gases. The French physicist Jacques Charles (1746–1823), who was the first person to fill a balloon with hydrogen gas and who made the first solo balloon flight, showed that the volume of a given amount of gas (at constant pressure) increases with the temperature of the gas. That is, the volume increases when the temperature increases. A plot of the volume of a given sample of gas (at constant pressure) versus its temperature (in Celsius degrees) gives a straight line. This type of relationship is called *linear,* and this behavior is shown for several gases in Fig. 13.7.

The solid lines in Fig. 13.7 are based on actual measurements of temperature and volume for the gases listed. As we cool the gases they eventually liquefy, so we cannot determine any experimental points below this temperature. However, when we extend each straight line (which is called *extrapolation* and is shown here by a dashed line), something very interesting happens. *All* of the lines extrapolate to zero volume at the same temperature: −273 °C. This suggests that −273 °C is the lowest possible temperature, because a negative volume is physically impossible. In fact, experiments have shown that matter cannot be cooled to temperatures lower than −273 °C. Therefore, this temperature is defined as **absolute zero** on the Kelvin scale. Temperatures such as 0.00000002 K have been obtained in the laboratory, but 0 K has never been reached.

When the volumes of the gases shown in Fig. 13.7 are plotted against temperature on the Kelvin scale rather than the Celsius scale, the plots shown in Fig. 13.8 result. These plots show that the volume of each gas is *directly proportional to the temperature* (in kelvins) and extrapolates to zero when the temperature is 0 K. Let's illustrate this statement with an example. Suppose we have 1 L of gas at 300 K. When we double the temperature of this gas to 600 K (without changing its pressure), the volume also

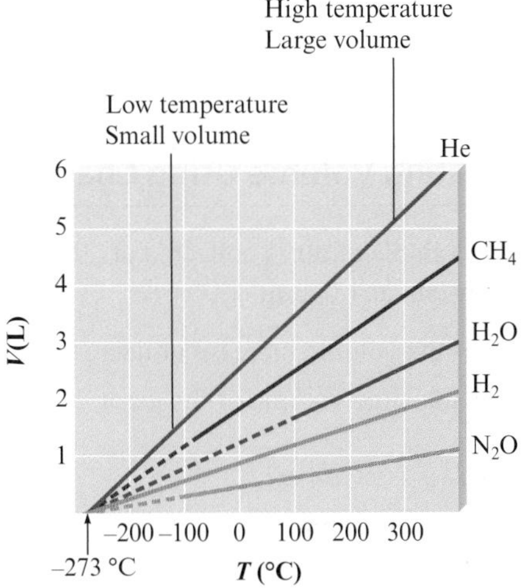

Figure 13.7 Plots of V (L) vs. T (°C) for several gases. Note that each sample of gas contains a different number of moles to spread out the plots.

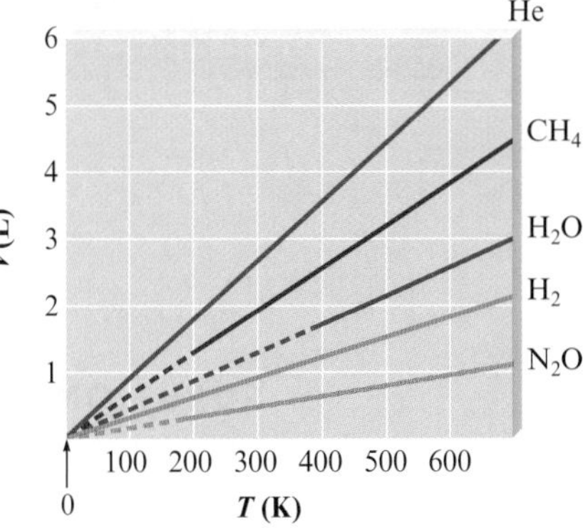

Figure 13.8 Plots of V vs. T as in Fig. 13.7, except that here the Kelvin scale is used for temperature.

doubles, to 2 L. Verify this type of behavior by looking carefully at the lines for various gases shown in Fig. 13.8.

The direct proportionality between volume and temperature (in kelvins) is represented by the equation known as **Charles's law:**

$$V = bT$$

where T is in kelvins and b is the proportionality constant. Charles's law holds for a given sample of gas at constant pressure. It tells us that (for a given amount of gas at a given pressure) the volume of the gas is directly proportional to the temperature on the Kelvin scale:

$$V = bT \quad \text{or} \quad \frac{V}{T} = b = \text{constant}$$

From Fig. 13.8 for Helium

V (L)	T (K)	b
0.7	100	0.01
1.7	200	0.01
2.7	300	0.01
3.7	400	0.01
5.7	600	0.01

Notice that in the second form, this equation states that the *ratio* of V to T (in kelvins) must be constant. (This is shown for helium in the margin.) Thus, when we triple the temperature (in kelvins) of a sample of gas, the volume of the gas triples as well.

$$\frac{V}{T} = \frac{3 \times V}{3 \times T} = b = \text{constant}$$

We can also write Charles's law in terms of V_1 and T_1 (the initial conditions) and V_2 and T_2 (the final conditions).

$$\frac{V_1}{T_1} = b \quad \text{and} \quad \frac{V_2}{T_2} = b$$

Thus

$$\frac{V_1}{T_1} = \frac{V_2}{T_2}$$

Charles's law in the form $V_1/T_1 = V_2/T_2$ applies only when both the amount of gas (moles) and the pressure are constant.

We will illustrate the use of this equation in Examples 13.4 and 13.5.

Critical Thinking

According to Charles's law, the volume of a gas is directly related to its temperature in Kelvin at constant pressure and number of moles. What if the volume of a gas was directly related to its temperature in Celsius at constant pressure and number of moles? How would the world be different?

Interactive Example 13.4 Calculating Volume Using Charles's Law, I

A 2.0-L sample of air is collected at 298 K and then cooled to 278 K. The pressure is held constant at 1.0 atm.

a. Does the volume increase or decrease?

b. Calculate the volume of the air at 278 K.

SOLUTION

Where Are We Going?

We want to determine if the volume will increase or decrease when the temperature is changed, and we want to calculate the new volume.

What Do We Know?

- We know the initial and final temperatures and the initial volume.
- The amount of gas and pressure are held constant.
- Charles's law: $\frac{V_1}{T_1} = \frac{V_2}{T_2}$.

How Do We Get There?

a. Because the gas is cooled, the volume of the gas must decrease:

$$\frac{V}{T} = \text{constant}$$

T is decreased, so *V* must decrease to maintain a constant ratio.

$\frac{V_1}{T_1} \Rightarrow \frac{V_2}{T_2}$

Temperature smaller, volume smaller

b. To calculate the new volume, V_2, we will use Charles's law in the form

$$\frac{V_1}{T_1} = \frac{V_2}{T_2}$$

We are given the following information:

Initial Conditions	Final Conditions
$T_1 = 298\text{ K}$	$T_2 = 278\text{ K}$
$V_1 = 2.0\text{ L}$	$V_2 = ?$

We want to solve the equation

$$\frac{V_1}{T_1} = \frac{V_2}{T_2}$$

for V_2. We can do this by multiplying both sides by T_2 and canceling.

$$T_2 \times \frac{V_1}{T_1} = \frac{V_2}{\cancel{T_2}} \times \cancel{T_2} = V_2$$

Thus

$$V_2 = T_2 \times \frac{V_1}{T_1} = 278\ \cancel{\text{K}} \times \frac{2.0\text{ L}}{298\ \cancel{\text{K}}} = 1.9\text{ L}$$

REALITY CHECK Because the temperature decreases, we expect the volume to decrease. The temperature decreased slightly, so we would expect the volume to decrease slightly. ■

Interactive Example 13.5

Calculating Volume Using Charles's Law, II

A sample of gas at 15 °C (at 1 atm) has a volume of 2.58 L. The temperature is then raised to 38 °C (at 1 atm).

a. Does the volume of the gas increase or decrease?

b. Calculate the new volume.

SOLUTION

Where Are We Going?

We want to determine if the volume will increase or decrease when the temperature is changed, and we want to calculate the new volume.

What Do We Know?

- We know the initial and final temperatures and the initial volume.
- The amount of gas and pressure are held constant.
- Charles's law: $\frac{V_1}{T_1} = \frac{V_2}{T_2}$.

How Do We Get There?

a. In this case we have a given sample (constant amount) of gas that is heated from 15 °C to 38 °C *while the pressure is held constant.* We know from Charles's law that the volume of a given sample of gas is directly proportional to the temperature (at constant pressure). So the increase in temperature will *increase* the volume; the new volume will be greater than 2.58 L.

b. To calculate the new volume, we use Charles's law in the form

$$\frac{V_1}{T_1} = \frac{V_2}{T_2}$$

We are given the following information:

Initial Conditions	**Final Conditions**
$T_1 = 15\ °C$	$T_2 = 38\ °C$
$V_1 = 2.58\ L$	$V_2 = ?$

As is often the case, the temperatures are given in Celsius degrees. However, for us to use Charles's law, the temperature *must be in kelvins.* Thus we must convert by adding 273 to each temperature.

Initial Conditions	**Final Conditions**
$T_1 = 15\ °C = 15 + 273$	$T_2 = 38\ °C = 38 + 273$
$= 288\ K$	$= 311\ K$
$V_1 = 2.58\ L$	$V_2 = ?$

Solving for V_2 gives

$$V_2 = V_1 \times \frac{T_2}{T_1} = 2.58\ L \left(\frac{311\ \cancel{K}}{288\ \cancel{K}}\right) = 2.79\ L$$

REALITY CHECK Because the temperature increases, we expect the volume to increase.

USGA photo by T. Casadevall

Researchers take samples from a steaming volcanic vent at Mount Baker in Washington.

SELF CHECK

Exercise 13.3 A child blows a bubble that contains air at 28 °C and has a volume of 23 cm^3 at 1 atm. As the bubble rises, it encounters a pocket of cold air (temperature 18 °C). If there is no change in pressure, will the bubble get larger or smaller as the air inside cools to 18 °C? Calculate the new volume of the bubble.

See Problems 13.29 and 13.30. ■

Notice from Example 13.5 that we adjust the volume of a gas for a temperature change by multiplying the original volume by the ratio of the Kelvin temperatures—final (T_2) over initial (T_1). Remember to check whether your answer makes sense. When the temperature increases (at constant pressure), the volume must increase, and vice versa.

Interactive Example 13.6 Calculating Temperature Using Charles's Law

In former times, gas volume was used as a way to measure temperature using devices called gas thermometers. Consider a gas that has a volume of 0.675 L at 35 °C and 1 atm pressure. What is the temperature (in units of °C) of a room where this gas has a volume of 0.535 L at 1 atm pressure?

SOLUTION

Where Are We Going?

We want to determine the new temperature of a gas given that the volume has decreased at constant pressure.

What Do We Know?

- We know the initial and final volumes and the initial temperature.
- The amount of gas and pressure are held constant.
- Charles's law: $\frac{V_1}{T_1} = \frac{V_2}{T_2}$.

How Do We Get There?

The information given in the problem is

Initial Conditions	Final Conditions
$T_1 = 35\ °C = 35 + 273 = 308\ K$	$T_2 = ?$
$V_1 = 0.675\ L$	$V_2 = 0.535\ L$
$P_1 = 1\ atm$	$P_2 = 1\ atm$

The pressure remains constant, so we can use Charles's law in the form

$$\frac{V_1}{T_1} = \frac{V_2}{T_2}$$

and solve for T_2. First we multiply both sides by T_2.

$$T_2 \times \frac{V_1}{T_1} = \frac{V_2}{\cancel{T_2}} \times \cancel{T_2} = V_2$$

Next we multiply both sides by T_1.

$$\cancel{T_1} \times T_2 \times \frac{V_1}{\cancel{T_1}} = T_1 \times V_2$$

This gives

$$T_2 \times V_1 = T_1 \times V_2$$

Now we divide both sides by V_1 (multiply by $1/V_1$),

$$\frac{1}{\cancel{V_1}} \times T_2 \times \cancel{V_1} = \frac{1}{V_1} \times T_1 \times V_2$$

and obtain

$$T_2 = T_1 \times \frac{V_2}{V_1}$$

We have now isolated T_2 on one side of the equation, and we can do the calculation.

$$T_2 = T_1 \times \frac{V_2}{V_1} = (308\text{ K}) \times \frac{0.535\ \cancel{\text{L}}}{0.675\ \cancel{\text{L}}} = 244\text{ K}$$

To convert from units of K to units of °C, we subtract 273 from the Kelvin temperature.

$$T_{°C} = T_K - 273 = 244 - 273 = -29\ °C$$

The room is very cold; the new temperature is −29 °C.

REALITY CHECK Because the volume is smaller, we expect the temperature to be lower. ■

13-4 Volume and Moles: Avogadro's Law

OBJECTIVE **To understand the law relating the volume and the number of moles of a sample of gas at constant temperature and pressure, and to do calculations involving this law.**

What is the relationship between the volume of a gas and the number of molecules present in the gas sample? Experiments show that when the number of moles of gas is doubled (at constant temperature and pressure), the volume doubles. In other words, the volume of a gas is directly proportional to the number of moles if temperature and pressure remain constant. Figure 13.9 illustrates this relationship, which can also be represented by the equation

$$V = an \qquad \text{or} \qquad \frac{V}{n} = a$$

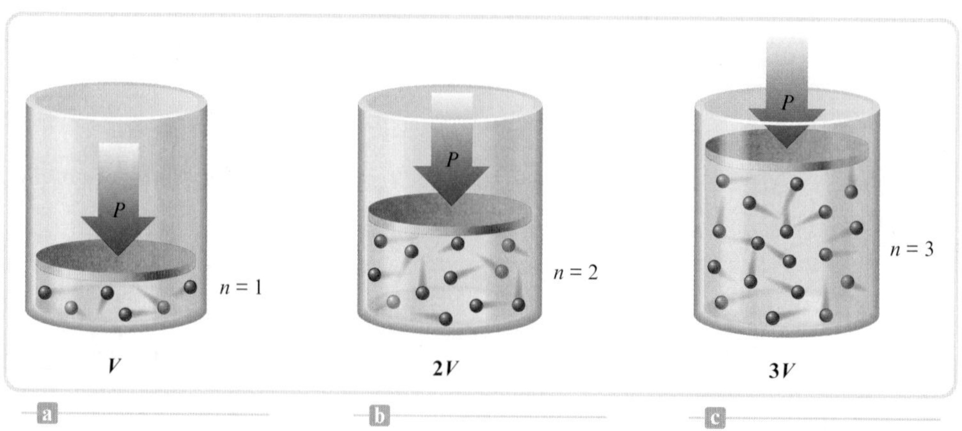

Figure 13.9 ▸ The relationship between volume V and number of moles n. As the number of moles is increased from 1 to 2 (a) to (b), the volume doubles. When the number of moles is tripled (c), the volume is also tripled. The temperature and pressure remain the same in these cases.

where V is the volume of the gas, n is the number of moles, and a is the proportionality constant. Note that this equation means that the ratio of V to n is constant as long as the temperature and pressure remain constant. Thus, when the number of moles of gas is increased by a factor of 5, the volume also increases by a factor of 5,

$$\frac{V}{n} = \frac{\cancel{5} \times V}{\cancel{5} \times n} = a = \text{constant}$$

and so on. In words, this equation means that *for a gas at constant temperature and pressure, the volume is directly proportional to the number of moles of gas.* This relationship is called **Avogadro's law** after the Italian scientist Amadeo Avogadro, who first postulated it in 1811.

For cases where the number of moles of gas is changed from an initial amount to another amount (at constant T and P), we can represent Avogadro's law as

$$\underset{\text{Initial amount}}{\frac{V_1}{n_1}} = a = \underset{\text{Final amount}}{\frac{V_2}{n_2}}$$

or

$$\frac{V_1}{n_1} = \frac{V_2}{n_2}$$

We will illustrate the use of this equation in Example 13.7.

Interactive Example 13.7

Using Avogadro's Law in Calculations

Suppose we have a 12.2-L sample containing 0.50 mole of oxygen gas, O_2, at a pressure of 1 atm and a temperature of 25 °C. If all of this O_2 is converted to ozone, O_3, at the same temperature and pressure, what will be the volume of the ozone formed?

SOLUTION

Where Are We Going?

We want to determine the volume of ozone (O_3) formed from 0.50 mole of O_2 given the volume of oxygen.

What Do We Know?

- We know the initial number of moles of oxygen and the volume of oxygen.
- The temperature and pressure are held constant.
- Avogadro's law: $\frac{V_1}{T_1} = \frac{V_2}{T_2}$.

What Information Do We Need?

- We need the balanced equation for the reaction to determine the number of moles of ozone formed.

How Do We Get There?

To do this problem we need to compare the moles of gas originally present to the moles of gas present after the reaction. We know that 0.50 mole of O_2 is present initially. To find out how many moles of O_3 will be present after the reaction, we need to use the balanced equation for the reaction.

$$3O_2(g) \rightarrow 2O_3(g)$$

We calculate the moles of O_3 produced by using the appropriate mole ratio from the balanced equation.

$$0.50 \cancel{\text{mol } O_2} \times \frac{2 \text{ mol } O_3}{3 \cancel{\text{mol } O_2}} = 0.33 \text{ mol } O_3$$

Avogadro's law states that

$$\frac{V_1}{n_1} = \frac{V_2}{n_2}$$

Math skill builder

$$\frac{V_1}{n_1} = \frac{V_2}{n_2}$$

$$n_2 \times \frac{V_1}{n_1} = \frac{V_2}{\cancel{n_2}} \times \cancel{n_2}$$

$$V_1 \times \frac{n_2}{n_1} = V_2$$

where V_1 is the volume of n_1 moles of O_2 gas and V_2 is the volume of n_2 moles of O_3 gas. In this case we have

Initial Conditions	Final Conditions
$n_1 = 0.50$ mol	$n_2 = 0.33$ mol
$V_1 = 12.2$ L	$V_2 = ?$

Solving Avogadro's law for V_2 gives

$$V_2 = V_1 \times \frac{n_2}{n_1} = 12.2 \text{ L}\left(\frac{0.33 \cancel{\text{mol}}}{0.50 \cancel{\text{mol}}}\right) = 8.1 \text{ L}$$

REALITY CHECK Note that the volume decreases, as it should, because fewer molecules are present in the gas after O_2 is converted to O_3.

SELF CHECK

Exercise 13.4 Consider two samples of nitrogen gas (composed of N_2 molecules). Sample 1 contains 1.5 moles of N_2 and has a volume of 36.7 L at 25 °C and 1 atm. Sample 2 has a volume of 16.5 L at 25 °C and 1 atm. Calculate the number of moles of N_2 in Sample 2.

See Problems 13.41 through 13.44. ■

13-5 The Ideal Gas Law

OBJECTIVE **To understand the ideal gas law and use it in calculations.**

We have considered three laws that describe the behavior of gases as it is revealed by experimental observations.

Constant n means a constant number of moles of gas.

Boyle's law: $PV = k$ or $V = \frac{k}{P}$ (at constant T and n)

Charles's law: $V = bT$ (at constant P and n)

Avogadro's law: $V = an$ (at constant T and P)

These relationships, which show how the volume of a gas depends on pressure, temperature, and number of moles of gas present, can be combined as follows:

$$V = R\left(\frac{Tn}{P}\right)$$

where R is the combined proportionality constant and is called the **universal gas constant.** When the pressure is expressed in atmospheres and the volume is in liters, R always has the value 0.08206 L atm/K mol. We can rearrange the above equation by multiplying both sides by P,

$$R = 0.08206 \frac{\text{L atm}}{\text{K mol}}$$

$$P \times V = \cancel{P} \times R\left(\frac{Tn}{\cancel{P}}\right)$$

to obtain the **ideal gas law** written in its usual form,

$$PV = nRT$$

The ideal gas law involves all the important characteristics of a gas: its pressure (P), volume (V), number of moles (n), and temperature (T). Knowledge of any three of these properties is enough to define completely the condition of the gas, because the fourth property can be determined from the ideal gas law.

It is important to recognize that the ideal gas law is based on experimental measurements of the properties of gases. A gas that obeys this equation is said to behave *ideally.* That is, this equation defines the behavior of an **ideal gas.** Most gases obey this equation closely at pressures of approximately 1 atm or lower, when the temperature is approximately 0 °C or higher. You should assume ideal gas behavior when working problems involving gases in this text.

The ideal gas law can be used to solve a variety of problems. Example 13.8 demonstrates one type, where you are asked to find one property characterizing the condition of a gas given the other three properties.

Interactive Example 13.8

Using the Ideal Gas Law in Calculations

A sample of hydrogen gas, H_2, has a volume of 8.56 L at a temperature of 0 °C and a pressure of 1.5 atm. Calculate the number of moles of H_2 present in this gas sample. (Assume that the gas behaves ideally.)

SOLUTION

Where Are We Going?

We want to determine the number of moles of hydrogen gas (H_2) present given conditions of temperature, pressure, and volume.

What Do We Know?

- We know the temperature, pressure, and volume of hydrogen gas.
- Ideal gas law: $PV = nRT$.

What Information Do We Need?

- $R = 0.08206$ L atm/mol K.

How Do We Get There?

In this problem we are given the pressure, volume, and temperature of the gas: $P = 1.5$ atm, $V = 8.56$ L, and $T = 0$ °C. Remember that the temperature must be changed to the Kelvin scale.

$$T = 0\ °\text{C} = 0 + 273 = 273\ \text{K}$$

We can calculate the number of moles of gas present by using the ideal gas law, $PV = nRT$. We solve for n by dividing both sides by RT:

$$\frac{PV}{RT} = n\frac{\cancel{RT}}{\cancel{RT}}$$

to give

$$\frac{PV}{RT} = n$$

Thus

$$n = \frac{PV}{RT} = \frac{(1.5\ \cancel{\text{atm}})(8.56\ \cancel{\text{L}})}{\left(0.08206\ \dfrac{\cancel{\text{L}}\ \cancel{\text{atm}}}{\cancel{\text{K}}\ \text{mol}}\right)(273\ \cancel{\text{K}})} = 0.57\ \text{mol}$$

SELF CHECK

Exercise 13.5 A weather balloon contains 1.10×10^5 moles of He and has a volume of 2.70×10^6 L at 1.00 atm pressure. Calculate the temperature of the helium in the balloon in kelvins and in Celsius degrees.

See Problems 13.53 through 13.60. ■

Interactive Example 13.9

Ideal Gas Law Calculations Involving Conversion of Units

What volume is occupied by 0.250 mole of carbon dioxide gas at 25 °C and 371 torr?

SOLUTION

Where Are We Going?

We want to determine the volume of carbon dioxide gas (CO_2) given the number of moles, pressure, and temperature.

What Do We Know?

- We know the number of moles, pressure, and temperature of the carbon dioxide.
- Ideal gas law: $PV = nRT$.

What Information Do We Need?

- $R = 0.08206$ L atm/mol K.

How Do We Get There?

We can use the ideal gas law to calculate the volume, but we must first convert pressure to atmospheres and temperature to the Kelvin scale.

Math skill builder

$PV = nRT$

$\frac{PV}{P} = \frac{nRT}{P}$

$V = \frac{nRT}{P}$

$$P = 371 \text{ torr} = 371 \cancel{\text{torr}} \times \frac{1.000 \text{ atm}}{760.0 \cancel{\text{torr}}} = 0.488 \text{ atm}$$

$$T = 25\ ^\circ\text{C} = 25 + 273 = 298 \text{ K}$$

We solve for V by dividing both sides of the ideal gas law ($PV = nRT$) by P.

$$V = \frac{nRT}{P} = \frac{(0.250 \cancel{\text{mol}})\left(0.08206 \frac{\text{L} \cancel{\text{atm}}}{\cancel{\text{K}} \cancel{\text{mol}}}\right)(298 \cancel{\text{K}})}{0.488 \cancel{\text{atm}}} = 12.5 \text{ L}$$

The volume of the sample of CO_2 is 12.5 L.

SELF CHECK

Exercise 13.6 Radon, a radioactive gas formed naturally in the soil, can cause lung cancer. It can pose a hazard to humans by seeping into houses, and there is concern about this problem in many areas. A 1.5-mole sample of radon gas has a volume of 21.0 L at 33 °C. What is the pressure of the gas?

See Problems 13.53 through 13.60. ■

Note that R has units of L atm/K mol. Accordingly, whenever we use the ideal gas law, we must express the volume in units of liters, the temperature in kelvins, and the pressure in atmospheres. When we are given data in other units, we must first convert to the appropriate units.

The ideal gas law can also be used to calculate the changes that will occur when the conditions of the gas are changed as illustrated in Example 13.10.

Interactive Example 13.10

Using the Ideal Gas Law Under Changing Conditions

Suppose we have a 0.240-mole sample of ammonia gas at 25 °C with a volume of 3.5 L at a pressure of 1.68 atm. The gas is compressed to a volume of 1.35 L at 25 °C. Use the ideal gas law to calculate the final pressure.

SOLUTION

Where Are We Going?

We want to use the ideal gas law equation to determine the pressure of ammonia gas given a change in volume.

What Do We Know?

- We know the initial number of moles, pressure, volume, and temperature of the ammonia.
- We know the new volume.
- Ideal gas law: $PV = nRT$.

How Do We Get There?

In this case we have a sample of ammonia gas in which the conditions are changed. We are given the following information:

Initial Conditions	**Final Conditions**
$V_1 = 3.5$ L	$V_2 = 1.35$ L
$P_1 = 1.68$ atm	$P_2 = ?$
$T_1 = 25\ °C = 25 + 273 = 298$ K	$T_2 = 25\ °C = 25 + 273 = 298$ K
$n_1 = 0.240$ mol	$n_2 = 0.240$ mol

Note that both n and T remain constant—only P and V change. Thus we could simply use Boyle's law ($P_1V_1 = P_2V_2$) to solve for P_2. However, we will use the ideal gas law to solve this problem in order to introduce the idea that one equation—the ideal gas equation—can be used to do almost any gas problem. The key idea here is that in using the ideal gas law to describe a change in conditions for a gas, we always *solve the ideal gas equation in such a way that the variables that change are on one side of the equals sign and the constant terms are on the other side.* That is, we start with the ideal gas equation in the conventional form ($PV = nRT$) and rearrange it so that all the terms that change are moved to one side and all the terms that do not change are moved to the other side. In this case the pressure and volume change, and the temperature and number of moles remain constant (as does R, by definition). So we write the ideal gas law as

$$\underbrace{PV}_{\text{Change}} = \underbrace{nRT}_{\text{Remain constant}}$$

Because n, R, and T remain the same in this case, we can write $P_1V_1 = nRT$ and $P_2V_2 = nRT$. Combining these gives

$$P_1V_1 = nRT = P_2V_2 \quad \text{or} \quad P_1V_1 = P_2V_2$$

and

$$P_2 = P_1 \times \frac{V_1}{V_2} = (1.68\ \text{atm})\left(\frac{3.5\ \cancel{\text{L}}}{1.35\ \cancel{\text{L}}}\right) = 4.4\ \text{atm}$$

REALITY CHECK Does this answer make sense? The volume was decreased (at constant temperature and constant number of moles), which means that the pressure should increase, as the calculation indicates.

SELF CHECK

Exercise 13.7 A sample of methane gas that has a volume of 3.8 L at 5 °C is heated to 86 °C at constant pressure. Calculate its new volume.

See Problems 13.61 and 13.62. ■

Note that in solving Example 13.10, we actually obtained Boyle's law ($P_1V_1 = P_2V_2$) from the ideal gas equation. You might well ask, "Why go to all this trouble?" The idea is to learn to use the ideal gas equation to solve all types of gas law problems. This way you will never have to ask yourself, "Is this a Boyle's law problem or a Charles's law problem?"

We continue to practice using the ideal gas law in Example 13.11. Remember, the key idea is to rearrange the equation so that the quantities that change are moved to one side of the equation and those that remain constant are moved to the other.

Interactive Example 13.11

Calculating Volume Changes Using the Ideal Gas Law

A sample of diborane gas, B_2H_6, a substance that bursts into flames when exposed to air, has a pressure of 0.454 atm at a temperature of −15 °C and a volume of 3.48 L. If conditions are changed so that the temperature is 36 °C and the pressure is 0.616 atm, what will be the new volume of the sample?

SOLUTION

Where Are We Going?

We want to use the ideal gas law equation to determine the volume of diborane gas.

What Do We Know?

- We know the initial pressure, volume, and temperature of the diborane gas.
- We know the new temperature and pressure.
- Ideal gas law: $PV = nRT$.

How Do We Get There?

We are given the following information:

Initial Conditions	Final Conditions
$P_1 = 0.454$ atm	$P_2 = 0.616$ atm
$V_1 = 3.48$ L	$V_2 = ?$
$T_1 = -15\ °C = 273 - 15 = 258$ K	$T_2 = 36\ °C = 273 + 36 = 309$ K

Math skill builder

$PV = nRT$

$\frac{PV}{T} = \frac{nRT}{T}$

$\frac{PV}{T} = nR$

Note that the value of n is not given. However, we know that n is constant (that is, $n_1 = n_2$) because no diborane gas is added or taken away. Thus, in this experiment, n is constant and P, V, and T change. Therefore, we rearrange the ideal gas equation ($PV = nRT$) by dividing both sides by T,

$$\underset{\text{Change}}{\frac{PV}{T}} = \underset{\text{Constant}}{nR}$$

which leads to the equation

$$\frac{P_1V_1}{T_1} = nR = \frac{P_2V_2}{T_2}$$

or

$$\frac{P_1V_1}{T_1} = \frac{P_2V_2}{T_2}$$

Snacks Need Chemistry, Too!

Have you ever wondered what makes popcorn pop? The popping is linked with the properties of gases. What happens when a gas is heated? Charles's law tells us that if the pressure is held constant, the volume of the gas must increase as the temperature is increased. But what happens if the gas being heated is trapped at a constant volume? We can see what happens by rearranging the ideal gas law ($PV = nRT$) as follows:

$$P = \left(\frac{nR}{V}\right)T$$

When n, R, and V are held constant, the pressure of a gas is directly proportional to the temperature. Thus, as the temperature of the trapped gas increases, its pressure also increases. This is exactly what happens inside a kernel of popcorn as it is heated. The moisture inside the kernel vaporized by the heat produces increasing pressure. The pressure finally becomes so great that the kernel breaks open, allowing the starch inside to expand to about 40 times its original size.

What's special about popcorn? Why does it pop while "regular" corn doesn't? William da Silva, a biologist at the University of Campinas in Brazil, has traced the "popability" of popcorn to its outer casing, called the pericarp. The molecules in the pericarp of popcorn, which are packed in a much more orderly way than in regular corn, transfer heat unusually quickly, producing a very fast pressure jump that pops the kernel. In addition, because the pericarp of popcorn is much thicker and stronger than that of regular corn, it can withstand more pressure, leading to a more explosive pop when the moment finally comes.

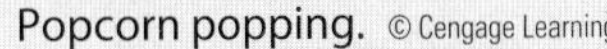

Popcorn popping. © Cengage Learning

We can now solve for V_2 by dividing both sides by P_2 and multiplying both sides by T_2.

$$\frac{1}{P_2} \times \frac{P_1V_1}{T_1} = \frac{\cancel{P_2}V_2}{T_2} \times \frac{1}{\cancel{P_2}} = \frac{V_2}{T_2}$$

$$T_2 \times \frac{P_1V_1}{P_2T_1} = \frac{V_2}{\cancel{T_2}} \times \cancel{T_2} = V_2$$

That is,

$$\frac{T_2P_1V_1}{P_2T_1} = V_2$$

It is sometimes convenient to think in terms of the ratios of the initial temperature and pressure and the final temperature and pressure. That is,

$$V_2 = \frac{T_2P_1V_1}{T_1P_2} = V_1 \times \frac{T_2}{T_1} \times \frac{P_1}{P_2}$$

Substituting the information given yields

$$V_2 = \frac{309\ \cancel{\text{K}}}{258\ \cancel{\text{K}}} \times \frac{0.454\ \cancel{\text{atm}}}{0.616\ \cancel{\text{atm}}} \times 3.48\ \text{L} = 3.07\ \text{L}$$

SELF CHECK

Exercise 13.8 A sample of argon gas with a volume of 11.0 L at a temperature of 13 °C and a pressure of 0.747 atm is heated to 56 °C and a pressure of 1.18 atm. Calculate the final volume.

See Problems 13.61 and 13.62. ■

The equation obtained in Example 13.11,

$$\frac{P_1V_1}{T_1} = \frac{P_2V_2}{T_2}$$

is often called the **combined gas law** equation. It holds when the amount of gas (moles) is held constant. While it may be convenient to remember this equation, it is not necessary because you can always use the ideal gas equation.

13-6 Dalton's Law of Partial Pressures

OBJECTIVE **To understand the relationship between the partial and total pressures of a gas mixture, and to use this relationship in calculations.**

Many important gases contain a mixture of components. One notable example is air. Scuba divers who are going deeper than 150 feet use another important mixture, helium and oxygen. Normal air is not used because the nitrogen present dissolves in the blood in large quantities as a result of the high pressures experienced by the diver under several hundred feet of water. When the diver returns too quickly to the surface, the nitrogen bubbles out of the blood just as soda fizzes when it's opened, and the diver gets "the bends"—a very painful and potentially fatal condition. Because helium gas is only sparingly soluble in blood, it does not cause this problem.

Studies of gaseous mixtures show that each component behaves independently of the others. In other words, a given amount of oxygen exerts the same pressure in a 1.0-L vessel whether it is alone or in the presence of nitrogen (as in the air) or helium.

Among the first scientists to study mixtures of gases was John Dalton. In 1803, Dalton summarized his observations in this statement: *For a mixture of gases in a container, the total pressure exerted is the sum of the partial pressures of the gases present. The* **partial pressure** *of a gas is the pressure that the gas would exert if it were alone in the container.* This statement, known as **Dalton's law of partial pressures,** can be expressed as follows for a mixture containing three gases:

$$P_{\text{total}} = P_1 + P_2 + P_3$$

where the subscripts refer to the individual gases (gas 1, gas 2, and gas 3). The pressures P_1, P_2, and P_3 are the partial pressures; that is, each gas is responsible for only part of the total pressure (Fig. 13.10).

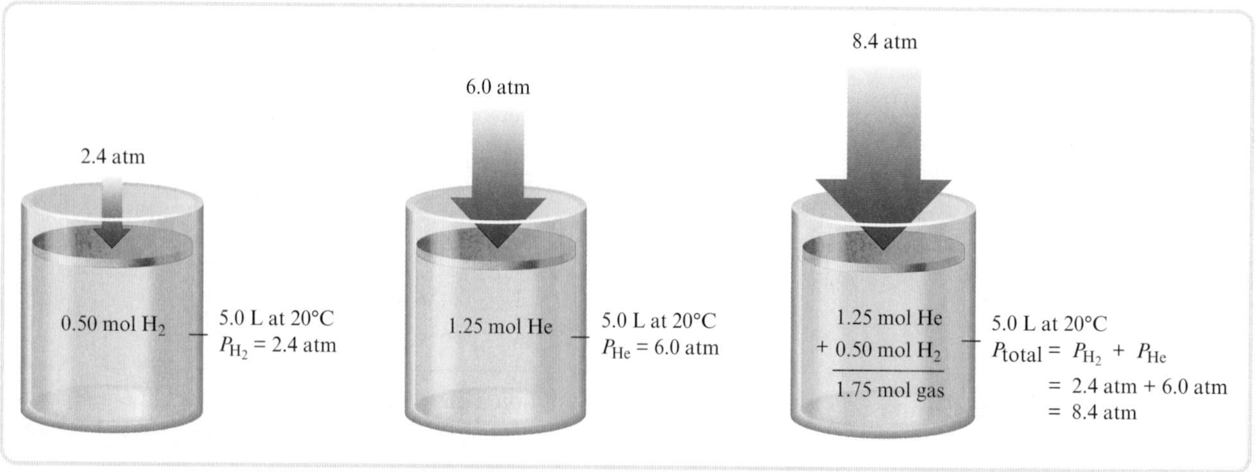

Figure 13.10 ▸ When two gases are present, the total pressure is the sum of the partial pressures of the gases.

Math skill builder

$$PV = nRT$$

$$\frac{PV}{V} = \frac{nRT}{V}$$

$$P = \frac{nRT}{V}$$

Assuming that each gas behaves ideally, we can calculate the partial pressure of each gas from the ideal gas law:

$$P_1 = \frac{n_1RT}{V},\ P_2 = \frac{n_2RT}{V},\ P_3 = \frac{n_3RT}{V}$$

The total pressure of the mixture, P_{total}, can be represented as

$$\begin{aligned} P_{total} = P_1 + P_2 + P_3 &= \frac{n_1RT}{V} + \frac{n_2RT}{V} + \frac{n_3RT}{V} \\ &= n_1\left(\frac{RT}{V}\right) + n_2\left(\frac{RT}{V}\right) + n_3\left(\frac{RT}{V}\right) \\ &= (n_1 + n_2 + n_3)\left(\frac{RT}{V}\right) \\ &= n_{total}\left(\frac{RT}{V}\right) \end{aligned}$$

where n_{total} is the sum of the numbers of moles of the gases in the mixture. Thus, for a mixture of ideal gases, it is the *total number of moles of particles* that is important, not the *identity* of the individual gas particles. This idea is illustrated in Fig. 13.11.

The fact that the pressure exerted by an ideal gas is affected by the *number* of gas particles and is independent of the *nature* of the gas particles tells us two important things about ideal gases:

1. The volume of the individual gas particle (atom or molecule) must not be very important.
2. The forces among the particles must not be very important.

If these factors were important, the pressure of the gas would depend on the nature of the individual particles. For example, an argon atom is much larger than a helium atom. Yet 1.75 moles of argon gas in a 5.0-L container at 20 °C exerts the same pressure as 1.75 moles of helium gas in a 5.0-L container at 20 °C.

The same idea applies to the forces among the particles. Although the forces among gas particles depend on the nature of the particles, this seems to have little influence on the behavior of an ideal gas. We will see that these observations strongly influence the model that we will construct to explain ideal gas behavior.

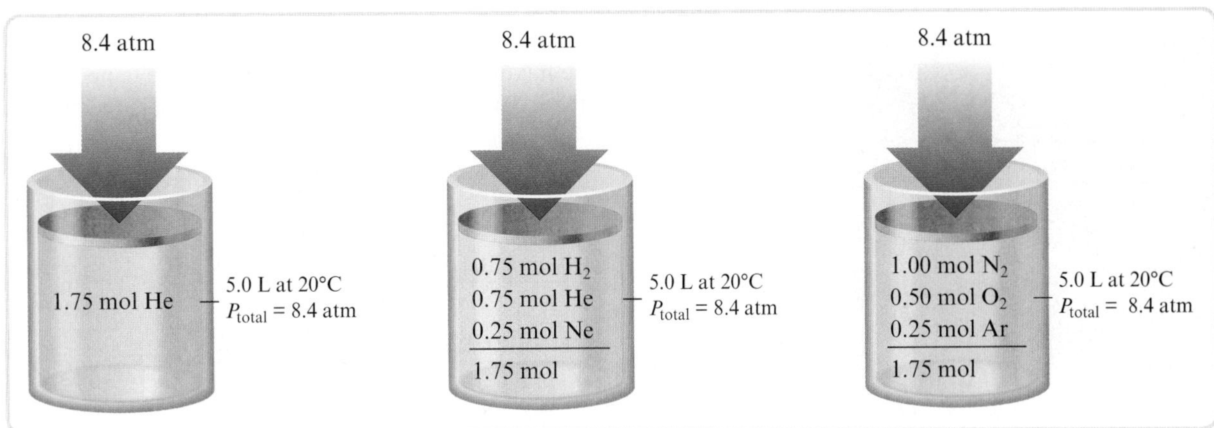

Figure 13.11 The total pressure of a mixture of gases depends on the number of moles of gas particles (atoms or molecules) present, not on the identities of the particles. Note that these three samples show the same total pressure because each contains 1.75 moles of gas. The detailed nature of the mixture is unimportant.

Interactive Example 13.12 Using Dalton's Law of Partial Pressures, I

Mixtures of helium and oxygen are used in the "air" tanks of underwater divers for deep dives. For a particular dive, 12 L of O_2 at 25 °C and 1.0 atm and 46 L of He at 25 °C and 1.0 atm were both pumped into a 5.0-L tank. Calculate the partial pressure of each gas and the total pressure in the tank at 25 °C.

SOLUTION

Where Are We Going?

We want to determine the partial pressure of helium and oxygen and the total pressure in the tank.

What Do We Know?

- We know the initial volume, pressure, and temperature of both gases.
- We know the final volume of the tank.
- The temperature remains constant.
- Ideal gas law: $PV = nRT$.
- Dalton's law of partial pressures: $P_{\text{total}} = P_1 + P_2 + \ldots$

What Information Do We Need?

- $R = 0.08206$ L atm/mol K.

How Do We Get There?

Math skill builder

$$PV = nRT$$

$$\frac{PV}{RT} = \frac{n\cancel{RT}}{\cancel{RT}}$$

$$\frac{PV}{RT} = n$$

Because the partial pressure of each gas depends on the moles of that gas present, we must first calculate the number of moles of each gas by using the ideal gas law in the form

$$n = \frac{PV}{RT}$$

From the above description we know that $P = 1.0$ atm, $V = 12$ L for O_2 and 46 L for He, and $T = 25 + 273 = 298$ K. Also, $R = 0.08206$ L atm/K mol (as always).

$$\text{Moles of } O_2 = n_{O_2} = \frac{(1.0 \cancel{\text{atm}})(12 \cancel{\text{L}})}{(0.08206 \cancel{\text{L}} \cancel{\text{atm}}/\cancel{\text{K}} \text{ mol})(298 \cancel{\text{K}})} = 0.49 \text{ mol}$$

$$\text{Moles of He} = n_{\text{He}} = \frac{(1.0 \cancel{\text{atm}})(46 \cancel{\text{L}})}{(0.08206 \cancel{\text{L}} \cancel{\text{atm}}/\cancel{\text{K}} \text{ mol})(298 \cancel{\text{K}})} = 1.9 \text{ mol}$$

Divers use a mixture of oxygen and helium in their breathing tanks when diving to depths greater than 150 feet.

The tank containing the mixture has a volume of 5.0 L, and the temperature is 25 °C (298 K). We can use these data and the ideal gas law to calculate the partial pressure of each gas.

$$P = \frac{nRT}{V}$$

$$P_{O_2} = \frac{(0.49 \cancel{\text{mol}})(0.08206 \cancel{\text{L}} \text{ atm}/\cancel{\text{K}} \cancel{\text{mol}})(298 \cancel{\text{K}})}{5.0 \cancel{\text{L}}} = 2.4 \text{ atm}$$

$$P_{\text{He}} = \frac{(1.9 \cancel{\text{mol}})(0.08206 \cancel{\text{L}} \text{ atm}/\cancel{\text{K}} \cancel{\text{mol}})(298 \cancel{\text{K}})}{5.0 \cancel{\text{L}}} = 9.3 \text{ atm}$$

The total pressure is the sum of the partial pressures.

$$P_{\text{total}} = P_{O_2} + P_{\text{He}} = 2.4 \text{ atm} + 9.3 \text{ atm} = 11.7 \text{ atm}$$

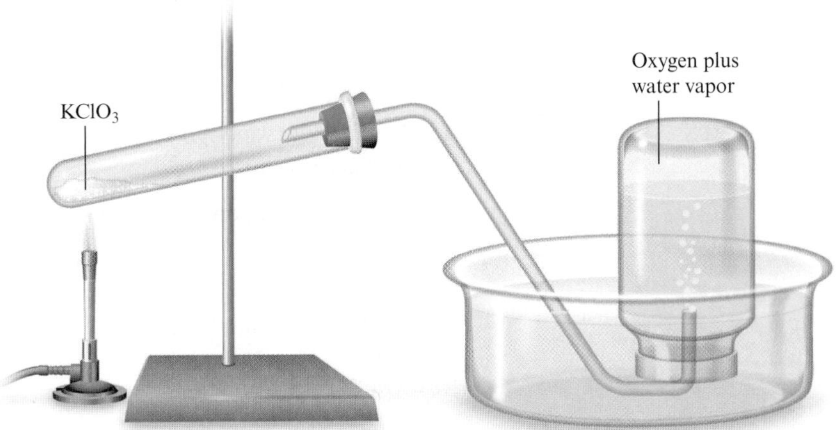

Figure 13.12 The production of oxygen by thermal decomposition of $KClO_3$.

REALITY CHECK The volume of each gas decreased, and the pressure of each gas increased. The partial pressure of helium is greater than that of oxygen, which makes sense because the initial temperatures and pressures of helium and oxygen were the same, but the initial volume of helium was much greater than that of oxygen.

SELF CHECK

Exercise 13.9 A 2.0-L flask contains a mixture of nitrogen gas and oxygen gas at 25 °C. The total pressure of the gaseous mixture is 0.91 atm, and the mixture is known to contain 0.050 mole of N_2. Calculate the partial pressure of oxygen and the moles of oxygen present.

See Problems 13.67 through 13.70. ■

Table 13.2 The Vapor Pressure of Water as a Function of Temperature

T (°C)	P (torr)
0.0	4.579
10.0	9.209
20.0	17.535
25.0	23.756
30.0	31.824
40.0	55.324
60.0	149.4
70.0	233.7
90.0	525.8

A mixture of gases occurs whenever a gas is collected by displacement of water. For example, Fig. 13.12 shows the collection of the oxygen gas that is produced by the decomposition of solid potassium chlorate. The gas is collected by bubbling it into a bottle that is initially filled with water. Thus the gas in the bottle is really a mixture of water vapor and oxygen. (Water vapor is present because molecules of water escape from the surface of the liquid and collect as a gas in the space above the liquid.) Therefore, the total pressure exerted by this mixture is the sum of the partial pressure of the gas being collected and the partial pressure of the water vapor. The partial pressure of the water vapor is called the vapor pressure of water. Because water molecules are more likely to escape from hot water than from cold water, the *vapor pressure* of water increases with temperature. This is shown by the values of vapor pressure at various temperatures in Table 13.2.

Interactive Example 13.13

Using Dalton's Law of Partial Pressures, II

A sample of solid potassium chlorate, $KClO_3$, was heated in a test tube (see Fig. 13.12) and decomposed according to the reaction

$$2KClO_3(s) \rightarrow 2KCl(s) + 3O_2(g)$$

The oxygen produced was collected by displacement of water at 22 °C. The resulting mixture of O_2 and H_2O vapor had a total pressure of 754 torr and a volume of 0.650 L. Calculate the partial pressure of O_2 in the gas collected and the number of moles of O_2 present. The vapor pressure of water at 22 °C is 21 torr.

SOLUTION

Where Are We Going?

We want to determine the partial pressure of oxygen collected by water displacement and the number of moles of O_2 present.

What Do We Know?

- We know the temperature, total pressure, and volume of gas collected by water displacement.
- We know the vapor pressure of water at this temperature.
- Ideal gas law: $PV = nRT$.
- Dalton's law of partial pressures: $P_{total} = P_1 + P_2 + \ldots$

What Information Do We Need?

- $R = 0.08206$ L atm/mol K.

How Do We Get There?

We know the total pressure (754 torr) and the partial pressure of water (vapor pressure = 21 torr). We can find the partial pressure of O_2 from Dalton's law of partial pressures:

$$P_{total} = P_{O_2} + P_{H_2O} = P_{O_2} + 21 \text{ torr} = 754 \text{ torr}$$

or

$$P_{O_2} + 21 \text{ torr} = 754 \text{ torr}$$

We can solve for P_{O_2} by subtracting 21 torr from both sides of the equation.

$$P_{O_2} = 754 \text{ torr} - 21 \text{ torr} = 733 \text{ torr}$$

Next we solve the ideal gas law for the number of moles of O_2.

$$n_{O_2} = \frac{P_{O_2}V}{RT}$$

In this case, $P_{O_2} = 733$ torr. We change the pressure to atmospheres as follows:

$$\frac{733 \cancel{\text{torr}}}{760 \cancel{\text{torr}}/\text{atm}} = 0.964 \text{ atm}$$

Math skill builder

$PV = nRT$

$\frac{PV}{RT} = \frac{n\cancel{RT}}{\cancel{RT}}$

$\frac{PV}{RT} = n$

Then,

$$V = 0.650 \text{ L}$$
$$T = 22\ °\text{C} = 22 + 273 = 295 \text{ K}$$
$$R = 0.08206 \text{ L atm/K mol}$$

so

$$n_{O_2} = \frac{(0.964 \cancel{\text{atm}})(0.650 \cancel{\text{L}})}{(0.08206 \cancel{\text{L}} \cancel{\text{atm}}/\cancel{\text{K}} \text{ mol})(295 \cancel{\text{K}})} = 2.59 \times 10^{-2} \text{ mol}$$

SELF CHECK

Exercise 13.10 Consider a sample of hydrogen gas collected over water at 25 °C where the vapor pressure of water is 24 torr. The volume occupied by the gaseous mixture is 0.500 L, and the total pressure is 0.950 atm. Calculate the partial pressure of H_2 and the number of moles of H_2 present.

See Problems 13.71 through 13.74. ■

13-7 Laws and Models: A Review

OBJECTIVE **To understand the relationship between laws and models (theories).**

In this chapter we have considered several properties of gases and have seen how the relationships among these properties can be expressed by various laws written in the form of mathematical equations. The most useful of these is the ideal gas equation, which relates all the important gas properties. However, under certain conditions gases do not obey the ideal gas equation. For example, at high pressures and/or low temperatures, the properties of gases deviate significantly from the predictions of the ideal gas equation. On the other hand, as the pressure is lowered and/or the temperature is increased, almost all gases show close agreement with the ideal gas equation. This means that an ideal gas is really a hypothetical substance. At low pressures and/or high temperatures, real gases *approach* the behavior expected for an ideal gas.

At this point we want to build a model (a theory) to explain *why* a gas behaves as it does. We want to answer the question, *What are the characteristics of the individual gas particles that cause a gas to behave as it does?* However, before we do this let's briefly review the scientific method. Recall that a law is a generalization about behavior that has been observed in many experiments. Laws are very useful; they allow us to predict the behavior of similar systems. For example, a chemist who prepares a new gaseous compound can assume that that substance will obey the ideal gas equation (at least at low P and/or high T).

However, laws do not tell us *why* nature behaves the way it does. Scientists try to answer this question by constructing theories (building models). The models in chemistry are speculations about how individual atoms or molecules (microscopic particles) cause the behavior of macroscopic systems (collections of atoms and molecules in large enough numbers so that we can observe them).

A model is considered successful if it explains known behavior and predicts correctly the results of future experiments. But a model can never be proved absolutely true. In fact, by its very nature *any model is an approximation* and is destined to be modified, at least in part. Models range from the simple (to predict approximate behavior) to the extraordinarily complex (to account precisely for observed behavior). In this text, we use relatively simple models that fit most experimental results.

13-8 The Kinetic Molecular Theory of Gases

OBJECTIVE **To understand the basic postulates of the kinetic molecular theory.**

A relatively simple model that attempts to explain the behavior of an ideal gas is the **kinetic molecular theory.** This model is based on speculations about the behavior of the individual particles (atoms or molecules) in a gas. The assumptions (postulates) of the kinetic molecular theory can be stated as follows:

Postulates of the Kinetic Molecular Theory of Gases

1. Gases consist of tiny particles (atoms or molecules).
2. These particles are so small, compared with the distances between them, that the volume (size) of the individual particles can be assumed to be negligible (zero).
3. The particles are in constant random motion, colliding with the walls of the container. These collisions with the walls cause the pressure exerted by the gas.
4. The particles are assumed not to attract or to repel each other.
5. The average kinetic energy of the gas particles is directly proportional to the Kelvin temperature of the gas.

The kinetic energy referred to in postulate 5 is the energy associated with the motion of a particle. Kinetic energy (KE) is given by the equation $KE = \frac{1}{2}mv^2$, where m is the mass of the particle and v is the velocity (speed) of the particle. The greater the mass or velocity of a particle, the greater its kinetic energy. Postulate 5 means that if a gas is heated to higher temperatures, the average speed of the particles increases; therefore, their kinetic energy increases.

Although real gases do not conform exactly to the five assumptions listed here, we will see in the next section that these postulates do indeed explain *ideal* gas behavior—behavior shown by real gases at high temperatures and/or low pressures.

Critical Thinking

You have learned the postulates of the kinetic molecular theory. What if we could not assume the fourth postulate to be true? How would this affect the measured pressure of a gas?

13-9 The Implications of the Kinetic Molecular Theory

OBJECTIVES

- **To understand the term *temperature.***
- **To learn how the kinetic molecular theory explains the gas laws.**

In this section we will discuss the *qualitative* relationships between the kinetic molecular (KM) theory and the properties of gases. That is, without going into the mathematical details, we will show how the kinetic molecular theory explains some of the observed properties of gases.

The Meaning of Temperature

In Chapter 2 we introduced temperature very practically as something we measure with a thermometer. We know that as the temperature of an object increases, the object feels "hotter" to the touch. But what does temperature really mean? How does matter change when it gets "hotter"? In Chapter 10 we introduced the idea that temperature is an index of molecular motion. The kinetic molecular theory allows us to further develop this concept. As postulate 5 of the KM theory states, the temperature of a gas reflects how rapidly, on average, its individual gas particles are moving. At high temperatures the particles move very fast and hit the walls of the container frequently, whereas at low temperatures the particles' motions are more sluggish and they collide with the walls of the container much less often. Therefore, temperature really is a measure of the motions of the gas particles. In fact, the Kelvin temperature of a gas is directly proportional to the average kinetic energy of the gas particles.

The Relationship Between Pressure and Temperature

To see how the meaning of temperature given above helps to explain gas behavior, picture a gas in a rigid container. As the gas is heated to a higher temperature, the particles move faster, hitting the walls more often. And, of course, the impacts become more forceful as the particles move faster. If the pressure is due to collisions with the walls, the gas pressure should increase as temperature is increased.

Is this what we observe when we measure the pressure of a gas as it is heated? Yes. A given sample of gas in a rigid container (if the volume is not changed) shows an increase in pressure as its temperature is increased.

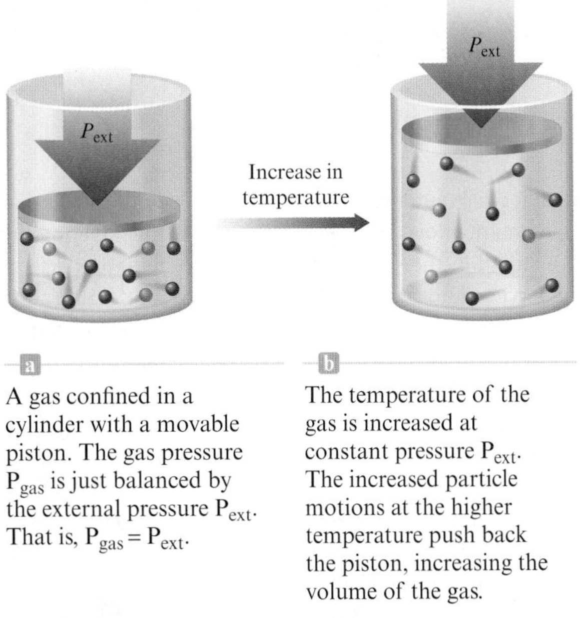

a

A gas confined in a cylinder with a movable piston. The gas pressure P_{gas} is just balanced by the external pressure P_{ext}. That is, $P_{gas} = P_{ext}$.

b

The temperature of the gas is increased at constant pressure P_{ext}. The increased particle motions at the higher temperature push back the piston, increasing the volume of the gas.

Figure 13.13

The Relationship Between Volume and Temperature

Now picture the gas in a container with a movable piston. As shown in Fig. 13.13(a), the gas pressure P_{gas} is just balanced by an external pressure P_{ext}. What happens when we heat the gas to a higher temperature? As the temperature increases, the particles move faster, causing the gas pressure to increase. As soon as the gas pressure P_{gas} becomes greater than P_{ext} (the pressure holding the piston), the piston moves up until $P_{gas} = P_{ext}$. Therefore, the KM model predicts that the volume of the gas will increase as we raise its temperature at a constant pressure [Fig. 13.13(b)]. This agrees with experimental observations (as summarized by Charles's law).

Example 13.14 Using the Kinetic Molecular Theory to Explain Gas Law Observations

Use the KM theory to predict what will happen to the pressure of a gas when its volume is decreased (n and T constant). Does this prediction agree with the experimental observations?

SOLUTION When we decrease the gas's volume (make the container smaller), the particles hit the walls more often because they do not have to travel so far between the walls. This would suggest an increase in pressure. This prediction on the basis of the model is in agreement with experimental observations of gas behavior (as summarized by Boyle's law). ■

In this section we have seen that the predictions of the kinetic molecular theory generally fit the behavior observed for gases. This makes it a useful and successful model.

CHEMISTRY IN FOCUS

The Chemistry of Air Bags

The inclusion of air bags in modern automobiles has led to a significant reduction in the number of injuries as a result of car crashes. Air bags are stored in the steering wheel and dashboard of all cars, and many autos now have additional air bags that protect the occupant's knees, head, and shoulders. In fact, some auto manufacturers now include air bags in the seat belts. Also, because deployment of an air bag can severely injure a child, all cars now have "smart" air bags that deploy with an inflation force that is proportional to the seat occupant's weight.

The term "air bag" is really a misnomer because air is not involved in the inflation process. Rather, an air bag inflates rapidly (in about 30 ms) due to the explosive production of N_2 gas. Originally, sodium azide, which decomposes to produce N_2,

$$2NaN_3(s) \rightarrow 2Na(s) + 3N_2(g)$$

was used, but it has now been replaced by less toxic materials.

The sensing devices that trigger the air bags must react very rapidly. For example, consider a car hitting a concrete bridge abutment. When this happens, an internal accelerometer sends a message to the control module that a collision possibly is occurring. The microprocessor then analyzes the measured deceleration from several accelerometers and door pressure sensors and decides whether air bag deployment is appropriate. All this happens within 8 to 40 ms of the initial impact.

Because an air bag must provide the appropriate cushioning effect, the bag begins to vent even as it is being filled. In fact, the maximum pressure in the bag is 5 pounds per square inch (psi), even in the middle of a collision event. Air bags represent a case where an explosive chemical reaction saves lives rather than the reverse.

Inflated air bags.

13-10 Gas Stoichiometry

OBJECTIVES

- **To understand the molar volume of an ideal gas.**
- **To learn the definition of STP.**
- **To use these concepts and the ideal gas equation.**

We have seen repeatedly in this chapter just how useful the ideal gas equation is. For example, if we know the pressure, volume, and temperature for a given sample of gas, we can calculate the number of moles present: $n = PV/RT$. This fact makes it possible to do stoichiometric calculations for reactions involving gases. We will illustrate this process in Example 13.15.

Interactive Example 13.15

Gas Stoichiometry: Calculating Volume

Calculate the volume of oxygen gas produced at 1.00 atm and 25 °C by the complete decomposition of 10.5 g of potassium chlorate. The balanced equation for the reaction is

$$2KClO_3(s) \rightarrow 2KCl(s) + 3O_2(g)$$

SOLUTION

Where Are We Going?

We want to determine the volume of oxygen gas collected by the decomposition of $KClO_3$.

What Do We Know?

- We know the temperature and pressure of oxygen gas.
- We know the mass of $KClO_3$.
- The balanced equation: $2KClO_3(s) \rightarrow 2KCl(s) + 3O_2(g)$.
- Ideal gas law: $PV = nRT$.

What Information Do We Need?

- $R = 0.08206$ L atm/mol K.
- We need the number of moles of oxygen gas.
- Molar mass of $KClO_3$.

How Do We Get There?

This is a stoichiometry problem very much like the type we considered in Chapter 9. The only difference is that in this case, we want to calculate the volume of a gaseous product rather than the number of grams. To do so, we can use the relationship between moles and volume given by the ideal gas law.

We'll summarize the steps required to do this problem in the following schematic:

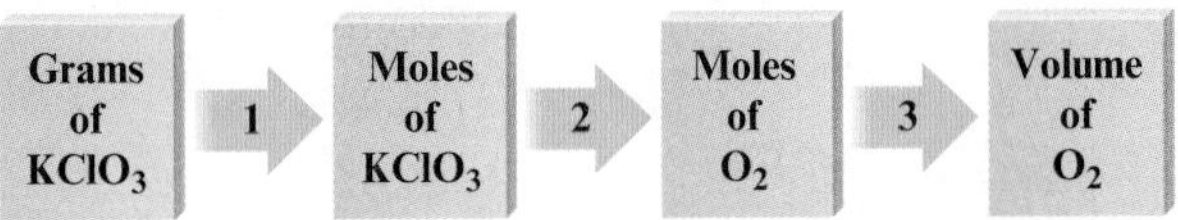

Math skill builder

$$\frac{10.5}{122.6} = 0.085644$$

0.085644 ⇨ 0.0856

Round off

$0.0856 = 8.56 \times 10^{-2}$

Step 1 To find the moles of $KClO_3$ in 10.5 g, we use the molar mass of $KClO_3$ (122.6 g).

$$10.5 \text{ g } \cancel{KClO_3} \times \frac{1 \text{ mol } KClO_3}{122.6 \text{ g } \cancel{KClO_3}} = 8.56 \times 10^{-2} \text{ mol } KClO_3$$

Step 2 To find the moles of O_2 produced, we use the mole ratio of O_2 to $KClO_3$ derived from the balanced equation.

$$8.56 \times 10^{-2} \cancel{\text{mol } KClO_3} \times \frac{3 \text{ mol } O_2}{2 \cancel{\text{mol } KClO_3}} = 1.28 \times 10^{-1} \text{ mol } O_2$$

Step 3 To find the volume of oxygen produced, we use the ideal gas law $PV = nRT$, where

$P = 1.00$ atm

$V = ?$

$n = 1.28 \times 10^{-1}$ mol, the moles of O_2 we calculated

$R = 0.08206$ L atm/K mol

$T = 25\ °C = 25 + 273 = 298$ K

Solving the ideal gas law for V gives

$$V = \frac{nRT}{P} = \frac{(1.28 \times 10^{-1} \cancel{\text{mol}})\left(0.08206 \frac{\text{L} \cancel{\text{atm}}}{\cancel{\text{K}} \cancel{\text{mol}}}\right)(298 \cancel{\text{K}})}{1.00 \cancel{\text{atm}}} = 3.13 \text{ L}$$

Thus 3.13 L of O_2 will be produced.

SELF CHECK

Exercise 13.11 Calculate the volume of hydrogen produced at 1.50 atm and 19 °C by the reaction of 26.5 g of zinc with excess hydrochloric acid according to the balanced equation

$$Zn(s) + 2HCl(aq) \rightarrow ZnCl_2(aq) + H_2(g)$$

See Problems 13.85 through 13.92. ■

In dealing with the stoichiometry of reactions involving gases, it is useful to define the volume occupied by 1 mole of a gas under certain specified conditions. For 1 mole of an ideal gas at 0 °C (273 K) and 1 atm, the volume of the gas given by the ideal gas law is

$$V = \frac{nRT}{P} = \frac{(1.00\ \cancel{mol})(0.08206\ L\ \cancel{atm}/\cancel{K}\ \cancel{mol})(273\ \cancel{K})}{1.00\ \cancel{atm}} = 22.4\ L$$

This volume of 22.4 L is called the **molar volume** of an ideal gas.

The conditions 0 °C and 1 atm are called **standard temperature and pressure (STP)**. Properties of gases are often given under these conditions. Remember, the molar volume of an ideal gas is 22.4 L *at STP*. That is, 22.4 L contains 1 mole of an ideal gas at STP.

Critical Thinking

What if STP was defined as normal room temperature (22 °C) and 1 atm? How would this affect the molar volume of an ideal gas? Include an explanation and a number.

Interactive Example 13.16

Gas Stoichiometry: Calculations Involving Gases at STP

A sample of nitrogen gas has a volume of 1.75 L at STP. How many moles of N_2 are present?

SOLUTION

Where Are We Going?

We want to determine the number of moles of nitrogen gas.

What Do We Know?

- The nitrogen gas has a volume of 1.75 L at STP.

What Information Do We Need?

- STP = 1.00 atm, 0 °C.
- At STP, 1 mole of an ideal gas occupies a volume of 22.4 L.

How Do We Get There?

We could solve this problem by using the ideal gas equation, but we can take a shortcut by using the molar volume of an ideal gas at STP. Because 1 mole of an ideal gas at STP has a volume of 22.4 L, a 1.75-L sample of N_2 at STP contains considerably less than 1 mole. We can find how many moles by using the equivalence statement

$$1.000\ mol = 22.4\ L\ (STP)$$

which leads to the conversion factor we need:

$$1.75\ \cancel{L\ N_2} \times \frac{1.000\ mol\ N_2}{22.4\ \cancel{L\ N_2}} = 7.81 \times 10^{-2}\ mol\ N_2$$

SELF CHECK

Exercise 13.12 Ammonia is commonly used as a fertilizer to provide a source of nitrogen for plants. A sample of $NH_3(g)$ occupies a volume of 5.00 L at 25 °C and 15.0 atm. What volume will this sample occupy at STP?

See Problems 13.95 through 13.98. ■

Standard conditions (STP) and molar volume are also useful in carrying out stoichiometric calculations on reactions involving gases, as shown in Example 13.17.

Interactive Example 13.17 Gas Stoichiometry: Reactions Involving Gases at STP

Quicklime, CaO, is produced by heating calcium carbonate, $CaCO_3$. Calculate the volume of CO_2 produced at STP from the decomposition of 152 g of $CaCO_3$ according to the reaction

$$CaCO_3(s) \rightarrow CaO(s) + CO_2(g)$$

SOLUTION

Where Are We Going?

We want to determine the volume of carbon dioxide produced from 152 g of $CaCO_3$.

What Do We Know?

- We know the temperature and pressure of carbon dioxide gas (STP).
- We know the mass of $CaCO_3$.
- The balanced equation: $CaCO_3(s) \rightarrow CaO(s) + CO_2(g)$.

What Information Do We Need?

- STP = 1.00 atm, 0 °C.
- At STP, 1 mole of an ideal gas occupies a volume of 22.4 L.
- We need the number of moles of carbon dioxide gas.
- Molar mass of $CaCO_3$.

How Do We Get There?

The strategy for solving this problem is summarized by the following schematic:

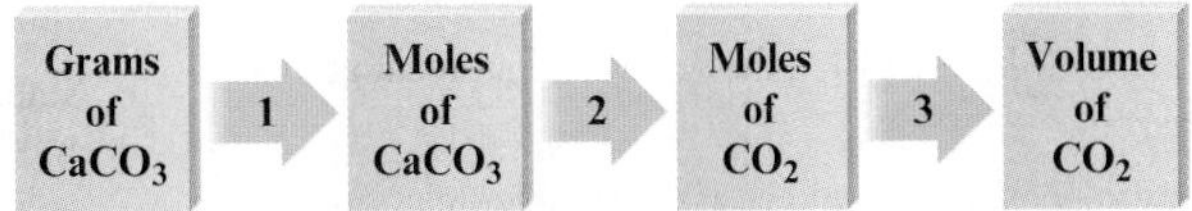

Step 1 Using the molar mass of $CaCO_3$ (100.1 g/mol), we calculate the number of moles of $CaCO_3$.

$$152 \text{ g } \cancel{CaCO_3} \times \frac{1 \text{ mol } CaCO_3}{100.1 \text{ g } \cancel{CaCO_3}} = 1.52 \text{ mol } CaCO_3$$

Step 2 Each mole of $CaCO_3$ produces 1 mole of CO_2, so 1.52 moles of CO_2 will be formed.

Step 3 We can convert the moles of CO_2 to volume by using the molar volume of an ideal gas, because the conditions are STP.

$$1.52 \text{ } \cancel{\text{mol } CO_2} \times \frac{22.4 \text{ L } CO_2}{1 \text{ } \cancel{\text{mol } CO_2}} = 34.1 \text{ L } CO_2$$

Thus the decomposition of 152 g of $CaCO_3$ produces 34.1 L of CO_2 at STP. ■

Note that the final step in Example 13.17 involves calculating the volume of gas from the number of moles. Because the conditions were specified as STP, we were able to use the molar volume of a gas at STP. If the conditions of a problem are different from STP, we must use the ideal gas law to compute the volume, as we did in Section 13-5.

CHAPTER 13 REVIEW

F directs you to the *Chemistry in Focus* feature in the chapter

Key Terms

barometer (13-1)
millimeters of mercury (mm Hg) (13-1)
torr (13-1)
standard atmosphere (13-1)
pascal (13-1)
Boyle's law (13-2)
absolute zero (13-3)
Charles's law (13-3)
Avogadro's law (13-4)
universal gas constant (13-5)
ideal gas law (13-5)
ideal gas (13-5)
combined gas law (13-5)
partial pressure (13-6)
Dalton's law of partial pressures (13-6)
kinetic molecular theory (13-8)
molar volume (13-10)
standard temperature and pressure (STP) (13-10)

For Review

- The common units for pressure are mm Hg (torr), atmosphere (atm), and pascal (Pa). The SI unit is the pascal.
- Boyle's law states that the volume of a given amount of gas at constant temperature varies inversely to its pressure. $PV = k$
- Charles's law states that the volume of a given amount of an ideal gas at constant pressure varies directly with its temperature (in kelvins). $V = bT$
    - At absolute zero (0 K; −273°C), the volume of an ideal gas extrapolates to zero.
- Avogadro's law states for an ideal gas at constant temperature and pressure, the volume varies directly with the number of moles of gas (n). $V = an$
- The ideal gas law describes the relationship among P, V, n, and T for an ideal gas. $PV = nRT$

$$R = 0.08206 \frac{\text{L atm}}{\text{K mol}}$$

  A gas that obeys this law exactly is called an ideal gas.
- From the ideal gas law we can obtain the combined gas law, which applies when n is constant.

$$\frac{P_1V_1}{T_1} = \frac{P_2V_2}{T_2}$$

- Dalton's law of partial pressures states the total pressure of a mixture of gases is equal to the sum of the individual (partial) pressures of the gases. $P_{total} = P_1 + P_2 + P_3 + \ldots$
- The kinetic molecular theory is a model based on the properties of individual gas components that explains the relationship of P, V, T, and n for an ideal gas.
- A law is a summary of experimental observation.
- A model (theory) is an attempt to explain observed behavior.
- The temperature of an ideal gas reflects the average kinetic energy of the gas particles.
- The pressure of a gas increases as its temperature increases because the gas particles speed up.
- The volume of a gas must increase because the gas particles speed up as a gas is heated to a higher temperature.
- Standard temperature and pressure (STP) is defined as $P = 1$ atm and $T = 273$ K (0°C).
- The volume of 1 mole of an ideal gas (the molar volume) is 22.4 L at STP.

All even-numbered Questions and Problems have answers in the back of this book and solutions in the *Student Solutions Guide*.

Active Learning Questions

These questions are designed to be considered by groups of students in class. Often these questions work well for introducing a particular topic in class.

1. As you increase the temperature of a gas in a sealed, rigid container, what happens to the density of the gas? Would the results be the same if you did the same experiment in a container with a movable piston at a constant external pressure? Explain.
2. A diagram in a chemistry book shows a magnified view of a flask of air.

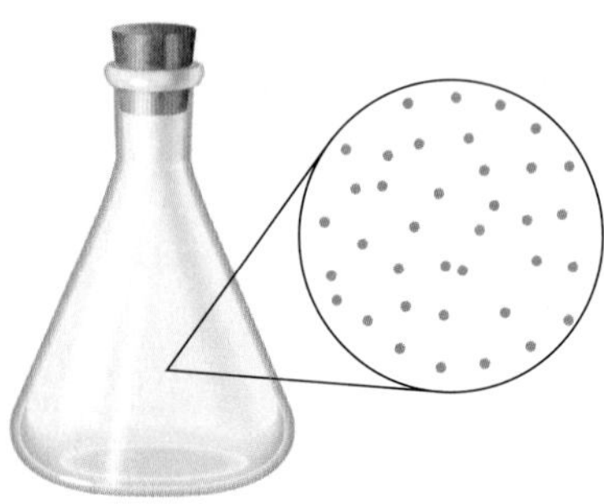

All even-numbered Questions and Problems have answers in the back of this book and solutions in the Student Solutions Guide.

What do you suppose is between the dots (which represent air molecules)?

a. air
b. dust
c. pollutants
d. oxygen
e. nothing

3. If you put a drinking straw in water, place your finger over the opening, and lift the straw out of the water, some water stays in the straw. Explain.

4. A chemistry student relates the following story: I noticed my tires were a bit low and went to the gas station. As I was filling the tires I thought about the kinetic molecular (KM) theory. I noticed the tires because the volume was low, and I realized that I was increasing both the pressure and volume of the tires. "Hmmm," I thought, "that goes against what I learned in chemistry, where I was told pressure and volume are inversely proportional." What is the fault of the logic of the chemistry student in this situation? Explain under what conditions pressure and volume are inversely related (draw pictures and use the KM theory).

5. Chemicals X and Y (both gases) react to form the gas XY, but it takes some time for the reaction to occur. Both X and Y are placed in a container with a piston (free to move), and you note the volume. As the reaction occurs, what happens to the volume of the container? Explain your answer.

6. Which statement best explains why a hot-air balloon rises when the air in the balloon is heated?

   a. According to Charles's law, the temperature of a gas is directly related to its volume. Thus the volume of the balloon increases, decreasing the density.
   b. Hot air rises inside the balloon, which lifts the balloon.
   c. The temperature of a gas is directly related to its pressure. The pressure therefore increases, which lifts the balloon.
   d. Some of the gas escapes from the bottom of the balloon, thus decreasing the mass of gas in the balloon. This decreases the density of the gas in the balloon, which lifts the balloon.
   e. Temperature is related to the velocity of the gas molecules. Thus the molecules are moving faster, hitting the balloon more, and lifting the balloon.

   For choices you did not pick, explain what you feel is wrong with them, and justify the choice you did pick.

7. If you release a helium balloon, it soars upward and eventually pops. Explain this behavior.

8. If you have any two gases in different containers that are the same size at the same pressure and same temperature, what is true about the moles of each gas? Why is this true?

9. Using postulates of the kinetic molecular theory, give a molecular interpretation of Boyle's law, Charles's law, and Dalton's law of partial pressures.

10. Rationalize the following observations.

    a. Aerosol cans will explode if heated.
    b. You can drink through a soda straw.
    c. A thin-walled can will collapse when the air inside is removed by a vacuum pump.
    d. Manufacturers produce different types of tennis balls for high and low altitudes.

11. Show how Boyle's law and Charles's law are special cases of the ideal gas law.

12. Look at the demonstration discussed in Fig. 13.1. How would this demonstration change if water was not added to the can? Explain.

13. How does Dalton's law of partial pressures help us with our model of ideal gases? That is, which postulates of the kinetic molecular theory does it support?

14. Draw molecular-level views that show the differences among solids, liquids, and gases.

15. Explain how increasing the number of moles of gas affects the pressure (assuming constant volume and temperature).

16. Explain how increasing the number of moles of gas affects the volume (assuming constant pressure and temperature).

17. Gases are said to exert pressure. Provide a molecular-level explanation for this.

18. Why is it incorrect to say that a sample of helium at 50 °C is twice as hot as a sample of helium at 25 °C?

19. We can use different units for pressure or volume, but we must use units of Kelvin for temperature. Why must we use the Kelvin temperature scale?

20. Estimate the mass of air at normal conditions that takes up the volume of your head. Provide support for your answer.

21. You are holding two balloons of the same volume. One balloon contains 1.0 g of helium. The other balloon contains neon. Calculate the mass of neon in the balloon.

22. You have helium gas in a two-bulbed container connected by a valve as shown below. Initially the valve is closed.

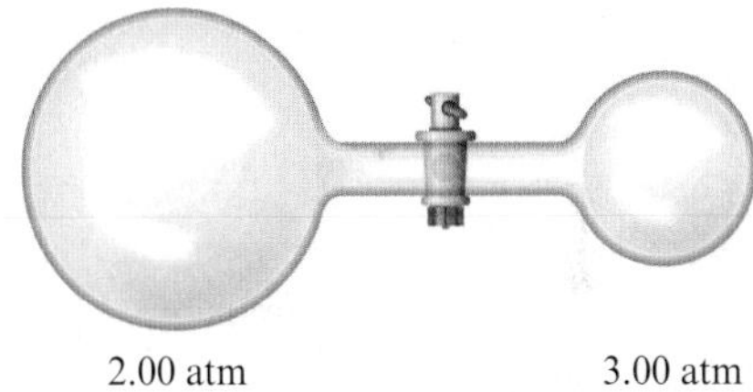

    a. When the valve is opened, will the total pressure in the apparatus be less than 5.00 atm, equal to 5.00 atm, or greater than 5.00 atm? Explain your answer.
    b. The left bulb has a volume of 9.00 L, and the right bulb has a volume of 3.00 L. Calculate the final pressure after the valve is opened.

23. Use the graphs below to answer the following questions.

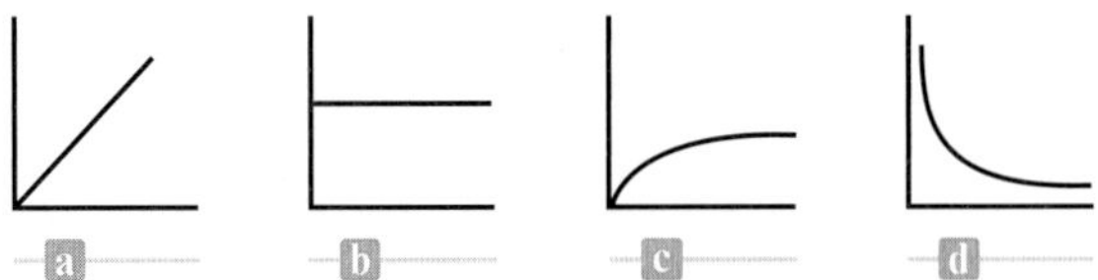

    a. Which of the above graphs best represents the relationship between the pressure and temperature (measured in kelvins) of 1 mole of an ideal gas?
    b. Which of the above graphs best represents the relationship between the pressure and volume of 1 mole of an ideal gas?
    c. Which of the above graphs best represents the relationship between the volume and temperature (measured in kelvins) of 1 mole of an ideal gas.

Questions and Problems

13-1 Pressure

Questions

1. The introduction to this chapter says that "we live immersed in a gaseous solution." What does that mean?
2. How are the three states of matter similar, and how do they differ?
3. Figure 13.1 shows an experiment that can be used effectively to demonstrate the pressure exerted by the atmosphere. Write an explanation of this experiment to a friend who has not yet taken any science courses to help him understand the concept of atmospheric pressure.
4. Describe a simple mercury barometer. How is such a barometer used to measure the pressure of the atmosphere?
5. If two gases that do not react with each other are placed in the same container, they will ________ completely with each other.
6. What are the common units used to measure pressure? Which unit is an experimental unit derived from the device used to measure atmospheric pressure?

Problems

7. Make the indicated pressure conversions.
   a. 45.2 kPa to atmospheres
   b. 755 mm Hg to atmospheres
   c. 802 torr to kilopascals
   d. 1.04 atm to millimeters of mercury
8. Make the indicated pressure conversions.
   a. 14.9 psi to atmospheres
   b. 795 torr to atmospheres
   c. 743 mm Hg to kilopascals
   d. 99,436 Pa to kilopascals
9. Make the indicated pressure conversions.
   a. 699 mm Hg to atmospheres
   b. 18.2 psi to mm Hg
   c. 862 mm Hg to torr
   d. 795 mm Hg to psi
10. Make the indicated pressure conversions.
   a. 17.3 psi to kilopascals
   b. 1.15 atm to psi
   c. 4.25 atm to mm Hg
   d. 224 psi to atmospheres
11. Make the indicated pressure conversions.
   a. 1.54×10^5 Pa to atmospheres
   b. 1.21 atm to pascals
   c. 97,345 Pa to mm Hg
   d. 1.32 kPa to pascals
12. Convert the following pressures into pascals.
   a. 774 torr
   b. 0.965 atm
   c. 112.5 kPa
   d. 801 mm Hg

13-2 Pressure and Volume: Boyle's Law

Questions

13. Pretend that you're talking to a friend who has not yet taken any science courses, and describe how you would explain Boyle's law to her.
14. In Fig. 13.4, when additional mercury is added to the right-hand arm of the J-shaped tube, the volume of the gas trapped above the mercury in the left-hand arm of the J-tube decreases. Explain.
15. The volume of a sample of ideal gas is inversely proportional to the ________ of the gas at constant temperature.
16. A mathematical expression that summarizes Boyle's law is ________.

Problems

17. For each of the following sets of pressure/volume data, calculate the new volume of the gas sample after the pressure change is made. Assume that the temperature and the amount of gas remain the same.
   a. $V = 125$ mL at 755 mm Hg; $V = ?$ mL at 780 mm Hg
   b. $V = 223$ mL at 1.08 atm; $V = ?$ mL at 0.951 atm
   c. $V = 3.02$ L at 103 kPa; $V = ?$ L at 121 kPa
18. For each of the following sets of pressure/volume data, calculate the new volume of the gas sample after the pressure change is made. Assume that the temperature and the amount of gas remain the same.
   a. $V = 375$ mL at 1.15 atm; $V = ?$ mL at 775 mm Hg
   b. $V = 195$ mL at 1.08 atm; $V = ?$ mL at 135 kPa
   c. $V = 6.75$ L at 131 kPa; $V = ?$ L at 765 mm Hg
19. For each of the following sets of pressure/volume data, calculate the missing quantity. Assume that the temperature and the amount of gas remain constant.
   a. $V = 19.3$ L at 102.1 kPa; $V = 10.0$ L at ? kPa
   b. $V = 25.7$ mL at 755 torr; $V = ?$ at 761 mm Hg
   c. $V = 51.2$ L at 1.05 atm; $V = ?$ at 112.2 kPa
20. For each of the following sets of pressure/volume data, calculate the missing quantity. Assume that the temperature and the amount of gas remain constant.
   a. $V = 53.2$ mL at 785 mm Hg; $V = ?$ mL at 700. mm Hg
   b. $V = 2.25$ L at 1.67 atm; $V = 2.00$ L at ? atm
   c. $V = 5.62$ L at 695 mm Hg; $V = ?$ L at 1.51 atm
21. What volume of gas would result if 225 mL of neon gas is compressed from 1.02 atm to 2.99 atm at constant temperature?
22. If the pressure on a 1.04-L sample of gas is doubled at constant temperature, what will be the new volume of the gas?
23. A sample of helium gas with a volume of 29.2 mL at 785 mm Hg is compressed at constant temperature until its volume is 15.1 mL. What will be the new pressure in the sample?
24. What pressure (in atmospheres) is required to compress 1.00 L of gas at 760. mm Hg pressure to a volume of 50.0 mL?

All even-numbered Questions and Problems have answers in the back of this book and solutions in the *Student Solutions Guide.*

13-3 Volume and Temperature: Charles's Law

Questions

25. Pretend that you're talking to a friend who has not yet taken any science courses, and describe how you would explain the concept of absolute zero to him.
26. Figures 13.7 and 13.8 show volume/temperature data for several samples of gases. Why do all the lines seem to extrapolate to the same point at −273 °C? Explain.
27. The volume of a sample of ideal gas is ________ proportional to its temperature (K) at constant pressure.
28. A mathematical expression that summarizes Charles's law is ________.

Problems

29. A sample of gas in a balloon has an initial temperature of 18 °C and a volume of 1340 L. If the temperature changes to 87 °C and there is no overall change of pressure or amount of gas, what is the new volume of the gas?
30. Suppose a 375-mL sample of neon gas at 78 °C is cooled to 22 °C at constant pressure. What will be the new volume of the neon sample?
31. For each of the following sets of volume/temperature data, calculate the missing quantity after the change is made. Assume that the pressure and the amount of gas remain the same.
    a. V = 2.03 L at 24 °C; V = 3.01 L at ? °C
    b. V = 127 mL at 273 K; V = ? mL at 373 K
    c. V = 49.7 mL at 34 °C; V = ? at 350 K
32. For each of the following sets of volume/temperature data, calculate the missing quantity. Assume that the pressure and the mass of gas remain constant.
    a. V = 25.0 L at 0 °C; V = 50.0 L at ? °C
    b. V = 247 mL at 25 °C; V = 255 mL at ? °C
    c. V = 1.00 mL at 2272 °C; V = ? at 25 °C
33. For each of the following sets of volume/temperature data, calculate the missing quantity after the change is made. Assume that the pressure and the amount of gas remain the same.
    a. V = 9.14 L at 24 °C; V = ? at 48 °C
    b. V = 24.9 mL at –12 °C; V = 49.9 mL at ? °C
    c. V = 925 mL at 25 K; V = ? at 273 K
34. For each of the following sets of volume/temperature data, calculate the missing quantity. Assume that the pressure and the mass of gas remain constant.
    a. $V = 2.01 \times 10^2$ L at 1150 °C; V = 5.00 L at ? °C
    b. V = 44.2 mL at 298 K; V = ? at 0 K
    c. V = 44.2 mL at 298 K; V = ? at 0 °C
35. Suppose 1.25 L of argon is cooled from 291 K to 78 K. What will be the new volume of the argon sample?
36. Suppose a 125-mL sample of argon is cooled from 450 K to 250 K at constant pressure. What will be the volume of the sample at the lower temperature?
37. If a 375-mL sample of neon gas is heated from 24 °C to 72 °C at constant pressure, what will be the volume of the sample at the higher temperature?
38. A sample of gas has a volume of 127 mL in a boiling water bath at 100 °C. Calculate the volume of the sample of gas at 10 °C intervals after the heat source is turned off and the gas sample begins to cool down to the temperature of the laboratory, 20 °C.

13-4 Volume and Moles: Avogadro's Law

Questions

39. At conditions of constant temperature and pressure, the volume of a sample of ideal gas is ________ proportional to the number of moles of gas present.
40. A mathematical expression that summarizes Avogadro's law is ________.

Problems

41. If 0.00901 mole of neon gas at a particular temperature and pressure occupies a volume of 242 mL, what volume would 0.00703 mole of neon occupy under the same conditions?
42. If 1.04 g of chlorine gas occupies a volume of 872 mL at a particular temperature and pressure, what volume will 2.08 g of chlorine gas occupy under the same conditions?
43. If 3.25 moles of argon gas occupies a volume of 100. L at a particular temperature and pressure, what volume does 14.15 moles of argon occupy under the same conditions?
44. If 2.71 g of argon gas occupies a volume of 4.21 L, what volume will 1.29 moles of argon occupy under the same conditions?

13-5 The Ideal Gas Law

Questions

45. What do we mean by an *ideal* gas?
46. Under what conditions do *real* gases behave most ideally?
47. Show how Boyle's gas law can be derived from the ideal gas law.
48. Show how Charles's gas law can be derived from the ideal gas law.

Problems

49. Given the following sets of values for three of the gas variables, calculate the unknown quantity.
    a. P = 782.4 mm Hg; V = ?; n = 0.1021 mol; T = 26.2 °C
    b. P = ? mm Hg; V = 27.5 mL; n = 0.007812 mol; T = 16.6 °C
    c. P = 1.045 atm; V = 45.2 mL; n = 0.002241 mol; T = ? °C
50. Given each of the following sets of values for an ideal gas, calculate the unknown quantity.
    a. P = 782 mm Hg; V = ?; n = 0.210 mol; T = 27 °C
    b. P = ? mm Hg; V = 644 mL; n = 0.0921 mol; T = 303 K
    c. P = 745 mm Hg; V = 11.2 L; n = 0.401 mol; T = ? K
51. What mass of neon gas is required to fill a 5.00-L container to a pressure of 1.02 atm at 25 °C?
52. Determine the pressure in a 125-L tank containing 56.2 kg of oxygen gas at 21 °C.
53. What volume will 2.04 g of helium gas occupy at 100. °C and 785 mm Hg pressure?

All even-numbered Questions and Problems have answers in the back of this book and solutions in the *Student Solutions Guide*.

54. At what temperature (in °C) will a 5.00-g sample of neon gas exert a pressure of 1.10 atm in a 7.00-L container?

55. What mass of helium gas is needed to pressurize a 100.0-L tank to 255 atm at 25 °C? What mass of oxygen gas would be needed to pressurize a similar tank to the same specifications?

56. Suppose that a 1.25-g sample of neon gas is confined in a 10.1-L container at 25 °C. What will be the pressure in the container? Suppose the temperature is then raised to 50 °C. What will the new pressure be after the temperature is increased?

57. At what temperature will a 1.0-g sample of neon gas exert a pressure of 500. torr in a 5.0-L container?

58. At what temperature would 4.25 g of oxygen gas, O_2, exert a pressure of 784 mm Hg in a 2.51-L container?

59. What pressure exists in a 200-L tank containing 5.0 kg of neon gas at 300. K?

60. Which flask will have the higher pressure: a 5.00-L flask containing 4.15 g of helium at 298 K, or a 10.0-L flask containing 56.2 g of argon at 303 K?

61. Suppose a 24.3-mL sample of helium gas at 25 °C and 1.01 atm is heated to 50. °C and compressed to a volume of 15.2 mL. What will be the pressure of the sample?

62. Suppose that 1.29 g of argon gas is confined to a volume of 2.41 L at 29 °C. What would be the pressure in the container? What would the pressure become if the temperature were raised to 42 °C without a change in volume?

63. What will the volume of the sample become if 459 mL of an ideal gas at 27 °C and 1.05 atm is cooled to 15 °C and 0.997 atm?

F 64. The "Chemistry in Focus" segment *Snacks Need Chemistry, Too!* discusses why popcorn "pops." You can estimate the pressure inside a kernel of popcorn at the time of popping by using the ideal gas law. Basically, you determine the mass of water released when the popcorn pops by measuring the mass of the popcorn both before and after popping. Assume that the difference in mass is the mass of water vapor lost on popping. Assume that the popcorn pops at the temperature of the cooking oil (225 °C) and that the volume of the "container" is the volume of the unpopped kernel. Although we are making several assumptions, we can at least get some idea of the magnitude of the pressure inside the kernel.

Assuming a total volume of 2.0 mL for 20 kernels and a mass of 0.250 g of water lost from them on popping, calculate the pressure inside the kernels just before they "pop."

13-6 Dalton's Law of Partial Pressures

Questions

65. Explain why the measured properties of a mixture of gases depend only on the total number of moles of particles, not on the identity of the individual gas particles. How is this observation summarized as a law?

66. We often collect small samples of gases in the laboratory by bubbling the gas into a bottle or flask containing water. Explain why the gas becomes saturated with water vapor and how we must take the presence of water vapor into account when calculating the properties of the gas sample.

Problems

67. If a gaseous mixture is made of 2.41 g of He and 2.79 g of Ne in an evacuated 1.04-L container at 25 °C, what will be the partial pressure of each gas and the total pressure in the container?

68. Suppose that 1.28 g of neon gas and 2.49 g of argon gas are confined in a 9.87-L container at 27 °C. What would be the pressure in the container?

69. A tank contains a mixture of 52.5 g of oxygen gas and 65.1 g of carbon dioxide gas at 27 °C. The total pressure in the tank is 9.21 atm. Calculate the partial pressure (in atm) of each gas in the mixture.

70. What mass of neon gas would be required to fill a 3.00-L flask to a pressure of 925 mm Hg at 26 °C? What mass of argon gas would be required to fill a similar flask to the same pressure at the same temperature?

71. A sample of oxygen gas is saturated with water vapor at 27 °C. The total pressure of the mixture is 772 torr, and the vapor pressure of water is 26.7 torr at 27 °C. What is the partial pressure of the oxygen gas?

72. Determine the partial pressure of each gas as shown in this figure. *Note:* The relative numbers of each type of gas are depicted in the figure.

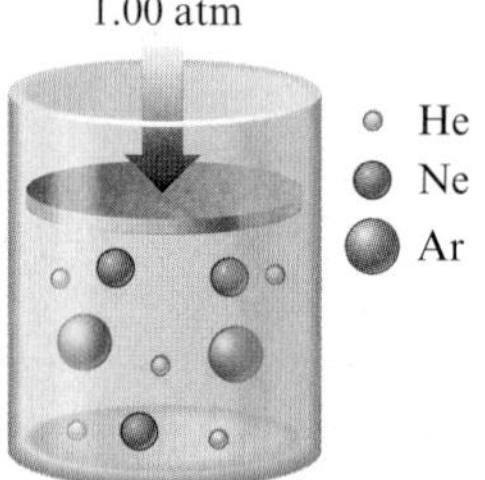

73. A 500.-mL sample of O_2 gas at 24 °C was prepared by decomposing a 3% aqueous solution of hydrogen peroxide, H_2O_2, in the presence of a small amount of manganese catalyst by the reaction

$$2H_2O_2(aq) \rightarrow 2H_2O(g) + O_2(g)$$

The oxygen thus prepared was collected by displacement of water. The total pressure of gas collected was 755 mm Hg. What is the partial pressure of O_2 in the mixture? How many moles of O_2 are in the mixture? (The vapor pressure of water at 24 °C is 23 mm Hg.)

74. Small quantities of hydrogen gas can be prepared in the laboratory by the addition of aqueous hydrochloric acid to metallic zinc.

$$Zn(s) + 2HCl(aq) \rightarrow ZnCl_2(aq) + H_2(g)$$

Typically, the hydrogen gas is bubbled through water for collection and becomes saturated with water vapor. Suppose 240. mL of hydrogen gas is collected at 30. °C and has a total pressure of 1.032 atm by this process. What is the partial pressure of hydrogen gas in the sample? How many moles of hydrogen gas are present in the sample? How many grams of zinc must have reacted to produce this quantity of hydrogen? (The vapor pressure of water is 32 torr at 30 °C.)

All even-numbered Questions and Problems have answers in the back of this book and solutions in the *Student Solutions Guide*.

13-7 Laws and Models: A Review

Questions

75. What is a scientific *law?* What is a *theory?* How do these concepts differ? Does a law explain a theory, or does a theory attempt to explain a law?

76. When is a scientific theory considered to be successful? Are all theories successful? Will a theory that has been successful in the past necessarily be successful in the future?

13-8 The Kinetic Molecular Theory of Gases

Questions

77. What do we assume about the volume of the actual molecules themselves in a sample of gas, compared to the bulk volume of the gas overall? Why?

78. Collisions of the molecules in a sample of gas with the walls of the container are responsible for the gas's observed ________.

79. Temperature is a measure of the average ________ of the molecules in a sample of gas.

80. The kinetic molecular theory of gases suggests that gas particles exert ________ attractive or repulsive forces on each other.

13-9 The Implications of the Kinetic Molecular Theory

Questions

81. How is the phenomenon of temperature explained on the basis of the kinetic molecular theory? What microscopic property of gas molecules is reflected in the temperature measured?

82. Explain, in terms of the kinetic molecular theory, how an increase in the temperature of a gas confined to a rigid container causes an increase in the pressure of the gas.

13-10 Gas Stoichiometry

Questions

83. What is the *molar volume* of a gas? Do all gases that behave ideally have the same molar volume?

84. What conditions are considered "standard temperature and pressure" (STP) for gases? Suggest a reason why these particular conditions might have been chosen for STP.

Problems

85. Calcium oxide can be used to "scrub" carbon dioxide from air.

$$CaO(s) + CO_2(g) \rightarrow CaCO_3(s)$$

What mass of CO_2 could be absorbed by 1.25 g of CaO? What volume would this CO_2 occupy at STP?

86. Consider the following reaction:

$$C(s) + O_2(g) \rightarrow CO_2(g)$$

What volume of oxygen gas at 25 °C and 1.02 atm would be required to react completely with 1.25 g of carbon?

87. Consider the following reaction for the combustion of octane, C_8H_{18}:

$$2C_8H_{18}(l) + 25O_2(g) \rightarrow 16CO_2(g) + 18H_2O(l)$$

What volume of oxygen gas at STP would be needed for the complete combustion of 10.0 g of octane?

88. Although we generally think of combustion reactions as involving oxygen gas, other rapid oxidation reactions are also referred to as combustions. For example, if magnesium metal is placed into chlorine gas, a rapid oxidation takes place, and magnesium chloride is produced.

$$Mg(s) + Cl_2(g) \rightarrow MgCl_2(s)$$

What volume of chlorine gas, measured at STP, is required to react completely with 1.02 g of magnesium?

89. Ammonia and gaseous hydrogen chloride combine to form ammonium chloride.

$$NH_3(g) + HCl(g) \rightarrow NH_4Cl(s)$$

If 4.21 L of $NH_3(g)$ at 27 °C and 1.02 atm is combined with 5.35 L of $HCl(g)$ at 26 °C and 0.998 atm, what mass of $NH_4Cl(s)$ will be produced? Which gas is the limiting reactant? Which gas is present in excess?

90. Calcium carbide, CaC_2, reacts with water to produce acetylene gas, C_2H_2.

$$CaC_2(s) + 2H_2O(l) \rightarrow C_2H_2(g) + Ca(OH)_2(s)$$

What volume of acetylene at 25 °C and 1.01 atm is generated by the complete reaction of 2.49 g of calcium carbide? What volume would this quantity of acetylene occupy at STP?

91. Many transition metal salts are hydrates: they contain a fixed number of water molecules bound per formula unit of the salt. For example, copper(II) sulfate most commonly exists as the pentahydrate, $CuSO_4 \cdot 5H_2O$. If 5.00 g of $CuSO_4 \cdot 5H_2O$ is heated strongly so as to drive off all of the waters of hydration as water vapor, what volume will this water vapor occupy at 350. °C and a pressure of 1.04 atm?

92. If water is added to magnesium nitride, ammonia gas is produced when the mixture is heated.

$$Mg_3N_2(s) + 3H_2O(l) \rightarrow 3MgO(s) + 2NH_3(g)$$

If 10.3 g of magnesium nitride is treated with water, what volume of ammonia gas would be collected at 24 °C and 752 mm Hg?

93. What volume does a mixture of 14.2 g of He and 21.6 g of H_2 occupy at 28 °C and 0.985 atm?

94. A 10.0-g sample of sodium metal reacts with 2.50 L of nitrogen gas at 0.976 atm and 28 °C to produce sodium nitride.
    a. Write a balanced chemical equation for this reaction. Do not include phases.
    b. How many grams of sodium nitride are produced in this reaction?

95. A sample of helium gas occupies a volume of 25.2 mL at 95 °C and a pressure of 892 mm Hg. Calculate the volume of the gas at STP.

96. The volume of a gas-filled balloon is 50.0 L at 20. °C and 742 torr. What volume will it occupy at STP?

97. A mixture contains 5.00 g *each* of O_2, N_2, CO_2, and Ne gas. Calculate the volume of this mixture at STP. Calculate the partial pressure of each gas in the mixture at STP.

98. A gaseous mixture contains 6.25 g of He and 4.97 g of Ne. What volume does the mixture occupy at STP? Calculate the partial pressure of each gas in the mixture at STP.

99. Consider the following *unbalanced* chemical equation for the combination reaction of sodium metal and chlorine gas:

$$Na(s) + Cl_2(g) \rightarrow NaCl(s)$$

What volume of chlorine gas, measured at STP, is necessary for the complete reaction of 4.81 g of sodium metal?

100. Welders commonly use an apparatus that contains a tank of acetylene (C_2H_2) gas and a tank of oxygen gas. When burned in pure oxygen, acetylene generates a large amount of heat.

$$2C_2H_2(g) + 5O_2(g) \rightarrow 2H_2O(g) + 4CO_2(g)$$

What volume of carbon dioxide gas at STP is produced if 1.00 g of acetylene is combusted completely?

101. During the making of steel, iron(II) oxide is reduced to metallic iron by treatment with carbon monoxide gas.

$$FeO(s) + CO(g) \rightarrow Fe(s) + CO_2(g)$$

Suppose 1.45 kg of Fe reacts. What volume of $CO(g)$ is required, and what volume of $CO_2(g)$ is produced, each measured at STP?

102. Consider the following reaction:

$$Zn(s) + 2HCl(aq) \rightarrow ZnCl_2(aq) + H_2(g)$$

What mass of zinc metal should be taken so as to produce 125 mL of H_2 measured at STP when reacted with excess hydrochloric acid?

Additional Problems

103. When doing any calculation involving gas samples, we must express the temperature in terms of the ________ temperature scale.

104. Two moles of ideal gas occupy a volume that is ________ the volume of 1 mole of ideal gas under the same temperature and pressure conditions.

105. Summarize the postulates of the kinetic molecular theory for gases. How does the kinetic molecular theory account for the observed properties of temperature and pressure?

106. Consider the flasks in the following diagrams.

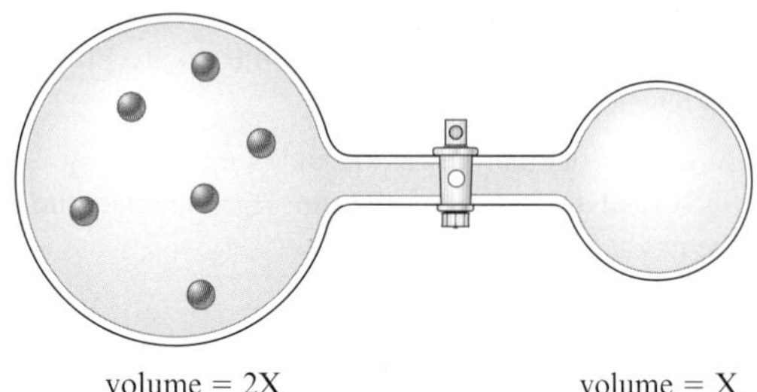

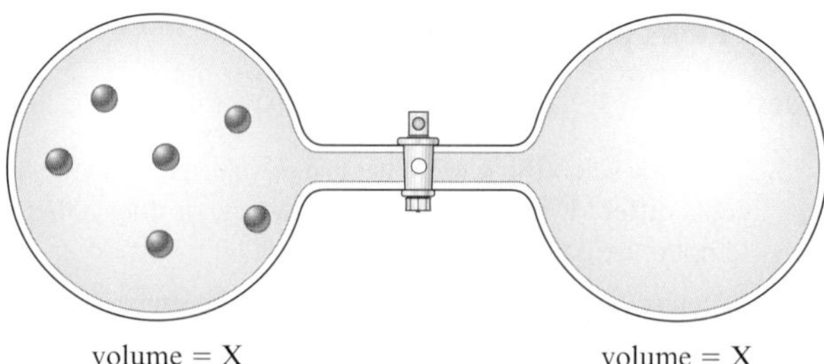

Assuming the connecting tube has negligible volume, draw what each diagram will look like after the stopcock between the two flasks is opened. Also, solve for the final pressure in each case, in terms of the original pressure. Assume temperature is constant.

107. For a mixture of gases in the same container, the total pressure exerted by the mixture of gases is the ________ of the pressures that those gases would exert if they were alone in the container under the same conditions.

108. A helium tank contains 25.2 L of helium at 8.40 atm pressure. Determine how many 1.50-L balloons at 755 mm Hg can be inflated with the gas in the tank, assuming that the tank will also have to contain He at 755 mm Hg after the balloons are filled (that is, it is not possible to empty the tank completely). The temperature is 25 °C in all cases.

109. As weather balloons rise from the earth's surface, the pressure of the atmosphere becomes less, tending to cause the volume of the balloons to expand. However, the temperature is much lower in the upper atmosphere than at sea level. Would this temperature effect tend to make such a balloon expand or contract? Weather balloons do, in fact, expand as they rise. What does this tell you?

110. When ammonium carbonate is heated, three gases are produced by its decomposition.

$$(NH_4)_2CO_3(s) \rightarrow 2NH_3(g) + CO_2(g) + H_2O(g)$$

What total volume of gas is produced, measured at 453 °C and 1.04 atm, if 52.0 g of ammonium carbonate is heated?

111. Carbon dioxide gas, in the dry state, may be produced by heating calcium carbonate.

$$CaCO_3(s) \rightarrow CaO(s) + CO_2(g)$$

What volume of CO_2, collected dry at 55 °C and a pressure of 774 torr, is produced by complete thermal decomposition of 10.0 g of $CaCO_3$?

112. Carbon dioxide gas, saturated with water vapor, can be produced by the addition of aqueous acid to calcium carbonate.

$$CaCO_3(s) + 2H^+(aq) \rightarrow Ca^{2+}(aq) + H_2O(l) + CO_2(g)$$

How many moles of $CO_2(g)$, collected at 60. °C and 774 torr total pressure, are produced by the complete reaction of 10.0 g of $CaCO_3$ with acid? What volume does this wet CO_2 occupy? What volume would the CO_2 occupy at 774 torr if a desiccant (a chemical drying agent) were added to remove the water? (The vapor pressure of water at 60. °C is 149.4 mm Hg.)

All even-numbered Questions and Problems have answers in the back of this book and solutions in the *Student Solutions Guide*.

113. Sulfur trioxide, SO_3, is produced in enormous quantities each year for use in the synthesis of sulfuric acid.

$$S(s) + O_2(g) \rightarrow SO_2(g)$$
$$2SO_2(g) + O_2(g) \rightarrow 2SO_3(g)$$

What volume of $O_2(g)$ at 350. °C and a pressure of 5.25 atm is needed to completely convert 5.00 g of sulfur to sulfur trioxide?

114. If you have any two gases in different, rigid containers that are the same size at the same pressure and same temperature, what is true about the number of moles of each gas? Choose the *best* answer.

a. The number of moles of each gas is *the same* because they are in the same size container. A direct relationship always exists between volume and moles; therefore, since the containers have the same volume, the moles must be the same.
b. The number of moles of each gas is *the same* because they are in the same size container at the same temperature and also exert the same pressure on the walls of each container.
c. The gas that has a *larger molar mass* will have a larger number of moles inside the container because the larger the mass of gas present in the sample, the more moles present (due to Avogadro's number).
d. The gas that has a *smaller molar mass* will have a larger number of moles inside the container because more particles must be present in order to collide with the walls of the container and exert the same pressure as the other gas.
e. It is *impossible* to draw a conclusion about the number of moles of each gas without knowing the actual identities of the two gases and the starting amounts present in each container.

115. If 10.0 g of liquid helium at 1.7 K is completely vaporized, what volume does the helium occupy at STP?

116. If 0.214 mole of argon gas occupies a volume of 652 mL at a particular temperature and pressure, what volume would 0.375 mole of argon occupy under the same conditions?

117. Convert the following pressures into mm Hg.

a. 0.903 atm
b. 2.1240×10^6 Pa
c. 445 kPa
d. 342 torr

118. Convert the following pressures into pascals.

a. 645 mm Hg
b. 221 kPa
c. 0.876 atm
d. 32 torr

119. For each of the following sets of pressure/volume data, calculate the missing quantity. Assume that the temperature and the amount of gas remain constant.

a. V = 123 L at 4.56 atm; V = ? at 1002 mm Hg
b. V = 634 mL at 25.2 mm Hg; V = 166 mL at ? atm
c. V = 443 L at 511 torr; V = ? at 1.05 kPa

120. For each of the following sets of pressure/volume data, calculate the missing quantity. Assume that the temperature and the amount of gas remain constant.

a. V = 255 mL at 1.00 mm Hg; V = ? at 2.00 torr
b. V = 1.3 L at 1.0 kPa; V = ? at 1.0 atm
c. V = 1.3 L at 1.0 kPa; V = ? at 1.0 mm Hg

121. A particular balloon is designed by its manufacturer to be inflated to a volume of no more than 2.5 L. If the balloon is filled with 2.0 L of helium at sea level, is released, and rises to an altitude at which the atmospheric pressure is only 500. mm Hg, will the balloon burst?

122. What pressure is needed to compress 1.52 L of air at 755 mm Hg to a volume of 450 mL (at constant temperature)?

123. An expandable vessel contains 729 mL of gas at 22 °C. What volume will the gas sample in the vessel have if it is placed in a boiling water bath (100. °C)?

124. For each of the following sets of volume/temperature data, calculate the missing quantity. Assume that the pressure and the amount of gas remain constant.

a. V = 100. mL at 74 °C; V = ? at −74 °C
b. V = 500. mL at 100 °C; V = 600. mL at ? °C
c. V = 10,000 L at 25 °C; V = ? at 0 K

125. For each of the following sets of volume/temperature data, calculate the missing quantity. Assume that the pressure and the amount of gas remain constant.

a. V = 22.4 L at 0 °C; V = 44.4 L at ? K
b. $V = 1.0 \times 10^{-3}$ mL at −272 °C; V = ? at 25 °C
c. V = 32.3 L at −40 °C; V = 1000. L at ? °C

126. A 75.2-mL sample of helium at 12 °C is heated to 192 °C. What is the new volume of the helium (assuming constant pressure)?

127. If 5.12 g of oxygen gas occupies a volume of 6.21 L at a certain temperature and pressure, what volume will 25.0 g of oxygen gas occupy under the same conditions?

128. You have a gas in a container fitted with a piston, and you change one of the conditions of the gas such that a change takes place, as shown below:

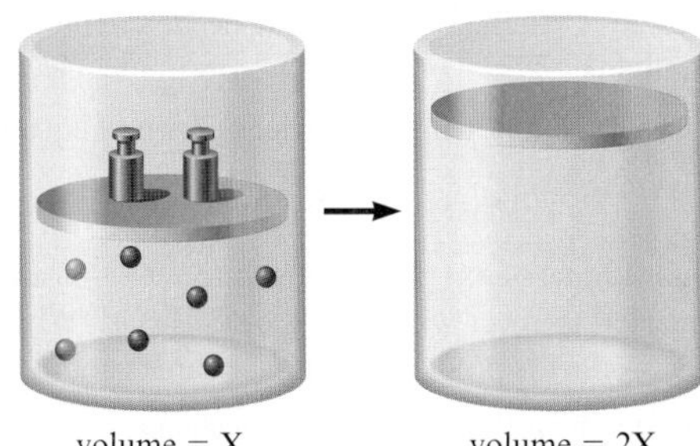

State three distinct changes you can make to accomplish this, and explain why each would work.

129. Given each of the following sets of values for three of the gas variables, calculate the unknown quantity.

a. P = 21.2 atm; V = 142 mL; n = 0.432 mol; T = ? K
b. P = ? atm; V = 1.23 mL; n = 0.000115 mol; T = 293 K
c. P = 755 mm Hg; V = ? mL; n = 0.473 mol; T = 131 °C

130. Given each of the following sets of values for three of the gas variables, calculate the unknown quantity.

a. $P = 1.034$ atm; $V = 21.2$ mL; $n = 0.00432$ mol; $T = ?$ K
b. $P = ?$ atm; $V = 1.73$ mL; $n = 0.000115$ mol; $T = 182$ K
c. $P = 1.23$ mm Hg; $V = ?$ L; $n = 0.773$ mol; $T = 152$ °C

131. What is the pressure inside a 10.0-L flask containing 14.2 g of N_2 at 26 °C?

132. Suppose three 100.-L tanks are to be filled separately with the gases CH_4, N_2, and CO_2, respectively. What mass of each gas is needed to produce a pressure of 120. atm in its tank at 27 °C?

133. At what temperature does 4.00 g of helium gas have a pressure of 1.00 atm in a 22.4-L vessel?

134. What is the pressure in a 100.-mL flask containing 55 mg of oxygen gas at 26 °C?

135. A weather balloon is filled with 1.0 L of helium at 23 °C and 1.0 atm. What volume does the balloon have when it has risen to a point in the atmosphere where the pressure is 220 torr and the temperature is −31 °C?

136. If 3.20 g of nitrogen gas occupies a volume of 1.71 L at 0 °C and a pressure of 1.50 atm, what would the volume become if 8.80 g of nitrogen gas were *added* at constant conditions of temperature and pressure?

137. If 1.0 mole of $N_2(g)$ is injected into a 5.0-L tank already containing 50. g of O_2 at 25 °C, what will be the total pressure in the tank?

138. A mixture at 33 °C contains H_2 at 325 torr, N_2 at 475 torr, and O_2 at 650. torr.

a. What is the total pressure of the gases in the system?
b. Which gas contains the greatest number of moles?

139. A flask of hydrogen gas is collected at 1.023 atm and 35 °C by displacement of water from the flask. The vapor pressure of water at 35 °C is 42.2 mm Hg. What is the partial pressure of hydrogen gas in the flask?

140. Consider the following chemical equation:

$$N_2(g) + 3H_2(g) \rightarrow 2NH_3(g)$$

What volumes of nitrogen gas and hydrogen gas, each measured at 11 °C and 0.998 atm, are needed to produce 5.00 g of ammonia?

141. Consider the following *unbalanced* chemical equation:

$$C_6H_{12}O_6(s) + O_2(g) \rightarrow CO_2(g) + H_2O(g)$$

What volume of oxygen gas, measured at 28 °C and 0.976 atm, is needed to react with 5.00 g of $C_6H_{12}O_6$? What volume of each product is produced under the same conditions?

142. Consider the following *unbalanced* chemical equation:

$$Cu_2S(s) + O_2(g) \rightarrow Cu_2O(s) + SO_2(g)$$

What volume of oxygen gas, measured at 27.5 °C and 0.998 atm, is required to react with 25 g of copper(I) sulfide? What volume of sulfur dioxide gas is produced under the same conditions?

143. When sodium bicarbonate, $NaHCO_3(s)$, is heated, sodium carbonate is produced, with the evolution of water vapor and carbon dioxide gas.

$$2NaHCO_3(s) \rightarrow Na_2CO_3(s) + H_2O(g) + CO_2(g)$$

What total volume of gas, measured at 29 °C and 769 torr, is produced when 1.00 g of $NaHCO_3(s)$ is completely converted to $Na_2CO_3(s)$?

144. What volume does 35 moles of N_2 occupy at STP?

145. A sample of oxygen gas has a volume of 125 L at 25 °C and a pressure of 0.987 atm. Calculate the volume of this oxygen sample at STP.

146. Consider the flasks in the following diagrams.

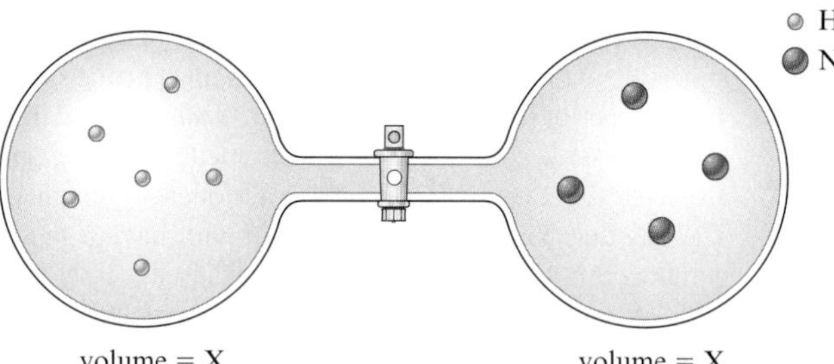

a. Which is greater, the initial pressure of helium or the initial pressure of neon? How much greater?
b. Assuming the connecting tube has negligible volume, draw what each diagram will look like after the stopcock between the two flasks is opened.
c. Solve for the final partial pressures of helium and neon in terms of their original pressures. Assume the temperature is constant.
d. Solve for the final pressure in terms of the original pressures of helium and neon. Assume temperature is constant.

147. What volume of CO_2 measured at STP is produced when 27.5 g of $CaCO_3$ is decomposed?

$$CaCO_3(s) \rightarrow CaO(s) + CO_2(g)$$

148. Concentrated hydrogen peroxide solutions are explosively decomposed by traces of transition metal ions (such as Mn or Fe):

$$2H_2O_2(aq) \rightarrow 2H_2O(l) + O_2(g)$$

What volume of pure $O_2(g)$, collected at 27 °C and 764 torr, would be generated by decomposition of 125 g of a 50.0% by mass hydrogen peroxide solution?

F **149.** The "Chemistry in Focus" segment *The Chemistry of Air Bags* discusses how the decomposition of sodium azide inflates the air bag. Use the balanced chemical equation in the segment to determine the mass of sodium azide required to inflate an air bag to 70.0 L at STP.

F **150.** The "Chemistry in Focus" segment *Breath Fingerprinting* discusses using breath analysis to diagnose diseases. The volume of the average human breath is approximately 500 mL, and carbon dioxide (CO_2) makes up 4% of what we exhale. Determine the mass of carbon dioxide exhaled in an average human breath.

All even-numbered Questions and Problems have answers in the back of this book and solutions in the *Student Solutions Guide.*

ChemWork Problems

These multiconcept problems (and additional ones) are found interactively online with the same type of assistance a student would get from an instructor.

151. Complete the following table for an ideal gas.

P (atm)	*V* (L)	*n* (mol)	*T*
6.74	______	2.00	155 °C
0.300	1.74	______	155 K
4.47	25.0	2.19	______ °C
______	2.25	10.5	93 °C

152. A glass vessel contains 28 g of nitrogen gas. Assuming ideal behavior, which of the processes listed below would double the pressure exerted on the walls of the vessel?

a. Adding 28 g of oxygen gas
b. Raising the temperature of the container from −73 °C to 127 °C
c. Adding enough mercury to fill one-half the container
d. Adding 32 g of oxygen gas
e. Raising the temperature of the container from 30. °C to 60. °C

153. A steel cylinder contains 150.0 moles of argon gas at a temperature of 25 °C and a pressure of 8.93 MPa. After some argon has been used, the pressure is 2.00 MPa at a temperature of 19 °C. What mass of argon remains in the cylinder?

154. A certain flexible weather balloon contains helium gas at a volume of 855 L. Initially, the balloon is at sea level where the temperature is 25 °C and the barometric pressure is 730 torr. The balloon then rises to an altitude of 6000 ft, where the pressure is 605 torr and the temperature is 15 °C. What is the change in volume of the balloon as it ascends from sea level to 6000 ft?

155. A large flask with a volume of 936 mL is evacuated and found to have a mass of 134.66 g. It is then filled to a pressure of 0.967 atm at 31 °C with a gas of unknown molar mass and then reweighed to give a new mass of 135.87 g. What is the molar mass of this gas?

156. A 20.0-L nickel container was charged with 0.859 atm of xenon gas and 1.37 atm of fluorine gas at 400 °C. The xenon and fluorine react to form xenon tetrafluoride. What mass of xenon tetrafluoride can be produced, assuming 100% yield?

157. Consider the *unbalanced* chemical equation:

$$CaSiO_3(s) + HF(g) \longrightarrow CaF_2(aq) + SiF_4(g) + H_2O(l)$$

Suppose a 32.9-g sample of $CaSiO_3$ is reacted with 31.8 L of HF at 27.0 °C and 1.00 atm. Assuming the reaction goes to completion, calculate the mass of SiF_4 and H_2O produced in the reaction.

158. Which of the following statements is(are) *true*?

a. If the number of moles of a gas is doubled, the volume will double, assuming the pressure and temperature of the gas remain constant.
b. If the temperature of a gas increases from 25 °C to 50 °C, the volume of the gas would double, assuming that the pressure and the number of moles of gas remain constant.
c. The device that measures atmospheric pressure is called a barometer.
d. If the volume of a gas decreases by one-half, then the pressure would double, assuming that the number of moles and the temperature of the gas remain constant.

All even-numbered Questions and Problems have answers in the back of this book and solutions in the *Student Solutions Guide.*

Liquids and Solids

14

Ice, the solid form of water, provides recreation for this ice climber. Gregg Epperson/Shutterstock.com

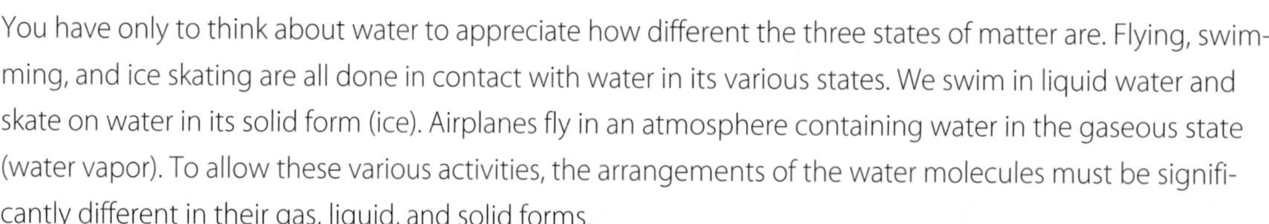

You have only to think about water to appreciate how different the three states of matter are. Flying, swimming, and ice skating are all done in contact with water in its various states. We swim in liquid water and skate on water in its solid form (ice). Airplanes fly in an atmosphere containing water in the gaseous state (water vapor). To allow these various activities, the arrangements of the water molecules must be significantly different in their gas, liquid, and solid forms.

In Chapter 13 we saw that the particles of a gas are far apart, are in rapid random motion, and have little effect on each other. Solids are obviously very different from gases. Gases have low densities, have high compressibilities, and completely fill a container. Solids have much greater densities than gases, are compressible only to a very slight extent, and are rigid; a solid maintains its shape regardless of its container. These properties indicate that the components of a solid are close together and exert large attractive forces on each other.

The properties of liquids lie somewhere between those of solids and of gases—but not midway between, as can be seen from some of the properties of the three states of water. For example, it takes about seven times more energy to change liquid water to steam (a gas) at 100 °C than to melt ice to form liquid water at 0 °C.

$$H_2O(s) \rightarrow H_2O(l) \qquad \text{energy required} \cong 6 \text{ kJ/mol}$$

$$H_2O(l) \rightarrow H_2O(g) \qquad \text{energy required} \cong 41 \text{ kJ/mol}$$

These values indicate that going from the liquid to the gaseous state involves a much greater change than going from the solid to the liquid. Therefore, we can conclude that the solid and liquid states are more similar than the liquid and gaseous states. This is also demonstrated by the densities of the three states of water (Table 14.1). Note that water in its gaseous state is about 2000 times less dense than in the solid and liquid states and that the latter two states have very similar densities.

We find in general that the liquid and solid states show many similarities and are strikingly different from the gaseous state (Fig. 14.1). The best way to picture the solid state is in terms of closely packed, highly ordered particles in contrast to the widely spaced, randomly arranged particles of a gas. The liquid state lies in between, but its properties indicate that it much more closely resembles the solid than the gaseous state. It is useful to picture a liquid in terms of particles that are generally quite close together, but with a more disordered arrangement than for the solid state and with some empty spaces. For most substances, the solid state has a higher density than the liquid, as Fig. 14.1 suggests. However, water is an exception to this rule. Ice has an unusual amount of empty space and so is less dense than liquid water, as indicated in Table 14.1.

In this chapter we will explore the important properties of liquids and solids. We will illustrate many of these properties by considering one of the earth's most important substances: water.

Wind surfers use liquid water for recreation. Dmitry Tsvetkov/Shutterstock.com

Table 14.1 ▸ Densities of the Three States of Water

State	Density (g/cm³)
solid (0 °C, 1 atm)	0.9168
liquid (25 °C, 1 atm)	0.9971
gas (100 °C, 1 atm)	5.88×10^{-4}

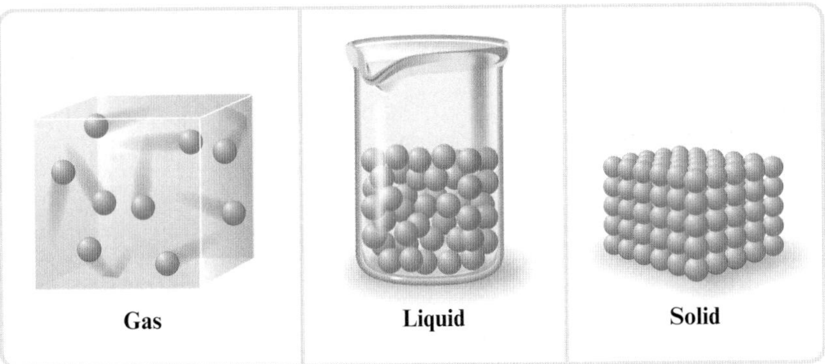

Figure 14.1 ▸ Representations of the gas, liquid, and solid states.

14-1 Water and Its Phase Changes

OBJECTIVE **To learn some of the important features of water.**

In the world around us we see many solids (soil, rocks, trees, concrete, and so on), and we are immersed in the gases of the atmosphere. But the liquid we most commonly see is water; it is virtually everywhere, covering about 70% of the earth's surface. Approximately 97% of the earth's water is found in the oceans, which are actually mixtures of water and huge quantities of dissolved salts.

Water is one of the most important substances on earth. It is crucial for sustaining the reactions within our bodies that keep us alive, but it also affects our lives in many indirect ways. The oceans help moderate the earth's temperature. Water cools automobile engines and nuclear power plants. Water provides a means of transportation on the earth's surface and acts as a medium for the growth of the myriad creatures we use as food, and much more.

The water we drink often has a taste because of the substances dissolved in it. It is not pure water.

Pure water is a colorless, tasteless substance that at 1 atm pressure freezes to form a solid at 0 °C and vaporizes completely to form a gas at 100 °C. This means that (at 1 atm pressure) the liquid range of water occurs between the temperatures 0 °C and 100 °C.

What happens when we heat liquid water? First the temperature of the water rises. Just as with gas molecules, the motions of the water molecules increase as it is heated. Eventually the temperature of the water reaches 100 °C; now bubbles develop in the interior of the liquid, float to the surface, and burst—the boiling point has been reached. An interesting thing happens at the boiling point: even though heating continues, the temperature stays at 100 °C until all the water has changed to vapor. Only when all of the water has changed to the gaseous state does the temperature begin to rise again. (We are now heating the vapor.) At 1 atm pressure, liquid water always changes to gaseous water at 100 °C, the **normal boiling point** for water.

The experiment just described is represented in Fig. 14.2, which is called the **heating/cooling curve** for water. Going from left to right on this graph means energy is being added (heating). Going from right to left on the graph means that energy is being removed (cooling).

When liquid water is cooled, the temperature decreases until it reaches 0 °C, where the liquid begins to freeze (Fig. 14.2). The temperature remains at 0 °C until all the liquid water has changed to ice and then begins to drop again as cooling continues. At 1 atm pressure, water freezes (or, in the opposite process, ice melts) at 0 °C. This is called the **normal freezing point** of water. Liquid and solid water can coexist indefi-

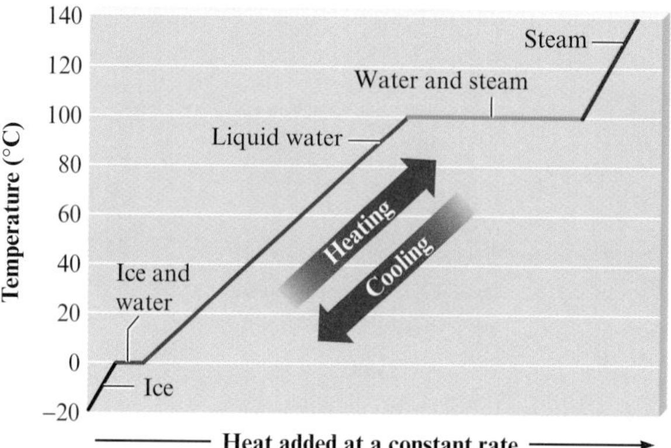

Figure 14.2 ▸ The heating/cooling curve for water heated or cooled at a constant rate. The plateau at the boiling point is longer than the plateau at the melting point, because it takes almost seven times as much energy (and thus seven times the heating time) to vaporize liquid water as to melt ice. Note that to make the diagram clear, the blue line is not drawn to scale. It actually takes more energy to melt ice and boil water than to heat water from 0 °C to 100 °C.

nitely if the temperature is held at 0 °C. However, at temperatures below 0 °C liquid water freezes, while at temperatures above 0 °C ice melts.

Interestingly, water expands when it freezes. That is, one gram of ice at 0 °C has a greater volume than one gram of liquid water at 0 °C. This has very important practical implications. For instance, water in a confined space can break its container when it freezes and expands. This accounts for the bursting of water pipes and engine blocks that are left unprotected in freezing weather.

The expansion of water when it freezes also explains why ice cubes float. Recall that density is defined as mass/volume. When one gram of liquid water freezes, its volume becomes greater (it expands). Therefore, the *density* of one gram of ice is less than the density of one gram of water, because in the case of ice we divide by a slightly larger volume. For example, at 0 °C the density of liquid water is

$$\frac{1.00\text{ g}}{1.00\text{ mL}} = 1.00\text{ g/mL}$$

and the density of ice is

$$\frac{1.00\text{ g}}{1.09\text{ mL}} = 0.917\text{ g/mL}$$

The lower density of ice also means that ice floats on the surface of lakes as they freeze, providing a layer of insulation that helps to prevent lakes and rivers from freezing solid in the winter. This means that aquatic life continues to have liquid water available through the winter.

14-2 Energy Requirements for the Changes of State

OBJECTIVES

- **To learn about interactions among water molecules.**
- **To understand and use heat of fusion and heat of vaporization.**

It is important to recognize that changes of state from solid to liquid and from liquid to gas are *physical* changes. No *chemical* bonds are broken in these processes. Ice, water, and steam all contain H_2O molecules. When water is boiled to form steam, water molecules are separated from each other (Fig. 14.3) but the individual molecules remain intact.

CHEMISTRY IN FOCUS

Whales Need Changes of State

Sperm whales are prodigious divers. They commonly dive a mile or more into the ocean, hovering at that depth in search of schools of squid or fish. To remain motionless at a given depth, the whale must have the same density as the surrounding water. Because the density of seawater increases with depth, the sperm whale has a system that automatically increases its density as it dives. This system involves the spermaceti organ found in the whale's head. Spermaceti is a waxy substance with the formula

$$CH_3{-}(CH_2)_{15}{-}O{-}\underset{\displaystyle \overset{\|}{O}}{C}{-}(CH_2)_{14}{-}CH_3$$

which is a liquid above 30 °C. At the ocean surface the spermaceti in the whale's head is a liquid, warmed by the flow of blood through the spermaceti organ. When the whale dives, this blood flow decreases and the colder water causes the spermaceti to begin freezing. Because solid spermaceti is more dense than the liquid state, the sperm whale's density increases as it dives, matching the increase in the water's density.* When the whale wants to resurface, blood flow through the spermaceti organ increases, remelting the spermaceti and making the whale more buoyant. So the sperm whale's sophisticated density-regulating mechanism is based on a simple change of state.

*For most substances, the solid state is more dense than the liquid state. Water is an important exception.

A sperm whale mother diving with her calf.

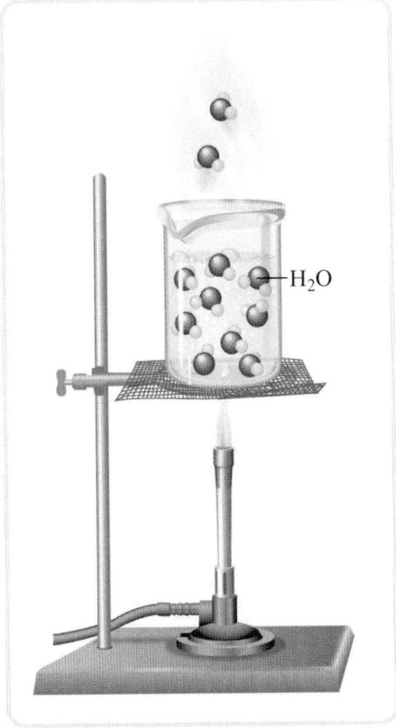

Figure 14.3 Both liquid water and gaseous water contain H_2O molecules. In liquid water the H_2O molecules are close together, whereas in the gaseous state the molecules are widely separated. The bubbles contain gaseous water.

The bonding forces that hold the atoms of a molecule together are called **intramolecular** (within the molecule) **forces.** The forces that occur among molecules that cause them to aggregate to form a solid or a liquid are called **intermolecular** (between the molecules) **forces.** These two types of forces are illustrated in Fig. 14.4.

It takes energy to melt ice and to vaporize water, because intermolecular forces between water molecules must be overcome. In ice the molecules are virtually locked in place, although they can vibrate about their positions. When energy is added, the vibrational motions increase, and the molecules eventually achieve the greater move-

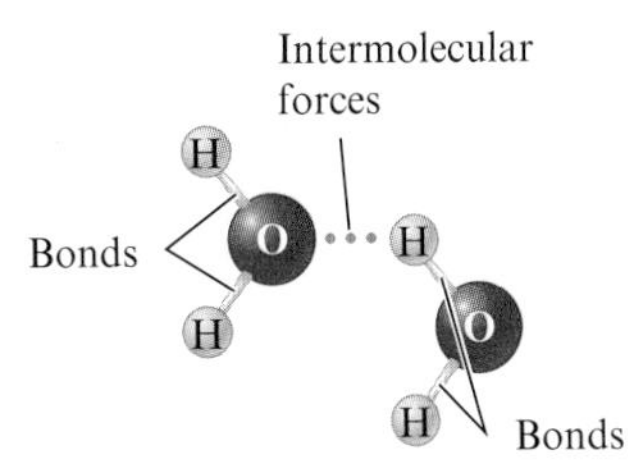

Figure 14.4 Intramolecular (bonding) forces exist between the atoms in a molecule and hold the molecule together. Intermolecular forces exist between molecules. These are the forces that cause water to condense to a liquid or form a solid at low enough temperatures. Intermolecular forces are typically much weaker than intramolecular forces.

Remember that temperature is a measure of the random motions (average kinetic energy) of the particles in a substance.

ment and disorder characteristic of liquid water. The ice has melted. As still more energy is added, the gaseous state is eventually reached, in which the individual molecules are far apart and interact relatively little. However, the gas still consists of water molecules. It would take *much* more energy to overcome the covalent bonds and decompose the water molecules into their component atoms.

The energy required to melt 1 mole of a substance is called the **molar heat of fusion.** For ice, the molar heat of fusion is 6.02 kJ/mol. The energy required to change 1 mole of liquid to its vapor is called the **molar heat of vaporization.** For water, the molar heat of vaporization is 40.6 kJ/mol at 100 °C. Notice in Fig. 14.2 that the plateau that corresponds to the vaporization of water is much longer than that for the melting of ice. This occurs because it takes much more energy (almost seven times as much) to vaporize a mole of water than to melt a mole of ice. This is consistent with our models of solids, liquids, and gases (Fig. 14.1). In liquids, the particles (molecules) are relatively close together, so most of the intermolecular forces are still present. However, when the molecules go from the liquid to the gaseous state, they must be moved far apart. To separate the molecules enough to form a gas, virtually all of the intermolecular forces must be overcome, and this requires large quantities of energy.

Interactive Example 14.1 Calculating Energy Changes: Solid to Liquid

Calculate the energy required to melt 8.5 g of ice at 0 °C. The molar heat of fusion for ice is 6.02 kJ/mol.

SOLUTION

Where Are We Going?

We want to determine the energy (in kJ) required to melt 8.5 g of ice at 0 °C.

What Do We Know?

- We have 8.5 g of ice (H_2O) at 0 °C.
- The molar heat of fusion of ice is 6.02 kJ/mol.

What Information Do We Need?

- We need to know the number of moles of ice in 8.5 g.

How Do We Get There?

The molar heat of fusion is the energy required to melt *1 mole* of ice. In this problem we have 8.5 g of solid water. We must find out how many moles of ice this mass represents. Because the molar mass of water is 16 + 2(1) = 18, we know that 1 mole of water has a mass of 18 g, so we can convert 8.5 g of H_2O to moles of H_2O.

$$8.5\ \cancel{\text{g H}_2\text{O}} \times \frac{1\ \text{mol H}_2\text{O}}{18\ \cancel{\text{g H}_2\text{O}}} = 0.47\ \text{mol H}_2\text{O}$$

Because 6.02 kJ of energy is required to melt a mole of solid water, our sample will take about half this amount (we have approximately half a mole of ice). To calculate the exact amount of energy required, we will use the equivalence statement

$$6.02\ \text{kJ required for 1 mol H}_2\text{O}$$

which leads to the conversion factor we need:

$$0.47\ \cancel{\text{mol H}_2\text{O}} \times \frac{6.02\ \text{kJ}}{\cancel{\text{mol H}_2\text{O}}} = 2.8\ \text{kJ}$$

This can be represented symbolically as

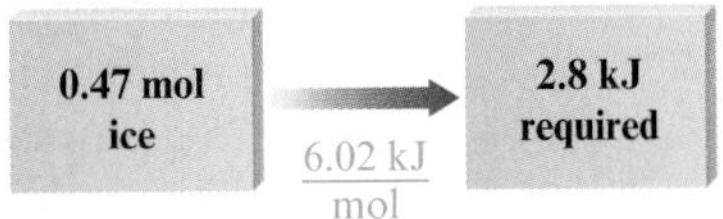

REALITY CHECK Because we have just under half of 1 mole of ice, our answer should be about half the molar heat of fusion of ice. The answer of 2.8 kJ is just under half of 6.02 kJ, so this answer makes sense. ■

Interactive Example 14.2 Calculating Energy Changes: Liquid to Gas

Specific heat capacity was discussed in Section 10-5.

Calculate the energy (in kJ) required to heat 25 g of liquid water from 25 °C to 100. °C and change it to steam at 100. °C. The specific heat capacity of liquid water is 4.18 J/g °C, and the molar heat of vaporization of water is 40.6 kJ/mol.

SOLUTION

Where Are We Going?

We want to determine the energy (in kJ) required to heat and vaporize a given quantity of water.

What Do We Know?

- We have 25 g of H_2O at 25 °C. The water will be heated to 100. °C and then vaporized at 100. °C
- The specific heat capacity of liquid water is 4.18 J/g °C.
- The molar vaporization of water is 40.6 kJ/mol.
- $Q = s \times m \times \Delta T$.

What Information Do We Need?

- We need to know the number of moles of water in 25 g.

How Do We Get There?

This problem can be split into two parts: (1) heating the water to its boiling point and (2) converting the liquid water to vapor at the boiling point.

Step 1 Heating to Boiling We must first supply energy to heat the liquid water from 25 °C to 100. °C. Because 4.18 J is required to heat *one* gram of water by *one* Celsius degree, we must multiply by both the mass of water (25 g) and the temperature change (100. °C − 25 °C = 75 °C),

Energy required (Q) = **Specific heat capacity (s)** × **Mass of water (m)** × **Temperature change (ΔT)**

which we can represent by the equation

$$Q = s \times m \times \Delta T$$

Thus

$$Q = 4.18 \frac{\text{J}}{\text{g} \,°\text{C}} \times 25\ \text{g} \times 75\ °\text{C} = 7.8 \times 10^3\ \text{J}$$

(Q: Energy required to heat 25 g of water from 25 °C to 100. °C; 4.18 J/g °C: Specific heat capacity; 25 g: Mass of water; 75 °C: Temperature change)

$$= 7.8 \times 10^3\ \text{J} \times \frac{1\ \text{kJ}}{1000\ \text{J}} = 7.8\ \text{kJ}$$

Step 2 Vaporization Now we must use the molar heat of vaporization to calculate the energy required to vaporize the 25 g of water at 100. °C. The heat of vaporization is given *per mole* rather than per gram, so we must first convert the 25 g of water to moles.

$$25\ \text{g}\ \cancel{H_2O} \times \frac{1\ \text{mol}\ H_2O}{18\ \text{g}\ \cancel{H_2O}} = 1.4\ \text{mol}\ H_2O$$

We can now calculate the energy required to vaporize the water.

$$\underbrace{\frac{40.6\ \text{kJ}}{\cancel{\text{mol}\ H_2O}}}_{\text{Molar heat of vaporization}} \times \underbrace{1.4\ \cancel{\text{mol}\ H_2O}}_{\text{Moles of water}} = 57\ \text{kJ}$$

The total energy is the sum of the two steps.

$$\underbrace{7.8\ \text{kJ}}_{\text{Heat from 25 °C to 100. °C}} + \underbrace{57\ \text{kJ}}_{\text{Change to vapor}} = 65\ \text{kJ}$$

SELF CHECK

Exercise 14.1 Calculate the total energy required to melt 15 g of ice at 0 °C, heat the water to 100. °C, and vaporize it to steam at 100. °C.

HINT Break the process into three steps and then take the sum.

See Problems 14.15 through 14.18. ■

14-3 Intermolecular Forces

OBJECTIVES

- **To learn about dipole–dipole attraction, hydrogen bonding, and London dispersion forces.**
- **To understand the effect of these forces on the properties of liquids.**

We have seen that covalent bonding forces within molecules arise from the sharing of electrons, but how do intermolecular forces arise? Actually several types of intermolecular forces exist. To illustrate one type, we will consider the forces that exist among water molecules.

The polarity of a molecule was discussed in Section 12-3.

As we saw in Chapter 12, water is a polar molecule—it has a dipole moment. When molecules with dipole moments are put together, they orient themselves to take advantage of their charge distributions. Molecules with dipole moments can attract each other by lining up so that the positive and negative ends are close to each other, as shown in Fig. 14.5(a). This is called a **dipole–dipole attraction.** In the liquid, the dipoles find the best compromise between attraction and repulsion, as shown in Fig. 14.5(b).

Dipole–dipole forces are typically only about 1% as strong as covalent or ionic bonds, and they become weaker as the distance between the dipoles increases. In the gas phase, where the molecules are usually very far apart, these forces are relatively unimportant.

See Section 12-2 for a discussion of electronegativity.

Particularly strong dipole-dipole forces occur between molecules in which hydrogen is bound to a highly electronegative atom, such as nitrogen, oxygen, or fluorine. Two factors account for the strengths of these interactions: the great polarity of the bond and the close approach of the dipoles, which is made possible by the very small size of the hydrogen atom. Because dipole–dipole attractions of this type are so unusu-

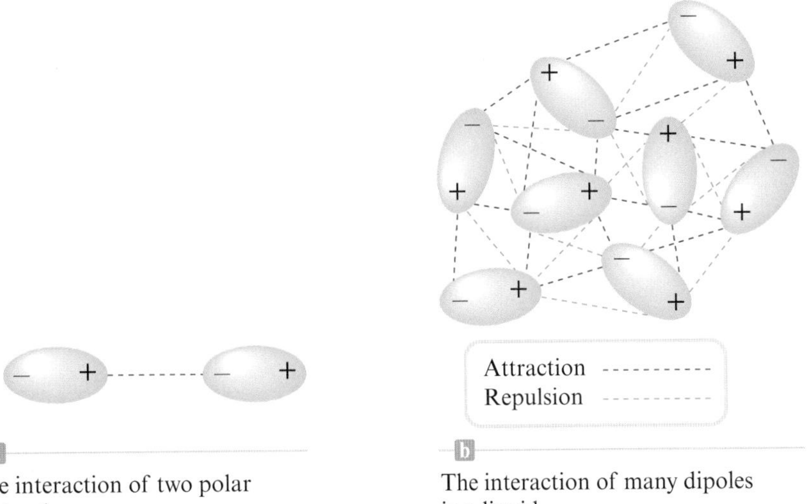

a The interaction of two polar molecules.

b The interaction of many dipoles in a liquid.

Figure 14.5

ally strong, they are given a special name—**hydrogen bonding.** Figure 14.6 illustrates hydrogen bonding among water molecules.

Hydrogen bonding has a very important effect on various physical properties. For example, the boiling points for the covalent compounds of hydrogen with the elements in Group 6 are given in Fig. 14.7. Note that the boiling point of water is much higher than would be expected from the trend shown by the other members of the series. Why? Because the especially large electronegativity value of the oxygen atom compared with that of the other group members causes the O—H bonds to be much more polar than the S—H, Se—H, or Te—H bonds. This leads to very strong hydrogen-bonding forces among the water molecules. An unusually large quantity of energy is required to overcome these interactions and separate the molecules to produce the gaseous state. That is, water molecules tend to remain together in the liquid state even at relatively high temperatures—hence the very high boiling point of water.

However, even molecules without dipole moments must exert forces on each other. We know this because all substances—even the noble gases—exist in the liquid and solid states at very low temperatures. There must be forces to hold the atoms or mol-

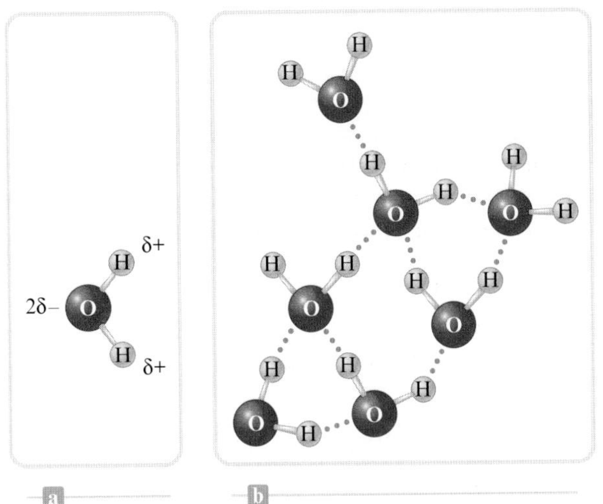

Figure 14.6 (a) The polar water molecule. (b) Hydrogen bonding among water molecules. The small size of the hydrogen atoms allows the molecules to get very close and thus to produce strong interactions.

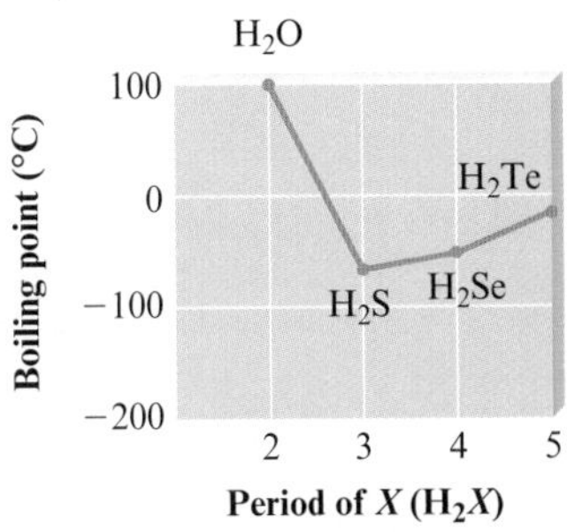

Figure 14.7 The boiling points of the covalent hydrides of elements in Group 6.

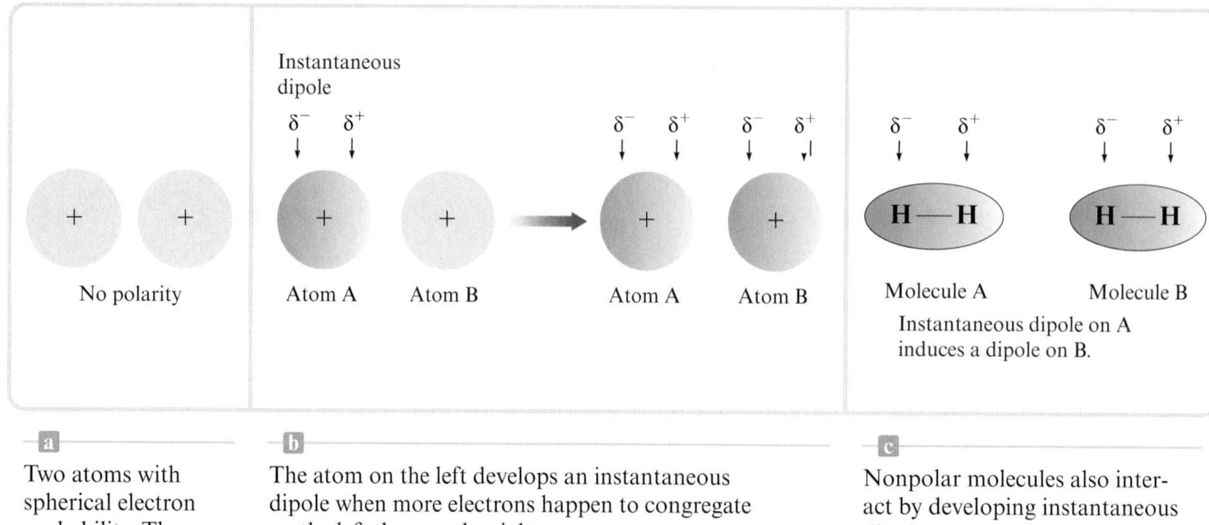

a Two atoms with spherical electron probability. These atoms have no polarity.

b The atom on the left develops an instantaneous dipole when more electrons happen to congregate on the left than on the right.

c Nonpolar molecules also interact by developing instantaneous dipoles.

Figure 14.8

ecules as close together as they are in these condensed states. The forces that exist among noble gas atoms and nonpolar molecules are called **London dispersion forces.** To understand the origin of these forces, consider a pair of noble gas atoms. Although we usually assume that the electrons of an atom are uniformly distributed about the nucleus [Fig. 14.8(a)], this is apparently not true at every instant. Atoms can develop a temporary dipolar arrangement of charge as the electrons move around the nucleus [Fig. 14.8(b)]. This *instantaneous dipole* can then *induce* a similar dipole in a neighboring atom, as shown in Fig. 14.8(b). The interatomic attraction thus formed is both weak and short-lived, but it can be very significant for large atoms and large molecules, as we will see.

Table 14.2 The Freezing Points of the Group 8 Elements

Element	Freezing Point (°C)
helium*	−272.0 (25 atm)
neon	−248.6
argon	−189.4
krypton	−157.3
xenon	−111.9

*Helium will not freeze unless the pressure is increased above 1 atm.

The motions of the atoms must be greatly slowed down before the weak London dispersion forces can lock the atoms into place to produce a solid. This explains, for instance, why the noble gas elements have such low freezing points (Table 14.2).

Nonpolar molecules such as H_2, N_2, and I_2, none of which has a permanent dipole moment, also attract each other by London dispersion forces [Fig. 14.8(c)]. London forces become more significant as the sizes of atoms or molecules increase. Larger size means there are more electrons available to form the dipoles.

Critical Thinking

You have learned the difference between intermolecular forces and intramolecular bonds. What if intermolecular forces were stronger than intramolecular bonds? What differences could you observe in the world?

14-4 Evaporation and Vapor Pressure

OBJECTIVE **To understand the relationship among vaporization, condensation, and vapor pressure.**

We all know that a liquid can evaporate from an open container. This is clear evidence that the molecules of a liquid can escape the liquid's surface and form a gas. This process, which is called **vaporization** or **evaporation,** requires energy to overcome the relatively strong intermolecular forces in the liquid.

Water is used to absorb heat from nuclear reactors. The water is then cooled in cooling towers before it is returned to the environment.

The fact that vaporization requires energy has great practical significance; in fact, one of the most important roles that water plays in our world is to act as a coolant. Because of the strong hydrogen bonding among its molecules in the liquid state, water has an unusually large heat of vaporization (41 kJ/mol). A significant portion of the sun's energy is spent evaporating water from the oceans, lakes, and rivers rather than warming the earth. The vaporization of water is also crucial to our body's temperature-control system, which relies on the evaporation of perspiration.

Vapor Pressure

Vapor, not *gas,* is the term we customarily use for the gaseous state of a substance that exists naturally as a solid or liquid at 25 °C and 1 atm.

When we place a given amount of liquid in a container and then close it, we observe that the amount of liquid at first decreases slightly but eventually becomes constant. The decrease occurs because there is a transfer of molecules from the liquid to the vapor phase (Fig. 14.9). However, as the number of vapor molecules increases, it becomes more and more likely that some of them will return to the liquid. The process by which vapor molecules form a liquid is called **condensation.** Eventually, the same number of molecules are leaving the liquid as are returning to it: the rate of condensation equals the rate of evaporation. *At this point no further change occurs in the amounts of liquid or vapor, because the two opposite processes exactly balance each other;* the system is at *equilibrium.* Note that this system is highly *dynamic* on the molecular level—molecules are constantly escaping from and entering the liquid. However, there is no *net* change because the two opposite processes just *balance* each other. As an analogy, consider two island cities connected by a bridge. Suppose the traffic flow on the bridge is the same in both directions. There is motion—we can see the cars

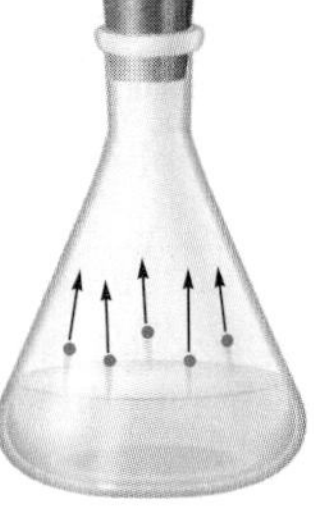

a Net evaporation occurs at first, so the amount of liquid decreases slightly.

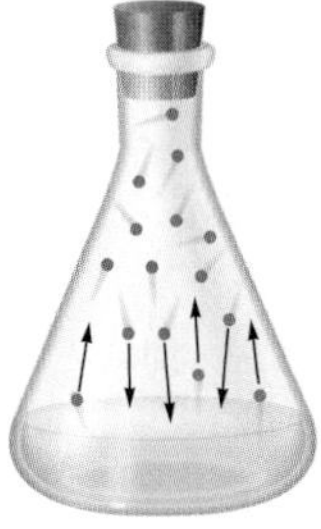

b As the number of vapor molecules increases, the rate of condensation increases. Finally the rate of condensation equals the rate of evaporation. The system is at equilibrium.

Figure 14.9 ▸ Behavior of a liquid in a closed container.

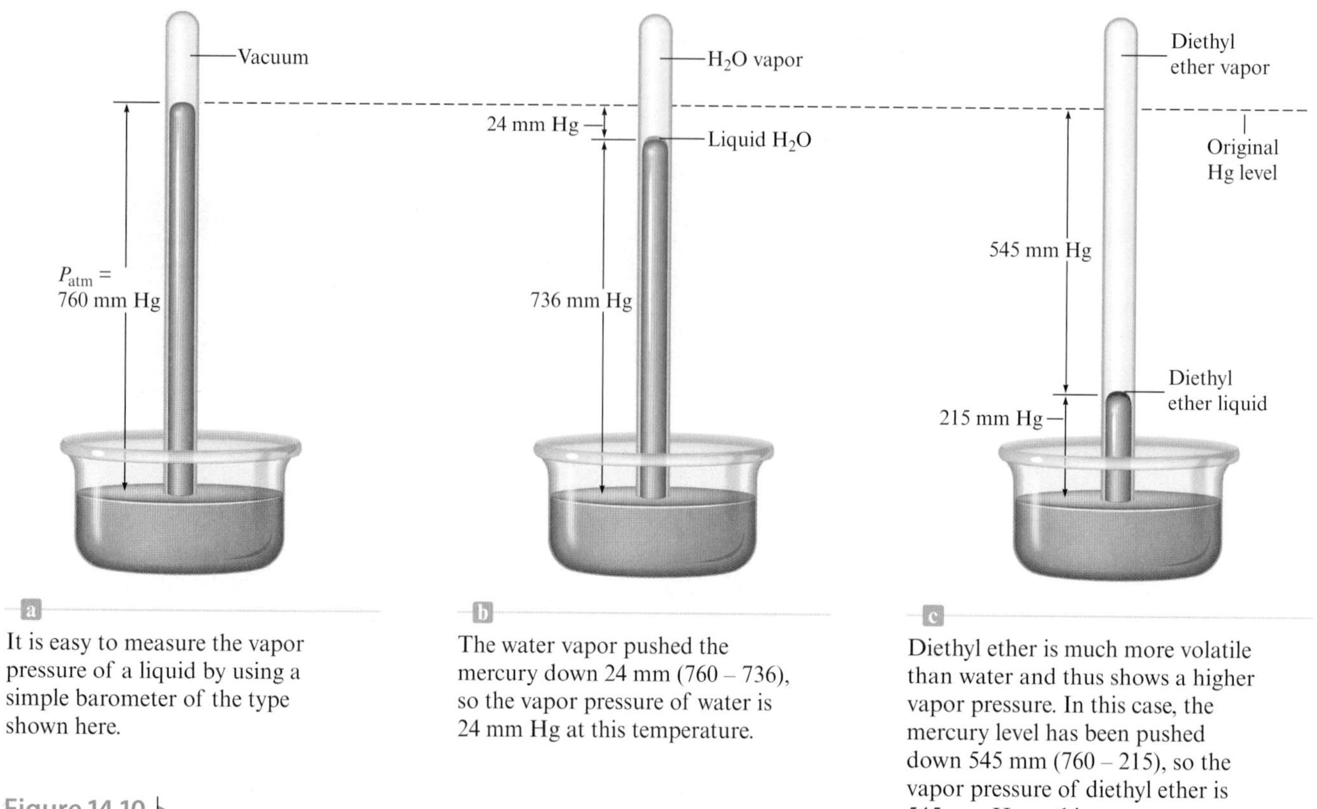

a It is easy to measure the vapor pressure of a liquid by using a simple barometer of the type shown here.

b The water vapor pushed the mercury down 24 mm (760 – 736), so the vapor pressure of water is 24 mm Hg at this temperature.

c Diethyl ether is much more volatile than water and thus shows a higher vapor pressure. In this case, the mercury level has been pushed down 545 mm (760 – 215), so the vapor pressure of diethyl ether is 545 mm Hg at this temperature.

Figure 14.10

traveling across the bridge—but the number of cars in each city is not changing because an equal number enter and leave each one. The result is no *net* change in the number of autos in each city: an equilibrium exists.

The pressure of the vapor present at equilibrium with its liquid is called the *equilibrium vapor pressure* or, more commonly, the **vapor pressure** of the liquid. A simple barometer can be used to measure the vapor pressure of a liquid, as shown in Fig. 14.10. Because mercury is so dense, any common liquid injected at the bottom of the column of mercury floats to the top, where it produces a vapor, and the pressure of this vapor pushes some mercury out of the tube. When the system reaches equilibrium, the vapor pressure can be determined from the change in the height of the mercury column.

In effect, we are using the space above the mercury in the tube as a closed container for each liquid. However, in this case as the liquid vaporizes, the vapor formed creates a pressure that pushes some mercury out of the tube and lowers the mercury level. The mercury level stops changing when the excess liquid floating on the mercury comes to equilibrium with the vapor. The change in the mercury level (in millimeters) from its initial position (before the liquid was injected) to its final position is equal to the vapor pressure of the liquid.

The vapor pressures of liquids vary widely (Fig. 14.10). Liquids with high vapor pressures are said to be *volatile*—they evaporate rapidly.

The vapor pressure of a liquid at a given temperature is determined by the *intermolecular forces* that act among the molecules. Liquids in which the intermolecular forces are large have relatively low vapor pressures, because such molecules need high energies to escape to the vapor phase. For example, although water is a much smaller molecule than diethyl ether, $C_2H_5—O—C_2H_5$, the strong hydrogen-bonding forces in water cause its vapor pressure to be much lower than that of ether (Fig. 14.10).

Interactive Example 14.3 Using Knowledge of Intermolecular Forces to Predict Vapor Pressure

Predict which substance in each of the following pairs will show the largest vapor pressure at a given temperature.

a. $H_2O(l)$, $CH_3OH(l)$

b. $CH_3OH(l)$, $CH_3CH_2CH_2CH_2OH(l)$

SOLUTION

a. Water contains two polar O—H bonds; methanol (CH_3OH) has only one. Therefore, the hydrogen bonding among H_2O molecules is expected to be much stronger than that among CH_3OH molecules. This gives water a lower vapor pressure than methanol.

b. Each of these molecules has one polar O—H bond. However, because $CH_3CH_2CH_2CH_2OH$ is a much larger molecule than CH_3OH, it has much greater London forces and thus is less likely to escape from its liquid. Thus $CH_3CH_2CH_2CH_2OH(l)$ has a lower vapor pressure than $CH_3OH(l)$. ■

14-5 The Solid State: Types of Solids

OBJECTIVE To learn about the various types of crystalline solids.

Solids play a very important role in our lives. The concrete we drive on, the trees that shade us, the windows we look through, the diamond in an engagement ring, and the plastic lenses in eyeglasses are all important solids. Most solids, such as wood, paper, and glass, contain mixtures of various components. However, some natural solids, such as diamonds and table salt, are nearly pure substances.

Many substances form **crystalline solids**—those with a regular arrangement of their components. This is illustrated by the partial structure of sodium chloride shown in Fig. 14.11. The highly ordered arrangement of the components in a crystalline solid produces beautiful, regularly shaped crystals such as those shown in Fig. 14.12.

There are many different types of crystalline solids. For example, both sugar and salt have beautiful crystals that we can easily see. However, although both dissolve readily in water, the properties of the resulting solutions are quite different. The salt solution readily conducts an electric current; the sugar solution does not. This behavior arises from the different natures of the components in these two solids. Common salt, NaCl, is an ionic solid that contains Na^+ and Cl^- ions. When solid sodium chloride dissolves in water, sodium ions and chloride ions are distributed throughout the resulting solution. These ions are free to move through the solution to conduct an electric

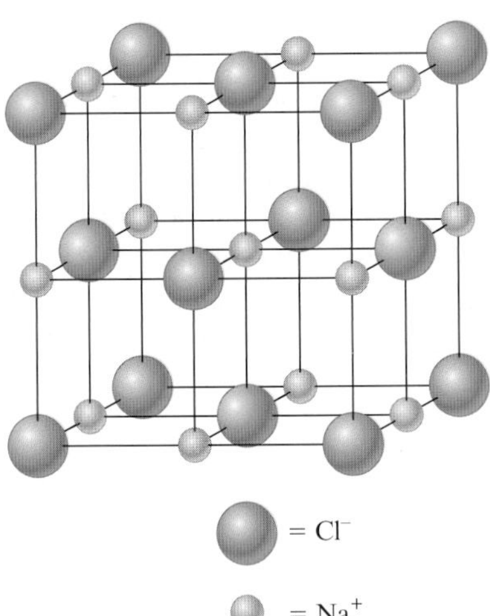

Figure 14.11 ▸ The regular arrangement of sodium and chloride ions in sodium chloride, a crystalline solid.

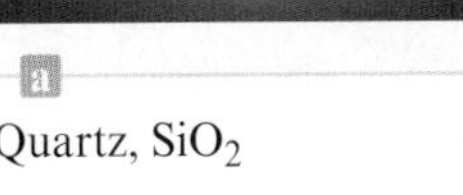

a Quartz, SiO_2

b Rock salt, NaCl

c Iron pyrite, FeS_2

© Cengage Learning

Figure 14.12 ▸ Several crystalline solids.

Gorilla Glass

When you think of glass, you probably think of a relatively fragile material—one that breaks readily when an errant baseball encounters a window. However, glass can be manufactured that at the same time is both thin and tough. An example of this type of glass is used on many of today's leading mobile electronic devices such as smartphones.

The main component of glass is silicon dioxide. Common glass is made by melting silicon dioxide and adding substances such as Na_2O, Al_2O_3, CaO, and B_2O_3. The specific additives used greatly influence the properties of the resulting glass. Corning® Gorilla Glass® is made by placing sheets of glass in a bath of molten salts containing K^+ ions at 400 °C. During this diffusive process some of the Na^+ ions in the regular glass are replaced by the larger K^+ ions, resulting in a superficial compressive stress layer. This compressive stress on the glass surface greatly strengthens the glass, making it resistant to breaking and scratching. Gorilla Glass for electronic devices is available in a variety of thicknesses, some as thin as 0.4 mm. It can be machined and shaped for a multiplicity of applications. To date, Gorilla Glass is used in more than one billion products worldwide, including smartphones, tablet computers, notebook computers, and TVs. There is little doubt that you touch Gorilla Glass many times during your daily activities.

Corning Inc.

current. Table sugar (sucrose), on the other hand, is composed of neutral molecules that are dispersed throughout the water when the solid dissolves. No ions are present, and the resulting solution does not conduct electricity. These examples illustrate two important types of crystalline solids: **ionic solids,** represented by sodium chloride; and **molecular solids,** represented by sucrose.

A third type of crystalline solid is represented by elements such as graphite and diamond (both pure carbon), boron, silicon, and all metals. These substances, which contain atoms of only one element covalently bonded to each other, are called **atomic solids.**

We have seen that crystalline solids can be grouped conveniently into three classes as shown in Fig. 14.13. Notice that the names of the three classes come from the com-

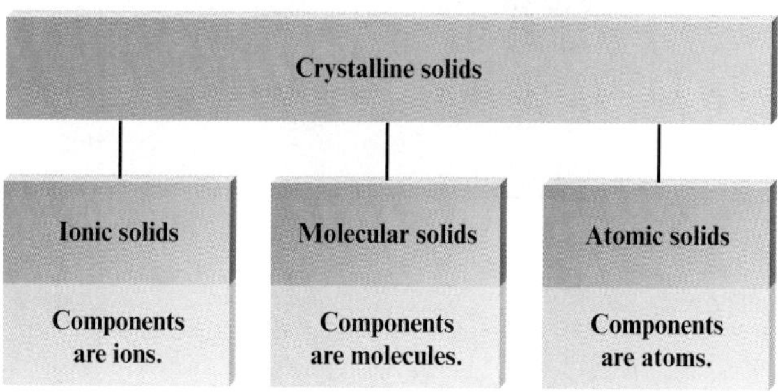

Figure 14.13 ▶ The classes of crystalline solids.

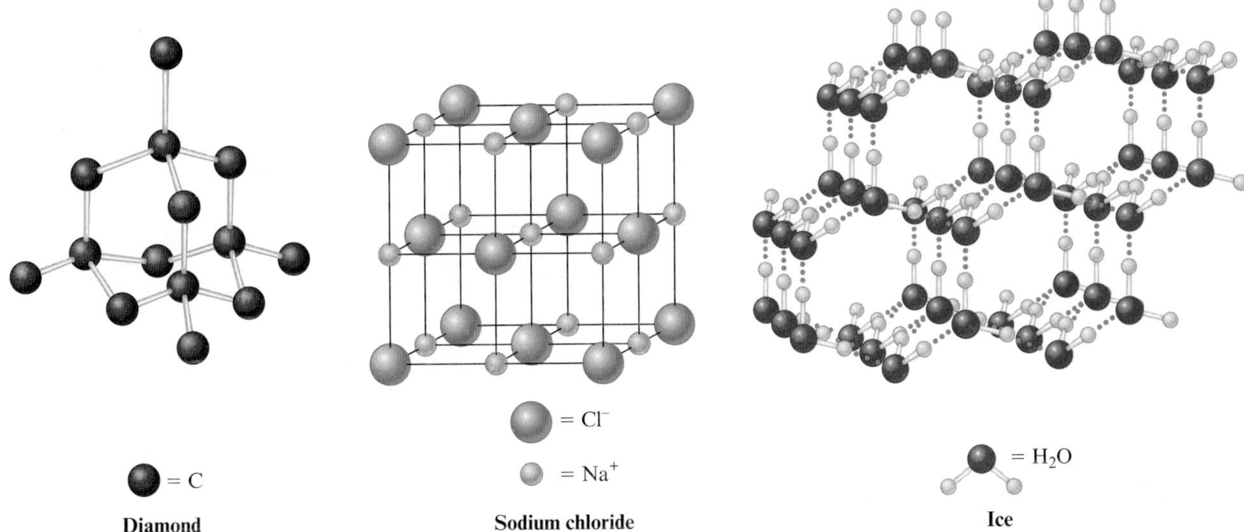

Figure 14.14 ▸ Examples of three types of crystalline solids. Only part of the structure is shown in each case. The structures continue in three dimensions with the same patterns.

ponents of the solid. An ionic solid contains ions, a molecular solid contains molecules, and an atomic solid contains atoms. Examples of the three types of solids are shown in Fig. 14.14.

The properties of a solid are determined primarily by the nature of the forces that hold the solid together. For example, although argon, copper, and diamond are all atomic solids (their components are atoms), they have strikingly different properties. Argon has a very low melting point (−189 °C), whereas diamond and copper melt at high temperatures (about 3500 °C and 1083 °C, respectively). Copper is an excellent conductor of electricity (it is widely used for electrical wires), whereas both argon and diamond are insulators. The shape of copper can easily be changed; it is both malleable (will form thin sheets) and ductile (can be pulled into a wire). Diamond, on the other hand, is the hardest natural substance known. The marked differences in properties among these three atomic solids are due to differences in bonding. We will explore the bonding in solids in the next section.

14-6 Bonding in Solids

OBJECTIVES

- **To understand the interparticle forces in crystalline solids.**
- **To learn about how the bonding in metals determines metallic properties.**

We have seen that crystalline solids can be divided into three classes, depending on the fundamental particle or unit of the solid. Ionic solids consist of oppositely charged ions packed together, molecular solids contain molecules, and atomic solids have atoms as their fundamental particles. Examples of the various types of solids are given in Table 14.3.

Ionic solids were also discussed in Section 12-5.

Ionic Solids

Ionic solids are stable substances with high melting points that are held together by the strong forces that exist between oppositely charged ions. The structures of ionic solids can be visualized best by thinking of the ions as spheres packed together as efficiently

Table 14.3 ▸ Examples of the Various Types of Solids

Type of Solid	Examples	Fundamental Unit(s)
ionic	sodium chloride, $NaCl(s)$	Na^+, Cl^- ions
ionic	ammonium nitrate, $NH_4NO_3(s)$	NH_4^+, NO_3^- ions
molecular	dry ice, $CO_2(s)$	CO_2 molecules
molecular	ice, $H_2O(s)$	H_2O molecules
atomic	diamond, $C(s)$	C atoms
atomic	iron, $Fe(s)$	Fe atoms
atomic	argon, $Ar(s)$	Ar atoms

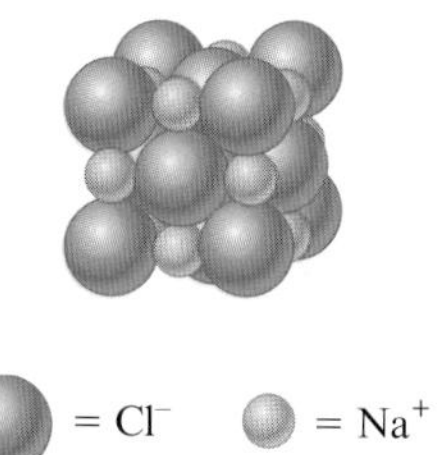

Figure 14.15 ▸ The packing of Cl^- and Na^+ ions in solid sodium chloride.

as possible. For example, in NaCl the larger Cl^- ions are packed together much like one would pack balls in a box. The smaller Na^+ ions occupy the small spaces ("holes") left among the spherical Cl^- ions, as represented in Fig. 14.15.

Molecular Solids

In a molecular solid the fundamental particle is a molecule. Examples of molecular solids include ice (contains H_2O molecules), dry ice (contains CO_2 molecules), sulfur (contains S_8 molecules), and white phosphorus (contains P_4 molecules). The latter two substances are shown in Fig. 14.16.

Molecular solids tend to melt at relatively low temperatures because the intermolecular forces that exist among the molecules are relatively weak. If the molecule has a dipole moment, dipole–dipole forces hold the solid together. In solids with nonpolar molecules, London dispersion forces hold the solid together.

Part of the structure of solid phosphorus is represented in Fig. 14.17. Note that the distances between P atoms in a given molecule are much shorter than the distances

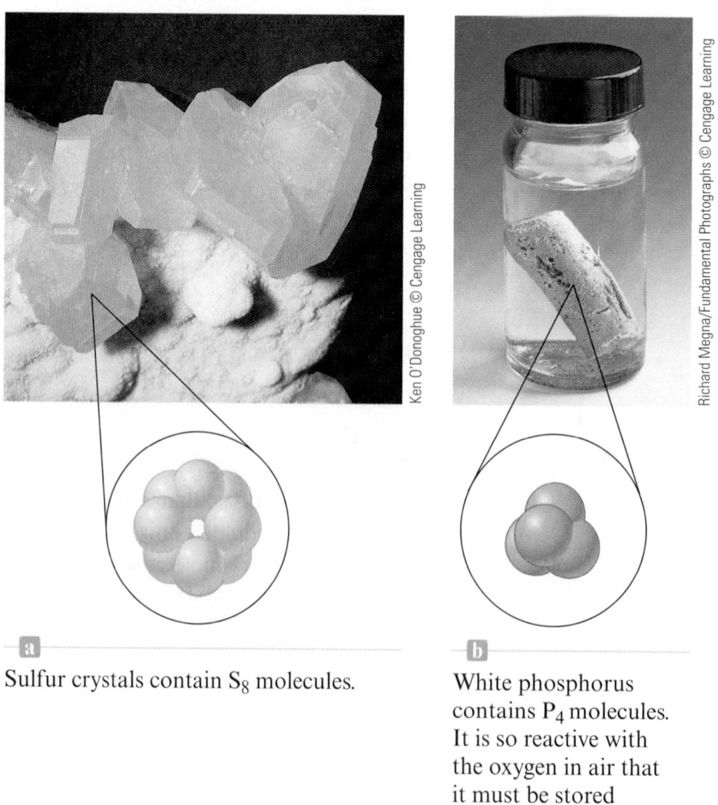

Figure 14.16 ▸

a Sulfur crystals contain S_8 molecules.

b White phosphorus contains P_4 molecules. It is so reactive with the oxygen in air that it must be stored under water.

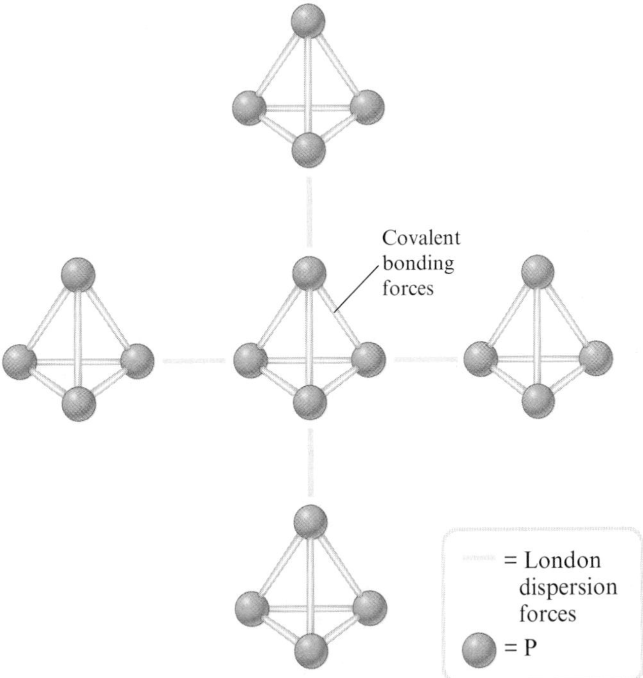

Figure 14.17 ▶ A representation of part of the structure of solid phosphorus, a molecular solid that contains P_4 molecules.

between the P_4 molecules. This is because the covalent bonds *between atoms* in the molecule are so much stronger than the London dispersion forces *between molecules.*

Atomic Solids

The properties of atomic solids vary greatly because of the different ways in which the fundamental particles, the atoms, can interact with each other. For example, the solids of the Group 8 elements have very low melting points (see Table 14.2), because these atoms, having filled valence orbitals, cannot form covalent bonds with each other. So the forces in these solids are the relatively weak London dispersion forces.

On the other hand, diamond, a form of solid carbon, is one of the hardest substances known and has an extremely high melting point (about 3500 °C). The incredible hardness of diamond arises from the very strong covalent carbon–carbon bonds in the crystal, which lead to a giant molecule. In fact, the entire crystal can be viewed as one huge molecule. A small part of the diamond structure is represented in Fig. 14.14. In diamond each carbon atom is bound covalently to four other carbon atoms to produce a very stable solid. Several other elements also form solids whereby the atoms join together covalently to form giant molecules. Silicon and boron are examples.

At this point you might be asking yourself, "Why aren't solids such as a crystal of diamond, which is a 'giant molecule,' classified as molecular solids?" The answer is that, by convention, a solid is classified as a molecular solid only if (like ice, dry ice, sulfur, and phosphorus) it contains small molecules. Substances like diamond that contain giant molecules are called network solids.

Bonding in Metals

Metals represent another type of atomic solid. Metals have familiar physical properties: they can be pulled into wires, can be hammered into sheets, and are efficient conductors of heat and electricity. However, although the shapes of most pure metals can be changed relatively easily, metals are also durable and have high melting points. These facts indicate that it is difficult to separate metal atoms but relatively easy to slide them past each other. In other words, the bonding in most metals is *strong* but *nondirectional.*

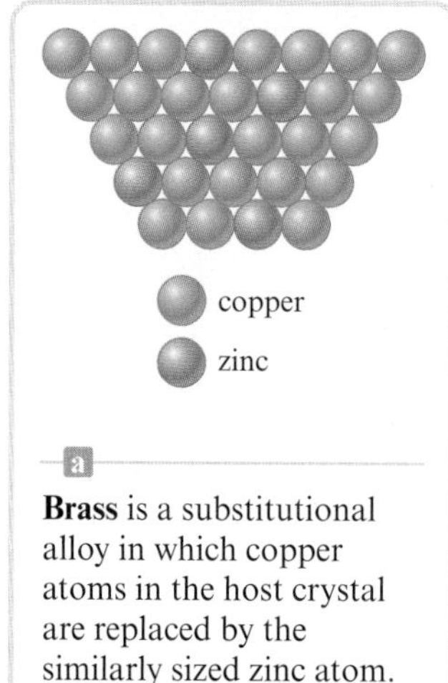

a

Brass is a substitutional alloy in which copper atoms in the host crystal are replaced by the similarly sized zinc atom.

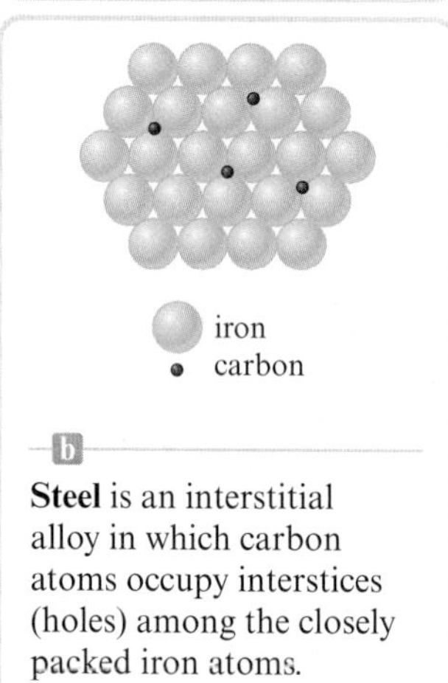

b

Steel is an interstitial alloy in which carbon atoms occupy interstices (holes) among the closely packed iron atoms.

Figure 14.18 ▸ Two types of alloys.

The simplest picture that explains these observations is the **electron sea model,** which pictures a regular array of metal atoms in a "sea" of valence electrons that are shared among the atoms in a nondirectional way and that are quite mobile in the metal crystal. The mobile electrons can conduct heat and electricity, and the atoms can be moved rather easily, as, for example, when the metal is hammered into a sheet or pulled into a wire.

Because of the nature of the metallic crystal, other elements can be introduced relatively easily to produce substances called alloys. An **alloy** is best defined as *a substance that contains a mixture of elements and has metallic properties.* There are two common types of alloys.

In a **substitutional alloy** some of the host metal atoms are *replaced* by other metal atoms of similar sizes. For example, in brass approximately one-third of the atoms in the host copper metal have been replaced by zinc atoms, as shown in Fig. 14.18(a). Sterling silver (93% silver and 7% copper) and pewter (85% tin, 7% copper, 6% bismuth, and 2% antimony) are other examples of substitutional alloys.

An **interstitial alloy** is formed when some of the interstices (holes) among the closely packed metal atoms are occupied by atoms much smaller than the host atoms, as shown in Fig. 14.18(b). Steel, the best-known interstitial alloy, contains carbon atoms in the "holes" of an iron crystal. The presence of interstitial atoms changes the properties of the host metal. Pure iron is relatively soft, ductile, and malleable because of the absence of strong directional bonding. The spherical metal atoms can be moved rather easily with respect to each other. However, when carbon, which forms strong directional bonds, is introduced into an iron crystal, the presence of the directional carbon–iron bonds makes the resulting alloy harder, stronger, and less ductile than pure iron. The amount of carbon directly affects the properties of steel. *Mild steels* (containing less than 0.2% carbon) are still ductile and malleable and are used for nails, cables, and chains. *Medium steels* (containing 0.2–0.6% carbon) are harder than mild steels and are used in rails and structural steel beams. *High-carbon steels* (containing 0.6–1.5% carbon) are tough and hard and are used for springs, tools, and cutlery.

Many types of steel contain other elements in addition to iron and carbon. Such steels are often called *alloy steels* and can be viewed as being mixed interstitial (carbon) and substitutional (other metals) alloys. An example is stainless steel, which has chromium and nickel atoms substituted for some of the iron atoms. The addition of these metals greatly increases the steel's resistance to corrosion.

T.J. Florian/Rainbow

A steel sculpture in Chicago.

Metal with a Memory

A distraught mother walks into the optical shop carrying her mangled pair of $400 eyeglasses. Her child had gotten into her purse, found her glasses, and twisted them into a pretzel. She hands them to the optometrist with little hope that they can be salvaged. The optometrist says not to worry and drops the glasses into a dish of warm water where the glasses magically spring back to their original shape. The optometrist hands the restored glasses to the woman and says there is no charge for repairing them.

How can the frames "remember" their original shape when placed in warm water? The answer is a nickel–titanium alloy called Nitinol that was developed in the late 1950s and early 1960s at the Naval Ordnance Laboratory in White Oak, Maryland, by William J. Buehler. (The name Nitinol comes from *Ni*ckel *Ti*tanium *N*aval *O*rdnance *L*aboratory.)

Nitinol has the amazing ability to remember a shape originally impressed in it. For example, note the accompanying photos. What causes Nitinol to behave this way? Although the details are too complicated to describe here, this phenomenon results from two different forms of solid Nitinol. When Nitinol is heated to a sufficiently high temperature, the Ni and Ti atoms arrange themselves in a way that leads to the most compact and regular pattern of the atoms—a form called austenite (A). When the alloy is cooled, its atoms rearrange slightly to a form called martensite (M). The shape desired (for example, the word *ICE*) is set into the alloy at a high temperature (A form), then the metal is cooled, causing it to assume the M form. In this process no visible change is noted. Then, if the image is deformed, it will magically return if the alloy is heated (hot water works fine) to a temperature that changes it back to the A form.

Nitinol has many medical applications, including hooks used by orthopedic surgeons to attach ligaments and tendons to bone and "baskets" to catch blood clots. In the latter case a length of Nitinol wire is shaped into a tiny basket and this shape is set at a high temperature. The wires forming the basket are then straightened so they can be inserted as a small bundle through a catheter. When the wires warm up in the blood, the basket shape springs back and acts as a filter to stop blood clots from moving to the heart.

One of the most promising consumer uses of Nitinol is for eyeglass frames. Nitinol is also now being used for braces to straighten crooked teeth.

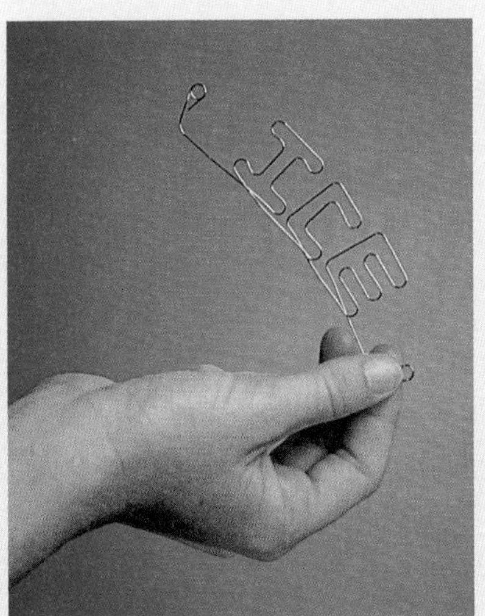

The word *ICE* is formed from Nitinol wire.

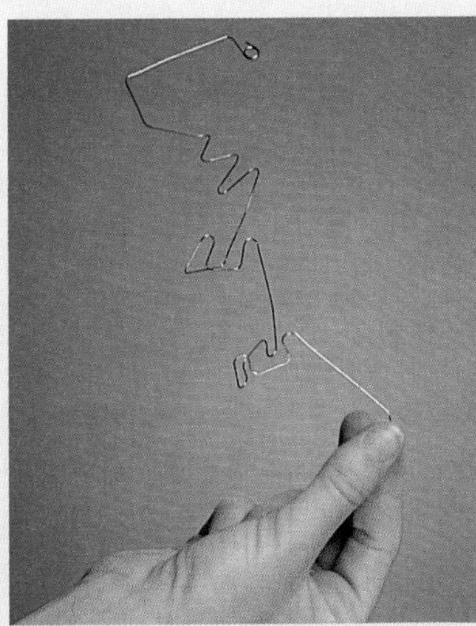

The wire is stretched to obliterate the word *ICE*.

The wire pops back to *ICE* when immersed in warm water.

CHEMISTRY IN FOCUS

Diamonds in the Ruff

Diamonds have been valued over the centuries for their luster and their durability. Natural diamonds are formed deep in the earth where the high temperatures and pressures favor this dense form of carbon. The diamonds are then slowly brought to the earth's surface by normal geologic processes, such as those operating in the kimberlite deposits in South Africa.

Synthetic diamonds.

Courtesy D.NEA Diamonds

Because of the relative rarity of natural diamonds, they are very costly. Thus there is a strong motivation to make synthetic diamonds. The first successful synthetic diamonds were made in 1955 at General Electric Research and Development Center. Believe it or not, the first material used to make synthetic diamonds was peanut butter. In fact, almost any source of carbon can be used to make synthetic diamonds. Although the early attempts to make synthetic diamonds produced black sand-like material, scientists have since learned to make beautiful, clear gem-quality diamonds that are virtually identical to natural diamonds.

Diamonds have long been used to express devotion between couples and are commonly used to formalize engagements. Recently a new use for synthetic diamonds has arisen. Some people are now using them to memorialize loved ones who have died. The idea of turning the cremated remains of a loved one into man-made diamonds started about a decade ago. Now the growth of this "industry" is being fueled by turning the ashes of deceased pets into diamonds. The company LifeGem in Elk Grove Village, Illinois, has produced more than 1000 animal diamonds in the past decade using the ashes from dogs, cats, birds, rabbits, horses, and one armadillo. The diamonds cost between \$1500 and \$15,000, the price depending on the color and size of the stone.

Producing a one-carat diamond requires less than a cup of ashes, and diamonds can be made from the hair of living pets. Many owners have the resulting diamonds set in rings or pendants, which can be worn daily to remind them of their lost pets. The wonders of chemistry now enable precious pets to become precious gems.

Interactive Example 14.4 Identifying Types of Crystalline Solids

Name the type of crystalline solid formed by each of the following substances:

a. ammonia
b. iron
c. cesium fluoride
d. argon
e. sulfur

SOLUTION

a. Solid ammonia contains NH_3 molecules, so it is a molecular solid.
b. Solid iron contains iron atoms as the fundamental particles. It is an atomic solid.
c. Solid cesium fluoride contains the Cs^+ and F^- ions. It is an ionic solid.
d. Solid argon contains argon atoms, which cannot form covalent bonds to each other. It is an atomic solid.
e. Sulfur contains S_8 molecules, so it is a molecular solid.

SELF CHECK

Exercise 14.2 Name the type of crystalline solid formed by each of the following substances:

a. sulfur trioxide
b. barium oxide
c. gold

See Problems 14.41 and 14.42. ■

CHAPTER 14 REVIEW

F directs you to the *Chemistry in Focus* feature in the chapter

Key Terms

normal boiling point (14-1)
heating/cooling curve (14-1)
normal freezing point (14-1)
intramolecular forces (14-2)
intermolecular forces (14-2)
molar heat of fusion (14-2)
molar heat of vaporization (14-2)
dipole–dipole attraction (14-3)
hydrogen bonding (14-3)
London dispersion forces (14-3)
vaporization (14-4)
evaporation (14-4)
condensation (14-4)
vapor pressure (14-4)
crystalline solids (14-5)
ionic solids (14-5)
molecular solids (14-5)
atomic solids (14-5)
electron sea model (14-6)
alloy (14-6)
substitutional alloy (14-6)
interstitial alloy (14-6)

For Review

- Phase changes can occur when energy enters or leaves a compound.
  - As energy enters a system, a substance can change from a solid to a liquid to a gas.

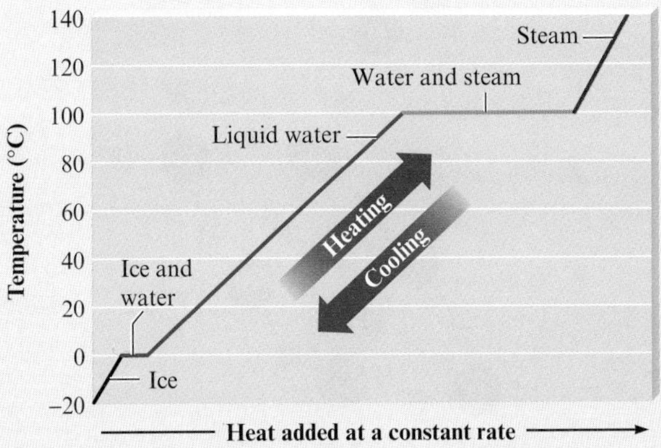

  - The molar heat of fusion is the energy required to melt 1 mole of a solid substance.
  - The normal melting point is the temperature at which a solid melts at a pressure of 1 atm.
  - The normal boiling point is the temperature at which a liquid boils at a pressure of 1 atm.
  - A substance with relatively large intermolecular forces requires relatively large amounts of energy to produce phase changes.
    - Water is an example of a substance with large intermolecular forces (hydrogen bonding).
- Intermolecular forces are the forces that occur among the molecules in a substance.
- The classes of intermolecular forces are
  - Dipole–dipole: due to the attractions among molecules that have dipole moments
  - Hydrogen bonding: particularly strong dipole–dipole forces that occur when hydrogen is bound to a highly electronegative atom (such as O, N, F)
  - London dispersion forces: forces that occur among nonpolar molecules (or atoms) when accidental dipoles develop in the electron "cloud"
- The equilibrium pressure of vapor over a liquid in a closed container is called the vapor pressure.
- The vapor pressure of a liquid is a balance between condensation and evaporation.
- Vapor pressure is relatively low for a substance (such as water) with large intermolecular forces.
- The boiling point of a liquid occurs at a temperature at which the vapor pressure of the liquid equals the atmospheric pressure.
- Types of solids
  - Ionic solid—the components are ions.
  - Molecular solid—the components are molecules.
  - Atomic solids—the components are atoms.

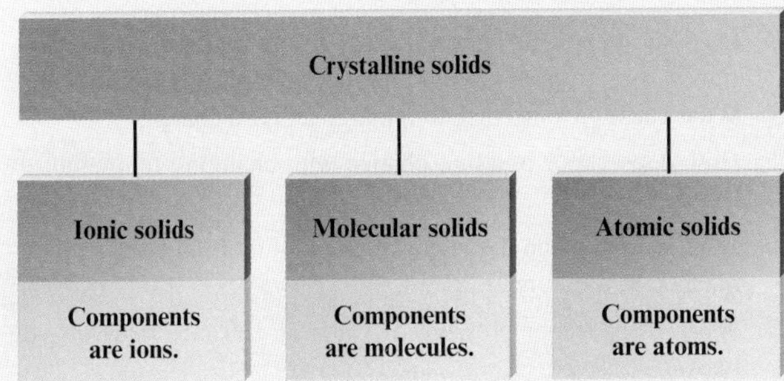

- Bonding in solids
  - Ionic solids are bound together by the attractions among the oppositely charged ions.
  - Molecular solids are bound together by the intermolecular forces among the molecules.
  - Atomic solids
    - The solids produced from noble gases have weak London dispersion forces.
    - Many atomic solids form "giant molecules" that are held together by covalent bonds.
- Metals are held together by nondirectional covalent bonds (called the electron sea model) among the closely packed atoms.
  - Metals form alloys of two types.
    - Substitutional: Different atoms are substituted for the host metal atoms.
    - Interstitial: Small atoms are introduced into the "holes" in the metallic structure.

Active Learning Questions

These questions are designed to be considered by groups of students in class. Often these questions work well for introducing a particular topic in class.

1. You seal a container half-filled with water. Which best describes what occurs in the container?
   a. Water evaporates until the air becomes saturated with water vapor; at this point, no more water evaporates.
   b. Water evaporates until the air becomes overly saturated (supersaturated) with water, and most of this water recondenses; this cycle continues until a certain amount of water vapor is present, and then the cycle ceases.
   c. The water does not evaporate because the container is sealed.
   d. Water evaporates, and then water evaporates and recondenses simultaneously and continuously.
   e. The water evaporates until it is eventually all in vapor form.

   Justify your choice and for choices you did not pick, explain what is wrong with them.
2. Explain the following: You add 100 mL of water to a 500-mL round-bottomed flask and heat the water until it is boiling. You remove the heat and stopper the flask, and the boiling stops. You then run cool water over the neck of the flask, and the boiling begins again. It seems as though you are boiling water by cooling it.
3. Is it possible for the dispersion forces in a particular substance to be stronger than hydrogen-bonding forces in another substance? Explain your answer.
4. Does the nature of intermolecular forces change when a substance goes from a solid to a liquid, or from a liquid to a gas? What causes a substance to undergo a phase change?
5. How does vapor pressure change with changing temperature? Explain.
6. What occurs when the vapor pressure of a liquid is equal to atmospheric pressure? Explain.
7. What is the vapor pressure of water at 100 °C? How do you know?
8. How do the following physical properties depend on the strength of intermolecular forces? Explain.
   a. melting point
   b. boiling point
   c. vapor pressure
9. Look at Fig. 14.2. Why doesn't temperature increase continuously over time? That is, why does the temperature stay constant for periods of time?
10. Which are stronger, intermolecular or intramolecular forces for a given molecule? What observation(s) have you made that supports this position? Explain.
11. Why does water evaporate at all?
12. Sketch a microscopic picture of water and distinguish between *intramolecular bonds* and *intermolecular forces.* Which correspond to the bonds we draw in Lewis structures?
13. Which has the stronger intermolecular forces: N_2 or H_2O? Explain.
14. Which gas would behave more ideally at the same conditions of pressure and temperature: CO or N_2? Why?
15. You have seen that the water molecule has a bent shape and therefore is a polar molecule. This accounts for many of water's interesting properties. What if the water molecule were linear? How would this affect the properties of water? How would life be different?
16. True or false? Methane (CH_4) is more likely to form stronger hydrogen bonding than is water because each methane molecule has twice as many hydrogen atoms. Provide a concise explanation of hydrogen bonding to go with your answer.
17. Why should it make sense that N_2 exists as a gas? Given your answer, how is it possible to make liquid nitrogen? Explain why lowering the temperature works.
18. White phosphorus and sulfur both are called molecular solids even though each is made of only phosphorus and sulfur, respectively. How can they be considered molecular solids? If this is true, why isn't diamond (which is made up only of carbon) a molecular solid?
19. Why is it incorrect to use the term "molecule of NaCl" but correct to use the term "molecule of H_2O"? Is the term "molecule of diamond" correct? Explain.
20. Which would you predict should be larger for a given substance: ΔH_{vap} or ΔH_{fus}? Explain why.
21. In the diagram below, which lines represent the hydrogen bonds?

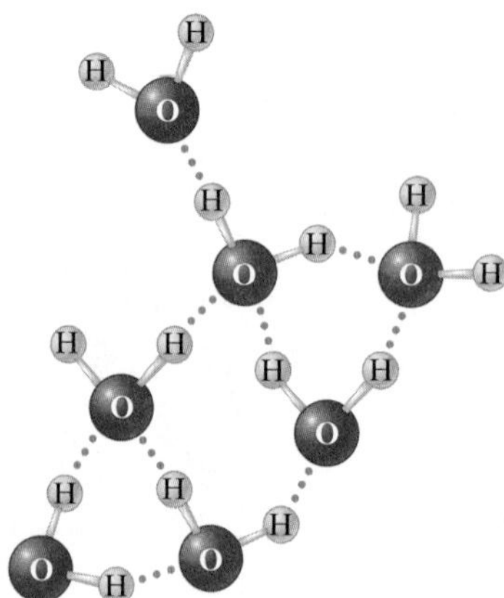

   a. The dotted lines between the hydrogen atoms of one water molecule and the oxygen atoms of a different water molecule.
   b. The solid lines between a hydrogen atom and oxygen atom in the same water molecule.
   c. Both the solid lines and dotted lines represent the hydrogen bonds.
   d. There are no hydrogen bonds represented in the diagram.

All even-numbered Questions and Problems have answers in the back of this book and solutions in the *Student Solutions Guide.*

22. Use the heating/cooling curve below to answer the following questions.

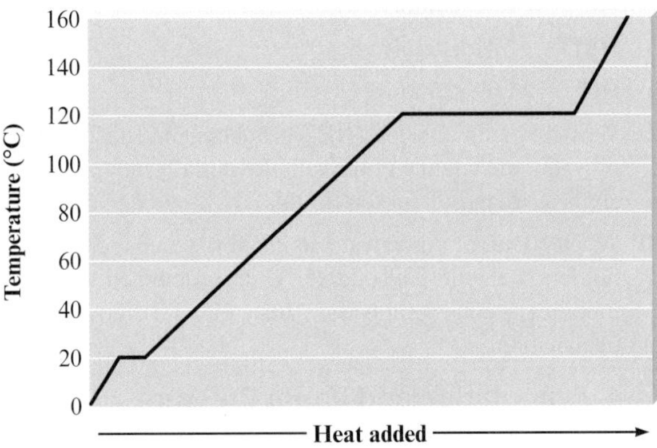

a. What is the freezing point of the liquid?
b. What is the boiling point of the liquid?
c. Which is greater: the head of fusion or the heat of vaporization? Explain.

23. Assume the two-dimensional structure of an ionic compound, M_xA_y, is

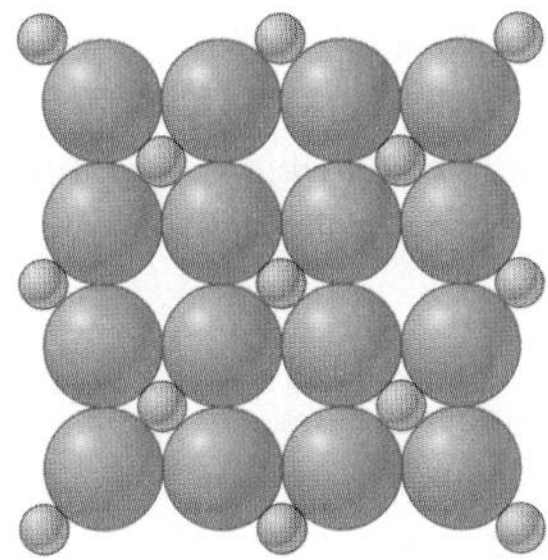

What is the empirical formula of this ionic compound?

Questions and Problems

14-1 Water and Its Phase Changes

Questions

1. Gases have (higher/lower) densities than liquids or solids.
2. Liquids and solids are (more/less) compressible than are gases.
3. What evidence do we have that the solid form of water is less dense than the liquid form of water at its freezing/melting point?
4. The enthalpy (ΔH) of *vaporization* of water is about seven times larger than water's enthalpy of *fusion* (41 kJ/mol vs. 6 kJ/mol). What does this tell us about the relative similarities among the solid, liquid, and gaseous states of water?
5. Consider a sample of ice being heated from −5 °C to +5 °C. Describe on both a macroscopic and a microscopic basis what happens to the ice as the temperature reaches 0 °C.
6. Sketch a heating/cooling curve for water, starting out at −20 °C and going up to 120 °C, applying heat to the sample at a constant rate. Mark on your sketch the portions of the curve that represent the melting of the solid and the boiling of the liquid.

14-2 Energy Requirements for the Changes of State

Questions

7. Are changes in state *physical* or *chemical* changes for molecular solids? Why?
8. Describe in detail the microscopic processes that take place when a solid melts and when a liquid boils. What kind of forces must be overcome? Are any chemical bonds broken during these processes?
9. Explain the difference between *intra*molecular and *inter*molecular forces.
10. The forces that connect two hydrogen atoms to an oxygen atom in a water molecule are (intermolecular/intramolecular), but the forces that hold water molecules close together in an ice cube are (intermolecular/intramolecular).
11. Discuss the similarities and differences between the arrangements of molecules and the forces between molecules in liquid water versus steam, and in liquid water versus ice.
12. What does the *molar heat of fusion* of a substance represent?

Problems

13. The following data have been collected for substance X. Construct a heating curve for substance X. (The drawing does not need to be absolutely to scale, but it should clearly show relative differences.)

normal melting point	−15 °C
molar heat of fusion	2.5 kJ/mol
normal boiling point	134 °C
molar heat of vaporization	55.3 kJ/mol

14. The molar heat of fusion of aluminum metal is 10.79 kJ/mol, whereas its heat of vaporization is 293.4 kJ/mol.
    a. Why is the heat of fusion of aluminum so much smaller than the heat of vaporization?
    b. What quantity of heat would be required to vaporize 1.00 g of aluminum at its normal boiling point?
    c. What quantity of heat would be evolved if 5.00 g of liquid aluminum freezes at its normal freezing point?
    d. What quantity of heat would be required to melt 0.105 mole of aluminum at its normal melting point?
15. The molar heat of fusion of benzene is 9.92 kJ/mol. Its molar heat of vaporization is 30.7 kJ/mol. Calculate the heat required to melt 8.25 g of benzene at its normal melting point. Calculate the heat required to vaporize 8.25 g of benzene at its normal boiling point. Why is the heat of vaporization more than three times the heat of fusion?
16. The molar heats of fusion and vaporization for silver are 11.3 kJ/mol and 250. kJ/mol, respectively. Silver's normal melting point is 962 °C, and its normal boiling point is 2212 °C. What quantity of heat is required to melt 12.5 g of silver at 962 °C? What quantity of heat is liberated when 4.59 g of silver vapor condenses at 2212 °C?

17. The molar heats of fusion and vaporization for water are 6.02 kJ/mol and 40.6 kJ/mol, respectively, and the specific heat capacity of liquid water is 4.18 J/g °C. What quantity of heat energy is required to melt 25.0 g of ice at 0 °C? What quantity of heat is required to vaporize 37.5 g of liquid water at 100. °C? What quantity of heat is required to warm 55.2 g of liquid water from 0 °C to 100. °C?

18. It requires 113 J to melt 1.00 g of sodium metal at its normal melting point of 98 °C. Calculate the *molar heat of fusion* of sodium.

14-3 Intermolecular Forces

Questions

19. Consider the iodine monochloride molecule, ICl. Because chlorine is more electronegative than iodine, this molecule is a dipole. How would you expect iodine monochloride molecules in the gaseous state to orient themselves with respect to each other as the sample is cooled and the molecules begin to aggregate? Sketch the orientation you would expect.

20. Dipole–dipole forces become ________ as the distance between the dipoles increases.

21. The text implies that hydrogen bonding is a special case of very strong dipole–dipole interactions possible among only certain atoms. What atoms in addition to hydrogen are necessary for hydrogen bonding? How does the small size of the hydrogen atom contribute to the unusual strength of the dipole–dipole forces involved in hydrogen bonding?

22. The normal boiling point of water is unusually high, compared to the boiling points of H_2S, H_2Se, and H_2Te. Explain this observation in terms of the *hydrogen bonding* that exists in water, but that does not exist in the other compounds.

23. Why are the dipole–dipole interactions between polar molecules *not* important in the vapor phase?

24. What are London dispersion forces, and how do they arise?

Problems

25. What type of intermolecular forces is active in the liquid state of each of the following substances?

    a. Ne
    b. CO
    c. CH_3OH
    d. Cl_2

26. Discuss the types of intermolecular forces acting in the liquid state of each of the following substances.

    a. Kr
    b. S_8
    c. NF_3
    d. H_2O

27. The boiling points of the noble gas elements are listed below. Comment on the trend in the boiling points. Why do the boiling points vary in this manner?

He	−272 °C	Kr	−152.3 °C
Ne	−245.9 °C	Xe	−107.1 °C
Ar	−185.7 °C	Rn	−61.8 °C

28. The heats of fusion of three substances are listed below. Explain the trend this list reflects.

HI	2.87 kJ/mol
HBr	2.41 kJ/mol
HCl	1.99 kJ/mol

29. When dry ammonia gas (NH_3) is bubbled into a 125-mL sample of water, the volume of the sample (initially, at least) *decreases* slightly. Suggest a reason for this.

30. When 50 mL of liquid water at 25 °C is added to 50 mL of ethanol (ethyl alcohol), also at 25 °C, the combined volume of the mixture is considerably *less* than 100 mL. Give a possible explanation.

14-4 Evaporation and Vapor Pressure

Questions

31. What is *evaporation*? What is *condensation*? Which of these processes is endothermic and which is exothermic?

32. If you've ever opened a bottle of rubbing alcohol or other solvent on a warm day, you may have heard a little "whoosh" as the vapor that had built up above the liquid escapes. Describe on a microscopic basis how a vapor pressure builds up in a closed container above a liquid. What processes in the container give rise to this phenomenon?

33. What do we mean by a *dynamic equilibrium?* Describe how the development of a vapor pressure above a liquid represents such an equilibrium.

34. Consider Fig. 14.10. Imagine you are talking to a friend who has not taken any science courses, and explain how the figure demonstrates the concept of vapor pressure and enables it to be measured.

Problems

35. Which substance in each pair would be expected to have a lower boiling point? Explain your reasoning.

    a. CH_3OH or $CH_3CH_2CH_2OH$
    b. CH_3CH_3 or CH_3CH_2OH
    c. H_2O or CH_4

36. Which substance in each pair would be expected to be more volatile at a particular temperature? Explain your reasoning.

    a. $H_2O(l)$ or $H_2S(l)$
    b. $H_2O(l)$ or $CH_3OH(l)$
    c. $CH_3OH(l)$ or $CH_3CH_2OH(l)$

37. Although water and ammonia differ in molar mass by only one unit, the boiling point of water is over 100 °C higher than that of ammonia. What forces in liquid water that do *not* exist in liquid ammonia could account for this observation?

38. Two molecules that contain the same number of each kind of atom but that have different molecular structures are said to be *isomers* of each other. For example, both ethyl alcohol and dimethyl ether (shown below) have the formula C_2H_6O and are isomers. Based on considerations of intermolecular forces, which substance would you expect to be more volatile? Which would you expect to have the higher boiling point? Explain.

dimethyl ether	ethyl alcohol
$CH_3—O—CH_3$	$CH_3—CH_2—OH$

All even-numbered Questions and Problems have answers in the back of this book and solutions in the *Student Solutions Guide*.

14-5 The Solid State: Types of Solids

Questions

39. What are crystalline solids? What kind of microscopic structure do such solids have? How is this microscopic structure reflected in the macroscopic appearance of such solids?

40. On the basis of the smaller units that make up the crystals, cite three types of crystalline solids. For each type of crystalline solid, give an example of a substance that forms that type of solid.

14-6 Bonding in Solids

Questions

41. How do *ionic* solids differ in structure from *molecular* solids? What are the fundamental particles in each? Give two examples of each type of solid and indicate the individual particles that make up the solids in each of your examples.

42. A common prank on college campuses is to switch the salt and sugar on dining hall tables, which is usually easy because the substances look so much alike. Yet, despite the similarity in their appearance, these two substances differ greatly in their properties, since one is a molecular solid and the other is an ionic solid. How do the properties differ and why?

43. Ionic solids are generally considerably harder than most molecular solids. Explain.

44. What types of forces exist between the individual particles in an ionic solid? Are these forces relatively strong or relatively weak?

45. The forces holding together a molecular solid are much (stronger/weaker) than the forces between particles in an ionic solid.

46. Explain the overall trend in melting points given below in terms of the forces among particles in the solids indicated.

Hydrogen, H_2	−259 °C
Ethyl alcohol, C_2H_5OH	−114 °C
Water, H_2O	0 °C
Sucrose, $C_{12}H_{22}O_{11}$	186 °C
Calcium chloride, $CaCl_2$	772 °C

47. What is a *network* solid? Give an example of a network solid and describe the bonding in such a solid. How does a network solid differ from a molecular solid?

48. Ionic solids do not conduct electricity in the solid state, but are strong conductors in the liquid state and when dissolved in water. Explain.

49. What is an *alloy?* Explain the differences in structure between substitutional and interstitial alloys. Give an example of each type.

F 50. The "Chemistry in Focus" segment *Metal with a Memory* discusses Nitinol, an alloy that "remembers" a shape originally impressed in it. Which elements compose Nitinol, and why is it classified as an alloy?

All even-numbered Questions and Problems have answers in the back of this book and solutions in the *Student Solutions Guide.*

Additional Problems

Matching

For Exercises 51–60 choose one of the following terms to match the definition or description given.

a. alloy
b. specific heat
c. crystalline solid
d. dipole–dipole attraction
e. equilibrium vapor pressure
f. intermolecular
g. intramolecular
h. ionic solids
i. London dispersion forces
j. molar heat of fusion
k. molar heat of vaporization
l. molecular solids
m. normal boiling point
n. semiconductor

51. boiling point at pressure of 1 atm
52. energy required to melt 1 mole of a substance
53. forces between atoms in a molecule
54. forces between molecules in a solid
55. instantaneous dipole forces for nonpolar molecules
56. lining up of opposite charges on adjacent polar molecules
57. maximum pressure of vapor that builds up in a closed container
58. mixture of elements having metallic properties overall
59. repeating arrangement of component species in a solid
60. solids that melt at relatively low temperatures

61. Given the densities and conditions of ice, liquid water, and steam listed in Table 14.1, calculate the volume of 1.0 g of water under each of these circumstances.

62. In carbon compounds a given group of atoms can often be arranged in more than one way. This means that more than one structure may be possible for the same atoms. For example, both the molecules diethyl ether and 1-butanol have the same number of each type of atom, but they have different structures and are said to be *isomers* of one another.

diethyl ether	$CH_3—CH_2—O—CH_2—CH_3$
1-butanol	$CH_3—CH_2—CH_2—CH_2—OH$

Which substance would you expect to have the larger vapor pressure? Why?

63. Which of the substances in each of the following sets would be expected to have the highest boiling point? Explain why.

a. Ga, KBr, O_2
b. Hg, NaCl, He
c. H_2, O_2, H_2O

64. Which of the substances below exhibit hydrogen bonding interactions?

a. CCl_2H_2
b. BeF_2
c. NO_3^-
d. HCN

65. When a person has a severe fever, one therapy to reduce the fever is an "alcohol rub." Explain how the evaporation of alcohol from the person's skin removes heat energy from the body.

66. Brass is an example of a/n ________ alloy, and steel is an example of a/n ________ alloy.

67. Some properties of potassium metal are summarized in the following table:

Normal melting point	63.5 °C
Normal boiling point	765.7 °C
Molar heat of fusion	2.334 kJ/mol
Molar heat of vaporization	79.87 kJ/mol
Specific heat of the solid	0.75 J/g °C

a. Calculate the quantity of heat required to heat 5.00 g of potassium from 25.3 °C to 45.2 °C.
b. Calculate the quantity of heat required to melt 1.35 moles of potassium at its normal melting point.
c. Calculate the quantity of heat required to vaporize 2.25 g of potassium at its normal boiling point.

68. What are some important uses of water, both in nature and in industry? What is the liquid range for water?

69. Describe, on both a microscopic and a macroscopic basis, what happens to a sample of water as it is cooled from room temperature to 50 °C below its normal freezing point.

70. Cake mixes and other packaged foods that require cooking often contain special directions for use at high elevations. Typically these directions indicate that the food should be cooked longer above 5000 ft. Explain why it takes longer to cook something at higher elevations.

71. Why is there no change in *intra*molecular forces when a solid is melted? Are intramolecular forces stronger or weaker than intermolecular forces?

72. What do we call the energies required, respectively, to melt and to vaporize 1 mole of a substance? Which of these energies is always larger for a given substance? Why?

73. The molar heat of vaporization of carbon disulfide, CS_2, is 28.4 kJ/mol at its normal boiling point of 46 °C. How much energy (heat) is required to vaporize 1.0 g of CS_2 at 46 °C? How much heat is evolved when 50. g of CS_2 is condensed from the vapor to the liquid form at 46 °C?

74. Which is stronger, a dipole–dipole attraction between two molecules or a covalent bond between two atoms within the same molecule? Explain.

75. For a liquid to boil, the intermolecular forces in the liquid must be overcome. Based on the types of intermolecular forces present, arrange the expected boiling points of the liquid states of the following substances in order from lowest to highest: NaCl(*l*), He(*l*), CO(*l*), H_2O(*l*).

76. What are *London dispersion forces* and how do they arise in a nonpolar molecule? Are London forces typically stronger or weaker than dipole–dipole attractions between polar molecules? Are London forces stronger or weaker than covalent bonds? Explain.

77. Discuss the types of intermolecular forces acting in the liquid state of each of the following substances.

a. N_2
b. NH_3
c. He
d. CO_2 (linear, nonpolar)

78. Discuss the types of intermolecular forces acting in the liquid state of each of the following substances.

a. Ar
b. H_2O
c. SeO_2
d. BF_3 (trigonal planar, nonpolar)

79. What do we mean when we say a liquid is *volatile?* Do volatile liquids have large or small vapor pressures? What types of intermolecular forces occur in highly volatile liquids?

80. Consider the molecules HF and HCl to answer the following questions. Explain your answers.

a. Which of the two molecules contains a *stronger* polar bond?
b. For which substance are the dipole–dipole interactions between the molecules *stronger*?
c. Starting with liquid HF and liquid HCl, as we heat these samples at the same rate, which would boil *first*?

81. Which type of solid is likely to have the highest melting point—an ionic solid, a molecular solid, or an atomic solid? Explain.

82. What types of intermolecular forces exist in a crystal of ice? How do these forces differ from the types of intermolecular forces that exist in a crystal of solid oxygen?

83. Discuss the *electron sea model* for metals. How does this model account for the fact that metals are very good conductors of electricity?

84. Water is unusual in that its solid form (ice) is less dense than its liquid form. Discuss some implications of this fact.

85. Describe in detail the microscopic processes that take place when a liquid boils. What kind of forces must be overcome? Are any chemical bonds broken during these processes?

86. Water at 100 °C (its normal boiling point) could certainly give you a bad burn if it were spilled on the skin, but steam at 100 °C could give you a much *worse* burn. Explain.

87. What is a *dipole–dipole attraction?* Give three examples of liquid substances in which you would expect dipole–dipole attractions to be large.

88. What is meant by *hydrogen bonding?* Give three examples of substances that would be expected to exhibit hydrogen bonding in the liquid state.

89. Although the noble gas elements are monatomic and could not give rise to dipole–dipole forces or hydrogen bonding, these elements still can be liquefied and solidified. Explain.

90. Describe, on a microscopic basis, the processes of *evaporation* and *condensation*. Which process requires an input of energy? Why?

F 91. The "Chemistry in Focus" segment *Gorilla Glass* discusses the glass currently used for products such as smartphones. Which addition to common glass is not an ionic solid?

All even-numbered Questions and Problems have answers in the back of this book and solutions in the *Student Solutions Guide*.

F 92. The "Chemistry in Focus" segment *Diamonds in the Ruff* discusses using the ashes of pets to produce diamonds. A diamond is an atomic solid. Why? Why are diamonds also referred to as network solids?

ChemWork Problems

These multiconcept problems (and additional ones) are found interactively online with the same type of assistance a student would get from an instructor.

93. Which of the following compound(s) exhibit only London dispersion intermolecular forces? Which compound(s) exhibit hydrogen-bonding forces? Considering only the compounds without hydrogen-bonding interactions, which compounds have dipole–dipole intermolecular forces?

a. SF_4
b. CO_2
c. CH_3CH_2OH
d. HF
e. ICl_5
f. XeF_4

94. Which of the following statements about intermolecular forces is(are) *true*?

a. London dispersion forces are the only type of intermolecular force that nonpolar molecules exhibit.
b. Molecules that have only London dispersion forces will always be gases at room temperature (25 °C).
c. The hydrogen-bonding forces in NH_3 are stronger than those in H_2O.
d. The molecules in $SO_2(g)$ exhibit dipole–dipole intermolecular interactions.
e. $CH_3CH_2CH_3$ has stronger London dispersion forces than does CH_4.

95. Identify the most important type of forces (ionic, hydrogen bonding, dipole–dipole, or London dispersion forces) among atoms or molecules present in the solids of each of the following substances.

Solid	Forces
$CF_3(CF_2CF_2)_nCF_3$	______
CO_2	______
NaI	______
NH_4Cl	______
$MgCl_2$	______

96. Rank the following compounds from lowest to highest boiling point.

a. $CH_3CH_2CH_2Cl$
b. CH_3CH_2Cl
c. $CH_3CH_2CH_2CH_2Cl$
d. CH_3Cl

97. Rank the following compounds from lowest to highest melting point.

a. CH_4
b. MgO
c. H_2O
d. H_2S

98. Which of the following statements is(are) *true*?

a. LiF will have a higher vapor pressure at 25 °C than H_2S.
b. HF will have a lower vapor pressure at −50 °C than HBr.
c. Cl_2 will have a higher boiling point than Ar.
d. HCl is more soluble in water than in CCl_4.
e. MgO will have a higher vapor pressure at 25 °C than CH_3CH_2OH.

All even-numbered Questions and Problems have answers in the back of this book and solutions in the *Student Solutions Guide*.

15 Solutions

CHAPTER 15

The salt water in the Monterey Bay Aquarium is an aqueous solution.

Most of the important chemistry that keeps plants, animals, and humans functioning occurs in aqueous solutions. Even the water that comes out of a tap is not pure water but a solution of various materials in water. For example, tap water may contain dissolved chlorine to disinfect it, dissolved minerals that make it "hard," and traces of many other substances that result from natural and human-initiated pollution. We encounter many other chemical solutions in our daily lives: air, shampoo, orange soda, coffee, gasoline, cough syrup, and many others.

A **solution** is a homogeneous mixture, a mixture in which the components are uniformly intermingled. This means that a sample from one part is the same as a sample from any other part. For example, the first sip of coffee is the same as the last sip.

The atmosphere that surrounds us is a gaseous solution containing $O_2(g)$, $N_2(g)$, and other gases randomly dispersed. Solutions can also be solids. For example, brass is a homogeneous mixture—a solution—of copper and zinc.

These examples illustrate that a solution can be a gas, a liquid, or a solid (Table 15.1). The substance present in the largest amount is called the **solvent,** and the other substance or substances are called **solutes.** For example, when we dissolve a teaspoon of sugar in a glass of water, the sugar is the solute and the water is the solvent.

Aqueous solutions are solutions with water as the solvent. Because they are so important, in this chapter we will concentrate on the properties of aqueous solutions.

Brass, a solid solution of copper and zinc, is used to make musical instruments and many other objects.
Pferd/Dreamstime.com

15-1 Solubility

OBJECTIVES

- **To understand the process of dissolving.**
- **To learn why certain components dissolve in water.**

What happens when you put a teaspoon of sugar in your iced tea and stir it, or when you add salt to water for cooking vegetables? Why do the sugar and salt "disappear" into the water? What does it mean when something dissolves—that is, when a solution forms?

We saw in Chapter 7 that when sodium chloride dissolves in water, the resulting solution conducts an electric current. This convinces us that the solution contains *ions* that can move (this is how the electric current is conducted). The dissolving of solid

Table 15.1 Various Types of Solutions

Example	State of Solution	Original State of Solute	State of Solvent
air, natural gas	gas	gas	gas
vodka in water, antifreeze in water	liquid	liquid	liquid
brass	solid	solid	solid
carbonated water (soda)	liquid	gas	liquid
seawater, sugar solution	liquid	solid	liquid

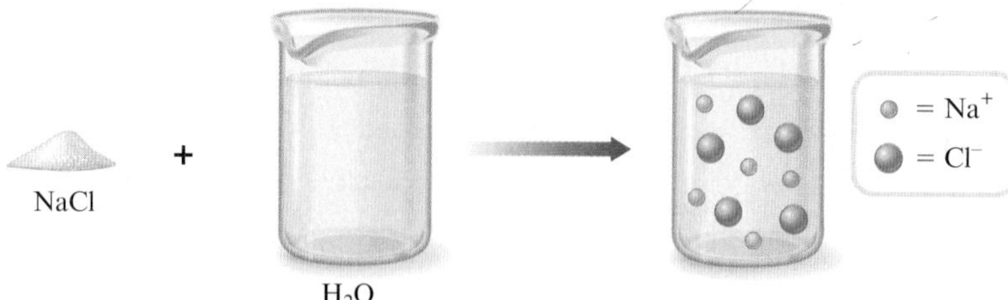

Figure 15.1 ▶ When solid sodium chloride dissolves, the ions are dispersed randomly throughout the solution.

sodium chloride in water is represented in Fig. 15.1. Notice that in the solid state the ions are packed closely together. However, when the solid dissolves, the ions are separated and dispersed throughout the solution. The strong ionic forces that hold the sodium chloride crystal together are overcome by the strong attractions between the ions and the polar water molecules. This process is represented in Fig. 15.2. Notice that each polar water molecule orients itself in a way to maximize its attraction with a Cl^- or Na^+ ion. The negative end of a water molecule is attracted to a Na^+ ion, while the positive end is attracted to a Cl^- ion. The strong forces holding the positive and negative ions in the solid are replaced by strong water–ion interactions, and the solid dissolves (the ions disperse).

It is important to remember that when an ionic substance (such as a salt) dissolves in water, it breaks up into *individual* cations (positive ions) and anions (negative ions), which are dispersed in the water. For instance, when ammonium nitrate, NH_4NO_3, dissolves in water, the resulting solution contains NH_4^+ and NO_3^- ions, which move around independently. This process can be represented as

$$NH_4NO_3(s) \xrightarrow{H_2O(l)} NH_4^+(aq) + NO_3^-(aq)$$

where (aq) indicates that the ions are surrounded by water molecules.

Water also dissolves many nonionic substances. Sugar is one example of a nonionic solute that is very soluble in water. Another example is ethanol, C_2H_5OH. Wine, beer, and mixed drinks are aqueous solutions of ethanol (and other substances). Why is ethanol so soluble in water? The answer lies in the structure of the ethanol molecule

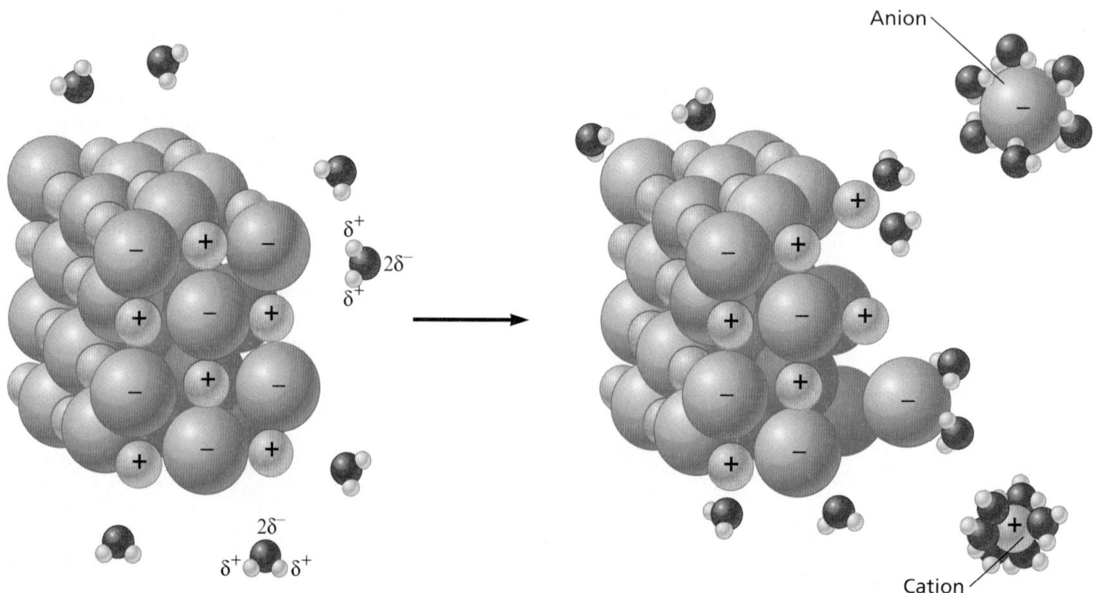

Figure 15.2 ▶ Polar water molecules interact with the positive and negative ions of a salt. These interactions replace the strong ionic forces holding the ions together in the undissolved solid, thus assisting in the dissolving process.

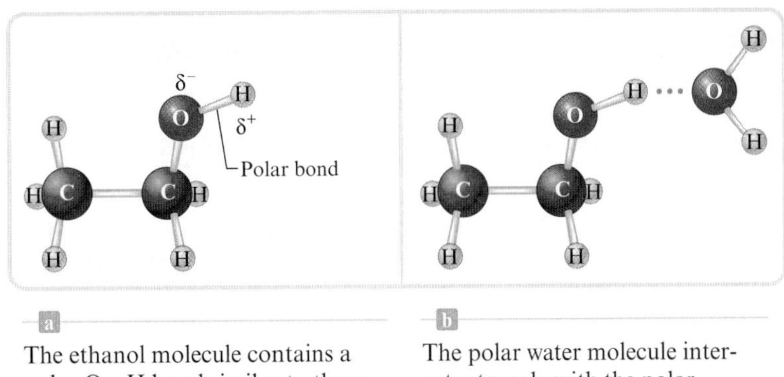

Figure 15.3 ▸ a The ethanol molecule contains a polar O—H bond similar to those in the water molecule. b The polar water molecule interacts strongly with the polar O—H bond in ethanol.

The oil from this tanker spreads out on the water as fire boats spray water to put out the fire.

[Fig. 15.3(a)]. The molecule contains a polar O—H bond like those in water, which makes it very compatible with water. Just as hydrogen bonds form among water molecules in pure water (Fig. 14.6), ethanol molecules can form hydrogen bonds with water molecules in a solution of the two. This is shown in Fig. 15.3(b).

The sugar molecule (common table sugar has the chemical name sucrose) is shown in Fig. 15.4. Notice that this molecule has many polar O—H groups, each of which can hydrogen-bond to a water molecule. Because of the attractions between sucrose and water molecules, solid sucrose is quite soluble in water.

Many substances do not dissolve in water. For example, when petroleum leaks from a damaged tanker, it does not disperse uniformly in the water (does not dissolve) but rather floats on the surface because its density is less than that of water. Petroleum is a mixture of molecules like the one shown in Fig. 15.5. Since carbon and hydrogen have very similar electronegativities, the bonding electrons are shared almost equally and the bonds are essentially nonpolar. The resulting molecule with its nonpolar bonds cannot form attractions to the polar water molecules and this prevents it from being soluble in water. This situation is represented in Fig. 15.6.

Notice in Fig. 15.6 that the water molecules in liquid water are associated with each other by hydrogen-bonding interactions. For a solute to dissolve in water, a "hole" must be made in the water structure for each solute particle. This will occur only if the

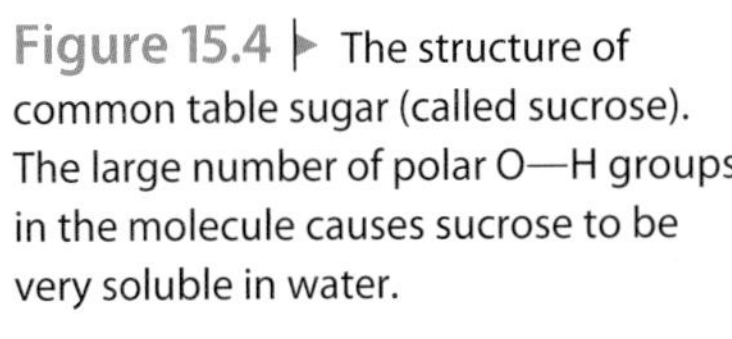

H CH₂OH O HO C C H H OH C C H OH H CH₂OH O C C H OH C CH₂OH O C C OH H

Figure 15.4 ▸ The structure of common table sugar (called sucrose). The large number of polar O—H groups in the molecule causes sucrose to be very soluble in water.

CH_3 CH_2 CH_2 CH_2 CH_2 CH_2 CH_2 CH_2 CH_2 CH_2 CH_2 CH_2 CH_2 CH_2 CH_2 CH_2 CH_2 CH_3

Figure 15.5 ▸ A molecule typical of those found in petroleum. The bonds are not polar.

Water, Water, Everywhere, But . . .

Although more than two thirds of the earth is covered by water, the earth is facing increasing water shortages as the global population grows. Why is that? The problem is that most of the earth's water is ocean water, which contains such high concentrations of dissolved minerals that the ocean water can't be consumed by humans and it kills crops. Humans need "fresh water" to sustain their lives. This fresh water is ultimately derived from rain, which supplies water to our lakes, rivers, and underground aquifers. However, as our populations are expanding, our supply of useable water is running short.

Because the earth has so much "salt water," the obvious answer to the water problem would seem to be removing minerals from ocean water, a process called "desalination." We know how to desalinate seawater. By forcing seawater through special membranes that trap the dissolved ions but allow water molecules to pass through, we can produce useable water. This is the most common method for producing drinking water in the Middle East and other arid regions. Worldwide more than 13,000 desalination plants produce more than 12 billion gallons of useable water every day.

Given the widespread use of desalination in the world, why is the process not used very much in the United States? The answer is simple: cost. Because the desalination process requires high-pressure pumps to force seawater through the special membranes, a great deal of electricity is needed. Also, the special membranes are very expensive. Thus, desalination costs currently are 30% higher than the costs of traditional sources of water. However, as water supplies are becoming tighter, major users of water are more willing to bear the higher costs of desalination. Water-challenged California is a good example. Poseidon Resources Corporation recently signed a contract to build a $300 million desalination plant in Carlsbad, California, a city just north of San Diego. This facility will be the largest in the Western Hemisphere, producing enough water for 100,000 homes (50 million gallons per day).

Poseidon is planning a second plant in Huntington Beach, California, and as many as 20 other similar projects are in the planning stages. This technology could go a long way toward satisfying our thirst in the future.

lost water–water interactions are replaced by similar water–solute interactions. In the case of sodium chloride, strong interactions occur between the polar water molecules and the Na^+ and Cl^- ions. This allows the sodium chloride to dissolve. In the case of ethanol or sucrose, hydrogen-bonding interactions can occur between the O—H groups on these molecules and water molecules, making these substances soluble as well. But

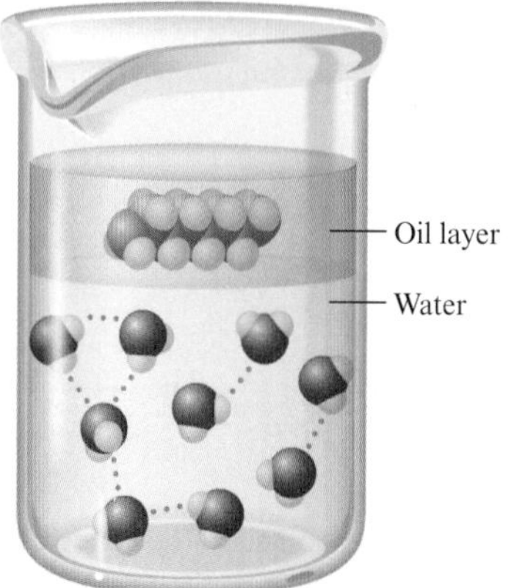

Figure 15.6 An oil layer floating on water. For a substance to dissolve, the water–water hydrogen bonds must be broken to make a "hole" for each solute particle. However, the water–water interactions will break only if they are replaced by similar strong interactions with the solute.

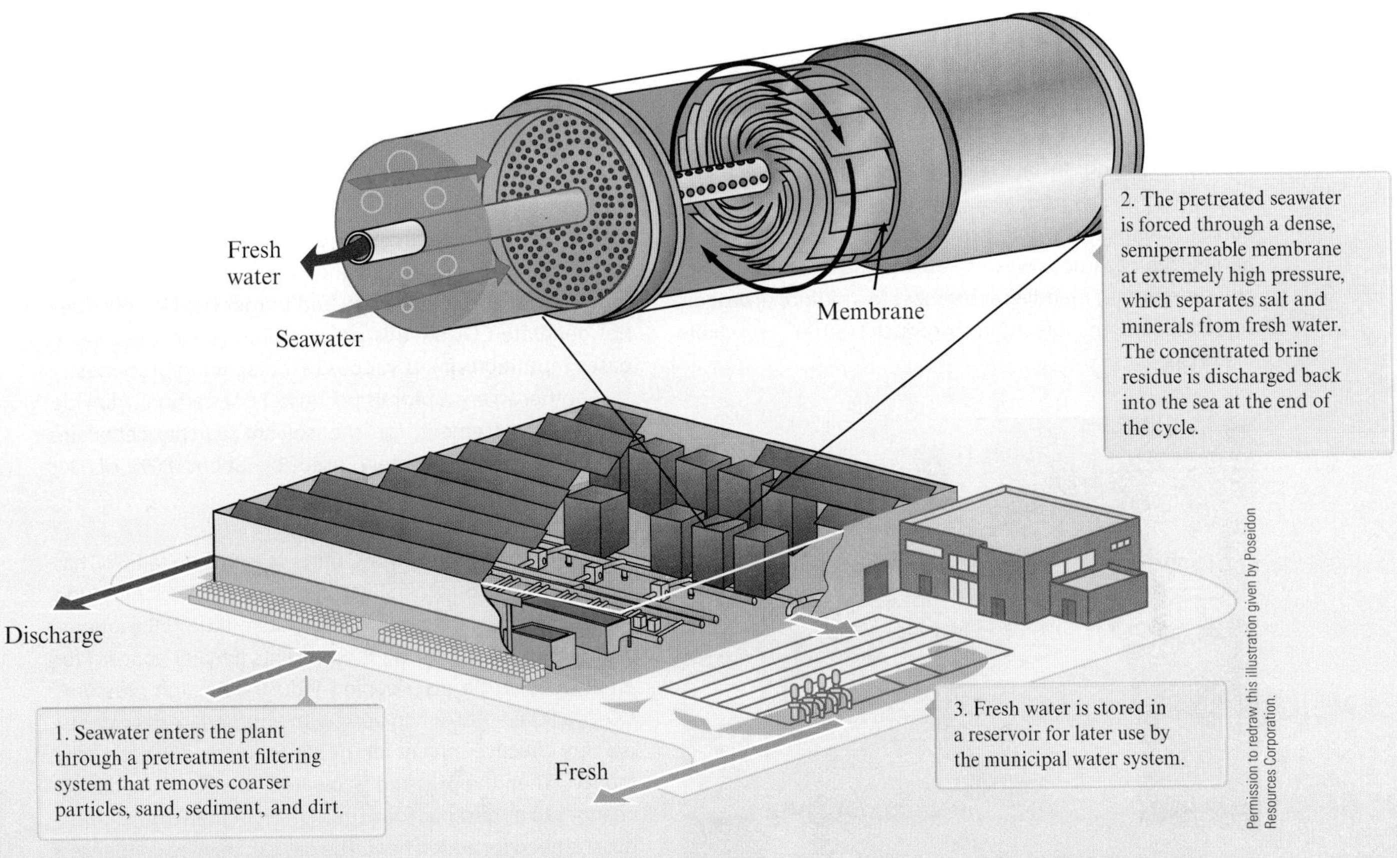

Fresh water

Seawater

Membrane

2. The pretreated seawater is forced through a dense, semipermeable membrane at extremely high pressure, which separates salt and minerals from fresh water. The concentrated brine residue is discharged back into the sea at the end of the cycle.

Discharge

1. Seawater enters the plant through a pretreatment filtering system that removes coarser particles, sand, sediment, and dirt.

Fresh

3. Fresh water is stored in a reservoir for later use by the municipal water system.

Permission to redraw this illustration given by Poseidon Resources Corporation.

A schematic diagram of the desalination plant in Carlsbad, California.

oil molecules are not soluble in water, because the many water–water interactions that would have to be broken to make "holes" for these large molecules are not replaced by favorable water–solute interactions.

These considerations account for the observed behavior that "*like dissolves like.*" In other words, we observe that a given solvent usually dissolves solutes that have polarities similar to its own. For example, water dissolves most polar solutes, because the solute–solvent interactions formed in the solution are similar to the water–water interactions present in the pure solvent. Likewise, nonpolar solvents dissolve nonpolar solutes. For example, dry-cleaning solvents used for removing grease stains from clothes are nonpolar liquids. "Grease" is composed of nonpolar molecules, so a nonpolar solvent is needed to remove a grease stain.

15-2 Solution Composition: An Introduction

OBJECTIVE **To learn qualitative terms associated with the concentration of a solution.**

Even for very soluble substances, there is a limit to how much solute can be dissolved in a given amount of solvent. For example, when you add sugar to a glass of water, the sugar rapidly disappears at first. However, as you continue to add more sugar, at some

Green Chemistry

Although some chemical industries have been culprits in the past for fouling the earth's environment, that situation is rapidly changing. In fact, a quiet revolution is sweeping through chemistry from academic labs to *Fortune* 500 companies. Chemistry is going green. *Green chemistry* means minimizing hazardous wastes, substituting water and other environmentally friendlier substances for traditional organic solvents, and manufacturing products out of recyclable materials.

Michael Newman/PhotoEdit

The dry-cleaning agent PERC is a health concern for workers in the dry-cleaning industry.

A good example of green chemistry is the increasing use of carbon dioxide, one of the by-products of the combustion of fossil fuels. For example, the Dow Chemical Company is now using CO_2 rather than chlorofluorocarbons (CFCs; substances known to catalyze the decomposition of protective stratospheric ozone) to put the "sponginess" into polystyrene egg cartons, meat trays, and burger boxes. Dow does not generate CO_2 for this process but instead uses waste gases captured from its various manufacturing processes.

Another very promising use of carbon dioxide is as a replacement for the solvent perchloroethylene (PERC; $Cl_2C{=}CCl_2$), now used by about 80% of dry cleaners in the United States. Chronic exposure to PERC has been linked to kidney and liver damage and cancer. Although PERC is not a hazard to the general public (little PERC adheres to dry-cleaned garments), it represents a major concern for employees in the dry-cleaning industry. At high pressures CO_2 is a liquid that, when used with appropriate detergents, is a very effective solvent for the soil found on dry-clean-only fabrics. When the pressure is lowered, the CO_2 immediately changes to its gaseous form, quickly drying the clothes without the need for added heat. The gas can then be condensed and reused for the next batch of clothes.

The good news is that green chemistry makes sense economically. When all of the costs are taken into account, green chemistry is usually cheaper chemistry as well. Everybody wins.

point the solid no longer dissolves but collects at the bottom of the glass. When a solution contains as much solute as will dissolve at that temperature, we say it is **saturated.** If a solid solute is added to a solution already saturated with that solute, the added solid does not dissolve. A solution that has *not* reached the limit of solute that will dissolve in it is said to be **unsaturated.** When more solute is added to an unsaturated solution, it dissolves.

Although a chemical compound always has the same composition, a solution is a mixture and the amounts of the substances present can vary in different solutions. For example, coffee can be strong or weak. Strong coffee has more coffee dissolved in a given amount of water than weak coffee. To describe a solution completely, we must specify the amounts of solvent and solute. We sometimes use the qualitative terms *concentrated* and *dilute* to describe a solution. A relatively large amount of solute is dissolved in a **concentrated** solution (strong coffee is concentrated). A relatively small amount of solute is dissolved in a **dilute** solution (weak coffee is dilute).

Although these qualitative terms serve a useful purpose, we often need to know the exact amount of solute present in a given amount of solution. In the next several sections, we will consider various ways to describe the composition of a solution.

15-3 Solution Composition: Mass Percent

OBJECTIVE **To understand the concentration term *mass percent* and learn how to calculate it.**

Describing the composition of a solution means specifying the amount of solute present in a given quantity of the solution. We typically give the amount of solute in terms of mass (number of grams) or in terms of moles. The quantity of solution is defined in terms of mass or volume.

One common way of describing a solution's composition is **mass percent** (sometimes called *weight percent*), which expresses the mass of solute present in a given mass of solution. The definition of mass percent is

$$\text{Mass percent} = \frac{\text{mass of solute}}{\text{mass of solution}} \times 100\%$$

$$= \frac{\text{grams of solute}}{\text{grams of solute} + \text{grams of solvent}} \times 100\%$$

For example, suppose a solution is prepared by dissolving 1.0 g of sodium chloride in 48 g of water. The solution has a mass of 49 g (48 g of H_2O plus 1.0 g of NaCl), and there is 1.0 g of solute (NaCl) present. The mass percent of solute, then, is

$$\frac{1.0 \text{ g solute}}{49 \text{ g solution}} \times 100\% = 0.020 \times 100\% = 2.0\% \text{ NaCl}$$

Interactive Example 15.1

Solution Composition: Calculating Mass Percent

A solution is prepared by mixing 1.00 g of ethanol, C_2H_5OH, with 100.0 g of water. Calculate the mass percent of ethanol in this solution.

SOLUTION

Where Are We Going?

We want to determine the mass percent of a given ethanol solution.

What Do We Know?

- We have 1.00 g ethanol (C_2H_5OH) in 100.0 g water (H_2O).
- $\text{Mass percent} = \frac{\text{mass of solute}}{\text{mass of solution}} \times 100\%.$

How Do We Get There?

In this case we have 1.00 g of solute (ethanol) and 100.0 g of solvent (water). We now apply the definition of mass percent.

$$\text{Mass percent } C_2H_5OH = \left(\frac{\text{grams of } C_2H_5OH}{\text{grams of solution}}\right) \times 100\%$$

$$= \left(\frac{1.00 \text{ g } C_2H_5OH}{100.0 \text{ g } H_2O + 1.00 \text{ g } C_2H_5OH}\right) \times 100\%$$

$$= \frac{1.00 \text{ g}}{101.0 \text{ g}} \times 100\%$$

$$= 0.990\% \text{ } C_2H_5OH$$

REALITY CHECK The percent mass is just under 1%, which makes sense because we have 1.00 g of ethanol in a bit more than 100.0 g of solution.

SELF CHECK

Exercise 15.1 A 135-g sample of seawater is evaporated to dryness, leaving 4.73 g of solid residue (the salts formerly dissolved in the seawater). Calculate the mass percent of solute present in the original seawater.

See Problems 15.15 and 15.16. ■

Interactive Example 15.2

Solution Composition: Determining Mass of Solute

Although milk is not a true solution (it is really a suspension of tiny globules of fat, protein, and other substrates in water), it does contain a dissolved sugar called lactose. Cow's milk typically contains 4.5% by mass of lactose, $C_{12}H_{22}O_{11}$. Calculate the mass of lactose present in 175 g of milk.

SOLUTION

Where Are We Going?

We want to determine the mass of lactose present in 175 g of milk.

What Do We Know?

- We have 175 g of milk.
- Milk contains 4.5% by mass of lactose, $C_{12}H_{22}O_{11}$.
- Mass percent $= \dfrac{\text{mass of solute}}{\text{mass of solution}} \times 100\%$.

How Do We Get There?

Using the definition of mass percent, we have

$$\text{Mass percent} = \frac{\text{grams of solute}}{\text{grams of solution}} \times 100\%$$

We now substitute the quantities we know:

$$\text{Mass percent} = \frac{\overbrace{\text{grams of solute}}^{\text{Mass of lactose}}}{\underbrace{175\text{ g}}_{\text{Mass of milk}}} \times 100\% = \overset{\text{Mass percent}}{4.5\%}$$

We now solve for grams of solute by multiplying both sides by 175 g,

$$\cancel{175\text{ g}} \times \frac{\text{grams of solute}}{\cancel{175\text{ g}}} \times 100\% = 4.5\% \times 175\text{ g}$$

and then dividing both sides by 100%,

$$\text{Grams of solute} \times \frac{\cancel{100\%}}{\cancel{100\%}} = \frac{4.5\%}{100\%} \times 175\text{ g}$$

to give

$$\text{Grams of solute} = 0.045 \times 175\text{ g} = 7.9\text{ g lactose}$$

SELF CHECK

Exercise 15.2 What mass of water must be added to 425 g of formaldehyde to prepare a 40.0% (by mass) solution of formaldehyde? This solution, called formalin, is used to preserve biological specimens.

HINT Substitute the known quantities into the definition for mass percent, and then solve for the unknown quantity (mass of solvent).

See Problems 15.17 and 15.18. ■

15-4 Solution Composition: Molarity

OBJECTIVES

- **To understand molarity.**
- **To learn to use molarity to calculate the number of moles of solute present.**

When a solution is described in terms of mass percent, the amount of solution is given in terms of its mass. However, it is often more convenient to measure the volume of a solution than to measure its mass. Because of this, chemists often describe a solution in terms of concentration. We define the *concentration* of a solution as the amount of solute in a *given volume* of solution. The most commonly used expression of concentration is **molarity (*M*).** Molarity describes the amount of solute in moles and the volume of the solution in liters. Molarity is *the number of moles of solute per volume of solution in liters.* That is

$$M = \text{molarity} = \frac{\text{moles of solute}}{\text{liters of solution}} = \frac{\text{mol}}{\text{L}}$$

A solution that is 1.0 molar (written as 1.0 M) contains 1.0 mole of solute per liter of solution.

Interactive Example 15.3 Solution Composition: Calculating Molarity, I

Calculate the molarity of a solution prepared by dissolving 11.5 g of solid NaOH in enough water to make 1.50 L of solution.

SOLUTION

Where Are We Going?

We want to determine the concentration (M) of a solution of NaOH.

What Do We Know?

- 11.5 g of NaOH is dissolved in 1.50 L of solution.
- $M = \dfrac{\text{moles of solute}}{\text{liters of solution}}$.

What Information Do We Need?

- We need to know the number of moles of NaOH in 11.5 g NaOH.

How Do We Get There?

We have the mass (in grams) of solute, so we need to convert the mass of solute to moles (using the molar mass of NaOH). Then we can divide the number of moles by the volume in liters.

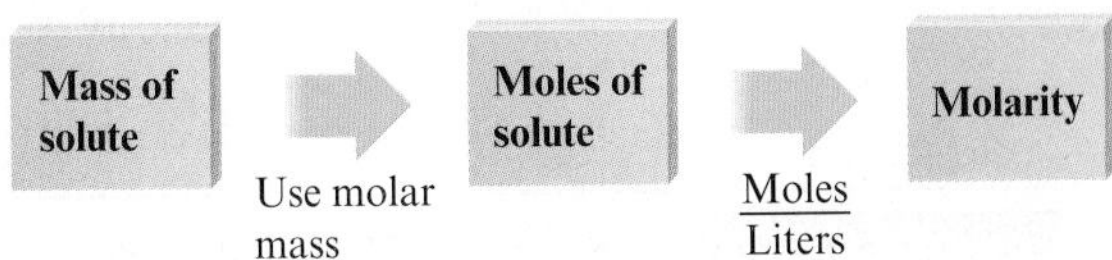

We compute the number of moles of solute, using the molar mass of NaOH (40.0 g).

$$11.5 \text{ g NaOH} \times \frac{1 \text{ mol NaOH}}{40.0 \text{ g NaOH}} = 0.288 \text{ mol NaOH}$$

Then we divide by the volume of the solution in liters.

$$\text{Molarity} = \frac{\text{moles of solute}}{\text{liters of solution}} = \frac{0.288 \text{ mol NaOH}}{1.50 \text{ L solution}} = 0.192\ M \text{ NaOH}$$ ■

Interactive Example 15.4

Solution Composition: Calculating Molarity, II

Calculate the molarity of a solution prepared by dissolving 1.56 g of gaseous HCl into enough water to make 26.8 mL of solution.

SOLUTION

Where Are We Going?

We want to determine the concentration (M) of a solution of HCl.

What Do We Know?

- 1.56 g of HCl is dissolved in 26.8 mL of solution.
- $M = \frac{\text{moles of solute}}{\text{liters of solution}}$.

What Information Do We Need?

- We need to know the number of moles of HCl in 1.56 g.
- We need to know the volume of the solution in liters.

How Do We Get There?

We must change 1.56 g of HCl to moles of HCl, and then we must change 26.8 mL to liters (because molarity is defined in terms of liters). First we calculate the number of moles of HCl (molar mass = 36.5 g).

$$1.56 \text{ g } \cancel{\text{HCl}} \times \frac{1 \text{ mol HCl}}{36.5 \text{ g } \cancel{\text{HCl}}} = 0.0427 \text{ mol HCl}$$

$$= 4.27 \times 10^{-2} \text{ mol HCl}$$

Next we change the volume of the solution from milliliters to liters, using the equivalence statement 1 L = 1000 mL, which gives the appropriate conversion factor.

$$26.8 \cancel{\text{mL}} \times \frac{1 \text{ L}}{1000 \cancel{\text{mL}}} = 0.0268 \text{ L}$$

$$= 2.68 \times 10^{-2} \text{ L}$$

Finally, we divide the moles of solute by the liters of solution.

$$\text{Molarity} = \frac{4.27 \times 10^{-2} \text{ mol HCl}}{2.68 \times 10^{-2} \text{ L solution}} = 1.59\ M \text{ HCl}$$

SELF CHECK

Exercise 15.3 Calculate the molarity of a solution prepared by dissolving 1.00 g of ethanol, C_2H_5OH, in enough water to give a final volume of 101 mL.

See Problems 15.37 through 15.42. ■

It is important to realize that the description of a solution's composition may not accurately reflect the true chemical nature of the solute as it is present in the dissolved state. Solute concentration is always written in terms of the form of the solute *before* it dissolves. For example, describing a solution as 1.0 *M* NaCl means that the solution was prepared by dissolving 1.0 mole of solid NaCl in enough water to make 1.0 L of solution; it does not mean that the solution contains 1.0 mole of NaCl units. Actually the solution contains 1.0 mole of Na^+ ions and 1.0 mole of Cl^- ions. That is, it contains 1.0 *M* Na^+ and 1.0 *M* Cl^-.

Interactive Example 15.5

Solution Composition: Calculating Ion Concentration from Molarity

Give the concentrations of all the ions in each of the following solutions:

a. 0.50 *M* $Co(NO_3)_2$

b. 1 *M* $FeCl_3$

A solution of cobalt(II) nitrate.

Tom Pantages

SOLUTION

Remember, ionic compounds separate into the component ions when they dissolve in water.

$Co(NO_3)_2$ → Co^{2+}, NO_3^-, NO_3^-

$FeCl_3$ → Fe^{3+}, Cl^-, Cl^-, Cl^-

a. When solid $Co(NO_3)_2$ dissolves, it produces ions as follows:

$$Co(NO_3)_2(s) \xrightarrow{H_2O(l)} Co^{2+}(aq) + 2NO_3^-(aq)$$

which we can represent as

$$1 \text{ mol } Co(NO_3)_2(s) \xrightarrow{H_2O(l)} 1 \text{ mol } Co^{2+}(aq) + 2 \text{ mol } NO_3^-(aq)$$

Therefore, a solution that is 0.50 *M* $Co(NO_3)_2$ contains 0.50 *M* Co^{2+} and (2×0.50) *M* NO_3^-, or 1.0 *M* NO_3^-.

b. When solid $FeCl_3$ dissolves, it produces ions as follows:

$$FeCl_3(s) \xrightarrow{H_2O(l)} Fe^{3+}(aq) + 3Cl^-(aq)$$

or

$$1 \text{ mol } FeCl_3(s) \xrightarrow{H_2O(l)} 1 \text{ mol } Fe^{3+}(aq) + 3 \text{ mol } Cl^-(aq)$$

A solution that is 1 *M* $FeCl_3$ contains 1 *M* Fe^{3+} ions and 3 *M* Cl^- ions.

SELF CHECK

Exercise 15.4 Give the concentrations of the ions in each of the following solutions:

a. 0.10 *M* Na_2CO_3

b. 0.010 *M* $Al_2(SO_4)_3$

See Problems 15.49 and 15.50. ■

Math skill builder

$$M = \frac{\text{moles of solute}}{\text{liters of solution}}$$

Liters × *M* ⇨ Moles of solute

Often we need to determine the number of moles of solute present in a given volume of a solution of known molarity. To do this, we use the definition of molarity. When we multiply the molarity of a solution by the volume (in liters), we get the moles of solute present in that sample:

$$\text{Liters of solution} \times \text{molarity} = \cancel{\text{liters of solution}} \times \frac{\text{moles of solute}}{\cancel{\text{liters of solution}}}$$

$$= \text{moles of solute}$$

Interactive Example 15.6 Solution Composition: Calculating Number of Moles from Molarity

How many moles of Ag^+ ions are present in 25 mL of a 0.75 *M* $AgNO_3$ solution?

SOLUTION

Where Are We Going?

We want to determine the number of moles of Ag^+ in a solution.

What Do We Know?

- We have 25 mL of 0.75 *M* $AgNO_3$.
- $M = \frac{\text{moles of solute}}{\text{liters of solution}}$.

How Do We Get There?

A 0.75 *M* $AgNO_3$ solution contains 0.75 *M* Ag^+ ions and 0.75 *M* NO_3^- ions. Next we must express the volume in liters. That is, we must convert from mL to L.

$$25 \cancel{\text{mL}} \times \frac{1 \text{ L}}{1000 \cancel{\text{mL}}} = 0.025 \text{ L} = 2.5 \times 10^{-2} \text{ L}$$

Now we multiply the volume times the molarity.

$$2.5 \times 10^{-2} \cancel{\text{L solution}} \times \frac{0.75 \text{ mol } Ag^+}{\cancel{\text{L solution}}} = 1.9 \times 10^{-2} \text{ mol } Ag^+$$

SELF CHECK

Exercise 15.5 Calculate the number of moles of Cl^- ions in 1.75 L of 1.0×10^{-3} *M* $AlCl_3$.

See Problems 15.49 and 15.50. ■

A **standard solution** is a solution whose *concentration is accurately known.* When the appropriate solute is available in pure form, a standard solution can be prepared by weighing out a sample of solute, transferring it completely to a *volumetric flask* (a flask of accurately known volume), and adding enough solvent to bring the volume up to the mark on the neck of the flask. This procedure is illustrated in Fig. 15.7.

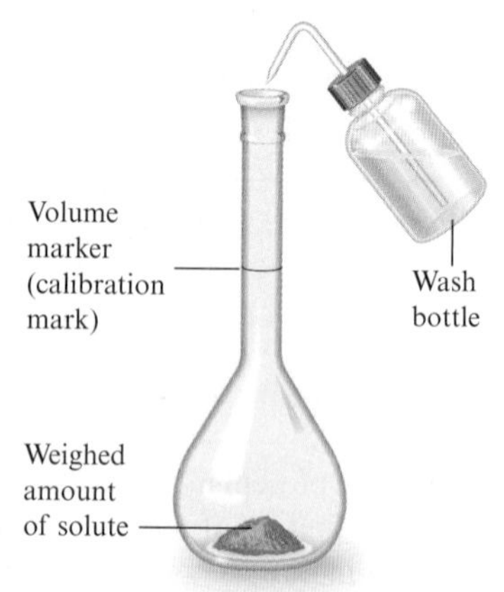

a Put a weighed amount of a substance (the solute) into the volumetric flask, and add a small quantity of water.

b Dissolve the solid in the water by gently swirling the flask (with the stopper in place).

c Add more water (with gentle swirling) until the level of the solution just reaches the mark etched on the neck of the flask. Then mix the solution thoroughly by inverting the flask several times.

Figure 15.7 ▸ Steps involved in the preparation of a standard aqueous solution.

Interactive Example 15.7

Solution Composition: Calculating Mass from Molarity

To analyze the alcohol content of a certain wine, a chemist needs 1.00 L of an aqueous 0.200 *M* $K_2Cr_2O_7$ (potassium dichromate) solution. How much solid $K_2Cr_2O_7$ (molar mass = 294.2 g/mol) must be weighed out to make this solution?

SOLUTION

Where Are We Going?

We want to determine the mass of $K_2Cr_2O_7$ needed to make a given solution.

What Do We Know?

- We want 1.00 L of 0.200 *M* $K_2Cr_2O_7$.
- The molar mass of $K_2Cr_2O_7$ is 294.2 g/mol.
- $M = \frac{\text{moles of solute}}{\text{liters of solution}}$.

How Do We Get There?

We need to calculate the number of grams of solute ($K_2Cr_2O_7$) present (and thus the mass needed to make the solution). First we determine the number of moles of $K_2Cr_2O_7$ present by multiplying the volume (in liters) by the molarity.

Liters × *M* ⇨ Moles of solute

$$1.00 \text{ L} \cancel{\text{ solution}} \times \frac{0.200 \text{ mol } K_2Cr_2O_7}{\text{L} \cancel{\text{ solution}}} = 0.200 \text{ mol } K_2Cr_2O_7$$

Then we convert the moles of $K_2Cr_2O_7$ to grams, using the molar mass of $K_2Cr_2O_7$ (294.2 g/mol).

$$0.200 \cancel{\text{ mol } K_2Cr_2O_7} \times \frac{294.2 \text{ g } K_2Cr_2O_7}{\cancel{\text{mol } K_2Cr_2O_7}} = 58.8 \text{ g } K_2Cr_2O_7$$

Therefore, to make 1.00 L of 0.200 *M* $K_2Cr_2O_7$, the chemist must weigh out 58.8 g of $K_2Cr_2O_7$ and dissolve it in enough water to make 1.00 L of solution. This is most easily done by using a 1.00-L volumetric flask (Fig. 15.7).

SELF CHECK

Exercise 15.6 Formalin is an aqueous solution of formaldehyde, HCHO, used as a preservative for biologic specimens. How many grams of formaldehyde must be used to prepare 2.5 L of 12.3 *M* formalin?

See Problems 15.51 and 15.52. ■

15-5 Dilution

OBJECTIVE **To learn to calculate the concentration of a solution made by diluting a stock solution.**

To save time and space in the laboratory, solutions that are routinely used are often purchased or prepared in concentrated form (called *stock solutions*). Water (or another solvent) is then added to achieve the molarity desired for a particular solution. The process of adding more solvent to a solution is called **dilution.** For example, the common laboratory acids are purchased as concentrated solutions and diluted with water as they are needed. A typical dilution calculation involves determining how much water must be added to an amount of stock solution to achieve a solution of the desired concentration. The key to doing these calculations is to remember that *only water is* *added in the dilution.* The amount of solute in the final, more dilute solution is the *same* as the amount of solute in the original, concentrated stock solution. That is,

Moles of solute after dilution = moles of solute before dilution

The number of moles of solute stays the same but more water is added, increasing the volume, so the molarity decreases.

$$M = \frac{\text{moles of solute}}{\text{volume (L)}}$$

Remains constant (moles of solute); Decreases (M); Increases (water added) (volume)

For example, suppose we want to prepare 500. mL of 1.00 M acetic acid, $HC_2H_3O_2$, from a 17.5 M stock solution of acetic acid. What volume of the stock solution is required?

The first step is to determine the number of moles of acetic acid needed in the final solution. We do this by multiplying the volume of the solution by its molarity.

$$\begin{matrix}\text{Volume of dilute} \\ \text{solution (liters)}\end{matrix} \times \begin{matrix}\text{molarity of} \\ \text{dilute solution}\end{matrix} = \begin{matrix}\text{moles of solute} \\ \text{present}\end{matrix}$$

The number of moles of solute present in the more dilute solution equals the number of moles of solute that must be present in the more concentrated (stock) solution, because this is the only source of acetic acid.

Because molarity is defined in terms of liters, we must first change 500. mL to liters and then multiply the volume (in liters) by the molarity.

Math skill builder
Liters × M ⇨ Moles of solute

$$\underset{V_{\text{dilute solution}}\text{ (in mL)}}{500.\ \cancel{\text{mL solution}}} \times \underset{\text{Convert mL to L}}{\frac{1\text{ L solution}}{1000\ \cancel{\text{mL solution}}}} = 0.500\text{ L solution}$$

$$0.500\ \cancel{\text{L solution}} \times \underset{M_{\text{dilute solution}}}{\frac{1.00\text{ mol }HC_2H_3O_2}{\cancel{\text{L solution}}}} = 0.500\text{ mol }HC_2H_3O_2$$

Now we need to find the volume of 17.5 M acetic acid that contains 0.500 mole of $HC_2H_3O_2$. We will call this unknown volume V. Because volume × molarity = moles, we have

$$V\text{ (in liters)} \times \frac{17.5\text{ mol }HC_2H_3O_2}{\text{L solution}} = 0.500\text{ mol }HC_2H_3O_2$$

Solving for V $\left(\text{by dividing both sides by } \frac{17.5\text{ mol}}{\text{L solution}}\right)$ gives

$$V = \frac{0.500\ \cancel{\text{mol }HC_2H_3O_2}}{\frac{17.5\ \cancel{\text{mol }HC_2H_3O_2}}{\text{L solution}}} = 0.0286\text{ L, or 28.6 mL, of solution}$$

Therefore, to make 500. mL of a 1.00 M acetic acid solution, we take 28.6 mL of 17.5 M acetic acid and dilute it to a total volume of 500. mL. This process is illustrated in Fig. 15.8. Because the moles of solute remain the same before and after dilution, we can write

$$\underbrace{M_1 \times V_1}_{\text{Initial Conditions}} = \text{moles of solute} = \underbrace{M_2 \times V_2}_{\text{Final Conditions}}$$

M_1: Molarity before dilution; V_1: Volume before dilution; M_2: Molarity after dilution; V_2: Volume after dilution

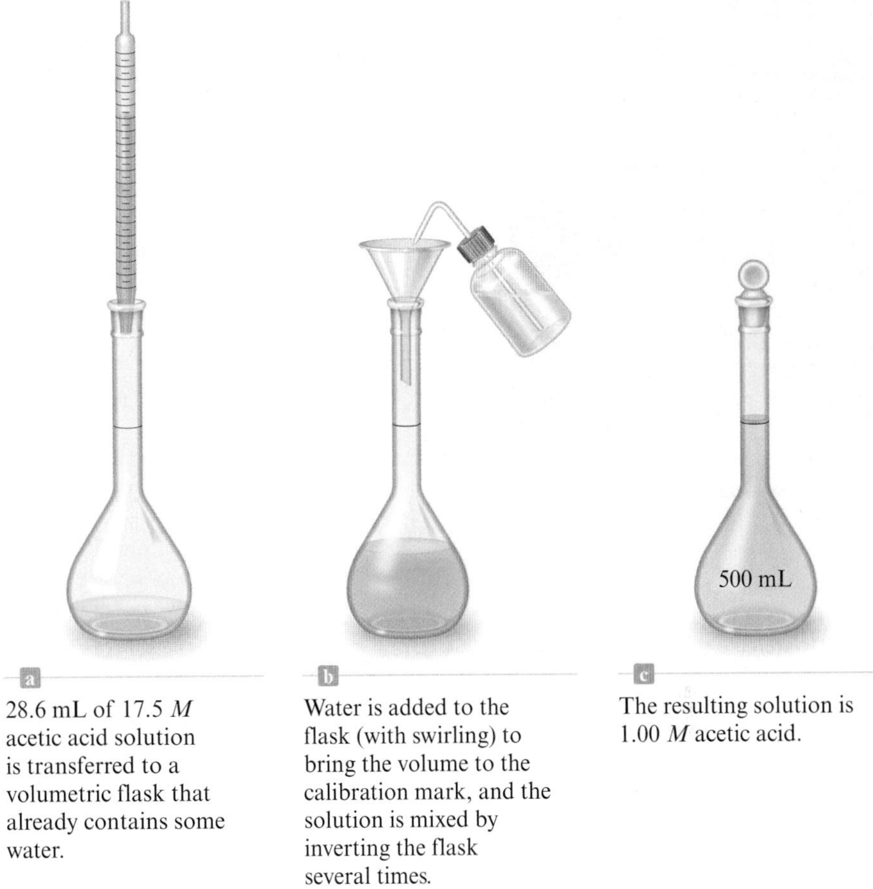

a 28.6 mL of 17.5 *M* acetic acid solution is transferred to a volumetric flask that already contains some water.

b Water is added to the flask (with swirling) to bring the volume to the calibration mark, and the solution is mixed by inverting the flask several times.

c The resulting solution is 1.00 *M* acetic acid.

Figure 15.8

We can check our calculations on acetic acid by showing that $M_1 \times V_1 = M_2 \times V_2$. In the above example, $M_1 = 17.5\ M$, $V_1 = 0.0286$ L, $V_2 = 0.500$ L, and $M_2 = 1.00\ M$, so

$$M_1 \times V_1 = 17.5 \frac{\text{mol}}{\text{L}} \times 0.0286 \text{ L} = 0.500 \text{ mol}$$

$$M_2 \times V_2 = 1.00 \frac{\text{mol}}{\text{L}} \times 0.500 \text{ L} = 0.500 \text{ mol}$$

and therefore

$$M_1 \times V_1 = M_2 \times V_2$$

This shows that the volume (V_2) we calculated is correct.

Interactive Example 15.8

Calculating Concentrations of Diluted Solutions

What volume of 16 *M* sulfuric acid must be used to prepare 1.5 L of a 0.10 *M* H_2SO_4 solution?

SOLUTION

Where Are We Going?

We want to determine the volume of sulfuric acid needed to prepare a given volume of a more dilute solution.

Charles D. Winters

Approximate dilutions can be carried out using a calibrated beaker. Here, concentrated sulfuric acid is being added to water to make a dilute solution.

What Do We Know?

Initial Conditions (concentrated)	Final Conditions (dilute)
$M_1 = 16 \frac{\text{mol}}{\text{L}}$	$M_2 = 0.10 \frac{\text{mol}}{\text{L}}$
$V_1 = ?$	$V_2 = 1.5\text{ L}$

$$\text{Moles of solute} = M_1 \times V_1 = M_2 \times V_2$$

How Do We Get There?

We can solve the equation

$$M_1 \times V_1 = M_2 \times V_2$$

for V_1 by dividing both sides by M_1,

$$\frac{M_1 \times V_1}{M_1} = \frac{M_2 \times V_2}{M_1}$$

to give

$$V_1 = \frac{M_2 \times V_2}{M_1}$$

Now we substitute the known values of M_2, V_2, and M_1.

$$V_1 = \frac{\left(0.10 \frac{\cancel{\text{mol}}}{\cancel{\text{L}}}\right)(1.5\text{ L})}{16 \frac{\cancel{\text{mol}}}{\cancel{\text{L}}}} = 9.4 \times 10^{-3}\text{ L}$$

$$9.4 \times 10^{-3}\ \cancel{\text{L}} \times \frac{1000\text{ mL}}{1\ \cancel{\text{L}}} = 9.4\text{ mL}$$

It is always best to add concentrated acid to water, not water to the acid. That way, if any splashing occurs accidentally, it is dilute acid that splashes.

Therefore, $V_1 = 9.4 \times 10^{-3}$ L, or 9.4 mL. To make 1.5 L of 0.10 M H_2SO_4 using 16 M H_2SO_4, we must take 9.4 mL of the concentrated acid and dilute it with water to a final volume of 1.5 L. The correct way to do this is to add the 9.4 mL of acid to about 1 L of water and then dilute to 1.5 L by adding more water.

SELF CHECK

Exercise 15.7 What volume of 12 M HCl must be taken to prepare 0.75 L of 0.25 M HCl?

See Problems 15.57 and 15.58. ■

15-6 Stoichiometry of Solution Reactions

OBJECTIVE **To understand the strategy for solving stoichiometric problems for solution reactions.**

Because so many important reactions occur in solution, it is important to be able to do stoichiometric calculations for solution reactions. The principles needed to perform these calculations are very similar to those developed in Chapter 9. It is helpful to think in terms of the following steps:

Steps for Solving Stoichiometric Problems Involving Solutions

Step 1 Write the balanced equation for the reaction. For reactions involving ions, it is best to write the net ionic equation.

Step 2 Calculate the moles of reactants.

Step 3 Determine which reactant is limiting.

Step 4 Calculate the moles of other reactants or products, as required.

Step 5 Convert to grams or other units, if required.

See Section 7-3 for a discussion of net ionic equations.

Interactive Example 15.9 Solution Stoichiometry: Calculating Mass of Reactants and Products

Calculate the mass of solid NaCl that must be added to 1.50 L of a 0.100 *M* $AgNO_3$ solution to precipitate all of the Ag^+ ions in the form of AgCl. Calculate the mass of AgCl formed.

SOLUTION

Where Are We Going?

We want to determine the mass of NaCl.

What Do We Know?

- We have 1.50 L of a 0.100 *M* $AgNO_3$ solution.

What Information Do We Need?

- We need the balanced equation between $AgNO_3$ and NaCl.
- We need the molar mass of NaCl.

How Do We Get There?

Step 1 *Write the balanced equation for the reaction.*
When added to the $AgNO_3$ solution (which contains Ag^+ and NO_3^- ions), the solid NaCl dissolves to yield Na^+ and Cl^- ions. Solid AgCl forms according to the following balanced net ionic reaction:

This reaction was discussed in Section 7-2.

$$Ag^+(aq) + Cl^-(aq) \rightarrow AgCl(s)$$

Step 2 *Calculate the moles of reactants.*
In this case we must add enough Cl^- ions to just react with all the Ag^+ ions present, so we must calculate the moles of Ag^+ ions present in 1.50 L of a 0.100 *M* $AgNO_3$ solution. (Remember that a 0.100 *M* $AgNO_3$ solution contains 0.100 *M* Ag^+ ions and 0.100 *M* NO_3^- ions.)

Math skill builder
Liters × *M* ⇨ Moles of solute

$$1.50\ \cancel{L} \times \frac{0.100\text{ mol }Ag^+}{\cancel{L}} = 0.150\text{ mol }Ag^+$$

Moles of Ag^+ present in 1.50 L of 0.100 *M* $AgNO_3$

Step 3 *Determine which reactant is limiting.*
In this situation we want to add just enough Cl^- to react with the Ag^+ present. That is, we want to precipitate *all* the Ag^+ in the solution. Thus the Ag^+ present determines the amount of Cl^- needed.

Richard Megna/Fundamental Photographs © Cengage Learning

When aqueous sodium chloride is added to a solution of silver nitrate, a white silver chloride precipitate forms.

Step 4 *Calculate the moles of Cl^- required.*
We have 0.150 mole of Ag^+ ions and, because one Ag^+ ion reacts with one Cl^- ion, we need 0.150 mole of Cl^-,

$$0.150 \cancel{\text{mol Ag}^+} \times \frac{1 \text{ mol Cl}^-}{1 \cancel{\text{mol Ag}^+}} = 0.150 \text{ mol Cl}^-$$

so 0.150 mole of AgCl will be formed.

$$0.150 \text{ mol Ag}^+ + 0.150 \text{ mol Cl}^- \rightarrow 0.150 \text{ mol AgCl}$$

Step 5 *Convert to grams of NaCl required.*
To produce 0.150 mole of Cl^-, we need 0.150 mole of NaCl. We calculate the mass of NaCl required as follows:

$$0.150 \cancel{\text{mol NaCl}} \times \frac{58.4 \text{ g NaCl}}{\cancel{\text{mol NaCl}}} = 8.76 \text{ g NaCl}$$

Moles → Times molar mass → Mass

The mass of AgCl formed is

$$0.150 \cancel{\text{mol AgCl}} \times \frac{143.3 \text{ g AgCl}}{\cancel{\text{mol AgCl}}} = 21.5 \text{ g AgCl}$$ ■

Interactive Example 15.10

Solution Stoichiometry: Determining Limiting Reactants and Calculating Mass of Products

See Section 7-2 for a discussion of this reaction.

When $Ba(NO_3)_2$ and K_2CrO_4 react in aqueous solution, the yellow solid $BaCrO_4$ is formed. Calculate the mass of $BaCrO_4$ that forms when 3.50×10^{-3} mole of solid $Ba(NO_3)_2$ is dissolved in 265 mL of 0.0100 *M* K_2CrO_4 solution.

SOLUTION

Richard Megna/Fundamental Photographs © Cengage Learning

Barium chromate precipitating.

Where Are We Going?

We want to determine the mass of $BaCrO_4$ that forms in a reaction of known amounts of solutions.

What Do We Know?

- We react 3.50×10^{-3} mole of $BaNO_3$ with 265 mL of 0.0100 *M* K_2CrO_4.

What Information Do We Need?

- We will need the balanced equation between $BaNO_3$ and K_2CrO_4.
- We will need the molar mass of $BaCrO_4$.

How Do We Get There?

Step 1 The original K_2CrO_4 solution contains the ions K^+ and CrO_4^{2-}. When the $Ba(NO_3)_2$ is dissolved in this solution, Ba^{2+} and NO_3^- ions are added. The Ba^{2+} and CrO_4^{2-} ions react to form solid $BaCrO_4$. The balanced net ionic equation is

$$Ba^{2+}(aq) + CrO_4^{2-}(aq) \rightarrow BaCrO_4(s)$$

Step 2 Next we determine the moles of reactants. We are told that 3.50×10^{-3} mole of $Ba(NO_3)_2$ is added to the K_2CrO_4 solution. Each formula unit of $Ba(NO_3)_2$ contains one Ba^{2+} ion, so 3.50×10^{-3} mole of $Ba(NO_3)_2$ gives 3.50×10^{-3} mole of Ba^{2+} ions in solution.

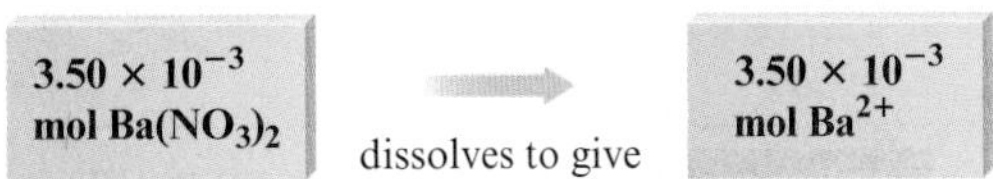

Because $V \times M$ = moles of solute, we can compute the moles of K_2CrO_4 in the solution from the volume and molarity of the original solution. First we must convert the volume of the solution (265 mL) to liters.

$$265 \text{ mL} \times \frac{1 \text{ L}}{1000 \text{ mL}} = 0.265 \text{ L}$$

Next we determine the number of moles of K_2CrO_4, using the molarity of the K_2CrO_4 solution (0.0100 M).

$$0.265 \text{ L} \times \frac{0.0100 \text{ mol } K_2CrO_4}{\text{L}} = 2.65 \times 10^{-3} \text{ mol } K_2CrO_4$$

We know that

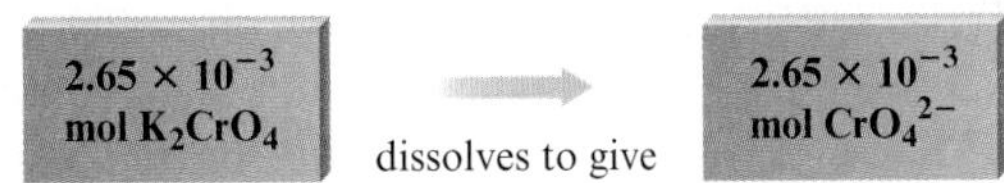

so the solution contains 2.65×10^{-3} mole of CrO_4^{2-} ions.

Step 3 The balanced equation tells us that one Ba^{2+} ion reacts with one CrO_4^{2-} ion. Because the number of moles of CrO_4^{2-} ions (2.65×10^{-3}) is smaller than the number of moles of Ba^{2+} ions (3.50×10^{-3}), the CrO_4^{2-} will run out first.

$$Ba^{2+}(aq) + CrO_4^{2-}(aq) \longrightarrow BaCrO_4(s)$$

3.50×10^{-3} mol | 2.65×10^{-3} mol

Smaller (runs out first)

Therefore, the CrO_4^{2-} is limiting.

Moles of CrO_4^{2-} → limits → Moles of $BaCrO_4$

Step 4 The 2.65×10^{-3} mole of CrO_4^{2-} ions will react with 2.65×10^{-3} mole of Ba^{2+} ions to form 2.65×10^{-3} mole of $BaCrO_4$.

2.65×10^{-3} mol Ba^{2+} + 2.65×10^{-3} mol CrO_4^{2-} → 2.65×10^{-3} mol $BaCrO_4(s)$

Step 5 The mass of $BaCrO_4$ formed is obtained from its molar mass (253.3 g) as follows:

$$2.65 \times 10^{-3} \text{ mol } BaCrO_4 \times \frac{253.3 \text{ g } BaCrO_4}{\text{mol } BaCrO_4} = 0.671 \text{ g } BaCrO_4$$

Critical Thinking

What if all ionic solids were soluble in water? How would this affect stoichiometry calculations for reactions in aqueous solution?

SELF CHECK

Exercise 15.8 When aqueous solutions of Na_2SO_4 and $Pb(NO_3)_2$ are mixed, $PbSO_4$ precipitates. Calculate the mass of $PbSO_4$ formed when 1.25 L of 0.0500 *M* $Pb(NO_3)_2$ and 2.00 L of 0.0250 *M* Na_2SO_4 are mixed.

HINT Calculate the moles of Pb^{2+} and SO_4^{2-} in the mixed solution, decide which ion is limiting, and calculate the moles of $PbSO_4$ formed.

See Problems 15.65 through 15.68. ■

15-7 Neutralization Reactions

OBJECTIVE **To learn how to do calculations involved in acid–base reactions.**

So far we have considered the stoichiometry of reactions in solution that result in the formation of a precipitate. Another common type of solution reaction occurs between an acid and a base. We introduced these reactions in Section 7-4. Recall from that discussion that an acid is a substance that furnishes H^+ ions. A strong acid, such as hydrochloric acid, HCl, dissociates (ionizes) completely in water.

$$HCl(aq) \rightarrow H^+(aq) + Cl^-(aq)$$

Strong bases are water-soluble metal hydroxides, which are completely dissociated in water. An example is NaOH, which dissolves in water to give Na^+ and OH^- ions.

$$NaOH(s) \xrightarrow{H_2O(l)} Na^+(aq) + OH^-(aq)$$

When a strong acid and a strong base react, the net ionic reaction is

$$H^+(aq) + OH^-(aq) \rightarrow H_2O(l)$$

An acid–base reaction is often called a **neutralization reaction.** When just enough strong base is added to react exactly with the strong acid in a solution, we say the acid has been *neutralized.* One product of this reaction is always water. The steps in dealing with the stoichiometry of any neutralization reaction are the same as those we followed in the previous section.

Interactive Example 15.11 Solution Stoichiometry: Calculating Volume in Neutralization Reactions

What volume of a 0.100 *M* HCl solution is needed to neutralize 25.0 mL of a 0.350 *M* NaOH solution?

SOLUTION

Where Are We Going?

We want to determine the volume of a given solution of HCl required to react with a known amount of NaOH.

What Do We Know?

- We have 25.0 mL of 0.350 *M* NaOH.
- The concentration of the HCl solution is 0.100 *M*.

What Information Do We Need?

- We need the balanced equation between HCl and NaOH.

How Do We Get There?

Step 1 *Write the balanced equation for the reaction.*

Hydrochloric acid is a strong acid, so all the HCl molecules dissociate to produce H^+ and Cl^- ions. Also, when the strong base NaOH dissolves, the solution contains Na^+ and OH^- ions. When these two solutions are mixed, the H^+ ions from the hydrochloric acid react with the OH^- ions from the sodium hydroxide solution to form water. The balanced net ionic equation for the reaction is

$$H^+(aq) + OH^-(aq) \rightarrow H_2O(l)$$

Step 2 *Calculate the moles of reactants.*

In this problem we are given a volume (25.0 mL) of 0.350 *M* NaOH, and we want to add just enough 0.100 *M* HCl to provide just enough H^+ ions to react with all the OH^-. Therefore, we must calculate the number of moles of OH^- ions in the 25.0-mL sample of 0.350 *M* NaOH. To do this, we first change the volume to liters and multiply by the molarity.

$$25.0\ \cancel{\text{mL NaOH}} \times \frac{1\ \cancel{\text{L}}}{1000\ \cancel{\text{mL}}} \times \frac{0.350\ \text{mol OH}^-}{\cancel{\text{L NaOH}}} = \underbrace{8.75 \times 10^{-3}\ \text{mol OH}^-}_{\text{Moles of OH}^-\text{ present in 25.0 mL of 0.350 } M \text{ NaOH}}$$

Step 3 *Determine which reactant is limiting.*

This problem requires the addition of just enough H^+ ions to react exactly with the OH^- ions present, so the number of moles of OH^- ions present determines the number of moles of H^+ that must be added. The OH^- ions are limiting.

Step 4 *Calculate the moles of H^+ required.*

The balanced equation tells us that the H^+ and OH^- ions react in a 1:1 ratio, so 8.75×10^{-3} mole of H^+ ions is required to neutralize (exactly react with) the 8.75×10^{-3} mole of OH^- ions present.

Step 5 *Calculate the volume of 0.100* M *HCl required.*

Next we must find the volume (V) of 0.100 *M* HCl required to furnish this amount of H^+ ions. Because the volume (in liters) times the molarity gives the number of moles, we have

$$\underbrace{V}_{\text{Unknown volume (in liters)}} \times \frac{0.100\ \text{mol H}^+}{\text{L}} = \underbrace{8.75 \times 10^{-3}\ \text{mol H}^+}_{\text{Moles of H}^+\text{ needed}}$$

Now we must solve for V by dividing both sides of the equation by 0.100.

$$V \times \frac{\cancel{0.100}\ \cancel{\text{mol H}^+}}{\cancel{0.100}\ \text{L}} = \frac{8.75 \times 10^{-3}\ \cancel{\text{mol H}^+}}{0.100}$$

$$V = 8.75 \times 10^{-2}\ \text{L}$$

Changing liters to milliliters, we have

$$V = 8.75 \times 10^{-2}\ \cancel{\text{L}} \times \frac{1000\ \text{mL}}{\cancel{\text{L}}} = 87.5\ \text{mL}$$

Therefore, 87.5 mL of 0.100 *M* HCl is required to neutralize 25.0 mL of 0.350 *M* NaOH.

SELF CHECK

Exercise 15.9 Calculate the volume of 0.10 *M* HNO_3 needed to neutralize 125 mL of 0.050 *M* KOH.

See Problems 15.69 through 15.74. ■

15-8 Solution Composition: Normality

OBJECTIVES

- To learn about normality and equivalent weight.
- To learn to use these concepts in stoichiometric calculations.

Normality is another unit of concentration that is sometimes used, especially when dealing with acids and bases. The use of normality focuses mainly on the H^+ and OH^- available in an acid–base reaction. Before we discuss normality, however, we need to define some terms. One **equivalent of an acid** is the *amount of that acid that can furnish 1 mole of* H^+ *ions.* Similarly, one **equivalent of a base** is defined as the *amount of that base that can furnish 1 mole of* OH^- *ions.* The **equivalent weight** of an acid or a base is the mass in grams of 1 equivalent (equiv) of that acid or base.

The common strong acids are HCl, HNO_3, and H_2SO_4. For HCl and HNO_3 each molecule of acid furnishes one H^+ ion, so 1 mole of HCl can furnish 1 mole of H^+ ions. This means that

Furnishes 1 mol H^+ ↓

$$1 \text{ mol HCl} = 1 \text{ equiv HCl}$$

$$\text{Molar mass (HCl)} = \text{equivalent weight (HCl)}$$

Likewise, for HNO_3,

$$1 \text{ mol } HNO_3 = 1 \text{ equiv } HNO_3$$

$$\text{Molar mass } (HNO_3) = \text{equivalent weight } (HNO_3)$$

However, H_2SO_4 can furnish *two* H^+ ions per molecule, so 1 mole of H_2SO_4 can furnish *two* moles of H^+. This means that

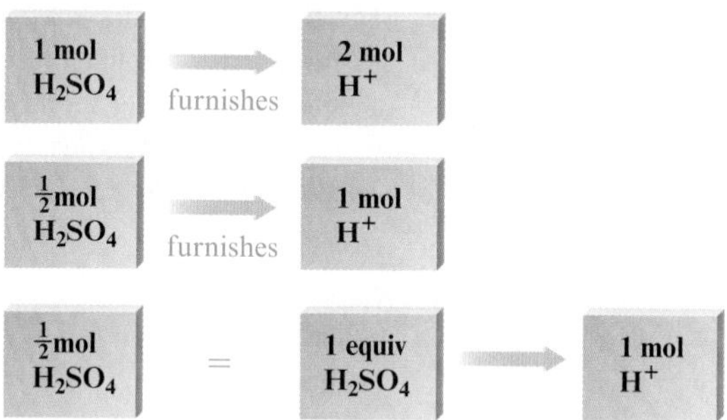

Because each mole of H_2SO_4 can furnish 2 moles of H^+, we need only to take $\frac{1}{2}$ mole of H_2SO_4 to get 1 equiv of H_2SO_4. Therefore,

$$\tfrac{1}{2} \text{ mol } H_2SO_4 = 1 \text{ equiv } H_2SO_4$$

and

$$\text{Equivalent weight } (H_2SO_4) = \tfrac{1}{2} \text{ molar mass } (H_2SO_4)$$
$$= \tfrac{1}{2}\,(98 \text{ g}) = 49 \text{ g}$$

The equivalent weight of H_2SO_4 is 49 g.

Table 15.2 ▸ The Molar Masses and Equivalent Weights of the Common Strong Acids and Bases

	Molar Mass (g)	Equivalent Weight (g)
Acid		
HCl	36.5	36.5
HNO_3	63.0	63.0
H_2SO_4	98.0	$49.0 = \frac{98.0}{2}$
Base		
NaOH	40.0	40.0
KOH	56.1	56.1

The common strong bases are NaOH and KOH. For NaOH and KOH, each formula unit furnishes one OH^- ion, so we can say

$$1 \text{ mol NaOH} = 1 \text{ equiv NaOH}$$
$$\text{Molar mass (NaOH)} = \text{equivalent weight (NaOH)}$$
$$1 \text{ mol KOH} = 1 \text{ equiv KOH}$$
$$\text{Molar mass (KOH)} = \text{equivalent weight (KOH)}$$

These ideas are summarized in Table 15.2.

Interactive Example 15.12

Solution Stoichiometry: Calculating Equivalent Weight

Phosphoric acid, H_3PO_4, can furnish three H^+ ions per molecule. Calculate the equivalent weight of H_3PO_4.

SOLUTION

Where Are We Going?

We want to determine the equivalent weight of phosphoric acid.

What Do We Know?

- The formula for phosphoric acid is H_3PO_4.
- The equivalent weight of an acid is the amount of acid that can furnish 1 mole of H^+ ions.

What Information Do We Need?

- We need to know the molar mass of H_3PO_4.

How Do We Get There?

The key point here involves how many protons (H^+ ions) each molecule of H_3PO_4 can furnish.

Because each H_3PO_4 can furnish three H^+ ions, 1 mole of H_3PO_4 can furnish 3 moles of H^+ ions:

So 1 equiv of H_3PO_4 (the amount that can furnish 1 mole of H^+) is one-third of a mole.

This means the equivalent weight of H_3PO_4 is one-third its molar mass.

Equivalent weight = $\frac{\text{Molar mass}}{3}$

$$\text{Equivalent weight } (H_3PO_4) = \frac{\text{molar mass } (H_3PO_4)}{3}$$

$$= \frac{98.0 \text{ g}}{3} = 32.7 \text{ g}$$ ■

Normality (N) is defined as the number of equivalents of solute per liter of solution.

$$\text{Normality} = N = \frac{\text{number of equivalents}}{1 \text{ liter of solution}} = \frac{\text{equivalents}}{\text{liter}} = \frac{\text{equiv}}{\text{L}}$$

This means that a 1 N solution contains 1 equivalent of solute per liter of solution. Notice that when we multiply the volume of a solution in liters by the normality, we get the number of equivalents.

Math skill builder
Liters × Normality ⇨ Equiv

$$N \times V = \frac{\text{equiv}}{\cancel{\text{L}}} \times \cancel{\text{L}} = \text{equiv}$$

Interactive Example 15.13

Solution Stoichiometry: Calculating Normality

A solution of sulfuric acid contains 86 g of H_2SO_4 per liter of solution. Calculate the normality of this solution.

SOLUTION

Where Are We Going?

We want to determine the normality of a given solution of H_2SO_4.

What Do We Know?

Whenever you need to calculate the concentration of a solution, first write the appropriate definition. Then decide how to calculate the quantities shown in the definition.

- We have 86 g of H_2SO_4 per liter of solution.
- $N = \frac{\text{equivalents}}{\text{L}}$.

What Information Do We Need?

- We need to know the molar mass of H_2SO_4.

How Do We Get There?

To find the number of equivalents present, we must calculate the number of equivalents represented by 86 g of H_2SO_4. To do this calculation, we focus on the definition of the equivalent: it is the amount of acid that furnishes 1 mole of H^+. Because H_2SO_4 can furnish two H^+ ions per molecule, 1 equiv of H_2SO_4 is $\frac{1}{2}$ mole of H_2SO_4, so

$$\text{Equivalent weight } (H_2SO_4) = \frac{\text{molar mass } (H_2SO_4)}{2}$$

$$= \frac{98.0 \text{ g}}{2} = 49.0 \text{ g}$$

We have 86 g of H_2SO_4.

$$86\ \text{g}\ \cancel{H_2SO_4} \times \frac{1\ \text{equiv}\ H_2SO_4}{49.0\ \text{g}\ \cancel{H_2SO_4}} = 1.8\ \text{equiv}\ H_2SO_4$$

$$N = \frac{\text{equiv}}{\text{L}} = \frac{1.8\ \text{equiv}\ H_2SO_4}{1.0\ \text{L}} = 1.8\ N\ H_2SO_4$$

REALITY CHECK We know that 86 g is more than 1 equiv of H_2SO_4 (49 g), so this answer makes sense.

SELF CHECK

Exercise 15.10 Calculate the normality of a solution containing 23.6 g of KOH in 755 mL of solution.

See Problems 15.79 and 15.80. ■

The main advantage of using equivalents is that 1 equiv of acid contains the same number of available H^+ ions as the number of OH^- ions present in 1 equiv of base. That is,

0.75 equiv base will react exactly with 0.75 equiv acid.

0.23 equiv base will react exactly with 0.23 equiv acid.

And so on.

In each of these cases, the *number of* H^+ ions furnished by the sample of acid is the same as the *number of* OH^- ions furnished by the sample of base. The point is that *n equivalents of any acid will exactly neutralize n equivalents of any base.*

Because we know that equal equivalents of acid and base are required for neutralization, we can say that

$$\text{equiv (acid)} = \text{equiv (base)}$$

That is,

$$N_{\text{acid}} \times V_{\text{acid}} = \text{equiv (acid)} = \text{equiv (base)} = N_{\text{base}} \times V_{\text{base}}$$

Therefore, for any neutralization reaction, the following relationship holds:

$$N_{\text{acid}} \times V_{\text{acid}} = N_{\text{base}} \times V_{\text{base}}$$

Interactive Example 15.14

Solution Stoichiometry: Using Normality in Calculations

What volume of a 0.075 *N* KOH solution is required to react exactly with 0.135 L of 0.45 *N* H_3PO_4?

SOLUTION

Where Are We Going?

We want to determine the volume of a given solution of KOH required to react with a known solution of H_3PO_4.

What Do We Know?

- We have 0.135 L of 0.45 *N* H_3PO_4.
- The concentration of the KOH solution is 0.075 *N*.
- We know $\text{equivalents}_{\text{acid}} = \text{equivalents}_{\text{base}}$.
- $N_{\text{acid}} \times V_{\text{acid}} = N_{\text{base}} \times V_{\text{base}}$

How Do We Get There?

We know that for neutralization, equiv (acid) = equiv (base), or

$$N_{acid} \times V_{acid} = N_{base} \times V_{base}$$

We want to calculate the volume of base, V_{base}, so we solve for V_{base} by dividing both sides by N_{base}.

$$\frac{N_{acid} \times V_{acid}}{N_{base}} = \frac{\cancel{N_{base}} \times V_{base}}{\cancel{N_{base}}} = V_{base}$$

Now we can substitute the given values $N_{acid} = 0.45\ N$, $V_{acid} = 0.135$ L, and $N_{base} = 0.075\ N$ into the equation.

$$V_{base} = \frac{N_{acid} \times V_{acid}}{N_{base}} = \frac{\left(0.45 \frac{\cancel{\text{equiv}}}{\cancel{\text{L}}}\right)(0.135 \text{ L})}{0.075 \frac{\cancel{\text{equiv}}}{\cancel{\text{L}}}} = 0.81 \text{ L}$$

This gives $V_{base} = 0.81$ L, so 0.81 L of 0.075 N KOH is required to react exactly with 0.135 L of 0.45 N H_3PO_4.

SELF CHECK

Exercise 15.11 What volume of 0.50 N H_2SO_4 is required to react exactly with 0.250 L of 0.80 N KOH?

See Problems 15.85 and 15.86. ■

CHAPTER 15 REVIEW

Ⓕ directs you to the *Chemistry in Focus* feature in the chapter

Key Terms

- solution (15)
- solvent (15)
- solutes (15)
- aqueous solutions (15)
- saturated (15-2)
- unsaturated (15-2)
- concentrated (15-2)
- dilute (15-2)
- mass percent (15-3)
- molarity (M) (15-4)
- standard solution (15-4)
- dilution (15-5)
- neutralization reaction (15-7)
- equivalent of an acid (15-8)
- equivalent of a base (15-8)
- equivalent weight (15-8)
- normality (N) (15-8)

For Review

- A solution is a homogeneous mixture of a solute dissolved in a solvent.
- Substances with similar polarities tend to dissolve with each other to form a solution.
- Water is a very polar substance and tends to dissolve ionic solids or other polar substances.
- Various terms are used to describe solutions:
  - Saturated—contains the maximum possible dissolved solid
  - Unsaturated—not saturated
  - Concentrated—contains a relatively large amount of solute
  - Dilute—contains a relatively small amount of solute
- Descriptions of solution composition:
  - Mass percent of solute $= \frac{\text{mass of solute}}{\text{mass of solution}} \times 100\%$
  - Molarity $= \frac{\text{moles of solute}}{\text{liters of solution}}$
  - Normality $= \frac{\text{equivalents of solute}}{\text{liters of solution}}$
- Dilution of a solution occurs when additional solvent is added to lower the concentration of a solution.
  - No solute is added so

  mol solute (before dilution) = mol solute (after dilution)
- A standard solution can be diluted to produce solutions with appropriate concentrations for various laboratory procedures.
- Moles of a dissolved reactant or product can be calculated from the known concentration and volume of the substance.

  Mol = (concentration)(volume)
- The properties of a solvent are affected by dissolving a solute.

Active Learning Questions

These questions are designed to be considered by groups of students in class. Often these questions work well for introducing a particular topic in class.

1. You have a solution of table salt in water. What happens to the salt concentration (increases, decreases, or stays the same) as the solution boils? Draw pictures to explain your answer.
2. Consider a sugar solution (solution A) with concentration x. You pour one-third of this solution into a beaker, and add an equivalent volume of water (solution B).
   a. What is the ratio of sugar in solutions A and B?
   b. Compare the volumes of solutions A and B.
   c. What is the ratio of the concentrations of sugar in solutions A and B?
3. You need to make 150.0 mL of a 0.10 M NaCl solution. You have solid NaCl, and your lab partner has a 2.5 M NaCl solution. Explain how you each independently make the solution you need.
4. You have two solutions containing solute A. To determine which solution has the highest concentration of A in molarity, which of the following must you know? (There may be more than one answer.)
   a. the mass in grams of A in each solution
   b. the molar mass of A
   c. the volume of water added to each solution
   d. the total volume of the solution

   Explain your answer.
5. Which of the following do you need to know to calculate the molarity of a salt solution? (There may be more than one answer.)
   a. the mass of salt added
   b. the molar mass of the salt
   c. the volume of water added
   d. the total volume of the solution

   Explain your answer.
6. Consider separate aqueous solutions of HCl and H_2SO_4 with the same concentrations in terms of molarity. You wish to neutralize an aqueous solution of NaOH. For which acid solution would you need to add more volume (in mL) to neutralize the base?
   a. The HCl solution.
   b. The H_2SO_4 solution.
   c. You need to know the acid concentrations to answer this question.
   d. You need to know the volume and concentration of the NaOH solution to answer this question.
   e. c and d

   Explain your answer.
7. Draw molecular-level pictures to differentiate between concentrated and dilute solutions.
8. Can one solution have a greater concentration than another in terms of weight percent, but a lower concentration in terms of molarity? Explain.
9. Explain why the formula $M_1V_1 = M_2V_2$ works when solving dilution problems.
10. You have equal masses of different solutes dissolved in equal volumes of solution. Which of the solutes listed below would make the solution with the highest concentration measured in molarity? Defend your answer.

    NaCl, $MgSO_4$, LiF, KNO_3
11. Which of the following solutions contains the greatest number of particles? Support your answer.
    a. 400.0 mL of 0.10 M sodium chloride
    b. 300.0 mL of 0.10 M calcium chloride
    c. 200.0 mL of 0.10 M iron(III) chloride
    d. 200.0 mL of 0.10 M potassium bromide
    e. 800.0 mL of 0.10 M sucrose (table sugar)
12. As with all quantitative problems in chemistry, make sure not to get "lost in the math." In particular, work on visualizing solutions at a molecular level. For example, consider the following.

    You have two separate beakers with aqueous solutions, one with 4 "units" of potassium sulfate and one with 3 "units" of barium nitrate.
    a. Draw molecular-level diagrams of both solutions.
    b. Draw a molecular-level diagram of the mixture of the two solutions before a reaction has taken place.
    c. Draw a molecular-level diagram of the product and solution formed after the reaction has taken place.
13. The figures below are molecular-level representations of four aqueous solutions of the same solute. Arrange the solutions from most to least concentrated.

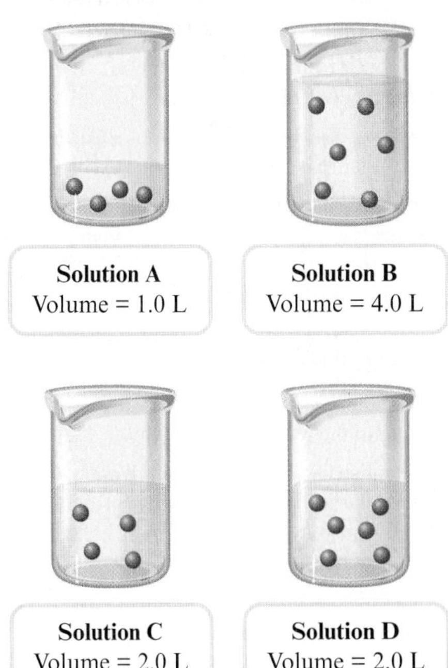

All even-numbered Questions and Problems have answers in the back of this book and solutions in the *Student Solutions Guide*.

14. The drawings below represent aqueous solutions. Solution A is 2.00 L of a 2.00 *M* aqueous solution of copper(II) nitrate. Solution B is 2.00 L of a 3.00 *M* aqueous solution of potassium hydroxide.

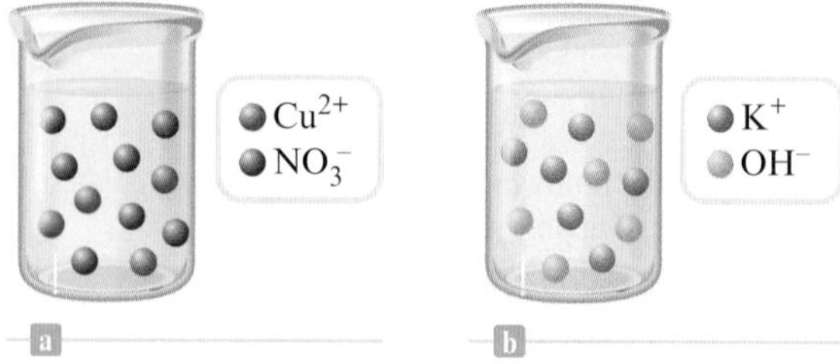

a. Draw a picture of the solution made by mixing solutions A and B together after the precipitation reaction takes place. Make sure this picture shows the correct relative volume compared to solutions A and B and the correct relative number of ions, along with the correct relative amount of solid formed.

b. Determine the concentrations (in *M*) of all ions left in solution (from part a) and the mass of solid formed.

Questions and Problems

15-1 Solubility

Questions

1. A solution is a *homogeneous mixture.* Can you give an example of a gaseous homogeneous mixture? A liquid homogeneous mixture? A solid homogeneous mixture?

2. How do the properties of a *non*homogeneous (heterogeneous) mixture differ from those of a solution? Give two examples of *non*homogeneous mixtures.

3. Suppose you dissolved a teaspoon of sugar in a glass of water. Which substance is the *solvent?* Which substance is the *solute?*

4. In a solution, the substance present in the largest amount is called the ________, whereas the other substances present are called the ________.

5. In Chapter 14, you learned that the bonding forces in ionic solids such as NaCl are very strong, yet many ionic solids dissolve readily in water. Explain.

6. An oil spill spreads out on the *surface* of water, rather than *dissolving* in the water. Explain why.

F 7. The "Chemistry in Focus" segment *Water, Water Everywhere, But . . .* discusses the desalinization of ocean water. Explain why many salts are soluble in water. Include molecular-level diagrams in your answer.

F 8. The "Chemistry in Focus" segment *Green Chemistry* discusses the use of gaseous carbon dioxide in place of CFCs and of liquid carbon dioxide in place of the dry-cleaning chemical PERC. Would you expect carbon dioxide to be very soluble in water? Explain your answer.

15-2 Solution Composition: An Introduction

Questions

9. What does it mean to say that a solution is *saturated* with a solute?

10. If additional solute is added to a(n) ________ solution, it will dissolve.

11. A solution is a homogeneous mixture and, unlike a compound, has ________ composition.

12. The label "concentrated H_2SO_4" on a bottle means that there is a relatively ________ amount of H_2SO_4 present in the solution.

15-3 Solution Composition: Mass Percent

Questions

13. How do we define the *mass percent* composition of a solution? Give an example of a solution, and explain the relative amounts of solute and solvent present in the solution in terms of the mass percent composition.

14. A solution that is 9% by mass glucose contains 9 g of glucose in every ________ g of solution.

Problems

15. Calculate the percent by mass of solute in each of the following solutions.
    a. 2.14 g of potassium chloride dissolved in 12.5 g of water
    b. 2.14 g of potassium chloride dissolved in 25.0 g of water
    c. 2.14 g of potassium chloride dissolved in 37.5 g of water
    d. 2.14 g of potassium chloride dissolved in 50.0 g of water

16. Calculate the percent by mass of solute in each of the following solutions.
    a. 5.00 g of calcium chloride dissolved in 95.0 g of water
    b. 1.00 g of calcium chloride dissolved in 19.0 g of water
    c. 15.0 g of calcium chloride dissolved in 285 g of water
    d. 2.00 mg of calcium chloride dissolved in 0.0380 g of water

17. Calculate the mass, in grams, of solute present in each of the following solutions.
    a. 375 g of 1.51% ammonium chloride solution
    b. 125 g of 2.91% sodium chloride solution
    c. 1.31 kg of 4.92% potassium nitrate solution
    d. 478 mg of 12.5% ammonium nitrate solution

18. Calculate how many grams of solute and solvent are needed to prepare the following solutions.
    a. 525 g of 3.91% iron(III) chloride solution
    b. 225 g of 11.9% sucrose solution
    c. 1.45 kg of 12.5% sodium chloride solution
    d. 635 g of 15.1% potassium nitrate solution

19. A sample of an iron alloy contains 92.1 g Fe, 2.59 g C, and 1.59 g Cr. Calculate the percent by mass of each component present in the alloy sample.

20. Consider the iron alloy described in Question 19. Suppose it is desired to prepare 1.00 kg of this alloy. What mass of each component would be necessary?

All even-numbered Questions and Problems have answers in the back of this book and solutions in the *Student Solutions Guide.*

All even-numbered Questions and Problems have answers in the back of this book and solutions in the *Student Solutions Guide*.

21. An aqueous solution is to be prepared that will be 7.51% by mass ammonium nitrate. What mass of NH_4NO_3 and what mass of water will be needed to prepare 1.25 kg of the solution?

22. If 67.1 g of $CaCl_2$ is added to 275 g of water, calculate the mass percent of $CaCl_2$ in the solution.

23. A solution is to be prepared that will be 4.50% by mass calcium chloride. To prepare 175 g of the solution, what mass of calcium chloride will be needed?

24. How many grams of KBr are contained in 125 g of a 6.25% (by mass) KBr solution?

25. What mass of each solute is present in 285 g of a solution that contains 5.00% by mass NaCl and 7.50% by mass Na_2CO_3?

26. Hydrogen peroxide solutions sold in drugstores as an antiseptic typically contain 3.0% of the active ingredient, H_2O_2. Hydrogen peroxide decomposes into water and oxygen gas when applied to a wound according to the balanced chemical equation

$$2H_2O_2(aq) \rightarrow 2H_2O(l) + O_2(g)$$

What approximate mass of hydrogen peroxide solution would be needed to produce 1.00 g of oxygen gas?

27. Sulfuric acid has a great affinity for water, and for this reason, the most concentrated form of sulfuric acid available is actually a 98.3% solution. The density of concentrated sulfuric acid is 1.84 g/mL. What mass of sulfuric acid is present in 1.00 L of the concentrated solution?

28. A solvent sold for use in the laboratory contains 0.95% of a stabilizing agent that prevents the solvent from reacting with the air. What mass of the stabilizing agent is present in 1.00 kg of the solvent?

15-4 Solution Composition: Molarity

Questions

29. A solution you used in last week's lab experiment was labeled "3 *M* HCl." Describe in words the composition of this solution.

30. A solution labeled "0.110 *M* $CaCl_2$" would contain ______ mole(s) of Ca^{2+} and ______ mole(s) of Cl^- in each liter of the solution.

31. What is a *standard* solution? Describe the steps involved in preparing a standard solution.

32. To prepare 500. mL of 1.02 *M* sugar solution, which of the following would you need?
   a. 500. mL of water and 1.02 mole of sugar
   b. 1.02 mole of sugar and enough water to make the total volume 500. mL
   c. 500. g of water and 1.02 mole of sugar
   d. 0.51 mole of sugar and enough water to make the total volume 500. mL

Problems

33. For each of the following solutions, the number of moles of solute is given, followed by the total volume of the solution prepared. Calculate the molarity of each solution.
   a. 0.521 mol NaCl; 125 mL
   b. 0.521 mol NaCl; 250. mL
   c. 0.521 mol NaCl; 500. mL
   d. 0.521 mol NaCl; 1.00 L

34. For each of the following solutions, the number of moles of solute is given, followed by the total volume of the solution prepared. Calculate the molarity of each solution.
   a. 0.754 mol KNO_3; 225 mL
   b. 0.0105 mol $CaCl_2$; 10.2 mL
   c. 3.15 mol NaCl; 5.00 L
   d. 0.499 mol NaBr; 100. mL

35. For each of the following solutions, the mass of solute is given, followed by the total volume of the solution prepared. Calculate the molarity of each solution.
   a. 3.51 g NaCl; 25 mL
   b. 3.51 g NaCl; 50. mL
   c. 3.51 g NaCl; 75 mL
   d. 3.51 g NaCl; 1.00 L

36. For each of the following solutions, the mass of solute is given, followed by the total volume of the solution prepared. Calculate the molarity of each solution.
   a. 5.59 g $CaCl_2$; 125 mL
   b. 2.34 g $CaCl_2$; 125 mL
   c. 8.73 g $CaCl_2$; 125 mL
   d. 11.5 g $CaCl_2$; 125 mL

37. A laboratory assistant needs to prepare 225 mL of 0.150 *M* $CaCl_2$ solution. How many grams of calcium chloride will she need?

38. What mass (in grams) of NH_4Cl is needed to prepare 450. mL of a 0.251 *M* NH_4Cl solution?

39. Standard solutions of calcium ion used to test for water hardness are prepared by dissolving pure calcium carbonate, $CaCO_3$, in dilute hydrochloric acid. A 1.745-g sample of $CaCO_3$ is placed in a 250.0-mL volumetric flask and dissolved in HCl. Then the solution is diluted to the calibration mark of the volumetric flask. Calculate the resulting molarity of calcium ion.

40. An alcoholic iodine solution ("tincture" of iodine) is prepared by dissolving 5.15 g of iodine crystals in enough alcohol to make a volume of 225 mL. Calculate the molarity of iodine in the solution.

41. If 42.5 g of NaOH is dissolved in water and diluted to a final volume of 225 mL, calculate the molarity of the solution.

42. Standard silver nitrate solutions are used in the analysis of samples containing chloride ion. How many grams of silver nitrate are needed to prepare 250. mL of a 0.100 *M* $AgNO_3$ solution?

43. How many *moles* of the indicated solute does each of the following solutions contain?
   a. 4.25 mL of 0.105 *M* $CaCl_2$ solution
   b. 11.3 mL of 0.405 *M* NaOH solution
   c. 1.25 L of 12.1 *M* HCl solution
   d. 27.5 mL of 1.98 *M* NaCl solution

44. How many *moles* of the indicated solute does each of the following solutions contain?
   a. 12.5 mL of 0.104 *M* HCl
   b. 27.3 mL of 0.223 *M* NaOH
   c. 36.8 mL of 0.501 *M* HNO_3
   d. 47.5 mL of 0.749 *M* KOH

45. What *mass* of the indicated solute does each of the following solutions contain?
   a. 2.50 L of 13.1 *M* HCl solution
   b. 15.6 mL of 0.155 *M* NaOH solution
   c. 135 mL of 2.01 *M* HNO_3 solution
   d. 4.21 L of 0.515 *M* $CaCl_2$ solution

46. What *mass* of the indicated solute does each of the following solutions contain?
   a. 17.8 mL of 0.119 M $CaCl_2$
   b. 27.6 mL of 0.288 M KCl
   c. 35.4 mL of 0.399 M $FeCl_3$
   d. 46.1 mL of 0.559 M KNO_3

47. What mass of NaOH pellets is required to prepare 3.5 L of 0.50 M NaOH solution?

48. What mass of solute is present in 225 mL of 0.355 M KBr solution?

49. Calculate the number of moles of the indicated ion present in each of the following solutions.
   a. Na^+ ion in 1.00 L of 0.251 M Na_2SO_4 solution
   b. Cl^- ion in 5.50 L of 0.10 M $FeCl_3$ solution
   c. NO_3^- ion in 100. mL of 0.55 M $Ba(NO_3)_2$ solution
   d. NH_4^+ ion in 250. mL of 0.350 M $(NH_4)_2SO_4$ solution

50. Calculate the number of moles of *each* ion present in each of the following solutions.
   a. 10.2 mL of 0.451 M $AlCl_3$ solution
   b. 5.51 L of 0.103 M Na_3PO_4 solution
   c. 1.75 mL of 1.25 M $CuCl_2$ solution
   d. 25.2 mL of 0.00157 M $Ca(OH)_2$ solution

51. An experiment calls for 125 mL of 0.105 M NaCl solution. What mass of NaCl is required? What mass of NaCl would be required for 1.00 L of the same solution?

52. Strong acid solutions may have their concentration determined by reaction with measured quantities of standard sodium carbonate solution. What mass of Na_2CO_3 is needed to prepare 250. mL of 0.0500 M Na_2CO_3 solution?

15-5 Dilution

Questions

53. When a concentrated stock solution is diluted to prepare a less concentrated reagent, the number of ________ is the same both before and after the dilution.

54. When the volume of a given solution is doubled (by adding water), the new concentration of solute is ________ the original concentration.

Problems

55. Calculate the new molarity if each of the following dilutions is made. Assume the volumes are additive.
   a. 55.0 mL of water is added to 25.0 mL of 0.119 M NaCl solution
   b. 125 mL of water is added to 45.3 mL of 0.701 M NaOH solution
   c. 550. mL of water is added to 125 mL of 3.01 M KOH solution
   d. 335 mL of water is added to 75.3 mL of 2.07 M $CaCl_2$ solution

56. Calculate the new molarity that results when 250. mL of water is added to each of the following solutions.
   a. 125 mL of 0.251 M HCl
   b. 445 mL of 0.499 M H_2SO_4
   c. 5.25 L of 0.101 M HNO_3
   d. 11.2 mL of 14.5 M $HC_2H_3O_2$

57. Many laboratories keep bottles of 3.0 M solutions of the common acids on hand. Given the following molarities of the concentrated acids, determine how many milliliters of each concentrated acid would be required to prepare 225 mL of a 3.0 M solution of the acid.

Acid	Molarity of Concentrated Reagent
HCl	12.1 M
HNO_3	15.9 M
H_2SO_4	18.0 M
$HC_2H_3O_2$	17.5 M
H_3PO_4	14.9 M

58. For convenience, one form of sodium hydroxide that is sold commercially is the saturated solution. This solution is 19.4 M, which is approximately 50% by mass sodium hydroxide. What volume of this solution would be needed to prepare 3.50 L of 3.00 M NaOH solution?

59. How would you prepare 275 mL of 0.350 M NaCl solution using an available 2.00 M solution?

60. Suppose 325 mL of 0.150 M NaOH is needed for your experiment. How would you prepare this if all that is available is a 1.01 M NaOH solution?

61. How much *water* must be added to 500. mL of 0.200 M HCl to produce a 0.150 M solution? (Assume that the volumes are additive.)

62. An experiment calls for 100. mL of 1.25 M HCl. All that is available in the lab is a bottle of concentrated HCl, whose label indicates that it is 12.1 M. How much of the concentrated HCl would be needed to prepare the desired solution?

15-6 Stoichiometry of Solution Reactions

Problems

63. The amount of nickel(II) present in an aqueous solution can be determined by precipitating the nickel with the organic chemical reagent dimethylglyoxime [$CH_3C(NOH)C(NOH)CH_3$, commonly abbreviated as "DMG"].

$$Ni^{2+}(aq) + 2DMG(aq) \rightarrow Ni(DMG)_2(s)$$

How many milliliters of 0.0703 M DMG solution is required to precipitate all the nickel(II) present in 10.0 mL of 0.103 M nickel(II) sulfate solution?

64. Generally only the carbonates of the Group 1 elements and the ammonium ion are soluble in water; most other carbonates are *insoluble*. How many milliliters of 0.125 M sodium carbonate solution would be needed to precipitate the calcium ion from 37.2 mL of 0.105 M $CaCl_2$ solution?

$$Na_2CO_3(aq) + CaCl_2(aq) \rightarrow CaCO_3(s) + 2NaCl(aq)$$

65. Many metal ions are precipitated from solution by the sulfide ion. As an example, consider treating a solution of copper(II) sulfate with sodium sulfide solution:

$$CuSO_4(aq) + Na_2S(aq) \rightarrow CuS(s) + Na_2SO_4(aq)$$

What volume of 0.105 M Na_2S solution would be required to precipitate all of the copper(II) ion from 27.5 mL of 0.121 M $CuSO_4$ solution?

All even-numbered Questions and Problems have answers in the back of this book and solutions in the *Student Solutions Guide*.

66. Calcium oxalate, CaC_2O_4, is very insoluble in water. What mass of sodium oxalate, $Na_2C_2O_4$, is required to precipitate the calcium ion from 37.5 mL of 0.104 *M* $CaCl_2$ solution?

67. When aqueous solutions of lead(II) ion are treated with potassium chromate solution, a bright yellow precipitate of lead(II) chromate, $PbCrO_4$, forms. How many grams of lead chromate form when a 1.00-g sample of $Pb(NO_3)_2$ is added to 25.0 mL of 1.00 *M* K_2CrO_4 solution?

68. Aluminum ion may be precipitated from aqueous solution by addition of hydroxide ion, forming $Al(OH)_3$. A large excess of hydroxide ion must not be added, however, because the precipitate of $Al(OH)_3$ will redissolve as a soluble compound containing aluminum ions and hydroxide ions begins to form. How many grams of solid NaOH should be added to 10.0 mL of 0.250 *M* $AlCl_3$ to just precipitate all the aluminum?

15-7 Neutralization Reactions

Problems

69. What volume of 0.502 *M* NaOH solution would be required to neutralize 27.2 mL of 0.491 *M* HNO_3 solution?

70. What volume of 0.995 *M* HCl solution could be neutralized by 125 mL of 3.01 *M* NaOH solution?

71. A sample of sodium hydrogen carbonate solid weighing 0.1015 g requires 47.21 mL of a hydrochloric acid solution to react completely.

$$HCl(aq) + NaHCO_3(s) \rightarrow NaCl(aq) + H_2O(l) + CO_2(g)$$

Calculate the molarity of the hydrochloric acid solution.

72. The total acidity in water samples can be determined by neutralization with standard sodium hydroxide solution. What is the total concentration of hydrogen ion, H^+, present in a water sample if 100. mL of the sample requires 7.2 mL of 2.5×10^{-3} *M* NaOH to be neutralized?

73. What volume of 1.00 *M* NaOH is required to neutralize each of the following solutions?
    a. 25.0 mL of 0.154 *M* acetic acid, $HC_2H_3O_2$
    b. 35.0 mL of 0.102 *M* hydrofluoric acid, HF
    c. 10.0 mL of 0.143 *M* phosphoric acid, H_3PO_4
    d. 35.0 mL of 0.220 *M* sulfuric acid, H_2SO_4

74. What volume of 0.101 *M* HNO_3 is required to neutralize each of the following solutions?
    a. 12.7 mL of 0.501 *M* NaOH
    b. 24.9 mL of 0.00491 *M* $Ba(OH)_2$
    c. 49.1 mL of 0.103 *M* NH_3
    d. 1.21 L of 0.102 *M* KOH

15-8 Solution Composition: Normality

Questions

75. One equivalent of an acid is the amount of the acid required to provide ________.

76. A solution that contains 1 equivalent of acid or base per liter is said to be a ________ solution.

77. Explain why the equivalent weight of H_2SO_4 is half the molar mass of this substance. How many hydrogen ions does each H_2SO_4 molecule produce when reacting with an excess of OH^- ions?

78. How many equivalents of hydroxide ion are needed to react with 1.53 equivalents of hydrogen ion? How did you know this when no balanced chemical equation was provided for the reaction?

Problems

79. For each of the following solutions, calculate the normality.
    a. 25.2 mL of 0.105 *M* HCl diluted with water to a total volume of 75.3 mL
    b. 0.253 *M* H_3PO_4
    c. 0.00103 *M* $Ca(OH)_2$

80. For each of the following solutions, the mass of solute taken is indicated, along with the total volume of solution prepared. Calculate the normality of each solution.
    a. 0.113 g NaOH; 10.2 mL
    b. 12.5 mg $Ca(OH)_2$; 100. mL
    c. 12.4 g H_2SO_4; 155 mL

81. Calculate the normality of each of the following solutions.
    a. 0.250 *M* HCl
    b. 0.105 *M* H_2SO_4
    c. 5.3×10^{-2} *M* H_3PO_4

82. Calculate the normality of each of the following solutions.
    a. 0.134 *M* NaOH
    b. 0.00521 *M* $Ca(OH)_2$
    c. 4.42 *M* H_3PO_4

83. A solution of phosphoric acid, H_3PO_4, is found to contain 35.2 g of H_3PO_4 per liter of solution. Calculate the molarity and normality of the solution.

84. A solution of the sparingly soluble base $Ca(OH)_2$ is prepared in a volumetric flask by dissolving 5.21 mg of $Ca(OH)_2$ to a total volume of 1000. mL. Calculate the molarity and normality of the solution.

85. How many milliliters of 0.50 *N* NaOH are required to neutralize exactly 15.0 mL of 0.35 *N* H_2SO_4?

86. What volume of 0.104 *N* H_2SO_4 is required to neutralize 15.2 mL of 0.152 *N* NaOH? What volume of 0.104 *M* H_2SO_4 is required to neutralize 15.2 mL of 0.152 *M* NaOH?

$$H_2SO_4(aq) + 2NaOH(aq) \rightarrow Na_2SO_4(aq) + 2H_2O(l)$$

87. What volume of 0.151 *N* NaOH is required to neutralize 24.2 mL of 0.125 *N* H_2SO_4? What volume of 0.151 *N* NaOH is required to neutralize 24.2 mL of 0.125 *M* H_2SO_4?

88. Suppose that 27.34 mL of standard 0.1021 *M* NaOH is required to neutralize 25.00 mL of an unknown H_2SO_4 solution. Calculate the molarity and the normality of the unknown solution.

Additional Problems

89. A mixture is prepared by mixing 50.0 g of ethanol, 50.0 g of water, and 5.0 g of sugar. What is the mass percent of each component in the mixture? How many grams of the mixture should one take in order to have 1.5 g of sugar? How many grams of the mixture should one take to have 10.0 g of ethanol?

All even-numbered Questions and Problems have answers in the back of this book and solutions in the *Student Solutions Guide*.

90. Explain the difference in meaning between the following two solutions: "50. g of NaCl dissolved in 1.0 L of water" and "50. g of NaCl dissolved in enough water to make 1.0 L of solution." For which solution can the molarity be calculated directly (using the molar mass of NaCl)?

91. Suppose 50.0 mL of 0.250 *M* $CoCl_2$ solution is added to 25.0 mL of 0.350 *M* $NiCl_2$ solution. Calculate the concentration, in moles per liter, of each of the ions present after mixing. Assume that the volumes are additive.

92. If 500. g of water is added to 75 g of 25% NaCl solution, what is the percent by mass of NaCl in the diluted solution?

93. Calculate the mass of AgCl formed, and the concentration of silver ion remaining in solution, when 10.0 g of solid $AgNO_3$ is added to 50. mL of 1.0×10^{-2} *M* NaCl solution. Assume there is no volume change upon addition of the solid.

94. Baking soda (sodium hydrogen carbonate, $NaHCO_3$) is often used to neutralize spills of acids on the benchtop in the laboratory. What mass of $NaHCO_3$ would be needed to neutralize a spill consisting of 25.2 mL of 6.01 *M* hydrochloric acid solution?

95. Many metal ions form insoluble sulfide compounds when a solution of the metal ion is treated with hydrogen sulfide gas. For example, nickel(II) precipitates nearly quantitatively as NiS when H_2S gas is bubbled through a nickel ion solution. How many milliliters of gaseous H_2S at STP are needed to precipitate all the nickel ion present in 10. mL of 0.050 *M* $NiCl_2$ solution?

96. Strictly speaking, the solvent is the component of a solution that is present in the largest amount on a *mole* basis. For solutions involving water, water is almost always the solvent because there tend to be many more water molecules present than molecules of any conceivable solute. To see why this is so, calculate the number of moles of water present in 1.0 L of water. Recall that the density of water is very nearly 1.0 g/mL under most conditions.

97. Aqueous ammonia is typically sold by chemical supply houses as the saturated solution, which has a concentration of 14.5 mol/L. What volume of NH_3 at STP is required to prepare 100. mL of concentrated ammonia solution?

98. What volume of hydrogen chloride gas at STP is required to prepare 500. mL of 0.100 *M* HCl solution?

99. What do we mean when we say that "like dissolves like"? Do two molecules have to be identical to be able to form a solution in one another?

100. The concentration of a solution of HCl is 33.1% by mass, and its density was measured to be 1.147 g/mL. How many milliliters of the HCl solution are required to obtain 10.0 g of HCl?

101. An experiment calls for 1.00 g of silver nitrate, but all that is available in the laboratory is a 0.50% solution of $AgNO_3$. Assuming the density of the silver nitrate solution to be very nearly that of water because it is so dilute, determine how many milliliters of the solution should be used.

102. You mix 225.0 mL of a 2.5 *M* HCl solution with 150.0 mL of a 0.75 *M* HCl solution. What is the molarity of the final solution?

103. A solution is 0.1% by mass calcium chloride. Therefore, 100. g of the solution contains ________ g of calcium chloride.

104. Calculate the mass, in grams, of NaCl present in each of the following solutions.

a. 11.5 g of 6.25% NaCl solution
b. 6.25 g of 11.5% NaCl solution
c. 54.3 g of 0.91% NaCl solution
d. 452 g of 12.3% NaCl solution

105. A 15.0% (by mass) NaCl solution is available. Determine what mass of the solution should be taken to obtain the following quantities of NaCl.

a. 10.0 g
b. 25.0 g
c. 100.0 g
d. 1.00 lb

106. A certain grade of steel is made by dissolving 5.0 g of carbon and 1.5 g of nickel per 100. g of molten iron. What is the mass percent of each component in the finished steel?

107. A sugar solution is prepared in such a way that it contains 10.% dextrose by mass. What quantity of this solution do we need to obtain 25 g of dextrose?

108. How many grams of Na_2CO_3 are contained in 500. g of a 5.5% by mass Na_2CO_3 solution?

109. What mass of KNO_3 is required to prepare 125 g of 1.5% KNO_3 solution?

110. A solution contains 7.5% by mass NaCl and 2.5% by mass KBr. What mass of *each* solute is contained in 125 g of the solution?

111. How many moles of each ion are present in 11.7 mL of 0.102 *M* Na_3PO_4 solution?

112. For each of the following solutions, the number of moles of solute is given, followed by the total volume of solution prepared. Calculate the molarity of each solution.

a. 0.50 mol KBr; 250 mL
b. 0.50 mol KBr; 500. mL
c. 0.50 mol KBr; 750 mL
d. 0.50 mol KBr; 1.0 L

113. For each of the following solutions, the mass of the solute is given, followed by the total volume of solution prepared. Calculate the molarity.

a. 5.0 g of $BaCl_2$; 2.5 L
b. 3.5 g of KBr; 75 mL
c. 21.5 g of Na_2CO_3; 175 mL
d. 55 g of $CaCl_2$; 1.2 L

114. If 125 g of sucrose, $C_{12}H_{22}O_{11}$, is dissolved in enough water to make 450. mL of solution, calculate the molarity.

115. Concentrated hydrochloric acid is made by pumping hydrogen chloride gas into distilled water. If concentrated HCl contains 439 g of HCl per liter, what is the molarity?

116. A large beaker contains 1.50 L of a 2.00 *M* iron(III) chloride solution.

a. How many moles of iron ions are in the solution? How many moles of chloride ions are in the solution?
b. You now add 0.500 L of a 4.00 *M* lead(II) nitrate solution to the beaker. Determine the mass of solid product formed (in grams).

117. How many *moles* of the indicated solute does each of the following solutions contain?

a. 1.5 L of 3.0 *M* H_2SO_4 solution
b. 35 mL of 5.4 *M* NaCl solution
c. 5.2 L of 18 *M* H_2SO_4 solution
d. 0.050 L of 1.1×10^{-3} *M* NaF solution

All even-numbered Questions and Problems have answers in the back of this book and solutions in the *Student Solutions Guide*.

118. How many *moles* and how many *grams* of the indicated solute does each of the following solutions contain?

a. 4.25 L of 0.105 *M* KCl solution
b. 15.1 mL of 0.225 *M* $NaNO_3$ solution
c. 25 mL of 3.0 *M* HCl
d. 100. mL of 0.505 *M* H_2SO_4

119. If 10. g of $AgNO_3$ is available, what volume of 0.25 *M* $AgNO_3$ solution can be prepared?

120. Calculate the number of moles of *each* ion present in each of the following solutions.

a. 1.25 L of 0.250 *M* Na_3PO_4 solution
b. 3.5 mL of 6.0 *M* H_2SO_4 solution
c. 25 mL of 0.15 *M* $AlCl_3$ solution
d. 1.50 L of 1.25 *M* $BaCl_2$ solution

121. Calcium carbonate, $CaCO_3$, can be obtained in a very pure state. Standard solutions of calcium ion are usually prepared by dissolving calcium carbonate in acid. What mass of $CaCO_3$ should be taken to prepare 500. mL of 0.0200 *M* calcium ion solution?

122. Calculate the new molarity when 150. mL of water is added to each of the following solutions.

a. 125 mL of 0.200 *M* HBr
b. 155 mL of 0.250 *M* $Ca(C_2H_3O_2)_2$
c. 0.500 L of 0.250 *M* H_3PO_4
d. 15 mL of 18.0 *M* H_2SO_4

123. How many milliliters of 18.0 *M* H_2SO_4 are required to prepare 35.0 mL of 0.250 *M* solution?

124. When 50. mL of 5.4 *M* NaCl is diluted to a final volume of 300. mL, what is the concentration of NaCl in the diluted solution?

125. When 10. L of water is added to 3.0 L of 6.0 *M* H_2SO_4, what is the molarity of the resulting solution? Assume the volumes are additive.

126. You pour 150.0 mL of a 0.250 *M* lead(II) nitrate solution into an empty 500-mL flask.

a. What is the concentration of nitrate ions in the solution?
b. What volume of 0.100 *M* sodium phosphate must be added to precipitate the lead(II) ions from the solution?

127. How many grams of $Ba(NO_3)_2$ are required to precipitate all the sulfate ion present in 15.3 mL of 0.139 *M* H_2SO_4 solution?

$$Ba(NO_3)_2(aq) + H_2SO_4(aq) \rightarrow BaSO_4(s) + 2HNO_3(aq)$$

128. What minimum volume of 16 *M* sulfuric acid must be used to prepare 750 mL of a 0.10 *M* H_2SO_4 solution?

129. What volume of 0.250 *M* HCl is required to neutralize each of the following solutions?

a. 25.0 mL of 0.103 *M* sodium hydroxide, NaOH
b. 50.0 mL of 0.00501 *M* calcium hydroxide, $Ca(OH)_2$
c. 20.0 mL of 0.226 *M* ammonia, NH_3
d. 15.0 mL of 0.0991 *M* potassium hydroxide, KOH

130. For each of the following solutions, the mass of solute taken is indicated, as well as the total volume of solution prepared. Calculate the normality of each solution.

a. 15.0 g of HCl; 500. mL
b. 49.0 g of H_2SO_4; 250. mL
c. 10.0 g of H_3PO_4; 100. mL

131. Calculate the normality of each of the following solutions.

a. 0.50 *M* acetic acid, $HC_2H_3O_2$
b. 0.00250 *M* sulfuric acid, H_2SO_4
c. 0.10 *M* potassium hydroxide, KOH

132. A sodium dihydrogen phosphate solution was prepared by dissolving 5.0 g of NaH_2PO_4 in enough water to make 500. mL of solution. What are the molarity and normality of the resulting solution?

133. How many milliliters of 0.105 *M* NaOH are required to neutralize exactly 14.2 mL of 0.141 *M* H_3PO_4?

134. If 27.5 mL of 3.5×10^{-2} *N* $Ca(OH)_2$ solution is needed to neutralize 10.0 mL of nitric acid solution of unknown concentration, what is the normality of the nitric acid?

Consider the following four solutions to answer Questions 135 and 136.

Solution 1: 100.0 g of cesium chloride dissolved in 1.0 L of water.
Solution 2: 100.0 g of sodium sulfate dissolved in 1.0 L of water.
Solution 3: 100.0 g of cesium fluoride dissolved in 1.0 L of water.
Solution 4: 100.0 g of sodium phosphate dissolved in 1.0 L of water.

135. Which of the solutions above has the *highest* concentration?

136. Which of the solutions above contains the *greatest* number of ions?

ChemWork Problems

These multiconcept problems (and additional ones) are found interactively online with the same type of assistance a student would get from an instructor.

137. Calculate the concentration of all ions present when 0.160 g of $MgCl_2$ is dissolved in enough water to make 100.0 mL of solution.

138. A solution is prepared by dissolving 0.6706 g of oxalic acid ($H_2C_2O_4$) in enough water to make 100.0 mL of solution. A 10.00-mL aliquot (portion) of this solution is then diluted to a final volume of 250.0 mL. What is the final molarity of the oxalic acid solution?

139. What volume of 0.100 *M* NaOH is required to precipitate all of the nickel(II) ions from 150.0 mL of a 0.249 *M* solution of $Ni(NO_3)_2$?

140. A 500.0-mL sample of 0.200 *M* sodium phosphate is mixed with 400.0 mL of 0.289 *M* barium chloride. What is the mass of the solid produced?

141. A 450.0-mL sample of a 0.257 *M* solution of silver nitrate is mixed with 400.0 mL of 0.200 *M* calcium chloride. What is the concentration of Cl^- in solution after the reaction is complete?

142. A 50.00-mL sample of aqueous $Ca(OH)_2$ requires 34.66 mL of a 0.944 *M* nitric acid for neutralization. Calculate the concentration (molarity) of the original solution of calcium hydroxide.

143. When organic compounds containing sulfur are burned, sulfur dioxide is produced. The amount of SO_2 formed can be determined by the reaction with hydrogen peroxide:

$$H_2O_2(aq) + SO_2(g) \longrightarrow H_2SO_4(aq)$$

The resulting sulfuric acid is then titrated with a standard NaOH solution. A 1.302-g sample of coal is burned, and the SO_2 is collected in a solution of hydrogen peroxide. It took 28.44 mL of a 0.1000 *M* NaOH solution to titrate the resulting sulfuric acid. Calculate the mass percent of sulfur in the coal sample. Sulfuric acid has two acidic hydrogens.

Questions

1. What are some of the general properties of gases that distinguish them from liquids and solids?
2. How does the pressure of the atmosphere arise? Sketch a representation of the device commonly used to measure the pressure of the atmosphere. Your textbook described a simple experiment to demonstrate the pressure of the atmosphere. Explain this experiment.
3. What is the SI unit of pressure? What units of pressure are commonly used in the United States? Why are these common units more convenient to use than the SI unit? Describe a *manometer* and explain how such a device can be used to measure the pressure of gas samples.
4. Your textbook gives several definitions and formulas for Boyle's law for gases. Write, in your *own* words, what this law really tells us about gases. Now write two mathematical expressions that describe Boyle's law. Do these two expressions tell us different things, or are they different representations of the same phenomena? Sketch the general shape of a graph of pressure versus volume for an ideal gas.
5. When using Boyle's law in solving problems in the textbook, you may have noticed that questions were often qualified by stating that "the temperature and amount of gas remain the same." Why was this qualification necessary?
6. What does Charles's law tell us about how the volume of a gas sample varies as the temperature of the sample is changed? How does this volume–temperature relationship *differ* from the volume–pressure relationship of Boyle's law? Give two mathematical expressions that describe Charles's law. For Charles's law to hold true, why must the pressure and amount of gas remain the same? Sketch the general shape of a graph of volume versus temperature (at constant pressure) for an ideal gas.
7. Explain how the concept of absolute zero came about through Charles's studies of gases. *Hint:* What would happen to the volume of a gas sample at absolute zero (if the gas did not liquefy first)? What temperature scale is defined with its lowest point as the absolute zero of temperature? What is absolute zero in Celsius degrees?
8. What does Avogadro's law tell us about the relationship between the volume of a sample of gas and the number of molecules the gas contains? Why must the temperature and pressure be held constant for valid comparisons using Avogadro's law? Does Avogadro's law describe a direct or an inverse relationship between the volume and the number of moles of gas?
9. What do we mean specifically by an *ideal* gas? Explain why the *ideal gas law* ($PV = nRT$) is actually a combination of Boyle's, Charles's, and Avogadro's gas laws. What is the numerical value and what are the specific units of the universal gas constant, R? Why is close attention to *units* especially important when doing ideal gas law calculations?
10. Dalton's law of partial pressures concerns the properties of mixtures of gases. What is meant by the *partial pressure* of an individual gas in a mixture? How does the *total pressure* of a gaseous mixture depend on the partial pressures of the individual gases in the mixture? How does Dalton's law help us realize that in an ideal gas sample, the volume of the individual molecules is insignificant compared with the bulk volume of the sample?
11. What happens to a gas sample when it is collected by displacement of, or by bubbling through, water? How is this taken into account when calculating the pressure of the gas?
12. Without consulting your textbook, list and explain the main postulates of the kinetic molecular theory for gases. How do these postulates help us account for the following bulk properties of a gas: the pressure of the gas and why the pressure of the gas increases with increased temperature; the fact that a gas fills its entire container; and the fact that the volume of a given sample of gas increases as its temperature is increased.
13. What does "STP" stand for? What conditions correspond to STP? What is the volume occupied by one mole of an ideal gas at STP?
14. In general, how do we envision the structures of solids and liquids? Explain how the densities and compressibilities of solids and liquids contrast with those properties of gaseous substances. How do we know that the structures of the solid and liquid states of a substance are more comparable to each other than to the properties of the substance in the gaseous state?
15. Describe some of the physical properties of water. Why is water one of the most important substances on earth?
16. Define the *normal* boiling point of water. Why does a sample of boiling water remain at the same temperature until all the water has been boiled? Define the normal freezing point of water. Sketch a representation of a heating/cooling curve for water, marking clearly the normal freezing and boiling points.
17. Are changes in state physical or chemical changes? Explain. What type of forces must be overcome to melt or vaporize a substance (are these forces *intra*molecular or *inter*molecular)? Define the *molar heat of fusion* and *molar heat of vaporization.* Why is the molar heat of vaporization of water so much larger than its molar heat of fusion? Why does the boiling point of a liquid vary with altitude?
18. What is a *dipole–dipole attraction?* How do the strengths of dipole–dipole forces compare with the strengths of typical covalent bonds? What is *hydrogen bonding?* What conditions are necessary for hydrogen bonding to exist in a substance or mixture? What experimental evidence do we have for hydrogen bonding?
19. Define *London dispersion forces.* Draw a picture showing how London forces arise. Are London forces relatively strong or relatively weak? Explain. Although London forces exist among all molecules, for what type of molecule are they the *only* major intermolecular force?

All even-numbered Questions and Problems have answers in the back of this book and solutions in the *Student Solutions Guide.*

20. Why does the process of *vaporization* require an input of energy? Why is it so important that water has a large heat of vaporization? What is *condensation?* Explain how the processes of vaporization and condensation represent an *equilibrium* in a closed container. Define the *equilibrium vapor pressure* of a liquid. Describe how this pressure arises in a closed container. Describe an experiment that demonstrates vapor pressure and enables us to measure the magnitude of that pressure. How is the magnitude of a liquid's vapor pressure related to the intermolecular forces in the liquid?

21. Define a *crystalline solid.* Describe in detail some important types of crystalline solids and name a substance that is an example of each type of solid. Explain how the particles are held together in each type of solid (the interparticle forces that exist). How do the interparticle forces in a solid influence the bulk physical properties of the solid?

22. Define the bonding that exists in metals and how this model explains some of the unique physical properties of metals. What are metal *alloys?* Identify the two main types of alloys, and describe how their structures differ. Give several examples of each type of alloy.

23. Define a *solution.* Describe how an ionic solute such as NaCl dissolves in water to form a solution. How are the strong bonding forces in a crystal of ionic solute overcome? Why do the ions in a solution not attract each other so strongly as to reconstitute the ionic solute? How does a molecular solid such as sugar dissolve in water? What forces between water molecules and the molecules of a molecular solid may help the solute dissolve? Why do some substances *not* dissolve in water at all?

24. Define a *saturated* solution. Does saturated mean the same thing as saying the solution is *concentrated?* Explain. Why does a solute dissolve only to a particular extent in water? How does formation of a saturated solution represent an equilibrium?

25. The concentration of a solution may be expressed in various ways. Suppose 5.00 g of NaCl were dissolved in 15.0 g of water, which resulted in 16.1 mL of solution after mixing. Explain how you would calculate the *mass percent* of NaCl and the *molarity* of NaCl.

26. When a solution is diluted by adding additional solvent, the *concentration* of solute changes but the *amount* of solute present does not change. Explain. Suppose 250. mL of water is added to 125 mL of 0.551 *M* NaCl solution. Explain *how* you would calculate the concentration of the solution after dilution.

27. What is one *equivalent* of an acid? What does an equivalent of a base represent? How is the equivalent weight of an acid or a base related to the substance's molar mass? Give an example of an acid and a base that have equivalent weights *equal* to their molar masses. Give an example of an acid and a base that have equivalent weights that are *not equal* to their molar masses. What is a *normal* solution of an acid or a base? How is the *normality* of an acid or a base solution related to its *molarity?* Give an example of a solution whose normality is equal to its molarity, and an example of a solution whose normality is *not* the same as its molarity.

Problems

28. a. If the pressure on a 125-mL sample of gas is increased from 755 mm Hg to 899 mm Hg at constant temperature, what will the volume of the sample become?
    b. If a sample of gas is compressed from an initial volume of 455 mL at 755 mm Hg to a final volume of 327 mL at constant temperature, what will be the new pressure in the gas sample?

29. a. If the temperature of a 255-mL sample of gas is increased from 35 °C to 55 °C at constant pressure, what will be the new volume of the gas sample?
    b. If a 325-mL sample of gas at 25 °C is immersed in liquid nitrogen at –196 °C, what will be the new volume of the gas sample?

30. Calculate the indicated quantity for each gas sample.
    a. The volume occupied by 1.15 g of helium gas at 25 °C and 1.01 atm pressure.
    b. The partial pressure of each gas if 2.27 g of H_2 and 1.03 g of He are confined to a 5.00-L container at 0 °C.
    c. The pressure existing in a 9.97-L tank containing 42.5 g of argon gas at 27 °C.

31. Chlorine gas, Cl_2, can be generated in small quantities by addition of concentrated hydrochloric acid to manganese(IV) oxide.

$$MnO_2(s) + 4HCl(aq) \rightarrow MnCl_2(aq) + 2H_2O(l) + Cl_2(g)$$

The chlorine gas is bubbled through water to dissolve any traces of HCl remaining and then is dried by bubbling through concentrated sulfuric acid.

After drying, what volume of Cl_2 gas at STP would be expected if 4.05 g of MnO_2 is treated with excess concentrated HCl?

32. When calcium carbonate is heated strongly, it evolves carbon dioxide gas.

$$CaCO_3(s) \rightarrow CaO(s) + CO_2(g)$$

If 1.25 g of $CaCO_3$ is heated, what mass of CO_2 would be produced? What volume would this quantity of CO_2 occupy at STP?

33. If an electric current is passed through molten sodium chloride, elemental chlorine gas is generated as the sodium chloride is decomposed.

$$2NaCl(l) \rightarrow 2Na(s) + Cl_2(g)$$

What volume of chlorine gas measured at 767 mm Hg at 25 °C would be generated by complete decomposition of 1.25 g of NaCl?

34. Calculate the indicated quantity for each solution.
    a. The percent by mass of solute when 2.05 g of NaCl is dissolved in 19.2 g of water.
    b. The mass of solute contained in 26.2 g of 10.5% $CaCl_2$ solution.
    c. The mass of NaCl required to prepare 225 g of 5.05% NaCl solution.

35. Calculate the indicated quantity for each solution.
   a. The mass of solute present in 235 mL of 0.251 *M* NaOH solution.
   b. The molarity of the solution when 0.293 mole of KNO_3 is dissolved in water to a final volume of 125 mL.
   c. The number of moles of HCl present in 5.05 L of 6.01 *M* solution.

36. Calculate the molarities of the solutions resulting when the indicated dilutions are made. Assume that the volumes are additive.
   a. 25 mL of water is added to 12.5 mL of 1.515 *M* NaOH solution.
   b. 75.0 mL of 0.252 *M* HCl is diluted to a volume of 225 mL.
   c. 52.1 mL fo 0.751 *M* HNO_3 is added to 250. mL of water.

37. Calculate the volume (in milliliters) of each of the following acid solutions that would be required to neutralize 36.2 mL of 0.259 *M* NaOH solution.
   a. 0.271 *M* HCl
   b. 0.119 *M* H_2SO_4
   c. 0.171 *M* H_3PO_4

38. If 125 mL of concentrated sulfuric acid solution (density 1.84 g/mL, 98.3% H_2SO_4 by mass) is diluted to a final volume of 3.01 L, calculate the following information.
   a. the mass of pure H_2SO_4 in the 125-mL sample.
   b. the molarity of the *concentrated* acid solution
   c. the molarity of the *dilute* acid solution
   d. the normality of the dilute acid solution
   e. the quantity of the dilute acid solution needed to neutralize 45.3 mL of 0.532 *M* NaOH solution

All even-numbered Questions and Problems have answers in the back of this book and solutions in the *Student Solutions Guide*.

Acids and Bases

16

CHAPTER 16

Gargoyles on the Notre Dame cathedral in Paris in need of restoration from decades of acid rain. Witold Skrypczak/SuperStock

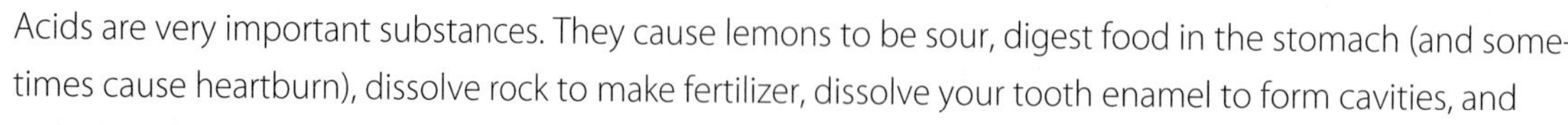

Acids are very important substances. They cause lemons to be sour, digest food in the stomach (and sometimes cause heartburn), dissolve rock to make fertilizer, dissolve your tooth enamel to form cavities, and clean the deposits out of your coffee maker. Acids are essential industrial chemicals. In fact, the chemical in first place in terms of the amount manufactured in the United States is sulfuric acid, H_2SO_4. Eighty *billion* pounds of this material are used every year in the manufacture of fertilizers, detergents, plastics, pharmaceuticals, storage batteries, and metals.

In this chapter we will consider the most important properties of acids and of their opposites, the bases.

A lemon tastes sour because it contains citric acid. Royalty-free Corbis

16-1 Acids and Bases

OBJECTIVE **To learn about two models of acids and bases and the relationship of conjugate acid–base pairs.**

Don't taste chemical reagents!

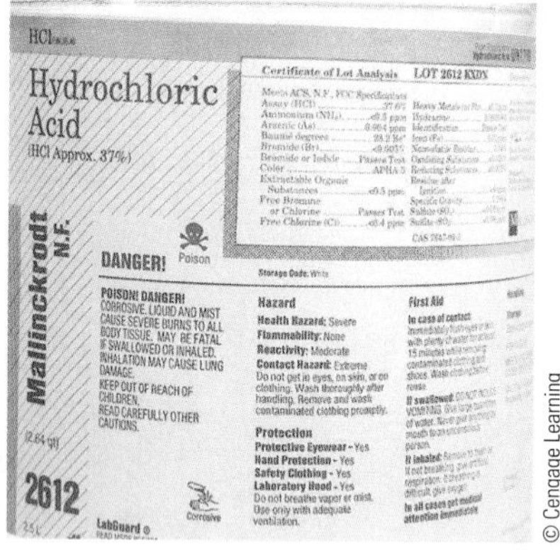

The label on a bottle of concentrated hydrochloric acid.

Acids were first recognized as substances that taste sour. Vinegar tastes sour because it is a dilute solution of acetic acid; citric acid is responsible for the sour taste of a lemon. Bases, sometimes called *alkalis,* are characterized by their bitter taste and slippery feel. Most hand soaps and commercial preparations for unclogging drains are highly basic.

The first person to recognize the essential nature of acids and bases was Svante Arrhenius. On the basis of his experiments with electrolytes, Arrhenius postulated that **acids** *produce hydrogen ions in aqueous solution,* whereas **bases** *produce hydroxide ions* (review Section 7-4).

For example, when hydrogen chloride gas is dissolved in water, each molecule produces ions as follows:

$$HCl(g) \xrightarrow{H_2O} H^+(aq) + Cl^-(aq)$$

This solution is the strong acid known as hydrochloric acid. On the other hand, when solid sodium hydroxide is dissolved in water, its ions separate producing a solution containing Na^+ and OH^- ions.

$$NaOH(s) \xrightarrow{H_2O} Na^+(aq) + OH^-(aq)$$

This solution is called a strong base.

Although the **Arrhenius concept of acids and bases** was a major step forward in understanding acid–base chemistry, this concept is limited because it allows for only one kind of base—the hydroxide ion. A more general definition of acids and bases was suggested by the Danish chemist Johannes Brønsted and the English chemist Thomas Lowry. In the **Brønsted–Lowry model,** *an acid is a proton (H^+) donor, and a base is a proton acceptor.* According to the Brønsted–Lowry model, the general reaction that occurs when an acid is dissolved in water can best be represented as an acid (HA) donating a proton to a water molecule to form a new acid (the **conjugate acid**) and a new base (the **conjugate base**).

Recall that (aq) means the substance is hydrated—it has water molecules clustered around it.

$$\underset{\text{Acid}}{HA(aq)} + \underset{\text{Base}}{H_2O(l)} \rightarrow \underset{\text{Conjugate acid}}{H_3O^+(aq)} + \underset{\text{Conjugate base}}{A^-(aq)}$$

This model emphasizes the significant role of the polar water molecule in pulling the proton from the acid. Note that the conjugate base is everything that remains of the acid molecule after a proton is lost. The conjugate acid is formed when the proton is transferred to the base. A **conjugate acid–base pair** consists of two substances related to each other by the donating and accepting of a *single proton.* In the above equation there are two conjugate acid–base pairs: HA (acid) and A^- (base), and H_2O (base) and H_3O^+ (acid). For example, when hydrogen chloride is dissolved in water it behaves as an acid.

Acid–conjugate base pair

$$HCl(aq) + H_2O(l) \rightarrow H_3O^+(aq) + Cl^-(aq)$$

Base–conjugate acid pair

In this case HCl is the acid that loses an H^+ ion to form Cl^-, its conjugate base. On the other hand, H_2O (behaving as a base) gains an H^+ ion to form H_3O^+ (the conjugate acid).

How can water act as a base? Remember that the oxygen of the water molecule has two unshared electron pairs, either of which can form a covalent bond with an H^+ ion. When gaseous HCl dissolves in water, the following reaction occurs:

$$\text{H—}\ddot{\text{O}}\text{:}(\text{—H}) + \overset{\delta+}{\text{H}}\text{—}\overset{\delta-}{\text{Cl}} \rightarrow \left[\text{H—}\ddot{\text{O}}\text{—H}(\text{—H})\right]^+ + \text{Cl}^-$$

Note that an H^+ ion is transferred from the HCl molecule to the water molecule to form H_3O^+, which is called the **hydronium ion.**

Interactive Example 16.1

Identifying Conjugate Acid–Base Pairs

Which of the following represent conjugate acid–base pairs?

a. HF, F^-

b. NH_4^+, NH_3

c. HCl, H_2O

SOLUTION

a. and b. HF, F^- and NH_4^+, NH_3 are conjugate acid–base pairs because the two species differ by one H^+.

$$HF \rightarrow H^+ + F^-$$

$$NH_4^+ \rightarrow H^+ + NH_3$$

c. HCl and H_2O are not a conjugate acid–base pair because they are not related by the removal or addition of one H^+. The conjugate base of HCl is Cl^-. The conjugate acid of H_2O is H_3O^+. ■

Interactive Example 16.2

Writing Conjugate Bases

Write the conjugate base for each of the following:

a. $HClO_4$ b. H_3PO_4 c. $CH_3NH_3^+$

SOLUTION To get the conjugate base for an acid, we must remove an H^+ ion.

a. $HClO_4 \rightarrow H^+ + ClO_4^-$
Acid Conjugate base

b. $H_3PO_4 \rightarrow H^+ + H_2PO_4^-$
Acid Conjugate base

c. $CH_3NH_3^+ \rightarrow H^+ + CH_3NH_2$
Acid Conjugate base

SELF CHECK

Exercise 16.1 Which of the following represent conjugate acid–base pairs?

a. H_2O, H_3O^+

b. OH^-, HNO_3

c. H_2SO_4, SO_4^{2-}

d. $HC_2H_3O_2$, $C_2H_3O_2^-$

See Problems 16.7 through 16.14. ■

16-2 Acid Strength

OBJECTIVES

- **To understand what acid strength means.**
- **To understand the relationship between acid strength and the strength of the conjugate base.**

We have seen that when an acid dissolves in water, a proton is transferred from the acid to water:

$$HA(aq) + H_2O(l) \rightarrow H_3O^+(aq) + A^-(aq)$$

In this reaction a new acid, H_3O^+ (called the conjugate acid), and a new base, A^- (the conjugate base), are formed. The conjugate acid and base can react with one another,

$$H_3O^+(aq) + A^-(aq) \rightarrow HA(aq) + H_2O(l)$$

to re-form the parent acid and a water molecule. Therefore, this reaction can occur "in both directions." The forward reaction is

$$HA(aq) + H_2O(l) \rightarrow H_3O^+(aq) + A^-(aq)$$

and the reverse reaction is

$$H_3O^+(aq) + A^-(aq) \rightarrow HA(aq) + H_2O(l)$$

Note that the products in the forward reaction are the reactants in the reverse reaction. We usually represent the situation in which the reaction can occur in both directions by double arrows:

$$HA(aq) + H_2O(l) \rightleftharpoons H_3O^+(aq) + A^-(aq)$$

This situation represents a competition for the H^+ ion between H_2O (in the forward reaction) and A^- (in the reverse reaction). If H_2O "wins" this competition—that is, if H_2O has a very high attraction for H^+ compared to A^-—then the solution will contain mostly H_3O^+ and A^-. We describe this situation by saying that the H_2O molecule is a much stronger base (more attraction for H^+) than A^-. In this case the forward reaction predominates:

$$HA(aq) + H_2O(l) \rightarrow H_3O^+(aq) + A^-(aq)$$

We say that the acid HA is *completely ionized* or *completely dissociated.* This situation represents a **strong acid.**

The opposite situation can also occur. Sometimes A^- "wins" the competition for the H^+ ion. In this case A^- is a much stronger base than H_2O and the reverse reaction predominates:

$$HA(aq) + H_2O(l) \leftarrow H_3O^+(aq) + A^-(aq)$$

Here, A^- has a much larger attraction for H^+ than does H_2O, and most of the HA molecules remain intact. This situation represents a **weak acid.**

We can determine what is actually going on in a solution by measuring its ability to conduct an electric current. Recall from Chapter 7 that a solution can conduct a current in proportion to the number of ions that are present (Fig. 7.2). When 1 mole of solid sodium chloride is dissolved in 1 L of water, the resulting solution is an excellent conductor of an electric current because the Na^+ and Cl^- ions separate completely. We call NaCl a strong electrolyte. Similarly, when 1 mole of hydrogen chloride is dissolved in 1 L of water, the resulting solution is an excellent conductor. Therefore, hydrogen chloride is also a strong electrolyte, which means that each HCl molecule must produce H^+ and Cl^- ions. This tells us that the forward reaction predominates:

A hydrochloric acid solution readily conducts electric current, as shown by the brightness of the bulb.

$$HCl(aq) + H_2O(l) \rightleftharpoons H_3O^+(aq) + Cl^-(aq)$$

(Accordingly, the arrow pointing right is longer than the arrow pointing left.) In solution there are virtually no HCl molecules, only H^+ and Cl^- ions. This shows that Cl^- is a very poor base compared to the H_2O molecule; it has virtually no ability to attract H^+ ions in water. This aqueous solution of hydrogen chloride (called *hydrochloric acid*) is a strong acid.

In general, the strength of an acid is defined by the position of its ionization (dissociation) reaction:

$$HA(aq) + H_2O(l) \rightleftharpoons H_3O^+(aq) + A^-(aq)$$

A strong acid is one for which *the forward reaction predominates.* This means that almost all the original HA is dissociated (ionized) [Fig. 16.1(a)]. There is an important

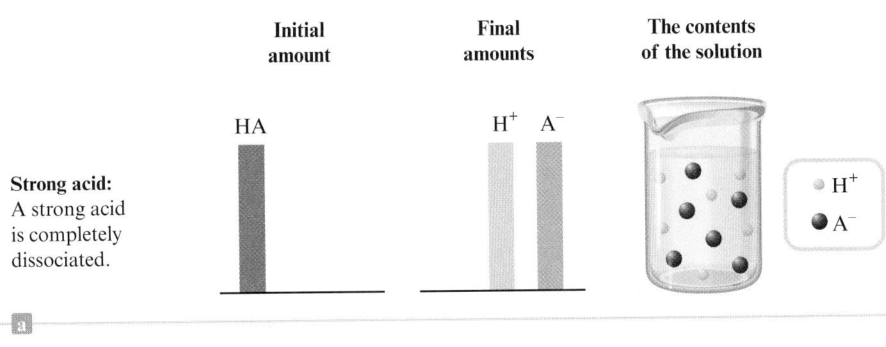

Figure 16.1 ▸ Graphical representation of the behavior of acids of different strengths in aqueous solution.

CHEMISTRY IN FOCUS

Carbonation—A Cool Trick

The sensations of taste and smell greatly affect our daily experience. For example, memories are often triggered by an odor that matches one that occurred when an event was originally stored in our memory banks. Likewise, the sense of taste has a powerful effect on our lives. For example, many people crave the intense sensation produced by the compounds found in chili peppers.

One sensation that is quite refreshing for most people is the effect of a chilled, carbonated beverage in the mouth. The sharp, tingling sensation experienced is not directly due to the bubbling of the dissolved carbon dioxide in the beverage. Rather, it arises because protons are produced as the CO_2 interacts with the water in the tissues of the mouth:

$$CO_2 + H_2O \rightleftharpoons H^+ + HCO_3^-$$

This reaction is speeded up by a biologic catalyst—an enzyme—called carbonic anhydrase. The acidification of the fluids in the nerve endings in the mouth leads to the sharp sensation produced by carbonated drinks.

Carbon dioxide also stimulates nerve sites that detect "coolness" in the mouth. In fact, researchers have identified a mutual enhancement between cooling and the presence of CO_2. Studies show that at a given concentration of CO_2, a colder drink feels more "pungent" than a warmer one. When tests were conducted on drinks in which the carbon dioxide concentration was varied, the results showed that a drink felt colder as the CO_2 concentration was increased, even though the drinks were all actually at the same temperature.

Thus a beverage can seem colder if it has a higher concentration of carbon dioxide. At the same time, cooling a carbonated beverage can intensify the tingling sensation caused by the acidity induced by the CO_2. This is truly a happy synergy.

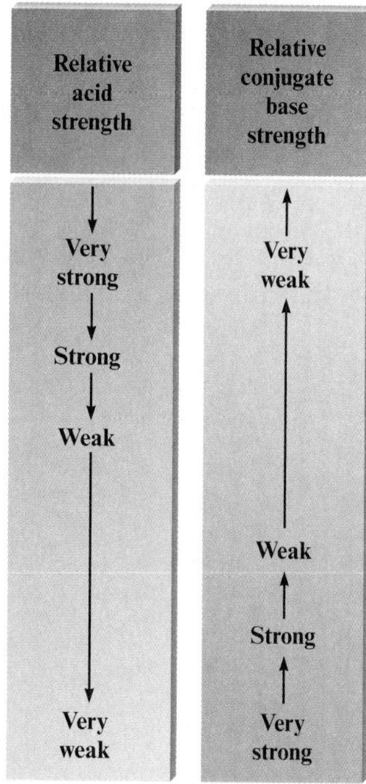

Figure 16.2 The relationship of acid strength and conjugate base strength for the dissociation reaction

$$\underset{\text{Acid}}{HA(aq)} + H_2O(l) \rightleftharpoons H_3O^+(aq) + \underset{\text{Conjugate base}}{A^-(aq)}$$

connection between the strength of an acid and that of its conjugate base. *A strong acid contains a relatively weak conjugate base*—one that has a low attraction for protons. A strong acid can be described as an acid whose conjugate base is a much weaker base than water (Fig. 16.2). In this case the water molecules win the competition for the H^+ ions.

In contrast to hydrochloric acid, when acetic acid, $HC_2H_3O_2$, is dissolved in water, the resulting solution conducts an electric current only weakly. That is, acetic acid is a weak electrolyte, which means that only a few ions are present. In other words, for the reaction

$$HC_2H_3O_2(aq) + H_2O(l) \rightleftharpoons H_3O^+(aq) + C_2H_3O_2^-(aq)$$

the reverse reaction predominates (thus the arrow pointing left is longer). In fact, measurements show that only about one in one hundred (1%) of the $HC_2H_3O_2$ molecules is dissociated (ionized) in a 0.1 M solution of acetic acid. Thus acetic acid is a weak acid. When acetic acid molecules are placed in water, almost all of the molecules remain undissociated. This tells us that the acetate ion, $C_2H_3O_2^-$, is an effective base—it very successfully attracts H^+ ions in water. This means that acetic acid remains largely in the form of $HC_2H_3O_2$ molecules in solution. A weak acid is one for which the *reverse reaction predominates.*

$$HA(aq) + H_2O(l) \leftarrow H_3O^+(aq) + A^-(aq)$$

Most of the acid originally placed in the solution is still present as HA at equilibrium. That is, a weak acid dissociates (ionizes) only to a very small extent in aqueous solution [Fig. 16.1(b)]. In contrast to a strong acid, a weak acid has a conjugate base that is a much stronger base than water. In this case a water molecule is not very successful in pulling an H^+ ion away from the conjugate base. *A weak acid contains a relatively strong conjugate base* (Fig. 16.2).

The various ways of describing the strength of an acid are summarized in Table 16.1.

CHEMISTRY IN FOCUS

Plants Fight Back

Plants sometimes do not seem to get much respect. We often think of them as rather dull life forms. We are used to animals communicating with each other, but we think of plants as mute. However, this perception is now changing. It is now becoming clear that plants communicate with other plants and also with insects. Ilya Roskin and his colleagues at Rutgers University, for example, have found that tobacco plants under attack by disease signal distress using the chemical salicylic acid, a precursor of aspirin. When a tobacco plant is infected with tobacco mosaic virus (TMV), which forms dark blisters on leaves and causes them to pucker and yellow, the sick plant produces large amounts of salicylic acid to alert its immune system to fight the virus. In addition, some of the salicylic acid is converted to methyl salicylate, a volatile compound that evaporates from the sick plant. Neighboring plants absorb this chemical and turn it back to salicylic acid, thus triggering their immune systems to protect them against the impending attack by TMV. Thus, as a tobacco plant gears up to fight an attack by TMV, it also warns its neighbors to be ready for this virus.

In another example of plant communication, a tobacco leaf under attack by a caterpillar emits a chemical signal that attracts a parasitic wasp that stings and kills the insect. Even more impressive is the ability of the plant to customize the emitted signal so that the wasp attracted will be the one that specializes in killing the particular caterpillar involved in the attack. The plant does this by changing the proportions of two chemicals emitted when a caterpillar chews on a leaf. Studies have shown that other plants, such as corn and cotton, also emit wasp-attracting chemicals when they face attack by caterpillars.

This research shows that plants can "speak up" to protect themselves. Scientists hope to learn to help them do this even more effectively.

Salicylic acid

Methyl salicylate

Agricultural Research Service/USDA

A wasp lays its eggs on a gypsy moth caterpillar on the leaf of a corn plant.

Critical Thinking

Vinegar contains acetic acid and is used in salad dressings. What if acetic acid was a strong acid instead of a weak acid? Would it be safe to use vinegar as a salad dressing?

Table 16.1 Ways to Describe Acid Strength

Property	Strong Acid	Weak Acid
the acid ionization (dissociation) reaction	forward reaction predominates	reverse reaction predominates
strength of the conjugate base compared with that of water	A^- is a much weaker base than H_2O	A^- is a much stronger base than H_2O

The common strong acids are sulfuric acid, $H_2SO_4(aq)$; hydrochloric acid, $HCl(aq)$; nitric acid, $HNO_3(aq)$; and perchloric acid, $HClO_4(aq)$. Sulfuric acid is actually a **diprotic acid,** an acid that can furnish two protons. The acid H_2SO_4 is a strong acid that is virtually 100% dissociated in water:

$$H_2SO_4(aq) \rightarrow H^+(aq) + HSO_4^-(aq)$$

An acetic acid solution conducts only a small amount of current as shown by the dimly lit bulb.

The HSO_4^- ion is also an acid but it is a weak acid:

$$HSO_4^-(aq) \rightleftharpoons H^+(aq) + SO_4^{2-}(aq)$$

Most of the HSO_4^- ions remain undissociated.

Most acids are **oxyacids,** in which the acidic hydrogen is attached to an oxygen atom (several oxyacids are shown below). The strong acids we have mentioned, except hydrochloric acid, are typical examples. **Organic acids,** those with a carbon-atom backbone, commonly contain the **carboxyl group:**

$$-C(=O)-O-H$$

Acids of this type are usually weak. An example is acetic acid, CH_3COOH, which is often written as $HC_2H_3O_2$.

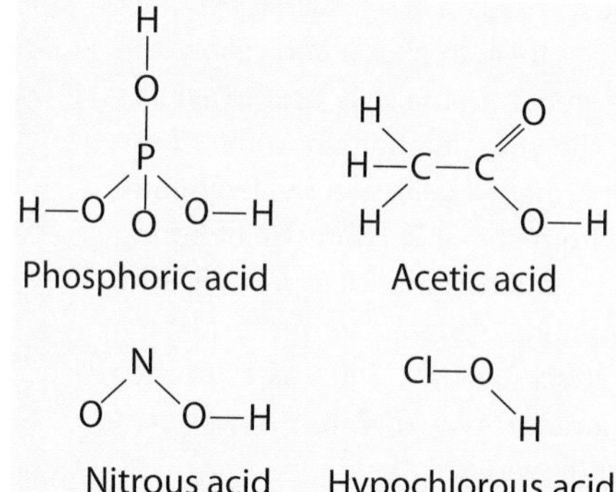

There are some important acids in which the acidic proton is attached to an atom other than oxygen. The most significant of these are the hydrohalic acids HX, where X represents a halogen atom. Examples are HCl(*aq*), a strong acid, and HF(*aq*), a weak acid.

16-3 Water as an Acid and a Base

OBJECTIVE **To learn about the ionization of water.**

A substance is said to be *amphoteric* if it can behave either as an acid or as a base. Water is the most common **amphoteric substance.** We can see this clearly in the ionization of water, which involves the transfer of a proton from one water molecule to another to produce a hydroxide ion and a hydronium ion.

$$H_2O(l) + H_2O(l) \rightleftharpoons H_3O^+(aq) + OH^-(aq)$$

In this reaction one water molecule acts as an acid by furnishing a proton, and the other acts as a base by accepting the proton. The forward reaction for this process does not occur to a very great extent. That is, in pure water only a tiny amount of H_3O^+ and OH^- exist. At 25 °C the actual concentrations are

$$[H_3O^+] = [OH^-] = 1.0 \times 10^{-7}\,M$$

Notice that in pure water the concentrations of $[H_3O^+]$ and $[OH^-]$ are equal because they are produced in equal numbers in the ionization reaction.

One of the most interesting and important things about water is that the mathematical *product* of the H_3O^+ and OH^- concentrations is always constant. We can find this constant by multiplying the concentrations of H_3O^+ and OH^- at 25 °C:

$$[H_3O^+][OH^-] = (1.0 \times 10^{-7})(1.0 \times 10^{-7}) = 1.0 \times 10^{-14}$$

We call this constant the **ion-product constant, K_w.** Thus at 25 °C

$$[H_3O^+][OH^-] = 1.0 \times 10^{-14} = K_w$$

To simplify the notation we often write H_3O^+ as just H^+. Thus we would write the K_w expression as follows:

$$[H^+][OH^-] = 1.0 \times 10^{-14} = K_w$$

The units are customarily omitted when the value of the constant is given and used.

It is important to recognize the meaning of K_w. In any aqueous solution at 25 °C, *no matter what it contains,* the product of $[H^+]$ and $[OH^-]$ must always equal 1.0×10^{-14}. This means that if the $[H^+]$ goes up, the $[OH^-]$ must go down so that the product of the two is still 1.0×10^{-14}. For example, if HCl gas is dissolved in water, increasing the $[H^+]$, the $[OH^-]$ must decrease.

There are three possible situations we might encounter in an aqueous solution. If we add an acid (an H^+ donor) to water, we get an *acidic solution.* In this case, because we have added a source of H^+, the $[H^+]$ will be greater than the $[OH^-]$. On the other hand, if we add a base (a source of OH^-) to water, the $[OH^-]$ will be greater than the $[H^+]$. This is a *basic solution.* Finally, we might have a situation in which $[H^+] = [OH^-]$. This is called a *neutral solution.* Pure water is automatically neutral, but we can also obtain a neutral solution by adding equal amounts of H^+ and OH^-. It is very important that you understand the definitions of neutral, acidic, and basic solutions. In summary:

Remember that H^+ represents H_3O^+.

1. In a **neutral solution,** $[H^+] = [OH^-]$
2. In an **acidic solution,** $[H^+] > [OH^-]$
3. In a **basic solution,** $[OH^-] > [H^+]$

In each case, however, $K_w = [H^+][OH^-] = 1.0 \times 10^{-14}$.

Interactive Example 16.3 Calculating Ion Concentrations in Water

Calculate $[H^+]$ or $[OH^-]$ as required for each of the following solutions at 25 °C, and state whether the solution is neutral, acidic, or basic.

a. $1.0 \times 10^{-5}\ M\ OH^-$ b. $1.0 \times 10^{-7}\ M\ OH^-$ c. $10.0\ M\ H^+$

SOLUTION

a. **Where Are We Going?**

We want to determine $[H^+]$ in a solution of given $[OH^-]$ at 25 °C.

What Do We Know?

- At 25 °C, $K_w = [H^+][OH^-] = 1.0 \times 10^{-14}$
- $[OH^-] = 1.0 \times 10^{-5}\ M$

How Do We Get There?

Math skill builder

$$K_w = [H^+][OH^-]$$

$$\frac{K_w}{[OH^-]} = [H^+]$$

We know that $K_w = [H^+][OH^-] = 1.0 \times 10^{-14}$. We need to calculate the $[H^+]$. However, the $[OH^-]$ is given—it is $1.0 \times 10^{-5}\ M$—so we will solve for $[H^+]$ by dividing both sides by $[OH^-]$.

$$[H^+] = \frac{1.0 \times 10^{-14}}{[OH^-]} = \frac{1.0 \times 10^{-14}}{1.0 \times 10^{-5}} = 1.0 \times 10^{-9}\ M$$

Because $[OH^-] = 1.0 \times 10^{-5}\ M$ is greater than $[H^+] = 1.0 \times 10^{-9}\ M$, the solution is basic. (Remember: The more negative the exponent, the smaller the number.)

b. **Where Are We Going?**

We want to determine $[H^+]$ in a solution of given $[OH^-]$ at 25 °C.

What Do We Know?

- At 25 °C, $K_w = [H^+][OH^-] = 1.0 \times 10^{-14}$
- $[OH^-] = 1.0 \times 10^{-7}\ M$

How Do We Get There?

Again the $[OH^-]$ is given, so we solve the K_w expression for $[H^+]$.

$$[H^+] = \frac{1.0 \times 10^{-14}}{[OH^-]} = \frac{1.0 \times 10^{-14}}{1.0 \times 10^{-7}} = 1.0 \times 10^{-7}\ M$$

Here $[H^+] = [OH^-] = 1.0 \times 10^{-7}\ M$, so the solution is neutral.

c. **Where Are We Going?**

We want to determine $[OH^-]$ in a solution of given $[H^+]$ at 25 °C.

What Do We Know?

- At 25 °C, $K_w = [H^+][OH^-] = 1.0 \times 10^{-14}$
- $[H^+] = 10.0\ M$

How Do We Get There?

In this case the $[H^+]$ is given, so we solve for $[OH^-]$.

Math skill builder

$$K_w = [H^+][OH^-]$$

$$\frac{K_w}{[H^+]} = [OH^-]$$

$$[OH^-] = \frac{1.0 \times 10^{-14}}{[H^+]} = \frac{1.0 \times 10^{-14}}{10.0} = 1.0 \times 10^{-15}\ M$$

Now we compare $[H^+] = 10.0\ M$ with $[OH^-] = 1.0 \times 10^{-15}\ M$. Because $[H^+]$ is greater than $[OH^-]$, the solution is acidic.

SELF CHECK

Exercise 16.2 Calculate $[H^+]$ in a solution in which $[OH^-] = 2.0 \times 10^{-2}\ M$. Is this solution acidic, neutral, or basic?

See Problems 16.31 through 16.34. ■

Example 16.4 Using the Ion-Product Constant in Calculations

Is it possible for an aqueous solution at 25 °C to have $[H^+] = 0.010\ M$ and $[OH^-] = 0.010\ M$?

SOLUTION

The concentration 0.010 M can also be expressed as $1.0 \times 10^{-2}\ M$. Thus, if $[H^+] = [OH^-] = 1.0 \times 10^{-2}\ M$, the product

$$[H^+][OH^-] = (1.0 \times 10^{-2})(1.0 \times 10^{-2}) = 1.0 \times 10^{-4}$$

This is not possible. The product of $[H^+]$ and $[OH^-]$ must always be 1.0×10^{-14} in water at 25 °C, so a solution could not have $[H^+] = [OH^-] = 0.010\ M$. If H^+ and OH^- are added to water in these amounts, they will react with each other to form H_2O,

$$H^+ + OH^- \rightarrow H_2O$$

until the product $[H^+][OH^-] = 1.0 \times 10^{-14}$.

This is a general result. When H^+ and OH^- are added to water in amounts such that the product of their concentrations is greater than 1.0×10^{-14}, they will react to form water until enough H^+ and OH^- are consumed so that $[H^+][OH^-] = 1.0 \times 10^{-14}$. ■

CHEMISTRY IN FOCUS

Airplane Rash

Because airplanes remain in service for many years, it is important to spot corrosion that might weaken the structure at an early stage. In the past, looking for minute signs of corrosion has been very tedious and labor-intensive, especially for large planes. This situation has changed, however, thanks to the paint system developed by Gerald S. Frankel and Jian Zhang of Ohio State University. The paint they created turns pink in areas that are beginning to corrode, making these areas easy to spot.

The secret to the paint's magic is phenolphthalein, the common acid–base indicator that turns pink in a basic solution. The corrosion of the aluminum skin of the airplane involves a reaction that forms OH^- ions, producing a basic area at the site of the corrosion that turns the phenolphthalein pink. Because this system is highly sensitive, corrosion can be corrected before it damages the plane.

Next time you fly, if the plane has pink spots you might want to wait for a later flight!

16-4 The pH Scale

OBJECTIVES

- **To understand pH and pOH.**
- **To learn to find pOH and pH for various solutions.**
- **To learn to use a calculator in these calculations.**

To express small numbers conveniently, chemists often use the "p scale," which is based on common logarithms (base 10 logs). In this system, if N represents some number, then

$$\mathrm{p}N = -\log N = (-1) \times \log N$$

That is, the p means to take the log of the number that follows and multiply the result by -1. For example, to express the number 1.0×10^{-7} on the p scale, we need to take the negative log of 1.0×10^{-7}.

$$\mathrm{p}(1.0 \times 10^{-7}) = -\log(1.0 \times 10^{-7}) = 7.00$$

Because the $[H^+]$ in an aqueous solution is typically quite small, using the p scale in the form of the **pH scale** provides a convenient way to represent solution acidity. The pH is defined as

$$\mathrm{pH} = -\log[H^+]$$

To obtain the pH value of a solution, we must compute the negative log of the $[H^+]$. In the case where $[H^+] = 1.0 \times 10^{-5}\ M$, the solution has a pH value of 5.00.

To represent pH to the appropriate number of significant figures, you need to know the following rule for logarithms: *the number of decimal places for a log must be equal to the number of significant figures in the original number.* Thus

2 significant figures

$$[H^+] = 1.0 \times 10^{-5}\ M$$

and

$$\mathrm{pH} = 5.00$$

2 decimal places

Interactive Example 16.5

Calculating pH

Calculate the pH value for each of the following solutions at 25 °C.

a. A solution in which $[H^+] = 1.0 \times 10^{-9}\ M$

b. A solution in which $[OH^-] = 1.0 \times 10^{-6}\ M$

SOLUTION

a. For this solution $[H^+] = 1.0 \times 10^{-9}$.

$$-\log(1.0 \times 10^{-9}) = 9.00$$

$$\text{pH} = 9.00$$

b. In this case we are given the $[OH^-]$. Thus we must first calculate $[H^+]$ from the K_w expression. We solve

$$K_w = [H^+][OH^-] = 1.0 \times 10^{-14}$$

for $[H^+]$ by dividing both sides by $[OH^-]$.

$$[H^+] = \frac{1.0 \times 10^{-14}}{[OH^-]} = \frac{1.0 \times 10^{-14}}{1.0 \times 10^{-6}} = 1.0 \times 10^{-8}$$

Now that we know the $[H^+]$, we can calculate the pH because $\text{pH} = -\log[H^+] = -\log(1.0 \times 10^{-8}) = 8.00$.

SELF CHECK

Exercise 16.3 Calculate the pH value for each of the following solutions at 25 °C.

a. A solution in which $[H^+] = 1.0 \times 10^{-3}\ M$

b. A solution in which $[OH^-] = 5.0 \times 10^{-5}\ M$

See Problems 16.41 through 16.44. ■

Table 16.2 ▸ The Relationship of the H^+ Concentration of a Solution to Its pH

$[H^+](M)$	pH
1.0×10^{-1}	1.00
1.0×10^{-2}	2.00
1.0×10^{-3}	3.00
1.0×10^{-4}	4.00
1.0×10^{-5}	5.00
1.0×10^{-6}	6.00
1.0×10^{-7}	7.00

Because the pH scale is a log scale based on 10, *the pH changes by 1 for every power-of-10 change in the $[H^+]$*. For example, a solution of pH 3 has an H^+ concentration of $10^{-3}\ M$, which is 10 times that of a solution of pH 4 ($[H^+] = 10^{-4}\ M$) and 100 times that of a solution of pH 5. This is illustrated in Table 16.2. Also note from Table 16.2 that *the pH decreases as the $[H^+]$ increases.* That is, a lower pH means a more acidic solution. The pH scale and the pH values for several common substances are shown in Fig. 16.3.

We often measure the pH of a solution by using a pH meter, an electronic device with a probe that can be inserted into a solution of unknown pH. A pH meter is shown in Fig. 16.4. Colored indicator paper is also commonly used to measure the pH of a solution when less accuracy is needed. A drop of the solution to be tested is placed on this special paper, which promptly turns to a color characteristic of a given pH (Fig. 16.5).

Log scales similar to the pH scale are used for representing other quantities. For example,

$$\text{pOH} = -\log[OH^-]$$

The symbol p means −log.

Therefore, in a solution in which

$$[OH^-] = 1.0 \times 10^{-12}\ M$$

the pOH is

$$-\log[OH^-] = -\log(1.0 \times 10^{-12}) = 12.00$$

	$[H^+]$	pH	
Basic	10^{-14}	14	1 *M* NaOH
	10^{-13}	13	
	10^{-12}	12	Ammonia (household cleaner)
	10^{-11}	11	
	10^{-10}	10	
	10^{-9}	9	
	10^{-8}	8	Blood
Neutral	10^{-7}	7	Pure water
			Milk
	10^{-6}	6	
	10^{-5}	5	
	10^{-4}	4	
	10^{-3}	3	Vinegar
			Lemon juice
Acidic	10^{-2}	2	Stomach acid
	10^{-1}	1	
	10^{0}	0	1 *M* HCl

Figure 16.3 ▸ The pH scale and pH values of some common substances.

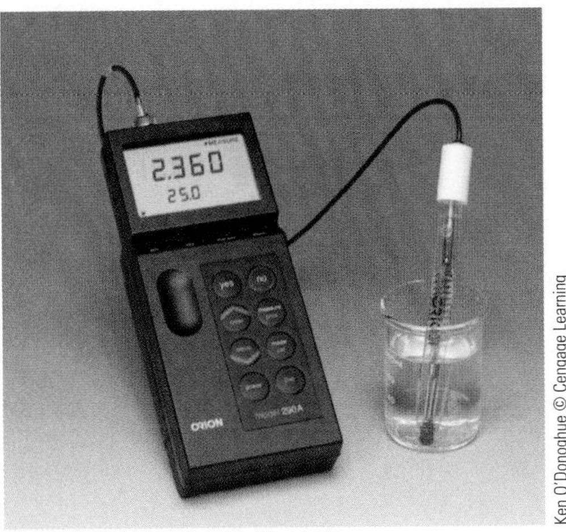

Figure 16.4 ▸ A pH meter. The electrodes on the right are placed in the solution with unknown pH. The difference between the $[H^+]$ in the solution sealed into one of the electrodes and the $[H^+]$ in the solution being analyzed is translated into an electrical potential and registered on the meter as a pH reading.

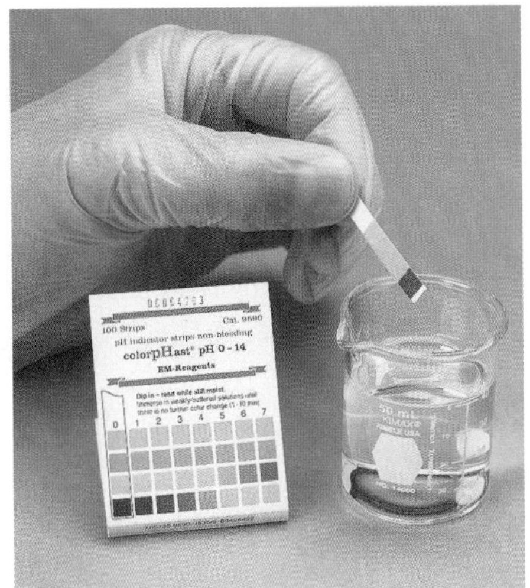

Figure 16.5 ▸ Indicator paper being used to measure the pH of a solution. The pH is determined by comparing the color that the solution turns the paper to the color chart.

Critical Thinking

What if an elected official decided to ban all products with a pH outside of the 6–8 range? How would this affect the products you could buy? Give some examples of products that would no longer be available.

Interactive Example 16.6

Calculating pH and pOH

Calculate the pH and pOH for each of the following solutions at 25 °C.

a. $1.0 \times 10^{-3}\ M\ OH^-$

b. $1.0\ M\ H^+$

SOLUTION

a. We are given the $[OH^-]$, so we can calculate the pOH value by taking $-\log[OH^-]$.

$$\text{pOH} = -\log[OH^-] = -\log(1.0 \times 10^{-3}) = 3.00$$

To calculate the pH, we must first solve the K_w expression for $[H^+]$.

$$[H^+] = \frac{K_w}{[OH^-]} = \frac{1.0 \times 10^{-14}}{1.0 \times 10^{-3}} = 1.0 \times 10^{-11}\ M$$

Now we compute the pH.

$$pH = -\log[H^+] = -\log(1.0 \times 10^{-11}) = 11.00$$

b. In this case we are given the $[H^+]$ and we can compute the pH.

$$pH = -\log[H^+] = -\log(1.0) = 0$$

We next solve the K_w expression for $[OH^-]$.

$$[OH^-] = \frac{K_w}{[H^+]} = \frac{1.0 \times 10^{-14}}{1.0} = 1.0 \times 10^{-14}\ M$$

Now we compute the pOH.

$$pOH = -\log[OH^-] = -\log(1.0 \times 10^{-14}) = 14.00 \ \blacksquare$$

We can obtain a convenient relationship between pH and pOH by starting with the K_w expression $[H^+][OH^-] = 1.0 \times 10^{-14}$ and taking the negative log of both sides.

$$-\log([H^+][OH^-]) = -\log(1.0 \times 10^{-14})$$

Scimat/Photo Researchers/Getty Images

Red blood cells can exist only over a narrow range of pH.

Because the log of a product equals the sum of the logs of the terms—that is, $\log(A \times B) = \log A + \log B$—we have

$$\underbrace{-\log[H^+]}_{pH} \underbrace{-\log[OH^-]}_{pOH} = -\log(1.0 \times 10^{-14}) = 14.00$$

which gives the equation

$$pH + pOH = 14.00$$

This means that once we know either the pH or the pOH for a solution, we can calculate the other. For example, if a solution has a pH of 6.00, the pOH is calculated as follows:

$$pH + pOH = 14.00$$
$$pOH = 14.00 - pH$$
$$pOH = 14.00 - 6.00 = 8.00$$

Interactive Example 16.7

Calculating pOH from pH

The pH of blood is about 7.4. What is the pOH of blood?

SOLUTION

$$pH + pOH = 14.00$$
$$pOH = 14.00 - pH$$
$$= 14.00 - 7.4$$
$$= 6.6$$

The pOH of blood is 6.6.

SELF CHECK

Exercise 16.4 A sample of rain in an area with severe air pollution has a pH of 3.5. What is the pOH of this rainwater?

See Problems 16.45 and 16.46. ■

It is also possible to find the $[H^+]$ or $[OH^-]$ from the pH or pOH. To find the $[H^+]$ from the pH, we must go back to the definition of pH:

$$pH = -\log[H^+]$$

or

$$-pH = \log[H^+]$$

To arrive at $[H^+]$ on the right-hand side of this equation we must "undo" the log operation. This is called taking the *antilog* or the *inverse* log.

$$\text{Inverse log }(-pH) = \text{inverse log }(\log[H^+])$$
$$\text{Inverse log }(-pH) = [H^+]$$

There are different methods for carrying out the inverse log operation on various calculators. One common method is the two-key (inv) (log) sequence. (Consult the user's manual for your calculator to find out how to do the antilog or inverse log operation.) The steps in going from pH to $[H^+]$ are as follows:

Math skill builder

This operation may involve a 10^x key on some calculators.

Charles D. Winters/Photo Researchers/Getty Images

Measuring the pH of glacial meltwater.

Steps for Calculating $[H^+]$ from pH

Step 1 Take the inverse log (antilog) of −pH to give $[H^+]$ by using the (inv) (log) keys in that order. (Your calculator may require different keys for this operation.)

Step 2 Press the minus [−] key.

Step 3 Enter the pH.

For practice, we will convert pH = 7.0 to $[H^+]$.

$$pH = 7.0$$
$$-pH = -7.0$$

The inverse log of −7.0 gives 1×10^{-7}.

$$[H^+] = 1 \times 10^{-7}\ M$$

This process is illustrated further in Example 16.8.

Interactive Example 16.8

Calculating $[H^+]$ from pH

The pH of a human blood sample was measured to be 7.41. What is the $[H^+]$ in this blood?

SOLUTION

$$pH = 7.41$$
$$-pH = -7.41$$
$$[H^+] = \text{inverse log of } -7.41 = 3.9 \times 10^{-8}$$
$$[H^+] = 3.9 \times 10^{-8}\ M$$

Notice that because the pH has two decimal places, we need two significant figures for $[H^+]$.

SELF CHECK

Exercise 16.5 The pH of rainwater in a polluted area was found to be 3.50. What is the $[H^+]$ for this rainwater?

See Problems 16.49 and 16.50. ■

A similar procedure is used to change from pOH to $[OH^-]$, as shown in Example 16.9.

CHEMISTRY IN FOCUS

Garden-Variety Acid–Base Indicators

What can flowers tell us about acids and bases? Actually, some flowers can tell us whether the soil they are growing in is acidic or basic. For example, in acidic soil, bigleaf hydrangea blossoms will be blue; in basic (alkaline) soil, the flowers will be red. What is the secret? The pigment in the flower is an acid–base indicator.

Generally, acid–base indicators are dyes that are weak acids. Because indicators are usually complex molecules, we often symbolize them as HIn. The reaction of the indicator with water can be written as

$$HIn(aq) + H_2O(l) \rightleftharpoons H_3O^+(aq) + In^-(aq)$$

To work as an acid–base indicator, the conjugate acid–base forms of these dyes must have different colors. The acidity level of the solution will determine whether the indicator is present mainly in its acidic form (HIn) or its basic form (In^-).

When placed in an acidic solution, most of the basic form of the indicator is converted to the acidic form by the reaction

$$In^-(aq) + H^+(aq) \rightarrow HIn(aq)$$

When placed in a basic solution, most of the acidic form of the indicator is converted to the basic form by the reaction

$$HIn(aq) + OH^-(aq) \rightarrow In^-(aq) + H_2O(l)$$

It turns out that many fruits, vegetables, and flowers can act as acid–base indicators. Red, blue, and purple plants often contain a class of chemicals called anthocyanins, which change color based on the acidity level of the surroundings. Perhaps the most famous of these plants is red cabbage. Red cabbage contains a mixture of anthocyanins and other pigments that allow it to be used as a "universal indicator." Red cabbage juice appears deep red at a

Interactive Example 16.9

Calculating $[OH^-]$ from pOH

The pOH of the water in a fish tank is found to be 6.59. What is the $[OH^-]$ for this water?

SOLUTION

We use the same steps as for converting pH to $[H^+]$, except that we use the pOH to calculate the $[OH^-]$.

$$pOH = 6.59$$

$$-pOH = -6.59$$

$$[OH^-] = \text{inverse log of } -6.59 = 2.6 \times 10^{-7}$$

$$[OH^-] = 2.6 \times 10^{-7}\ M$$

Note that two significant figures are required.

SELF CHECK

Exercise 16.6 The pOH of a liquid drain cleaner was found to be 10.50. What is the $[OH^-]$ for this cleaner?

See Problems 16.51 and 16.52. ■

16-5 Calculating the pH of Strong Acid Solutions

OBJECTIVE **To learn to calculate the pH of solutions of strong acids.**

In this section we will learn to calculate the pH for a solution containing a strong acid of known concentration. For example, if we know a solution contains 1.0 *M* HCl, how can we find the pH of the solution? To answer this question we must know that when pH of 1–2, purple at a pH of 4, blue at a pH of 8, and green at a pH of 11.

Other natural indicators include the skins of beets (which change from red to purple in very basic solutions), blueberries (which change from blue to red in acidic solutions), and a wide variety of flower petals, including delphiniums, geraniums, morning glories, and, of course, hydrangeas.

Royalty-free Corbis

Neil Holmes/Holmes Garden Photos/Alamy

HCl dissolves in water, each molecule dissociates (ionizes) into H^+ and Cl^- ions. That is, we must know that HCl is a strong acid. Thus, although the label on the bottle says 1.0 *M* HCl, the solution contains virtually no HCl molecules. A 1.0 *M* HCl solution contains H^+ and Cl^- ions rather than HCl molecules. Typically, container labels indicate the substance(s) used to make up the solution but do not necessarily describe the solution components after dissolution. In this case,

$$1.0\ M\ \text{HCl} \rightarrow 1.0\ M\ \text{H}^+ \text{ and } 1.0\ M\ \text{Cl}^-$$

Therefore, the $[H^+]$ in the solution is 1.0 *M*. The pH is then

$$\text{pH} = -\log[\text{H}^+] = -\log(1.0) = 0$$

Interactive Example 16.10 Calculating the pH of Strong Acid Solutions

Calculate the pH of 0.10 *M* HNO_3.

SOLUTION HNO_3 is a strong acid, so the ions in solution are H^+ and NO_3^-. In this case,

$$0.10\ M\ \text{HNO}_3 \rightarrow 0.10\ M\ \text{H}^+ \text{ and } 0.10\ M\ \text{NO}_3^-$$

Thus

$$[\text{H}^+] = 0.10\ M \quad \text{and} \quad \text{pH} = -\log(0.10) = 1.00$$

SELF CHECK

Exercise 16.7 Calculate the pH of a solution of 5.0×10^{-3} *M* HCl.

See Problems 16.57 and 16.58. ■

16-6 Buffered Solutions

OBJECTIVE To understand the general characteristics of buffered solutions.

A **buffered solution** is one that resists a change in its pH even when a strong acid or base is added to it. For example, when 0.01 mole of HCl is added to 1 L of pure water, the pH changes from its initial value of 7 to 2, a change of 5 pH units. However, when 0.01 mole of HCl is added to a solution containing both 0.1 *M* acetic acid ($HC_2H_3O_2$) and 0.1 *M* sodium acetate ($NaC_2H_3O_2$), the pH changes from an initial value of 4.74 to 4.66, a change of only 0.08 pH unit. The latter solution is buffered—it undergoes only a very slight change in pH when a strong acid or base is added to it.

Buffered solutions are vitally important to living organisms whose cells can survive only in a very narrow pH range. Many goldfish have died because their owners did not realize the importance of buffering the aquarium water at an appropriate pH. For humans to survive, the pH of the blood must be maintained between 7.35 and 7.45. This narrow range is maintained by several different buffering systems.

For goldfish to survive, the pH of the water must be carefully controlled.

A solution is buffered by the *presence of a weak acid and its conjugate base.* An example of a buffered solution is an aqueous solution that contains acetic acid and sodium acetate. The sodium acetate is a salt that furnishes acetate ions (the conjugate base of acetic acid) when it dissolves. To see how this system acts as a buffer, we must recognize that the species present in this solution are

$$HC_2H_3O_2, \quad Na^+, \quad C_2H_3O_2^-$$

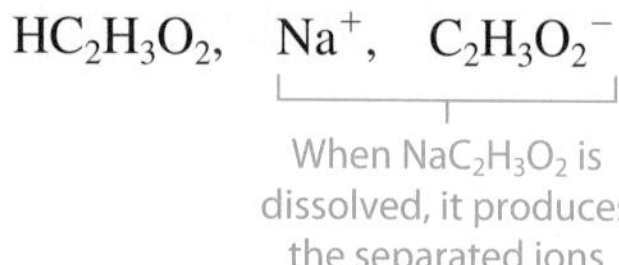

What happens in this solution when a strong acid such as HCl is added? In pure water, the H^+ ions from the HCl would accumulate, thus lowering the pH.

$$HCl \xrightarrow{100\%} H^+ + Cl^-$$

However, this buffered solution contains $C_2H_3O_2^-$ ions, which are basic. That is, $C_2H_3O_2^-$ has a strong affinity for H^+, as evidenced by the fact that $HC_2H_3O_2$ is a weak acid. This means that the $C_2H_3O_2^-$ and H^+ ions do not exist together in large numbers. Because the $C_2H_3O_2^-$ ion has a high affinity for H^+, these two combine to form $HC_2H_3O_2$ molecules. Thus the H^+ from the added HCl does not accumulate in solution but reacts with the $C_2H_3O_2^-$ as follows:

$$H^+(aq) + C_2H_3O_2^-(aq) \rightarrow HC_2H_3O_2(aq)$$

Next consider what happens when a strong base such as sodium hydroxide is added to the buffered solution. If this base were added to pure water, the OH^- ions from the solid would accumulate and greatly change (raise) the pH.

$$NaOH \xrightarrow{100\%} Na^+ + OH^-$$

However, in the buffered solution the OH^- ion, which has a *very strong* affinity for H^+, reacts with $HC_2H_3O_2$ molecules as follows:

$$HC_2H_3O_2(aq) + OH^-(aq) \rightarrow H_2O(l) + C_2H_3O_2^-(aq)$$

This happens because, although $C_2H_3O_2^-$ has a strong affinity for H^+, OH^- has a much stronger affinity for H^+ and thus can remove H^+ ions from acetic acid molecules.

Note that the buffering materials dissolved in the solution prevent added H^+ or OH^- from building up in the solution. Any added H^+ is trapped by $C_2H_3O_2^-$ to form $HC_2H_3O_2$. Any added OH^- reacts with $HC_2H_3O_2$ to form H_2O and $C_2H_3O_2^-$.

The general properties of a buffered solution are summarized in Table 16.3.

Table 16.3 ▸ The Characteristics of a Buffer

1. The solution contains a weak acid HA and its conjugate base A^-.
2. The buffer resists changes in pH by reacting with any added H^+ or OH^- so that these ions do not accumulate.
3. Any added H^+ reacts with the base A^-.
$$H^+(aq) + A^-(aq) \rightarrow HA(aq)$$
4. Any added OH^- reacts with the weak acid HA.
$$OH^-(aq) + HA(aq) \rightarrow H_2O(l) + A^-(aq)$$

CHAPTER 16 REVIEW

F directs you to the *Chemistry in Focus* feature in the chapter

Key Terms

acids (16-1)
bases (16-1)
Arrhenius concept of acids and bases (16-1)
Brønsted–Lowry model (16-1)
conjugate acid (16-1)
conjugate base (16-1)
conjugate acid–base pair (16-1)
hydronium ion (16-1)
strong acid (16-2)
weak acid (16-2)
diprotic acid (16-2)
oxyacids (16-2)
organic acids (16-2)
carboxyl group (16-2)
amphoteric substance (16-3)
ion-product constant, K_w (16-3)
neutral solution (16-3)
acidic solution (16-3)
basic solution (16-3)
pH scale (16-4)
buffered solution (16-6)

For Review

- The Arrhenius model
  - Acids produce H^+ in aqueous solution.
  - Bases produce OH^- in aqueous solution.
- Brønsted–Lowry model
  - Acid → proton donor
  - Base → proton acceptor
  - $HA(aq) + H_2O(l) \rightleftharpoons H_3O^+(aq) + A^-(aq)$
- Acid strength
  - A strong acid is completely dissociated (ionized).
    - $HA(aq) + H_2O(l) \rightarrow H_3O^+(aq) + A^-(aq)$
    - Has a weak conjugate base
  - A weak acid is dissociated (ionized) only to a very slight extent.
    - Has a strong conjugate base
- Water is an acid and a base (amphoteric).
- $H_2O(l) + H_2O(l) \rightleftharpoons H_3O^+(aq) + OH^-(aq)$
  - $K_w = [H_3O^+][OH^-] = 1.0 \times 10^{-14}$ at 25 °C

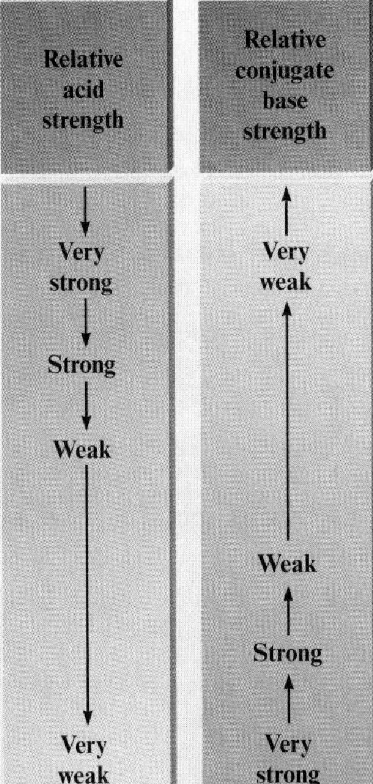

- The acidity of a solution is expressed in terms of $[H^+]$.
- $pH = -\log [H^+]$
  - Lower pH means greater acidity ($[H^+]$).
- $pOH = -\log [OH^-]$
- For a neutral solution $[H^+] = [OH^-]$.
- For an acidic solution $[H^+] > [OH^-]$.
- For a basic solution $[H^+] < [OH^-]$.

- The pH of a solution can be measured by
  - Indicators
  - pH meter
- For a strong acid the concentration of H^+ in the solution equals the initial acid concentration.
- A buffered solution is a solution that resists a change in its pH when an acid or base is added.
- Buffered solutions contain a weak acid and its conjugate base.

The Characteristics of a Buffer

1. The solution contains a weak acid HA and its conjugate base A^-.
2. The buffer resists changes in pH by reacting with any added H^+ or OH^- so that these ions do not accumulate.
3. Any added H^+ reacts with the base A^-.
$$H^+(aq) + A^-(aq) \rightarrow HA(aq)$$
4. Any added OH^- reacts with the weak acid HA.
$$OH^-(aq) + HA(aq) \rightarrow H_2O(l) + A^-(aq)$$

Active Learning Questions

These questions are designed to be considered by groups of students in class. Often these questions work well for introducing a particular topic in class.

1. You are asked for the H^+ concentration in a solution of $NaOH(aq)$. Because sodium hydroxide is a strong base, can we say there is no H^+, since having H^+ would imply that the solution is acidic?
2. Explain why Cl^- does not affect the pH of an aqueous solution.
3. Write the general reaction for an acid acting in water. What is the base in this case? The conjugate acid? The conjugate base?
4. Differentiate among the terms *concentrated, dilute, weak,* and *strong* in describing acids. Use molecular-level pictures to support your answer.
5. What is meant by "pH"? True or false: A strong acid always has a lower pH than a weak acid does. Explain.
6. Consider two separate solutions: one containing a weak acid, HA, and one containing HCl. Assume that you start with 10 molecules of each.
   a. Draw a molecular-level picture of what each solution looks like.
   b. Arrange the following from strongest to weakest base: Cl^-, H_2O, A^-. Explain.
7. Why is the pH of water at 25 °C equal to 7.00?
8. Can the pH of a solution be negative? Explain.
9. Stanley's grade-point average (GPA) is 3.28. What is Stanley's p(GPA)?
10. A friend asks the following: "Consider a buffered solution made up of the weak acid HA and its salt NaA. If a strong base like NaOH is added, the HA reacts with the OH^- to make A^-. Thus, the amount of acid (HA) is decreased, and the amount of base (A^-) is increased. Analogously, adding HCl to the buffered solution forms more of the acid (HA) by reacting with the base (A^-). How can we claim that a buffered solution resists changes in the pH of the solution?" How would you explain buffering to your friend?
11. Mixing together aqueous solutions of acetic acid and sodium hydroxide can make a buffered solution. Explain.
12. Could a buffered solution be made by mixing aqueous solutions of HCl and NaOH? Explain.
13. Consider the equation: $HA(aq) + H_2O \rightleftharpoons H_3O^+(aq) + A^-(aq)$.
    a. If water is a better base than A^-, which way will equilibrium lie?
    b. If water is a better base than A^-, does this mean that HA is a strong or a weak acid?
    c. If water is a better base than A^-, is the value for K_a greater or less than 1?
14. Choose the answer that best completes the following statement and defend your answer. When 100.0 mL of water is added to 100.0 mL of 1.00 M HCl,
    a. the pH decreases because the solution is diluted.
    b. the pH does not change because water is neutral.
    c. the pH is doubled because the volume is now doubled.
    d. the pH increases because the concentration of H^+ decreases.
    e. the solution is completely neutralized.
15. You mix a solution of a strong acid with a pH of 4 and an equal volume of a strong acid solution with a pH of 6. Is the final pH less than 4, between 4 and 5, 5, between 5 and 6, or greater than 6? Explain.
16. The following figures are molecular-level representations of acid solutions. Label each as a strong acid or a weak acid.

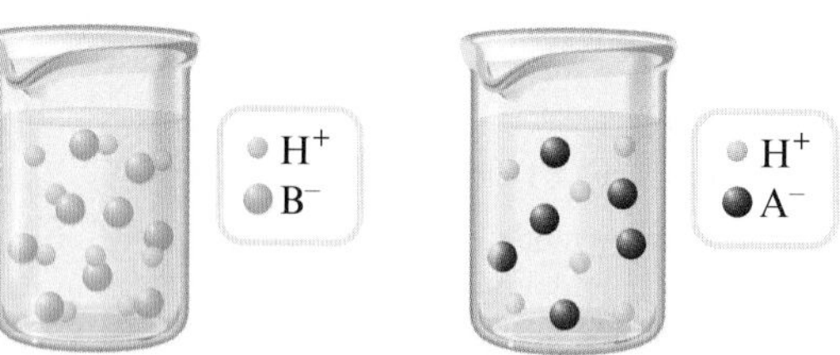

17. Answer the following questions concerning buffered solutions.
    a. Explain what a buffered solution does.
    b. Describe the substances that make up a buffered solution.
    c. Explain how a buffered solution works.

All even-numbered Questions and Problems have answers in the back of this book and solutions in the *Student Solutions Guide*.

Questions and Problems

16-1 Acids and Bases

Questions

1. What are some physical properties that historically led chemists to classify various substances as acids and bases?
2. Write an equation showing how $HCl(g)$ behaves as an Arrhenius acid when dissolved in water. Write an equation showing how $NaOH(s)$ behaves as an Arrhenius base when dissolved in water.
3. According to the Brønsted-Lowry model, an acid is a "proton donor" and a base is a "proton acceptor." Explain.
4. How do the components of a conjugate acid–base pair differ from one another? Give an example of a conjugate acid–base pair to illustrate your answer.
5. Given the general equation illustrating the reaction of the acid HA in water,

$$HA(aq) + H_2O(l) \rightarrow H_3O^+(aq) + A^-(aq)$$

explain why water is considered a *base* in the Brønsted-Lowry model.

6. According to Arrhenius, ________ produce hydrogen ions in aqueous solution, whereas ________ produce hydroxide ions.

Problems

7. Which of the following do *not* represent a conjugate acid–base pair? For those pairs that are not conjugate acid–base pairs, write the correct conjugate acid–base pair for each species in the pair.
   a. HI, I^-
   b. $HClO, HClO_2$
   c. H_3PO_4, PO_4^{3-}
   d. H_2CO_3, CO_3^{2-}
8. Which of the following do *not* represent a conjugate acid–base pair? For those pairs that are not conjugate acid–base pairs, write the correct conjugate acid–base pair for each species in the pair.
   a. H_2SO_4, SO_4^{2-}
   b. $H_2PO_4^-, HPO_4^{2-}$
   c. $HClO_4, Cl^-$
   d. NH_4^+, NH_2^-
9. In each of the following chemical equations, identify the conjugate acid–base pairs.
   a. $HF(aq) + H_2O(l) \rightleftharpoons F^-(aq) + H_3O^+(aq)$
   b. $CN^-(aq) + H_2O(l) \rightleftharpoons HCN(aq) + OH^-(aq)$
   c. $HCO_3^-(aq) + H_2O(l) \rightleftharpoons H_2CO_3(aq) + OH^-(aq)$
10. In each of the following chemical equations, identify the conjugate acid–base pairs.
   a. $NH_3(aq) + H_2O(l) \rightleftharpoons NH_4^+(aq) + OH^-(aq)$
   b. $PO_4^{3-}(aq) + H_2O(l) \rightleftharpoons HPO_4^{2-}(aq) + OH^-(aq)$
   c. $C_2H_3O_2^-(aq) + H_2O(l) \rightleftharpoons HC_2H_3O_2(aq) + OH^-(aq)$
11. Write the conjugate *acid* for each of the following bases:
   a. PO_4^{3-}
   b. IO_3^-
   c. NO_3^-
   d. NH_2^-
12. Write the conjugate *acid* for each of the following bases.
   a. ClO^-
   b. Cl^-
   c. ClO_3^-
   d. ClO_4^-
13. Write the conjugate *base* for each of the following acids:
   a. H_2S
   b. HS^-
   c. NH_3
   d. H_2SO_3
14. Write the conjugate *base* for each of the following acids.
   a. $HBrO$
   b. HNO_2
   c. HSO_3^-
   d. $CH_3NH_3^+$
15. Write a chemical equation showing how each of the following species can behave as indicated when dissolved in water.
   a. HSO_3^- as an acid
   b. CO_3^{2-} as a base
   c. $H_2PO_4^-$ as an acid
   d. $C_2H_3O_2^-$ as a base
16. Write a chemical equation showing how each of the following species can behave as indicated when dissolved in water.
   a. O^{2-} as a base
   b. NH_3 as a base
   c. HSO_4^- as an acid
   d. HNO_2 as an acid

16-2 Acid Strength

Questions

17. What does it mean to say that an acid is *strong* in aqueous solution? What does this reveal about the ability of the acid's anion to attract protons?
18. What does it mean to say that an acid is *weak* in aqueous solution? What does this reveal about the ability of the acid's anion to attract protons?
19. How is the strength of an acid related to the fact that a competition for protons exists in aqueous solution between water molecules and the anion of the acid?
20. A strong acid has a weak conjugate base, whereas a weak acid has a relatively strong conjugate base. Explain.
21. Write the formula for the *hydronium* ion. Write an equation for the formation of the hydronium ion when an acid is dissolved in water.
22. Name four strong acids. For each of these, write the equation showing the acid dissociating in water.

All even-numbered Questions and Problems have answers in the back of this book and solutions in the *Student Solutions Guide*.

23. Organic acids contain the carboxyl group

```
      O
     //
  —C
     \
      O—H
```

Using acetic acid, CH_3—COOH, and propionic acid, CH_3CH_2—COOH, write equations showing how the carboxyl group enables these substances to behave as weak acids when dissolved in water.

24. What is an *oxyacid?* Write the formulas of three acids that are oxyacids. Write the formulas of three acids that are *not* oxyacids.

25. Which of the following acids have relatively *strong* conjugate bases?

a. HCN
b. H_2S
c. $HBrO_4$
d. HNO_3

26. The "Chemistry in Focus" segment *Plants Fight Back* discusses how tobacco plants under attack by disease produce salicylic acid. Examine the structure of salicylic acid and predict whether it behaves as a monoprotic or a diprotic acid.

16-3 Water as an Acid and a Base

Questions

27. Water is the most common *amphoteric* substance, which means that, depending on the circumstances, water can behave either as an acid or as a base. Using HF as an example of an acid and NH_3 as an example of a base, write equations for these substances reacting with water, in which water behaves as a base and as an acid, respectively.

28. Anions containing hydrogen (for example, HCO_3^- and $H_2PO_4^{2-}$) show amphoteric behavior when reacting with other acids or bases. Write equations illustrating the amphoterism of these anions.

29. What is meant by the *ion-product constant* for water, K_w? What does this constant signify? Write an equation for the chemical reaction from which the constant is derived.

30. What happens to the hydroxide ion concentration in aqueous solutions when we increase the hydrogen ion concentration by adding an acid? What happens to the hydrogen ion concentration in aqueous solutions when we increase the hydroxide ion concentration by adding a base? Explain.

Problems

31. Calculate the $[H^+]$ in each of the following solutions, and indicate whether the solution is acidic or basic.

a. $[OH^-] = 2.32 \times 10^{-4}\ M$
b. $[OH^-] = 8.99 \times 10^{-10}\ M$
c. $[OH^-] = 4.34 \times 10^{-6}\ M$
d. $[OH^-] = 6.22 \times 10^{-12}\ M$

32. Calculate the $[H^+]$ in each of the following solutions, and indicate whether the solution is acidic, basic, or neutral.

a. $[OH^-] = 3.99 \times 10^{-5}\ M$
b. $[OH^-] = 2.91 \times 10^{-9}\ M$
c. $[OH^-] = 7.23 \times 10^{-2}\ M$
d. $[OH^-] = 9.11 \times 10^{-7}\ M$

33. Calculate the $[OH^-]$ in each of the following solutions, and indicate whether the solution is acidic or basic.

a. $[H^+] = 4.01 \times 10^{-4}\ M$
b. $[H^+] = 7.22 \times 10^{-6}\ M$
c. $[H^+] = 8.05 \times 10^{-7}\ M$
d. $[H^+] = 5.43 \times 10^{-9}\ M$

34. Calculate the $[OH^-]$ in each of the following solutions, and indicate whether the solution is acidic or basic.

a. $[H^+] = 1.02 \times 10^{-7}\ M$
b. $[H^+] = 9.77 \times 10^{-8}\ M$
c. $[H^+] = 3.41 \times 10^{-3}\ M$
d. $[H^+] = 4.79 \times 10^{-11}\ M$

35. For each pair of concentrations, tell which represents the more acidic solution.

a. $[H^+] = 1.2 \times 10^{-3}\ M$ or $[H^+] = 4.5 \times 10^{-4}\ M$
b. $[H^+] = 2.6 \times 10^{-6}\ M$ or $[H^+] = 4.3 \times 10^{-8}\ M$
c. $[H^+] = 0.000010\ M$ or $[H^+] = 0.0000010\ M$

36. For each pair of concentrations, tell which represents the more *basic* solution.

a. $[H^+] = 3.99 \times 10^{-6}\ M$ or $[OH^-] = 6.03 \times 10^{-4}\ M$
b. $[H^+] = 1.79 \times 10^{-5}\ M$ or $[OH^-] = 4.21 \times 10^{-6}\ M$
c. $[H^+] = 7.81 \times 10^{-3}\ M$ or $[OH^-] = 8.04 \times 10^{-4}\ M$

16-4 The pH Scale

Questions

37. Why do scientists tend to express the acidity of a solution in terms of its pH, rather than in terms of the molarity of hydrogen ion present? How is pH defined mathematically?

38. Using Fig. 16.3, list the approximate pH value of five "everyday" solutions. How do the familiar properties (such as the sour taste for acids) of these solutions correspond to their indicated pH?

39. For a hydrogen ion concentration of $2.33 \times 10^{-6}\ M$, how many *decimal places* should we give when expressing the pH of the solution?

40. The "Chemistry in Focus" segment *Garden-Variety Acid–Base Indicators* discusses acid–base indicators found in nature. What colors are exhibited by red cabbage juice under acid conditions? Under basic conditions?

Problems

41. Calculate the pH corresponding to each of the hydrogen ion concentrations given below, and indicate whether each solution is acidic or basic.

a. $[H^+] = 4.02 \times 10^{-3}\ M$
b. $[H^+] = 8.99 \times 10^{-7}\ M$
c. $[H^+] = 2.39 \times 10^{-6}\ M$
d. $[H^+] = 1.89 \times 10^{-10}\ M$

All even-numbered Questions and Problems have answers in the back of this book and solutions in the *Student Solutions Guide*.

42. Calculate the pH corresponding to each of the hydrogen ion concentrations given below, and indicate whether each solution is acidic, basic, or neutral.

a. $[H^+] = 0.00100\ M$
b. $[H^+] = 2.19 \times 10^{-4}\ M$
c. $[H^+] = 9.18 \times 10^{-11}\ M$
d. $[H^+] = 4.71 \times 10^{-7}\ M$

43. Calculate the pH corresponding to each of the hydroxide ion concentrations given below, and indicate whether each solution is acidic or basic.

a. $[OH^-] = 4.73 \times 10^{-4}\ M$
b. $[OH^-] = 5.99 \times 10^{-1}\ M$
c. $[OH^-] = 2.87 \times 10^{-8}\ M$
d. $[OH^-] = 6.39 \times 10^{-3}\ M$

44. Calculate the pH corresponding to each of the hydroxide ion concentrations given below, and indicate whether each solution is acidic or basic.

a. $[OH^-] = 8.63 \times 10^{-3}\ M$
b. $[OH^-] = 7.44 \times 10^{-6}\ M$
c. $[OH^-] = 9.35 \times 10^{-9}\ M$
d. $[OH^-] = 1.21 \times 10^{-11}\ M$

45. Calculate the pH corresponding to each of the pOH values listed, and indicate whether each solution is acidic, basic, or neutral.

a. pOH = 4.32
b. pOH = 8.90
c. pOH = 1.81
d. pOH = 13.1

46. Calculate the pOH value corresponding to each of the pH values listed, and tell whether each solution is acidic or basic.

a. pH = 9.78
b. pH = 4.01
c. pH = 2.79
d. pH = 11.21

47. For each hydrogen ion concentration listed, calculate the pH of the solution as well as the concentration of hydroxide ion in the solution. Indicate whether each solution is acidic or basic.

a. $[H^+] = 4.76 \times 10^{-8}\ M$
b. $[H^+] = 8.92 \times 10^{-3}\ M$
c. $[H^+] = 7.00 \times 10^{-5}\ M$
d. $[H^+] = 1.25 \times 10^{-12}\ M$

48. For each hydrogen ion concentration listed, calculate the pH of the solution as well as the concentration of hydroxide ion in the solution. Indicate whether each solution is acidic or basic.

a. $[H^+] = 1.91 \times 10^{-2}\ M$
b. $[H^+] = 4.83 \times 10^{-7}\ M$
c. $[H^+] = 8.92 \times 10^{-11}\ M$
d. $[H^+] = 6.14 \times 10^{-5}\ M$

49. Calculate the hydrogen ion concentration, in moles per liter, for solutions with each of the following pH values.

a. pH = 9.01
b. pH = 6.89
c. pH = 1.02
d. pH = 7.00

50. Calculate the hydrogen ion concentration, in moles per liter, for solutions with each of the following pH values.

a. pH = 1.04
b. pH = 13.1
c. pH = 5.99
d. pH = 8.62

51. Calculate the hydrogen ion concentration, in moles per liter, for solutions with each of the following pOH values.

a. pOH = 4.95
b. pOH = 7.00
c. pOH = 12.94
d. pOH = 1.02

52. Calculate the hydrogen ion concentration, in moles per liter, for solutions with each of the following pH or pOH values.

a. pOH = 4.99
b. pH = 7.74
c. pOH = 10.74
d. pH = 2.25

53. Calculate the pH of each of the following solutions from the information given.

a. $[H^+] = 4.78 \times 10^{-2}\ M$
b. pOH = 4.56
c. $[OH^-] = 9.74 \times 10^{-3}\ M$
d. $[H^+] = 1.24 \times 10^{-8}\ M$

54. Calculate the pH of each of the following solutions from the information given.

a. $[H^+] = 4.39 \times 10^{-6}\ M$
b. pOH = 10.36
c. $[OH^-] = 9.37 \times 10^{-9}\ M$
d. $[H^+] = 3.31 \times 10^{-1}\ M$

16-5 Calculating the pH of Strong Acid Solutions

Questions

55. When 1 mole of gaseous hydrogen chloride is dissolved in enough water to make 1 L of solution, approximately how many HCl molecules remain in the solution? Explain.

56. A bottle of acid solution is labeled "3 M HNO_3." What are the substances that are actually present in the solution? Are any HNO_3 molecules present? Why or why not?

Problems

57. Calculate the hydrogen ion concentration and the pH of each of the following solutions of strong acids.

a. $1.04 \times 10^{-4}\ M$ HCl
b. 0.00301 M HNO_3
c. $5.41 \times 10^{-4}\ M$ $HClO_4$
d. $6.42 \times 10^{-2}\ M$ HNO_3

58. Calculate the pH of each of the following solutions of strong acids.

a. $1.21 \times 10^{-3}\ M$ HNO_3
b. 0.000199 M $HClO_4$
c. $5.01 \times 10^{-5}\ M$ HCl
d. 0.00104 M HBr

16-6 Buffered Solutions

Questions

59. What characteristic properties do buffered solutions possess?

60. What two components make up a buffered solution? Give an example of a combination that would serve as a buffered solution.

61. Which component of a buffered solution is capable of combining with an added strong acid? Using your example from Exercise 60, show how this component would react with added HCl.

62. Which component of a buffered solution consumes added strong base? Using your example from Exercise 60, show how this component would react with added NaOH.

Problems

63. Which of the following combinations would act as buffered solutions?
    a. HCl and NaCl
    b. CH_3COOH and KCH_3COO
    c. H_2S and NaHS
    d. H_2S and Na_2S

64. A buffered solution is prepared containing acetic acid, $HC_2H_3O_2$, and sodium acetate, $NaC_2H_3O_2$, both at 0.5 *M*. Write a chemical equation showing how this buffered solution would resist a decrease in its pH if a few drops of aqueous strong acid HCl solution were added to it. Write a chemical equation showing how this buffered solution would resist an increase in its pH if a few drops of aqueous strong base NaOH solution were added to it.

Additional Problems

65. The concepts of acid–base equilibria were developed in this chapter for aqueous solutions (in aqueous solutions, water is the solvent and is intimately involved in the equilibria). However, the Brønsted–Lowry acid–base theory can be extended easily to other solvents. One such solvent that has been investigated in depth is liquid ammonia, NH_3.
    a. Write a chemical equation indicating how HCl behaves as an acid in liquid ammonia.
    b. Write a chemical equation indicating how OH^- behaves as a base in liquid ammonia.

66. *Strong bases* are bases that completely ionize in water to produce hydroxide ion, OH^-. The strong bases include the hydroxides of the Group 1 elements. For example, if 1.0 mole of NaOH is dissolved per liter, the concentration of OH^- ion is 1.0 *M*. Calculate the $[OH^-]$, pOH, and pH for each of the following strong base solutions.
    a. 0.10 *M* NaOH
    b. 2.0×10^{-4} *M* KOH
    c. 6.2×10^{-3} *M* CsOH
    d. 0.0001 *M* NaOH

67. Which of the following conditions indicate an *acidic* solution?
    a. pH = 3.04
    b. $[H^+] > 1.0 \times 10^{-7}\ M$
    c. pOH = 4.51
    d. $[OH^-] = 3.21 \times 10^{-12}\ M$

68. Which of the following conditions indicate a *basic* solution?
    a. pOH = 11.21
    b. pH = 9.42
    c. $[OH^-] > [H^+]$
    d. $[OH^-] > 1.0 \times 10^{-7}\ M$

69. Buffered solutions are mixtures of a weak acid and its conjugate base. Explain why a mixture of a *strong* acid and its conjugate base (such as HCl and Cl^-) is not buffered.

70. Which of the following acids have relatively strong conjugate bases?
    a. CH_3COOH ($HC_2H_3O_2$)
    b. HF
    c. H_2S
    d. HCl

71. Is it possible for a solution to have $[H^+] = 0.002\ M$ and $[OH^-] = 5.2 \times 10^{-6}\ M$ at 25 °C? Explain.

72. Despite HCl's being a strong acid, the pH of $1.00 \times 10^{-7}\ M$ HCl is *not* exactly 7.00. Can you suggest a reason why?

73. According to Arrhenius, bases are species that produce ________ ions in aqueous solution.

74. According to the Brønsted–Lowry model, a base is a species that ________ protons.

75. A conjugate acid–base pair consists of two substances related by the donating and accepting of a(n) ________.

76. Acetate ion, $C_2H_3O_2^-$, has a stronger affinity for protons than does water. Therefore, when dissolved in water, acetate ion behaves as a(n) ________.

77. An acid such as HCl that strongly conducts an electric current when dissolved in water is said to be a(n) ________ acid.

78. Draw the structure of the carboxyl group, —COOH. Show how a molecule containing the carboxyl group behaves as an acid when dissolved in water.

79. Because of ________, even pure water contains measurable quantities of H^+ and OH^-.

80. The ion-product constant for water, K_w, has the value ________ at 25 °C.

81. The number of ________ in the logarithm of a number is equal to the number of significant figures in the number.

82. A solution with pH = 9 has a (higher/lower) hydrogen ion concentration than a solution with pOH = 9.

83. A 0.20 *M* HCl solution contains ________ *M* hydrogen ion and ________ *M* chloride ion concentrations.

84. A buffered solution is one that resists a change in ________ when either a strong acid or a strong base is added to it.

85. A(n) ________ solution contains a conjugate acid–base pair and through this is able to resist changes in its pH.

All even-numbered Questions and Problems have answers in the back of this book and solutions in the *Student Solutions Guide*.

86. When sodium hydroxide, NaOH, is added dropwise to a buffered solution, the ________ component of the buffer consumes the added hydroxide ion.

87. When hydrochloric acid, HCl, is added dropwise to a buffered solution, the ________ component of the buffer consumes the added hydrogen ion.

88. The following are representations of acid–base reactions:

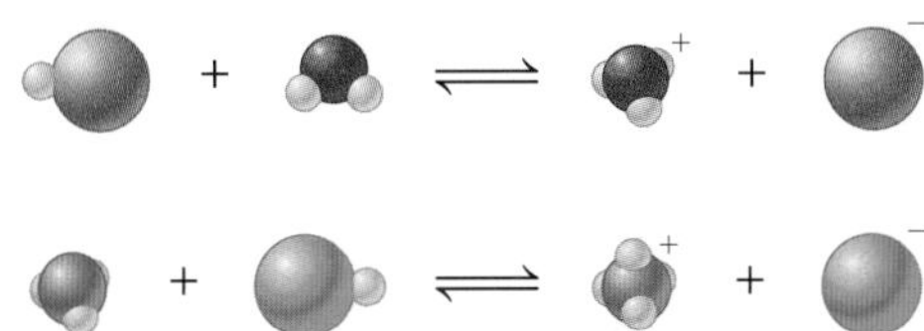

a. Label each of the species in both equations as an acid or base, and explain.

b. For those species that are acids, which labels apply: Arrhenius acid, Brønsted-Lowry acid? What about the bases?

89. In each of the following chemical equations, identify the conjugate acid–base pairs.

a. $CH_3NH_2 + H_2O \rightleftharpoons CH_3NH_3^+ + OH^-$

b. $CH_3COOH + NH_3 \rightleftharpoons CH_3COO^- + NH_4^+$

c. $HF + NH_3 \rightleftharpoons F^- + NH_4^+$

90. Write the conjugate *acid* for each of the following.

a. NH_3

b. NH_2^-

c. H_2O

d. OH^-

91. Write the conjugate *base* for each of the following.

a. H_3PO_4

b. HCO_3^-

c. HF

d. H_2SO_4

92. Which of the following combinations would act as buffered solutions?

a. HCN and NaCN

b. H_3PO_4 and K_3PO_4

c. HF and KF

d. $HC_3H_5O_2$ and $NaC_3H_5O_2$

93. Which of the following bases have relatively *strong* conjugate acids?

a. F^-

b. Cl^-

c. HSO_4^-

d. NO_3^-

94. Calculate $[H^+]$ in each of the following solutions, and indicate whether the solution is acidic, basic, or neutral.

a. $[OH^-] = 4.22 \times 10^{-3}\ M$

b. $[OH^-] = 1.01 \times 10^{-13}\ M$

c. $[OH^-] = 3.05 \times 10^{-7}\ M$

d. $[OH^-] = 6.02 \times 10^{-6}\ M$

95. Calculate $[OH^-]$ in each of the following solutions, and indicate whether the solution is acidic, basic, or neutral.

a. $[H^+] = 4.21 \times 10^{-7}\ M$

b. $[H^+] = 0.00035\ M$

c. $[H^+] = 0.00000010\ M$

d. $[H^+] = 9.9 \times 10^{-6}\ M$

96. Calculate the pH corresponding to each of the pOH values listed, and indicate whether each solution is acidic, basic, or neutral.

a. pOH = 4.32

b. pOH = 8.90

c. pOH = 1.81

d. pOH = 13.1

97. Calculate the pH of each of the solutions indicated below. Tell whether the solution is acidic, basic, or neutral.

a. $[H^+] = 1.49 \times 10^{-3}\ M$

b. $[OH^-] = 6.54 \times 10^{-4}\ M$

c. $[H^+] = 9.81 \times 10^{-9}\ M$

d. $[OH^-] = 7.45 \times 10^{-10}\ M$

98. Calculate the pH corresponding to each of the hydroxide ion concentrations given below, and indicate whether each solution is acidic, basic, or neutral.

a. $[OH^-] = 1.4 \times 10^{-6}\ M$

b. $[OH^-] = 9.35 \times 10^{-9}\ M$

c. $[OH^-] = 2.21 \times 10^{-1}\ M$

d. $[OH^-] = 7.98 \times 10^{-12}\ M$

99. Calculate the pOH corresponding to each of the pH values listed, and indicate whether each solution is acidic, basic, or neutral.

a. pH = 1.02

b. pH = 13.4

c. pH = 9.03

d. pH = 7.20

100. For each hydrogen or hydroxide ion concentration listed, calculate the concentration of the complementary ion and the pH and pOH of the solution.

a. $[H^+] = 5.72 \times 10^{-4}\ M$

b. $[OH^-] = 8.91 \times 10^{-5}\ M$

c. $[H^+] = 2.87 \times 10^{-12}\ M$

d. $[OH^-] = 7.22 \times 10^{-8}\ M$

101. Calculate the hydrogen ion concentration, in moles per liter, for solutions with each of the following pH values.

a. pH = 8.34

b. pH = 5.90

c. pH = 2.65

d. pH = 12.6

102. Calculate the hydrogen ion concentration, in moles per liter, for solutions with each of the following pH or pOH values.

a. pH = 5.41

b. pOH = 12.04

c. pH = 11.91

d. pOH = 3.89

All even-numbered Questions and Problems have answers in the back of this book and solutions in the *Student Solutions Guide.*

103. Calculate the hydrogen ion concentration, in moles per liter, for solutions with each of the following pH or pOH values.

a. pOH = 0.90
b. pH = 0.90
c. pOH = 10.3
d. pH = 5.33

104. Calculate the hydrogen ion concentration and the pH of each of the following solutions of strong acids.

a. 1.4×10^{-3} *M* $HClO_4$
b. 3.0×10^{-5} *M* HCl
c. 5.0×10^{-2} *M* HNO_3
d. 0.0010 *M* HCl

105. Write the formulas for *three* combinations of weak acid and salt that would act as buffered solutions. For each of your combinations, write chemical equations showing how the components of the buffered solution would consume added acid and base.

ChemWork Problems

These multiconcept problems (and additional ones) are found interactively online with the same type of assistance a student would get from an instructor.

106. Choose pairs in which the species listed *first* is the conjugate *base* of the species listed *second.*

a. S^{2-}, HS^-
b. H^+, OH^-
c. HBr, Br^-
d. NO_2^-, HNO_2

107. Complete the table for each of the following solutions:

	$[H^+]$	pH	pOH	$[OH^-]$
0.0070 *M* HNO_3	______	______	______	______
3.0 *M* KOH	______	______	______	______

108. Consider 0.25 *M* solutions of the following salts: NaCl, RbOCl, KI, $Ba(ClO_4)_2$, and NH_4NO_3. For each salt, indicate whether the solution is acidic, basic, or neutral.

All even-numbered Questions and Problems have answers in the back of this book and solutions in the *Student Solutions Guide.*

Appendix

Using Your Calculator

In this section we will review how to use your calculator to perform common mathematical operations. This discussion assumes that your calculator uses the algebraic operating system, the system used by most brands.

One very important principle to keep in mind as you use your calculator is that it is not a substitute for your brain. Keep thinking as you do the calculations. Keep asking yourself, "Does the answer make sense?"

Addition, Subtraction, Multiplication, and Division

Performing these operations on a pair of numbers always involves the following steps:

1. Enter the first number, using the numbered keys and the decimal (.) key if needed.
2. Enter the operation to be performed.
3. Enter the second number.
4. Press the "equals" key to display the answer.

For example, the operation

$$15.1 + 0.32$$

is carried out as follows:

Press	Display
15.1	15.1
+	15.1
.32	0.32
=	15.42

The answer given by the display is 15.42. If this is the final result of a calculation, you should round it off to the correct number of significant figures (15.4), as discussed in Section 2-5. If this number is to be used in further calculations, use it exactly as it appears on the display. Round off only the final answer in the calculation.

Do the following operations for practice. The detailed procedures are given below.

a. $1.5 + 32.86$ b. $23.5 - 0.41$

c. 0.33×153 d. $\frac{9.3}{0.56}$ or $9.3 \div 0.56$

Procedures

a. Press	Display	b. Press	Display
1.5	1.5	23.5	23.5
+	1.5	−	23.5
32.86	32.86	.41	0.41
=	34.36	=	23.09
Rounded:	34.4	Rounded:	23.1

c. Press	Display	d. Press	Display
.33	0.33	9.3	9.3
×	0.33	÷	9.3
153	153	.56	0.56
=	50.49	=	16.607143
Rounded:	50.	Rounded:	17

Squares, Square Roots, Reciprocals, and Logs

Now we will consider four additional operations that we often need to solve chemistry problems.

The *squaring* of a number is done with a key labeled X^2. The *square root* key is usually labeled $\sqrt{X}$. To take the *reciprocal* of a number, you need the 1/X key. The *logarithm* of a number is determined by using a key labeled log or logX.

To perform these operations, take the following steps:

1. Enter the number.
2. Press the appropriate function key.
3. The answer is displayed automatically.

For example, let's calculate the square root of 235.

Press	Display
235	235
$\sqrt{X}$	15.32971
Rounded:	15.3

We can obtain the log of 23 as follows:

Press	Display
23	23
log	1.3617278
Rounded:	1.36

Often a key on a calculator serves two functions. In this case, the first function is listed on the key and the second is shown on the calculator just above the key. For example, on some calculators the top row of keys appears as follows:

	1/X	X^2		
2nd	R/S	$\sqrt{X}$	off	on/C

To make the calculator square a number, we must use 2nd and then $\sqrt{X}$; pressing 2nd tells the calculator we want the function that is listed *above* the key. Thus we can obtain the square of 11.56 on this calculator as follows:

Press	Display
11.56	11.56
2nd then $\sqrt{X}$	133.6336
Rounded:	133.6

We obtain the reciprocal of 384 (1/384) on this calculator as follows:

Press	Display
384	384
2nd then R/S	0.0026042
Rounded:	0.00260

Your calculator may be different. See the user's manual if you are having trouble with these operations.

Chain Calculations

In solving problems you often have to perform a series of calculations—a calculation chain. This is generally quite easy if you key in the chain as you read the numbers and operations in order. For example, to perform the calculation

$$\frac{14.68 + 1.58 - 0.87}{0.0850}$$

you should use the appropriate keys as you read it to yourself:

14.68 plus 1.58 equals; minus .87 equals;
divided by 0.0850 equals

The details follow.

Press	Display
14.68	14.68
+	14.68
1.58	1.58
=	16.26
−	16.26
.87	0.87
=	15.39
÷	15.39
.0850	0.0850
=	181.05882
Rounded:	181

Note that you must press (=) after every operation to keep the calculation "up to date."

For more practice, consider the calculation

$$(0.360)(298) + \frac{(14.8)(16.0)}{1.50}$$

Here you are adding two numbers, but each must be obtained by the indicated calculations. One procedure is to calculate each number first and then add them. The first term is

$$(0.360)(298) = 107.28$$

The second term,

$$\frac{(14.8)(16.0)}{1.50}$$

can be computed easily by reading it to yourself. It "reads"

14.8 times 16.0 equals; divided by 1.50 equals

and is summarized as follows:

Press	Display
14.8	14.8
×	14.8
16.0	16.0
=	236.8
÷	236.8
1.50	1.50
=	157.86667

Now we can keep this last number on the calculator and add it to 107.28 from the first calculation.

Press	Display
+	157.86667
107.28	107.28
=	265.14667
Rounded:	265

To summarize,

$$(0.360)(298) + \frac{(14.8)(16.0)}{1.50}$$

becomes

$$107.28 + 157.86667$$

and the sum is 265.14667 or, rounded to the correct number of significant figures, 265. There are other ways to do this calculation, but this is the safest way (assuming you are careful).

A common type of chain calculation involves a number of terms multiplied together in the numerator and the denominator, as in

$$\frac{(323)(.0821)(1.46)}{(4.05)(76)}$$

There are many possible sequences by which this calculation can be carried out, but the following seems the most natural.

323 times .0821 equals; times 1.46 equals;
divided by 4.05 equals; divided by 76 equals

This sequence is summarized as follows:

Press	Display
323	323
×	323
.0821	0.0821
=	26.5183
×	26.5183
1.46	1.46
=	38.716718
÷	38.716718
4.05	4.05
=	9.5596835
÷	9.5596835
76	76
=	0.1257853

The answer is 0.1257853, which, when rounded to the correct number of significant figures, is 0.13. Note that when two or more numbers are multiplied in the denominator, you must divide by *each* one.

Here are some additional chain calculations (with solutions) to give you more practice.

a. $15 - (0.750)(243)$

b. $\dfrac{(13.1)(43.5)}{(1.8)(63)}$

c. $\dfrac{(85.8)(0.142)}{(16.46)(18.0)} + \dfrac{(131)(0.0156)}{10.17}$

d. $(18.1)(0.051) - \dfrac{(325)(1.87)}{(14.0)(3.81)} + \dfrac{1.56 - 0.43}{1.33}$

Solutions

a. $15 - 182 = -167$

b. 5.0

c. $0.0411 + 0.201 = 0.242$

d. $0.92 - 11.4 + 0.850 = -9.6$

In performing chain calculations, take the following steps in the order listed.

1. Perform any additions and subtractions that appear inside parentheses.
2. Complete the multiplications and divisions of individual terms.
3. Add and subtract individual terms as required.

Basic Algebra

In solving chemistry problems you will use, over and over again, relatively few mathematical procedures. In this section we review the few algebraic manipulations that you will need.

Solving an Equation

In the course of solving a chemistry problem, we often construct an algebraic equation that includes the unknown quantity (the thing we want to calculate). An example is

$$(1.5)V = (0.23)(0.08206)(298)$$

We need to "solve this equation for V." That is, we need to isolate V on one side of the equals sign with all the numbers on the other side. How can we do this? The key idea in solving an algebraic equation is that *doing the same thing on both sides of the equals sign* does not change the equality. That is, it is always "legal" to do the same thing to both sides of the equation. Here we want to solve for V, so we must get the number 1.5 on the other side of the equals sign. We can do this by dividing *both sides* by 1.5.

$$\frac{(1.5)V}{1.5} = \frac{(0.23)(0.08206)(298)}{1.5}$$

Now the 1.5 in the denominator on the left cancels the 1.5 in the numerator:

$$\frac{(\cancel{1.5})V}{\cancel{1.5}} = \frac{(0.23)(0.08206)(298)}{1.5}$$

to give

$$V = \frac{(0.23)(0.08206)(298)}{1.5}$$

Using the procedures in "Using Your Calculator" for chain calculations, we can now obtain the value for V with a calculator.

$$V = 3.7$$

Sometimes it is necessary to solve an equation that consists of symbols. For example, consider the equation

$$\frac{P_1V_1}{T_1} = \frac{P_2V_2}{T_2}$$

Let's assume we want to solve for T_2. That is, we want to isolate T_2 on one side of the equation. There are several possible ways to proceed, keeping in mind that we always do the same thing on both sides of the equals sign. First we multiply both sides by T_2.

$$T_2 \times \frac{P_1V_1}{T_1} = \frac{P_2V_2}{\cancel{T_2}} \times \cancel{T_2}$$

This cancels T_2 on the right. Next we multiply both sides by T_1.

$$T_2 \times \frac{P_1V_1}{\cancel{T_1}} \times \cancel{T_1} = P_2V_2T_1$$

This cancels T_1 on the left. Now we divide both sides by P_1V_1.

$$T_2 \times \frac{\cancel{P_1V_1}}{\cancel{P_1V_1}} = \frac{P_2V_2T_1}{P_1V_1}$$

This yields the desired equation,

$$T_2 = \frac{P_2V_2T_1}{P_1V_1}$$

For practice, solve each of the following equations for the variable indicated.

a. $PV = k$; solve for P

b. $1.5x + 6 = 3$; solve for x

c. $PV = nRT$; solve for n

d. $\frac{P_1V_1}{T_1} = \frac{P_2V_2}{T_2}$; solve for V_2

e. $\frac{°F - 32}{°C} = \frac{9}{5}$; solve for °C

f. $\frac{°F - 32}{°C} = \frac{9}{5}$; solve for °F

Solutions

a. $\frac{PV}{V} = \frac{k}{V}$

$P = \frac{k}{V}$

b. $1.5x + 6 - 6 = 3 - 6$

$1.5x = -3$

$\frac{1.5x}{1.5} = \frac{-3}{1.5}$

$x = -\frac{3}{1.5} = -2$

c. $\frac{PV}{RT} = \frac{nRT}{RT}$

$\frac{PV}{RT} = n$

d. $\frac{P_1V_1}{T_1} \times T_2 = \frac{P_2V_2}{T_2} \times T_2$

$\frac{P_1V_1T_2}{T_1P_2} = \frac{P_2V_2}{P_2}$

$\frac{P_1V_1T_2}{T_1P_2} = V_2$

e. $\frac{°F - 32}{°C} \times °C = \frac{9}{5}°C$

$\frac{5}{9}(°F - 32) = \frac{5}{9} \times \frac{9}{5}°C$

$\frac{5}{9}(°F - 32) = °C$

f. $\frac{°F - 32}{°C} \times °C = \frac{9}{5}°C$

$°F - 32 + 32 = \frac{9}{5}°C + 32$

$°F = \frac{9}{5}°C + 32$

Scientific (Exponential) Notation

The numbers we must work with in scientific measurements are often very large or very small; thus it is convenient to express them using powers of 10. For example, the number 1,300,000 can be expressed as 1.3×10^6, which means multiply 1.3 by 10 six times, or

$$1.3 \times 10^6 = 1.3 \times \underbrace{10 \times 10 \times 10 \times 10 \times 10 \times 10}_{10^6 = 1 \text{ million}}$$

A number written in scientific notation always has the form:

A number (between 1 and 10) times
the appropriate power of 10

To represent a large number such as 20,500 in scientific notation, we must move the decimal point in such a way as to achieve a number between 1 and 10 and then multiply the result by a power of 10 to compensate for moving the decimal point. In this case, we must move the decimal point four places to the left.

2 0 5 0 0
4 3 2 1

to give a number between 1 and 10:

2.05

where we retain only the significant figures (the number 20,500 has three significant figures). To compensate for moving the decimal point four places to the left, we must multiply by 10^4. Thus

$$20{,}500 = 2.05 \times 10^4$$

As another example, the number 1985 can be expressed as 1.985×10^3. To end up with the number 1.985, which is between 1 and 10, we had to move the decimal point three places to the left. To compensate for that, we must multiply by 10^3. Some other examples are given in the accompanying list.

Number	Exponential Notation
5.6	5.6×10^0 or 5.6×1
39	3.9×10^1
943	9.43×10^2
1126	1.126×10^3

So far, we have considered numbers greater than 1. How do we represent a number such as 0.0034 in exponential notation? First, to achieve a number between 1 and 10, we start with 0.0034 and move the decimal point three places to the right.

0 . 0 0 3 4
1 2 3

This yields 3.4. Then, to compensate for moving the decimal point to the right, we must multiply by a power of 10 with a negative exponent—in this case, 10^{-3}. Thus

$$0.0034 = 3.4 \times 10^{-3}$$

In a similar way, the number 0.00000014 can be written as 1.4×10^{-7}, because going from 0.00000014 to 1.4 requires that we move the decimal point seven places to the right.

Mathematical Operations with Exponentials

We next consider how various mathematical operations are performed using exponential numbers. First we cover the various rules for these operations; then we consider how to perform them on your calculator.

Multiplication and Division

When two numbers expressed in exponential notation are multiplied, the initial numbers are multiplied and the exponents of 10 are *added.*

$$(M \times 10^m)(N \times 10^n) = (MN) \times 10^{m+n}$$

For example (to two significant figures, as required),

$$(3.2 \times 10^4)(2.8 \times 10^3) = 9.0 \times 10^7$$

When the numbers are multiplied, if a result greater than 10 is obtained for the initial number, the decimal point is moved one place to the left and the exponent of 10 is increased by 1.

$$\begin{aligned}(5.8 \times 10^2)(4.3 \times 10^8) &= 24.9 \times 10^{10}\\ &= 2.49 \times 10^{11}\\ &= 2.5 \times 10^{11} \quad \text{(two significant figures)}\end{aligned}$$

Division of two numbers expressed in exponential notation involves normal division of the initial numbers and *subtraction* of the exponent of the divisor from that of the dividend. For example,

$$\frac{4.8 \times 10^8}{\underbrace{2.1 \times 10^3}_{\text{Divisor}}} = \frac{4.8}{2.1} \times 10^{(8-3)} = 2.3 \times 10^5$$

If the initial number resulting from the division is less than 1, the decimal point is moved one place to the right and the exponent of 10 is decreased by 1. For example,

$$\begin{aligned}\frac{6.4 \times 10^3}{8.3 \times 10^5} = \frac{6.4}{8.3} \times 10^{(3-5)} &= 0.77 \times 10^{-2}\\ &= 7.7 \times 10^{-3}\end{aligned}$$

Addition and Subtraction

In order for us to add or subtract numbers expressed in exponential notation, *the exponents of the numbers must be the same.* For example, to add 1.31×10^5 and 4.2×10^4, we must rewrite one number so that the exponents of both are the same. The number 1.31×10^5 can be written 13.1×10^4: decreasing the exponent by 1 compensates for moving the decimal point one place to the right. Now we can add the numbers.

$$\begin{array}{r} 13.1 \times 10^4 \\ +\ 4.2 \times 10^4 \\ \hline 17.3 \times 10^4 \end{array}$$

In correct exponential notation, the result is expressed as 1.73×10^5.

To perform addition or subtraction with numbers expressed in exponential notation, we add or subtract only the initial numbers. The exponent of the result is the same as the exponents of the numbers being added or subtracted. To subtract 1.8×10^2 from 8.99×10^3, we first convert 1.8×10^2 to 0.18×10^3 so that both numbers have the same exponent. Then we subtract.

$$\begin{array}{r} 8.99 \times 10^3 \\ -0.18 \times 10^3 \\ \hline 8.81 \times 10^3 \end{array}$$

Powers and Roots

When a number expressed in exponential notation is taken to some power, the initial number is taken to the appropriate power and the exponent of 10 is *multiplied* by that power.

$$(N \times 10^n)^m = N^m \times 10^{m \times n}$$

For example,

$$\begin{aligned}(7.5 \times 10^2)^2 &= (7.5)^2 \times 10^{2 \times 2}\\ &= 56. \times 10^4\\ &= 5.6 \times 10^5\end{aligned}$$

When a root is taken of a number expressed in exponential notation, the root of the initial number is taken and the exponent of 10 is divided by the number representing the root. For example, we take the square root of a number as follows:

$$\sqrt{N \times 10^n} = (N \times 10^n)^{1/2} = \sqrt{N} \times 10^{n/2}$$

For example,

$$\begin{aligned}(2.9 \times 10^6)^{1/2} &= \sqrt{2.9} \times 10^{6/2}\\ &= 1.7 \times 10^3\end{aligned}$$

Using a Calculator to Perform Mathematical Operations on Exponentials

In dealing with exponential numbers, you must first learn to enter them into your calculator. First the number is keyed in and then the exponent. There is a special key that must be pressed just before the exponent is entered. This key is often labeled EE or exp. For example, the number 1.56×10^6 is entered as follows:

Press	Display
1.56	1.56
EE or exp	1.56 00
6	1.56 06

To enter a number with a negative exponent, use the change-of-sign key +/− after entering the exponent number. For example, the number 7.54×10^{-3} is entered as follows:

Press	Display
7.54	7.54
EE or exp	7.54 00
3	7.54 03
+/−	7.54 −03

Once a number with an exponent is entered into your calculator, the mathematical operations are performed exactly the same as with a "regular" number. For example, the numbers 1.0×10^3 and 1.0×10^2 are multiplied as follows:

Press	Display
1.0	1.0
EE or exp	1.0 00
3	1.0 03
×	1 03
1.0	1.0
EE or exp	1.0 00
2	1.0 02
=	1 05

The answer is correctly represented as 1.0×10^5.

The numbers 1.50×10^5 and 1.1×10^4 are added as follows:

Press	Display
1.5	1.50
EE or exp	1.50 00
5	1.50 05
+	1.5 05
1.1	1.1
EE or exp	1.1 00
4	1.1 04
=	1.61 05

The answer is correctly represented as 1.61×10^5. Note that when exponential numbers are added, the calculator automatically takes into account any difference in exponents.

To take the power, root, or reciprocal of an exponential number, enter the number first, then press the appropriate key or keys. For example, the square root of 5.6×10^3 is obtained as follows:

Press	Display	
5.6	5.6	
EE or exp	5.6	00
3	5.6	03
$\sqrt{X}$	7.4833148	01

The answer is correctly represented as 7.5×10^1.

Practice by performing the following operations that involve exponential numbers. The answers follow the exercises.

a. $7.9 \times 10^2 \times 4.3 \times 10^4$

b. $\dfrac{5.4 \times 10^3}{4.6 \times 10^5}$

c. $1.7 \times 10^2 + 1.63 \times 10^3$

d. $4.3 \times 10^{-3} + 1 \times 10^{-4}$

e. $(8.6 \times 10^{-6})^2$

f. $\dfrac{1}{8.3 \times 10^2}$

g. $\log(1.0 \times 10^{-7})$

h. $-\log(1.3 \times 10^{-5})$

i. $\sqrt{6.7 \times 10^9}$

Solutions

a. 3.4×10^7

b. 1.2×10^{-2}

c. 1.80×10^3

d. 4.4×10^{-3}

e. 7.4×10^{-11}

f. 1.2×10^{-3}

g. -7.00

h. 4.89

i. 8.2×10^4

Graphing Functions

In interpreting the results of a scientific experiment, it is often useful to make a graph. If possible, the function to be graphed should be in a form that gives a straight line. The equation for a straight line (a *linear equation*) can be represented in the general form

$$y = mx + b$$

where y is the *dependent variable*, x is the *independent variable*, m is the *slope*, and b is the *intercept* with the y axis.

To illustrate the characteristics of a linear equation, the function $y = 3x + 4$ is plotted in Fig. A.1. For this equation $m = 3$ and $b = 4$. Note that the y intercept occurs when $x = 0$. In this case the y intercept is 4, as can be seen from the equation ($b = 4$).

The slope of a straight line is defined as the ratio of the rate of change in y to that in x:

$$m = \text{slope} = \frac{\Delta y}{\Delta x}$$

For the equation $y = 3x + 4$, y changes three times as fast as x (because x has a coefficient of 3). Thus the slope in this case is 3. This can be verified from the graph. For the triangle shown in Fig. A.1,

$$\Delta y = 15 - 16 = 36 \qquad \text{and} \qquad \Delta x = 15 - 3 = 12$$

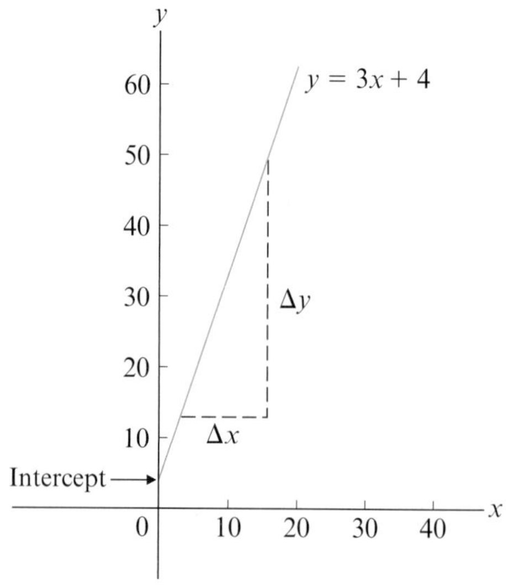

Figure A.1 Graph of the linear equation $y = 3x + 4$.

Thus

$$\text{Slope} = \frac{\Delta y}{\Delta x} = \frac{36}{12} = 3$$

This example illustrates a general method for obtaining the slope of a line from the graph of that line. Simply draw a triangle with one side parallel to the y axis and the other side parallel to the x axis, as shown in Fig. A.1. Then determine the lengths of the sides to get Δy and Δx, respectively, and compute the ratio $\Delta y/\Delta x$.

SI Units and Conversion Factors

These conversion factors are given with more significant figures than those typically used in the body of the text.

Length ▸ SI Unit: Meter (m)

1 meter	=	1.0936 yards
1 centimeter	=	0.39370 inch
1 inch	=	2.54 centimeters (exactly)
1 kilometer	=	0.62137 mile
1 mile	=	5280. feet
	=	1.6093 kilometers

Mass ▸ SI Unit: Kilogram (kg)

1 kilogram	=	1000 grams
	=	2.2046 pounds
1 pound	=	453.59 grams
	=	0.45359 kilogram
	=	16 ounces
1 atomic mass unit	=	1.66057×10^{-27} kilograms

Volume ▸ SI Unit: Cubic Meter (m^3)

1 liter	=	10^{-3} m^3
	=	1 dm^3
	=	1.0567 quarts
1 gallon	=	4 quarts
	=	8 pints
	=	3.7854 liters
1 quart	=	32 fluid ounces
	=	0.94635 liter

Pressure ▸ SI Unit: Pascal (Pa)

1 atmosphere	=	101.325 kilopascals
	=	760. torr (mm Hg)
	=	14.70 pounds per square inch

Energy ▸ SI Unit: Joule (J)

1 joule	=	0.23901 calorie
1 calorie	=	4.184 joules

Solutions to Self Check Exercises

Chapter 2

Self Check Exercise 2.1

$357 = 3.57 \times 10^2$
$0.0055 = 5.5 \times 10^{-3}$

Self Check Exercise 2.2

a. Three significant figures. The leading zeros (to the left of the 1) do not count, but the trailing zeros do.
b. Five significant figures. The one captive zero and the two trailing zeros all count.
c. This is an exact number obtained by counting the cars. It has an unlimited number of significant figures.

Self Check Exercise 2.3

a. $12.6 \times \underset{\text{Limiting}}{0.53} = 6.678 = 6.7$

b. $12.6 \times \underset{\text{Limiting}}{0.53} = 6.7$; $\quad \begin{array}{r} 6.7 \text{ Limiting} \\ -4.59 \\ \hline 2.11 \end{array} = 2.1$

c. $\begin{array}{r} 25.36 \\ -4.15 \\ \hline 21.21 \end{array} \quad \dfrac{21.21}{2.317} = 9.15408 = 9.154$

Self Check Exercise 2.4

$$0.750 \text{ L} \times \frac{1.06 \text{ qt}}{1 \text{ L}} = 0.795 \text{ qt}$$

Self Check Exercise 2.5

$$225 \frac{\text{mi}}{\text{h}} \times \frac{1760 \text{ yd}}{1 \text{ mi}} \times \frac{1 \text{ m}}{1.094 \text{ yd}} \times \frac{1 \text{ km}}{1000 \text{ m}} = 362 \frac{\text{km}}{\text{h}}$$

Self Check Exercise 2.6

The best way to solve this problem is to convert 172 K to Celsius degrees. To do this we will use the formula $T_{°C} = T_K - 273$.

In this case

$$T_{°C} = T_K - 273 = 172 - 273 = -101$$

So 172 K = −101 °C, which is a lower temperature than −75 °C. Thus 172 K is colder than −75 °C.

Self Check Exercise 2.7

The problem is 41 °C = ? °F.

Using the formula

$$T_{°F} = 1.80(T_{°C}) + 32$$

we have

$$T_{°F} = ?\ °F = 1.80(41) + 32 = 74 + 32 = 106$$

That is, 41 °C = 106 °F.

Self Check Exercise 2.8

This problem can be stated as 239 °F = ? °C.

Using the formula

$$T_{°C} = \frac{T_{°F} - 32}{1.80}$$

we have in this case

$$T_{°C} = ?\ °C = \frac{239 - 32}{1.80} = \frac{207}{1.80} = 115$$

That is, 239 °F = 115 °C.

Self Check Exercise 2.9

We obtain the density of the cleaner by dividing its mass by its volume.

$$\text{Density} = \frac{\text{mass}}{\text{volume}} = \frac{28.1 \text{ g}}{35.8 \text{ mL}} = 0.785 \text{ g/mL}$$

This density identifies the liquid as isopropyl alcohol.

Chapter 3

Self Check Exercise 3.1

Items (a) and (c) are physical properties. When the solid gallium melts, it forms liquid gallium. There is no change in composition. Items (b) and (d) reflect the ability to change composition and are thus chemical properties. Statement (b) means that platinum does not react with oxygen to form some new substance. Statement (d) means that copper does react in the air to form a new substance, which is green.

Self Check Exercise 3.2

a. Milk turns sour because new substances are formed. This is a chemical change.
b. Melting the wax is a physical change (a change of state). When the wax burns, new substances are formed. This is a chemical change.

Self Check Exercise 3.3

a. Maple syrup is a homogeneous mixture of sugar and other substances dispersed uniformly in water.
b. Helium and oxygen form a homogeneous mixture.
c. Oil and vinegar salad dressing is a heterogeneous mixture. (Note the two distinct layers the next time you look at a bottle of dressing.)
d. Common salt is a pure substance (sodium chloride), so it always has the same composition. (Note that other substances such as iodine are often added to commercial preparations of table salt, which is mostly sodium chloride. Thus commercial table salt is a homogeneous mixture.)

Chapter 4

Self Check Exercise 4.1

a. P_4O_{10} b. UF_6 c. $AlCl_3$

Self Check Exercise 4.2

In the symbol $^{90}_{38}Sr$, the number 38 is the atomic number, which represents the number of protons in the nucleus of a strontium atom. Because the atom is neutral overall, it must also have 38 electrons. The number 90 (the mass number) represents the number of protons plus the number of neutrons. Thus the number of neutrons is $A - Z = 90 - 38 = 52$.

Self Check Exercise 4.3

The atom $^{201}_{80}Hg$ has 80 protons, 80 electrons, and $201 - 80 =$ 121 neutrons.

Self Check Exercise 4.4

The atomic number for phosphorus is 15 and the mass number is $15 + 17 = 32$. Thus the symbol for the atom is $^{32}_{15}P$.

Self Check Exercise 4.5

Element	Symbol	Atomic Number	Metal or Nonmetal	Family Name
a. argon	Ar	18	nonmetal	noble gas
b. chlorine	Cl	17	nonmetal	halogen
c. barium	Ba	56	metal	alkaline earth metal
d. cesium	Cs	55	metal	alkali metal

Self Check Exercise 4.6

a. KI $(1+) + (1-) = 0$
b. Mg_3N_2 $3(2+) + 2(3-) = (6+) + (6-) = 0$
c. Al_2O_3 $2(3+) + 3(2-) = 0$

Chapter 5

Self Check Exercise 5.1

a. rubidium oxide
b. strontium iodide
c. potassium sulfide

Self Check Exercise 5.2

a. The compound $PbBr_2$ must contain Pb^{2+}—named lead(II)—to balance the charges of the two Br^- ions. Thus the name is lead(II) bromide. The compound $PbBr_4$ must contain Pb^{4+}—named lead(IV)—to balance the charges of the four Br^- ions. The name is therefore lead(IV) bromide.
b. The compound FeS contains the S^{2-} ion (sulfide) and thus the iron cation present must be Fe^{2+}, iron(II). The name is iron(II) sulfide. The compound Fe_2S_3 contains three S^{2-} ions and two iron cations of unknown charge. We can determine the iron charge from the following:

$$\underset{\text{Iron charge}}{2(?+)} + \underset{S^{2-}\text{ charge}}{3(2-)} = 0$$

In this case, ? must represent 3 because

$$2(3+) + 3(2-) = 0$$

Thus Fe_2S_3 contains Fe^{3+} and S^{2-}, and its name is iron(III) sulfide.
c. The compound $AlBr_3$ contains Al^{3+} and Br^-. Because aluminum forms only one ion (Al^{3+}), no Roman numeral is required. The name is aluminum bromide.
d. The compound Na_2S contains Na^+ and S^{2-} ions. The name is sodium sulfide. (Because sodium forms only Na^+, no Roman numeral is needed.)
e. The compound $CoCl_3$ contains three Cl^- ions. Thus the cobalt cation must be Co^{3+}, which is named cobalt(III) because cobalt is a transition metal and can form more than one type of cation. Thus the name of $CoCl_3$ is cobalt(III) chloride.

Self Check Exercise 5.3

Compound	Individual Names	Prefixes	Name
a. CCl_4	carbon	none	carbon
	chloride	*tetra-*	tetrachloride
b. NO_2	nitrogen	none	nitrogen
	oxide	*di-*	dioxide
c. IF_5	iodine	none	iodine
	fluoride	*penta-*	pentafluoride

Self Check Exercise 5.4

a. silicon dioxide
b. dioxygen difluoride
c. xenon hexafluoride

Self Check Exercise 5.5

a. chlorine trifluoride
b. vanadium(V) fluoride
c. copper(I) chloride
d. manganese(IV) oxide
e. magnesium oxide
f. water

Self Check Exercise 5.6

a. calcium hydroxide
b. sodium phosphate
c. potassium permanganate
d. ammonium dichromate
e. cobalt(II) perchlorate (Perchlorate has a 1− charge, so the cation must be Co^{2+} to balance the two ClO_4^- ions.)
f. potassium chlorate
g. copper(II) nitrite (This compound contains two NO_2^- (nitrite) ions and thus must contain a Cu^{2+} cation.)

Self Check Exercise 5.7

Compound **Name**

a. $NaHCO_3$ sodium hydrogen carbonate
Contains Na^+ and HCO_3^-; often called sodium bicarbonate (common name).
b. $BaSO_4$ barium sulfate
Contains Ba^{2+} and SO_4^{2-}.
c. $CsClO_4$ cesium perchlorate
Contains Cs^+ and ClO_4^-.
d. BrF_5 bromine pentafluoride
Both nonmetals (Type III binary).
e. NaBr sodium bromide
Contains Na^+ and Br^- (Type I binary).
f. KOCl potassium hypochlorite
Contains K^+ and OCl^-.

g. $Zn_3(PO_4)_2$ zinc(II) phosphate
Contains Zn^{2+} and PO_4^{3-}; Zn is a transition metal and officially requires a Roman numeral. However, because Zn forms only the Zn^{2+} cation, the II is usually left out. Thus the name of the compound is usually given as zinc phosphate.

Self Check Exercise 5.8

Name	Chemical Formula
a. ammonium sulfate	$(NH_4)_2SO_4$
b. vanadium(V) fluoride	VF_5
c. disulfur dichloride	S_2Cl_2
d. rubidium peroxide	Rb_2O_2
e. aluminum oxide	Al_2O_3

a. ammonium sulfate $(NH_4)_2SO_4$
Two ammonium ions (NH_4^+) are required for each sulfate ion (SO_4^{2-}) to achieve charge balance.

b. vanadium(V) fluoride VF_5
The compound contains V^{5+} ions and requires five F^- ions for charge balance.

c. disulfur dichloride S_2Cl_2
The prefix *di-* indicates two of each atom.

d. rubidium peroxide Rb_2O_2
Because rubidium is in Group 1, it forms only 1+ ions. Thus two Rb^+ ions are needed to balance the 2− charge on the peroxide ion (O_2^{2-}).

e. aluminum oxide Al_2O_3
Aluminum forms only 3+ ions. Two Al^{3+} ions are required to balance the charge on three O^{2-} ions.

Chapter 6

Self Check Exercise 6.1

a. $Mg(s) + H_2O(l) \rightarrow Mg(OH)_2(s) + H_2(g)$
Note that magnesium (which is in Group 2) always forms the Mg^{2+} cation and thus requires two OH^- anions for a zero net charge.

b. Ammonium dichromate contains the polyatomic ions NH_4^+ and $Cr_2O_7^{2-}$ (you should have these memorized). Because NH_4^+ has a 1+ charge, two NH_4^+ cations are required for each $Cr_2O_7^{2-}$, with it 2− charge, to give the formula $(NH_4)_2Cr_2O_7$. Chromium(III) oxide contains Cr^{3+} ions—signified by chromium(III)—and O^{2-} (the oxide ion). To achieve a net charge of zero, the solid must contain two Cr^{3+} ions for every three O^{2-} ions, so the formula is Cr_2O_3. Nitrogen gas contains diatomic molecules and is written $N_2(g)$, and gaseous water is written $H_2O(g)$. Thus the unbalanced equation for the decomposition of ammonium dichromate is

$$(NH_4)_2Cr_2O_7(s) \rightarrow Cr_2O_3(s) + N_2(g) + H_2O(g)$$

c. Gaseous ammonia, $NH_3(g)$, and gaseous oxygen, $O_2(g)$, react to form nitrogen monoxide gas, $NO(g)$, plus gaseous water, $H_2O(g)$. The unbalanced equation is

$$NH_3(g) + O_2(g) \rightarrow NO(g) + H_2O(g)$$

Self Check Exercise 6.2

Step 1 The reactants are propane, $C_3H_8(g)$, and oxygen, $O_2(g)$; the products are carbon dioxide, $CO_2(g)$, and water, $H_2O(g)$. All are in the gaseous state.

Step 2 The unbalanced equation for the reaction is

$$C_3H_8(g) + O_2(g) \rightarrow CO_2(g) + H_2O(g)$$

Step 3 We start with C_3H_8 because it is the most complicated molecule. C_3H_8 contains three carbon atoms per molecule, so a coefficient of 3 is needed for CO_2.

$$C_3H_8(g) + O_2(g) \rightarrow 3CO_2(g) + H_2O(g)$$

Also, each C_3H_8 molecule contains eight hydrogen atoms, so a coefficient of 4 is required for H_2O.

$$C_3H_8(g) + O_2(g) \rightarrow 3CO_2(g) + 4H_2O(g)$$

The final element to be balanced is oxygen. Note that the left side of the equation now has two oxygen atoms, and the right side has ten. We can balance the oxygen by using a coefficient of 5 for O_2.

$$C_3H_8(g) + 5O_2(g) \rightarrow 3CO_2(g) + 4H_2O(g)$$

Step 4 *Check:*

$$3\text{ C, }8\text{ H, }10\text{ O} \rightarrow 3\text{ C, }8\text{ H, }10\text{ O}$$

Reactant atoms — Product atoms

We cannot divide all coefficients by a given integer to give smaller integer coefficients.

Self Check Exercise 6.3

a. $NH_4NO_2(s) \rightarrow N_2(g) + H_2O(g)$ (unbalanced)
$NH_4NO_2(s) \rightarrow N_2(g) + 2H_2O(g)$ (balanced)

b. $NO(g) \rightarrow N_2O(g) + NO_2(g)$ (unbalanced)
$3NO(g) \rightarrow N_2O(g) + NO_2(g)$ (balanced)

c. $HNO_3(l) \rightarrow NO_2(g) + H_2O(l) + O_2(g)$ (unbalanced)
$4HNO_3(l) \rightarrow 4NO_2(g) + 2H_2O(l) + O_2(g)$ (balanced)

Chapter 7

Self Check Exercise 7.1

a. The ions present are

$$Ba^{2+}(aq) + 2NO_3^-(aq) + Na^+(aq) + Cl^-(aq) \rightarrow$$

Ions in $Ba(NO_3)_2(aq)$ — Ions in $NaCl(aq)$

Exchanging the anions gives the possible solid products $BaCl_2$ and $NaNO_3$. Using Table 7.1, we see that both substances are very soluble (rules 1, 2, and 3). Thus no solid forms.

b. The ions present in the mixed solution before any reaction occurs are

$$2Na^+(aq) + S^{2-}(aq) + Cu^{2+}(aq) + 2NO_3^-(aq) \rightarrow$$

Ions in $Na_2S(aq)$ — Ions in $Cu(NO_3)_2(aq)$

Exchanging the anions gives the possible solid products CuS and $NaNO_3$. According to rules 1 and 2 in Table 7.1, $NaNO_3$ is soluble, and by rule 6, CuS should be insoluble. Thus CuS will precipitate. The balanced equation is

$$Na_2S(aq) + Cu(NO_3)_2(aq) \rightarrow CuS(s) + 2NaNO_3(aq)$$

c. The ions present are

$$\underbrace{NH_4^+(aq) + Cl^-(aq)}_{\text{Ions in } NH_4Cl(aq)} + \underbrace{Pb^{2+}(aq) + 2NO_3^-(aq)}_{\text{Ions in } Pb(NO_3)_2(aq)} \rightarrow$$

Exchanging the anions gives the possible solid products NH_4NO_3 and $PbCl_2$. NH_4NO_3 is soluble (rules 1 and 2) and $PbCl_2$ is insoluble (rule 3). Thus $PbCl_2$ will precipitate. The balanced equation is

$$2NH_4Cl(aq) + Pb(NO_3)_2(aq) \rightarrow PbCl_2(s) + 2NH_4NO_3(aq)$$

Self Check Exercise 7.2

a. *Molecular equation:*
$Na_2S(aq) + Cu(NO_3)_2(aq) \rightarrow CuS(s) + 2NaNO_3(aq)$
Complete ionic equation:
$2Na^+(aq) + S^{2-}(aq) + Cu^{2+}(aq) + 2NO_3^-(aq) \rightarrow CuS(s) + 2Na^+(aq) + 2NO_3^-(aq)$
Net ionic equation:
$S^{2-}(aq) + Cu^{2+}(aq) \rightarrow CuS(s)$

b. *Molecular equation:*
$2NH_4Cl(aq) + Pb(NO_3)_2(aq) \rightarrow PbCl_2(s) + 2NH_4NO_3(aq)$
Complete ionic equation:
$2NH_4^+(aq) + 2Cl^-(aq) + Pb^{2+}(aq) + 2NO_3^-(aq) \rightarrow PbCl_2(s) + 2NH_4^+(aq) + 2NO_3^-(aq)$
Net ionic equation:
$2Cl^-(aq) + Pb^{2+}(aq) \rightarrow PbCl_2(s)$

Self Check Exercise 7.3

a. The compound NaBr contains the ions Na^+ and Br^-. Thus each sodium atom loses one electron ($Na \rightarrow Na^+ + e^-$), and each bromine atom gains one electron ($Br + e^- \rightarrow Br^-$).

$$Na + Na + Br - Br \rightarrow (Na^+Br^-) + (Na^+Br^-)$$

e^- e^-

b. The compound CaO contains the Ca^{2+} and O^{2-} ions. Thus each calcium atom loses two electrons ($Ca \rightarrow Ca^{2+} + 2e^-$), and each oxygen atom gains two electrons ($O + 2e^- \rightarrow O^{2-}$).

$$Ca + Ca + O - O \rightarrow (Ca^{2+}O^{2-}) + (Ca^{2+}O^{2-})$$

$2e^-$ $2e^-$

Self Check Exercise 7.4

a. oxidation–reduction reaction; combustion reaction
b. synthesis reaction; oxidation–reduction reaction; combustion reaction
c. synthesis reaction; oxidation–reduction reaction
d. decomposition reaction; oxidation–reduction reaction
e. precipitation reaction (and double displacement)
f. synthesis reaction; oxidation–reduction reaction
g. acid–base reaction (and double displacement)
h. combustion reaction; oxidation–reduction reaction

Chapter 8

Self Check Exercise 8.1

The average mass of nitrogen is 14.01 amu. The appropriate equivalence statement is 1 N atom = 14.01 amu, which yields the conversion factor we need:

$$\underset{\text{(exact)}}{23\ \cancel{\text{N atoms}}} \times \frac{14.01\ \text{amu}}{\cancel{\text{N atom}}} = 322.2\ \text{amu}$$

Self Check Exercise 8.2

The average mass of oxygen is 16.00 amu, which gives the equivalence statement 1 O atom = 16.00 amu. The number of oxygen atoms present is

$$288\ \cancel{\text{amu}} \times \frac{1\ \text{O atom}}{16.00\ \cancel{\text{amu}}} = 18.0\ \text{O atoms}$$

Self Check Exercise 8.3

Note that the sample of 5.00×10^{20} atoms of chromium is less than 1 mole (6.022×10^{23} atoms) of chromium. What fraction of a mole it represents can be determined as follows:

$$5.00 \times 10^{20}\ \cancel{\text{atoms Cr}} \times \frac{1\ \text{mol Cr}}{6.022 \times 10^{23}\ \cancel{\text{atoms Cr}}} = 8.30 \times 10^{-4}\ \text{mol Cr}$$

Because the mass of 1 mole of chromium atoms is 52.00 g, the mass of 5.00×10^{20} atoms can be determined as follows:

$$8.30 \times 10^{-4}\ \cancel{\text{mol Cr}} \times \frac{52.00\ \text{g Cr}}{1\ \cancel{\text{mol Cr}}} = 4.32 \times 10^{-2}\ \text{g Cr}$$

Self Check Exercise 8.4

Each molecule of C_2H_3Cl contains two carbon atoms, three hydrogen atoms, and one chlorine atom, so 1 mole of C_2H_3Cl molecules contains 2 moles of C atoms, 3 moles of H atoms, and 1 mole of Cl atoms.

Mass of 2 mol C atoms: $2 \times 12.01 = 24.02$ g
Mass of 3 mol H atoms: $3 \times 1.008 = 3.024$ g
Mass of 1 mol Cl atoms: $1 \times 35.45 = 35.45$ g
62.494 g

The molar mass of C_2H_3Cl is 62.49 g (rounding to the correct number of significant figures).

Self Check Exercise 8.5

The formula for sodium sulfate is Na_2SO_4. One mole of Na_2SO_4 contains 2 moles of sodium ions and 1 mole of sulfate ions.

1 mole of Na_2SO_4 $\rightarrow$ 1 mole of [Na^+, Na^+, SO_4^{2-}] (2 mol Na^+; 1 mol SO_4^{2-})

Mass of 2 mol Na^+ $= 2 \times 22.99 = 45.98$ g
Mass of 1 mol SO_4^{2-} $= 32.07 + 4(16.00) = 96.07$ g
Mass of 1 mol Na_2SO_4 $= 142.05$ g

The molar mass for sodium sulfate is 142.05 g.

A sample of sodium sulfate with a mass of 300.0 g represents more than 1 mol. (Compare 300.0 g to the molar mass of Na_2SO_4.) We calculate the number of moles of Na_2SO_4 present in 300.0 g as follows:

$$300.0\ \text{g}\ \cancel{\text{Na}_2\text{SO}_4} \times \frac{1\ \text{mol Na}_2\text{SO}_4}{142.05\ \text{g}\ \cancel{\text{Na}_2\text{SO}_4}} = 2.112\ \text{mol Na}_2\text{SO}_4$$

Self Check Exercise 8.6

First we must compute the mass of 1 mole of C_2F_4 molecules (the molar mass). Because 1 mole of C_2F_4 contains 2 moles of C atoms and 4 moles of F atoms, we have

$$2\ \cancel{\text{mol}}\ \text{C} \times \frac{12.01\ \text{g}}{\cancel{\text{mol}}} = 24.02\ \text{g C}$$

$$4\ \cancel{\text{mol}}\ \text{F} \times \frac{19.00\ \text{g}}{\cancel{\text{mol}}} = 76.00\ \text{g F}$$

Mass of 1 mole of C_2F_4: 100.02 g = molar mass

Using the equivalence statement 100.02 g C_2F_4 = 1 mole C_2F_4, we calculate the moles of C_2F_4 units in 135 g of Teflon.

$$135\ \text{g}\ \cancel{\text{C}_2\text{F}_4}\ \text{units} \times \frac{1\ \text{mol C}_2\text{F}_4}{100.02\ \text{g}\ \cancel{\text{C}_2\text{F}_4}} = 1.35\ \text{mol C}_2\text{F}_4\ \text{units}$$

Next, using the equivalence statement 1 mol = 6.022×10^{23} units, we calculate the number of C_2F_4 units in 135 mol of Teflon.

$$135\ \cancel{\text{mol}}\ \text{C}_2\text{F}_4 \times \frac{6.022 \times 10^{23}\ \text{units}}{1\ \cancel{\text{mol}}} = 8.13 \times 10^{23}\ \text{C}_2\text{F}_4\ \text{units}$$

Self Check Exercise 8.7

The molar mass of penicillin F is computed as follows:

$$\text{C: } 14\ \cancel{\text{mol}} \times 12.01\ \frac{\text{g}}{\cancel{\text{mol}}} = 168.1\ \text{g}$$

$$\text{H: } 20\ \cancel{\text{mol}} \times 1.008\ \frac{\text{g}}{\cancel{\text{mol}}} = 20.16\ \text{g}$$

$$\text{N: } 2\ \cancel{\text{mol}} \times 14.01\ \frac{\text{g}}{\cancel{\text{mol}}} = 28.02\ \text{g}$$

$$\text{S: } 1\ \cancel{\text{mol}} \times 32.07\ \frac{\text{g}}{\cancel{\text{mol}}} = 32.07\ \text{g}$$

$$\text{O: } 4\ \cancel{\text{mol}} \times 16.00\ \frac{\text{g}}{\cancel{\text{mol}}} = 64.00\ \text{g}$$

Mass of 1 mole of $C_{14}H_{20}N_2SO_4$ = 312.39 g = 312.4 g

$$\text{Mass percent of C} = \frac{168.1\ \text{g C}}{312.4\ \text{g C}_{14}\text{H}_{20}\text{N}_2\text{SO}_4} \times 100\% = 53.81\%$$

$$\text{Mass percent of H} = \frac{20.16\ \text{g H}}{312.4\ \text{g C}_{14}\text{H}_{20}\text{N}_2\text{SO}_4} \times 100\% = 6.453\%$$

$$\text{Mass percent of N} = \frac{28.02\ \text{g N}}{312.4\ \text{g C}_{14}\text{H}_{20}\text{N}_2\text{SO}_4} \times 100\% = 8.969\%$$

$$\text{Mass percent of S} = \frac{32.07\ \text{g S}}{312.4\ \text{g C}_{14}\text{H}_{20}\text{N}_2\text{SO}_4} \times 100\% = 10.27\%$$

$$\text{Mass percent of O} = \frac{64.00\ \text{g O}}{312.4\ \text{g C}_{14}\text{H}_{20}\text{N}_2\text{SO}_4} \times 100\% = 20.49\%$$

Check: The percentages add up to 99.99%.

Self Check Exercise 8.8

Step 1 0.6884 g lead and 0.2356 g chlorine

Step 2
$$0.6884\ \text{g}\ \cancel{\text{Pb}} \times \frac{1\ \text{mol Pb}}{207.2\ \text{g}\ \cancel{\text{Pb}}} = 0.003322\ \text{mol Pb}$$

$$0.2356\ \text{g}\ \cancel{\text{Cl}} \times \frac{1\ \text{mol Cl}}{35.45\ \text{g}\ \cancel{\text{Cl}}} = 0.006646\ \text{mol Cl}$$

Step 3
$$\frac{0.003322\ \text{mol Pb}}{0.003322} = 1.000\ \text{mol Pb}$$

$$\frac{0.006646\ \text{mol Cl}}{0.003322} = 2.001\ \text{mol Cl}$$

These numbers are very close to integers, so step 4 is unnecessary. The empirical formula is $PbCl_2$.

Self Check Exercise 8.9

Step 1 0.8007 g C, 0.9333 g N, 0.2016 g H, and 2.133 g O

Step 2
$$0.8007\ \text{g}\ \cancel{\text{C}} \times \frac{1\ \text{mol C}}{12.01\ \text{g}\ \cancel{\text{C}}} = 0.06667\ \text{mol C}$$

$$0.9333\ \text{g}\ \cancel{\text{N}} \times \frac{1\ \text{mol N}}{14.01\ \text{g}\ \cancel{\text{N}}} = 0.06662\ \text{mol N}$$

$$0.2016\ \text{g}\ \cancel{\text{H}} \times \frac{1\ \text{mol H}}{1.008\ \text{g}\ \cancel{\text{H}}} = 0.2000\ \text{mol H}$$

$$2.133\ \text{g}\ \cancel{\text{O}} \times \frac{1\ \text{mol O}}{16.00\ \text{g}\ \cancel{\text{O}}} = 0.1333\ \text{mol O}$$

Step 3
$$\frac{0.06667\ \text{mol C}}{0.06667} = 1.001\ \text{mol C}$$

$$\frac{0.06662\ \text{mol N}}{0.06667} = 1.000\ \text{mol N}$$

$$\frac{0.2000\ \text{mol H}}{0.06662} = 3.002\ \text{mol H}$$

$$\frac{0.1333\ \text{mol O}}{0.06662} = 2.001\ \text{mol O}$$

The empirical formula is CNH_3O_2.

Self Check Exercise 8.10

Step 1 In 100.00 g of Nylon-6 the masses of elements present are 63.68 g C, 12.38 g N, 9.80 g H, and 14.14 g O.

Step 2
$$63.68\ \text{g}\ \cancel{\text{C}} \times \frac{1\ \text{mol C}}{12.01\ \text{g}\ \cancel{\text{C}}} = 5.302\ \text{mol C}$$

$$12.38\ \text{g}\ \cancel{\text{N}} \times \frac{1\ \text{mol N}}{14.01\ \text{g}\ \cancel{\text{N}}} = 0.8837\ \text{mol N}$$

$$9.80\ \text{g}\ \cancel{\text{H}} \times \frac{1\ \text{mol H}}{1.008\ \text{g}\ \cancel{\text{H}}} = 9.72\ \text{mol H}$$

$$14.14\ \text{g}\ \cancel{\text{O}} \times \frac{1\ \text{mol O}}{16.00\ \text{g}\ \cancel{\text{O}}} = 0.8838\ \text{mol O}$$

Step 3 $\frac{5.302 \text{ mol C}}{0.8836} = 6.000 \text{ mol C}$

$$\frac{0.8837 \text{ mol N}}{0.8837} = 1.000 \text{ mol N}$$

$$\frac{9.72 \text{ mol H}}{0.8837} = 11.0 \text{ mol H}$$

$$\frac{0.8838 \text{ mol O}}{0.8837} = 1.000 \text{ mol O}$$

The empirical formula for Nylon-6 is $C_6NH_{11}O$.

Self Check Exercise 8.11

Step 1 First we convert the mass percents to mass in grams. In 100.0 g of the compound, there are 71.65 g of chlorine, 24.27 g of carbon, and 4.07 g of hydrogen.

Step 2 We use these masses to compute the moles of atoms present.

$$71.65 \text{ g Cl} \times \frac{1 \text{ mol Cl}}{35.45 \text{ g Cl}} = 2.021 \text{ mol Cl}$$

$$24.27 \text{ g C} \times \frac{1 \text{ mol C}}{12.01 \text{ g C}} = 2.021 \text{ mol C}$$

$$4.07 \text{ g H} \times \frac{1 \text{ mol H}}{1.008 \text{ g H}} = 4.04 \text{ mol H}$$

Step 3 Dividing each mole value by 2.021 (the smallest number of moles present), we obtain the empirical formula $ClCH_2$.

To determine the molecular formula, we must compare the empirical formula mass to the molar mass. The empirical formula mass is 49.48.

Cl: 35.45
C: 12.01
2 H: $2 \times (1.008)$
$ClCH_2$: 49.48 = empirical formula mass

The molar mass is known to be 98.96. We know that

$$\text{Molar mass} = n \times (\text{empirical formula mass})$$

So we can obtain the value of n as follows:

$$\frac{\text{Molar mass}}{\text{Empirical formula mass}} = \frac{98.96}{49.48} = 2$$

$$\text{Molecular formula} = (ClCH_2)_2 = Cl_2C_2H_4$$

This substance is composed of molecules with the formula $Cl_2C_2H_4$.

Chapter 9

Self Check Exercise 9.1

The problem can be stated as follows:

$$4.30 \text{ mol } C_3H_8 \underset{\text{yields}}{\rightarrow} ? \text{ mol } CO_2$$

From the balanced equation

$$C_3H_8(g) + 5O_2(g) \rightarrow 3CO_2(g) + 4H_2O(g)$$

we derive the equivalence statement

$$1 \text{ mol } C_3H_8 = 3 \text{ mol } CO_2$$

The appropriate conversion factor (moles of C_3H_8 must cancel) is $3 \text{ mol } CO_2/1 \text{ mol } C_3H_8$, and the calculation is

$$4.30 \text{ mol } C_3H_8 \times \frac{3 \text{ mol } CO_2}{1 \text{ mol } C_3H_8} = 12.9 \text{ mol } CO_2$$

Thus we can say

$$4.30 \text{ mol } C_3H_8 \text{ yields } 12.9 \text{ mol } CO_2$$

Self Check Exercise 9.2

The problem can be sketched as follows:

$$C_3H_8(g) + 5O_2(g) \rightarrow 3CO_2(g) + 4H_2O(g)$$

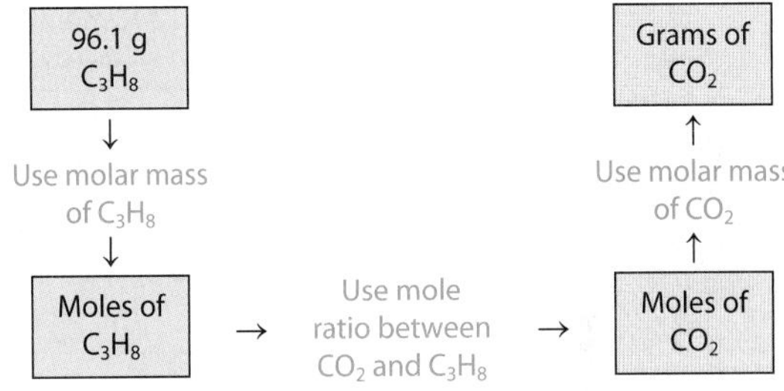

We have already done the first step in Example 9.4.

96.1 g C_3H_8 → $\frac{1 \text{ mol}}{44.09 \text{ g}}$ → 2.18 mol C_3H_8

To find out how many moles of CO_2 can be produced from 2.18 moles of C_3H_8, we see from the balanced equation that 3 moles of CO_2 is produced for each mole of C_3H_8 reacted. The mole ratio we need is $3 \text{ mol } CO_2/1 \text{ mol } C_3H_8$. The conversion is therefore

$$2.18 \text{ mol } C_3H_8 \times \frac{3 \text{ mol } CO_2}{1 \text{ mol } C_3H_8} = 6.54 \text{ mol } CO_2$$

Next, using the molar mass of CO_2, which is $12.01 + 32.00 = 44.01$ g, we calculate the mass of CO_2 produced.

$$6.54 \text{ mol } CO_2 \times \frac{44.01 \text{ g } CO_2}{1 \text{ mol } CO_2} = 288 \text{ g } CO_2$$

The sequence of steps we took to find the mass of carbon dioxide produced from 96.1 g of propane is summarized in the following diagram.

96.1 g C_3H_8 → $\frac{1 \text{ mol } C_3H_8}{44.09 \text{ g } C_3H_8}$ → 2.18 mol C_3H_8

2.18 mol C_3H_8 → $\frac{3 \text{ mol } CO_2}{1 \text{ mol } C_3H_8}$ → 6.54 mol CO_2

6.54 mol CO_2 → $\frac{44.01 \text{ g } CO_2}{1 \text{ mol } CO_2}$ → 288 g CO_2

Mass   Moles

Self Check Exercise 9.3

We sketch the problem as follows:

$$C_3H_8(g) + 5O_2(g) \rightarrow 3CO_2(g) + 4H_2O(g)$$

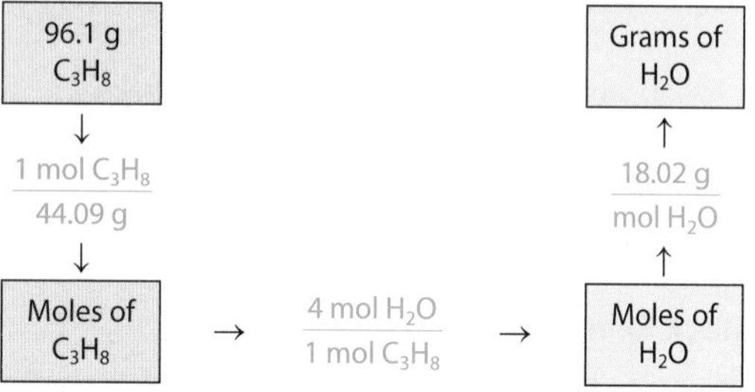

Then we do the calculations.

$$96.1\ \text{g } C_3H_8 \rightarrow \frac{1\ \text{mol } C_3H_8}{44.09\ \text{g}} \rightarrow 2.18\ \text{mol } C_3H_8$$

$$2.18\ \text{mol } C_3H_8 \rightarrow \frac{4\ \text{mol } H_2O}{1\ \text{mol } C_3H_8} \rightarrow 8.72\ \text{mol } H_2O$$

$$8.72\ \text{mol } H_2O \rightarrow \frac{18.02\ \text{g}}{\text{mol } H_2O} \rightarrow 157\ \text{g } H_2O$$

Therefore, 157 g of H_2O is produced from 96.1 g C_3H_8.

Self Check Exercise 9.4

a. We first write the balanced equation.

$$SiO_2(s) + 4HF(aq) \rightarrow SiF_4(g) + 2H_2O(l)$$

The map of the steps required is

$$SiO_2(s) + 4HF(aq) \rightarrow SiF_4(g) + 2H_2O(l)$$

We convert 5.68 g of SiO_2 to moles as follows:

$$5.68\ \cancel{\text{g } SiO_2} \times \frac{1\ \text{mol } SiO_2}{60.09\ \cancel{\text{g } SiO_2}} = 9.45 \times 10^{-2}\ \text{mol } SiO_2$$

Using the balanced equation, we obtain the appropriate mole ratio and convert to moles of HF.

$$9.45 \times 10^{-2}\ \cancel{\text{mol } SiO_2} \times \frac{4\ \text{mol HF}}{1\ \cancel{\text{mol } SiO_2}} = 3.78 \times 10^{-1}\ \text{mol HF}$$

Finally, we calculate the mass of HF by using its molar mass.

$$3.78 \times 10^{-1}\ \cancel{\text{mol HF}} \times \frac{20.01\ \text{g HF}}{\cancel{\text{mol HF}}} = 7.56\ \text{g HF}$$

b. The map for this problem is

$$SiO_2(s) + 4HF(aq) \rightarrow SiF_4(g) + 2H_2O(l)$$

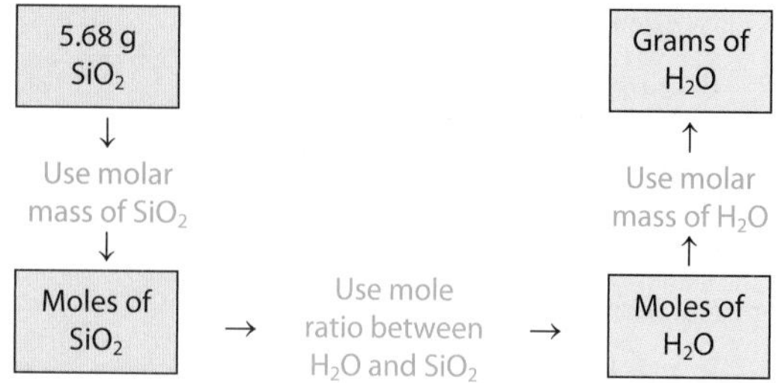

We have already accomplished the first conversion in part a. Using the balanced equation, we obtain moles of H_2O as follows:

$$9.45 \times 10^{-2}\ \cancel{\text{mol } SiO_2} \times \frac{2\ \text{mol } H_2O}{1\ \cancel{\text{mol } SiO_2}} = 1.89 \times 10^{-1}\ \text{mol } H_2O$$

The mass of water formed is

$$1.89 \times 10^{-1}\ \cancel{\text{mol } H_2O} \times \frac{18.02\ \text{g } H_2O}{\cancel{\text{mol } H_2O}} = 3.41\ \text{g } H_2O$$

Self Check Exercise 9.5

In this problem, we know the mass of the product to be formed by the reaction

$$CO(g) + 2H_2(g) \rightarrow CH_3OH(l)$$

and we want to find the masses of reactants needed. The procedure is the same one we have been following. We must first convert the mass of CH_3OH to moles, then use the balanced equation to obtain moles of H_2 and CO needed, and then convert these moles to masses. Using the molar mass of CH_3OH (32.04 g/mol), we convert to moles of CH_3OH.

First we convert kilograms to grams.

$$6.0\ \cancel{\text{kg}}\ CH_3OH \times \frac{1000\ \text{g}}{\cancel{\text{kg}}} = 6.0 \times 10^3\ \text{g } CH_3OH$$

Next we convert 6.0×10^3 g CH_3OH to moles of CH_3OH, using the conversion factor 1 mol CH_3OH/32.04 g CH_3OH.

$$6.0 \times 10^3\ \cancel{\text{g } CH_3OH} \times \frac{1\ \text{mol } CH_3OH}{32.04\ \cancel{\text{g } CH_3OH}} = 1.9 \times 10^2\ \text{mol } CH_3OH$$

Then we have two questions to answer:

? mol of H_2 $\longrightarrow$ (required to produce) 1.9×10^2 mol CH_3OH

? mol of CO $\longrightarrow$ (required to produce) 1.9×10^2 mol CH_3OH

To answer these questions, we use the balanced equation

$$CO(g) + 2H_2(g) \rightarrow CH_3OH(l)$$

to obtain mole ratios between the reactants and the products. In the balanced equation the coefficients for both CO and CH_3OH are 1, so we can write the equivalence statement

$$1\ \text{mol CO} = 1\ \text{mol } CH_3OH$$

Using the mole ratio 1 mol CO/1 mol CH_3OH, we can now convert from moles of CH_3OH to moles of CO.

$$1.9 \times 10^2 \text{ mol CH}_3\text{OH} \times \frac{1 \text{ mol CO}}{1 \text{ mol CH}_3\text{OH}} = 1.9 \times 10^2 \text{ mol CO}$$

To calculate the moles of H_2 required, we construct the equivalence statement between CH_3OH and H_2, using the coefficients in the balanced equation.

$$2 \text{ mol H}_2 = 1 \text{ mol CH}_3\text{OH}$$

Using the mole ratio 2 mol H_2/1 mol CH_3OH, we can convert moles of CH_3OH to moles of H_2.

$$1.9 \times 10^2 \text{ mol CH}_3\text{OH} \times \frac{2 \text{ mol H}_2}{1 \text{ mol CH}_3\text{OH}} = 3.8 \times 10^2 \text{ mol H}_2$$

We now have the moles of reactants required to produce 6.0 kg of CH_3OH. Since we need the masses of reactants, we must use the molar masses to convert from moles to mass.

$$1.9 \times 10^2 \text{ mol CO} \times \frac{28.01 \text{ g CO}}{1 \text{ mol CO}} = 5.3 \times 10^3 \text{ g CO}$$

$$3.8 \times 10^2 \text{ mol H}_2 \times \frac{2.016 \text{ g H}_2}{1 \text{ mol H}_2} = 7.7 \times 10^2 \text{ H}_2$$

Therefore, we need 5.3×10^3 g CO to react with 7.7×10^2 g H_2 to form 6.0×10^3 g (6.0 kg) of CH_3OH. This whole process is mapped in the following diagram.

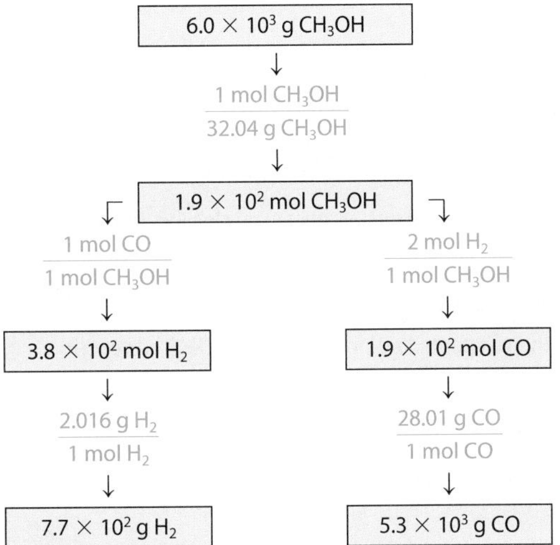

Self Check Exercise 9.6

Step 1 The balanced equation for the reaction is

$$6\text{Li}(s) + \text{N}_2(g) \rightarrow 2\text{Li}_3\text{N}(s)$$

Step 2 To determine the limiting reactant, we must convert the masses of lithium (atomic mass = 6.941 g) and nitrogen (molar mass = 28.02 g) to moles.

$$56.0 \text{ g Li} \times \frac{1 \text{ mol Li}}{6.941 \text{ g Li}} = 8.07 \text{ mol Li}$$

$$56.0 \text{ g N}_2 \times \frac{1 \text{ mol N}_2}{28.02 \text{ g N}_2} = 2.00 \text{ mol N}_2$$

Step 3 Using the mole ratio from the balanced equation, we can calculate the moles of lithium required to react with 2.00 moles of nitrogen.

$$2.00 \text{ mol N}_2 \times \frac{6 \text{ mol Li}}{1 \text{ mol N}_2} = 12.0 \text{ mol Li}$$

Therefore, 12.0 moles of Li is required to react with 2.00 moles of N_2. However, we have only 8.07 mol of Li, so lithium is limiting. It will be consumed before the nitrogen runs out.

Step 4 Because lithium is the limiting reactant, we must use the 8.07 moles of Li to determine how many moles of Li_3N can be formed.

$$8.07 \text{ mol Li} \times \frac{2 \text{ mol Li}_3\text{N}}{6 \text{ mol Li}} = 2.69 \text{ mol Li}_3\text{N}$$

Step 5 We can now use the molar mass of Li_3N (34.83 g) to calculate the mass of Li_3N formed.

$$2.69 \text{ mol Li}_3\text{N} \times \frac{34.83 \text{ g Li}_3\text{N}}{1 \text{ mol Li}_3\text{N}} = 93.7 \text{ g Li}_3\text{N}$$

Self Check Exercise 9.7

a. **Step 1** The balanced equation is

$$\text{TiCl}_4(g) + \text{O}_2(g) \rightarrow \text{TiO}_2(s) + 2\text{Cl}_2(g)$$

Step 2 The numbers of moles of reactants are

$$6.71 \times 10^3 \text{ g TiCl}_4 \times \frac{1 \text{ mol TiCl}_4}{189.68 \text{ g TiCl}_4} = 3.54 \times 10^1 \text{ mol TiCl}_4$$

$$2.45 \times 10^3 \text{ g O}_2 \times \frac{1 \text{ mol O}_2}{32.00 \text{ g O}_2} = 7.66 \times 10^1 \text{ mol O}_2$$

Step 3 In the balanced equation both $TiCl_4$ and O_2 have coefficients of 1, so

$$1 \text{ mol TiCl}_4 = 1 \text{ mol O}_2$$

and

$$3.54 \times 10^1 \text{ mol TiCl}_4 \times \frac{1 \text{ mol O}_2}{1 \text{ mol TiCl}_4} = 3.54 \times 10^1 \text{ mol O}_2 \text{ required}$$

We have 7.66×10^1 moles of O_2, so the O_2 is in excess and the $TiCl_4$ is limiting. This makes sense. $TiCl_4$ and O_2 react in a 1:1 mole ratio, so the $TiCl_4$ is limiting because fewer moles of $TiCl_4$ are present than moles of O_2.

Step 4 We will now use the moles of $TiCl_4$ (the limiting reactant) to determine the moles of TiO_2 that would form if the reaction produced 100% of the expected yield (the theoretical yield).

$$3.54 \times 10^1 \text{ mol TiCl}_4 \times \frac{1 \text{ mol TiO}_2}{1 \text{ mol TiCl}_4} = 3.54 \times 10^1 \text{ mol TiO}_2$$

The mass of TiO_2 expected for 100% yield is

$$3.54 \times 10^1 \text{ mol TiO}_2 \times \frac{79.88 \text{ g TiO}_2}{1 \text{ mol TiO}_2} = 2.83 \times 10^3 \text{ g TiO}_2$$

This amount represents the theoretical yield.

b. Because the reaction is said to give only a 75.0% yield of TiO_2, we use the definition of percent yield,

$$\frac{\text{Actual yield}}{\text{Theoretical yield}} \times 100\% = \%\ \text{yield}$$

to write the equation

$$\frac{\text{Actual yield}}{2.83 \times 10^3\ \text{g}\ TiO_2} \times 100\% = 75.0\%\ \text{yield}$$

We now want to solve for the actual yield. First we divide both sides by 100%.

$$\frac{\text{Actual yield}}{2.83 \times 10^3\ \text{g}\ TiO_2} \times \frac{100\%}{100\%} = \frac{75.0}{100} = 0.750$$

Then we multiply both sides by 2.83×10^3 g TiO_2.

$$2.83 \times 10^3\ \text{g}\ TiO_2 \times \frac{\text{Actual yield}}{2.83 \times 10^3\ \text{g}\ TiO_2} = 0.750 \times 2.83 \times 10^3\ \text{g}\ TiO_2$$

$$\text{Actual yield} = 0.750 \times 2.83 \times 10^3\ \text{g}\ TiO_2 = 2.12 \times 10^3\ \text{g}\ TiO_2$$

Thus 2.12×10^3 g of $TiO_2(s)$ is actually obtained in this reaction.

Chapter 10

Self Check Exercise 10.1

The conversion factor needed is $\frac{1\ \text{cal}}{4.184\ \text{J}}$, and the conversion is

$$28.4\ \cancel{\text{J}} \times \frac{1\ \text{cal}}{4.184\ \cancel{\text{J}}} = 6.79\ \text{cal}$$

Self Check Exercise 10.2

We know that it takes 4.184 J of energy to change the temperature of each gram of water by 1 °C, so we must multiply 4.184 by the mass of water (454 g) and the temperature change (98.6 °C − 5.4 °C = 93.2 °C).

$$4.184 \frac{\text{J}}{\cancel{\text{g}}\ \cancel{°\text{C}}} \times 454\ \cancel{\text{g}} \times 93.2\ \cancel{°\text{C}} = 1.77 \times 10^5\ \text{J}$$

Self Check Exercise 10.3

From Table 10.1, the specific heat capacity for solid gold is 0.13 J/g °C. Because it takes 0.13 J to change the temperature of *one* gram of gold by *one* Celsius degree, we must multiply 0.13 by the sample size (5.63 g) and the change in temperature (32 °C − 21 °C = 11 °C).

$$0.13 \frac{\text{J}}{\text{g}\ °\text{C}} \times 5.63\ \text{g} \times 11\ °\text{C} = 8.1\ \text{J}$$

We can change this energy to units of calories as follows:

$$8.1\ \text{J} \times \frac{1\ \text{cal}}{4.184\ \text{J}} = 1.9\ \text{cal}$$

Self Check Exercise 10.4

Table 10.1 lists the specific heat capacities of several metals. We want to calculate the specific heat capacity (s) for this metal and then use Table 10.1 to identify the metal. Using the equation

$$Q = s \times m \times \Delta T$$

we can solve for s by dividing both sides by m (the mass of the sample) and by ΔT:

$$\frac{Q}{m \times \Delta T} = s$$

In this case,

Q = energy (heat) required = 10.1 J

m = 2.8 g

ΔT = temperature change = 36 °C − 21 °C = 15 °C

so

$$s = \frac{Q}{m \times \Delta T} = \frac{10.1\ \text{J}}{(2.8\ \text{g})(15\ °\text{C})} = 0.24\ \text{J/g}\ °\text{C}$$

Table 10.1 shows that silver has a specific heat capacity of 0.24 J/g °C. The metal is silver.

Self Check Exercise 10.5

We are told that 1652 kJ of energy is *released* when 4 moles of Fe reacts. We first need to determine what number of moles 1.00 g Fe represents.

$$1.00\ \text{g Fe} \times \frac{1\ \text{mol}}{55.85\ \text{g}} = 1.79 \times 10^{-2}\ \text{mol Fe}$$

$$1.79 \times 10^{-2}\ \text{mol Fe} \times \frac{1652\ \text{kJ}}{4\ \text{mol Fe}} = 7.39\ \text{kJ}$$

Thus 7.39 kJ of energy (as heat) is released when 1.00 g of iron reacts.

Self Check Exercise 10.6

Noting the reactants and products in the desired reaction

$$S(s) + O_2(g) \rightarrow SO_2(g)$$

We need to reverse the second equation and multiply it by $\frac{1}{2}$. This reverses the sign and cuts the amount of energy by a factor of 2.

$$\tfrac{1}{2}[2SO_3(g) \rightarrow 2SO_2(g) + O_2(g)] \qquad \Delta H = \frac{198.2\ \text{kJ}}{2}$$

or

$$SO_3(g) \rightarrow SO_2(g) + \tfrac{1}{2}O_2(g) \qquad \Delta H = 99.1\ \text{kJ}$$

Now we add this reaction to the first reaction.

$$S(s) + \tfrac{3}{2}O_2(g) \rightarrow SO_3(g) \qquad \Delta H = -395.2\ \text{kJ}$$

$$SO_3(g) \rightarrow SO_2(g) + \tfrac{1}{2}O_2(g) \qquad \Delta H = 99.1\ \text{kJ}$$

$$S(s) + O_2(g) \rightarrow SO_2(g) \qquad \Delta H = -296.1\ \text{kJ}$$

Chapter 11

Self Check Exercise 11.1

a. Circular pathways for electrons in the Bohr model.
b. Three-dimensional probability maps that represent the likelihood that the electron will occupy a given point in space.
c. The surface that contains 90% of the total electron probability.
d. A set of orbitals of a given type of orbital within a principal energy level. For example, there are three sublevels in principal energy level 3 (*s*, *p*, *d*).

Self Check Exercise 11.2

Element	Electron Configuration	Orbital Diagram 1s	2s	2p	3s	3p
Al	$1s^22s^22p^63s^23p^1$ $[Ne]3s^23p^1$	↑↓	↑↓	↑↓ ↑↓ ↑↓	↑↓	↑ _ _
Si	$[Ne]3s^23p^2$	↑↓	↑↓	↑↓ ↑↓ ↑↓	↑↓	↑ ↑ _
P	$[Ne]3s^23p^3$	↑↓	↑↓	↑↓ ↑↓ ↑↓	↑↓	↑ ↑ ↑
S	$[Ne]3s^23p^4$	↑↓	↑↓	↑↓ ↑↓ ↑↓	↑↓	↑↓ ↑ ↑
Cl	$[Ne]3s^23p^5$	↑↓	↑↓	↑↓ ↑↓ ↑↓	↑↓	↑↓ ↑↓ ↑
Ar	$[Ne]3s^23p^6$	↑↓	↑↓	↑↓ ↑↓ ↑↓	↑↓	↑↓ ↑↓ ↑↓

Self Check Exercise 11.3

F: $1s^22s^22p^5$ or $[He]2s^22p^5$
Si: $1s^22s^22p^63s^23p^2$ or $[Ne]3s^23p^2$
Cs: $1s^22s^22p^63s^23p^64s^23d^{10}4p^65s^24d^{10}5p^66s^1$ or $[Xe]6s^1$
Pb: $1s^22s^22p^63s^23p^64s^23d^{10}4p^65s^24d^{10}5p^66s^24f^{14}5d^{10}6p^2$ or $[Xe]6s^24f^{14}5d^{10}6p^2$
I: $1s^22s^22p^63s^23p^64s^23d^{10}4p^65s^24d^{10}5p^5$ or $[Kr]5s^24d^{10}5p^5$

Silicon (Si): In Group 4 and Period 3, it is the second of the "3*p* elements." The configuration is $1s^22s^22p^63s^23p^2$, or $[Ne]3s^23p^2$.
Cesium (Cs): In Group 1 and Period 6, it is the first of the "6*s* elements." The configuration is $1s^22s^22p^63s^23p^64s^23d^{10}4p^65s^24d^{10}5p^66s^1$, or $[Xe]6s^1$.
Lead (Pb): In Group 4 and Period 6, it is the second of the "6*p* elements." The configuration is $[Xe]6s^24f^{14}5d^{10}6p^2$.
Iodine (I): In Group 7 and Period 5, it is the fifth of the "5*p* elements." The configuration is $[Kr]5s^24d^{10}5p^5$.

Chapter 12

Self Check Exercise 12.1

Using the electronegativity values given in Fig. 12.3, we choose the bond in which the atoms exhibit the largest difference in electronegativity. (Electronegativity values are shown in parentheses.)

a.	H—C (2.1)(2.5)	>	H—P (2.1)(2.1)	c.	S—O (2.5)(3.5)	>	N—O (3.0)(3.5)
b.	O—I (3.5)(2.5)	>	O—F (3.5)(4.0)	d.	N—H (3.0)(2.1)	>	Si—H (1.8)(2.1)

Self Check Exercise 12.2

H has one electron, and Cl has seven valence electrons. This gives a total of eight valence electrons. We first draw in the bonding pair:

H—Cl, which could be drawn as H : Cl

We have six electrons yet to place. The H already has two electrons, so we place three lone pairs around the chlorine to satisfy the octet rule.

H—C̤̈l: or H:C̤̈l:

Self Check Exercise 12.3

Step 1 O_3: 3(6) = 18 valence electrons
Step 2 O—O—O
Step 3 Ö=Ö—Ö: and :Ö—Ö=Ö
This molecule shows resonance (it has two valid Lewis structures).

Self Check Exercise 12.4

See Table A.1 for the answer to Self Check Exercise 12.4.

Self Check Exercise 12.5

a. NH_4^+
The Lewis structure is [H—N(—H)(—H)—H]⁺

(See Self Check Exercise 12.4.) There are four pairs of electrons around the nitrogen. This requires a tetrahedral arrangement of electron pairs. The NH_4^+ ion has a tetrahedral molecular structure (row 3 in Table A.1), because all electron pairs are shared.

b. SO_4^{2-}
The Lewis structure is [:Ö—S(—Ö:)(—Ö:)—Ö:]²⁻

(See Self Check Exercise 12.4.) The four electron pairs around the sulfur require a tetrahedral arrangement. The SO_4^{2-} has a tetrahedral molecular structure (row 3 in Table A.1).

c. NF_3
The Lewis structure is :F̤̈—N̈(—F̤̈:)—F̤̈:

(See Self Check Exercise 12.4.) The four pairs of electrons on the nitrogen require a tetrahedral arrangement. In this case, only three of the pairs are shared with the fluorine atoms, leaving one lone pair. Thus the molecular structure is a trigonal pyramid (row 4 in Table A.1).

d. H_2S
The Lewis structure is H—S̤̈—H

(See Self Check Exercise 12.4.) The four pairs of electrons around the sulfur require a tetrahedral arrangement. In this case, two pairs are shared with hydrogen atoms, leaving two lone pairs. Thus the molecular structure is bent or V-shaped (row 5 in Table A.1).

e. ClO_3^-
The Lewis structure is [:Ö—C̤̈l(—Ö:)—Ö:]⁻

(See Self Check Exercise 12.4.) The four pairs of electrons require a tetrahedral arrangement. In this case, three pairs are shared with oxygen atoms, leaving one lone pair. Thus the molecular structure is a trigonal pyramid (row 4 in Table A.1).

Molecule or Ion	Total Valence Electrons	Draw Single Bonds	Calculate Number of Electrons Remaining	Use Remaining Electrons to Achieve Noble Gas Configurations	Check	
					Atoms	Electrons
a. NF_3	$5 + 3(7) = 26$	F—N(—F)—F	$26 - 6 = 20$	:F̤—N̈(—:F̤:)—F̤:	N F	8 8
b. O_2	$2(6) = 12$	O—O	$12 - 2 = 10$	Ö=Ö	O	8
c. CO	$4 + 6 = 10$	C—O	$10 - 2 = 8$	:C≡O:	C O	8 8
d. PH_3	$5 + 3(1) = 8$	H—P(—H)—H	$8 - 6 = 2$	H—P̈(—H)—H	P H	8 2
e. H_2S	$2(1) + 6 = 8$	H—S—H	$8 - 4 = 4$	H—S̤̈—H	S H	8 2
f. SO_4^{2-}	$6 + 4(6) + 2 = 32$	O—S(—O)(—O)—O	$32 - 8 = 24$	$[$:Ö—S(—:Ö:)(—:Ö:)—Ö:$]^{2-}$	S O	8 8
g. NH_4^+	$5 + 4(1) - 1 = 8$	H—N(—H)(—H)—H	$8 - 8 = 0$	$[$H—N(—H)(—H)—H$]^{+}$	N H	8 2
h. ClO_3^-	$7 + 3(6) + 1 = 26$	O—Cl(—O)—O	$26 - 6 = 20$	$[$:Ö—C̈l(—:Ö:)—Ö:$]^{-}$	Cl O	8 8
i. SO_2	$6 + 2(6) = 18$	O—S—O	$18 - 4 = 14$	Ö=S̈—Ö: and :Ö—S̈=Ö	S O	8 8

Table A.1 ▸ Answer to Self Check Exercise 12.4

f. BeF_2
The Lewis structure is :F̤—Be—F̤:
The two electron pairs on beryllium require a linear arrangement. Because both pairs are shared by fluorine atoms, the molecular structure is also linear (row 1 in Table A.1).

Chapter 13

Self Check Exercise 13.1

We know that 1.000 atm = 760.0 mm Hg. So

$$525 \text{ mm Hg} \times \frac{1.000 \text{ atm}}{760.0 \text{ mm Hg}} = 0.691 \text{ atm}$$

Self Check Exercise 13.2

Initial Conditions	Final Conditions
$P_1 = 635$ torr	$P_2 = 785$ torr
$V_1 = 1.51$ L	$V_2 = ?$

Solving Boyle's law ($P_1V_1 = P_2V_2$) for V_2 gives

$$V_2 = V_1 \times \frac{P_1}{P_2}$$

$$= 1.51 \text{ L} \times \frac{635 \text{ torr}}{785 \text{ torr}} = 1.22 \text{ L}$$

Note that the volume decreased, as the increase in pressure led us to expect.

Self Check Exercise 13.3

Because the temperature of the gas inside the bubble decreases (at constant pressure), the bubble gets smaller. The conditions are

Initial Conditions

$T_1 = 28\ °C = 28 + 273 = 301$ K

$V_1 = 23\ cm^3$

Final Conditions

$T_2 = 18\ °C = 18 + 273 = 291$ K

$V_2 = ?$

Solving Charles's law,

$$\frac{V_1}{T_1} = \frac{V_2}{T_2}$$

for V_2 gives

$$V_2 = V_1 \times \frac{T_2}{T_1} = 23 \text{ cm}^3 \times \frac{291 \text{ K}}{301 \text{ K}} = 22 \text{ cm}^3$$

Self Check Exercise 13.4

Because the temperature and pressure of the two samples are the same, we can use Avogadro's law in the form

$$\frac{V_1}{n_1} = \frac{V_2}{n_2}$$

The following information is given:

Sample 1	**Sample 2**
$V_1 = 36.7$ L	$V_2 = 16.5$ L
$n_1 = 1.5$ mol	$n_2 = ?$

We can now solve Avogadro's law for the value of n_2 (the moles of N_2 in sample 2):

$$n_2 = n_1 \times \frac{V_2}{V_1} = 1.5 \text{ mol} \times \frac{16.5 \text{ L}}{36.7 \text{ L}} = 0.67 \text{ mol}$$

Here n_2 is smaller than n_1, which makes sense in view of the fact that V_2 is smaller than V_1.

Note: We isolate n_2 from Avogadro's law as given above by multiplying both sides of the equation by n_2 and then by n_1/V_1,

$$\left(n_2 \times \frac{n_1}{V_1}\right)\frac{V_1}{n_1} = \left(n_2 \times \frac{n_1}{V_1}\right)\frac{V_2}{n_2}$$

to give $n_2 = n_1 \times V_2/V_1$.

Self Check Exercise 13.5

We are given the following information:

$$P = 1.00 \text{ atm}$$
$$V = 2.70 \times 10^6 \text{ L}$$
$$n = 1.10 \times 10^5 \text{ mol}$$

We solve for T by dividing both sides of the ideal gas law by nR:

$$\frac{PV}{nR} = \frac{nRT}{nR}$$

to give

$$T = \frac{PV}{nR} = \frac{(1.00 \text{ atm})(2.70 \times 10^6 \text{ L})}{(1.10 \times 10^5 \text{ mol})\left(0.08206 \dfrac{\text{L atm}}{\text{K mol}}\right)}$$

$$= 299 \text{ K}$$

The temperature of the helium is 299 K, or 299 − 273 = 26 °C.

Self Check Exercise 13.6

We are given the following information about the radon sample:

$$n = 1.5 \text{ mol}$$
$$V = 21.0 \text{ L}$$
$$T = 33 \text{ °C} = 33 + 273 = 306 \text{ K}$$
$$P = ?$$

We solve the ideal gas law ($PV = nRT$) for P by dividing both sides of the equation by V:

$$P = \frac{nRT}{V} = \frac{(1.5 \text{ mol})\left(0.08206 \dfrac{\text{L atm}}{\text{K mol}}\right)(306 \text{ K})}{21.0 \text{ L}}$$

$$= 1.8 \text{ atm}$$

Self Check Exercise 13.7

To solve this problem, we take the ideal gas law and separate those quantities that change from those that remain constant (on opposite sides of the equation). In this case, volume and temperature change, and number of moles and pressure (and, of course, R) remain constant. So $PV = nRT$ becomes $V/T = nR/P$, which leads to

$$\frac{V_1}{T_1} = \frac{nR}{P} \quad \text{and} \quad \frac{V_2}{T_2} = \frac{nR}{P}$$

Combining these gives

$$\frac{V_1}{T_1} = \frac{nR}{P} = \frac{V_2}{T_2} \quad \text{or} \quad \frac{V_1}{T_1} = \frac{V_2}{T_2}$$

We are given

Initial Conditions

$T_1 = 5 \text{ °C} = 5 + 273 = 278$ K

$V_1 = 3.8$ L

Final Conditions

$T_2 = 86 \text{ °C} = 86 + 273 = 359$ K

$V_2 = ?$

Thus

$$V_2 = \frac{T_2V_1}{T_1} = \frac{(359 \text{ K})(3.8 \text{ L})}{278 \text{ K}} = 4.9 \text{ L}$$

Check: Is the answer sensible? In this case, the temperature was increased (at constant pressure), so the volume should increase. The answer makes sense.

Note that this problem could be described as a "Charles's law problem." The real advantage of using the ideal gas law is that you need to remember only *one* equation to do virtually any problem involving gases.

Self Check Exercise 13.8

We are given the following information:

Initial Conditions

$P_1 = 0.747$ atm

$T_1 = 13 \text{ °C} = 13 + 273 = 286$ K

$V_1 = 11.0$ L

Final Conditions

$P_2 = 1.18$ atm

$T_2 = 56 \text{ °C} = 56 + 273 = 329$ K

$V_2 = ?$

In this case, the number of moles remains constant. Thus we can say

$$\frac{P_1V_1}{T_1} = nR \quad \text{and} \quad \frac{P_2V_2}{T_2} = nR$$

or

$$\frac{P_1V_1}{T_1} = \frac{P_2V_2}{T_2}$$

Solving for V_2 gives

$$V_2 = V_1 \times \frac{T_2}{T_1} \times \frac{P_1}{P_2} = (11.0\text{ L})\left(\frac{329\text{ K}}{286\text{ K}}\right)\left(\frac{0.747\text{ atm}}{1.18\text{ atm}}\right)$$
$$= 8.01\text{ L}$$

Self Check Exercise 13.9

As usual when dealing with gases, we can use the ideal gas equation $PV = nRT$. First consider the information given:

$$P = 0.91\text{ atm} = P_{\text{total}}$$
$$V = 2.0\text{ L}$$
$$T = 25\text{ °C} = 25 + 273 = 298\text{ K}$$

Given this information, we can calculate the number of moles of gas in the mixture: $n_{\text{total}} = n_{N_2} + n_{O_2}$. Solving for n in the ideal gas equation gives

$$n_{\text{total}} = \frac{P_{\text{total}}V}{RT} = \frac{(0.91\text{ atm})(2.0\text{ L})}{\left(0.08206\frac{\text{L atm}}{\text{K mol}}\right)(298\text{ K})} = 0.074\text{ mol}$$

We also know that 0.050 mole of N_2 is present. Because

$$n_{\text{total}} = \underset{\underset{(0.050\text{ mol})}{\uparrow}}{n_{N_2}} + n_{O_2} = 0.074\text{ mol}$$

we can calculate the moles of O_2 present.

$$0.050\text{ mol} + n_{O_2} = 0.074\text{ mol}$$
$$n_{O_2} = 0.074\text{ mol} - 0.050\text{ mol} = 0.024\text{ mol}$$

Now that we know the moles of oxygen present, we can calculate the partial pressure of oxygen from the ideal gas equation.

$$P_{O_2} = \frac{n_{O_2}RT}{V} = \frac{(0.024\text{ mol})\left(0.08206\frac{\text{L atm}}{\text{K mol}}\right)(298\text{ K})}{2.0\text{ L}}$$
$$= 0.29\text{ atm}$$

Although it is not requested, note that the partial pressure of the N_2 must be 0.62 atm, because

$$\underbrace{0.62\text{ atm}}_{P_{N_2}} + \underbrace{0.29\text{ atm}}_{P_{O_2}} = \underbrace{0.91\text{ atm}}_{P_{\text{total}}}$$

Self Check Exercise 13.10

The volume is 0.500 L, the temperature is 25 °C (or 25 + 273 = 298 K), and the total pressure is given as 0.950 atm. Of this total pressure, 24 torr is due to the water vapor. We can calculate the partial pressure of the H_2 because we know that

$$P_{\text{total}} = P_{H_2} + \underset{\underset{24\text{ torr}}{\uparrow}}{P_{H_2O}} = 0.950\text{ atm}$$

Before we carry out the calculation, however, we must convert the pressures to the same units. Converting P_{H_2O} to atmospheres gives

$$24\text{ torr} \times \frac{1.000\text{ atm}}{760.0\text{ torr}} = 0.032\text{ atm}$$

Thus

$$P_{\text{total}} = P_{H_2} + P_{H_2O} = 0.950\text{ atm} = P_{H_2} + 0.032\text{ atm}$$

and

$$P_{H_2} = 0.950\text{ atm} - 0.032\text{ atm} = 0.918\text{ atm}$$

Now that we know the partial pressure of the hydrogen gas, we can use the ideal gas equation to calculate the moles of H_2.

$$n_{H_2} = \frac{P_{H_2}V}{RT} = \frac{(0.918\text{ atm})(0.500\text{ L})}{\left(0.08206\frac{\text{L atm}}{\text{K mol}}\right)(298\text{ K})}$$
$$= 0.0188\text{ mol} = 1.88 \times 10^{-2}\text{ mol}$$

The sample of gas contains 1.88×10^{-2} mole of H_2, which exerts a partial pressure of 0.918 atm.

Self Check Exercise 13.11

We will solve this problem by taking the following steps:

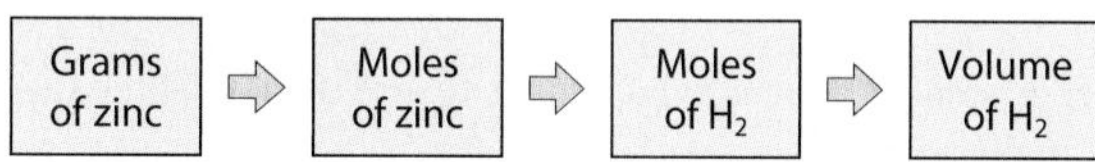

Step 1 Using the atomic mass of zinc (65.38), we calculate the moles of zinc in 26.5 g.

$$26.5\text{ g Zn} \times \frac{1\text{ mol Zn}}{65.38\text{ g Zn}} = 0.405\text{ mol Zn}$$

Step 2 Using the balanced equation, we next calculate the moles of H_2 produced.

$$0.405\text{ mol Zn} \times \frac{1\text{ mol H}_2}{1\text{ mol Zn}} = 0.405\text{ mol H}_2$$

Step 3 Now that we know the moles of H_2, we can compute the volume of H_2 by using the ideal gas law, where

$$P = 1.50\text{ atm}$$
$$V = ?$$
$$n = 0.405\text{ mol}$$
$$R = 0.08206\text{ L atm/K mol}$$
$$T = 19\text{ °C} = 19 + 273 = 292\text{ K}$$

$$V = \frac{nRT}{P} = \frac{(0.405\text{ mol})\left(0.08206\frac{\text{L atm}}{\text{K mol}}\right)(292\text{ K})}{1.50\text{ atm}}$$
$$= 6.47\text{ L of H}_2$$

Self Check Exercise 13.12

Although there are several possible ways to do this problem, the most convenient method involves using the molar volume at STP. First we use the ideal gas equation to calculate the moles of NH_3 present:

$$n = \frac{PV}{RT}$$

where $P = 15.0$ atm, $V = 5.00$ L, and $T = 25\ °C + 273 = 298$ K.

$$n = \frac{(15.0\ \cancel{\text{atm}})(5.00\ \cancel{\text{L}})}{\left(0.08206\ \frac{\cancel{\text{L}}\ \cancel{\text{atm}}}{\cancel{\text{K}}\ \text{mol}}\right)(298\ \cancel{\text{K}})} = 3.07\ \text{mol}$$

We know that at STP each mole of gas occupies 22.4 L. Therefore, 3.07 mol has the volume

$$3.07\ \cancel{\text{mol}} \times \frac{22.4\ \text{L}}{1\ \cancel{\text{mol}}} = 68.8\ \text{L}$$

The volume of the ammonia at STP is 68.8 L.

Chapter 14

Self Check Exercise 14.1

Energy to melt the ice:

$$15\ \cancel{\text{g H}_2\text{O}} \times \frac{1\ \text{mol H}_2\text{O}}{18\ \cancel{\text{g H}_2\text{O}}} = 0.83\ \text{mol H}_2\text{O}$$

$$0.83\ \cancel{\text{mol H}_2\text{O}} \times 6.02\ \frac{\text{kJ}}{\cancel{\text{mol H}_2\text{O}}} = 5.0\ \text{kJ}$$

Energy to heat the water from 0 °C to 100 °C:

$$4.18\ \frac{\text{J}}{\cancel{\text{g}}\ \cancel{°\text{C}}} \times 15\ \cancel{\text{g}} \times 100\ \cancel{°\text{C}} = 6300\ \text{J}$$

$$6300\ \cancel{\text{J}} \times \frac{1\ \text{kJ}}{1000\ \cancel{\text{J}}} = 6.3\ \text{kJ}$$

Energy to vaporize the water at 100 °C:

$$0.83\ \cancel{\text{mol H}_2\text{O}} \times 40.6\ \frac{\text{kJ}}{\cancel{\text{mol H}_2\text{O}}} = 34\ \text{kJ}$$

Total energy required:

$$5.0\ \text{kJ} + 6.3\ \text{kJ} + 34\ \text{kJ} = 45\ \text{kJ}$$

Self Check Exercise 14.2

a. Contains SO_3 molecules—a molecular solid.
b. Contains Ba^{2+} and O^{2-} ions—an ionic solid.
c. Contains Au atoms—an atomic solid.

Chapter 15

Self Check Exercise 15.1

$$\text{Mass percent} = \frac{\text{mass of solute}}{\text{mass of solution}} \times 100\%$$

For this sample, the mass of solution is 135 g and the mass of the solute is 4.73 g, so

$$\text{Mass percent} = \frac{4.73\ \text{g solute}}{135\ \text{g solution}} \times 100\%$$
$$= 3.50\%$$

Self Check Exercise 15.2

Using the definition of mass percent, we have

$$\frac{\text{Mass of solute}}{\text{Mass of solution}} = \frac{\text{grams of solute}}{\text{grams of solute} + \text{grams of solvent}} \times 100\% = 40.0\%$$

There are 425 grams of solute (formaldehyde). Substituting, we have

$$\frac{425\ \text{g}}{425\ \text{g} + \text{grams of solvent}} \times 100\% = 40.0\%$$

We must now solve for grams of solvent (water). This will take some patience, but we can do it if we proceed step by step. First we divide both sides by 100%.

$$\frac{425\ \text{g}}{425\ \text{g} + \text{grams of solvent}} \times \frac{\cancel{100\%}}{\cancel{100\%}} = \frac{40.0\%}{100\%} = 0.400$$

Now we have

$$\frac{425\ \text{g}}{425\ \text{g} + \text{grams of solvent}} = 0.400$$

Next we multiply both sides by (425 g + grams of solvent).

$$\cancel{(425\ \text{g} + \text{grams of solvent})} \times \frac{425\ \text{g}}{\cancel{425\ \text{g} + \text{grams of solvent}}}$$
$$= 0.400 \times (425\ \text{g} + \text{grams of solvent})$$

This gives

$$425\ \text{g} = 0.400 \times (425\ \text{g} + \text{grams of solvent})$$

Carrying out the multiplication gives

$$425\ \text{g} = 170.\ \text{g} + 0.400\ (\text{grams of solvent})$$

Now we subtract 170. g from both sides,

$$425\ \text{g} - 170.\ \text{g} = \cancel{170.\ \text{g}} - \cancel{170.\ \text{g}} + 0.400\ (\text{grams of solvent})$$
$$255\ \text{g} = 0.400\ (\text{grams of solvent})$$

and divide both sides by 0.400.

$$\frac{255\ \text{g}}{0.400} = \frac{\cancel{0.400}}{\cancel{0.400}}(\text{grams of solvent})$$

We finally have the answer:

$$\frac{255\ \text{g}}{0.400} = 638\ \text{g} = \text{grams of solvent}$$
$$= \text{mass of water needed}$$

Self Check Exercise 15.3

The moles of ethanol can be obtained from its molar mass (46.1).

$$1.00\ \cancel{\text{g C}_2\text{H}_5\text{OH}} \times \frac{1\ \text{mol C}_2\text{H}_5\text{OH}}{46.1\ \cancel{\text{g C}_2\text{H}_5\text{OH}}} = 2.17 \times 10^{-2}\ \text{mol C}_2\text{H}_5\text{OH}$$

$$\text{Volume in liters} = 101\ \cancel{\text{mL}} \times \frac{1\ \text{L}}{1000\ \cancel{\text{mL}}} = 0.101\ \text{L}$$

$$\text{Molarity of C}_2\text{H}_5\text{OH} = \frac{\text{moles of C}_2\text{H}_5\text{OH}}{\text{liters of solution}}$$
$$= \frac{2.17 \times 10^{-2}\ \text{mol}}{0.101\ \text{L}}$$
$$= 0.215\ M$$

Self Check Exercise 15.4

When Na_2CO_3 and $Al_2(SO_4)_3$ dissolve in water, they produce ions as follows:

$$Na_2CO_3(s) \xrightarrow{H_2O(l)} 2Na^+(aq) + CO_3^{2-}(aq)$$

$$Al_2(SO_4)_3(s) \xrightarrow{H_2O(l)} 2Al^{3+}(aq) + 3SO_4^{2-}(aq)$$

Therefore, in a 0.10 M Na_2CO_3 solution, the concentration of Na^+ ions is $2 \times 0.10\ M = 0.20\ M$ and the concentration of CO_3^{2-} ions is 0.10 M. In a 0.010 M $Al_2(SO_4)_3$ solution, the concentration of Al^{3+} ions is $2 \times 0.010\ M = 0.020\ M$ and the concentration of SO_4^{2-} ions is $3 \times 0.010\ M = 0.030\ M$.

Self Check Exercise 15.5

When solid $AlCl_3$ dissolves, it produces ions as follows:

$$AlCl_3(s) \xrightarrow{H_2O(l)} Al^{3+}(aq) + 3Cl^-(aq)$$

so a $1.0 \times 10^{-3}\ M$ $AlCl_3$ solution contains $1.0 \times 10^{-3}\ M$ Al^{3+} ions and $3.0 \times 10^{-3}\ M$ Cl^- ions.

To calculate the moles of Cl^- ions in 1.75 L of the $1.0 \times 10^{-3}\ M$ $AlCl_3$ solution, we must multiply the volume by the molarity.

$$1.75 \text{ L solution} \times 3.0 \times 10^{-3}\ M\ Cl^-$$

$$= 1.75 \text{ L solution} \times \frac{3.0 \times 10^{-3} \text{ mol } Cl^-}{\text{L solution}}$$

$$= 5.25 \times 10^{-3} \text{ mol } Cl^- = 5.3 \times 10^{-3} \text{ mol } Cl^-$$

Self Check Exercise 15.6

We must first determine the number of moles of formaldehyde in 2.5 L of 12.3 M formalin. Remember that volume of solution (in liters) times molarity gives moles of solute. In this case, the volume of solution is 2.5 L and the molarity is 12.3 moles of HCHO per liter of solution.

$$2.5 \text{ L solution} \times \frac{12.3 \text{ mol HCHO}}{\text{L solution}} = 31 \text{ mol HCHO}$$

Next, using the molar mass of HCHO (30.0 g), we convert 31 moles of HCHO to grams.

$$31 \text{ mol HCHO} \times \frac{30.0 \text{ g HCHO}}{1 \text{ mol HCHO}} = 9.3 \times 10^2 \text{ g HCHO}$$

Therefore, 2.5 L of 12.3 M formalin contains 9.3×10^2 g of formaldehyde. We must weigh out 930 g of formaldehyde and dissolve it in enough water to make 2.5 L of solution.

Self Check Exercise 15.7

We are given the following information:

$$M_1 = 12 \frac{\text{mol}}{\text{L}} \qquad M_2 = 0.25 \frac{\text{mol}}{\text{L}}$$

$$V_1 = ? \text{ (what we need to find)} \qquad V_2 = 0.75 \text{ L}$$

Using the fact that the moles of solute do not change upon dilution, we know that

$$M_1 \times V_1 = M_2 \times V_2$$

Solving for V_1 by dividing both sides by M_1 gives

$$V_1 = \frac{M_2 \times V_2}{M_1} = \frac{0.25 \frac{\text{mol}}{\text{L}} \times 0.75 \text{ L}}{12 \frac{\text{mol}}{\text{L}}}$$

and

$$V_1 = 0.016 \text{ L} = 16 \text{ mL}$$

Self Check Exercise 15.8

Step 1 When the aqueous solutions of Na_2SO_4 (containing Na^+ and SO_4^{2-} ions) and $Pb(NO_3)_2$ (containing Pb^{2+} and NO_3^- ions) are mixed, solid $PbSO_4$ is formed.

$$Pb^{2+}(aq) + SO_4^{2-}(aq) \rightarrow PbSO_4(s)$$

Step 2 We must first determine whether Pb^{2+} or SO_4^{2-} is the limiting reactant by calculating the moles of Pb^{2+} and SO_4^{2-} ions present. Because 0.0500 M $Pb(NO_3)_2$ contains 0.0500 M Pb^{2+} ions, we can calculate the moles of Pb^{2+} ions in 1.25 L of this solution as follows:

$$1.25 \text{ L} \times \frac{0.0500 \text{ mol } Pb^{2+}}{\text{L}} = 0.0625 \text{ mol } Pb^{2+}$$

The 0.0250 M Na_2SO_4 solution contains 0.0250 M SO_4^{2-} ions, and the number of moles of SO_4^{2-} ions in 2.00 L of this solution is

$$2.00 \text{ L} \times \frac{0.0250 \text{ mol } SO_4^{2-}}{\text{L}} = 0.0500 \text{ mol } SO_4^{2-}$$

Step 3 Pb^{2+} and SO_4^{2-} react in a 1:1 ratio, so the amount of SO_4^{2-} ions is limiting because SO_4^{2-} is present in the smaller number of moles.

Step 4 The Pb^{2+} ions are present in excess, and only 0.0500 mole of solid $PbSO_4$ will be formed.

Step 5 We calculate the mass of $PbSO_4$ by using the molar mass of $PbSO_4$ (303.3 g).

$$0.0500 \text{ mol } PbSO_4 \times \frac{303.3 \text{ g } PbSO_4}{1 \text{ mol } PbSO_4} = 15.2 \text{ g } PbSO_4$$

Self Check Exercise 15.9

Step 1 Because nitric acid is a strong acid, the nitric acid solution contains H^+ and NO_3^- ions. The KOH solution contains K^+ and OH^- ions. When these solutions are mixed, the H^+ and OH^- react to form water.

$$H^+(aq) + OH^-(aq) \rightarrow H_2O(l)$$

Step 2 The number of moles of OH^- present in 125 mL of 0.050 M KOH is

$$125 \text{ mL} \times \frac{1 \text{ L}}{1000 \text{ mL}} \times \frac{0.050 \text{ mol } OH^-}{\text{L}} = 6.3 \times 10^{-3} \text{ mol } OH^-$$

Step 3 H^+ and OH^- react in a 1:1 ratio, so we need 6.3×10^{-3} mole of H^+ from the 0.100 M HNO_3.

Step 4 6.3×10^{-3} mole of OH^- requires 6.3×10^{-3} mole of H^+ to form 6.3×10^{-3} mole of H_2O.
Therefore,

$$V \times \frac{0.100 \text{ mol } H^+}{\text{L}} = 6.3 \times 10^{-3} \text{ mol } H^+$$

where V represents the volume in liters of 0.100 M HNO_3 required. Solving for V, we have

$$V = \frac{6.3 \times 10^{-3} \cancel{\text{mol H}^+}}{\dfrac{0.100 \cancel{\text{mol H}^+}}{\text{L}}} = 6.3 \times 10^{-2}\ \text{L}$$

$$= 6.3 \times 10^{-2}\ \cancel{\text{L}} \times \frac{1000\ \text{mL}}{\cancel{\text{L}}} = 63\ \text{mL}$$

Self Check Exercise 15.10

From the definition of normality, N = equiv/L, we need to calculate (1) the equivalents of KOH and (2) the volume of the solution in liters. To find the number of equivalents, we use the equivalent weight of KOH, which is 56.1 g (see Table 15.2).

$$23.6\ \cancel{\text{g KOH}} \times \frac{1\ \text{equiv KOH}}{56.1\ \cancel{\text{g KOH}}} = 0.421\ \text{equiv KOH}$$

Next we convert the volume to liters.

$$755\ \cancel{\text{mL}} \times \frac{1\ \text{L}}{1000\ \cancel{\text{mL}}} = 0.755\ \text{L}$$

Finally, we substitute these values into the equation that defines normality.

$$\text{Normality} = \frac{\text{equiv}}{\text{L}} = \frac{0.421\ \text{equiv}}{0.755\ \text{L}} = 0.558\ N$$

Self Check Exercise 15.11

To solve this problem, we use the relationship

$$N_{\text{acid}} \times V_{\text{acid}} = N_{\text{base}} \times V_{\text{base}}$$

where

$$N_{\text{acid}} = 0.50\ \frac{\text{equiv}}{\text{L}}$$

$$V_{\text{acid}} = ?$$

$$N_{\text{base}} = 0.80\ \frac{\text{equiv}}{\text{L}}$$

$$V_{\text{base}} = 0.250\ \text{L}$$

We solve the equation

$$N_{\text{acid}} \times V_{\text{acid}} = N_{\text{base}} \times V_{\text{base}}$$

for V_{acid} by dividing both sides by N_{acid}.

$$\frac{N_{\text{acid}} \times V_{\text{acid}}}{N_{\text{acid}}} = \frac{N_{\text{base}} \times V_{\text{base}}}{N_{\text{acid}}}$$

$$V_{\text{acid}} = \frac{N_{\text{base}} \times V_{\text{base}}}{N_{\text{acid}}} = \frac{\left(0.80\ \dfrac{\cancel{\text{equiv}}}{\cancel{\text{L}}}\right) \times (0.250\ \text{L})}{0.50\ \dfrac{\cancel{\text{equiv}}}{\cancel{\text{L}}}}$$

$$V_{\text{acid}} = 0.40\ \text{L}$$

Therefore, 0.40 L of 0.50 N H_2SO_4 is required to neutralize 0.250 L of 0.80 N KOH.

Chapter 16

Self Check Exercise 16.1

The conjugate acid–base pairs are

H_2O,	H_3O^+
Base	Conjugate acid

and

$HC_2H_3O_2$,	$C_2H_3O_2^-$
Acid	Conjugate base

The members of both pairs differ by one H^+.

Self Check Exercise 16.2

Because $[H^+][OH^-] = 1.0 \times 10^{-14}$, we can solve for $[H^+]$.

$$[H^-] = \frac{1.0 \times 10^{-14}}{[OH^-]} = \frac{1.0 \times 10^{-14}}{2.0 \times 10^{-2}} = 5.0 \times 10^{-13}\ M$$

This solution is basic: $[OH^-] = 2.0 \times 10^{-2}\ M$ is greater than $[H^+] = 5.0 \times 10^{-13}\ M$.

Self Check Exercise 16.3

a. Because $[H^+] = 1.0 \times 10^{-3}\ M$, we get pH = 3.00 because $\text{pH} = -\log[H^+] = -\log[1.0 \times 10^{-3}] = 3.00$.
b. Because $[OH^-] = 5.0 \times 10^{-5}\ M$, we can find $[H^+]$ from the K_w expression.

$$[H^+] = \frac{K_w}{[OH^-]} = \frac{1.0 \times 10^{-14}}{5.0 \times 10^{-5}} = 2.0 \times 10^{-10}\ M$$

$$\text{pH} = -\log[H^+] = -\log[2.0 \times 10^{-10}] = 9.70$$

Self Check Exercise 16.4

$$\text{pOH} + \text{pH} = 14.00$$
$$\text{pOH} = 14.00 - \text{pH} = 14.00 - 3.5$$
$$\text{pOH} = 10.5$$

Self Check Exercise 16.5

Step 1 pH = 3.50
Step 2 −pH = −3.50
Step 3 [inv] [log] $-3.50 = 3.2 \times 10^{-4}$
$[H^+] = 3.2 \times 10^{-4}\ M$

Self Check Exercise 16.6

Step 1 pOH = 10.50
Step 2 −pOH = −10.50
Step 3 [inv] [log] $-10.50 = 3.2 \times 10^{-11}$
$[OH^-] = 3.2 \times 10^{-11}\ M$

Self Check Exercise 16.7

Because HCl is a strong acid, it is completely dissociated:

$$5.0 \times 10^{-3}\ M\ \text{HCl} \rightarrow 5.0 \times 10^{-3}\ M\ H^+ \text{ and } 5.0 \times 10^{-3}\ M\ Cl^-$$

so $[H^+] = 5.0 \times 10^{-3}\ M$.

$$\text{pH} = -\log(5.0 \times 10^{-3}) = 2.30$$

Answers to Even-Numbered End-of-Chapter Questions and Exercises

Chapter 1

2. The answer depends on the student's experiences.
4. Answers will depend on the student's responses.
6. Answers will depend on the student's choices.
8. Recognize the problem and state it clearly; propose possible solutions or explanations; decide which solution/explanation is best through experiments.
10. Answers will depend on student responses. A quantitative observation must include a number, such as "There are three windows in this room." A qualitative observation could include something like "The chair is blue."
12. False. Theories can be refined and changed because they are interpretations. They represent possible explanations of why nature behaves in a particular way. Theories are refined by performing experiments and making new observations, not by proving the existing observations as false (which is something that can be witnessed and recorded).
14. When a scientist formulates a hypothesis, he or she wants it to be proved correct. Financial success and prestige are dependent on publishing and producing such ideas.
16. Chemistry is the study of very real interactions among different samples of matter. When we first begin to study chemistry, we try to be as general and nonspecific as possible, trying to learn the basic principles for application to many situations in the future. At the beginning of study, the solution to a problem is not as important as learning how to recognize and interpret the problem, and how to propose reasonable, testable hypotheses.
18. A good student will learn the background and fundamentals of the subject from the classes and textbook; will develop the ability to recognize and solve problems and to extend what was learned in the classroom to "real" situations; will learn to make careful observations; and will be able to communicate effectively. Whereas some academic subjects may emphasize use of one or more of these skills, chemistry makes extensive use of all of them.

Chapter 2

2. "Scientific notation" means we have to put the decimal point after the first significant figure, and then express the order of magnitude of the number as a power of 10. So we want to put the decimal point after the first 2:

$$2421 \rightarrow 2.421 \times 10^{\text{to some power}}$$

To be able to move the decimal point three places to the left in going from 2421 to 2.421 means you will need a power of 10^3 after the number, where the exponent 3 shows that you moved the decimal point three places to the left:

$$2421 \rightarrow 2.421 \times 10^{\text{to some power}} = 2.421 \times 10^3$$

4. (a) 10^6; (b) 10^{-2}; (c) 10^{-4}; (d) 10^9
6. (a) negative; (b) zero; (c) negative; (d) positive
8. (a) 2789; (b) 0.002789; (c) 93,000,000; (d) 42.89; (e) 99,990; (f) 0.00009999
10. (a) three places to the left; (b) one place to the left; (c) five places to the right; (d) one place to the left; (e) two places to the right; (f) two places to the left
12. (a) 6244; (b) 0.09117; (c) 82.99; (d) 0.0001771; (e) 545.1; (f) 0.00002934
14. (a) 3.1×10^3; (b) 1×10^6; (c) 1 or 1×10^0; (d) 1.8×10^{-5}; (e) 1×10^7; (f) 1.00×10^6; (g) 1.00×10^{-7}; (h) 1×10^1
16. (a) kilo; (b) milli; (c) nano; (d) mega; (e) deci; (f) micro
18. about ¼ pound **20.** about an inch **22.** 161 km
24. the woman **26.** (a) inch; (b) yard; (c) mile **28.** b
30. Typically we read the scale on measuring devices to 0.1 unit of the smallest scale division on the device. We estimate this final significant figure, which makes the final significant figure in the measurement uncertain.
32. The scale of the ruler is marked to the nearest tenth of a centimeter. Writing 2.850 would imply that the scale was marked to the nearest hundredth of a centimeter (and that the zero in the thousandths place had been estimated).
34. (a) probably only two; (b) infinite (definition); (c) infinite (definition); (d) probably one; (e) three (the race is defined to be 500. miles)
36. It is better to round off only the final answer and to carry through extra digits in intermediate calculations. If there are enough steps to the calculation, rounding off in each step may lead to a cumulative error in the final answer.
38. (a) 1.57×10^6; (b) 2.77×10^{-3}; (c) 7.76×10^{-2}; (d) 1.17×10^{-3}
40. (a) 3.42×10^{-4}; (b) 1.034×10^4; (c) 1.7992×10^1; (d) 3.37×10^5
42. 170. mL; 18 mL limits the precision to the ones place.
44. three; b, c, and d; a contains two significant figures
46. none
48. (a) 2.3; (b) 9.1×10^2; (c) 1.323×10^3; (d) 6.63×10^{-13}
50. (a) one; (b) four; (c) two; (d) three
52. (a) 2.045; (b) 3.8×10^3; (c) 5.19×10^{-5}; (d) 3.8418×10^{-7}
54. an infinite number, a definition
56. $\frac{5280 \text{ ft}}{1 \text{ mi}}; \frac{1 \text{ mi}}{5280 \text{ ft}}$ **58.** $\frac{1 \text{ lb}}{\$1.75}$
60. (a) 2.44 yd; (b) 42.2 m; (c) 115 in; (d) 2238 cm; (e) 648.1 mi; (f) 716.9 km; (g) 0.0362 km; (h) 5.01×10^4 cm
62. (a) 0.2543 kg; (b) 2.75×10^3 g; (c) 6.06 lb; (d) 97.0 oz; (e) 1.177 lb; (f) 794 g; (g) 2.5×10^2 g; (h) 1.62 oz
64. 4117 km **66.** 1×10^{-8} cm; 4×10^{-9} in.; 0.1 nm
68. freezing/melting **70.** 373
72. Fahrenheit (F)
74. (a) 144 K; (b) 72 K; (c) 664 °F; (d) −101 °C
76. (a) 173 °F; (b) 104 °F; (c) −459 °F; (d) 90. °F
78. (a) 2 °C; (b) 28 °C; (c) −5.8 °F (−6 °F); (d) −40 °C (−40 is where both temperature scales have the same value)
80. g/cm^3 (g/mL) **82.** 100 $in.^3$
84. Density is a characteristic property of a pure substance.
86. silver
88. (a) 20.1 g/cm^3; (b) 1.05 g/cm^3; (c) 0.907 g/cm^3; (d) 1.30 g/cm^3
90. 4140 g; 0.408 L **92.** float **94.** 14.6 mL
96. (a) 966 g; (b) 394 g; (c) 567 g; (d) 135 g
98. (a) 301,100,000,000,000,000,000,000; (b) 5,091,000,000; (c) 720; (d) 123,400; (e) 0.000432002; (f) 0.03001; (g) 0.00000029901; (h) 0.42

100. e

102. (a) 5.07×10^4 kryll; (b) 0.12 blim; (c) 3.70×10^{-5} blim^2

104. $\frac{45 \text{ mi}}{\text{hr}} \times \frac{1.61 \text{ km}}{1 \text{ mi}} \times \frac{1000 \text{ m}}{1 \text{ km}} \times \frac{1 \text{ hr}}{3600 \text{ s}} = 20.$ m/s

106. Because \$1 = 1.44 euros and 1 kg = 2.2 lb, the peaches will cost \$0.87/lb.

108. °X = 1.26 °C + 14

110. 3.50 g/L (3.50×10^{-3} g/cm^3) **112.** 959 g

114. (a) negative; (b) negative; (c) positive; (d) zero; (e) negative

116. (a) 2, positive; (b) 11, negative; (c) 3, positive; (d) 5, negative; (e) 5, positive; (f) 0, zero; (g) 1, negative; (h) 7, negative

118. (a) 1, positive; (b) 3, negative; (c) 0, zero; (d) 3, positive; (e) 9, negative

120. (a) 0.0000298; (b) 4,358,000,000; (c) 0.0000019928; (d) 602,000,000,000,000,000,000,000; (e) 0.101; (f) 0.00787; (g) 98,700,000; (h) 378.99; (i) 0.1093; (j) 2.9004; (k) 0.00039; (l) 0.00000001904

122. (a) 1×10^{-2}; (b) 1×10^2; (c) 5.5×10^{-2}; (d) 3.1×10^9; (e) 1×10^3; (f) 1×10^8; (g) 2.9×10^2; (h) 3.453×10^4

124. The student's answer depends on the glassware used. Two possible answers are:

Using a buret where 0 mL starts at the top:

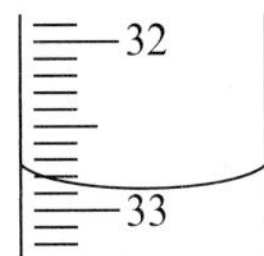

Using a graduated cylinder where 0 mL starts at the bottom:

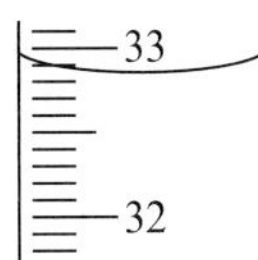

126. 1 L = 1 dm^3 = 1000 cm^3 = 1000 mL **128.** 0.105 m

130. They weigh the same. **132.** 5×10^{11} nm **134.** 0.830 m

136. (a) 0.000426; (b) 4.02×10^{-5}; (c) 5.99×10^6; (d) 400.; (e) 0.00600

138. (a) 2149.6; (b) 5.37×10^3; (c) 3.83×10^{-2}; (d) -8.64×10^5

140. (a) 7.6166×10^6; (b) 7.24×10^3; (c) 1.92×10^{-5}; (d) 2.4482×10^{-3}

142. $\frac{1 \text{ yr}}{12 \text{ mo}}; \frac{12 \text{ mo}}{1 \text{ yr}}$

144. (a) 25.7 kg; (b) 3.38 gal; (c) 0.132 qt; (d) 1.09×10^4 mL; (e) 2.03×10^3 g; (f) 0.58 qt

146. 2.4 metric tons

148. (a) 352 K; (b) −18 °C; (c) −43 °C; (d) 257 °F

150. 78.2 g **152.** 0.59 g/cm^3

154. (a) 23 °F; (b) 32 °F; (c) −321 °F; (d) −459 °F; (e) 187 °F; (f) −459 °F

156. (a) 100 km $\times \frac{1 \text{ mile}}{1.6093 \text{ km}}$ = 62 miles, or about 60 miles, taking significant figures into account.

(b) 22,300 kg $\times \frac{2.2046 \text{ lbs}}{1 \text{ kg}}$ = 49,200 lbs. of fuel was needed; 22,300 lbs. were added, so 26,900 additional pounds were needed.

158. $\frac{10^{-8} \text{ g}}{\text{L}} \times \frac{3.7854 \text{ L}}{1 \text{ gallon}} \times \frac{1 \text{ lb.}}{453.59 \text{ g}} \approx 8 \times 10^{-11}$ lb/gal

160. 2, 0.51; 3, 29.1; 3, 8.61; 3, 1.89; 4, 134.6; 3, 14.4

162. 16.9 m/s

164. liquid

166. 24 g/cm^3

Chapter 3

2. forces among the particles in the matter

4. solids **6.** gaseous **8.** stronger

10. Because gases are mostly empty space, they can be *compressed* easily to smaller volumes. In solids and liquids, most of the sample's bulk volume is filled with the molecules, leaving little empty space.

12. chemical change **14.** malleable; ductile **16.** d

18. (a) physical; (b) chemical; (c) chemical; (d) chemical; (e) physical; (f) physical; (g) chemical; (h) physical; (i) physical; (j) physical; (k) chemical

20. Compounds consist of two or more elements combined together chemically in a fixed composition, no matter what their source may be. For example, water on Earth consists of molecules containing one oxygen atom and two hydrogen atoms. Water on Mars (or any other planet) has the same composition.

22. compounds **24.** He, F_2, S_8

26. no; heating causes a reaction to form iron(II) sulfide, a pure substance

28. Heterogeneous mixtures: salad dressing, jelly beans, the change in my pocket; solutions: window cleaner, shampoo, rubbing alcohol

30. (a) mixture; (b) mixture; (c) mixture; (d) pure substance

32. Concrete is a mixture. It consists of sand, gravel, water, and cement (which consists of limestone, clay, shale, and gypsum). The composition of concrete can vary.

34. Consider a mixture of salt (sodium chloride) and sand. Salt is soluble in water; sand is not. The mixture is added to water and stirred to dissolve the salt, and is then filtered. The salt solution passes through the filter; the sand remains on the filter. The water can then be evaporated from the salt.

36. Each component of the mixture retains its own identity during the separation.

38. (a) compound, pure substance; (b) element, pure substance; (c) homogeneous mixture

40. physical

42. False; no reaction has taken place. The substances are merely separating, not changing into different substances. This is an example of a heterogeneous mixture.

44. b **46.** physical **48.** pure substance; compound; element

50. (a) heterogeneous; (b) homogeneous; (c) heterogeneous; (d) heterogeneous; (e) homogeneous

52. Answers depend on student responses.

54. false

56. O_2 and P_4 are both still elements, even though the ordinary forms of these elements consist of molecules containing more than one atom (but all atoms in each respective molecule are the same). P_2O_5 is a compound, because it is made up of two or more different elements (not all the atoms in the P_2O_5 molecule are the same).

58. Assuming there is enough water present in the mixture to have dissolved all the salt, filter the mixture to separate out the sand from the mixture. Then distill the filtrate (consisting of salt and water), which will boil off the water, leaving the salt.

60. The most obvious difference is the physical states: water is a liquid under room conditions, hydrogen and oxygen are both gases. Hydrogen is flammable. Oxygen supports combustion. Water does neither.

62. (a) false; (b) false; (c) true; (d) false; (e) true

64. a, d

Chapter 4

2. Robert Boyle **4.** oxygen; carbon; hydrogen

6. (a) Trace elements are elements that are present in tiny amounts. Trace elements in the body, while present in small amounts, are essential. (b) Answers will vary. For example, chromium assists in the metabolism of sugars and cobalt is present in vitamin B_{12}.

8. Answer depends on student choices/examples.

10. (a) copper; (b) cobalt; (c) calcium; (d) carbon; (e) chromium; (f) cesium; (g) chlorine; (h) cadmium
12. silicon; Ni; silver; K; calcium
14. B: barium, Ba; berkelium, Bk; beryllium, Be; bismuth, Bi; bohrium, Bh; boron, B; bromine, Br
N: neodymium, Nd; neon, Ne; neptunium, Np; nickel, Ni; niobium, Nb; nitrogen, N; nobelium, No
P: palladium, Pd; phosphorus, P; platinum, Pt; plutonium, Pu; polonium, Po; potassium, K; praseodymium, Pr; promethium, Pm; protactinium, Pa
S: samarium, Sm; scandium, Sc; seaborgium, Sg; selenium, Se; silicon, Si; silver, Ag; sodium, Na; strontium, Sr; sulfur, S
16. (a) Elements are made of tiny particles called atoms; (b) All atoms of a given element are identical; (c) The atoms of a given element are different from those of any other element; (d) A given compound always has the same numbers and types of atoms; (e) Atoms are neither created nor destroyed in chemical processes. A chemical reaction simply changes the way the atoms are grouped together.
18. According to Dalton, all atoms of the same element are *identical;* in particular, every atom of a given element has the same *mass* as every other atom of that element. If a given compound always contains the *same relative numbers* of atoms of each kind, and those atoms always have the same *masses,* then the compound made from those elements always contains the same relative masses of its elements.
20. (a) CO_2; (b) CO; (c) $CaCO_3$; (d) H_2SO_4; (e) $BaCl_2$; (f) Al_2S_3
22. False; Rutherford's bombardment experiments with metal foil suggested that the alpha particles were being deflected by coming near a *dense, positively charged* atomic nucleus.
24. protons **26.** neutron; electron **28.** electrons
30. False; the mass number represents the total number of protons and neutrons in the nucleus.
32. Neutrons are uncharged and contribute only to the mass.
34. Atoms of the same element (atoms with the same number of protons in the nucleus) may have different numbers of neutrons, and so will have different masses.
36.

Z	Symbol	Name
32	Ge	germanium
30	Zn	zinc
24	Cr	chromium
74	W	tungsten
38	Sr	strontium
27	Co	cobalt
4	Be	beryllium

38. (a) ${}^{54}_{26}Fe$; (b) ${}^{56}_{26}Fe$; (c) ${}^{57}_{26}Fe$; (d) ${}^{14}_{7}N$; (e) ${}^{15}_{7}N$; (f) ${}^{15}_{7}N$
40. Researchers have found that the concentrations of hydrogen-2 (deuterium) and oxygen-18 in drinking water vary significantly from region to region in the United States. By collecting hair samples around the country, they have also found that 86% of the variations in the hair samples' hydrogen and oxygen isotopes result from the isotopic composition of the local water.
42.

Name	Symbol	Atomic Number	Mass Number	Number of Neutrons
oxygen	${}^{17}_{8}O$	8	17	9
oxygen	${}^{17}_{8}O$	8	17	9
neon	${}^{20}_{10}Ne$	10	20	10
iron	${}^{56}_{26}Fe$	26	56	30
plutonium	${}^{244}_{94}Pu$	94	244	150
mercury	${}^{202}_{80}Hg$	80	202	122
cobalt	${}^{59}_{27}Co$	27	59	32
nickel	${}^{56}_{28}Ni$	28	56	28
fluorine	${}^{19}_{9}F$	9	19	10
chromium	${}^{50}_{24}Cr$	24	50	26

44. vertical; groups **46.** true
48. nonmetallic gaseous elements: oxygen, nitrogen, fluorine, chlorine, hydrogen, and the noble gases; There are no metallic gaseous elements at room conditions
50. metalloids or semimetals
52. (a) fluorine, chlorine, bromine, iodine, astatine; (b) lithium, sodium, potassium, rubidium, cesium, francium; (c) beryllium, magnesium, calcium, strontium, barium, radium; (d) helium, neon, argon, krypton, xenon, radon
54. Arsenic is a metalloid. Other elements in the same group (5A) include nitrogen (N), phosphorus (P), antinomy (Sb), and bismuth (Bi).
56. Most elements are too reactive to be found in the uncombined form in nature and are found only in compounds.
58. These elements are found uncombined in nature and do not readily react with other elements. Although these elements were once thought to form no compounds, this now has been shown to be untrue.
60. diatomic gases: H_2, N_2, O_2, F_2, Cl_2; monatomic gases: He, Ne, Kr, Xe, Rn, Ar
62. chlorine **64.** carbon **66.** electrons **68.** 2− **70.** *-ide*
72. False; N^{3-} contains 7 protons and 10 electrons; P^{3-} contains 15 protons and 18 electrons.
74. number of protons = 26; number of electrons = 23; number of neutrons = 30
76. (a) two electrons gained; (b) three electrons gained; (c) three electrons lost; (d) two electrons lost; (e) one electron lost; (f) two electrons lost
78. (a) P^{3-}; (b) Ra^{2+}; (c) At^-; (d) no ion; (e) Cs^+; (f) Se^{2-}
80. Sodium chloride is an ionic compound, consisting of Na^+ and Cl^- ions. When NaCl is dissolved in water, these ions are set free and can move independently to conduct the electric current.
82. The total number of positive charges must equal the total number of negative charges so that the crystals of an ionic compound have *no net charge.* A macroscopic sample of compound ordinarily has no net charge.
84. (a) CsI, BaI_2, AlI_3; (b) Cs_2O, BaO, Al_2O_3; (c) Cs_3P, Ba_3P_2, AlP; (d) Cs_2Se, BaSe, Al_2Se_3; (e) CsH, BaH_2, AlH_3
86. (a) 7, halogens; (b) 8, noble gases; (c) 2, alkaline earth elements; (d) 2, alkaline earth elements; (e) 4; (f) 6; (g) 8, noble gases; (h) 1, alkali metals
88. d
90. Most of an atom's mass is concentrated in the nucleus: the *protons* and *neutrons* that constitute the nucleus have similar masses and are each nearly 2000 times more massive than electrons. The chemical properties of an atom depend on the number and location of the *electrons* it possesses. Electrons are found in the outer regions of the atom and are involved in interactions between atoms.
92. $C_6H_{12}O_6$
94. (a) 29 electrons, 34 neutrons, 29 electrons; (b) 35 protons, 45 neutrons, 35 electrons; (c) 12 protons, 12 neutrons, 12 electrons
96. The chief use of gold in ancient times was as *ornamentation,* whether in statuary or in jewelry. Gold possesses an especially beautiful luster; since it is relatively soft and malleable, it can be worked finely by artisans. Among the metals, gold is inert to attack by most substances in the environment.
98. ${}^{81}_{35}Br^-$ **100.** a, b, c, d **102.** Cu^{2+}; mass number = 63
104. (a) CO_2; (b) $AlCl_3$; (c) $HClO_4$; (d) SCl_6
106. (a) ${}^{13}_{6}C$; (b) ${}^{13}_{6}C$; (c) ${}^{13}_{6}C$; (d) ${}^{44}_{19}K$; (e) ${}^{41}_{20}Ca$; (f) ${}^{35}_{19}K$
108.

Symbol	Number of Protons	Number of Neutrons	Mass Number
${}^{41}_{20}Ca$	20	21	41
${}^{55}_{25}Mn$	25	30	55
${}^{109}_{47}Ag$	47	62	109
${}^{45}_{21}Sc$	21	24	45

110. Cu-63: 29 protons, 29 electrons, 34 neutrons, ${}^{63}_{29}Cu$
Cu-65: 29 protons, 29 electrons, 36 neutrons, ${}^{65}_{29}Cu$

112. Sn; Be; H; Cl; Ra; Xe; Zn; O

114.

Atom	G or L	Ion
O	G	O^{2-}
Mg	L	Mg^{2+}
Rb	L	Rb^+
Br	G	Br^-
Cl	G	Cl^-

116.

#p	#n	#e
50	70	50
12	13	10
26	30	24
34	45	34
17	18	17
29	34	29

Chapter 5

2. A binary chemical compound contains only two elements; the major types are ionic (compounds of a metal and a nonmetal) and nonionic or molecular (compounds between two nonmetals). Answers depend on student responses.

4. cation (positive ion)

6. Sodium chloride consists of Na^+ ions and Cl^- ions in an extended crystal lattice array. No discrete NaCl pairs are present.

8. Roman numeral

10. (a) lithium iodide; (b) magnesium fluoride; (c) strontium oxide; (d) aluminum bromide; (e) calcium sulfide; (f) sodium oxide

12. (a) correct; (b) incorrect, silver oxide; (c) incorrect, lithium oxide; (d) correct; (e) incorrect, cesium sulfide

14. (a) iron(III) iodide; (b) manganese(II) chloride; (c) mercury(II) oxide; (d) copper(I) sulfide; (e) cobalt(II) oxide; (f) tin(IV) bromide

16. (a) cobaltous chloride; (b) chromic bromide; (c) plumbous oxide; (d) stannic oxide; (e) cobaltic oxide; (f) ferric chloride

18. (a) germanium tetrahydride; (b) dinitrogen tetrabromide; (c) diphosphorus pentoxide; (d) carbon dioxide; (e) ammonia; (f) silicon dioxide

20. Na_2O: sodium oxide; N_2O: dinitrogen monoxide. For Na_2O, the compound contains a metal and a nonmetal in which the charges must balance. When forming this compound, Na always forms a 1+ charge and oxygen always forms a 2− charge. Therefore, the prefixes are not needed. For N_2O, the compound contains only nonmetals and the charges do not have to balance. Therefore, prefixes are needed to tell us how many of each atom are present.

22. (a) radium chloride, ionic; (b) selenium dichloride, nonionic; (c) phosphorus trichloride, nonionic; (d) sodium phosphide, ionic; (e) manganese(II) fluoride, ionic; (f) zinc oxide, ionic

24. An oxyanion is a polyatomic ion containing a given element and one or more oxygen atoms. The oxyanions of chlorine and bromine are given below:

Oxyanion	Name	Oxyanion	Name
ClO^-	hypochlorite	BrO^-	hypobromite
ClO_2^-	chlorite	BrO_2^-	bromite
ClO_3^-	chlorate	BrO_3^-	bromate
ClO_4^-	perchlorate	BrO_4^-	perbromate

26. *hypo-* (fewest); *per-* (most)

28. IO^-, hypoiodite; IO_2^-, iodite; IO_3^-, iodate; IO_4^-, periodate

30. (a) Cl^-; (b) ClO^-; (c) ClO_3^-; (d) ClO_4^-

32. CN^-, cyanide; CO_3^{2-}, carbonate; HCO_3^-, hydrogen carbonate; $C_2H_3O_2^-$, acetate

34. (a) ammonium ion; (b) dihydrogen phosphate ion; (c) sulfate ion; (d) hydrogen sulfite ion (bisulfite ion); (e) perchlorate ion; (f) iodate ion

36. (a) sodium permanganate; (b) aluminum phosphate; (c) chromium(II) carbonate, chromous carbonate; (d) calcium hypochlorite; (e) barium carbonate; (f) calcium chromate

38. oxygen

40. (a) hypochlorous acid; (b) sulfurous acid; (c) bromic acid; (d) hypoiodous acid; (e) perbromic acid; (f) hydrosulfuric acid; (g) hydroselenic acid; (h) phosphorous acid

42. (a) MgF_2; (b) FeI_3; (c) HgS; (d) Ba_3N_2; (e) $PbCl_2$; (f) SnF_4; (g) Ag_2O; (h) K_2Se

44. (a) N_2O; (b) NO_2; (c) N_2O_4; (d) SF_6; (e) PBr_3; (f) CI_4; (g) OCl_2

46. (a) $NH_4C_2H_3O_2$; (b) $Fe(OH)_2$; (c) $Co_2(CO_3)_3$; (d) $BaCr_2O_7$; (e) $PbSO_4$; (f) KH_2PO_4; (g) Li_2O_2; (h) $Zn(ClO_3)_2$

48. (a) HCN; (b) HNO_3; (c) H_2SO_4; (d) H_3PO_4; (e) HClO or HOCl; (f) HBr; (g) $HBrO_2$; (h) HF

50. (a) $Ca(HSO_4)_2$; (b) $Zn_3(PO_4)_2$; (c) $Fe(ClO_4)_3$; (d) $Co(OH)_3$; (e) K_2CrO_4; (f) $Al(H_2PO_4)_3$; (g) $LiHCO_3$; (h) $Mn(C_2H_3O_2)_2$; (i) $MgHPO_4$; (j) $CsClO_2$; (k) BaO_2; (l) $NiCO_3$

52. A moist paste of NaCl would contain Na^+ and Cl^- ions in solution and would serve as a *conductor* of electrical impulses.

54. $H \rightarrow H^+$ (hydrogen ion) + e^-; $H + e^- \rightarrow H^-$ (hydride ion)

56. ClO_4^-, $HClO_4$; IO_3^-, HIO_3; ClO^-, HClO; BrO_2^-, $HBrO_2$; ClO_2^-, $HClO_2$

58. (a) gold(III) bromide (auric bromide); (b) cobalt(III) cyanide (cobaltic cyanide); (c) magnesium hydrogen phosphate; (d) diboron hexahydride (common name diborane); (e) ammonia; (f) silver(I) sulfate (usually called silver sulfate); (g) beryllium hydroxide

60. b

62. (a) $M(C_2H_3O_2)_2$; (b) $M(MnO_4)_2$; (c) MO; (d) $MHPO_4$; (e) $M(OH)_2$; (f) $M(NO_2)_2$

64. (a) Mn^{2+}; (b) Cl^-; 18 electrons; (c) $MnCl_2$; manganese(II) chloride

66.

$Ca(NO_3)_2$	$CaSO_4$	$Ca(HSO_4)_2$	$Ca(H_2PO_4)_2$	CaO	$CaCl_2$
$Sr(NO_3)_2$	$SrSO_4$	$Sr(HSO_4)_2$	$Sr(H_2PO_4)_2$	SrO	$SrCl_2$
NH_4NO_3	$(NH_4)_2SO_4$	NH_4HSO_4	$NH_4H_2PO_4$	$(NH_4)_2O$	NH_4Cl
$Al(NO_3)_3$	$Al_2(SO_4)_3$	$Al(HSO_4)_3$	$Al(H_2PO_4)_3$	Al_2O_3	$AlCl_3$
$Fe(NO_3)_3$	$Fe_2(SO_4)_3$	$Fe(HSO_4)_3$	$Fe(H_2PO_4)_3$	Fe_2O_3	$FeCl_3$
$Ni(NO_3)_2$	$NiSO_4$	$Ni(HSO_4)_2$	$Ni(H_2PO_4)_2$	NiO	$NiCl_2$
$AgNO_3$	Ag_2SO_4	$AgHSO_4$	AgH_2PO_4	Ag_2O	AgCl
$Au(NO_3)_3$	$Au_2(SO_4)_3$	$Au(HSO_4)_3$	$Au(H_2PO_4)_3$	Au_2O_3	$AuCl_3$
KNO_3	K_2SO_4	$KHSO_4$	KH_2PO_4	K_2O	KCl
$Hg(NO_3)_2$	$HgSO_4$	$Hg(HSO_4)_2$	$Hg(H_2PO_4)_2$	HgO	$HgCl_2$
$Ba(NO_3)_2$	$BaSO_4$	$Ba(HSO_4)_2$	$Ba(H_2PO_4)_2$	BaO	$BaCl_2$

68. $(NH_4)_3PO_4$ **70.** F_2, Cl_2 (gas); Br_2 (liquid); I_2, At_2 (solid)

72. 1+ **74.** 2−

76. (a) $Al(13e) \rightarrow Al^{3+}(10e) + 3e^-$; (b) $S(16e) + 2e^- \rightarrow S^{2-}(18e)$; (c) $Cu(29e) \rightarrow Cu^+(28e) + e^-$; (d) $F(9e) + e^- \rightarrow F^-(10e)$; (e) $Zn(30e) \rightarrow Zn^{2+}(28e) + 2e^-$; (f) $P(15e) + 3e^- \rightarrow P^{3-}(18e)$

78. (a) Na_2S; (b) KCl; (c) BaO; (d) MgSe; (e) $CuBr_2$; (f) AlI_3; (g) Al_2O_3; (h) Ca_3N_2

80. (a) silver(I) oxide or just silver oxide; (b) correct; (c) iron(III) oxide; (d) plumbic oxide; (e) correct

82. (a) stannous chloride; (b) ferrous oxide; (c) stannic oxide; (d) plumbous sulfide; (e) cobaltic sulfide; (f) chromous chloride

84. (a) iron(III) acetate; (b) bromine monofluoride; (c) potassium peroxide; (d) silicon tetrabromide; (e) copper(II) permanganate; (f) calcium chromate

86. a

88. (a) carbonate; (b) chlorate; (c) sulfate; (d) phosphate; (e) perchlorate; (f) permanganate

90. $SrBr_2$

92. (a) NaH_2PO_4; (b) $LiClO_4$; (c) $Cu(HCO_3)_2$; (d) $KC_2H_3O_2$; (e) BaO_2; (f) Cs_2SO_3

94.

Atom	G or L	Ion
K	L	K^+
Cs	L	Cs^+
Br	G	Br^-
S	G	S^{2-}
Se	G	Se^{2-}

96. cobalt(II) nitrite; arsenic pentafluoride; lithium cyanide; potassium sulfite; lithium nitride; lead(II) chromate

98. b, d

Chapter 6

2. Most of these products contain a peroxide, which decomposes and releases oxygen gas.

4. Bubbling takes place as the hydrogen peroxide chemically decomposes into water and oxygen gas.

6. The appearance of the black color actually signals the breakdown of starches and sugars in the bread to elemental carbon. You may also see steam coming from the bread (water produced by the breakdown of the carbohydrates).

8. (a) N_2, H_2; the reactants are on the left side of the arrow. (b) NH_3; the products are on the right side of the arrow.

10. Balancing an equation ensures that no atoms are created or destroyed during the reaction. The total mass after the reaction must be the same as the total mass before the reaction.

12. gaseous **14.** $CaCO_3(s) \rightarrow CaO(s) + CO_2(g)$

16. $N_2H_4(l) \rightarrow N_2(g) + H_2(g)$ **18.** $Ag_2O(s) \rightarrow Ag(s) + O_2(g)$

20. $CaCO_3(s) + HCl(aq) \rightarrow CaCl_2(aq) + H_2O(l) + CO_2(g)$

22. $SiO_2(s) + C(s) \rightarrow Si(s) + CO(g)$

24. $Zn(s) + HCl(aq) \rightarrow H_2(g) + ZnCl_2(aq)$

26. $SO_2(g) + H_2O(l) \rightarrow H_2SO_3(aq)$; $SO_3(g) + H_2O(l) \rightarrow H_2SO_4(aq)$

28. $NO(g) + O_3(g) \rightarrow NO_2(g) + O_2(g)$

30. $P_4(s) + O_2(g) \rightarrow P_2O_5(s)$ **32.** $Xe(g) + F_2(g) \rightarrow XeF_4(s)$

34. $NH_3(g) + O_2(g) \rightarrow HNO_3(aq) + H_2O(l)$

36. To balance a chemical equation we must have the same number of each type of atom on both sides of the equation. In addition, we must balance the equation we are given, that is, we are not to change the nature of the substances.

For example, the equation $2H_2O_2(aq) \rightarrow 2H_2O(l) + O_2(g)$ can be represented as

The equation $H_2O_2(aq) \rightarrow H_2(g) + O_2(g)$ can be represented as

38. $2K(s) + 2H_2O(l) \rightarrow H_2(g) + 2KOH(aq)$

40. (a) $Na_2SO_4(aq) + CaCl_2(aq) \rightarrow CaSO_4(s) + 2NaCl(aq)$;
(b) $3Fe(s) + 4H_2O(g) \rightarrow Fe_3O_4(s) + 4H_2(g)$;
(c) $Ca(OH)_2(aq) + 2HCl(aq) \rightarrow CaCl_2(aq) + 2H_2O(l)$;
(d) $Br_2(g) + 2H_2O(l) + SO_2(g) \rightarrow 2HBr(aq) + H_2SO_4(aq)$;
(e) $3NaOH(s) + H_3PO_4(aq) \rightarrow Na_3PO_4(aq) + 3H_2O(l)$;
(f) $2NaNO_3(s) \rightarrow 2NaNO_2(s) + O_2(g)$;
(g) $2Na_2O_2(s) + 2H_2O(l) \rightarrow 4NaOH(aq) + O_2(g)$;
(h) $4Si(s) + S_8(s) \rightarrow 2Si_2S_4(s)$

42. (a) $4NaCl(s) + 2SO_2(g) + 2H_2O(g) + O_2(g) \rightarrow 2Na_2SO_4(s) + 4HCl(g)$; (b) $3Br_2(l) + I_2(s) \rightarrow 2IBr_3(s)$;
(c) $Ca(s) + 2H_2O(g) \rightarrow Ca(OH)_2(aq) + H_2(g)$;
(d) $2BF_3(g) + 3H_2O(g) \rightarrow B_2O_3(s) + 6HF(g)$;
(e) $SO_2(g) + 2Cl_2(g) \rightarrow SOCl_2(l) + Cl_2O(g)$;
(f) $Li_2O(s) + H_2O(l) \rightarrow 2LiOH(aq)$;
(g) $Mg(s) + CuO(s) \rightarrow MgO(s) + Cu(l)$;
(h) $Fe_3O_4(s) + 4H_2(g) \rightarrow 3Fe(l) + 4H_2O(g)$

44. (a) $Ba(NO_3)_2(aq) + Na_2CrO_4(aq) \rightarrow BaCrO_4(s) + 2NaNO_3(aq)$;
(b) $PbCl_2(aq) + K_2SO_4(aq) \rightarrow PbSO_4(s) + 2KCl(aq)$;
(c) $C_2H_5OH(l) + 3O_2(g) \rightarrow 2CO_2(g) + 3H_2O(l)$;
(d) $CaC_2(s) + 2H_2O(l) \rightarrow Ca(OH)_2(s) + C_2H_2(g)$;
(e) $Sr(s) + 2HNO_3(aq) \rightarrow Sr(NO_3)_2(aq) + H_2(g)$;
(f) $BaO_2(s) + H_2SO_4(aq) \rightarrow BaSO_4(s) + H_2O_2(aq)$;
(g) $2AsI_3(s) \rightarrow 2As(s) + 3I_2(s)$; (h) $2CuSO_4(aq) + 4KI(s) \rightarrow 2CuI(s) + I_2(s) + 2K_2SO_4(aq)$

46. a **48.** whole numbers

50. $2Al_2O_3(s) + 3C(s) \rightarrow 4Al(s) + 3CO_2(g)$

52. True; coefficients can be fractions when balancing a chemical equation because the coefficients represent a ratio of the moles needed for the reaction to occur. As a result, moles can be fractions, because they represent an amount. The key is to make sure the atoms are conserved from reactants to products. Take note that the accepted convention is that the "best" balanced equation is the one with the smallest integers (although not required).

54. $BaO_2(s) + H_2O(l) \rightarrow BaO(s) + H_2O_2(aq)$

56. $2KClO_3(s) \rightarrow 2KCl(s) + 3O_2(g)$

58. $4FeO(s) + O_2(g) \rightarrow 2Fe_2O_3(s)$

60. $3LiAlH_4(s) + AlCl_3(s) \rightarrow 4AlH_3(s) + 3LiCl(s)$

62. $Fe(s) + S(s) \rightarrow FeS(s)$

64. $K_2CrO_4(aq) + BaCl_2(aq) \rightarrow BaCrO_4(s) + 2KCl(aq)$

66. $2NaCl(aq) + 2H_2O(l) \rightarrow Cl_2(g) + H_2(g) + 2NaOH(aq, s)$
$2NaBr(aq) + 2H_2O(l) \rightarrow Br_2(l) + H_2(g) + 2NaOH(aq, s)$
$2NaI(aq) + 2H_2O(l) \rightarrow I_2(s) + H_2(g) + 2NaOH(aq, s)$

68. e **70.** $CuO(s) + H_2SO_4(aq) \rightarrow CuSO_4(aq) + H_2O(l)$

72. a **74.** $2Na_2S_2O_3(aq) + I_2(aq) \rightarrow Na_2S_4O_6(aq) + 2NaI(aq)$

76. Answers will vary, but the following balanced equation should be reported: $4NH_3(g) + 3O_2(g) \rightarrow 2N_2(g) + 6H_2O(g)$

78. $4Fe(s) + 3O_2(g) \rightarrow 2Fe_2O_3(s)$; $2PbO_2(s) \rightarrow 2PbO(s) + O_2(g)$; $2H_2O_2(l) \rightarrow O_2(g) + 2H_2O(l)$

Chapter 7

2. Driving forces are types of *changes* in a system that pull a reaction in the *direction of product formation;* driving forces include formation of a *solid,* formation of *water,* formation of a *gas,* and transfer of electrons.

4. A reactant in aqueous solution is indicated with (*aq*); formation of a solid is indicated with (*s*).

6. There are twice as many chloride ions as magnesium ions.

8. The simplest evidence is that solutions of ionic substances conduct electricity.

10. a

12. (a) soluble, Rule 1; (b) soluble, Rule 2; (c) insoluble, Rule 4; (d) insoluble, Rule 5; (e) soluble, Rule 2; (f) insoluble, Rule 3; (g) soluble, Rule 2; (h) insoluble, Rule 6

14. (a) Rule 5; (b) Rule 6; (c) Rule 6; (d) Rule 3; (e) Rule 4

16. (a) $MnCO_3$, Rule 6; (b) $CaSO_4$, Rule 4; (c) Hg_2Cl_2, Rule 3; (d) no precipitate, most sodium and nitrate salts are soluble; (e) $Ni(OH)_2$, Rule 5; (f) $BaSO_4$, Rule 4

18. (a) $Na_2CO_3(aq) + CuSO_4(aq) \rightarrow Na_2SO_4(aq) + \underline{CuCO_3(s)}$
(b) $HCl(aq) + AgC_2H_3O_2(aq) \rightarrow HC_2H_3O_2(aq) + \underline{AgCl(s)}$
(c) no precipitate
(d) $3(NH_4)_2S(aq) + 2FeCl_3(aq) \rightarrow 6NH_4Cl(aq) + \underline{Fe_2S_3(s)}$
(e) $H_2SO_4(aq) + Pb(NO_3)_2(aq) \rightarrow 2HNO_3(aq) + \underline{PbSO_4(s)}$
(f) $2K_3PO_4(aq) + 3CaCl_2(aq) \rightarrow 6KCl(aq) + \underline{Ca_3(PO_4)_2(s)}$

20. (a) $CaCl_2(aq) + 2AgNO_3(aq) \rightarrow Ca(NO_3)_2(aq) + 2AgCl(s)$;
(b) $2AgNO_3(aq) + K_2CrO_4(aq) \rightarrow Ag_2CrO_4(s) + 2KNO_3(aq)$;
(c) $BaCl_2(aq) + K_2SO_4(aq) \rightarrow BaSO_4(s) + 2KCl(aq)$

22. The precipitate is lead(II) phosphate. The balanced equation is: $2Na_3PO_4(aq) + 3Pb(NO_3)_2(aq) \rightarrow Pb_3(PO_4)_2(s) + 6NaNO_3(aq)$

24. e

26. molecular: $K_2SO_4(aq) + Pb(NO_3)_2(aq) \rightarrow 2KNO_3(aq) + PbSO_4(s)$
complete ionic: $2K^+(aq) + SO_4^{2-}(aq) + Pb^{2+}(aq) + 2NO_3^-(aq) \rightarrow 2K^+(aq) + 2NO_3^-(aq) + PbSO_4(s)$
net ionic: $SO_4^{2-}(aq) + Pb^{2+}(aq) \rightarrow PbSO_4(s)$

28. $Ag^+(aq) + Cl^-(aq) \rightarrow AgCl(s)$; $Pb^{2+}(aq) + 2Cl^-(aq) \rightarrow PbCl_2(s)$; $Hg_2^{2+}(aq) + 2Cl^-(aq) \rightarrow Hg_2Cl_2(s)$

30. $Co^{2+}(aq) + S^{2-}(aq) \rightarrow CoS(s)$; $2Co^{3+}(aq) + 3S^{2-}(aq) \rightarrow Co_2S_3(s)$; $Fe^{2+}(aq) + S^{2-}(aq) \rightarrow FeS(s)$; $2Fe^{3+}(aq) + 3S^{2-}(aq) \rightarrow Fe_2S_3(s)$

32. The strong bases are those hydroxide compounds that dissociate fully when dissolved in water. The strong bases that are highly soluble in water (NaOH, KOH) are also strong electrolytes.

34. acids: HCl (hydrochloric), HNO_3 (nitric), H_2SO_4 (sulfuric); bases: hydroxides of Group 1A elements: NaOH, KOH, RbOH, CsOH

36. A salt is the ionic product remaining in solution when an acid neutralizes a base. For example, in the reaction $HCl(aq) + NaOH(aq) \rightarrow NaCl(aq) + H_2O(l)$, sodium chloride is the salt produced by the neutralization reaction.

38. $HBr(aq) \rightarrow H^+(aq) + Br^-(aq)$; $HClO_4(aq) \rightarrow H^+(aq) + ClO_4^-(aq)$

40. (a) $H_2SO_4(aq) + 2KOH(aq) \rightarrow K_2SO_4(aq) + 2H_2O(l)$
(b) $HNO_3(aq) + NaOH(aq) \rightarrow NaNO_3(aq) + H_2O(l)$
(c) $2HCl(aq) + Ca(OH)_2(aq) \rightarrow CaCl_2(aq) + 2H_2O(l)$
(d) $2HClO_4(aq) + Ba(OH)_2(aq) \rightarrow Ba(ClO_4)_2(aq) + 2H_2O(l)$

42. Answer depends on student choice of example: $2Na(s) + Cl_2(g) \rightarrow 2NaCl(s)$ is an example.

44. The aluminum atoms lose three electrons to become Al^{3+} ions; Fe^{3+} ions gain three electrons to become Fe atoms.

46. Each magnesium atom would lose two electrons. Each oxygen atom would gain two electrons (so the O_2 molecule would gain four electrons). Two magnesium atoms would be required to react with each O_2 molecule. Magnesium ions are charged 2+, oxide ions are charged 2−.

48. $AlBr_3$ is made up of Al^{3+} ions and Br^- ions. Aluminum atoms each lose three electrons, and bromine atoms each gain one electron (Br_2 gains two electrons).

50. (a) $P_4(s) + 5O_2(g) \rightarrow P_4O_{10}(s)$;
(b) $MgO(s) + C(s) \rightarrow Mg(s) + CO(g)$;
(c) $Sr(s) + 2H_2O(l) \rightarrow Sr(OH)_2(aq) + H_2(g)$;
(d) $Co(s) + 2HCl(aq) \rightarrow CoCl_2(aq) + H_2(g)$

52. The reaction includes aluminum metal as a reactant and products that contain aluminum ions. For this reaction, electrons must be transferred. That is, to make an aluminum cation, electrons must be removed from the metal. An oxidation reduction is one that involves transfers of electrons.

54. (a) oxidation–reduction; (b) oxidation–reduction; (c) acid–base; (d) acid–base, precipitation; (e) precipitation; (f) precipitation; (g) oxidation–reduction; (h) oxidation–reduction; (i) acid–base

56. oxidation–reduction

58. A decomposition reaction is one in which a given compound is broken down into simpler compounds or constituent elements. The reactions $CaCO_3(s) \rightarrow CaO(s) + CO_2(g)$ and $2HgO(s) \rightarrow 2Hg(l) + O_2(g)$ represent decomposition reactions. Such reactions often may be classified in other ways. For example, the reaction of $HgO(s)$ is also an oxidation–reduction reaction.

60. (a) $C_3H_8(g) + 5O_2(g) \rightarrow 3CO_2(g) + 4H_2O(g)$;
(b) $C_2H_4(g) + 3O_2(g) \rightarrow 2CO_2(g) + 2H_2O(g)$;
(c) $2C_8H_{18}(l) + 25O_2(g) \rightarrow 16CO_2(g) + 18H_2O(g)$

62. Answer depends on student selection.

64. (a) $8Fe(s) + S_8(s) \rightarrow 8FeS(s)$; (b) $4Co(s) + 3O_2(g) \rightarrow 2Co_2O_3(s)$; (c) $Cl_2O_7(g) + H_2O(l) \rightarrow 2HClO_4(aq)$

66. (a) $2Al(s) + 3Br_2(l) \rightarrow 2AlBr_3(s)$
(b) $Zn(s) + 2HClO_4(aq) \rightarrow Zn(ClO_4)_2(aq) + H_2(g)$
(c) $3Na(s) + P(s) \rightarrow Na_3P(s)$
(d) $CH_4(g) + 4Cl_2(g) \rightarrow CCl_4(l) + 4HCl(g)$
(e) $Cu(s) + 2AgNO_3(aq) \rightarrow Cu(NO_3)_2(aq) + 2Ag(s)$

68. c only

70. (a) $HNO_3(aq) + KOH(aq) \rightarrow H_2O(l) + \underline{KNO_3}(aq)$;
(b) $H_2SO_4(aq) + Ba(OH)_2(aq) \rightarrow \underline{BaSO_4}(s) + 2H_2O(l)$;
(c) $HClO_4(aq) + NaOH(aq) \rightarrow H_2O(l) + \underline{NaClO_4}(aq)$;
(d) $2HCl(aq) + Ca(OH)_2(aq) \rightarrow \underline{CaCl_2}(aq) + H_2O(l)$

72. (a) $2AgNO_3(aq) + H_2SO_4(aq) \rightarrow Ag_2SO_4(s) + 2HNO_3(aq)$;
(b) $Ca(NO_3)_2(aq) + H_2SO_4(aq) \rightarrow CaSO_4(s) + 2HNO_3(aq)$;
(c) $Pb(NO_3)_2(aq) + H_2SO_4(aq) \rightarrow PbSO_4(s) + 2HNO_3(aq)$

74. (a) $H_2SO_4(aq) + 2NaOH(aq) \rightarrow 2H_2O(l) + Na_2SO_4(aq)$;
(b) $HNO_3(aq) + RbOH(aq) \rightarrow H_2O(l) + RbNO_3(aq)$;
(c) $HClO_4(aq) + KOH(aq) \rightarrow 2H_2O(l) + KClO_4(aq)$;
(d) $HCl(aq) + KOH(aq) \rightarrow H_2O(l) + KCl(aq)$

76. molecular: $Na_2SO_4(aq) + CaCl_2(aq) \rightarrow CaSO_4(s) + 2NaCl(aq)$
complete ionic: $2Na^+(aq) + SO_4^{2-}(aq) + Ca^{2+}(aq) + 2Cl^-(aq) \rightarrow CaSO_4(s) + 2Na^+(aq) + 2Cl^-(aq)$
net ionic: $SO_4^{2-}(aq) + Ca^{2+}(aq) \rightarrow CaSO_4(s)$

78. Aluminum atoms lose three electrons to become Al^{3+} ions; iodine atoms gain one electron each to become I^- ions.

80. (a) $2Na(s) + O_2(g) \rightarrow Na_2O_2(s)$;
(b) $Fe(s) + H_2SO_4(aq) \rightarrow FeSO_4(aq) + H_2(g)$;
(c) $2Al_2O_3(s) \rightarrow 4Al(s) + 3O_2(g)$;
(d) $2Fe(s) + 3Br_2(l) \rightarrow 2FeBr_3(s)$;
(e) $Zn(s) + 2HNO_3(aq) \rightarrow Zn(NO_3)_2(aq) + H_2(g)$

82. a, b, c

84. (a) $2NaHCO_3(s) \rightarrow Na_2CO_3(s) + H_2O(g) + CO_2(g)$;
(b) $2NaClO_3(s) \rightarrow 2NaCl(s) + 3O_2(g)$;
(c) $2HgO(s) \rightarrow 2Hg(l) + O_2(g)$;
(d) $C_{12}H_{22}O_{11}(s) \rightarrow 12C(s) + 11H_2O(g)$;
(e) $2H_2O_2(l) \rightarrow 2H_2O(l) + O_2(g)$

86. $6Na + N_2 \rightarrow 2Na_3N$

88. (a) one; (b) one; (c) two; (d) two; (e) three

90. False; the balanced molecular equation is: $Ba(OH)_2(aq) + H_2SO_4(aq) \rightarrow BaSO_4(s) + 2H_2O(l)$. The complete ionic equation is: $Ba^{2+}(aq) + 2OH^-(aq) + 2H^+(aq) + SO_4^{2-}(aq) \rightarrow BaSO_4(s) + 2H_2O(l)$. The net ionic equation includes all species that take part in the chemical reaction. The OH^- and H^+ ions form water, so they are also included in the net ionic equation. Thus the complete ionic equation and net ionic equation are the same.

92. $3Na_2CrO_4(aq) + 2AlBr_3(aq) \rightarrow Al_2(CrO_4)_3(s) + 6NaBr(aq)$

94. $PbCl_2$; $PbSO_4$; $Pb_3(PO_4)_2$; AgCl; Ag_3PO_4

96. $PbSO_4$; AgCl; none

Chapter 8

2. The empirical formula is the lowest whole-number ratio of atoms in the compound. The graphic of PVDF shows four of each type of atom (carbon, hydrogen, and fluorine), so the empirical formula is CHF.

4. The average atomic mass takes into account the various isotopes of an element and the relative abundances in which those isotopes are found.

6. (a) one; (b) five; (c) ten; (d) 50; (e) ten

8. A sample containing 54 Br atoms would weigh 4315 amu; 5672.9 amu of Br would represent 71 Br atoms.

10. 118.7 **12.** 1.674×10^{24} atoms Ag, 176.7 g Cu **14.** 177 g

16. 5.90×10^{-23} g **18.** 2.0 moles of carbon

20. (a) 1.53 mol S; (b) 3.59×10^5 mol Pb; (c) 9.22×10^{-5} mol Cl; (d) 0.578 mol Li; (e) 1.574 mol Cu; (f) 9.43×10^{-4} mol Sr

22. (a) 0.221 g; (b) 0.0676 g; (c) 3.64×10^3 g; (d) 1.84×10^{-5} g; (e) 86.4 g; (f) 7.47×10^{-3} g

24. (a) 1.16×10^{-20} g; (b) 6.98×10^3 amu; (c) 2.24 mol; (d) 6.98×10^3 g; (e) 1.35×10^{24} atoms; (f) 7.53×10^{25} atoms

26. The molar mass is calculated by summing the individual atomic masses of the atoms in the formula. In the compound CH_4, the atomic mass of carbon and the atomic mass of four hydrogens are summed (giving a molar mass of 16.042 g/mol).

28. (a) potassium hydrogen carbonate; 100.12 g; (b) mercury(I) chloride, mercurous chloride; 472.1 g; (c) hydrogen peroxide; 34.02 g; (d) beryllium chloride; 79.91 g; (e) aluminum sulfate; 342.2 g; (f) potassium chlorate; 122.55 g
30. (a) $Ba(ClO_4)_2$, 336.2 g; (b) $MgSO_4$, 120.38 g; (c) $PbCl_2$, 278.1 g; (d) $Cu(NO_3)_2$, 187.57 g; (e) $SnCl_4$, 260.5 g
32. (a) 0.463 mol; (b) 11.3 mol; (c) 7.18×10^{-3} mol; (d) 2.36×10^{-7} mol; (e) 0.362 mol; (f) 0.0129 mol
34. (a) 2.64×10^{-5} mol; (b) 38.1 mol; (c) 7.76×10^{-6} mol; (d) 3.49×10^{-2} mol; (e) 2.09×10^{-3} mol; (f) 2.69×10^{-2} mol
36. (a) 49.2 mg; (b) 7.44×10^4 kg; (c) 59.1 g; (d) 3.27 μg; (e) 4.00 g; (f) 521 g
38. (a) 77.6 g; (b) 177 g; (c) 6.09×10^{-3} g; (d) 0.220 g; (e) 1.26×10^3 g; (f) 3.78×10^{-2} g
40. (a) 2.13×10^{24} molecules; (b) 3.33×10^{22} molecules; (c) 1.58×10^{18} molecules; (d) 2.69×10^{19} molecules; (e) 3.93×10^{19} molecules
42. (a) 0.0141 mol S; (b) 0.0159 mol S; (c) 0.0258 mol S; (d) 0.0254 mol S
44. The percent composition of each element in the compound does not change, because a compound consists of the same percent composition by mass regardless of the starting amount in the sample.
46. (a) 80.34% Zn; 19.66% O; (b) 58.91% Na; 41.09% S; (c) 41.68% Mg; 54.86% O; 3.456% H; (d) 5.926% H; 94.06% O; (e) 95.20% Ca; 4.789% H; (f) 83.01% K; 16.99% O
48. (a) 28.45% Cu; (b) 44.30% Cu; (c) 44.06% Fe; (d) 34.43% Fe; (e) 18.84% Co; (f) 13.40% Co; (g) 88.12% Sn; (h) 78.77% Sn
50. (a) 57.12%; (b) 36.81%; (c) 39.17%; (d) 22.27%; (e) 43.19%
52. (a) 47.06% S^{2-}; (b) 63.89% Cl^-; (c) 10.44% O^{2-}; (d) 62.08% SO_4^{2-}
54. The empirical formula indicates the smallest whole-number ratio of the number and type of atoms present in a molecule. For example, NO_2 and N_2O_4 both have two oxygen atoms for every nitrogen atom and therefore have the same empirical formula.
56. a, c 58. NCl_3 60. BH_3 62. $SnCl_4$ 64. Co_2S_3
66. AlF_3 68. Li_2O 70. Li_3N
72. $C_3H_6O_2$ 74. PCl_3, PCl_5
76. molecular formula = $C_6H_{12}O_6$; empirical formula = CH_2O
78. C_6H_6 80. $C_4H_{10}O_2$
82. The molecular formula is C_3H_8, which is the same as the empirical formula.
84. 5.00 g Al, 0.185 mol, 1.12×10^{23} atoms; 0.140 g Fe, 0.00250 mole, 1.51×10^{21} atoms; 2.7×10^2 g Cu, 4.3 mol, 2.6×10^{24} atoms; 0.00250 g Mg, 1.03×10^{-4} mol, 6.19×10^{19} atoms; 0.062 g Na, 2.7×10^{-3} mol, 1.6×10^{21} atoms; 3.95×10^{-18} g U, 1.66×10^{-20} mol, 1.00×10^4 atoms
86. 24.8% X, 17.4% Y, 57.8% Z. If the molecular formula were actually $X_4Y_2Z_6$, the percent composition would be the same: the *relative* mass of each element present would not change. The molecular formula is always a whole-number multiple of the empirical formula.
88. Cu_2O, CuO
90. (a) 2.82×10^{23} H atoms, 1.41×10^{23} O atoms; (b) 9.32×10^{22} C atoms, 1.86×10^{23} O atoms; (c) 1.02×10^{19} C atoms and H atoms; (d) 1.63×10^{25} C atoms, 2.99×10^{25} H atoms, 1.50×10^{25} O atoms
92. (a) 4.141 g C, 52.96% C, 2.076×10^{23} C atoms; (b) 0.0305 g C, 42.88% C, 1.53×10^{21} C atoms; (c) 14.4 g C, 76.6% C, 7.23×10^{23} C atoms
94. All are true (a, b, c, d, e) 96. a
98. 2.554×10^{-22} g 100. a 102. $C_7H_6O_3$
104. The average mass takes into account not only the exact masses of the isotopes of an element, but also the relative abundance of the isotopes in nature.
106. 8.61×10^{11} sodium atoms; 6.92×10^{24} amu
108. (a) 2.0×10^2 g K; (b) 0.0612 g Hg; (c) 1.27×10^{-3} g Mn; (d) 325 g P; (e) 2.7×10^6 g Fe; (f) 868 g Li; (g) 0.2290 g F
110. (a) 151.9 g; (b) 454.4 g; (c) 150.7 g; (d) 129.8 g; (e) 187.6 g
112. (a) 0.311 mol; (b) 0.270 mol; (c) 0.0501 mol; (d) 2.8 mol; (e) 6.2 mol
114. $CuCO_3$, Na_3PO_4, P_4O_{10}
116. (a) 1.15×10^{22} molecules; (b) 2.08×10^{24} molecules; (c) 4.95×10^{22} molecules; (d) 2.18×10^{22} molecules; (e) 6.32×10^{20} formula units (substance is ionic)
118. 5.97×10^{22} atoms
120. (d) The percent mass of an element in a compound is independent of the amount of compound present.
122. 124.9 g NaOH 124. sulfur (S) 126. $BaCl_2$
128. (a) 1355.37 g/mol; (b) 1.8×10^{-4} mol; (c) 810 g; (d) 5.3×10^{25} H atoms; (e) 2.3×10^{-14} g; (f) 2.3×10^{-21} g
130. (a) 1.9×10^{22}; (b) 2.6×10^{22}; (c) 5.01×10^{23}
132. 60.00% C, 4.476% H, 35.53% O
134. empirical formula = CH_2O; molecular formula = $C_6H_{12}O_6$

Chapter 9

2. The coefficients of this balanced chemical equation indicate the relative numbers of molecules (or moles) of each reactant that combine as well as the number of molecules (or moles) of the product formed.
4. e
6. (a) $3MnO_2(s) + 4Al(s) \rightarrow 3Mn(s) + 2Al_2O_3(s)$.
Three formula units (or three moles) of manganese(IV) oxide react with four atoms (or four moles) of aluminum, producing three atoms (or three moles) of manganese and two formula units (or two moles) of aluminum oxide.
(b) $B_2O_3(s) + 3CaF_2(s) \rightarrow 2BF_3(g) + 3CaO(s)$.
One molecule (or one mole) of diboron trioxide reacts with three formula units (or three moles) of calcium fluoride, producing two molecules (or two moles) of boron trifluoride and three formula units (or three moles) of calcium oxide.
(c) $3NO_2(g) + H_2O(l) \rightarrow 2HNO_3(aq) + NO(g)$.
Three molecules (or three moles) of nitrogen dioxide react with one molecule (or one mole) of water, producing two molecules (or two moles) of nitric acid and one molecule (or one mole) of nitrogen monoxide.
(d) $C_6H_6(g) + 3H_2(g) \rightarrow C_6H_{12}(g)$.
One molecule (or one mole) of benzene (C_6H_6) reacts with three molecules (or three moles) of hydrogen gas, producing one molecule (or one mole) of cyclohexane (C_6H_{12}).
8. Balanced chemical equations tell us in what molar ratios substances combine to form products, not in what mass proportions they combine. How could a total of 3 g of reactants produce 2 g of product?
10. $\frac{2 \text{ mol } O_2}{1 \text{ mol } CH_4}; \frac{1 \text{ mol } CO_2}{1 \text{ mol } CH_4}; \frac{2 \text{ mol } H_2O}{1 \text{ mol } CH_4}$
12. (a) 0.125 mol Fe, 0.0625 mol CO_2; (b) 0.125 mol KCl, 0.0625 mol I_2; (c) 0.500 mol H_3BO_3, 0.125 mol Na_2SO_4; (d) 0.125 mol $Ca(OH)_2$, 0.125 mol C_2H_2
14. (a) 0.50 mol NH_4Cl (27 g); (b) 0.125 mol CS_2 (9.5 g), 0.25 mol H_2S (8.5 g); (c) 0.50 mol H_3PO_3 (41 g), 1.5 mol HCl (55 g); (d) 0.50 mol $NaHCO_3$ (42 g)
16. (a) 0.469 mol O_2; (b) 0.938 mol Se; (c) 0.625 mol CH_3CHO; (d) 1.25 mol Fe
18. Stoichiometry is the process of using a chemical equation to calculate the relative masses of reactants and products involved in a reaction.
20. (a) 1.86×10^{-4} mol Ag; (b) 6.63×10^{-4} mol $(NH_4)_2S$; (c) 2.59×10^{-7} mol U; (d) 81.6 mol SO_2; (e) 1.12 mol $Fe(NO_3)_3$
22. (a) 44.8 g; (b) 0.0529 g; (c) 0.319 g; (d) 4.31×10^7 g; (e) 0.0182 g
24. (a) 0.0310 mol; (b) 0.00555 mol; (c) 0.00475 mol; (d) 0.139 mol
26. 7.87 g of H_2O 28. 1.52 g C_2H_2 30. 0.959 g Na_2CO_3
32. 2.68 g ethyl alcohol 34. 4.704 g O_2
36. 8.62 kg Hg 38. 0.501 g C

40. Gas mileage is about 19 miles per gallon.
Balanced equation:
$2C_8H_{18} + 25O_2 \rightarrow 16CO_2 + 18H_2O$

$$1 \text{ gallon of gasoline} \times \frac{3.7854 \text{ L}}{1 \text{ gallon}} \times \frac{1000 \text{ mL}}{1 \text{ L}} \times \frac{0.75 \text{ g } C_8H_{18}}{1 \text{ mL}} \times \frac{1 \text{ mol } C_8H_{18}}{114.224 \text{ g } C_8H_{18}} \times \frac{16 \text{ mol } CO_2}{2 \text{ mol } C_8H_{18}} \times \frac{44.01 \text{ g } CO_2}{1 \text{ mol } CO_2} \times \frac{1 \text{ lb } CO_2}{453.59 \text{ g } CO_2} \times \frac{1 \text{ mile}}{1 \text{ lb } CO_2} = 19.29 \text{ miles traveled}$$

42. To determine the limiting reactant, first calculate the number of moles of each reactant present. Then determine how these numbers of moles correspond to the stoichiometric ratio indicated by the balanced chemical equation for the reaction. For each reactant, use the stoichiometric ratios from the balanced chemical equation to calculate how much of the *other* reactants would be required to react completely.

44. a

46. (a) H_2SO_4 is limiting, 4.90 g SO_2, 0.918 g H_2O; (b) H_2SO_4 is limiting, 6.30 g $Mn(SO_4)_2$, 0.918 g H_2O; (c) O_2 is limiting, 6.67 g SO_2, 1.88 g H_2O; (d) $AgNO_3$ is limiting, 3.18 g Ag, 2.09 g $Al(NO_3)_3$

48. (a) O_2 is limiting, 0.458 g CO_2; (b) CO_2 is limiting, 0.409 g H_2O; (c) MnO_2 is limiting, 0.207 g H_2O; (d) I_2 is limiting, 1.28 g ICl

50. (a) HCl is the limiting reactant; 18.3 g $AlCl_3$; 0.415 g H_2; (b) NaOH is the limiting reactant; 19.9 g Na_2CO_3; 3.38 g H_2O; (c) $Pb(NO_3)_2$ is the limiting reactant; 12.6 g $PbCl_2$; 5.71 g HNO_3; (d) I_2 is the limiting reactant; 19.6 g KI

52. CuO **54.** 1.79 g Fe_2O_3

56. Sodium sulfate is the limiting reactant; calcium chloride is present in excess.

58. 0.67 kg SiC

60. If the reaction occurs in a solvent, the product may have a substantial solubility in the solvent; the reaction may come to equilibrium before the full yield of product is achieved; loss of product may occur through operator error.

62. 3.71 g in theory; 76.5% yield

64. $2LiOH(s) + CO_2(g) \rightarrow Li_2CO_3(s) + H_2O(g)$. 142 g of CO_2 can be ultimately absorbed; 102 g is 71.8% of the canister's capacity.

66. theoretical, 72.4 g Cu; percent, 62.6% **68.** 28.6 g $NaHCO_3$

70. $C_6H_{12}O_6 + 6O_2 \rightarrow 6CO_2 + 6H_2O$; 1.47 g CO_2 **72.** at least 325 mg

74. (a) $UO_2(s) + 4HF(aq) \rightarrow UF_4(aq) + 2H_2O(l)$. One formula unit of uranium(IV) oxide combines with four molecules of hydrofluoric acid, producing one uranium(IV) fluoride molecule and two water molecules. One mole of uranium(IV) oxide combines with four moles of hydrofluoric acid to produce one mole of uranium(IV) fluoride and two moles of water;
(b) $2NaC_2H_3O_2(aq) + H_2SO_4(aq) \rightarrow Na_2SO_4(aq) + 2HC_2H_3O_2(aq)$. Two molecules (formula units) of sodium acetate react exactly with one molecule of sulfuric acid, producing one molecule (formula unit) of sodium acetate and two molecules of acetic acid. Two moles of sodium acetate combine with one mole of sulfuric acid, producing one mole of sodium sulfate and two moles of acetic acid;
(c) $Mg(s) + 2HCl(aq) \rightarrow MgCl_2(aq) + H_2(g)$. One magnesium atom reacts with two hydrochloric acid molecules (formula units) to produce one molecule (formula unit) of magnesium chloride and one molecule of hydrogen gas. One mole of magnesium combines with two moles of hydrochloric acid, producing one mole of magnesium chloride and one mole of gaseous hydrogen;
(d) $B_2O_3(s) + 3H_2O(l) \rightarrow 2B(OH)_3(s)$. One molecule (formula unit) of diboron trioxide reacts exactly with three molecules of water, producing two molecules of boron trihydroxide (boric acid). One mole of diboron trioxide combines with three moles of water to produce two moles of boron trihydroxide (boric acid).

76. for O_2, 5 mol O_2/1 mol C_3H_8; for CO_2, 3 mol CO_2/1 mol C_3H_8; for H_2O, 4 mol H_2O/1 mol C_3H_8

78. (a) 0.0588 mol NH_4Cl; (b) 0.0178 mol $CaCO_3$; (c) 0.0217 mol Na_2O; (d) 0.0323 mol PCl_3

80. (a) 3.2×10^2 g HNO_3; (b) 0.0612 g Hg; (c) 4.49×10^{-3} g K_2CrO_4; (d) 1.40×10^3 g $AlCl_3$; (e) 7.2×10^6 g SF_6; (f) 2.13×10^3 g NH_3; (g) 0.9397 g Na_2O_2

82. 1.9×10^2 kg SO_3 **84.** 0.667 g O_2 **86.** 0.0771 g H_2

88. (a) Br_2 is limiting reactant, 6.4 g NaBr; (b) $CuSO_4$ is limiting reactant, 5.1 g $ZnSO_4$, 2.0 g Cu; (c) NH_4Cl is limiting reactant, 1.6 g NH_3, 1.7 g H_2O, 5.5 g NaCl; (d) Fe_2O_3 is limiting reactant, 3.5 g Fe, 4.1 g CO_2

90. 0.624 mol N_2, 17.5 g N_2; 1.25 mol H_2O, 22.5 g H_2O

92. 5.0 g **94.** 68.12%

96. (a) 36.73 g Fe_2O_3; (b) 12.41 g Al; (c) 23.45 g Al_2O_3

98. 28.2 g $NaNH_2$; 0.728 g H_2

100. (a) 660. g C_3H_3N; (b) 672 g H_2O; (c) 0 g propylene, 289 g ammonia, 405 g oxygen

Chapter 10

2. Potential energy is energy due to position or composition. A stone at the top of a hill possesses potential energy because the stone may eventually roll down the hill. A gallon of gasoline possesses potential energy because heat will be released when the gasoline is burned.

4. constant

6. Ball A initially possesses potential energy by virtue of its position at the top of the hill. As ball A rolls down the hill, its potential energy is converted to kinetic energy and frictional (heat) energy. When ball A reaches the bottom of the hill and hits ball B, it transfers its kinetic energy to ball B. Ball A then has only the potential energy corresponding to its new position.

8. The hot tea is at a higher temperature, which means the particles in the hot tea have higher average kinetic energies. When the tea spills on the skin, energy flows from the hot tea to the skin, until the tea and skin are at the same temperature. This sudden inflow of energy causes the burn.

10. Temperature is the concept by which we express the thermal energy contained in a sample. We cannot measure the motions of the particles/kinetic energy in a sample of matter directly. We know, however, that if two objects are at different temperatures, the one with the higher temperature has molecules that have higher average kinetic energies than the molecules of the object at the lower temperature.

12. When the chemical system evolves energy, the energy evolved from the reacting chemicals is transferred to the surroundings.

14. (a) endothermic; (b) exothermic; (c) exothermic; (d) endothermic

16. internal **18.** losing **20.** -21 kJ

22. (a) $\frac{1 \text{ J}}{4.184 \text{ cal}}$; (b) $\frac{4.184 \text{ cal}}{1 \text{ J}}$; (c) $\frac{1 \text{ kcal}}{1000 \text{ cal}}$; (d) $\frac{1000 \text{ J}}{1 \text{ kJ}}$

24. 6540 J = 6.54 kJ

26. (a) 8.254 kcal; (b) 0.0415 kcal; (c) 8.231 kcal; (d) 752.9 kcal

28. (a) 32,820 J; 32.82 kJ; (b) 1.90×10^5 J; 190. kJ; (c) 2.600×10^5 J; 260.0 kJ; (d) 1.800×10^5 J; 180.0 kJ

30. (a) 190.9 kJ; (b) 17.43 kcal; (c) 657.5 cal; (d) 2.394×10^4 J

32. 5.8×10^2 J (two significant figures) **34.** 29 °C

36. exothermic **38.** 14.6 kJ (~15 kJ to one significant figure)

40. calorimeter **42.** (a) -9.23 kJ; (b) -148 kJ; (c) $+296$ kJ/mol

44. (a) -29.5 kJ; (b) $\Delta H = -1360$ kJ; (c) 453 kJ/mol H_2O

46. -220 kJ **48.** 226 kJ

50. Once everything in the universe is at the same temperature, no further thermodynamic work can be done. Even though the total energy of the universe will be the same, the energy will have been dispersed evenly, making it effectively useless.

52. Concentrated sources of energy, such as petroleum, are being used so as to disperse the energy they contain, making that energy unavailable for further human use.

54. Petroleum consists mainly of hydrocarbons, which are molecules containing chains of carbon atoms with hydrogen atoms attached to the chains. The fractions are based on the number of carbon atoms in the chains: for example, gasoline is a mixture of hydrocarbons with 5–10 carbon atoms in the chains, whereas asphalt is a mixture of hydrocarbons with 25 or more carbon atoms in the chains. Different fractions have different physical properties and uses, but all can be combusted to produce energy. See Table 10.3.
56. Tetraethyl lead was used as an additive for gasoline to promote smoother running of engines. It is no longer widely used because of concerns about the lead being released into the environment as the leaded gasoline burns.
58. The greenhouse effect is a warming effect due to the presence of gases in the atmosphere that absorb infrared radiation that has reached the earth from the sun; the gases do not allow the energy to pass back into space. A limited greenhouse effect is desirable because it moderates the temperature changes in the atmosphere that would otherwise be more drastic between daytime when the sun is shining and nighttime. Having too high a concentration of greenhouse gases, however, will elevate the temperature of the earth too much, affecting climate, crops, the polar ice caps, the temperatures of the oceans, and so on. Carbon dioxide produced by combustion reactions is our greatest concern as a greenhouse gas.
60. The second law of thermodynamics says that the entropy of the universe is always increasing. Energy spread and matter spread lead to greater entropy (greater disorder) in the universe.
62. Formation of a solid precipitate represents a concentration of matter.
64. The molecules in liquid water are moving around freely and are therefore more "disordered" than when the molecules are held rigidly in a solid lattice in ice. The entropy increases during melting.
66. -55.5 kJ **68.** 7.65 kcal
70. 2.0×10^2 J (two significant figures) **72.** 3.8×10^5 J
74. 62.5 °C = 63 °C **76.** 0.711 J/g °C
78. $w = +168$ kJ **80.** -252 kJ
82. $\frac{1}{2}C + F \rightarrow A + B + D \quad \Delta H = 47.0$ kJ
84. c **86.** 5.1×10^2 J **88.** 3.97×10^3 kJ

Chapter 11

2. Rutherford was not able to determine where the electrons were in the atom or what they were doing.
4. The different forms of electromagnetic radiation all exhibit the same wavelike behavior and are propagated through space at the same speed (the "speed of light"). The types of electromagnetic radiation differ in their frequency (and wavelength) and in the resulting amount of energy carried per photon.
6. The frequency of electromagnetic radiation represents how many waves pass a given location per second. The speed of electromagnetic radiation represents how fast the waves propagate through space. The frequency and the speed are not the same.
8. The greenhouse gases do not absorb light in the visible wavelengths. Therefore, this light passes through the atmosphere and warms the earth, keeping the earth much warmer than it would be without these gases. As we are increasing our use of fossil fuels, the level of CO_2 in the atmosphere is increasing gradually but significantly. An increase in the level of CO_2 will warm the earth further, eventually changing the weather patterns on the earth's surface and melting the polar ice caps.
10. exactly equal to **12.** It is emitted as a photon. **14.** absorbs
16. When excited hydrogen atoms emit their excess energy, the photons of radiation emitted always have exactly the same wavelength and energy. This means that the hydrogen atom possesses only certain allowed energy states and that the photons emitted correspond to the electron changing from one of these allowed energy states to another allowed energy state. The energy of the photon emitted corresponds to the energy difference between the allowed states. If the hydrogen atom did *not* possess discrete energy levels, then the photons emitted would have random wavelengths and energies.
18. They are identical.
20. Energy is emitted at wavelengths corresponding to specific transitions for the electron among the energy levels of hydrogen.
22. The electron moves to an orbit farther from the nucleus of the atom.
24. Bohr's theory explained the experimentally observed line spectrum of hydrogen exactly. The theory was discarded because the calculated properties did not correspond closely to experimental measurements for atoms other than hydrogen.
26. An orbit refers to a definite, exact circular pathway around the nucleus, in which Bohr postulated that an electron would be found. An orbital represents a region of space in which there is a high probability of finding the electron.
28. The firefly analogy is intended to demonstrate the concept of a probability map for electron density. In the wave mechanical model of the atom, we cannot say specifically where the electron is in the atom; we can say only where there is a high probability of finding the electron. The analogy is to imagine a time-exposure photograph of a firefly in a closed room. Most of the time, the firefly will be found near the center of the room.
30. b
32. two-lobed ("dumbbell"-shaped); lower in energy and closer to the nucleus; similar shape
34. excited
36.

Value of n	Possible Subshells
1	1*s*
2	2*s*, 2*p*
3	3*s*, 3*p*, 3*d*
4	4*s*, 4*p*, 4*d*, 4*f*

38. Electrons have an intrinsic spin (they spin on their own axes). Geometrically, there are only two senses possible for spin (clockwise or counterclockwise). This means only two electrons can occupy an orbital, with the opposite sense or direction of spin. This idea is called the Pauli exclusion principle.
40. increases **42.** paired (opposite spin)
44. (a) not possible; (b) possible; (c) not possible; (d) possible
46. For a hydrogen atom in its ground state, the electron is in the 1*s* orbital. The 1*s* orbital has the lowest energy of all hydrogen orbitals.
48. similar type of orbitals being filled in the same way; chemical properties of the members of the group are similar
50. (a) silicon; (b) beryllium; (c) neon; (d) argon
52. (a) selenium; (b) scandium; (c) sulfur; (d) iodine
54. (a)

1*s*	2*s*	2*p*	3*s*	3*p*
[↑↓]	[↑↓]	[↑↓][↑↓][↑↓]	[↑↓]	[↑][ ][ ]

(b)

1*s*	2*s*	2*p*	3*s*	3*p*
[↑↓]	[↑↓]	[↑↓][↑↓][↑↓]	[↑↓]	[↑][↑][↑]

(c)

1*s*	2*s*	2*p*	3*s*	3*p*	4*s*	3*d*	4*p*
[↑↓]	[↑↓]	[↑↓][↑↓][↑↓]	[↑↓]	[↑↓][↑↓][↑↓]	[↑↓]	[↑↓][↑↓][↑↓][↑↓][↑↓]	[↑↓][↑↓][↑]

(d)

1*s*	2*s*	2*p*	3*s*	3*p*
[↑↓]	[↑↓]	[↑↓][↑↓][↑↓]	[↑↓]	[↑↓][↑↓][↑↓]

56. Specific answers depend on student choice of elements. Any Group 1 element would have one valence electron. Any Group 3 element would have three valence electrons. Any Group 5 element would have five valence electrons. Any Group 7 element would have seven valence electrons.
58. The properties of Rb and Sr suggest that they are members of Groups 1 and 2, respectively, and so must be filling the 5*s* orbital. The 5*s* orbital is lower in energy than (and fills before) the 4*d* orbitals.
60. (a) aluminum; (b) potassium; (c) bromine; (d) tin
62. (a) 1 valence electron; (b) 5 valence electrons (*d* electrons are not counted as valence electrons); (c) 3 valence electrons; (d) 2 valence electrons (*d* electrons are not counted as valence electrons)
64. (a) 6; (b) 8; (c) 10; (d) 0

66. (a) $[Rn]7s^26d^15f^3$; (b) $[Ar]4s^23d^5$; (c) $[Xe]6s^24f^{14}5d^{10}$; (d) $[Rn]7s^1$

68. $[Rn]7s^25f^{14}6d^5$

70. The metallic elements lose electrons and form positive ions (cations); the nonmetallic elements gain electrons and form negative ions (anions).

72. All exist as diatomic molecules (F_2, Cl_2, Br_2, I_2); are nonmetals; have relatively high electronegativities; and form 1− ions in reacting with metallic elements.

74. Elements at the left of a period (horizontal row) lose electrons most readily; at the left of a period (given principal energy level) the nuclear charge is the smallest and the electrons are least tightly held.

76. The elements of a given period (horizontal row) have valence electrons in the same subshells, but nuclear charge increases across a period going from left to right. Atoms at the left side have smaller nuclear charges and bind their valence electrons less tightly.

78. When substances absorb energy, the electrons become excited (move to higher energy levels). Upon returning to the ground state, energy is released, some of which is in the visible spectrum. Because we see colors, this tells us only certain wavelengths of light are released, which means that only certain transitions are allowed. This is what is meant by quantized energy levels. If all wavelengths of light were emitted, we would see white light.

80. (a) Li; (b) Ca; (c) Cl; (d) S

82. (a) Na; (b) S; (c) N; (d) F

84. speed of light **86.** photons **88.** quantized

90. orbital **92.** transition metal **94.** spins

96. (a) $1s^22s^22p^63s^23p^64s^1$; $[Ar]4s^1$;

1s	2s	2p	3s	3p	4s
↑↓	↑↓	↑↓ ↑↓ ↑↓	↑↓	↑↓ ↑↓ ↑↓	↑

(b) $1s^22s^22p^63s^23p^64s^23d^2$; $[Ar]4s^23d^2$;

1s	2s	2p	3s	3p	4s
↑↓	↑↓	↑↓ ↑↓ ↑↓	↑↓	↑↓ ↑↓ ↑↓	↑↓

3d				
↑	↑			

(c) $1s^22s^22p^63s^23p^2$; $[Ne]3s^23p^2$;

1s	2s	2p	3s	3p
↑↓	↑↓	↑↓ ↑↓ ↑↓	↑↓	↑ ↑

(d) $1s^22s^22p^63s^23p^64s^23d^6$; $[Ar]4s^23d^6$;

1s	2s	2p	3s	3p	4s
↑↓	↑↓	↑↓ ↑↓ ↑↓	↑↓	↑↓ ↑↓ ↑↓	↑↓

3d				
↑↓	↑	↑	↑	↑

(e) $1s^22s^22p^63s^23p^64s^23d^{10}$; $[Ar]4s^23d^{10}$;

1s	2s	2p	3s	3p	4s
↑↓	↑↓	↑↓ ↑↓ ↑↓	↑↓	↑↓ ↑↓ ↑↓	↑↓

3d				
↑↓	↑↓	↑↓	↑↓	↑↓

98. (a) ns^2; (b) ns^2np^5; (c) ns^2np^4; (d) ns^1; (e) ns^2np^4

100. (a) 2.7×10^{-12} m; (b) 4.4×10^{-34} m; (c) 2×10^{-35} m; the wavelengths for the ball and the person are infinitesimally small, whereas the wavelength for the electron is nearly the same order of magnitude as the diameter of a typical atom.

102. Light is emitted from the hydrogen atom only at certain fixed wavelengths. If the energy levels of hydrogen were continuous, a hydrogen atom would emit energy at all possible wavelengths.

104. 3

106. (a) $[Ne]3s^23p^4$; 16 electrons in this ground state atom, which corresponds to sulfur. (b) $[He]2s^12p^4$; 7 electrons in this excited state atom, which corresponds to nitrogen. One electron was excited from the 2s to 2p. (c) $[Ar]4s^23d^{10}4p^5$; 35 electrons in this ground state ion. A charge of −1 means an electron was gained, thus making the ion Se^-, which corresponds to 34 protons.

108. (a) $1s^22s^22p^63s^23p^64s^23d^{10}4p^5$
(b) $1s^22s^22p^63s^23p^64s^23d^{10}4p^65s^24d^{10}5p^6$
(c) $1s^22s^22p^63s^23p^64s^23d^{10}4p^65s^24d^{10}5p^66s^2$
(d) $1s^22s^22p^63s^23p^64s^23d^{10}4p^4$

110. (a) five (2s, 2p); (b) seven (3s, 3p); (c) one (3s); (d) three (3s, 3p)

112. (a) ns^2np^3; (b) ns^1; (c) ns^2np^5; (d) ns^2np^4; (e) ns^2

114. a < c < b **116.** metals, low; nonmetals, high

118. (a) Ca; (b) P; (c) K **120.** b, c, e

122. (a) Te; (b) Ge; (c) F **124.** B; C; Al

126.

	Symbol	IE	AR
$1s^22s^22p^63s^2$	Mg	0.738	160
$1s^22s^22p^63s^23p^4$	S	0.999	104
$1s^22s^22p^63s^23p^64s^2$	Ca	0.590	197

Chapter 12

2. The *bond energy* represents the energy required to break a chemical bond.

4. A covalent bond represents the *sharing* of electrons by nuclei.

6. In H_2 and HF, the bonding is covalent in nature, with an electron pair being shared between the atoms. In H_2, the two atoms are identical and so the sharing is equal; in HF, the two atoms are different and so the bonding is polar covalent. Both of these are in marked contrast to the situation in NaF: NaF is an ionic compound, and an electron is completely transferred from sodium to fluorine, thereby producing the separate ions.

8. A bond is polar if the centers of positive and negative charge do not coincide at the same point. The bond has a negative end and a positive end. Any molecule in which the atoms in the bonds are not identical will have polar bonds (although the molecule as a whole may not be polar). Two simple examples are HCl and HF.

10. The difference in electronegativity between the atoms in the bond

12. (a) At is most electronegative, Cs is least electronegative; (b) Sr is most electronegative, Ba and Ra have the same electronegativities; (c) O is most electronegative, Rb is least electronegative

14. (a) covalent; (b) polar covalent; (c) ionic **16.** d

18. (a) O—Br; (b) N—F; (c) P—O; (d) H—O

20. (a) Na—N; (b) K—P; (c) Na—Cl; (d) Mg—Cl

22. The presence of strong bond dipoles and a large overall dipole moment makes water a very polar substance. Properties of water that are dependent on its dipole moment involve its freezing point, melting point, vapor pressure, and ability to dissolve many substances.

24. (a) H; (b) Cl; (c) I

26. (a) $^{\delta-}S \rightarrow P^{\delta+}$; (b) $^{\delta+}S \rightarrow F^{\delta-}$; (c) $^{\delta+}S \rightarrow Cl^{\delta-}$; (d) $^{\delta+}S \rightarrow Br^{\delta-}$

28. (a) $^{\delta+}H \rightarrow C^{\delta-}$; (b) $^{\delta+}N \rightarrow O^{\delta-}$; (c) $^{\delta+}S \rightarrow N^{\delta-}$; (d) $^{\delta+}C \rightarrow N^{\delta-}$

30. preceding

32. Atoms in covalent molecules gain a configuration like that of a noble gas by sharing one or more pairs of electrons between atoms: such shared pairs of electrons "belong" to each of the bonding atoms at the same time. In ionic bonding, one atom completely donates one or more electrons to another atom, and then the resulting ions behave independently of one another (they are not "attached" to one another, although they are mutually attracted).

34. (a) Br^- [Kr]; (b) Cs^+ [Xe]; (c) P^{3-} [Ar]; (d) S^{2-} [Ar]

36. (a) Cl^-, S^{2-}, P^{3-}; (b) F^-, O^{2-}, N^{3-}; (c) Br^-, Se^{2-}, As^{3-}; (d) I^-, Te^{2-}

38. (a) $AlBr_3$; (b) Al_2O_3; (c) AlP; (d) AlH_3

40. Answers depend on student choice of examples.

42. An ionic solid such as NaCl consists of an array of alternating positively and negatively charged ions: that is, each positive ion has as its nearest neighbors a group of negative ions, and each negative ion has a group of positive ions surrounding it. In most ionic solids, the ions are packed as tightly as possible.

44. In forming an anion, an atom gains additional electrons in its outermost (valence) shell. Having additional electrons in the valence shell increases the repulsive forces between electrons, and the outermost shell becomes larger to accommodate this.

46. (a) Mg is larger than Mg^{2+}. Positive ions are always smaller than the atoms from which they are formed. (b) K^+ is larger than Ca^{2+}. Both species contain the same number of electrons, but the nuclear charge of K^+ is less than the nuclear charge of Ca^{2+} (19 protons versus 20 protons), thus the electrons are not drawn as closely to the nucleus. (c) Br^- is larger than Rb^+. Both species contain the same number of electrons, but the nuclear charge of Br^- is less (35 protons versus 37 protons), thus the electrons are not drawn as closely to the nucleus. (d) Se^{2-} is larger than Se. Negative ions tend to be larger than the neutral form of that atom, because the negative ions contain a larger number of electrons in the valence shell.

48. (a) I; (b) F^-; (c) F^-

50. When atoms form covalent bonds, they try to attain a valence-electron configuration similar to that of the following noble gas element. When the elements in the first few horizontal rows of the periodic table form covalent bonds, they attempt to achieve the configurations of the noble gases helium (two valence electrons, duet rule) and neon and argon (eight valence electrons, octet rule).

52. These elements attain a total of eight valence electrons, giving the valence-electron configurations of the noble gases Ne and Ar.

54. Two atoms in a molecule are connected by a triple bond if the atoms share three pairs of electrons (six electrons) to complete their outermost shells. A simple molecule containing a triple bond is acetylene, C_2H_2 (H:C:::C:H).

56. (a) Mg:; (b) :Br:; (c) :S:; (d) :Si

58. (a) 24; (b) 16; (c) 20; (d) 17

60. (a) H–S–H (bent, S with two lone pairs) (b) SiF_4: Si bonded to four :F: (c) $H_2C{=}CH_2$ (d) $H_3C{-}CH(H){-}CH_3$ (propane)

62. (a) :Cl–P–Cl: with :Cl: on P (b) H–C(Cl)(Cl)–Cl ($CHCl_3$) (c) Cl–CH₂–CH₂–Cl drawn as :Cl–C(H)(H)–C(H)(H)–Cl: (d) H–N(H)–N(H)–H

64. :O≡C–Ö: ⟷ Ö=C=Ö ⟷ :Ö–C≡O:

66. (a) ClO_3^- (:Ö–Cl–Ö: with :Ö: on Cl)$^-$

(b) O_2^{2-} (:Ö–Ö:)$^{2-}$

(c) $C_2H_3O_2^-$ (H–C(H)(H)–C(=O)–Ö:)$^-$ ↔ (H–C(H)(H)–C(–Ö:)=O)$^-$

68. (a) $[CO_3]^{2-}$: three resonance structures, each with one C=O double bond and two C–O single bonds

(b) $[NH_4]^+$: N bonded to four H

(c) [:Cl–O:]$^-$

70. The geometric structure of NH_3 is that of a trigonal pyramid. The nitrogen atom of NH_3 is surrounded by four electron pairs (three are bonding, one is a lone pair). The H—N—H bond angle is somewhat less than 109.5° (because of the presence of the lone pair).

72. SiF_4 has a tetrahedral geometric structure; eight pairs of electrons on Si; ~109.5°

74. The general molecular structure of a molecule is determined by how many electron pairs surround the central atom in the molecule, and by which of those pairs are used for bonding to the other atoms of the molecule.

76. Geometry shows that only two points in space are needed to indicate a straight line. A diatomic molecule represents two points in space.

78. In NF_3, the nitrogen atom has *four* pairs of valence electrons; in BF_3, only *three* pairs of valence electrons surround the boron atom. The nonbonding pair on nitrogen in NF_3 pushes the three F atoms out of the plane of the N atom.

80. (a) four pairs of electrons in a (distorted) tetrahedral arrangement; (b) four pairs of electrons in a (slightly distorted) tetrahedral arrangement; (c) four pairs of electrons in a tetrahedral arrangement

82. (a) tetrahedral; (b) trigonal pyramidal; (c) bent or V-shaped

84. (a) basically tetrahedral arrangement of the oxygens around the phosphorus; (b) tetrahedral; (c) trigonal pyramid

86. (a) approximately tetrahedral (a little less than 109.5°); (b) approximately tetrahedral (a little less than 109.5°); (c) tetrahedral (109.5°); (d) trigonal planar (120°) because of the double bond

88. (a) bent or V-shaped; 120°; (b) trigonal planar; 120°; (c) bent or V-shaped; 120°; (d) linear; 180°

90. double

92. C is the correct answer. N—N contains equal electron sharing, so it is nonpolar covalent.

94. The bond energy is the energy required to break the bond.

96. (a) Be; (b) N; (c) F **98.** a, c

100. (a) O; (b) Br; (c) I **102.** d

104. (a) Na_2Se; (b) RbF; (c) K_2Te; (d) BaSe; (e) KAt; (f) FrCl

106. (a) Na^+; (b) Al^{3+}; (c) F^-; (d) Na^+

108. (a) 24; (b) 32; (c) 32; (d) 32

110. (a) N_2H_4 H–N(H)–N(H)–H

(b) C_2H_6 H–C(H)(H)–C(H)(H)–H

(c) NCl_3 :Cl–N–Cl: with :Cl: on N

(d) $SiCl_4$:Cl–Si–Cl: with :Cl: above and below Si

112. (a) NO_3^- $\left[\ddot{O}{=}N{-}\ddot{O}{:}\ \text{with}\ {:}\ddot{O}{:}\ \text{on N}\right]^- \leftrightarrow \left[{:}\ddot{O}{-}N{-}\ddot{O}{:}\ \text{with}\ {:}O{:}\ \text{(double bond) on N}\right]^- \leftrightarrow \left[{:}\ddot{O}{-}N{=}\ddot{O}\ \text{with}\ {:}\ddot{O}{:}\ \text{on N}\right]^-$

(b) CO_3^{2-} $\left[\ddot{O}{=}C{-}\ddot{O}{:}\ \text{with}\ {:}\ddot{O}{:}\ \text{on C}\right]^{2-} \leftrightarrow \left[{:}\ddot{O}{-}C{=}\ddot{O}\ \text{with}\ {:}\ddot{O}{:}\ \text{on C}\right]^{2-} \leftrightarrow \left[{:}\ddot{O}{-}C{-}\ddot{O}{:}\ \text{with}\ {:}O{:}\ \text{(double bond) on C}\right]^{2-}$

(c) NH_4^+ $\left[H{-}N{-}H\ \text{with H above and H below N}\right]^+$

114. (a) four pairs arranged tetrahedrally; (b) four pairs arranged tetrahedrally; (c) three pairs arranged trigonally (planar)
116. (a) trigonal pyramid; (b) nonlinear (V-shaped); (c) tetrahedral
118. (a) nonlinear (V-shaped); (b) trigonal planar; (c) basically trigonal planar around C, distorted somewhat by H; (d) linear
120. Ionic compounds tend to be hard, crystalline substances with relatively high melting and boiling points. Covalently bonded substances tend to be gases, liquids, or relatively soft solids, with much lower melting and boiling points.
122. (a) nonpolar covalent; (b) ionic; (c) ionic; (d) polar covalent; (e) polar covalent; (f) nonpolar covalent; (g) nonpolar covalent; (h) ionic
124. O—F, P—Cl, P—F, Si—F
126. Na^+: $1s^22s^22p^6$; K^+: $1s^22s^22p^63s^23p^6$; Li^+: $1s^2$; Cs^+: $1s^22s^22p^63s^23p^64s^23d^{10}4p^65s^24d^{10}5p^6$
128.

Formula	Compound Name	Molecular Structure
CO_2	carbon dioxide	linear
NH_3	ammonia	trigonal pyramidal
SO_3	sulfur trioxide	trigonal planar
H_2O	water	bent or V-shaped
ClO_4^-	perchlorate ion	tetrahedral

Chapter 13

2. Solids are rigid and incompressible and have definite shapes and volumes. Liquids are less rigid than solids; although they have definite volumes, liquids take the shape of their containers. Gases have no fixed volume or shape; they take the volume and shape of their container and are affected more by changes in their pressure and temperature than are solids or liquids.
4. A mercury barometer consists of a tube filled with mercury that is then inverted over a reservoir of mercury, the surface of which is open to the atmosphere. The pressure of the atmosphere is reflected in the height to which the column of mercury in the tube is supported.
6. Pressure units include mm Hg, torr, pascals, and psi. The unit "mm Hg" is derived from the barometer, because in a traditional mercury barometer, we measure the height of the mercury column (in millimeters) above the reservoir of mercury.
8. (a) 1.01 atm; (b) 1.05 atm; (c) 99.1 kPa; (d) 99.436 kPa
10. (a) 119 kPa; (b) 16.9 psi; (c) 3.23×10^3 mm Hg; (d) 15.2 atm
12. (a) 1.03×10^5 Pa; (b) 9.78×10^4 Pa; (c) 1.125×10^5 Pa; (d) 1.07×10^5 Pa
14. Additional mercury increases the pressure on the gas sample, causing the volume of the gas upon which the pressure is exerted to decrease (Boyle's law).
16. $PV = k$; $P_1V_1 = P_2V_2$
18. (a) 423 mL; (b) 158 mL; (c) 8.67 L
20. (a) 59.7 mL; (b) 1.88 atm; (c) 3.40 L
22. 0.520 L **24.** 20.0 atm
26. Charles's law indicates that an ideal gas decreases by 1/273 of its volume for every Celsius degree its temperature is lowered. This means an ideal gas would approach a volume of zero at −273 °C.
28. $V = bT$; $V_1/T_1 = V_2/T_2$ **30.** 315 mL
32. (a) 273 °C; (b) 35 °C; (c) 0.117 mL
34. (a) 35.4 K = −238 °C; (b) 0 mL (absolute zero; a real gas would condense to a solid or liquid); (c) 40.5 mL
36. 69.4 mL (69 mL to two significant figures)
38. 90 °C, 124 mL; 80 °C, 120. mL; 70 °C, 117 mL; 60 °C, 113 mL; 50 °C, 110. mL; 40 °C, 107 mL; 30 °C, 103 mL; 20 °C, 99.8 mL
40. $V = an$; $V_1/n_1 = V_2/n_2$
42. 1744 mL (1.74×10^3 mL) **44.** 80.1 L
46. Real gases behave most ideally at relatively high temperatures and relatively low pressures. We usually assume that a real gas's behavior approaches ideal behavior if the temperature is over 0 °C (273 K) and the pressure is 1 atm or lower.
48. For an ideal gas, $PV = nRT$ is true under any conditions. Consider a particular sample of gas (n remains constant) at a particular fixed pressure (P remains constant). Suppose that at temperature T_1 the volume of the gas sample is V_1. For this set of conditions, the ideal gas equation would be given by $PV_1 = nRT_1$. If the temperature of the gas sample changes to a new temperature, T_2, then the volume of the gas sample changes to a new volume, V_2. For this new set of conditions, the ideal gas equation would be given by $PV_2 = nRT_2$. If we make a *ratio* of these two expressions for the ideal gas equation for this gas sample, and cancel out terms that are constant for this situation (P, n, and R), we get

$$\frac{PV_1}{PV_2} = \frac{nRT_1}{nRT_2}, \text{ or } \frac{V_1}{V_2} = \frac{T_1}{T_2},$$

which can be rearranged to the familiar form of Charles's law,

$$\frac{V_1}{T_1} = \frac{V_2}{T_2}.$$

50. (a) 5.02 L; (b) 3.56 atm = 2.70×10^3 mm Hg; (c) 334 K
52. 339 atm **54.** 106 °C
56. 0.150 atm; 0.163 atm **58.** 238 K/−35 °C
60. The helium (5.07 atm) is at a higher pressure than the argon (3.50 atm).
62. 0.332 atm; 0.346 atm **64.** ~283 atm (2.8×10^2 atm)
66. As a gas is bubbled through water, the bubbles of gas become saturated with water vapor, thus forming a gaseous mixture. The total pressure for a sample of gas that has been collected by bubbling through water is made up of two components: the pressure of the sample gas and the pressure of water vapor. The partial pressure of the gas equals the total pressure of the sample minus the vapor pressure of water.
68. 0.314 atm **70.** 3.00 g Ne; 5.94 g Ar
72. $P_{Ar} = 2(0.100\text{ atm}) = 0.20$ atm; $P_{Ne} = 3(0.100\text{ atm}) = 0.30$ atm; $P_{He} = 5(0.100\text{ atm}) = 0.50$ atm
74. $P_{hydrogen} = 0.990$ atm; 9.55×10^{-3} mol H_2; 0.625 g Zn
76. A theory is successful if it explains known experimental observations. Theories that have been successful in the past may not be successful in the future (for example, as technology evolves, more sophisticated experiments may be possible in the future).
78. pressure **80.** no
82. If the temperature of a sample of gas is increased, the average kinetic energy of the particles of gas increases. This means that the speeds of the particles increase. If the particles have a higher speed, they hit the walls of the container more frequently and with greater force, thereby increasing the pressure.
84. STP = 0 °C, 1 atm pressure. These conditions were chosen because they are easy to attain and reproduce *experimentally*. The barometric pressure within a laboratory will usually be near 1 atm, and 0 °C can be attained with a simple ice bath.

86. 2.50 L O_2 **88.** 0.940 L
90. 0.941 L; 0.870 L **92.** 5.03 L (dry volume)
94. (a) $6Na + N_2 \rightarrow 2Na_3N$; (b) 12.0 g Na_3N
96. 45.5 L **98.** 40.5 L; $P_{He} = 0.864$ atm; $P_{Ne} = 0.136$ atm
100. 1.72 L **102.** 0.365 g **104.** twice
106. In both cases, the gas particles will distribute uniformly throughout both flasks.
In case 1, $P_{final} = (2/3)P_{initial}$.
In case 2, $P_{final} = (1/2)P_{initial}$.
108. 125 balloons **110.** 124 L
112. 0.0999 mol CO_2; 3.32 L wet; 2.68 L dry
114. b **116.** 1.14 L
118. (a) 8.60×10^4 Pa; (b) 2.21×10^5 Pa; (c) 8.88×10^4 Pa; (d) 4.3×10^3 Pa
120. (a) 128 mL; (b) 1.3×10^{-2} L; (c) 9.8 L
122. 2.55×10^3 mm Hg
124. (a) 57.3 mL; (b) 448 K = 175 °C; (c) zero (absolute zero; a real gas would condense to a solid or liquid)
126. 123 mL
128. Three changes you can make to double the volume are to:
(1) *increase the temperature (double the temperature in the kelvin scale).* If the temperature is increased, the gas particles have more kinetic energy and will hit the piston with more force (and more pressure). Therefore, the piston will move up until the pressure inside the container is the same as outside the container (causing the volume to increase).
(2) *add moles of gas to the container (double the amount).* By adding moles of gas to the container, gas particles will hit the walls of the container more frequently (and thus exert more pressure). The piston will move up until the pressure inside the container is the same as outside the container (causing the volume to increase).
(3) *decrease the pressure outside the container (by half).* By decreasing the pressure outside the container, the pressure inside becomes greater than the pressure outside. The gas particles inside will push the piston up until the pressure inside the container is the same as outside the container (causing the volume to increase).
130. (a) 61.8 K; (b) 0.993 atm; (c) 1.66×10^4 L
132. 487 mol gas needed; 7.79 kg CH_4; 13.6 kg N_2; 21.4 kg CO_2
134. 0.42 atm **136.** 6.41 L
138. (a) P_{Total} = 325 torr + 475 torr + 650. torr = 1450. torr
(b) Since the volume and temperature are constant, there is a direct relationship between the pressure and number of moles. The gas with the highest pressure (which is O_2) must contain the greatest number of moles and collide with the walls more frequently.
140. 3.43 L N_2; 10.3 L H_2
142. 5.8 L O_2; 3.9 L SO_2 **144.** 7.8×10^2 L
146. (a) The pressure of helium is 1.5 times as great as the pressure of neon. (b) When the valve is opened, the gases disperse uniformly throughout the entire apparatus.
(c) $P_{f(Ne)} = \frac{1}{2}P_{o(Ne)}$ and $P_{f(He)} = \frac{1}{2}P_{o(He)}$
(d) The original pressure of helium is 1.5 times the original pressure of neon.

$$P_f = \frac{1}{2}P_{o(Ne)} + \frac{1}{2}P_{o(He)} = \frac{1}{2}P_{o(Ne)} + \frac{1.5}{2}P_{o(Ne)} = \frac{2.5}{2}P_{o(Ne)} = 1.25P_{o(Ne)}$$

148. 22.4 L O_2 **150.** 0.04 g CO_2 **152.** b; c; d
154. $\Delta V = 142$ L **156.** 51.4 g XeF_4 **158.** a; c; d

Chapter 14

2. less
4. Because it requires so much more energy to vaporize water than to melt ice, this suggests that the gaseous state is significantly different from the liquid state, but that the liquid and solid states are relatively similar.
6. See Fig. 14.2.
8. When a solid is heated, the molecules begin to vibrate/move more quickly. When enough energy has been added to overcome the intermolecular forces that hold the molecules in a crystal lattice, the solid melts. As the liquid is heated, the molecules begin to move more quickly and more randomly. When enough energy has been added, molecules having sufficient kinetic energy will begin to escape from the surface of the liquid. Once the pressure of vapor coming from the liquid is equal to the pressure above the liquid, the liquid boils. Only intermolecular forces need to be overcome in this process: no chemical bonds are broken.
10. intramolecular; intermolecular
12. The quantity of energy that must be applied to melt 1 mole of the substance.
14. (a) In going from a liquid to a gas, a considerably larger amount of the heat being applied has to be converted to the kinetic energy of the atoms escaping from the liquid; (b) 10.9 kJ; (c) −2.00 kJ (heat is evolved); (d) 1.13 kJ
16. 2.44 kJ; −10.6 kJ (heat is evolved)
18. 2.60 kJ/mol **20.** weaker
22. The hydrogen bonding that can exist when H is bonded to O (or N or F) is an additional intermolecular force, which means additional energy must be added to separate the molecules during boiling.
24. London dispersion forces are instantaneous dipole forces that arise when the electron cloud of an atom is momentarily distorted by a nearby dipole, temporarily separating the centers of positive and negative charge in the atom.
26. (a) London dispersion forces; (b) London dispersion forces; (c) dipole–dipole forces, London dispersion forces; (d) hydrogen bonding (H bonded to O), London dispersion forces
28. An increase in the heat of fusion is observed for an increase in the size of the halogen atom (the electron cloud of a larger atom is more easily polarized by a neighboring dipole, thus giving larger London dispersion forces).
30. For a homogeneous mixture to form, the forces between molecules of the two substances being mixed must be at least *comparable in magnitude* to the intermolecular forces within each separate substance. In the case of a water–ethanol mixture, the forces that exist when water and ethanol are mixed are stronger than water–water or ethanol–ethanol forces in the separate substances. Ethanol and water molecules can approach one another more closely in the mixture than either substance's molecules could approach a like molecule in the separate substances. Strong hydrogen bonding occurs in both ethanol and water.
32. When a liquid is placed into a closed container, a dynamic equilibrium is set up, in which vaporization of the liquid and condensation of the vapor are occurring at the same rate. Once the equilibrium has been achieved, there is a net concentration of molecules in the vapor state, which gives rise to the observed vapor pressure.
34. A liquid is injected at the bottom of the column of mercury and rises to the surface of the mercury, where the liquid evaporates into the vacuum above the mercury column. As the liquid evaporates, the pressure of the vapor increases in the space above the mercury and presses down on the mercury. The level of mercury therefore drops, and the amount by which the mercury level drops (in mm Hg) is equivalent to the vapor pressure of the liquid.
36. (a) H_2S. H_2O exhibits hydrogen bonding, and H_2S does not.
(b) CH_3OH. H_2O exhibits stronger hydrogen bonding than CH_3OH, because there are two locations where hydrogen bonding is possible on water. (c) CH_3OH. Both are capable of hydrogen bonding, but generally lighter molecules are more volatile than heavier molecules.
38. Both substances have the same molar mass. Ethyl alcohol contains a hydrogen atom directly bonded to an oxygen atom, however. Therefore, hydrogen bonding can exist in ethyl alcohol, whereas only weak dipole–dipole forces exist in dimethyl ether. Dimethyl ether is more volatile; ethyl alcohol has a higher boiling point.

40. *Ionic* solids have positive and negative ions as their fundamental particles; a simple example is sodium chloride, in which Na^+ and Cl^- ions are held together by strong electrostatic forces. *Molecular* solids have molecules as their fundamental particles, with the molecules being held together in the crystal by dipole–dipole forces, hydrogen-bonding forces, or London dispersion forces (depending on the identity of the substance); simple examples of molecular solids include ice (H_2O) and ordinary table sugar (sucrose). *Atomic* solids have simple atoms as their fundamental particles, with the atoms being held together in the crystal by either covalent bonding (as in graphite or diamond) or metallic bonding (as in copper or other metals).
42. sugar: molecular solid, relatively "soft," melts at a relatively low temperature, dissolves as molecules, does not conduct electricity when dissolved or melted; salt: ionic solid, relatively "hard," melts at a high temperature, dissolves as positively and negatively charged ions, conducts electricity when dissolved or melted.
44. Strong electrostatic forces exist between oppositely charged ions in ionic solids.
46. In liquid hydrogen, the only intermolecular forces are weak London dispersion forces. In ethyl alcohol and water, hydrogen bonding is possible, but the hydrogen bonding forces are weaker in ethyl alcohol because of the influence of the remainder of the molecule. In sucrose, hydrogen bonding also is possible but now at several places in the molecule, leading to stronger forces. In calcium chloride, there exists an ionic crystal lattice with even stronger forces between the particles.
48. Although ions exist in both the solid and liquid states, in the solid state the ions are rigidly held in place in the crystal lattice and cannot move so as to conduct an electric current.
50. Nitinol is an alloy of nickel and titanium. When nickel and titanium are heated to a sufficiently high temperature during the production of Nitinol, the atoms arrange themselves in a compact and regular pattern of the atoms.
52. j 54. f 56. d 58. a 60. l
62. Diethyl ether has the larger vapor pressure. No hydrogen bonding is possible because the O atom does not have a hydrogen atom attached. Hydrogen bonding can occur *only* when a hydrogen atom is *directly* attached to a strongly electronegative atom (such as N, O, or F). Hydrogen bonding *is* possible in 1-butanol (1-butano contains an —OH group).
64. None of the substances listed exhibit hydrogen bonding interactions.
66. substitutional; interstitial
68. Water is the solvent in which cellular processes take place in living creatures. Water in the oceans moderates the earth's temperature. Water is used in industry as a cooling agent, and it serves as a means of transportation. The liquid range is 0 °C to 100 °C at 1 atm pressure.
70. At higher altitudes, the boiling points of liquids are lower because there is a lower atmospheric pressure above the liquid. The temperature at which food cooks is determined by the temperature to which the water in the food can be heated before it escapes as steam. Thus, food cooks at a lower temperature at high elevations where the boiling point of water is lowered.
72. Heat of fusion (melt); heat of vaporization (boil). The heat of vaporization is always larger, because virtually all of the intermolecular forces must be overcome to form a gas. In a liquid, considerable intermolecular forces remain. Going from a solid to a liquid requires less energy than going from a liquid to a gas.
74. Dipole–dipole interactions are typically 1% as strong as a covalent bond. Dipole–dipole interactions represent electrostatic attractions between portions of molecules that carry only a *partial* positive or negative charge, and such forces require the molecules that are interacting to come *near* each other.
76. London dispersion forces are relatively weak forces that arise among noble gas atoms and in nonpolar molecules. London forces arise from *instantaneous dipoles* that develop when one atom (or molecule) momentarily distorts the electron cloud of another atom (or molecule). London forces are typically weaker than either permanent dipole–dipole forces or covalent bonds.
78. (a) London dispersion forces (nonpolar atoms); (b) hydrogen bonding (H attached to O), London dispersion forces; (c) dipole–dipole (polar molecules), London dispersion forces; (d) London dispersion forces (nonpolar molecules)
80. (a) HF contains a stronger polar bond compared to HCl because of the greater electronegativity difference between its two atoms; therefore, it has more unequal sharing of electrons. (b) HF contains stronger dipole–dipole interactions because the molecule itself is more polar. It has more charge separation within the molecule, which leads to stronger dipole–dipole interactions between the molecules. (c) HCl would boil first because the intermolecular forces are not as strong as in HF. HCl exhibits dipole–dipole interactions, but HF exhibits hydrogen bonding (which is a stronger form of dipole–dipole interactions). It would take less energy to disturb the dipole–dipole interactions in HCl and make it go to the gas phase. (HF would require more energy and thus has a higher boiling point.)
82. Strong *hydrogen-bonding* forces are present in an ice crystal, while only the much weaker *London forces* exist in the crystal of a nonpolar substance like oxygen.
84. Ice floats on liquid water; water expands when it is frozen.
86. Although they are at the same *temperature*, steam at 100 °C contains a larger amount of *energy* than hot water, equal to the heat of vaporization of water.
88. Hydrogen bonding is a special case of dipole–dipole interactions that occur among molecules containing hydrogen atoms bonded to highly electronegative atoms such as fluorine, oxygen, or nitrogen. The bonds are very polar, and the small size of the hydrogen atom (compared to other atoms) allows the dipoles to approach each other very closely. Examples: H_2O, NH_3, HF.
90. Evaporation and condensation are opposite processes. Evaporation is an endothermic process; condensation is an exothermic process. Evaporation requires an input of energy to provide the increased kinetic energy possessed by the molecules when they are in the gaseous state. It occurs when the molecules in a liquid are moving fast enough and randomly enough that molecules are able to escape from the surface of the liquid and enter the vapor phase.
92. Diamonds are made of only one element (carbon). The very strong covalent bonds among the carbon atoms in diamond lead to a giant molecule, and these types of substances are referred to as network solids.
94. a, d, and e are true.
96. CH_3Cl, CH_3CH_2Cl, $CH_3CH_2CH_2Cl$, $CH_3CH_2CH_2CH_2Cl$
98. b, c, and d are true.

Chapter 15

2. A *non*homogeneous mixture may differ in composition in various places in the mixture, whereas a solution (a homogeneous mixture) has the same composition throughout. Examples of nonhomogeneous mixtures include spaghetti sauce, a jar of jelly beans, and a mixture of salt and sugar.
4. solvent; solutes
6. "Like dissolves like." The hydrocarbons in oil have intermolecular forces that are very different from those in water, so the oil spreads out rather than dissolving in the water.
8. Carbon dioxide is somewhat soluble in water, especially if pressurized (otherwise, the soda you may be drinking while studying chemistry would be "flat"). Carbon dioxide's solubility in water is approximately 1.5 g/L at 25 °C under a pressure of approximately 1 atm. The carbon dioxide molecule overall is nonpolar, because the two individual C—O bond dipoles cancel each other due to the linearity of the molecule. However, these bond dipoles are able to interact with water, making CO_2 more soluble in water than nonpolar molecules such as O_2 or N_2, which do not possess individual bond dipoles.

10. unsaturated **12.** large **14.** 100.

16. (a) 5.00%; (b) 5.00%; (c) 5.00%; (d) 5.00%

18. (a) 20.5 g $FeCl_3$; 504.5 g (505 g) water; (b) 26.8 g sucrose; 198.2 g (198 g) water; (c) 181.3 g (181 g) NaCl; 1268.7 (1.27×10^3) g water; (d) 95.9 g KNO_3; 539.1 (539) g water

20. 957 g Fe; 26.9 g C; 16.5 g Cr **22.** 19.6% $CaCl_2$

24. 7.81 g KBr **26.** approximately 71 g **28.** 9.5 g

30. 0.110 mol; 0.220 mol **32.** b

34. (a) 3.35 *M*; (b) 1.03 *M*; (c) 0.630 *M*; (d) 4.99 *M*

36. (a) 0.403 *M*; (b) 0.169 *M*; (c) 0.629 *M*; (d) 0.829 *M*

38. 6.04 g NH_4Cl **40.** 0.0902 *M* **42.** 4.25 g $AgNO_3$

44. (a) 0.00130 mol; (b) 0.00609 mol; (c) 0.0184 mol; (d) 0.0356 mol

46. (a) 0.235 g; (b) 0.593 g; (c) 2.29 g; (d) 2.61 g

48. 9.51 g

50. (a) 4.60×10^{-3} mol Al^{3+}, 1.38×10^{-2} mol Cl^-; (b) 1.70 mol Na^+, 0.568 mol PO_4^{3-}; (c) 2.19×10^{-3} mol Cu^{2+}, 4.38×10^{-3} mol Cl^-; (d) 3.96×10^{-5} mol Ca^{2+}, 7.91×10^{-5} mol OH^-

52. 1.33 g **54.** half

56. (a) 0.0837 *M*; (b) 0.320 *M*; (c) 0.0964 *M*; (d) 0.622 *M*

58. 0.541 L (541 mL)

60. Dilute 48.3 mL of the 1.01 *M* solution to a final volume of 325 mL.

62. 10.3 mL **64.** 31.2 mL **66.** 0.523 g

68. 0.300 g **70.** 378 mL **72.** 1.8×10^{-4} *M*

74. (a) 63.0 mL; (b) 2.42 mL; (c) 50.1 mL; (d) 1.22 L

76. 1 *N*

78. 1.53 equivalents OH^- ion. By definition, one equivalent of OH^- ion exactly neutralizes one equivalent of H^+ ion.

80. (a) 0.277 *N*; (b) 3.37×10^{-3} *N*; (c) 1.63 *N*

82. (a) 0.134 *N*; (b) 0.0104 *N*; (c) 13.3 *N*

84. 7.03×10^{-5} *M*, 1.41×10^{-4} *N*

86. 22.2 mL, 11.1 mL **88.** 0.05583 *M*, 0.1117 *N*

90. Molarity is defined as the number of moles of solute contained in 1 liter of *total* solution volume (solute plus solvent after mixing). In the first example, the total volume after mixing is *not* known and the molarity cannot be calculated. In the second example, the final volume after mixing is known and the molarity can be calculated simply.

92. 3.3% **94.** 12.7 g $NaHCO_3$ **96.** 56 mol

98. 1.12 L HCl at STP **100.** 26.3 mL **102.** 1.8 *M* HCl

104. (a) 0.719 g NaCl; (b) 0.719 g NaCl; (c) 0.49 g NaCl; (d) 55.6 g NaCl

106. 4.7% C, 1.4% Ni, 93.9% Fe

108. 28 g Na_2CO_3 **110.** 9.4 g NaCl, 3.1 g KBr

112. (a) 2.0 *M*; (b) 1.0 *M*; (c) 0.67 *M*; (d) 0.50 *M*

114. 0.812 *M*

116. (a) 3.00 moles of Fe^{3+}, 9.00 moles of Cl^-; (b) 556 g $PbCl_2$

118. (a) 0.446 mol, 33.3 g; (b) 0.00340 mol, 0.289 g;(c) 0.075 mol, 2.7 g; (d) 0.0505 mol, 4.95 g

120. (a) 0.938 mol Na^+, 0.313 mol PO_4^{3-}; (b) 0.042 mol H^+, 0.021 mol SO_4^{2-}; (c) 0.0038 mol Al^{3+}, 0.011 mol Cl^-; (d) 1.88 mol Ba^{2+}, 3.75 mol Cl^-

122. (a) 0.0909 *M*; (b) 0.127 *M*; (c) 0.192 *M*; (d) 1.6 *M*

124. 0.90 *M*

126. (a) 0.500 *M*; (b) 250. mL

128. 4.7 mL

130. (a) 0.822 *N* HCl; (b) 4.00 *N* H_2SO_4; (c) 3.06 *N* H_3PO_4

132. 0.083 *M* NaH_2PO_4, 0.17 *N* NaH_2PO_4

134. 9.6×10^{-2} *N* HNO_3

136. Solution 4 contains the greatest number of ions.

138. 2.979×10^{-3} *M* **140.** 23.2 g $Ba_3(PO_4)_2$

142. 0.327 *M* $Ca(OH)_2$

Chapter 16

2. $HCl(g) \xrightarrow{H_2O} H^+(aq) + Cl^-(aq)$; $NaOH(s) \xrightarrow{H_2O} Na^+(aq) + OH^-(aq)$

4. A conjugate acid–base pair differs by one hydrogen ion, H^+. For example, $HC_2H_3O_2$ (acetic acid) differs from its conjugate base, $C_2H_3O_2^-$ (acetate ion), by a single H^+ ion.

$$HC_2H_3O_2(aq) \rightleftharpoons C_2H_3O_2^-(aq) + H^+(aq)$$

6. acids; bases

8. (a) not a conjugate pair; H_2SO_4, HSO_4^-; HSO_4^-, SO_4^{2-}; (b) conjugate pair; (c) not a conjugate pair; $HClO_4$, ClO_4^-; HCl, Cl^-; (d) not a conjugate pair; NH_4^+, NH_3; NH_3, NH_2^-

10. (a) NH_3 (base), NH_4^+ (acid); H_2O (acid), OH^- (base); (b) PO_4^{3-} (base), H_2O (acid); HPO_4^{2-} (acid), OH^- (base); (c) $C_2H_3O_2^-$ (base), H_2O (acid); $HC_2H_3O_2$ (acid), OH^- (base)

12. (a) HClO; (b) HCl; (c) $HClO_3$; (d) $HClO_4$

14. (a) BrO^-; (b) NO_2^-; (c) SO_3^{2-}; (d) CH_3NH_2

16. (a) $O^{2-}(aq) + H_2O(l) \rightleftharpoons OH^-(aq) + OH^-(aq)$; (b) $NH_3(aq) + H_2O(l) \rightleftharpoons NH_4^+(aq) + OH^-(aq)$; (c) $HSO_4^-(aq) + H_2O(l) \rightleftharpoons SO_4^{2-}(aq) + H_3O^+(aq)$; (d) $HNO_2(aq) + H_2O(l) \rightleftharpoons NO_2^-(aq) + H_3O^+(aq)$

18. If an acid is weak in aqueous solution, it does not easily transfer protons to water (and does not fully ionize). If an acid does not lose protons easily, then the acid's anion must strongly attract protons.

20. A strong acid loses its protons easily and fully ionizes in water; the acid's conjugate base is poor at attracting and holding protons and is a relatively weak base. A weak acid resists loss of its protons and does not ionize to a great extent in water; the acid's conjugate base attracts and holds protons tightly and is a relatively strong base.

22. H_2SO_4 (sulfuric): $H_2SO_4 + H_2O \rightarrow HSO_4^- + H_3O^+$; HCl (hydrochloric): $HCl + H_2O \rightarrow Cl^- + H_3O^+$; HNO_3 (nitric): $HNO_3 + H_2O \rightarrow NO_3^- + H_3O^+$; $HClO_4$ (perchloric): $HClO_4 + H_2O \rightarrow ClO_4^- + H_3O^+$

24. An oxyacid is an acid containing a particular element that is bonded to one or more oxygen atoms. HNO_3, H_2SO_4, and $HClO_4$ are oxyacids. HCl, HF, and HBr are not oxyacids.

26. Salicylic acid is a monoprotic acid: only the hydrogen of the carboxyl group ionizes.

28. HCO_3^- can behave as an acid if it reacts with a substance that more strongly gains protons than does HCO_3^- itself. For example, HCO_3^- would behave as an acid when reacting with hydroxide ion (a much stronger base): $HCO_3^-(aq) + OH^-(aq) \rightarrow CO_3^{2-}(aq) + H_2O(l)$. On the other hand, HCO_3^- would behave as a base when reacted with a substance that more readily loses protons than does HCO_3^- itself. For example, HCO_3^- would behave as a base when reacting with hydrochloric acid (a much stronger acid): $HCO_3^-(aq) + HCl(aq) \rightarrow H_2CO_3(aq) + Cl^-(aq)$. $H_2PO_4^- + OH^- \rightarrow HPO_4^{2-} + H_2O$ and $H_2PO_4^- + H_3O^+ \rightarrow H_3PO_4 + H_2O$.

30. The concentrations of H^+ and OH^- ions in water and in dilute aqueous solutions are *not* independent of one another. Rather, they are related by the ion product equilibrium constant, K_w. $K_w = [H^+(aq)][OH^-(aq)] = 1.00 \times 10^{-14}$ at 25 °C. If the concentration of one ion is *increased* by addition of a reagent producing H^+ or OH^-, then the concentration of the complementary ion will *decrease* so that the constant's value will hold true. If an acid is added to a solution, the concentration of hydroxide ion in the solution will decrease. Similarly, if a base is added to a solution, then the concentration of hydrogen ion will decrease.

32. (a) $[H^+] = 2.5 \times 10^{-10}$ *M*; basic; (b) $[H^+] = 3.4 \times 10^{-6}$ *M*; acidic; (c) $[H^+] = 1.4 \times 10^{-13}$ *M*; basic; (d) $[H^+] = 1.1 \times 10^{-8}$ *M*; basic

34. (a) $[OH^-] = 9.8 \times 10^{-8}$ *M*; acidic; (b) $[OH^-] = 1.02 \times 10^{-7}$ *M* (1.0×10^{-7} *M*); basic; (c) $[OH^-] = 2.9 \times 10^{-12}$ *M*; acidic; (d) $[OH^-] = 2.1 \times 10^{-4}$ *M*; basic

36. (a) $[OH^-] = 6.03 \times 10^{-4}\ M$; (b) $[OH^-] = 4.21 \times 10^{-6}\ M$; (c) $[OH^-] = 8.04 \times 10^{-4}\ M$
38. Answers will depend on student choices.
40. pH 1–2, deep red; pH 4, purple; pH 8, blue; pH 11, green
42. (a) 3.000 (acidic); (b) 3.660 (acidic); (c) 10.037 (basic); (d) 6.327 (acidic)
44. (a) pH = 11.94 (basic); (b) pH = 8.87 (basic); (c) pH = 5.97 (acidic); (d) pH = 3.08 (acidic)
46. (a) 4.22 (basic); (b) 9.99 (acidic); (c) 11.21 (acidic); (d) 2.79 (basic)
48. (a) pH = 1.719, $[OH^-] = 5.2 \times 10^{-13}\ M$; (b) pH = 6.316, $[OH^-] = 2.1 \times 10^{-8}\ M$; (c) pH = 10.050, $[OH^-] = 1.1 \times 10^{-4}\ M$; (d) pH = 4.212, $[OH^-] = 1.6 \times 10^{-10}\ M$
50. (a) 0.091 M; (b) $8 \times 10^{-14}\ M$; (c) $1.0 \times 10^{-6}\ M$; (d) $2.4 \times 10^{-9}\ M$
52. (a) $9.8 \times 10^{-10}\ M$; (b) $1.8 \times 10^{-8}\ M$; (c) $5.5 \times 10^{-4}\ M$; (d) $5.6 \times 10^{-3}\ M$
54. (a) 5.358; (b) 3.64; (c) 5.97; (d) 0.480
56. The solution contains water molecules, H_3O^+ ions (protons), and NO_3^- ions. Because HNO_3 is a strong acid that is completely ionized in water, no HNO_3 molecules are present.
58. (a) pH = 2.917; (b) pH = 3.701; (c) pH = 4.300; (d) pH = 2.983
60. A buffered solution consists of a mixture of a weak acid and its conjugate base; one example of a buffered solution is a mixture of acetic acid ($HC_2H_3O_2$) and sodium acetate ($NaC_2H_3O_2$).
62. The weak acid component of a buffered solution is capable of reacting with added strong base. For example, using the buffered solution given as an example in Exercise 60, acetic acid would consume added sodium hydroxide as follows: $HC_2H_3O_2(aq) + NaOH(aq) \rightarrow NaC_2H_3O_2(aq) + H_2O(l)$. Acetic acid *neutralizes* the added NaOH and prevents it from affecting the overall pH of the solution.
64. HCl: $H_3O^+ + C_2H_3O_2^- \rightarrow HC_2H_3O_2 + H_2O$; NaOH: $OH^- + HC_2H_3O_2 \rightarrow C_2H_3O_2^- + H_2O$
66. (a) $[OH^-(aq)] = 0.10\ M$, pOH = 1.00, pH = 13.00; (b) $[OH^-(aq)] = 2.0 \times 10^{-4}\ M$, pOH = 3.70, pH = 10.30; (c) $[OH^-(aq)] = 6.2 \times 10^{-3}\ M$, pOH = 2.21, pH = 11.79; (d) $[OH^-(aq)] = 0.0001\ M$, pOH = 4.0, pH = 10.0
68. b, c, d
70. (a) CH_3COO^- is a relatively strong base; (b) F^- is a relatively strong base; (c) HS^- is a relatively strong base; (d) Cl^- is a very weak base
72. Having a concentration as small as $10^{-7}\ M$ for HCl means that the contribution to the total hydrogen ion concentration from the dissociation of water must also be considered in determining the pH of the solution.
74. accepts 76. base
78. $-C(=O)OH$; $CH_3COOH + H_2O \rightleftharpoons C_2H_3O_2^- + H_3O^+$
80. $1.0 \times 10^{-14}\ M$ 82. lower 84. pH 86. weak acid
88. (a) Equation 1: (acid1) + (base1) → (conjugate acid1) + (conjugate base1)
Equation 2: (base2) + (acid2) → (conjugate acid2) + (conjugate base2)
The acids are the proton donors, and the bases are the proton acceptors. By looking at which species is positively charged or negatively charged in the products, it's possible to determine which reactant is the proton donor and which is the proton acceptor.
(b) An Arrhenius acid produces hydrogen ions. An Arrhenius base produces hydroxide ions. Therefore, acid1 is considered an Arrhenius acid. A Brønsted–Lowry acid is a proton donor, and a Brønsted–Lowry base is a proton acceptor. Thus, acid1 and acid2 are both Brønsted–Lowry acids, and base1 and base2 are both Brønsted–Lowry bases.
90. (a) NH_4^+; (b) NH_3; (c) H_3O^+; (d) H_2O
92. (a) A buffer: HCN and NaCN are conjugates. (b) Not a buffer: PO_4^{3-} (of K_3PO_4) is not the conjugate base of H_3PO_4. (c) A buffer: HF and KF are conjugates. (d) A buffer: $HC_3H_5O_2$ and $NaC_3H_5O_2$ are conjugates.
94. (a) $[H^+(aq)] = 2.4 \times 10^{-12}\ M$, solution is basic; (b) $[H^+(aq)] = 9.9 \times 10^{-2}\ M$, solution is acidic; (c) $[H^+(aq)] = 3.3 \times 10^{-8}\ M$, solution is basic; (d) $[H^+(aq)] = 1.7 \times 10^{-9}\ M$, solution is basic
96. (a) 9.68 (basic); (b) 5.10 (acidic); (c) 12.19 (basic); (d) 0.9 (acidic)
98. (a) pH = 8.15; solution is basic; (b) pH = 5.97; solution is acidic; (c) pH = 13.34; solution is basic; (d) pH = 2.90; solution is acidic
100. (a) $[OH^-(aq)] = 1.8 \times 10^{-11}\ M$, pH = 3.24, pOH = 10.76; (b) $[H^+(aq)] = 1.1 \times 10^{-10}\ M$, pH = 9.95, pOH = 4.05; (c) $[OH^-(aq)] = 3.5 \times 10^{-3}\ M$, pH = 11.54, pH = 2.46; (d) $[H^+(aq)] = 1.4 \times 10^{-7}\ M$, pH = 6.86, pOH = 7.14
102. (a) $[H^+] = 3.9 \times 10^{-6}\ M$; (b) $[H^+] = 1.1 \times 10^{-2}\ M$; (c) $[H^+] = 1.2 \times 10^{-12}\ M$; (d) $[H^+] = 7.8 \times 10^{-11}\ M$
104. (a) $[H^+(aq)] = 1.4 \times 10^{-3}\ M$, pH = 2.85; (b) $[H^+(aq)] = 3.0 \times 10^{-5}\ M$, pH = 4.52; (c) $[H^+(aq)] = 5.0 \times 10^{-2}\ M$, pH = 1.30; (d) $[H^+(aq)] = 0.0010\ M$, pH = 3.00
106. a and d
108. NaCl: neutral; RbOCl: basic; KI: neutral; $Ba(ClO_4)_2$: neutral; NH_4NO_3: acidic

Answers to Even-Numbered Cumulative Review Exercises

Chapters 1–3

2. After having covered three chapters in this book, you should have adopted an "active" approach to your study of chemistry. You can't just sit and take notes in class, or just review the solved examples in the textbook. You must learn to *interpret* problems and reduce them to simple mathematical relationships.
4. Some courses, particularly those in your major field, have obvious and immediate utility. Other courses—chemistry included—provide general *background* knowledge that will prove useful in understanding your own major, or other subjects related to your major.
6. Whenever a scientific measurement is made, we always employ the instrument or measuring device to the limits of its precision. This usually means that we *estimate* the last significant figure of the measurement. An example of the uncertainty in the last significant figure is given for measuring the length of a pin in the text in Fig. 2.5. Scientists appreciate the limits of experimental techniques and instruments and always *assume* that the last digit in a number representing a measurement has been estimated. Because instruments or measuring devices always have a limit to their precision, uncertainty cannot be completely excluded from measurements.
8. Dimensional analysis is a method of problem solving that pays particular attention to the units of measurements and uses these units as if they were algebraic symbols that multiply, divide, and cancel. Consider the following example: One dozen eggs costs $1.25. Suppose we want to know how much one egg costs, and also how much three dozen eggs will cost. To solve these problems, we need two equivalence statements:

   1 dozen eggs = 12 eggs

   1 dozen eggs = $1.25

   The calculations are

   $$\frac{\$1.25}{12\text{ eggs}} = \$0.104 = \$0.10$$

   as the cost of one egg and

   $$\frac{\$1.25}{1\text{ dozen}} \times 3\text{ dozen} = \$3.75$$

   as the cost of three dozen eggs. See Section 2-6 of the text for how we construct conversion factors from equivalence statements.
10. Scientists say that matter is anything that "has mass and occupies space." Matter is the "stuff" of which everything is made. It can be classified and subdivided in many ways, depending on what we are trying to demonstrate. All the types of matter we have studied are made of atoms. They differ in whether these atoms are all of one element, or are of more than one element, and also in whether these atoms are in physical mixtures or chemical combinations.
    Matter can also be classified according to its physical state (solid, liquid, or gas). In addition, it can be classified as a pure substance (one type of molecule) or a mixture (more than one type of molecule).
12. An element is a fundamental substance that cannot be broken down into simpler substances by chemical methods. An element consists of atoms of only one type. Compounds, on the other hand, *can* be broken down into simpler substances. For example, both sulfur and oxygen are *elements*. When sulfur and oxygen are placed together and heated, the *compound* sulfur dioxide (SO_2) forms. Each molecule of sulfur dioxide contains one sulfur atom and two oxygen atoms. On a mass basis, SO_2 always consists of 50% each, by mass, sulfur and oxygen—that is, sulfur dioxide has a constant composition. Sulfur dioxide from any source would have the same composition (or it wouldn't be sulfur dioxide!).
14. (a) 8.917×10^{-4}; (b) 0.0002795; (c) 4913; (d) 8.51×10^{7}; (e) 1.219×10^{2}; (f) 3.396×10^{-9}
16. (a) two; (b) two; (c) three; (d) three; (e) one; (f) two; (g) two; (h) three
18. (a) 0.785 g/mL; (b) 2.03 L; (c) 1.06 kg; (d) 9.33 cm^3; (e) 2.0×10^{2} g

Chapters 4–5

2. Although you don't have to memorize all the elements, you should at least be able to give the symbol or name for the most common elements (listed in Table 4.3).
4. The main postulates of Dalton's theory are: (1) elements are made up of tiny particles called atoms; (2) all atoms of a given element are identical; (3) although all atoms of a given element are identical, these atoms are different from the atoms of all other elements; (4) atoms of one element can combine with atoms of another element to form a compound that will always have the same relative numbers and types of atoms for its composition; and (5) atoms are merely rearranged into new groupings during an ordinary chemical reaction, and no atom is ever destroyed and no new atom is ever created during such a reaction.
6. The expression "nuclear atom" indicates that the atom has a dense center of positive charge (nucleus) around which the electrons move through primarily empty space. Rutherford's experiment involved shooting a beam of α particles at a thin sheet of metal foil. According to the "plum pudding" model of the atom, these positively charged α particles should have passed through the foil. Rutherford detected that a small number of α particles bounced backward to the source of α particles or were deflected from the foil at large angles. Rutherford realized that his observations could be explained if the atoms of the metal foil had a small, dense, positively charged nucleus, with a significant amount of empty space between nuclei. The empty space between nuclei would allow most of the α particles to pass through the foil. If an α particle were to hit a nucleus head-on, it would be deflected backward. If a positively charged α particle passed *near* a positively charged nucleus, then the α particle would be deflected by the repulsive forces. Rutherford's experiment disproved the "plum pudding" model, which envisioned the atom as a uniform sphere of positive charge, with enough negatively charged electrons scattered throughout to balance out the positive charge.
8. Isotopes represent atoms of the same element that have different atomic masses. Isotopes result from the different numbers of neutrons in the nuclei of atoms of a given element. They have the same atomic number (number of protons in the nucleus) but have different mass numbers (total number of protons and neutrons in the nucleus). The different isotopes of an atom are indicated by the form ${}^{A}_{Z}\text{X}$, in which Z represents the atomic number and A the mass number of element X. For example, ${}^{13}_{6}\text{C}$ represents a nuclide of carbon with atomic number 6 (6 protons in the nucleus) and mass number 13 (6 protons plus 7 neutrons in the nucleus). The various isotopes of an element have identical *chemical* properties. The *physical* properties of the isotopes of an element may differ slightly because of the small difference in mass.

10. Most elements are too reactive to be found in nature in other than the combined form. Gold, silver, platinum, and some of the gaseous elements (such as O_2, N_2, He, and Ar) are found in the elemental form.
12. Ionic compounds typically are hard, crystalline solids with high melting and boiling points. The ability of aqueous solutions of ionic substances to conduct electricity means that ionic substances consist of positively and negatively charged particles (ions). A sample of an ionic substance has no net electrical charge because the total number of positive charges is *balanced* by an equal number of negative charges. An ionic compound could not consist of only cations or only anions because a net charge of zero cannot be obtained when all ions have the same charge. Also, ions of like charge will repel each other.
14. When naming ionic compounds, the positive ion (cation) is named first. For simple binary Type I ionic compounds, the ending *-ide* is added to the root name of the negative ion (anion). For example, the name for K_2S would be "potassium sulf*ide*"—potassium is the cation, sulfide is the anion. Type II compounds, which involve elements that form more than one stable ion, are named by either of two systems: the Roman numeral system (which is preferred by most chemists) and the *-ous*/*-ic* system. For example, iron can react with oxygen to form either of two stable oxides, FeO or Fe_2O_3. Under the Roman numeral system, FeO would be named iron(II) oxide to show that it contains Fe^{2+} ions; Fe_2O_3 would be named iron(III) oxide to indicate that it contains Fe^{3+} ions. Under the *-ous*/*-ic* system, FeO is named ferr*ous* oxide and Fe_2O_3 is called ferr*ic* oxide. Type II compounds usually involve transition metals and nonmetals.
16. A polyatomic ion is an ion containing more than one atom. Some common polyatomic ions are listed in Table 5.4. Parentheses are used in writing formulas containing polyatomic ions to indicate how many polyatomic ions are present. For example, the correct formula for calcium phosphate is $Ca_3(PO_4)_2$, which indicates that three calcium ions are combined for every two phosphate ions. If we did *not* write the parentheses around the formula for the phosphate ion (that is, if we wrote Ca_3PO_{42}), people might think that 42 oxygen atoms were present!
18. Acids are substances that produce protons (H^+ ions) when dissolved in water. For acids that do *not* contain oxygen, the prefix *hydro-* and the suffix *-ic* are used with the root name of the element present in the acid (for example: HCl, *hydro*chlor*ic* acid; H_2S, *hydro*sulfur*ic* acid; HF, *hydro*fluor*ic* acid). For acids whose anions contain oxygen, a series of prefixes and suffixes is used with the name of the central atom in the anion: these prefixes and suffixes indicate the relative (not actual) number of oxygen atoms present in the anion. Most elements that form oxyanions form *two* such anions—for example, sulfur forms sulf*ite* ion (SO_3^{2-}) and sulf*ate* ion (SO_4^{2-}). For an element that forms two oxyanions, the acid containing the anions will have the ending *-ous* if the *-ite* anion is involved and the ending *-ic* if the *-ate* anion is present. For example, H_2SO_3 is sulfur*ous* acid and H_2SO_4 is sulfur*ic* acid. The Group 7 elements each form *four* oxyanions/oxyacids. The prefix *hypo-* is used for the oxyacid that contains fewer oxygen atoms than the *-ite* anion, and the prefix *per-* is used for the oxyacid that contains more oxygen atoms than the *-ate* anion. For example,

Acid	Name	Anion	Name
HBrO	*hypo*brom*ous* acid	BrO^-	hypobromite
$HBrO_2$	brom*ous* acid	BrO_2^-	bromite
$HBrO_3$	brom*ic* acid	BrO_3^-	bromate
$HBrO_4$	*per*brom*ic* acid	BrO_4^-	perbromate

20. Elements in the same family have the same electron configuration and tend to undergo similar chemical reactions with other groups. For example, Li, Na, K, Rb, and Cs all react with elemental chlorine gas, Cl_2, to form an ionic compound of general formula M^+Cl^-.
22. (a) 8, 8, 9; (b) 92, 92, 143; (c) 17, 17, 20; (d) 1, 1, 2; (e) 2, 2, 2; (f) 50, 50, 69; (g) 54, 54, 70; h) 30, 30, 34
24. (a) 12 protons, 10 electrons; (b) 26 protons, 24 electrons; (c) 26 protons, 23 electrons; (d) 9 protons, 10 electrons; (e) 28 protons, 26 electrons; (f) 30 protons; 28 electrons; (g) 27 protons, 24 electrons; (h) 7 protons, 10 electrons; (i) 16 protons, 18 electrons; (j) 37 protons, 36 electrons; (k) 34 protons, 36 electrons; (l) 19 protons, 18 electrons
26. (a) CuI; (b) $CoCl_2$; (c) Ag_2S; (d) Hg_2Br_2; (e) HgO; (f) Cr_2S_3; (g) PbO_2; (h) K_3N; (i) SnF_2; (j) Fe_2O_3
28. (a) NH_4^+, ammonium ion; (b) SO_3^{2-}, sulfite ion; (c) NO_3^-, nitrate ion; (d) SO_4^{2-}, sulfate ion; (e) NO_2^-, nitrite ion; (f) CN^-, cyanide ion; (g) OH^-, hydroxide ion; (h) ClO_4^-, perchlorate ion; (i) ClO^-, hypochlorite ion; (j) PO_4^{3-}, phosphate ion
30. (a) xenon dioxide; (b) iodine pentachloride; (c) phosphorus trichloride; (d) carbon monoxide; (e) oxygen diflouride; (f) diphosphorus pentoxide; (g) arsenic triiodide; (h) sulfur trioxide

Chapters 6–7

2. A chemical equation indicates the substances necessary for a given chemical reaction, and the substances produced by that chemical reaction. The substances to the left of the arrow are called the *reactants;* those to the right of the arrow are called the *products.* A *balanced* equation indicates the relative numbers of molecules in the reaction.
4. Never change the *subscripts* of a *formula:* changing the subscripts changes the *identity* of a substance and makes the equation invalid. When balancing a chemical equation, we adjust only the *coefficients* in front of a formula: changing a coefficient changes the *number* of molecules being used in the reaction, *without* changing the *identity* of the substance.
6. A precipitation reaction is one in which a solid is produced when two aqueous solutions are combined. The driving force in such a reaction is the formation of the solid, thus removing ions from the solution. Examples depend on student input.
8. Nearly all compounds containing the nitrate, sodium, potassium, and ammonium ions are soluble in water. Most salts containing the chloride and sulfate ions are soluble in water, with specific exceptions (see Table 7.1). Most compounds containing the hydroxide, sulfide, carbonate, and phosphate ions are *not* soluble in water (unless the compound also contains Na^+, K^+, or NH_4^+). For example, suppose we combine barium chloride and sulfuric acid solutions:

$$BaCl_2(aq) + H_2SO_4(aq) \rightarrow BaSO_4(s) + 2HCl(aq)$$
$$Ba^{2+}(aq) + SO_4^{2-}(aq) \rightarrow BaSO_4(s) \text{ [net ionic reaction]}$$

Because barium sulfate is not soluble in water, a precipitate of $BaSO_4(s)$ forms.
10. Acids (such as the acetic acid found in vinegar) were first noted primarily because of their sour taste, whereas bases were first characterized by their bitter taste and slippery feel on the skin. Acids and bases neutralize each other, forming water: $H^+(aq) + OH^-(aq) \rightarrow H_2O(l)$. Strong acids and bases ionize *fully* when dissolved in water, which means they are also strong electrolytes.
Strong acids: HCl, HNO_3, and H_2SO_4
Strong bases: Group 1 hydroxides (for example, NaOH and KOH)
12. Oxidation–reduction reactions; oxidation; reduction; No: if one species is going to lose electrons, there must be another species present capable of gaining them; Examples depend on student input.
14. In a synthesis reaction, elements or simple compounds react to produce more complex substances. For example,

$$N_2(g) + 3H_2(g) \rightarrow 2NH_3(g)$$
$$NaOH(aq) + CO_2(g) \rightarrow NaHCO_3(s)$$

Decomposition reactions represent the breakdown of complex substances into simpler substances. For example, $2H_2O_2(aq) \rightarrow 2H_2O(l) + O_2(g)$. Synthesis and decomposition reactions are often oxidation–reduction reactions, although not always. For example, the synthesis reaction between NaOH and CO_2 does *not* represent oxidation–reduction.

16. (a) $C(s) + O_2(g) \rightarrow CO_2(g)$; (b) $2C(s) + O_2(g) \rightarrow 2CO(g)$; (c) $2Li(l) + 2C(s) \rightarrow Li_2C_2(s)$; (d) $FeO(s) + C(s) \rightarrow Fe(l) + CO(g)$; (e) $C(s) + 2F_2(g) \rightarrow CF_4(g)$
18. (a) $Ba(NO_3)_2(aq) + K_2CrO_4(aq) \rightarrow BaCrO_4(s) + 2KNO_3(aq)$; (b) $NaOH(aq) + HC_2H_3O_2(aq) \rightarrow H_2O(l) + NaC_2H_3O_2(aq)$ (then evaporate the water from the solution); (c) $AgNO_3(aq) + NaCl(aq) \rightarrow AgCl(s) + NaNO_3(aq)$; (d) $Pb(NO_3)_2(aq) + H_2SO_4(aq) \rightarrow PbSO_4(s) + 2HNO_3(aq)$; (e) $2NaOH(aq) + H_2SO_4(aq) \rightarrow Na_2SO_4(aq) + 2H_2O(l)$ (then evaporate the water from the solution); (f) $Ba(NO_3)_2(aq) + 2Na_2CO_3(aq) \rightarrow BaCO_3(s) + 2NaNO_3(aq)$
20. (a) $FeO(s) + 2HNO_3(aq) \rightarrow Fe(NO_3)_2(aq) + H_2O(l)$; acid–base; double-displacement; (b) $2Mg(s) + 2CO_2(g) + O_2(g) \rightarrow 2MgCO_3(s)$; synthesis; oxidation–reduction; (c) $2NaOH(s) + CuSO_4(aq) \rightarrow Cu(OH)_5(s) + Na_2SO_4(aq)$; precipitation; double-displacement; (d) $HI(aq) + KOH(aq) \rightarrow KI(aq) + H_2O(l)$; acid–base; double-displacement; (e) $C_3H_8(g) + 5O_2(g) \rightarrow 3CO_2(g) + 4H_2O(g)$; combustion; oxidation–reduction; (f) $Co(NH_3)_6Cl_2(s) \rightarrow CoCl_2(s) + 6NH_3(g)$; decomposition; (g) $2HCl(aq) + Pb(C_2H_3O_2)_2(aq) \rightarrow 2HC_2H_3O_2(aq) + PbCl_2(aq)$; precipitation; double-displacement; (h) $C_{12}H_{22}O_{11}(s) \rightarrow 12C(s) + 11H_2O(g)$; decomposition; oxidation–reduction; (i) $2Al(s) + 6HNO_3(aq) \rightarrow 2Al(NO_3)_3(aq) + 3H_2(g)$; oxidation–reduction; single-displacement; (j) $4B(s) + 3O_2(g) \rightarrow 2B_2O_3(s)$; synthesis; oxidation–reduction
22. Answer will depend on student examples.
24. (a) no reaction (all combinations are soluble)
(b) $Ca^{2+}(aq) + SO_4^{2-}(aq) \rightarrow CaSO_4(s)$
(c) $Pb^{2+}(aq) + S^{2-}(aq) \rightarrow PbS(s)$
(d) $2Fe^{3+}(aq) + 3CO_3^{2-}(aq) \rightarrow Fe_2(CO_3)_3(s)$
(e) $Hg_2^{2+}(aq) + 2Cl^-(aq) \rightarrow Hg_2Cl_2(s)$
(f) $Ag^+(aq) + Cl^-(aq) \rightarrow AgCl(s)$
(g) $3Ca^{2+}(aq) + 2PO_4^{3-}(aq) \rightarrow Ca_3(PO_4)_2(s)$
(h) no reaction (all combinations are soluble)

Chapters 8–9

2. On a microscopic basis, one mole of a substance represents Avogadro's number (6.022×10^{23}) of individual units (atoms or molecules) of the substance. On a macroscopic basis, one mole of a substance represents the amount of substance present when the molar mass of the substance in grams is taken. Chemists have chosen these definitions so that a simple relationship will exist between measurable amounts of substances (grams) and the actual number of atoms or molecules present, and so that the number of particles present in samples of *different* substances can easily be compared.
4. The molar mass of a compound is the mass in grams of one mole of the compound and is calculated by summing the average atomic masses of all the atoms present in a molecule of the compound. For example, for H_3PO_4: molar mass $H_3PO_4 = 3(1.008\text{ g}) + 1(30.97\text{ g}) + 4(16.00\text{ g}) = 97.99\text{ g}$.
6. The *empirical* formula of a compound represents the *relative* number of atoms of each type present in a molecule of the compound, whereas the *molecular* formula represents the *actual* number of atoms of each type present in a real molecule. For example, both acetylene (molecular formula C_2H_2) and benzene (molecular formula C_6H_6) have the same relative number of carbon and hydrogen atoms, and thus have the same empirical formula (CH). The molar mass of the compound must be determined before calculating the actual molecular formula. Since real molecules cannot contain fractional *parts* of atoms, the molecular formula is always a *whole-number multiple* of the empirical formula.
8. Answer depends on student examples chosen for Exercise 7.
10. for O_2: $\frac{5\text{ mol }O_2}{1\text{ mol }C_3H_8}$; $0.55\text{ mol }C_3H_8 \times \frac{5\text{ mol }O_2}{1\text{ mol }C_3H_8} = 2.8\ (2.75)\text{ mol }O_2$
for CO_2: $\frac{3\text{ mol }CO_2}{1\text{ mol }C_3H_8}$; $0.55\text{ mol }C_3H_8 \times \frac{3\text{ mol }CO_2}{1\text{ mol }C_3H_8} = 1.7\ (1.65)\text{ mol }CO_2$
for H_2O: $\frac{4\text{ mol }H_2O}{1\text{ mol }C_3H_8}$; $0.55\text{ mol }C_3H_8 \times \frac{4\text{ mol }H_2O}{1\text{ mol }C_3H_8} = 2.2\text{ mol }H_2O$
12. When arbitrary amounts of reactants are used, one reactant will be present, stoichiometrically, in the least amount: this substance is called the *limiting reactant.* It *limits* the amount of product that can form in the experiment, because once this substance has reacted completely, the reaction must *stop.* The other reactants in the experiment are present *in excess,* which means that a portion of these reactants will be present *unchanged* after the reaction ends.
14. The *theoretical yield* for an experiment is the mass of product calculated assuming the limiting reactant for the experiment is completely consumed. The *actual yield* for an experiment is the mass of product actually collected by the experimenter. Any experiment is restricted by the skills of the experimenter and by the inherent limitations of the experimental method: for these reasons, the actual yield is often less than the theoretical yield. Although one would expect that the actual yield should never exceed the theoretical yield, in real experiments, sometimes this happens. However, an actual yield greater than a theoretical yield usually means that something is *wrong* in either the experiment (for example, impurities may be present) or the calculations.
16. (a) 92.26% C; (b) 32.37% Na; (c) 15.77% C; (d) 20.24% Al; (e) 88.82% Cu; (f) 79.89% Cu; (g) 71.06% Co; (h) 40.00% C
18. (a) 53.0 g $SiCl_4$, 3.75 g C; (b) 20.0 g LiOH; (c) 12.8 g NaOH, 2.56 g O_2; (d) 9.84 g Sn, 2.99 g H_2O
20. 11.7 g CO; 18.3 g CO_2

Chapters 10–12

2. Temperature is a measure of the random motions of the components of a substance; in other words, temperature is a measure of the average kinetic energy of the particles in a sample. The molecules in warm water must be moving faster than the molecules in cold water (the molecules have the same mass, so if the temperature is higher, the average velocity of the particles must be higher in the warm water). Heat is the energy that flows because of a difference in temperature.
4. Thermodynamics is the study of energy and energy changes. The first law of thermodynamics is the law of conservation of energy: the energy of the universe is constant. Energy cannot be created or destroyed, only transferred from one place to another or from one form to another. The internal energy of a system, *E,* represents the total of the kinetic and potential energies of all particles in a system. A flow of heat may be produced when there is a change in internal energy in the system, but it is not correct to say that the system "contains" the heat: part of the internal energy is *converted* to heat energy during the process (under other conditions, the change in internal energy might be expressed as work rather than a heat flow).
6. The enthalpy change represents the heat energy that flows (at constant pressure) on a molar basis when a reaction occurs. The enthalpy change is a state function (which we make great use of in Hess's law calculations). Enthalpy changes are typically measured in insulated reaction vessels called calorimeters (a simple calorimeter is shown in Fig. 10.6).
8. Consider petroleum. A gallon of gasoline contains concentrated, stored energy. We can use that energy to make a car move, but when we do, the energy stored in the gasoline is dispersed throughout the environment. Although the energy is still there (it is conserved), it is no longer in a concentrated, useful form. Thus, although the energy content of the universe remains constant, the energy that is now stored in concentrated forms in oil, coal, wood, and other sources is gradually being dispersed to the universe, where it can do no work.

10. A driving force is an effect that tends to make a process occur. Two important driving forces are dispersion of energy during a process or dispersion of matter during a process (energy spread and matter spread). For example, a log burns in a fireplace because the energy contained in the log is dispersed to the universe when it burns. If we put a teaspoon of sugar into a glass of water, the dissolving of the sugar is a favorable process because the matter of the sugar is dispersed when it dissolves. Entropy is a measure of the randomness or disorder in a system. The entropy of the universe is constantly increasing because of matter spread and energy spread. A spontaneous process is one that occurs without outside intervention: the spontaneity of a reaction depends on the energy spread and matter spread if the reaction takes place. A reaction that disperses energy and also disperses matter will always be spontaneous. Reactions that require an input of energy may still be spontaneous if the matter spread is large enough.
12. (a) 464 kJ; (b) 69.3 kJ; (c) 1.40 mol (22.5 g)
14. An atom in its *ground state* is in its lowest possible energy state. When an atom possesses more energy than in its ground state, the atom is in an *excited state.* An atom is promoted from its ground state to an excited state by absorbing energy; when the atom returns from an excited state to its ground state it emits the excess energy as electromagnetic radiation. Atoms do not gain or emit radiation randomly, but rather do so only in discrete bundles of radiation called *photons.* The photons of radiation emitted by atoms are characterized by the wavelength (color) of the radiation: longer-wavelength photons carry less energy than shorter-wavelength photons. The energy of a photon emitted by an atom corresponds *exactly* to the difference in energy between two allowed energy states in an atom.
16. Bohr pictured the electron moving in certain circular orbits around the nucleus, with each orbit being associated with a specific energy (resulting from the attraction between the nucleus and the electron and from the kinetic energy of the electron). Bohr assumed that when an atom absorbs energy, the electron moves from its ground state ($n = 1$) to an orbit farther away from the nucleus ($n = 2, 3, 4, \ldots$). Bohr postulated that when an excited atom returns to its ground state, the atom emits the excess energy as radiation. Because the Bohr orbits are located at fixed distances from the nucleus and from each other, when an electron moves from one fixed orbit to another, the energy change is of a definite amount, which corresponds to the emission of a photon with a particular characteristic wavelength and energy. When the simple Bohr model for the atom was applied to the emission spectra of other elements, however, the theory could not predict or explain the observed emission spectra of these elements.
18. The lowest-energy hydrogen atomic orbital is called the $1s$ orbital. The $1s$ orbital is spherical in shape (the electron density around the nucleus is uniform in all directions). The orbital does *not* have a sharp edge (it appears fuzzy) because the probability of finding the electron gradually decreases as distance from the nucleus increases. The orbital does *not* represent just a spherical surface on which the electron moves (this would be similar to Bohr's original theory)—instead, the $1s$ orbital represents a probability map of electron density around the nucleus for the first principal energy level.
20. The third principal energy level of hydrogen is divided into three sublevels: the $3s$, $3p$, and $3d$ sublevels. The $3s$ subshell consists of the single $3s$ orbital, which is spherical in shape. The $3p$ subshell consists of a set of three equal-energy $3p$ orbitals: each of these $3p$ orbitals has the same shape ("dumbbell"), but each of the $3p$ orbitals is oriented in a different direction in space. The $3d$ subshell consists of a set of five $3d$ orbitals with shapes as indicated in Fig. 11.28, which are oriented in different directions around the nucleus. The fourth principal energy level of hydrogen is divided into four sublevels: the $4s$, $4p$, $4d$, and $4f$ orbitals. The $4s$ subshell consists of the single $4s$ orbital. The $4p$ subshell consists of a set of three $4p$ orbitals. The $4d$ subshell consists of a set of five $4d$ orbitals. The shapes of the $4s$, $4p$, and $4d$ orbitals are the *same* as the shapes of the orbitals of the third principal energy level—the orbitals of the fourth principal energy level are *larger* and *farther from the nucleus* than the orbitals of the third level, however. The fourth principal energy level also contains a $4f$ subshell consisting of seven $4f$ orbitals (the shapes of the $4f$ orbitals are beyond the scope of this text).
22. Atoms have a series of *principal energy levels* indexed by the letter n. The $n = 1$ level is closest to the nucleus, and the energies of the levels increase as the value of n (and distance from the nucleus) increases. Each principal energy level is divided into *sublevels* (sets of orbitals) of different characteristic shapes designated by the letters *s, p, d,* and *f.* Each s subshell consists of a single s orbital; each p subshell consists of a set of three p orbitals; each d subshell consists of a set of five d orbitals; and so on. An orbital can be empty or it can contain one or two electrons, but never more than two electrons (if an orbital contains two electrons, then the electrons must have opposite spins). The shape of an orbital represents a probability map for finding electrons—it does not represent a trajectory or pathway for electron movements.
24. The valence electrons are the electrons in an atom's outermost shell. The valence electrons are those most likely to be involved in chemical reactions because they are at the outside edge of the atom.
26. The general periodic table you drew for Question 25 should resemble that found in Fig. 11.31. From the column and row location of an element, you should be able to determine its valence configuration. For example, the element in the third horizontal row of the second vertical column has $3s^2$ as its valence configuration. The element in the seventh vertical column of the second horizontal row has valence configuration $2s^22p^5$.
28. The ionization energy of an atom represents the energy required to remove an electron from the atom in the gas phase. Moving from top to bottom in a vertical group on the periodic table, the ionization energies decrease. The ionization energies increase when going from left to right within a horizontal row within the periodic table. The relative sizes of atoms also vary systematically with the location of an element on the periodic table. Within a given vertical group, the atoms become progressively larger when going from the top of the group to the bottom. Moving from left to right within a horizontal row on the periodic table, the atoms become progressively smaller.
30. To form an ionic compound, a metallic element reacts with a nonmetallic element, with the metallic element losing electrons to form a positive ion and the nonmetallic element gaining electrons to form a negative ion. The aggregate form of such a compound consists of a crystal lattice of alternating positively and negatively charged ions: a given positive ion is attracted by surrounding negatively charged ions, and a given negative ion is attracted by surrounding positively charged ions. Similar electrostatic attractions exist in three dimensions throughout the crystal of the ionic solid, leading to a very stable system (with very high melting and boiling points, for example). As evidence for the existence of ionic bonding, ionic solids do not conduct electricity (the ions are rigidly held), but melts or solutions of such substances do conduct electric current. For example, when sodium metal and chlorine gas react, a typical ionic substance (sodium chloride) results: $2Na(s) + Cl_2(g) \rightarrow 2Na^+Cl^-(s)$.
32. Electronegativity represents the relative ability of an atom in a molecule to attract shared electrons to itself. The larger the difference in electronegativity between two atoms joined in a bond, the more polar is the bond. Examples depend on student choice of elements.
34. It has been observed over many, many experiments that when an active metal like sodium or magnesium reacts with a nonmetal, the sodium atoms always form Na^+ ions and the magnesium atoms always form Mg^{2+} ions. It has also been observed that when nonmetallic elements like nitrogen, oxygen, or fluorine form simple ions, the ions are always N^{3-}, O^{2-}, and F^-, respectively. Observing that these elements always form the same ions and those ions all contain eight electrons in the outermost shell, scientists speculated that a species that has an octet of electrons (like the noble gas neon) must be very fundamentally stable. The *repeated* observation that so many elements, when reacting, tend to attain an electron configuration that is isoelectronic with a noble gas led chemists to speculate that *all* elements try to attain such a configuration for their outermost shells. Covalently and polar covalently bonded molecules also strive to attain pseudo–noble gas electron configurations. For a covalently bonded molecule like F_2, each F atom provides one electron of the pair of electrons that constitutes the covalent bond. Each

F atom feels also the influence of the other F atom's electron in the shared pair, and each F atom effectively fills its outermost shell.

36. Bonding between atoms to form a molecule involves only the outermost electrons of the atoms, so only these *valence* electrons are shown in the Lewis structures of molecules. The most important requisite for the formation of a stable compound is that each atom of a molecule attain a noble gas electron configuration. In Lewis structures, arrange the bonding and nonbonding valence electrons to try to complete the octet (or duet) for as many atoms as possible.

38. You could choose practically any molecules for your discussion. Let's illustrate the method for ammonia, NH_3. First, count the total number of valence electrons available in the molecule (without regard to their source). For NH_3, since nitrogen is in Group 5, one nitrogen atom would contribute five valence electrons. Since hydrogen atoms have only one electron each, the three hydrogen atoms provide an additional three valence electrons, for a total of eight valence electrons overall. Next, write down the symbols for the atoms in the molecule, and use one pair of electrons (represented by a line) to form a bond between each pair of bound atoms.

```
H—N—H
  |
  H
```

These three bonds use six of the eight valence electrons. Because each hydrogen already has its duet and the nitrogen atom has only six electrons around it so far, the final two valence electrons must represent a lone pair on the nitrogen.

```
  ..
H—N—H
  |
  H
```

40. Boron and beryllium compounds sometimes do not fit the octet rule. For example, in BF_3, the boron atom has only six valence electrons in its outermost shell, whereas in BeF_2, the beryllium atom has only four electrons in its outermost shell. Other exceptions to the octet rule include any molecule with an odd number of valence electrons (such as NO or NO_2).

42.

Number of Valence Pairs	Bond Angle	Examples
2	180°	BeF_2, BeH_2
3	120°	BCl_3
4	109.5°	CH_4, CCl_4, GeF_4

44. (a) $[Kr]5s^2$; (b) $[Ne]3s^23p^1$; (c) $[Ne]3s^23p^5$; (d) $[Ar]4s^1$; (e) $[Ne]3s^23p^4$; (f) $[Ar]4s^23d^{10}4p^3$

46.

```
  ..
H—O—H
  ..
```

4 electron pairs tetrahedrally oriented on O; nonlinear (bent, V-shaped) geometry; H—O—H bond angle slightly less than 109.5° because of lone pairs

```
  ..
H—P—H
  |
  H
```

4 electron pairs tetrahedrally oriented on P; trigonal pyramidal geometry; H—P—H bond angles slightly less than 109.5° because of lone pair

```
       ..
      :Br:
  ..   |   ..
 :Br—C—Br:
  ..   |   ..
      :Br:
       ..
```

4 electron pairs tetrahedrally oriented on C; overall tetrahedral geometry; Br—C—Br bond angles 109.5°

```
 [      ..      ]-
 [     :O:      ]
 [  ..  |   ..  ]
 [ :O—Cl—O:     ]
 [  ..  |   ..  ]
 [     :O:      ]
 [      ..      ]
```

4 electron pairs tetrahedrally oriented on Cl; overall tetrahedral geometry; O—Cl—O bond angles 109.5°

```
 :F:
    \     ..
     B—F:
    /     ..
 :F:
```

3 electron pairs trigonally oriented on B (exception to octet rule); overall trigonal geometry; F—B—F bond angles 120°

```
 ..       ..
:F—Be—F:
 ..       ..
```

2 electron pairs linearly oriented on Be (exception to octet rule); overall linear geometry; F—Be—F bond angle 180°

Chapters 13–15

2. The pressure of the atmosphere represents the mass of the gases in the atmosphere pressing down on the surface of the earth. The device most commonly used to measure the pressure of the atmosphere is the mercury barometer, shown in Fig. 13.2. A simple experiment to demonstrate the pressure of the atmosphere is shown in Fig. 13.1.

4. Boyle's law says that the volume of a gas sample will decrease if you squeeze it harder (at constant temperature, for a fixed amount of gas). Two mathematical statements of Boyle's law are

$$P \times V = \text{constant}$$

$$P_1 \times V_1 = P_2 \times V_2$$

These two mathematical formulas say the same thing: if the pressure on a sample of gas is increased, the volume of the sample will decrease. A graph of Boyle's law data is given as Fig. 13.5: this type of graph ($xy = k$) is known to mathematicians as a *hyperbola*.

6. Charles's law says that if you heat a sample of gas, the volume of the sample will increase (assuming the pressure and amount of gas remain the same). When the temperature is given in kelvins, Charles's law expresses a *direct* proportionality (if you *increase* T, then V *increases*), whereas Boyle's law expresses an *inverse* proportionality (if you *increase* P, then V *decreases*). Two mathematical statements of Charles's law are $V = bT$ and $(V_1/T_1) = (V_2/T_2)$. With this second formulation, we can determine volume–temperature information for a given gas sample under two sets of conditions. Charles's law holds true only if the amount of gas remains the same (the volume of a gas sample would increase if more gas were present) and also if the pressure remains the same (a change in pressure also changes the volume of a gas sample). A graph of volume versus temperature (at constant pressure) for an ideal gas is a straight line with an intercept at –273 °C (see Fig. 13.7).

8. Avogadro's law says that the volume of a sample of gas is directly proportional to the number of moles (or molecules) of gas present (at constant temperature and pressure). Avogadro's law holds true only for gas samples compared under the same conditions of temperature and pressure. Avogadro's law expresses a direct proportionality: the more gas in a sample, the larger the sample's volume.

10. The "partial" pressure of an individual gas in a mixture of gases represents the pressure the gas would exert in the same container at the same temperature if it were the *only* gas present. The *total* pressure in a mixture of gases is the sum of the individual partial pressures of the gases present in the mixture. The fact that the partial pressures of the gases in a mixture are additive suggests that the total pressure in a container is a function of the *number* of molecules present, and not of the identity of the molecules or of any other property (such as the molecules' inherent atomic size).

12. The main postulates of the kinetic molecular theory for gases are: (a) gases consist of tiny particles (atoms or molecules), and the size of these particles themselves is negligible compared with the bulk volume of a gas sample; (b) the particles in a gas are in constant random motion, colliding with each other and with the walls of the container; (c) the particles in a gas sample do not assert any attractive or repulsive forces on one another; and (d) the average kinetic energy of the gas particles is directly related to the absolute temperature of the gas sample. The pressure exerted by a gas results from the molecules colliding with (and pushing on) the walls of the container; the pressure increases with temperature because, at a higher temperature, the molecules move faster and hit the walls of the container with greater force. A gas fills the volume available to it because the molecules in a gas are in constant *random* motion: the randomness of the molecules' motion means that they eventually will move out into the available volume until the distribution of molecules is uniform; at constant pressure, the volume of a gas sample increases as the temperature is increased because with each collision having greater force, the container must expand so that the molecules are farther apart if the pressure is to remain constant.

14. The molecules are much closer together in solids and liquids than in gaseous substances and interact with each other to a much greater extent. Solids and liquids have much greater densities than do gases, and are much less compressible, because so little room exists between the molecules in the solid and liquid states (the volume of a solid or liquid is not affected very much by temperature or pressure). We know that the solid and liquid states of a substance are similar to each other in structure, since it typically takes only a few kilojoules of energy to melt 1 mole of a solid, whereas it may take 10 times more energy to convert a liquid to the vapor state.

16. The *normal* boiling point of water—that is, water's boiling point at a pressure of exactly 760 mm Hg—is 100 °C. Water remains at 100 °C while boiling, because the additional energy added to the sample is used to overcome attractive forces among the water molecules as they go from the condensed, liquid state to the gaseous state. The normal (760 mm Hg) freezing point of water is exactly 0 °C. A cooling curve for water is given in Fig. 14.2.

18. Dipole–dipole forces arise when molecules with permanent dipole moments try to orient themselves so that the positive end of one polar molecule can attract the negative end of another polar molecule. Dipole–dipole forces are not nearly as strong as ionic or covalent bonding forces (only about 1% as strong as covalent bonding forces) since electrostatic attraction is related to the *magnitude* of the charges of the attracting species and drops off rapidly with distance. Hydrogen bonding is an especially strong dipole–dipole attractive force that can exist when hydrogen atoms are directly bonded to the most electronegative atoms (N, O, and F). Because the hydrogen atom is so small, dipoles involving N—H, O—H, and F—H bonds can approach each other much more closely than can other dipoles; because the magnitude of dipole–dipole forces is related to distance, unusually strong attractive forces can exist. The much higher boiling point of water than that of the other covalent hydrogen compounds of the Group 6 elements is evidence for the special strength of hydrogen bonding.

20. The vaporization of a liquid requires an input of energy to overcome the intermolecular forces that exist between the molecules in the liquid state. The large heat of vaporization of water is essential to life since much of the excess energy striking the earth from the sun is dissipated in vaporizing water. Condensation refers to the process by which molecules in the vapor state form a liquid. In a closed container containing a liquid with some empty space above the liquid, an equilibrium occurs between vaporization and condensation. When the liquid is first placed in the container, the liquid phase begins to evaporate into the empty space. As the number of molecules in the vapor phase increases, however, some of these molecules begin to reenter the liquid phase. Eventually, each time a molecule of liquid somewhere in the container enters the vapor phase, another molecule of vapor reenters the liquid phase. No further net change occurs in the amount of liquid phase. The pressure of the vapor in such an equilibrium situation is characteristic for the liquid at each temperature. A simple experiment to determine the vapor pressure of a liquid is shown in Fig. 14.10. Typically, liquids with strong intermolecular forces have smaller vapor pressures (they have more difficulty in evaporating) than do liquids with very weak intermolecular forces.

22. The *electron sea model* explains many properties of metallic elements. This model pictures a regular array of metal atoms set in a "sea" of mobile valence electrons. The electrons can move easily throughout the metal to conduct heat or electricity, and the lattice of atoms and cations can be deformed with little effort, allowing the metal to be hammered into a sheet or stretched into wire. An alloy is a material that contains a mixture of elements that overall has metallic properties. *Substitutional* alloys consist of a host metal in which some of the atoms in the metal's crystalline structure are replaced by atoms of other metallic elements. For example, sterling silver is an alloy in which some silver atoms have been replaced by copper atoms. An *interstitial alloy* is formed when other, smaller atoms enter the interstices (holes) between atoms in the host metal's crystal structure. Steel is an interstitial alloy in which carbon atoms enter the interstices of a crystal of iron atoms.

24. A saturated solution contains as much solute as can dissolve at a particular temperature. Saying that a solution is *saturated* does not *necessarily* mean that the solute is present at a high concentration—for example, magnesium hydroxide dissolves only to a very small extent before the solution is saturated. A saturated solution is in equilibrium with undissolved solute: as molecules of solute dissolve from the solid in one place in the solution, dissolved molecules rejoin the solid phase in another part of the solution. Once the rates of dissolving and solid formation become equal, no further net change occurs in the concentration of the solution and the solution is saturated.

26. Adding more solvent to a solution to dilute the solution does *not* change the number of moles of solute present, but changes only the *volume* in which the solute is dispersed. If molarity is used to describe the solution's concentration, then the number of *liters* is changed when solvent is added and the number of *moles per liter* (the molarity) changes, but the actual number of *moles* of solute does *not* change. For example, 125 mL of 0.551 *M* NaCl contains 0.0689 mole of NaCl. The solution will *still* contain 0.0689 mole of NaCl after 250 mL of water is added to it. The volume and the concentration will change, but the number of moles of solute in the solution will *not* change. The 0.0689 mole of NaCl, divided by the total volume of the diluted solution in liters, gives the new molarity (0.184 *M*).

28. (a) 105 mL; (b) 1.05×10^3 mm Hg

30. (a) 6.96 L; (b) $P_{\text{hydrogen}} = 5.05$ atm; $P_{\text{helium}} = 1.15$ atm; (c) 2.63 atm

32. 0.550 g CO_2; 0.280 L CO_2 at STP

34. (a) 9.65% NaCl; (b) 2.75 g $CaCl_2$; (c) 11.4 g NaCl

36. (a) 0.505 *M*; (b) 0.0840 *M*; (c) 0.130 *M*

38. (a) 226 g; (b) 18.4 *M*; (c) 0.764 *M*; (d) 1.53 *N*; (e) 15.8 mL

Index and Glossary